AUTOMOTIVE ENGINES

Other Books and Instructional Materials
by William H. Crouse and *Donald L. Anglin

Automotive Chassis and Body*
  Workbook for Automotive Chassis and Body*
Automotive Electrical Equipment
  Workbook for Automotive Electrical Equipment*
Automotive Engines*
  Workbook for Automotive Engines*
Automotive Fuel, Lubricating, and Cooling Systems*
  Workbook for Automotive Fuel, Lubricating, and Cooling Systems*
Automotive Transmissions and Power Trains*
  Workbook for Automotive Transmissions and Power Trains*
Automotive Service Business: Operation and Management
Automotive Emission Control
Automotive Engine Design
Workbook for Automotive Service and Trouble Diagnosis
Workbook for Automotive Tools
Automotive Mechanics
  Study Guide for Automotive Mechanics
  Testbook for Automotive Mechanics*
  Workbook for Automotive Mechanics*
  Automotive Troubleshooting Cards
The Auto Book
  Auto Shop Workbook*
  Auto Study Guide
  Auto Test Book*
  Auto Cassette Series
General Power Mechanics (with Robert Worthington and Morton Margules*)
Small Engines: Operation and Maintenance
  Workbook for Small Engines: Operation and Maintenance

Automotive Transparencies by
William H. Crouse and Jay D. Helsel

Automotive Brakes
Automotive Electrical Systems
Automotive Engine Systems
Automotive Transmissions and Power Trains
Automotive Steering Systems
Automotive Suspension Systems
Engines and Fuel Systems

AUTOMOTIVE ENGINES

CONSTRUCTION, OPERATION, AND MAINTENANCE

William H. Crouse
Donald L. Anglin

Fifth Edition

McGRAW-HILL BOOK COMPANY
Gregg Division
NEW YORK
ST. LOUIS
DALLAS
SAN FRANCISCO
AUCKLAND
DÜSSELDORF
JOHANNESBURG
KUALA LUMPUR
LONDON
MEXICO
MONTREAL
NEW DELHI
PANAMA
PARIS
SÃO PAULO
SINGAPORE
SYDNEY
TOKYO
TORONTO

About the Authors

William H. Crouse

Behind William H. Crouse's clear technical writing is a background of sound mechanical engineering training as well as a variety of practical industrial experience. After finishing high school, he spent a year working in a tinplate mill. Summers, while still in school, he worked in General Motors plants, and for three years he worked in the Delco-Remy Division shops. Later he became Director of Field Education in the Delco-Remy Division of General Motors Corporation, which gave him an opportunity to develop and use his natural writing talent in the preparation of service bulletins and educational literature.

During the war years, he wrote a number of technical manuals for the Armed Forces. After the war, he became Editor of Technical Education Books for the McGraw-Hill Book Company. He has contributed numerous articles to automotive and engineering magazines and has written many outstanding books. He was the first Editor-in-Chief of the 15-volume McGraw-Hill Encyclopedia of Science and Technology.

William H. Crouse's outstanding work in the automotive field has earned for him membership in the Society of Automotive Engineers and in the American Society of Engineering Education.

Donald L. Anglin

Trained in the automotive and diesel service field, Donald L. Anglin has worked both as a mechanic and as a service manager. He has taught automotive courses and has also worked as curriculum supervisor and school administrator for an automotive trade school. Interested in all types of vehicle performance, he has served as a racing-car mechanic and as a consultant to truck fleets on maintenance problems.

Currently he serves as editorial assistant to William H. Crouse, visiting automotive instructors and service shops. Together they have coauthored magazine articles on automotive education and several books in the McGraw-Hill Automotive Technology Series.

Donald L. Anglin is a Certified General Automotive Mechanic and holds many other licenses and certificates in automotive education, service, and related areas. His work in the automotive service field has earned for him membership in the American Society of Mechanical Engineers and the Society of Automotive Engineers. In addition, he is an automotive instructor at Piedmont Virginia Community College, Charlottesville, Virginia.

Library of Congress Cataloging in Publication Data

Crouse, William Harry, (date)
Automotive engines.

(McGraw-Hill automotive technology series)
Includes index.
1. Automobiles—Motors. I. Anglin, Donald L., joint author. II. Title.
TL210.C7 1976 629.2′5 75-6627
ISBN 0-07-014602-0

AUTOMOTIVE ENGINES

6 7 8 9 0 WCWC 8 3 2 1 0

The editors for this book were Ardelle Cleverdon and Susan Berkowitz, the designer was Dennis G. Purdy, and the production supervisor was Rena Shindelman. The cover illustrator was Shelley Freshman. It was set in Melior by York Graphic Services, Inc.
Printed and bound by Webcrafters, Inc.

CONTENTS

PREFACE

This is the fifth edition of *Automotive Engines*. The book has undergone many changes in its five editions. These changes parallel the many changes in the automotive engine. Engines have been improved year after year to make them more economical, more powerful, more flexible, and longer lasting. Now, new forces have aimed the thrust of engine-design work toward reduction of atmospheric pollution. This effort has meant the introduction of various emission controls. It has intensified the search for alternative power plants and the development of various modifications of the present piston engine.

This new edition of *Automotive Engines* covers all these new developments. However, despite the present furor over alternative power plants, electric cars, turbine engines, automobiles powered by hydrogen fuel or spinning rotors, and various exotic engines, engineers generally agree that the piston engine will be with us for a long time. It seems that no other engine has the potential to take over. Thus, automotive-piston-engine technicans and mechanics will be needed for many years to come.

This edition has been largely rewritten, and much new material and many new illustrations have been added. The new developments covered include electronic and high-energy ignition systems, the most modern engine-testing instruments, stratified charge, the Honda system, automotive emission controls (two chapters), overhead camshafts, catalytic converters, and so on. There are two chapters on the Wankel engine (one on servicing), a new chapter on engine tune-up, an updated trouble-diagnosis chapter, new material on the Stirling engine, new developments in engine-servicing techniques (such as the use of coated valves to fight valve-seat recession), and much more. A feature of the new edition is the introduction of the metric system of measurements. When a United States Customary measurement is used, it is usually followed by its metric equivalent in brackets, for example, 0.002 inch [0.0508 mm]. Tables of equivalents are also included.

A new edition of the *Workbook for Automotive Engines* has been prepared. It includes the basic engine-service jobs as proposed in the latest recommendations of the Motor Vehicle Manufacturers Association–American Vocational Association Industry Planning Council. Taken together, *Automotive Engines* and the *Workbook for Automotive Engines* supply the student with the background information and "hands-on" experience needed to become a qualified and certified automotive engine mechanic.

To assist the automotive instructor, the *Instructor's Planning Guide for Automotive Engines* is available from McGraw-Hill. The instructor's guide was prepared to help the automotive instructor do the best possible job of teaching by most effectively utilizing the textbook, workbook, and other related instructional materials. The instructor's guide contains suggestions on student motivation, classroom instruction and related shop activities, the automotive curriculum, and much more. It also includes the answer key for the tests at the end of each jobsheet in the *Workbook for Automotive Engines*.

Also in the instructor's guide is a list of various related textbooks and ancillary instructional materials available from McGraw-Hill. Used singly or together, these items form a comprehensive student learning and activity package. They provide the student with meaningful learning experiences and help the student develop job competencies in automotive engines and related fields. The instructor's guide explains how the various available materials can be used, either singly or in combination, to satisfy any teaching requirement.

WILLIAM H. CROUSE
DONALD L. ANGLIN

TO THE STUDENT

Automotive Engines is one of eight books in the McGraw-Hill Automotive Technology Series. These books cover in detail the construction, operation, and maintenance of automotive vehicles. They are designed to give you the complete background of information you need to become successful in the automotive service business. The books satisfy the recommendations of the Motor Vehicle Manufacturers Association–American Vocational Association Industry Planning Council. The books also meet the requirements for automotive mechanics certification and state vocational educational programs, and recommendations for automotive trade apprenticeship training. Furthermore, the comprehensive coverage of the subject matter makes the books valuable additions to the library of anyone interested in automotive engineering, manufacturing, sales, service, and operation.

Meeting the Standards

The eight books in the McGraw-Hill Automotive Technology Series meet the standards of the Motor Vehicle Manufacturers Association (MVMA) for associate degrees in automotive servicing and automotive service management. These standards are described in the MVMA booklet "Community College Guide for Associate Degree Programs in Auto and Truck Service and Management." The books also cover the subjects recommended by the American National Standards Institute in their detailed standard D18.1–1972, "American National Standard for Training of Automotive Mechanics for Passenger Cars and Light Trucks."

In addition, the books cover the subject matter tested by the National Institute for Automotive Service Excellence (NIASE). The tests given by NIASE are used for certifying general automotive mechanics and automotive technicians working in specific areas of specialization under the NIASE voluntary mechanic testing and certification program.

Getting Practical Experience

At the same time that you study the books in the McGraw-Hill Automotive Technology Series, you should be getting practical experience in the shop. You should handle automotive parts, automotive tools, and automotive servicing equipment, and you should perform actual servicing jobs. This is what is meant by getting practical experience. To assist you in your shop work, there are workbooks for each book in the series. For example, the *Workbook for Automotive Engines* includes the jobs that cover every basic servicing procedure on automotive engines. If you do every job covered in the workbook, you will have had "hands-on" experience with all engine-servicing work.

If you are taking an automotive mechanics course in school, you will have an instructor to guide you in your classroom and shop activities. But even if you are not taking a course, the workbook can act as an instructor. It tells you, step by step, how to do the various servicing jobs. Perhaps you can meet others who are taking a school course in automotive mechanics. You can talk over any problems you have with them. A local garage or service station is a good source of information. If you can get acquainted with the automotive mechanics there, you will find they have a great deal of practical information. Watch them at their work if you can. Make notes of important points for filing in your notebook.

Service Publications

While you are in the service shop, study the various publications received at the shop. Automobile manufacturers, as well as suppliers of parts, accessories, and tools, publish shop manuals, service bulletins, and parts catalogs. All these help service personnel do better jobs. In addition, numerous automotive magazines are published which deal with problems and methods of automotive service. All these publications will be of great value to you; study them carefully.

These activities will help you obtain practical experience in automotive mechanics. Sooner or later this experience, plus the knowledge that you have gained in studying the books in the McGraw-Hill Automotive Technology Series, will permit you to step into the automotive shop on a full-time basis. Or, if you are already in the shop, you will be equipped to step up to a better and a more responsible job.

Checking Up on Yourself

You can check up on your progress in your studies by answering the questions given every few pages in the book. There are two types of tests, progress quizzes and chapter checkups, the answers to which are given at the back of the book. Each progress quiz should be taken just after you have completed the pages preceding it. These quizzes allow you to check yourself as you finish a lesson. On the other hand, the chapter checkup may cover several lessons, since it is a review test of the entire chapter. Because it is a review test, you should review the entire chapter by rereading it or at least glancing through it to check important points before trying the test. If any of the questions stump you, reread the pages in the book that will give you the answer. This sort of review is valuable; it will help you to remember the information you need when you work in an automotive shop.

Keeping a Notebook

Keeping a notebook is a valuable part of your training. Start it now, at the beginning of your studies of automotive engines. Your notebook will help you in many ways: It will be a record of your progress; it will become a storehouse of valuable information you will refer to time after time; it will help you learn; and it will help you organize your training program so that it will do you the most good.

When you study a lesson in the book, have your notebook open in front of you. Start with a fresh notebook page at the beginning of each lesson. Write the lesson or textbook page number and date at the top of the page. As you read your lesson, jot down the important points.

In the shop, use a small scratch pad or cards to jot down important points. You can transfer your notes to your notebook later.

You can also make sketches in your notebook showing wiring or hose diagrams, fuel circuits, and so on. Save articles and illustrations from technical and hot-rod magazines. File them in your notebook. Also, save instruction sheets that come with service parts. Piston-ring sets, for example, have instruction sheets explaining how to install the rings. Glue or tape these to sheets of paper and file them in your notebook.

As you can see, your notebook will become a valued possession—a permanent record of all you have learned about automotive engines.

Glossary and Index

A glossary (a definition list) of automotive terms is given in the back of the book. Whenever you have any doubt about the meaning of a term or what purpose some automotive part has, you should refer to this list. Also, there is an index at the back of the book. This index will steer you to the page in the book where you will find the information you are seeking.

And now, good luck to you. You are studying a fascinating, complex, and admirable machine—the automobile. Your studies can lead you to success in the automotive field, a field where opportunities are nearly unlimited.

ACKNOWLEDGMENTS

During the preparation of this fifth edition of *Automotive Engines,* the authors were given invaluable aid and inspiration by many people in the automotive industry and in the field of education. The authors gratefully acknowledge their indebtedness and offer their sincere thanks to these people. All cooperated with the aim of providing accurate and complete information that would be useful in training automotive mechanics.

Special thanks are owed to the following organizations for information and illustrations that they supplied: AC Spark Plug Division of General Motors Corporation; American Motors Corporation; Automotive Rebuilders, Inc.; Autoscan, Inc.; Black and Decker Manufacturing Company; Bohn Aluminum and Brass Company, Division of Universal American Corporation; Buick Motor Division of General Motors Corporation; Cadillac Motor Car Division of General Motors Corporation; Caterpiller Tractor Company; Chevrolet Motor Division of General Motors Corporation; Chrysler Corporation; Clayton Manufacturing Company; Cummins Engine Company Inc.; DAF of Holland; Dana Corporation; Detroit Diesel Allison Division of General Motors Corporation; Delco-Remy Division of General Motors Corporation; Dow Corning Corporation; Federal-Mogul Corporation; Ford Motor Company; Ford Motor Company of Germany; Ford of Britain; General Motors Corporation; Grant Piston Rings; Hall Manufacturing Company; Harrison Radiator Division of General Motors Corporation; Hillman Motor Car Company, Limited; Honda; Interindustry Emission Control Program; Johnson Bronze Company; Johnson Motors; K-D Manufacturing Company; Kent-Moore Inc.; Lawnboy Division of Outboard Marine Corporation; Lord Manufacturing Company; Los Angeles County Air Pollution Control District; McCord Replacement Products Division of McCord Corporation; Mercedes-Benz, Daimler-Benz Aktiengesellshaft; MG Car Company, Limited; Muskegan Piston Ring Company; NSU of Germany; Oldsmobile Division of General Motors Corporation; Perkins Engines, Limited; Pontiac Division of General Motors Corporation; Renault; Replacement Parts Division of TRW Inc., Rottler Boring Bar Company; SAAB of Sweden; Sealed Power Company; Service Parts Division of Dana Corporation; Snap-on Tools Corporation; Sun Electric Corporation; Sunnen Products Company; Texaco, Inc.; Toyo Kogyo Company, Limited; Toyota Motor Sales, Limited; United Tool Processes; Universal Testproducts, Inc.; Volkswagen; and Young and Rubican. To all these organizations and the people who represent them, sincere thanks.

WILLIAM H. CROUSE
DONALD L. ANGLIN

chapter 1

AUTOMOTIVE ENGINES

In this chapter, we introduce the various kinds of engines used in automotive vehicles. We also discuss the accessory systems needed to keep them running. One accessory system is the fuel system; another is the cooling system. In later chapters, we will discuss in detail engine and engine-accessory construction, operation, and servicing.

⊘ **1-1 Engine Types** There are two major types of engines used in automobiles today. One is the *piston engine,* in which pistons move up and down, or reciprocate. The other is the *rotary engine,* in which a rotor spins. Piston engines are by far the most common. This is the kind of engine you see in Chevrolets, Fords, and Plymouths, for example. Figure 1-1 shows a piston engine removed from the car. Figure 1-2 shows, in outline view, where the engine is located in the automobile. Figure 1-3 shows a piston engine partly cut away so you can see the interior parts. Figure 1-4 shows a piston. Chapters 7 to 9 describe all the interior parts of the piston engine and explain how they work together to produce power.

The rotary engine has rotors that spin. There are two kinds, the Wankel and the turbine. There are practically no turbine engines in automobiles today. But the Wankel engine is operating in hundreds of thousands of cars. The most widely known Wankel engine is the Mazda, made by Toyo Kogyo of Japan. Figure 1-5 shows a Mazda Wankel engine. Figure 1-6 shows a rotor from this engine. Turbine and Wankel engines are described in detail in Chap. 10.

Another type of automotive engine is the *diesel engine.* None of the automobiles made in the United States has a diesel engine. But many automobiles made in other countries use diesels. Mercedes-Benz of Germany, for instance, has made hundreds of thousands of automobiles with diesel engines. Many of these have been imported into the United States. The diesel is also a piston engine, but it operates on a different principle from the gasoline engine used in American cars. Diesel engines are described in Chap. 6.

Other types of automotive power plants include steam engines, Stirling engines, spinning flywheels, and electric motors. These are described in later chapters. First, however, we want to introduce you to the four basic systems that the reciprocating piston engine and the Wankel, or rotary, engine must have. These are:

Fuel system
Ignition system
Lubricating system
Cooling system

The *fuel system* mixes gasoline with air to make a combustible mixture. This mixture burns in the engine cylinders to produce high pressure. The high pressure forces the pistons to move or the rotor to spin. The movement is carried to shafts so that shafts turn and the wheels rotate to move the car.

The *ignition system* supplies a constant stream of sparks to the engine. These sparks set fire to, or *ignite,* the air-fuel mixture. The mixture then burns to produce power.

The *lubricating system* keeps all moving parts

Fig. 1-1. Piston engine removed from the automobile. *(Chrysler Corporation)*

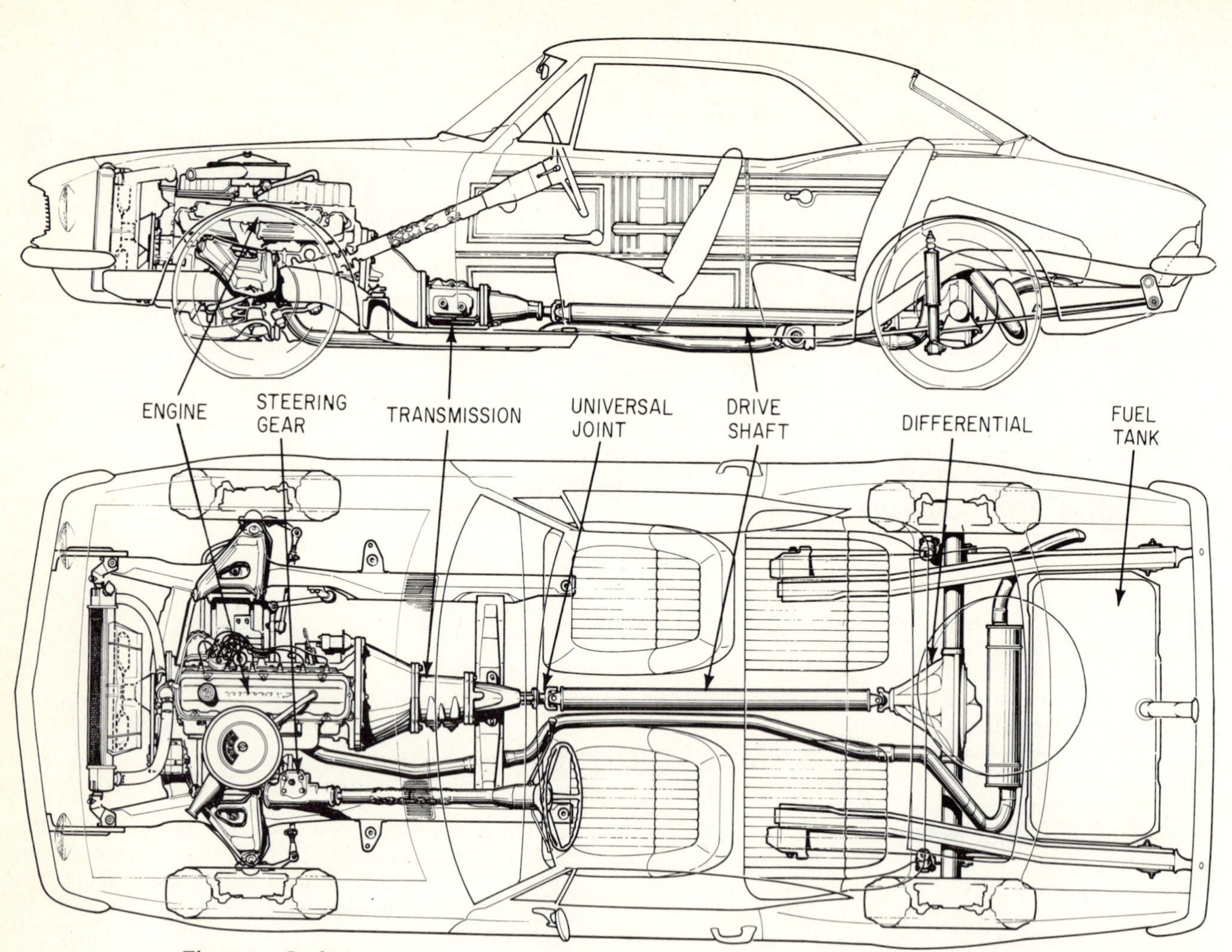

Fig. 1-2. Outline views of an automobile showing essential parts. (*Chevrolet Motor Division of General Motors Corporation*)

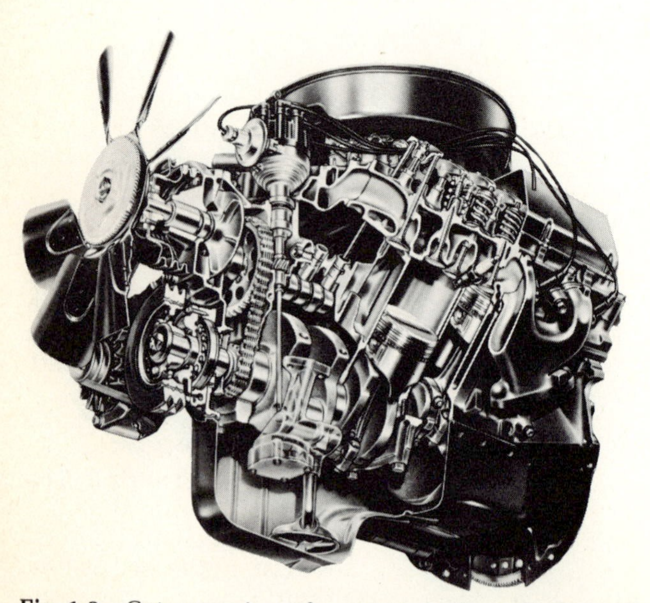

Fig. 1-3. Cutaway view of a piston engine. (*Ford Motor Company*)

Fig. 1-4. Piston from piston engine. (*Piedmont Virginia Community College*)

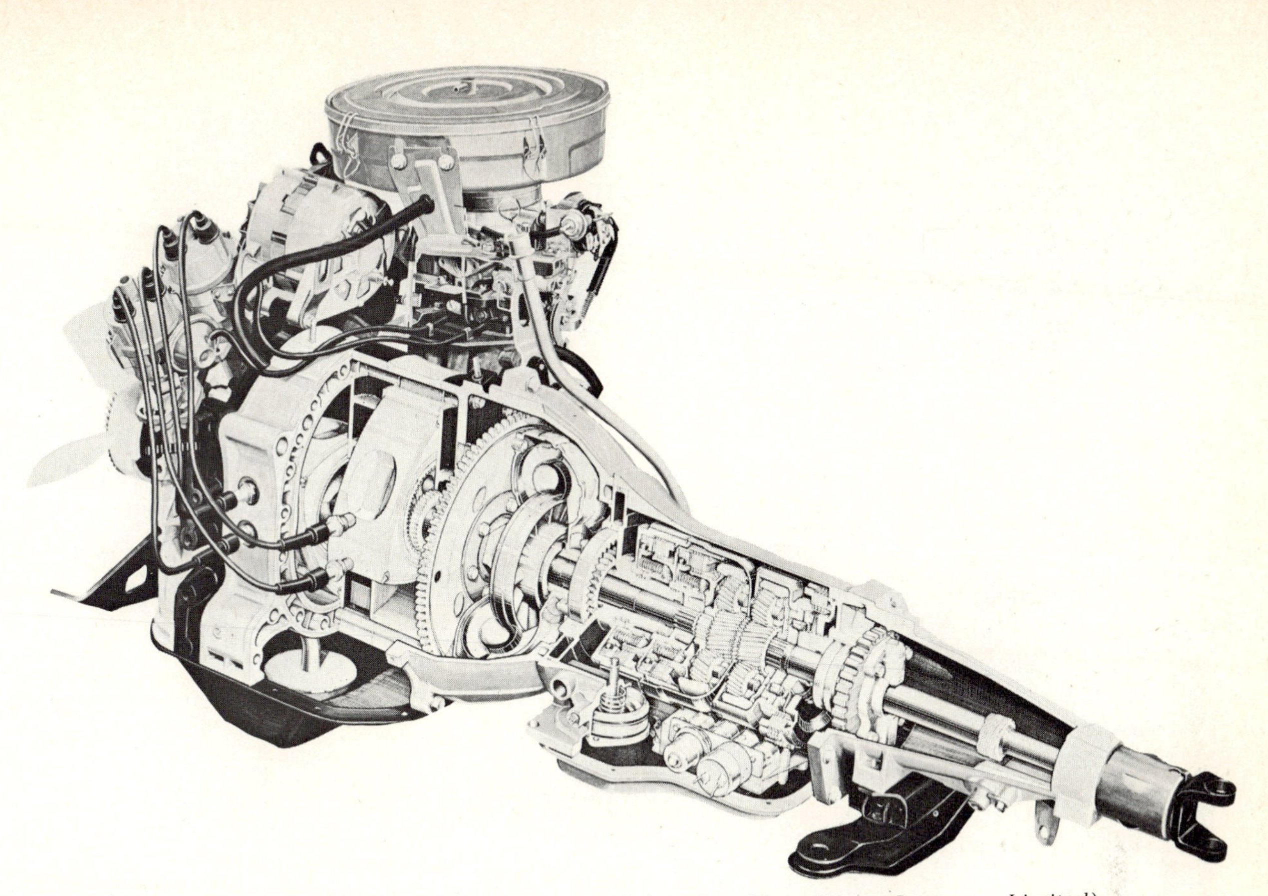

Fig. 1-5. Cutaway view of a Mazda Wankel engine. (*Toyo Kogyo Company, Limited*)

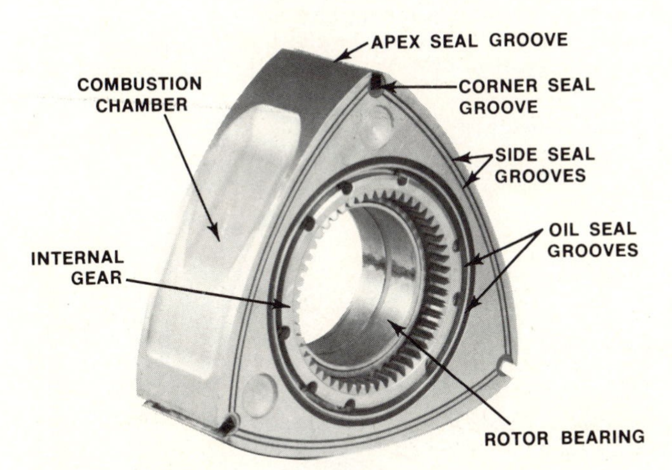

Fig. 1-6. Rotor from a Wankel engine. (*Toyo Kogyo Company, Limited*)

in the engine coated with oil so that they will move easily without undue wear.

The *cooling system*, in most engines, circulates a mixture of water and antifreeze between the engine and a radiator. This mixture is called the *coolant*. The coolant carries heat away from the engine, thus preventing the engine from overheating.

All these systems are described briefly in this chapter and are discussed in detail in later chapters.

⊘ 1-2 Fuel System Figure 1-7 shows a fuel system on an automobile chassis. The fuel system includes a fuel tank, fuel lines, a fuel pump, and a carburetor.

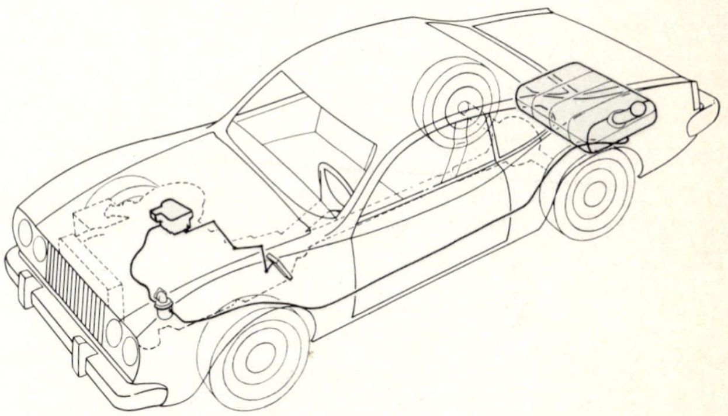

Fig. 1-7. Simplified view of a fuel system.

1. *FUEL TANK* The fuel tank (Fig. 1-8) is simply a sheet-metal tank with an opening through which fuel can be pumped into the tank. It also has an opening through which fuel is removed from the tank and sent to the carburetor.

2. *FUEL PUMP* The fuel pump draws fuel from the tank and sends it to the carburetor. There are two types of fuel pumps, mechanical and electrical. The *mechanical fuel pump* (Fig. 1-9) is mounted on the engine. It has an airtight, flexible diaphragm attached by linkage to a rocker arm. The rocker arm rests on an eccentric (an offset collar) on the camshaft. As the camshaft rotates, the rocker arm rocks.

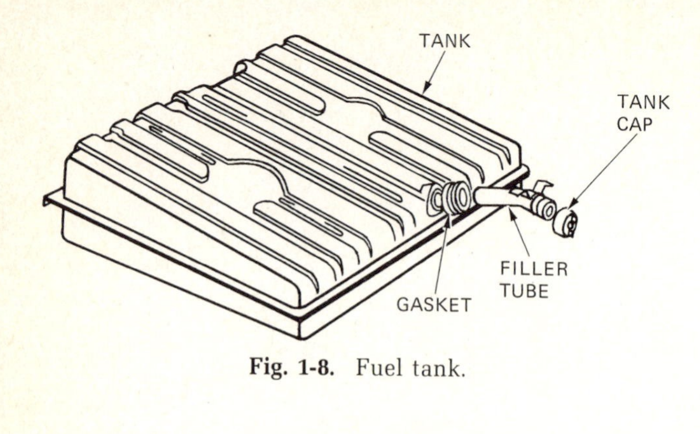

Fig. 1-8. Fuel tank.

Fig. 1-9. Mechanical fuel pump. (*AC-Delco Division of General Motors Corporation*)

This motion forces the diaphragm to move up and down. The up-and-down motion of the diaphragm pumps fuel from the fuel tank to the carburetor.

The *electric fuel pump* uses a plunger, bellows, or rotor to do the pumping. The electricity comes from the battery or the alternator or generator. One type of electric fuel pump is mounted inside the fuel tank. Fuel pumps are discussed in greater detail in Chap. 13.

3. *CARBURETOR* The carburetor (Fig. 1-10) is a mixing device. It mixes fuel from the fuel pump with air and sends the mixture to the engine cylinders. In operation, the carburetor throws a fine spray of fuel into the air passing through it. The fuel vaporizes and mixes with the air to form an air-fuel mixture that will burn.

Carburetors have been made in many sizes and shapes, but basically all carburetors are the same. Each carburetor has a reservoir—the *float bowl*—in which a small amount of fuel can be stored. Also, each carburetor has a series of passages through which the fuel flows on its way from the float bowl to the discharge nozzles. These passages make up the *carburetor systems*—the idle, low-speed, high-speed, and choke systems. These systems will be discussed in Chap. 14.

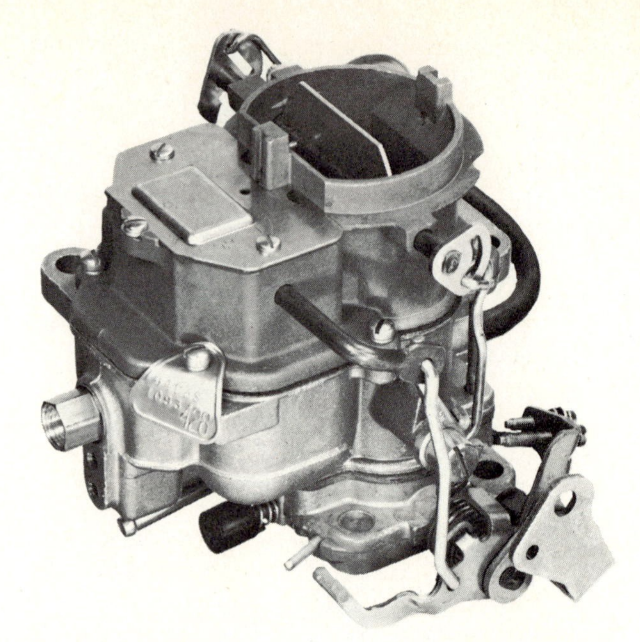

Fig. 1-10. Carburetor. (*Carter Carburetor Division of ACF Industries*)

⊘ 1-3 Ignition System The engine requires a system that will set fire to, or *ignite*, the air-fuel mixture that the fuel system delivers to the engine cylinders. The ignition system does this job. Figure 1-11 is a simplified drawing of an ignition system. It consists of an ignition switch, an ignition coil, a distributor, spark plugs, and wiring. The *battery*, or *generator* or *alternator*, delivers low-voltage current to the ignition system. The ignition coil and distributor turn this current into high-voltage surges. These surges flow through the wiring to the spark plugs. There, the surges produce electric sparks. The sparks ignite the air-fuel mixture in the engine cylinders. The burning air-fuel mixture produces high temper-

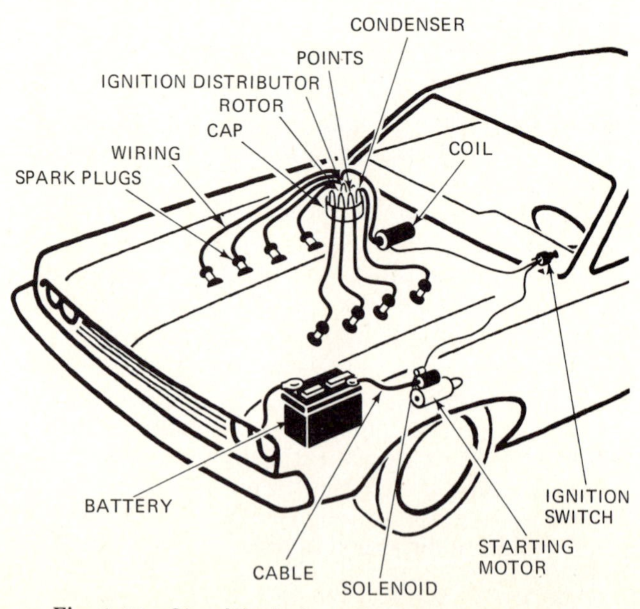

Fig. 1-11. Simplified view of an ignition system.

atures and pressure. The pressure pushes the pistons down so the pistons move. This movement causes shafts and thus the wheels to turn.

⊘ 1-4 Lubricating System The engine has many moving metal parts. These parts must be coated with oil so that they will not come in contact with each other and will slide or rotate easily. If the metal parts rubbed against each other, they would wear rapidly and the engine would last only a few miles. The parts move on thin layers of oil supplied by the engine lubricating system. The lubricating system includes an oil pan in which a reserve of oil is kept, an oil pump that sends oil from the oil pan up to the engine, and a series of oil passages that carry oil to all moving parts.

⊘ 1-5 Cooling System Fire produces heat. And there is plenty of fire in the running engine. The burning of the air-fuel mixture produces this heat. Some of this heat leaves with the hot exhaust gas (the gas that is left after the air-fuel mixture burns). But a great deal of heat remains in the engine. If the heat stayed in the engine, the engine would soon be ruined. The cooling system removes this heat.

Figure 1-12 shows an engine cooling system. Its principle of operation is simple. There are openings, called *water jackets,* surrounding the engine cylinders. Coolant (water mixed with antifreeze) circulates through these water jackets, forced through by the water pump. As the coolant passes through, it gets hot; that is, it takes some of the heat from the engine. Then the hot coolant flows through the radiator where it loses heat. So the cooling system is continually removing heat from the engine.

⊘ 1-6 What's Coming in Later Chapters In order for you to understand just how the engine works, you will need to know some basic principles. So in later chapters we shall answer such questions as:

What is heat?
What is a chemical reaction?
What makes the piston move?

Also, we shall explain how the power gets from the engine to the wheels so that the car moves. And we shall discuss engine construction in detail and explain how all the parts fit and work together. The accessory systems will be discussed, too. There will be chapters on engine and accessory service, emission controls, and tune-up. When you have finished this book and have done the related shop work, you should qualify as an automotive mechanic.

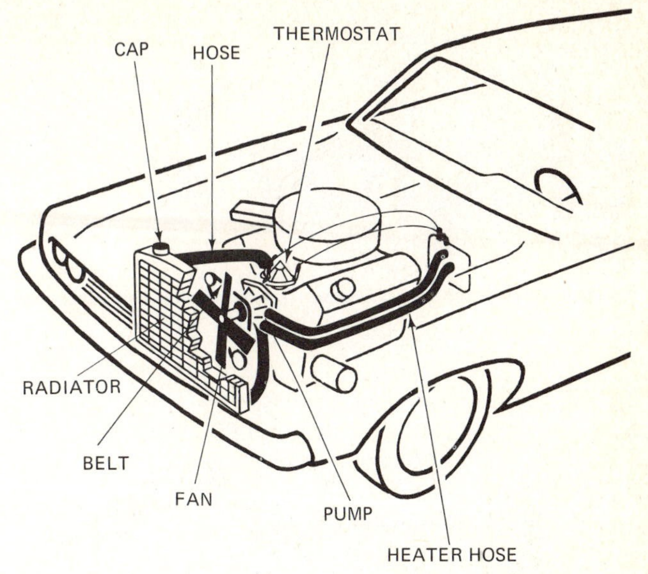

Fig. 1-12. Simplified view of an engine cooling system.

CHAPTER 1 CHECKUP

NOTE: Since the following is a chapter review test, you should review the chapter before taking the test.

At the end of each chapter in this book, there are quizzes that will test your memory and understanding of what you have been studying. In addition, within the longer chapters there are "Progress Quizzes" that will test you on portions of chapters. The purpose of these tests is to give you the chance to check up on yourself. That is, you may or may not feel that you have mastered what you have been studying. Taking the tests lets you know where you stand. If you can answer all the questions, you have done an outstanding job of studying. If any of the questions stump you, go back into the chapter and reread the pages that will give you the answer. This way, you can be sure that you are making steady progress on your way to becoming an engine expert.

Completing the Sentences The sentences that follow are incomplete. After each sentence there are several words or phrases, only one of which will correctly complete the sentence. Write each sentence in your notebook, selecting the proper word or phrase to complete it correctly.[1]

1. The two major types of engine used in automobiles today are the: (*a*) piston engine and reciprocating engine, (*b*) Wankel engine and rotary engine, (*c*) reciprocating engine and rotary engine.
2. The diesel engine is a: (*a*) rotary engine, (*b*) reciprocating engine.
3. The major parts of the fuel system are the carburetor: (*a*) engine, and fuel pump, (*b*) fuel pump, and fuel tank, (*c*) fuel lines, and fuel pump.
4. The major parts of the ignition system are the ignition switch, ignition coil, wiring: (*a*) spark plugs, and engine, (*b*) spark plugs, and distributor, (*c*) condenser, and spark plugs.
5. The major parts of the cooling system are the coolant, pump: (*a*) water jackets, and radiator, (*b*) heater, and radiator, (*c*) water, and radiator.

[1] Answers are given at the back of the book.

chapter 2

AUTOMOTIVE COMPONENTS

This chapter describes the various component parts of the automobile, aside from the engine, including such items as the frame, power train, and car body. This material will serve as background information so that you will have a better understanding of the relationship between the engine and other automotive components.

⊘ 2-1 Components of the Automobile Before we begin our studies of the engine, let us first look at the complete automobile (Fig. 2-1), examine its component parts, and find out how these parts work together with the engine. You will probably find that much of the information is already familiar to you. However, as you read the next few pages, you may find some new and interesting relationships among automobile components that might not have occurred to you.

The automobile can be said to consist of five basic components.

1. The engine, or power plant, which is the source of power.
2. The frame, which supports the engine, wheels, and body.
3. The power train, which carries power from the engine to the car wheels. The power train includes the clutch, transmission, drive shaft (propeller shaft), differential, and wheel axles.
4. The car body.
5. The car-body accessories, which include lights, heater, radio, windshield wipers, convertible-top raiser, and so on.

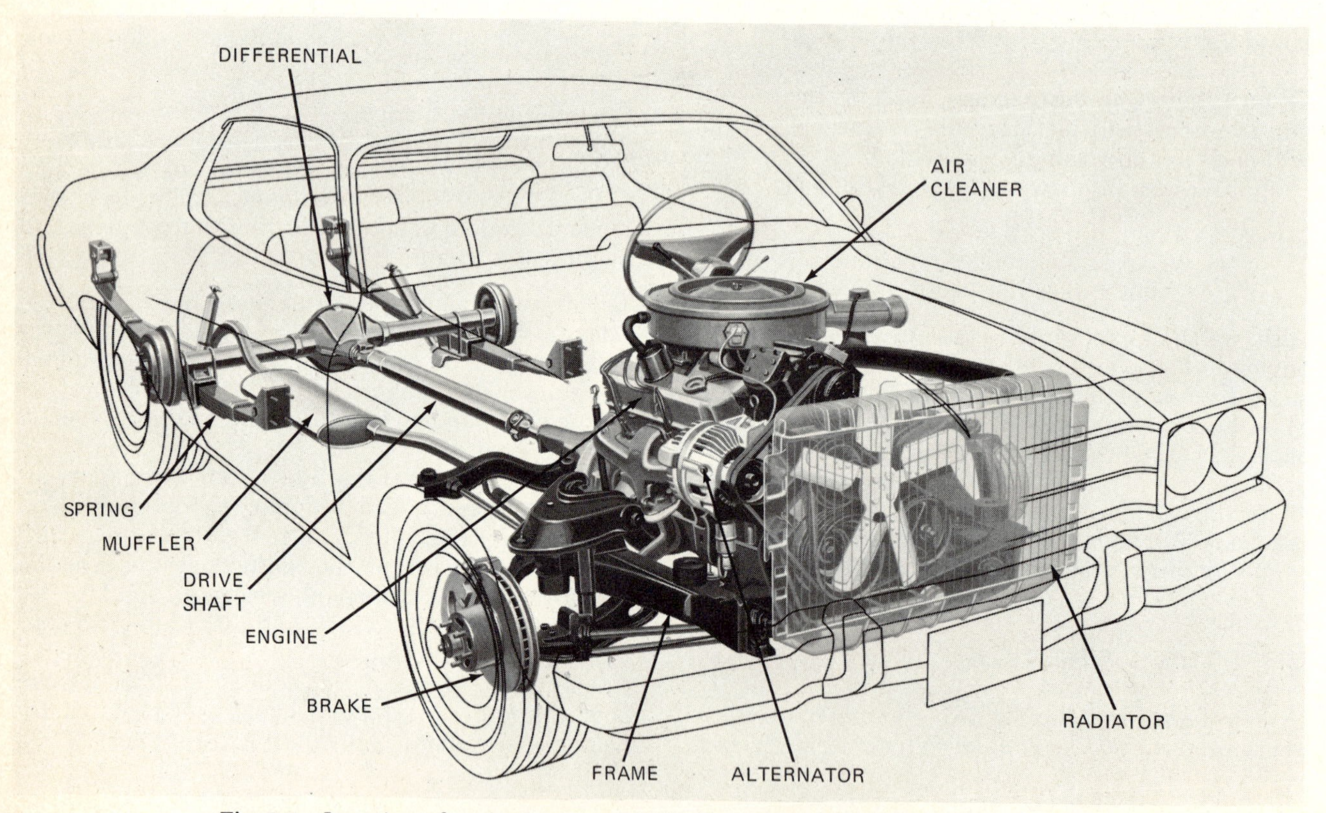

Fig. 2-1. Location of major components in the automobile. (*Young and Rubican*)

Let us look at each of these further. Remember, the discussions of automobile components that are presented in this chapter are "get-acquainted" descriptions. Later chapters in the book will describe the construction, operation, maintenance, and servicing of engines. Other books in the McGraw-Hill Automotive Technology Series discuss the other automobile components in similar detail.

⊘ 2-2 Frame and Chassis The automobile frame is made of steel members welded together. It supports the engine, car body, and other component parts. The frame with the parts shown in Fig. 2-2 attached is called the *chassis*. This includes all the essential mechanical parts that make the car run and control its operation. The engine is mounted on the frame with rubber pads which absorb vibration and keep it from passing on up to the passengers.

⊘ 2-3 Springs The wheels are attached to the frame through springs. The springs support the weight of the vehicle and also permit the wheels to move up and down as the wheels meet irregularities in the road. Figure 2-3 shows springs at front and rear wheels.

⊘ 2-4 Shock Absorbers Springs alone will not give a smooth ride. They will continue to oscillate, or expand and contract, after the car has moved over a bump or hole in the road. To prevent this excessive activity, shock absorbers (Fig. 2-4) are used. The typical shock absorber includes a piston fitted in a cylinder. When the wheels move up and down, the piston moves in the cylinder. This forces a fluid to flow through small openings, or restricting orifices. The result is a damping effect which prevents excessive or repeated movements of the wheels after the car has moved over a hole or bump in the road.

⊘ 2-5 Steering System The steering system permits the front wheels to be pivoted to the right or left so that the car can be steered. Figure 2-3 shows how a front wheel is supported, and Fig. 2-5 is a simplified drawing of a steering system. When the steering wheel is turned, gears in the steering-gear box cause the tie rods to move, thus causing the wheels to pivot to the left or right.

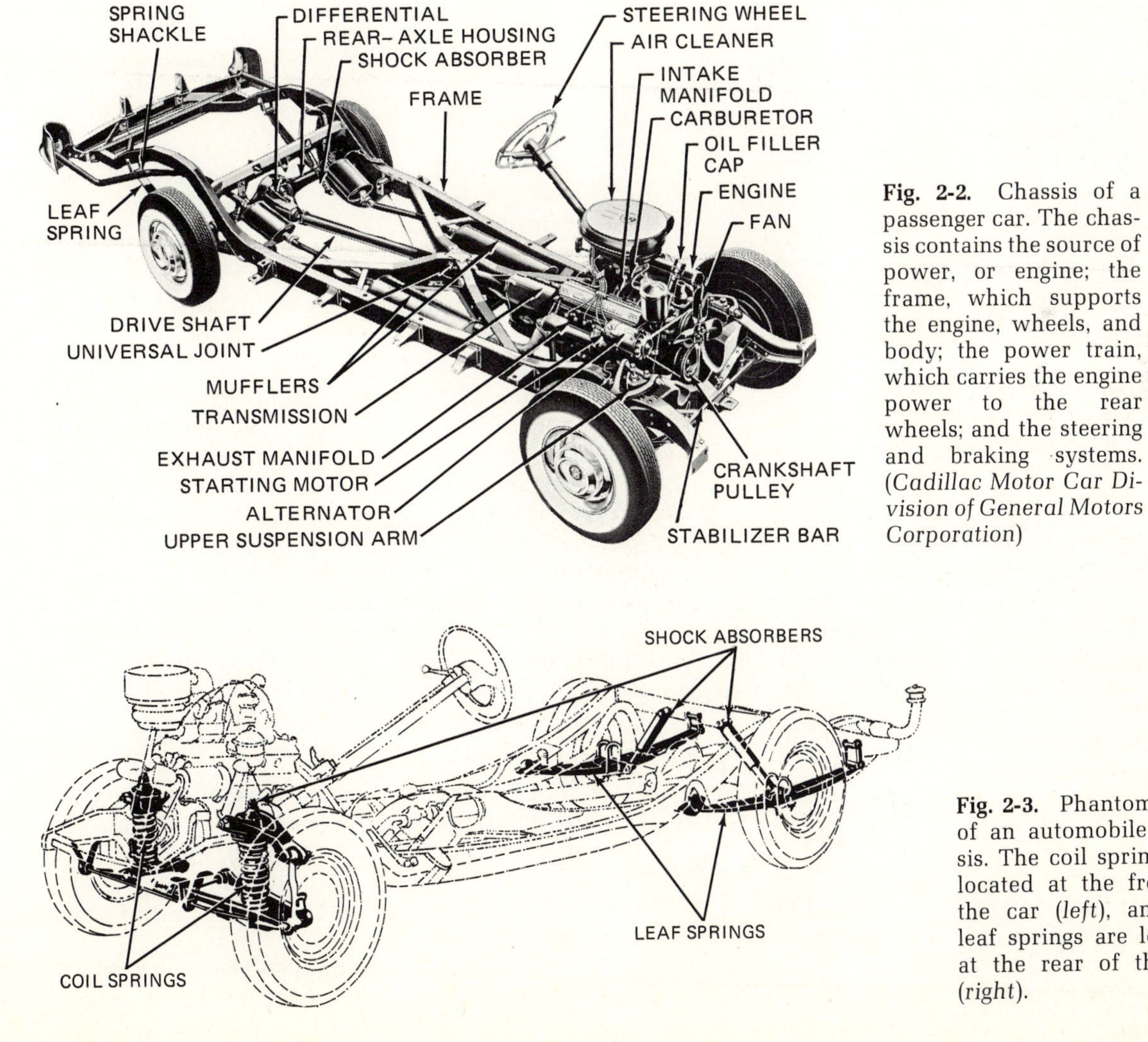

Fig. 2-2. Chassis of a passenger car. The chassis contains the source of power, or engine; the frame, which supports the engine, wheels, and body; the power train, which carries the engine power to the rear wheels; and the steering and braking systems. (*Cadillac Motor Car Division of General Motors Corporation*)

Fig. 2-3. Phantom view of an automobile chassis. The coil springs are located at the front of the car (*left*), and the leaf springs are located at the rear of the car (*right*).

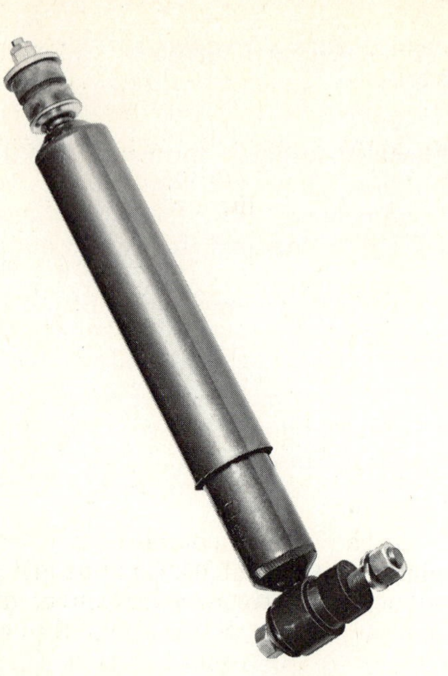

Fig. 2-4. Direct-acting shock absorber.

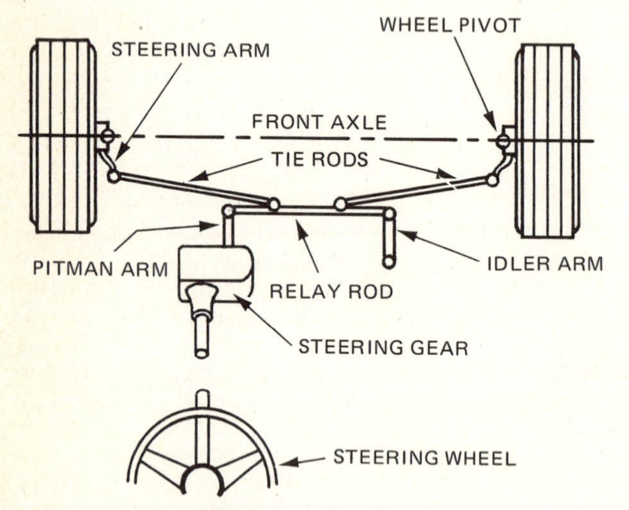

Fig. 2-5. Simplified drawing of a steering system.

⊘ **2-6 Brakes** Brakes are necessary to slow or stop the car. A typical brake system is shown in Fig. 2-6. When the brake pedal is pushed, a piston is forced to move in the master cylinder. This pushes hydraulic brake fluid through tubes to the wheel cylinders. The pressure of the brake fluid forces pistons to move, pushing the brake shoes against drums or disks on the wheels. The friction of the stationary brake shoes against the revolving drums or disks causes the wheels to slow down or stop rotating, thus braking the car.

⊘ **2-7 Tires** The tires support the weight of the car and also transmit the driving or braking force from the wheels to the road. Tires consist of a casing of interwoven fiber cords on which a rubber covering,

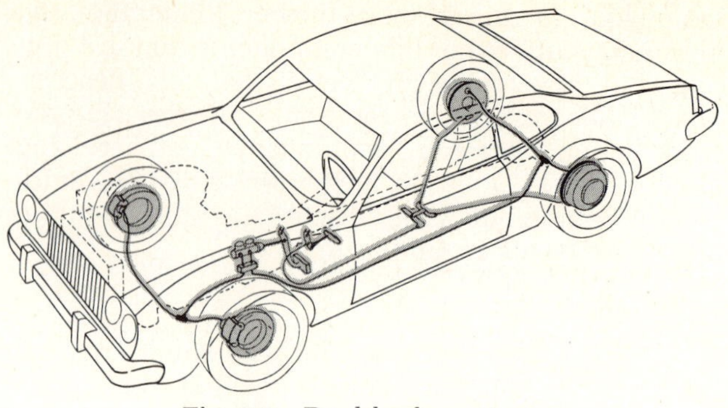

Fig. 2-6. Dual-brake system.

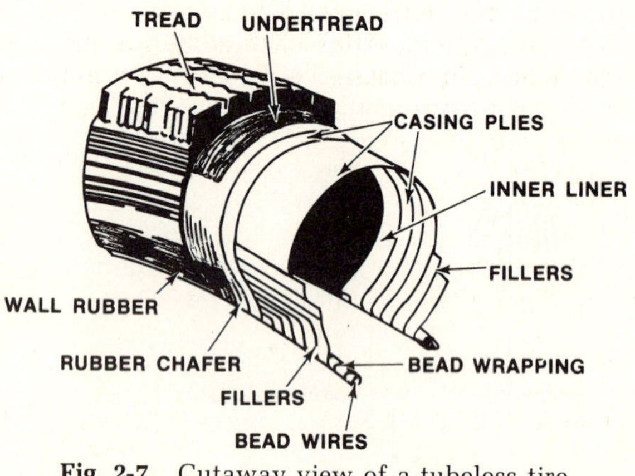

Fig. 2-7. Cutaway view of a tubeless tire.

with rubber tread, is applied (Fig. 2-7). Some tires have fiber-glass, steel mesh, rayon, or nylon reinforcing in the casing.

Check Your Progress

Progress Quiz 2-1 You have now moved into the actual study of the automobile. The automotive component fundamentals you have just been reading about are of special interest to anyone working in the automotive field. Find out how well you remember what you have been reading by taking the quiz that follows. Don't be discouraged if any of the questions stump you. Just reread the past few pages and then try the questions again. Remember, even the best students usually reread their lessons several times. The combination of taking the quiz and rereading the preceding pages will help you fix essential facts firmly in your mind.

Correcting Parts Lists The purpose of this exercise is to enable you to spot the unrelated part in a list. For example, in the list "shoe, pants, shirt, milk, tie, coat," you would see that "milk" does not belong because it is the only thing named that is not something you wear.

In each of the following lists, you will find one item that does not belong. Write each list in your notebook, but *do not write* the item that does not belong.

1. Parts of the chassis include the frame, engine, wheels, body, power train, brakes, steering system.
2. Parts of the suspension include the shock absorbers, coil springs, leaf springs, steering wheel.
3. Parts in the shock absorber include the piston, cylinder tube, brake shoe, valve.
4. Parts in the brake system include the master brake cylinder, wheel cylinder, engine support, brake shoes, brake drums.
5. Parts of the car include the power plant, frame, steering system, transmission, power train, car body.

Completing the Sentences The sentences that follow are incomplete. After each sentence there are several words or phrases, only one of which will correctly complete the sentence. Write each sentence in your notebook, selecting the proper word or phrase to complete it correctly.

1. The title of this book is *Automotive:* (*a*) *Mechanics,* (*b*) *Chassis,* (*c*) *Engines,* (*d*) *Electricity.*
2. The power plant, or engine, is the source of: (*a*) electricity, (*b*) power, (*c*) gravity, (*d*) effort.
3. The engine and body are attached to the: (*a*) brake system, (*b*) frame, (*c*) steering, (*d*) power train.
4. When the engine, wheels, power train, brakes, and steering system are installed on the frame, the assembly is called the: (*a*) automobile, (*b*) chassis, (*c*) car, (*d*) framework.

⊘ 2-8 Power Train The *power train* carries the power that the engine produces to the car wheels. It consists of the clutch (on cars with a manual transmission), transmission, drive shaft (propeller shaft), differential, and wheel axles (Figs. 2-1 and 2-8).

⊘ 2-9 Clutch The clutch shown in Fig. 2-9 is the type used with manual transmissions (not automatic). Its purpose is to permit the driver to couple or uncouple the engine and the transmission. When the clutch is in the coupling (or normal running) position, power flows through it from the engine to the transmission. If the transmission is in gear (see ⊘ 2-10), then power flows on through to the car wheels so that the car moves. Essentially, then, the clutch permits the driver to uncouple the engine temporarily so that the gears can be shifted from one forward-gear position to another (or into reverse or neutral). It is necessary to interrupt the flow of power (by uncoupling) before gears are shifted. Otherwise, gear shifting would be extremely difficult if not impossible.

The clutch (Fig. 2-9) contains a friction disk (or driven plate) about 1 foot [0.305 m (meter)] in diameter. It also contains a spring arrangement and a pressure plate for pressing this disk tightly against the smooth face of the flywheel. The friction disk is splined to the clutch shaft. The splines consist of two sets of teeth, an internal set on the hub of the friction disk and a matching external set on the clutch shaft. These splines permit the friction disk to slide back and forth along the shaft but force the disk and the shaft to rotate together. External splines can be seen on the shaft in Fig. 2-9.

The flywheel, which is attached to the end of the engine crankshaft, rotates when the engine is running. When the clutch is engaged or applied (that is, in the coupling position), the friction disk is held tightly against the flywheel (by the clutch springs and pressure plate) so that it must rotate with the flywheel (see Fig. 2-10). This rotary motion is carried through the friction disk and clutch shaft to the transmission.

To disengage or release (uncouple) the clutch, the clutch pedal is pushed down by the foot. This

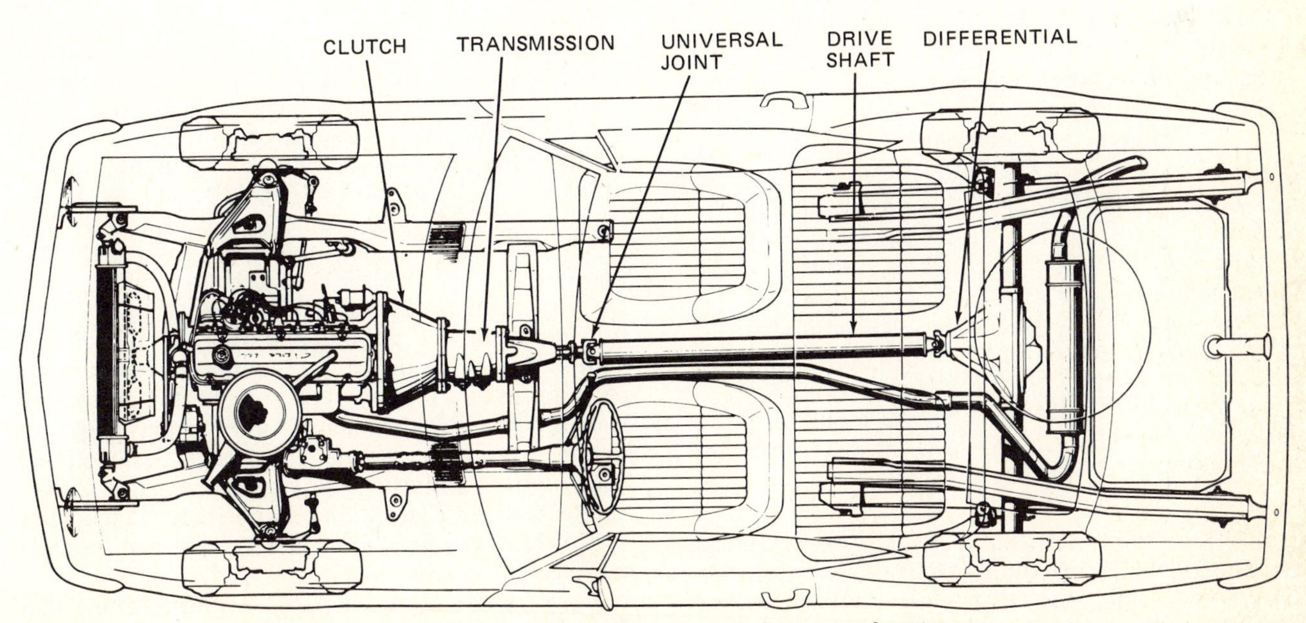

Fig. 2-8. Power train of an automobile, showing location of major components.

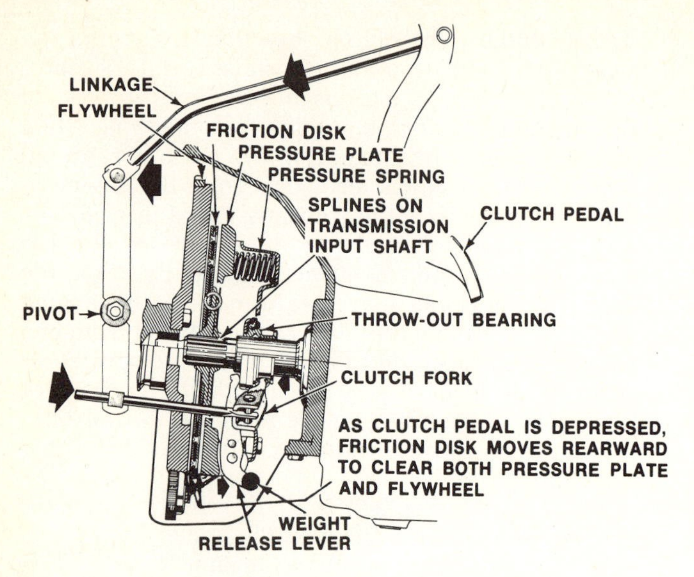

Fig. 2-9. Sectional view of a clutch, with the linkage to the clutch pedal. (*Buick Motor Division of General Motors Corporation*)

causes the clutch fork to pivot so that the clutch throw-out-bearing is forced inward. As the throw-out bearing moves inward, it operates release levers. The release levers take up the spring pressure and lift the pressure plate away from the friction disk. The friction disk is no longer pressed against the flywheel face, and the engine can run independently of the power train. Releasing the clutch pedal permits the clutch fork to release the throw-out bearing so that the springs once again cause the pressure plate to force the friction disk against the flywheel face. The two again revolve together.

⊘ 2-10 Transmission The transmission provides a means of altering the *gear ratio* between the engine and car wheels. In the usual manual three-speed transmission (not automatic) used in passenger cars, the three possible forward-gear ratios cause the engine crankshaft to turn about four, eight, or twelve times for each car-wheel revolution. There is also a reverse gear in the transmission. In addition, there is a neutral position in which no power flows through the transmission.

The different gear ratios are necessary since the engine does not develop much power at low engine speeds. It must be turning at a fairly high speed for it to deliver enough power to start the car moving. Thus, on first starting and after clutch disengagement, the transmission is placed in *low gear*. Now, after the clutch is engaged, the engine crankshaft turns 12 times for each wheel revolution. Engine and car speed are now increased until the car is moving 5 to 10 miles per hour [8.047 to 16.093 km/h (kilometers per hour)]. At this point, the engine may be turning 2,000 rpm (revolutions per minute). The clutch is then disengaged and the engine speed reduced so that gears may be shifted in the transmission. Gears are then shifted to second and the clutch reengaged. The ratio is now about 8:1 (eight to one), or eight crankshaft revolutions to one wheel revolution. When engine speed is again increased, the car speed will be increased. For example, 2,000 engine rpm would give a car speed of possibly 20 miles per hour [32.187 km/h]. Next the gears are shifted into high, the clutch being disengaged and reengaged for this operation. This gives a ratio of about 4:1.

In automatic transmissions, the various gear ratios between the engine and the car wheels are achieved automatically; that is, the driver does not shift gears. Automatic controls in the automatic transmission select or supply the proper gear ratio to suit driving conditions. Such transmissions make use of a fluid coupling or a torque converter, as well as mechanical, hydraulic, and possibly electric controls. These transmissions are discussed in detail in another book in the McGraw-Hill Automotive Technology Series (*Automotive Transmissions and Power Trains*).

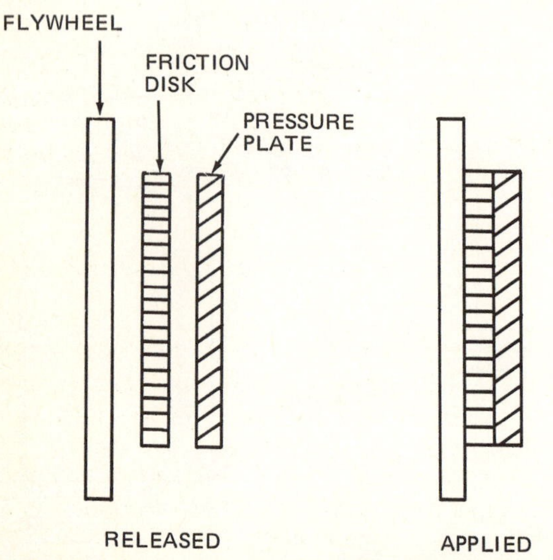

Fig. 2-10. Simplified drawing of clutch parts, showing the released and applied positions of the friction disk and pressure plate.

⊘ **2-11 Gears** Before we discuss transmissions, let us look at gear construction. When two meshing gears have the same number of teeth (Fig. 2-11), both gears will turn at the same speed. But when one gear has more teeth than the other (Fig. 2-12), then the smaller gear will turn faster than the larger gear. For instance, when a 24-tooth gear is meshed with a 12-tooth gear, the smaller gear will turn twice as fast as the larger one. The gear ratio between the two gears is 24:12, or 2:1. When a 12-tooth gear is meshed with a 36-tooth gear, the smaller gear will turn three times every time the larger gear turns once. The gear ratio is 36:12, or 3:1. Thus, if the larger gear is turning at 1,500 rpm, the smaller gear will be turning at 4,500 rpm.

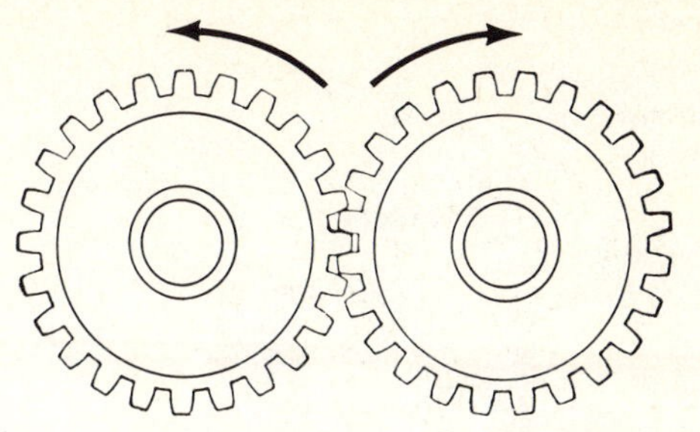

Fig. 2-11. Two meshing gears with the same number of teeth.

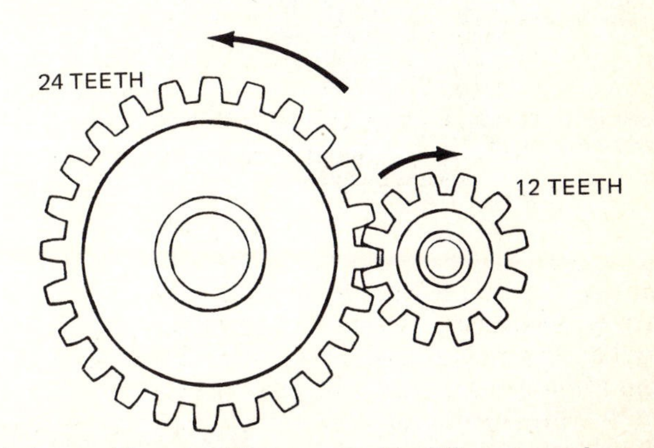

Fig. 2-12. Two meshing gears with different numbers of teeth. The smaller gear with fewer teeth turns faster than the larger gear.

⊘ **2-12 Operation of Transmission** Essentially, the typical standard passenger-car transmission consists of three shafts and eight gears of various sizes. A simplified version of a standard transmission is shown in Fig. 2-13. Four of the gears are rigidly attached to the countershaft. One of the countershaft gears is permanently meshed with the clutch gear on the end of the clutch shaft. When the engine is running and the clutch is engaged, the countershaft is driven by the clutch gear. It turns in a direction opposite, or counter, to the rotation of the clutch gear (that is why it is called a *countershaft*). With the gears in neutral (Fig. 2-13) and the car stationary, the transmission main shaft does not turn. The transmission shaft is connected, through the drive shaft, to the final drive (see Figs. 2-1 and 2-8). The two gears on the transmission main shaft may be shifted back and forth along the splines on the shaft by operation of the gearshift lever in the driving compartment. The splines permit endwise (axial) movement of the gears but cause the gears and shaft to rotate together. (Note that a floorboard shift lever is shown in Figs. 2-13 to 2-17. This type of lever is shown because it illustrates more clearly the lever action in shifting gears. The transmission action is the same, regardless of whether a floorboard or a steering-column type of shift lever is used.)

1. *LOW GEAR* When the shift lever is operated to place the gears in low (Fig. 2-14), the lower end of the shift lever enters a slot in the low-and-reverse shifter yoke. The shifter yoke is moved forward by the lever movement. This moves the large gear on the transmission main shaft (the low-and-reverse gear) along the shaft until it meshes with the small gear on the countershaft. The clutch is disengaged for this operation, and so the clutch shaft and countershaft stop rotating. When the clutch is again engaged, the transmission main shaft is driven through the countershaft low gear and the low-and-reverse gear. Because the various gears are of different sizes, the countershaft turns more slowly than the clutch shaft. The transmission main shaft turns more slowly than the countershaft. This gives a gear reduction of about 3:1; that is, the clutch shaft turns about three times to turn the transmission main shaft

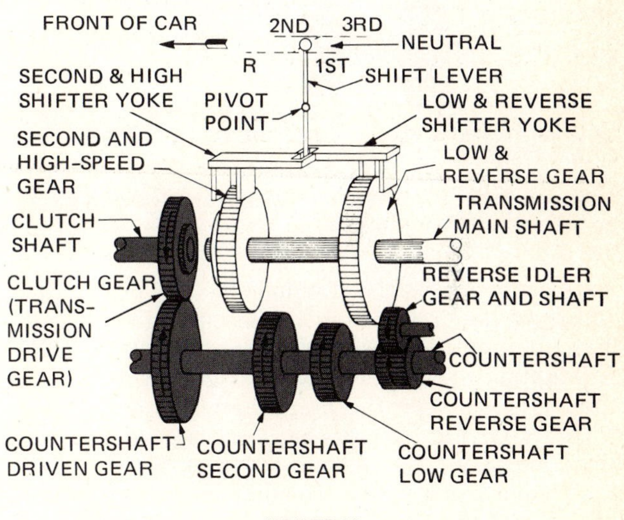

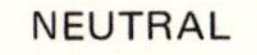

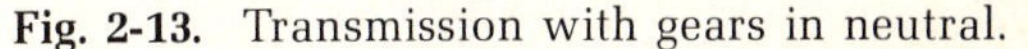

Fig. 2-13. Transmission with gears in neutral.

once. Further gear reduction in the final drive (or differential) at the rear axles produces an overall gear reduction of about 12:1 between the clutch shaft (or engine crankshaft) and the wheels.

2. *SECOND GEAR* When the clutch is operated and the gearshift lever is moved to second (Fig. 2-15), the low-and-reverse gear on the transmission main shaft is de-meshed from the countershaft low gear. The second-and-high-speed gear is moved into mesh with the countershaft second gear. Since these two

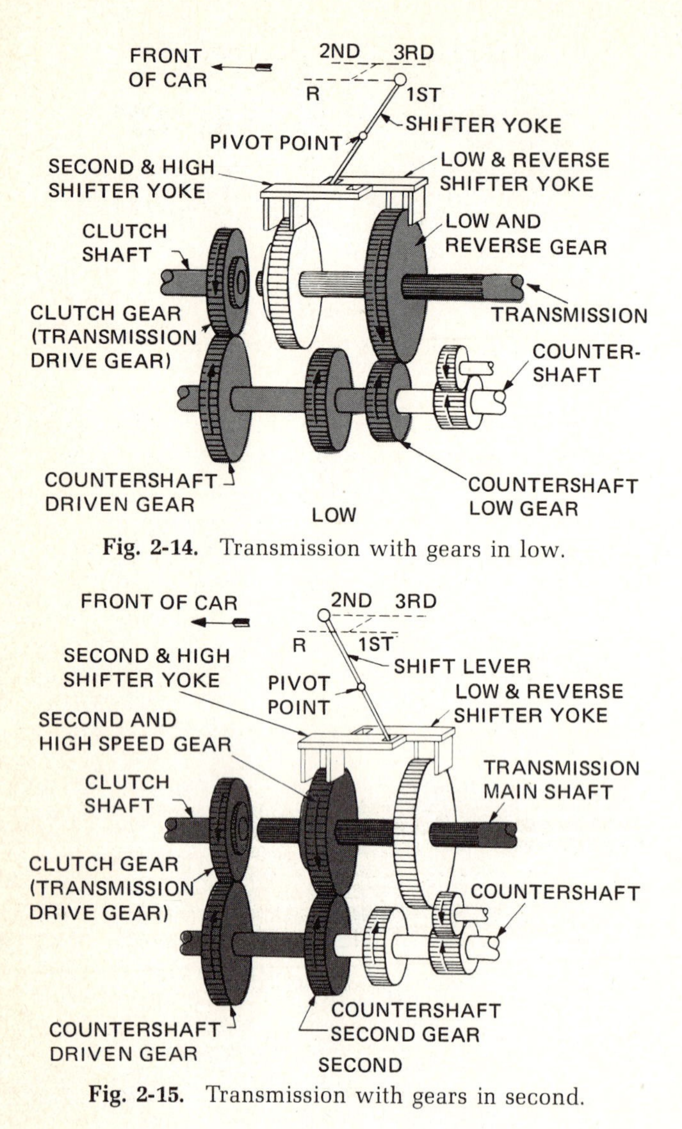

Fig. 2-14. Transmission with gears in low.

Fig. 2-15. Transmission with gears in second.

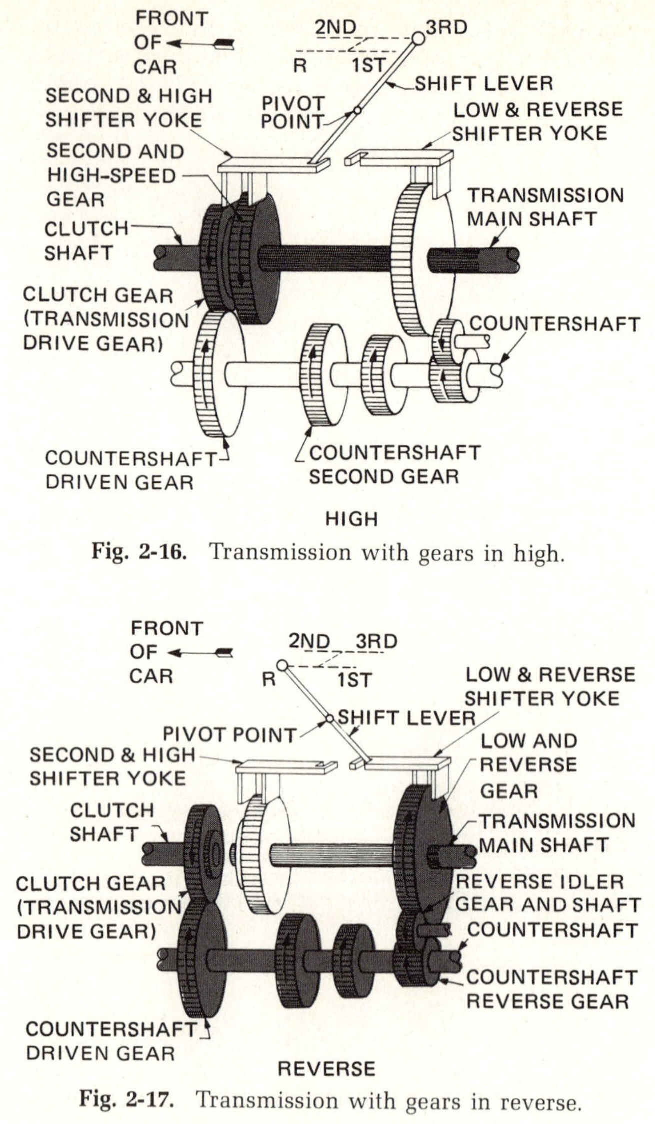

Fig. 2-16. Transmission with gears in high.

Fig. 2-17. Transmission with gears in reverse.

gears are more nearly of equal size, the gear reduction from the clutch shaft to the transmission main shaft is about 2:1. With the gear reduction in the differential, this gives an overall gear reduction of about 8:1 between the crankshaft and the wheels.

3. *HIGH GEAR* When the gears are shifted into high (Fig. 2-16), the second-and-high-speed gear is de-meshed from the countershaft second gear. The second-and-high-speed gear is then moved on along the transmission main shaft until teeth on the side of this gear mesh with teeth on the side of the clutch gear. These two gears now turn together so that the transmission main shaft turns at the same speed as the clutch shaft. There is now a 1:1 gear ratio between the two shafts. The differential gear reduction gives a gear ratio of about 4:1 between the crankshaft and the wheels.

4. *REVERSE GEAR* When the gears are placed in reverse (Fig. 2-17), the low-and-reverse gear is moved into mesh with the reverse idler gear. The reverse idler gear is always in mesh with the countershaft reverse gear. Interposing the idler gear between the countershaft reverse gear and the low-and-reverse gear causes the low-and-reverse gear to be rotated in the reverse direction. Now, when the clutch is engaged, the car will move backward.

⊘ 2-13 Other Transmissions Although the previous discussion outlines the basic principles of all transmissions, somewhat more complex transmissions are used in modern cars, trucks, and buses. These transmissions may use helical gears instead of the plain spur gears shown in the illustrations. Also synchromesh devices are used to simplify gear shifting. These devices synchronize the gears that are about to be meshed so that the meshing teeth move at the same speed and therefore mesh without clashing.

Many cars now have automatic transmissions which use torque converters and gearing systems quite different from the one described previously. Automatic transmissions, along with conventional transmissions, are discussed in another book in the

McGraw-Hill Automotive Technology Series (*Automotive Transmissions and Power Trains*).

⊘ 2-14 Drive Shaft The drive shaft, or propeller shaft (Figs. 2-1 and 2-8), carries the power from the transmission to the differential at the rear-wheel axles. The drive shaft is more than a simple line shaft; it is connected at one end to the rigidly mounted transmission and at the other end to the differential, which moves up and down with wheel-spring movement. Two separate actions are produced by this movement. First, the distance between the transmission and the differential changes as the differential moves up and down. As the springs compress and the differential moves up, the distance increases. As the springs expand and the differential moves down, the distance decreases. Second, the driving angle changes as the differential moves up and down.

1. *SLIP JOINT* Since the drive shaft tends to shorten and lengthen with the up-and-down differential movement, a device must be included to permit this action. The device used is called a *slip joint*. It is located either at the front or at the rear of the drive shaft. A typical slip joint is shown in Fig. 2-18. It consists of the externally splined slip yoke and the internally splined front shaft. When the slip joint is assembled, the yoke enters the shaft. The two must turn together because of the splines. However, the yoke can slip in the shaft to shorten or lengthen the effective length of the drive shaft.

2. *UNIVERSAL JOINTS* To take care of the differences in the angle of drive as the differential moves up and down, the drive shaft has one or more universal joints. A universal joint is shown in assembled view in Fig. 2-18. A simple universal joint is shown in Fig. 2-19. It is essentially a double-hinged joint through which the driving shaft can transmit power to the driven shaft, even though the two shafts are somewhat out of line with each other.

Each of the two shafts has a Y-shaped yoke on the end. Between the yokes is a center member shaped like a cross. The four arms of this cross member are assembled in bearings in the ends of the shaft yokes. The bearings permit the cross-member arms to turn in the yokes. The driving shaft causes the cross member to rotate with it. At the same time the cross member causes the driven member to rotate.

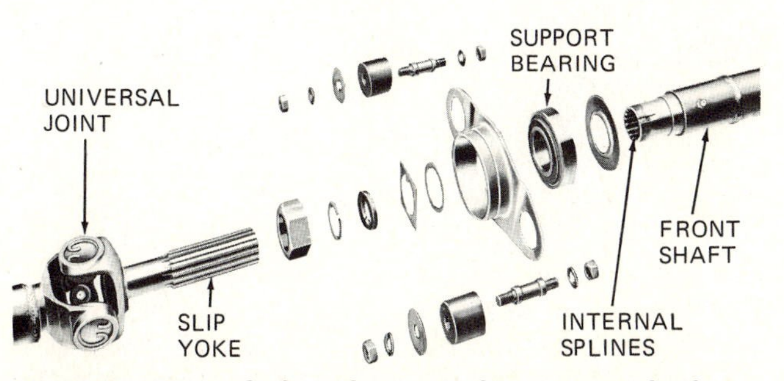

Fig. 2-18. Drive shaft and support bearing, partly disassembled so that the slip joint can be seen. External splines are on the universal-joint yoke, and internal splines are in the shaft.

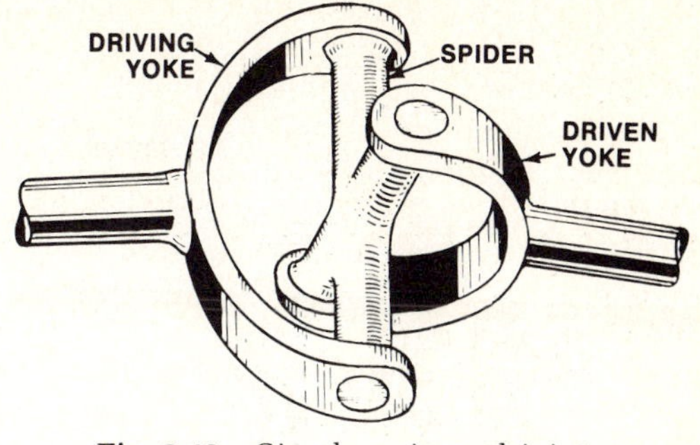

Fig. 2-19. Simple universal joint.

⊘ 2-15 Differential If the car were driven in a straight line without turning, no differential would be necessary. However, when the car rounds a turn, the outer wheel must travel farther than the inner wheel. For example, when a right-angle turn is made with the inner wheel turning on a 20-foot [6.096 m] radius, this wheel will travel about 31 feet [9.449 m] (Fig. 2-20). The outer wheel, being nearly 5 feet [1.524 m] from the inner wheel, turns on a radius of about $24\frac{2}{3}$ feet [7.518 m]. It therefore travels nearly 39 feet [11.887 m]. This is 8 feet [2.438 m] more than the distance the inner wheel travels.

If the drive line were geared rigidly to both wheels so that both wheels turned together, then each wheel would have to skid an average of 4 feet [1.219 m] to make the turn just described. On this basis tires would not last long. Furthermore, the skidding would make the car very hard to handle on turns; the car would probably go completely out of control. The differential prevents these difficulties by allowing the wheels to rotate different amounts when turns are made.

To study the construction and operation of the differential, let us build, in effect, a simple differ-

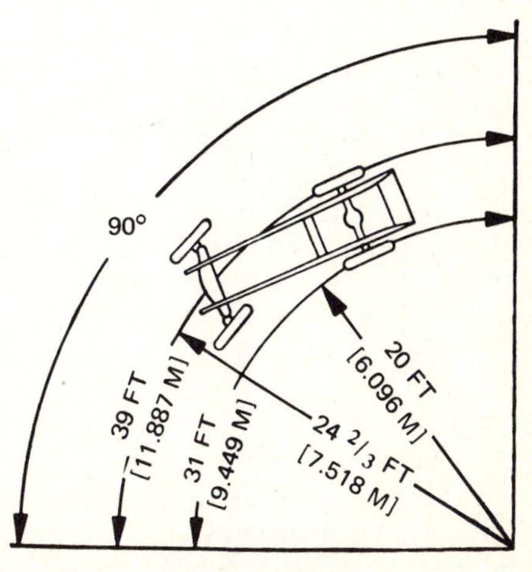

Fig. 2-20. Difference of rear-wheel travel as a car makes a 90-degree turn with the inner rear wheel on a 20-foot [6.096 m] radius.

ential. Figure 2-21 shows the basic parts of the differential. In Fig. 2-22, we put the parts together to build the complete differential. The two rear wheels are mounted on separate axles. Each axle has a bevel gear, called a *differential side gear*, that drives it (Fig. 2-22*a*). There is a differential case assembled to one axle which encases the two differential side gears (Fig. 2-22*a*). The differential case has a bearing that permits it to turn independently on the axle. Inside the case there is a shaft that supports two bevel gears, called *differential pinion gears* (Fig. 2-22*b*). These gears mesh with the two axle bevel gears. Thus when the differential case is rotated, both axle gears rotate and both wheels turn. The differential case is rotated through a ring gear attached to it (Fig. 2-22*c*), which is driven by the drive pinion on the end of the drive line (Fig. 2-22*d*). The drive line turns the drive pinion; the drive pinion turns the ring gear and the differential case.

If one of the wheel axles, with its differential side gear, were held stationary as the differential case rotated, then the differential pinion gear would also rotate as it "ran around" on the stationary bevel gear. It would be forced to do this because it is mounted on a shaft that is carried around when the differential case rotates. As the differential pinion gear rotates, it carries rotary motion to the other axle bevel gear, causing the axle bevel gear and the wheel to rotate.

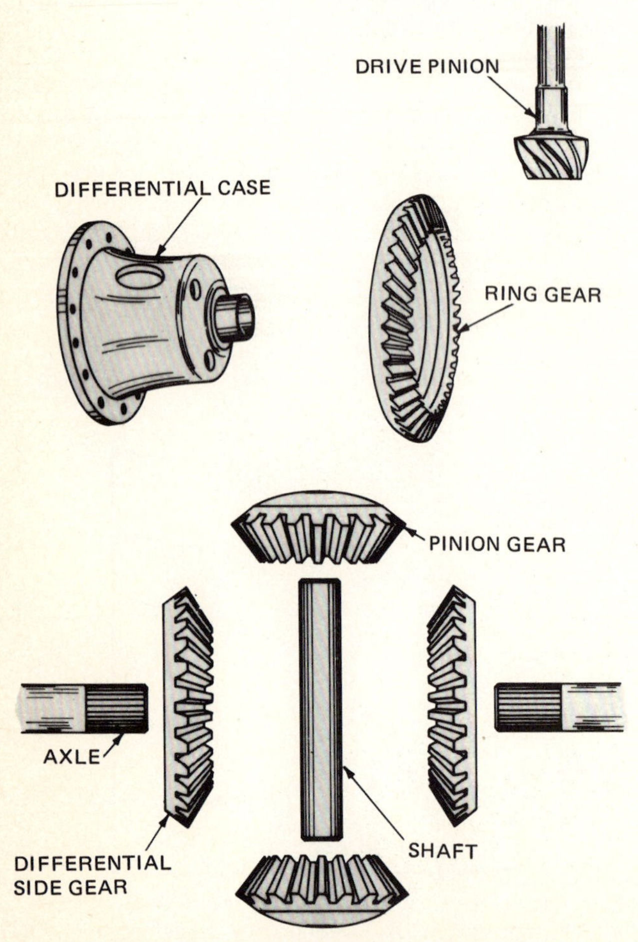

Fig. 2-21. Basic parts of a differential.

In actual operation, when the car is moving on a straight road, the ring gear, differential case, differential pinion gears, and the two differential side gears rotate without any relative motion. They turn as a unit. However, when the car goes around a curve, the differential pinion gear rotates on its shaft. This action permits the outer wheel to turn more rapidly than the inner wheel.

The actual differential is more complicated than the simplified unit shown in Fig. 2-22. An actual differential is shown in cutaway view in Fig. 2-23. Driving power enters through the drive pinion on the end of the drive shaft. The drive pinion is meshed with the ring gear. The differential pinion-gear shaft is mounted in the differential case, and two differential pinion gears are mounted on this shaft. There is a differential side gear on each axle shaft. The other parts shown in the figure support these parts. Shims and adjustment nuts permit bearing adjustment.

Check Your Progress

Progress Quiz 2-2 When a mechanic in an automotive shop has a battery on charge, it must be checked periodically to see how it is taking the charge. Likewise, in this book, you must stop periodically to check your progress and see how you are "taking the charge" of information. The following questions will help you find out how well you remember what you have just read. You may have some difficulty answering the questions. But don't be discouraged. Just reread the past few pages and then try the questions again. Most good students reread their lessons several times. Rereading the pages and rechecking the questions will help you learn how to pick out the important facts to remember. This is good practice. Before you finish the book you will find your ability to remember essential information greatly improved. This ability will be of immense help to you in your work.

Correcting Parts List The purpose of this exercise is to enable you to spot the unrelated part in a list. For example, in the list "shoes, pants, shirt, milk, tie, coat," you would see that "milk" does not belong because it is the only thing named that is not something you wear.

In each of the following lists, you will find one item that does not belong. Write each list in your notebook, but *do not write* the item that does not belong.

1. Parts in the power train include the clutch, steering wheel, transmission, drive shaft, differential.
2. Parts in the clutch include the friction disk, flywheel, pistons, clutch pedal, throw-out bearing.
3. Parts in the transmission include the clutch gear, tires, countershaft, reverse idler gear, main shaft.

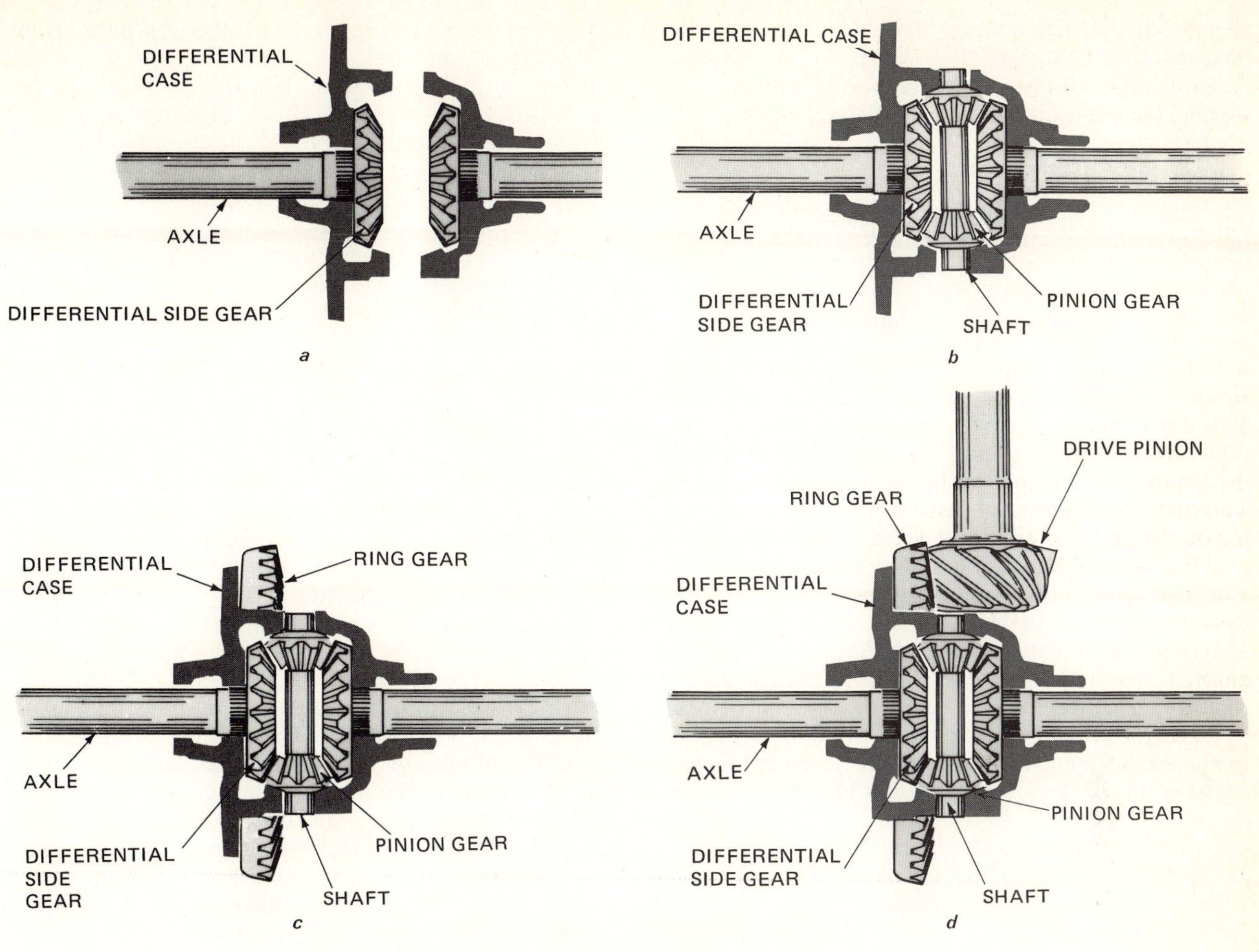

Fig. 2-22. Step-by-step assembly of the differential.

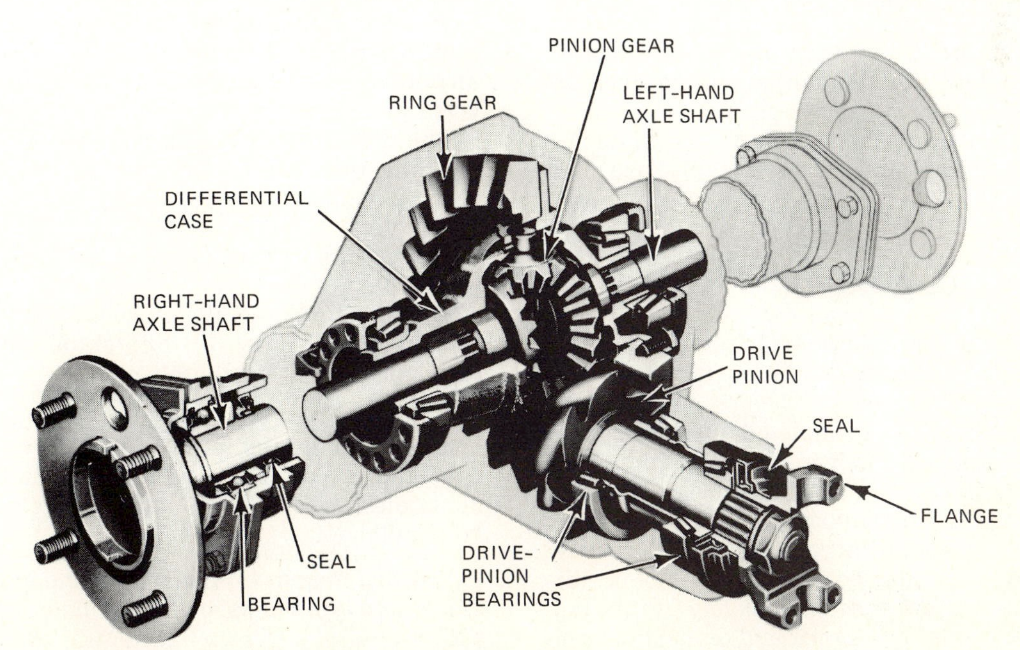

Fig. 2-23. Cutaway view of a differential and rear axle. (*Ford Motor Company*)

4. Parts in the drive shaft include the pitman arm, universal joint, slip joint.
5. Parts in the differential include the drive pinion, shaft, ring gear, clutch, differential side gear.

Completing the Sentences The sentences that follow are incomplete. After each sentence there are several words or phrases, only one of which will correctly complete the sentence. Write each sentence in your notebook, selecting the proper word or phrase to complete it correctly.

1. The power train contains the clutch, transmission, drive shaft, and: (*a*) frame, (*b*) chassis, (*c*) differential, (*d*) body.
2. There is a double-faced friction disk splined to a shaft in the: (*a*) transmission, (*b*) differential, (*c*) drive shaft, (*d*) clutch.
3. Two meshed gears have a gear ratio of 4:1. This means that while the large gear is turning five times, the small gear will turn: (*a*) $\frac{4}{5}$ times, (*b*) $\frac{5}{4}$ times, (*c*) 10 times, (*d*) 20 times.
4. In the transmission, the countershaft drive is meshed with the: (*a*) reverse idler gear, (*b*) clutch gear, (*c*) second-and-low gear, (*d*) high-speed gear.
5. In low gear, the countershaft low gear is meshed with the: (*a*) second-and-high-speed gear, (*b*) low-and-reverse gear, (*c*) second-and-low gear, (*d*) reverse idler gear.
6. In high gear, the transmission main shaft turns at the same speed as the: (*a*) countershaft, (*b*) reverse-idler-gear shaft, (*c*) clutch shaft.
7. To take care of the difference in driving angle as the differential moves up and down, the drive shaft has one or more: (*a*) slip joints, (*b*) shaft joints, (*c*) universal joints.
8. To take care of the lengthening and shortening of the drive shaft with differential movement, the drive shaft has a: (*a*) slip joint, (*b*) shaft joint, (*c*) universal joint.
9. In the differential, the drive pinion meshes with the: (*a*) ring gear, (*b*) differential pinion gear, (*c*) differential side gear, (*d*) drive gear.
10. The differential pinion gear meshes with the: (*a*) ring gear, (*b*) differential side gears, (*c*) drive gear, (*d*) pinion gear.

CHAPTER 2 CHECKUP

NOTE: Since the following is a chapter review test, you should review the chapter before taking the test.

Now that you have completed another chapter in the book, you will want to test your knowledge of the subjects covered in the chapter. The questions that follow have two purposes. One is to test your knowledge. The second purpose is to help you review the chapter. It may be that you will not be able to answer all the questions immediately. If this happens, turn back into the chapter and reread the pages that will give you the answer. For instance, under "Listing Parts" you are asked to list the parts in the differential, besides bearings, that are in motion when the car is moving. If you cannot remember them all, turn back to the illustration of a differential in the chapter and refer to it when writing your list. The act of writing down the names of the parts will help you remember them.

NOTE: Write your answers in your notebook. Then later, when you finish the book, you will find your notebook filled with valuable information to which you can quickly refer.

Completing the Sentences The sentences that follow are incomplete. After each sentence there are several words or phrases, only one of which will correctly complete the sentence. Write each sentence in your notebook, selecting the proper word or phrase to complete it correctly.

1. The major part that must be added to make the chassis a complete automobile is the: (*a*) engine, (*b*) frame, (*c*) body, (*d*) wheels, (*e*) brakes.
2. The purpose of the shock absorbers is to: (*a*) damp spring oscillations, (*b*) improve the rigidity of spring mountings, (*c*) strengthen the frame, (*d*) permit the use of coil springs instead of leaf springs.
3. The pitman arm in the steering gear is linked to the front wheels through the: (*a*) worm, (*b*) steering wheel, (*c*) tie rods, (*d*) steering shaft.
4. The brake shoes are curved to conform to the inner diameter of the: (*a*) wheel, (*b*) tire, (*c*) brake drum, (*d*) master cylinder, (*e*) pedal.
5. The power train transmits power from the engine to the: (*a*) crankshaft, (*b*) rear wheels, (*c*) front wheels, (*d*) steering gear, (*e*) power plant.
6. The clutch part that is between the pressure plate and the engine flywheel is called the: (*a*) throw-out bearing, (*b*) clutch fork, (*c*) friction disk, (*d*) pedal.
7. The gear that is always in mesh with the countershaft drive gear in the transmission is called the: (*a*) second-and-high-speed gear, (*b*) clutch gear, (*c*) reverse idler gear, (*d*) low-and-reverse gear.
8. The drive shaft has one or more: (*a*) idler gears, (*b*) universal joints, (*c*) synchromeshes, (*d*) ring gears.
9. The total number of pinions and gears in the standard differential is: (*a*) two, (*b*) three, (*c*) six, (*d*) nine.
10. In the differential, the ring gear is attached to the: (*a*) differential case, (*b*) differential side gears, (*c*) rear axles, (*d*) drive shaft.

Listing Parts In the following, you are asked to list parts that make up the various automotive components discussed in the chapter. Write these lists in your notebook.

1. List five major components of the chassis.
2. List two types of springs used in automobile chassis.
3. List five parts in a steering system.

4. List five parts used in a hydraulic brake system.
5. List the four major components of the power train.
6. List five parts in a clutch.
7. List six gears used in a typical transmission.
8. List two types of joints used in a drive shaft.
9. List the parts in the differential, aside from bearings, that are in motion when the car is moving.
10. List several body and body-accessory parts.

Purpose and Operation of Components In the following, you are asked to write the purpose and operation of certain components of the automobile discussed in this chapter. If you have any difficulty in writing your explanations, turn back into the chapter and reread the pages that will give you the answer. Then write your explanation. Don't copy; try to tell it in your own words. This is a good way to fix the explanation firmly in your mind. Write in your notebook.

1. What is the purpose of the car frame?
2. What is the purpose of the car springs?
3. What is the purpose of shock absorbers?
4. Briefly, how does a shock absorber work?
5. How does the steering system work?
6. How does the hydraulic brake system work?
7. What happens in the clutch when the clutch pedal is depressed?
8. What happens in the transmission when the gears are shifted into second speed?
9. What is the purpose of the universal joints in the drive shaft?
10. What happens in the differential when the car turns a corner?

SUGGESTIONS FOR FURTHER STUDY

If you would like to do some further studying of the engine chassis, brakes, differential, transmission, clutch, and so forth, there are several things you can do. First, you can read *Automotive Chassis and Body* and *Automotive Transmissions and Power Trains,* which are other books in the McGraw-Hill Automotive Technology Series. Also, you can inspect your own and your friends' cars. We do not suggest that you get out your tool kit and start tearing them down, however. You are not quite ready for that yet. You can go into your school automotive shop or to a friendly service shop where repair work on all these parts is done. By watching what goes on in a service shop you will learn a great deal about the various automobile components, how they are put together, and how the parts look.

You may be able to borrow shop repair manuals from car-dealer service shops or from your school automotive-shop library. These manuals are sometimes available from the car manufacturers for a price. Your school automotive shop may have cutaway parts on exhibit which are used as teaching aids. Studying these displays, if your school shop supplies them, will help you understand how the various parts are constructed and how they work.

chapter 3

MEASUREMENTS: U.S. CUSTOMARY AND METRIC

In any engine work, you will be taking measurements of one kind or another. Engine specifications are given in terms of measurements—length, diameter, speed, power, torque, and so on. In this chapter, we introduce you to the measurements that are made with a ruler, feeler gauge, or micrometer. We talk about the system of measurements currently used in the United States, the U.S. Customary System, and the system used in other countries, the metric system. All imported cars are made using metric measurements, and the United States is gradually adopting the metric system. So, if you want to work on foreign cars and expect to stay in the automotive servicing business, you will need to know about the metric system.

⊘ 3-1 Linear Measurements Linear measurements are measurements that are taken in a straight line. If you are using the familiar U.S. Customary System, you will be taking linear measurements in feet (ft), inches (in), and fractions of an inch. But if you are using the metric system, you will be taking linear measurements in meters [m], centimeters [cm], or millimeters [mm]. We will explain these units of measure in following pages.

⊘ 3-2 The Metric System As we mentioned previously, all imported automobiles, as well as other imported machines, are built according to specifications based on the *metric system* of measurement. That is, measurements on these cars are made in centimeters and millimeters. When you work on these cars, you must use tools that are marked in metric system units. So let us find out about the metric system.

First, let's look at the system of measurements currently used in the United States. This system is called the *U.S. Customary System* (USCS). Measurements in this system include those listed in the following table.

Length	Liquids	Weight
12 inches = 1 foot	16 ounces = 1 pint	16 ounces = 1 pound
3 feet = 1 yard	2 pints = 1 quart	2,000 pounds = 1 ton
5,280 feet = 1 mile	4 quarts = 1 gallon	

These figures may be a little confusing. USCS measurements seem to have no rhyme or reason.

The metric system, however, is scientific. It is based on the decimal system, that is, on multiples of 10. For example, 1 meter (which is 39.37 inches) is 10 decimeters, or 100 centimeters, or 1,000 millimeters.

In the metric system, volume is measured in liters. One liter equals 1.057 quarts. The conversion table in Fig. 3-1 gives you the metric measurements that you will need to work on imported cars, along with their USCS equivalents. With this table, you can convert metric measurements into USCS measurements or vice versa.

In most service manuals for imported cars, the manufacturers give specifications in both the metric and USCS measurements. Thus the stroke of a certain engine would be given as 73.7 mm [2.902 inches]. Specified torque for tightening main-bearing cap bolts would be given as 4.5 to 5.5 kg-m (kilogram-meters) [32.5 to 39.5 pound-feet].

NOTE: If you are going to work on imported cars, you will need a set of metric wrenches because these cars have some nuts and bolts that domestic wrenches will not fit.

In this book, we will usually give both the USCS and metric measurements whenever specifications are presented. As a rule, the USCS unit will be given first, followed by its metric equivalent in brackets. For example: 1 inch [25.4 mm].

⊘ 3-3 Ruler The ruler, also called the *scale,* is used to measure on flat surfaces (Fig. 3-2). The ruler is marked off in inches and fractions of an inch (or,

LENGTH

1 in (inch)	= 25.4 mm (millimeters)	or 0.0254 m (meter)
1 cm	= 0.39 in	or 0.03281 ft (foot)
1 mm (millimeter)	= 0.039 in	or 0.003281 ft
1 ft	= 304.8 mm	or 0.3048 m
1 mi (mile)	= 1.609 km (kilometers)	or 1,609 m
1 km	= 0.62 mi	or 3,281 ft

VOLUME/CAPACITY

1 in^3 (cubic inch)	= 16.39 cm^3 (cubic centimeters)	or 0.01 l (liter)
1 cm^3	= 0.061 in^3	or 0.001 l
1 l	= 61.02 in^3	or 1.057 qt (quarts)
1 gal (gallon)	= 4 qt	or 3.780 l

WEIGHT

1 kg (kilogram)	= 2.2 lb (pounds)	= 35.2 oz (ounces)
1 lb	= 0.454 kg	

Here are the metric measurements, taken from the complete metric system table, that you will work with most often.

LENGTH

1 km	= 1,000 m	= 100,000 cm
1 m	= 100 cm	= 1,000 mm

VOLUME/CAPACITY

1 kl (kiloliter)	= 1,000 l	= 100,000 cl (centiliters)
1 l	= 1.000 cm^3	= 1,000 ml (milliliters)

WEIGHT

1 kg	= 1,000 g (grams)	= 100,000 cg (centigrams)

Fig. 3-1. Conversion table: U.S. Customary System versus metric system.

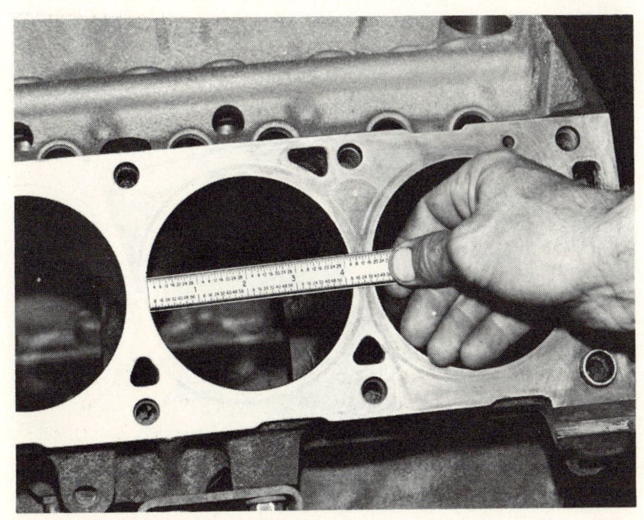

Fig. 3-2. The bore of a cylinder is its diameter, or the distance across the cylinder. A ruler, or scale, can be used to measure the bore.

in the metric system, in centimeters and millimeters). Some scales have metric units on one side and USCS units on the other side (Fig. 3-3). Although the ruler can be used for some automotive service jobs,

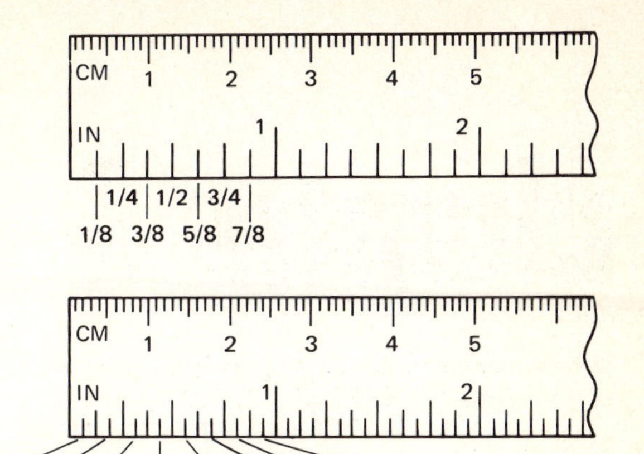

Fig. 3-3. Ruler, or steel scale, marked in both inches (U.S. Customary System) and centimeters (metric system).

other jobs require measuring tools that are more accurate.

⊘ 3-4 Feeler Gauge Feeler gauges are strips or blades of hardened and tempered steel or other metal. They are supplied in sets which contain blades of various thickness (Fig. 3-4). Each blade is marked to indicate its thickness in thousandths of an inch (or in millimeters). The blade marked "4," for example, is 0.004 inch thick. These blades are very thin. For example, it would take 250 blades marked "4" stacked one on top of another to measure a full inch (250 × 0.004 inch = 1.000 inch [25.4 mm]).

Some feeler-gauge sets have blades marked in both inches and millimeters. These are very handy since they can be used on both domestic and imported cars.

Feeler gauges are used to measure small clearances, say, between a piston and an engine cylinder, or between a brake shoe and a brake drum.

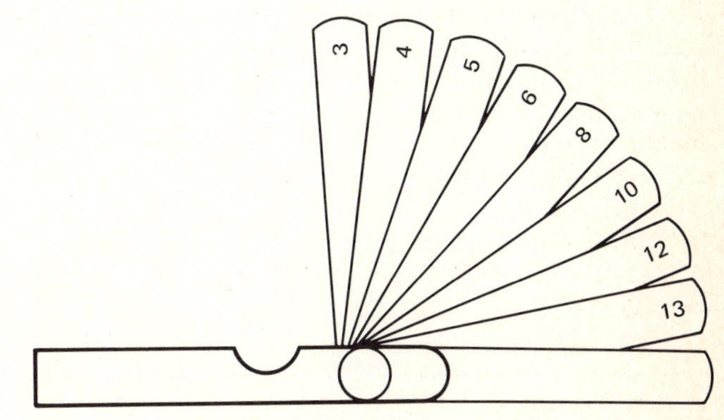

Fig. 3-4. Set of feeler gauges.

⊘ 3-5 Stepped Feeler Gauges Stepped feeler gauges are blades that have two thicknesses; the tip of the blade is thinner than the rest of the blade. In Fig. 3-5, the blade marked "10-12" has a tip that

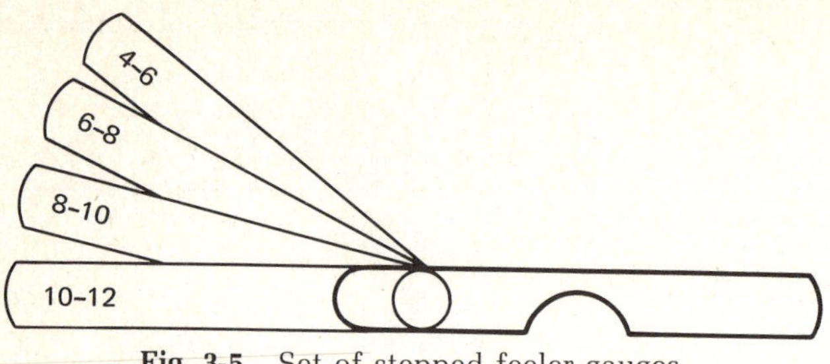

Fig. 3-5. Set of stepped feeler gauges.

is 0.010 inch [0.254 mm] thick while the rest of the blade is 0.012 inch [0.305 mm] thick. This type of gauge, often called a "go no-go" gauge, is handy for some jobs, such as adjusting valve-tappet clearances. For example, suppose the specification tells you the proper valve-tappet clearance is between 0.006 and 0.008 inch [0.152 and 0.203 mm]. You should make the adjustment with a 6–8 feeler gauge. When the adjustment is right, the 0.006-inch [0.152 mm] tip fits the clearance, but the 0.008-inch [0.203 mm] part of the blade does not fit.

⊘ 3-6 Wire Feeler Gauges Wire feeler gauges (Fig. 3-6) are round steel wires of various diameters. They are used to check small gaps, openings, and so on.

⊘ 3-7 Micrometers There are two types of micrometers: outside and inside micrometers. The *outside micrometer* (Fig. 3-7), commonly called the "mike," is a tool that can measure the thickness and diameter of solid objects in thousandths of an inch. It is a precision instrument and must be treated with care! Study Fig. 3-7 carefully and identify the thimble, hub, spindle, frame, and anvil. In the cutaway view of the mike (Fig. 3-7) note that there are screw threads and a screw nut inside the thimble. By turning the thimble, you can move the spindle toward or away from the anvil.

Here's how to use the outside micrometer, or mike. Suppose you want to measure the diameter of a rod. Hold the rod between the end of the spindle and the anvil, as shown in Fig. 3-8. Turn the thimble gently to move the end of the spindle toward the rod. The rod should be touched lightly on one side by the anvil and on the other side by the end of the spindle.

CAUTION: Never turn the spindle tight against anything. If you do, you can ruin the mike. Turn the thimble only until the spindle touches.

The mike in Fig. 3-8 has a small knob on the end of the thimble called a *ratchet stop*. When the spindle approaches the object being measured, turn the ratchet stop instead of the thimble. Then, when the spindle comes into contact with the object, the ratchet continues to turn without putting any pressure on the object. When you use the ratchet stop, you cannot damage the mike by turning the thimble too tightly.

NOTE: The outside micrometer shown in Fig. 3-7 is a 1-inch [25.4 mm] mike. It measures anything up to 1 inch in thickness or diameter. Other mikes are

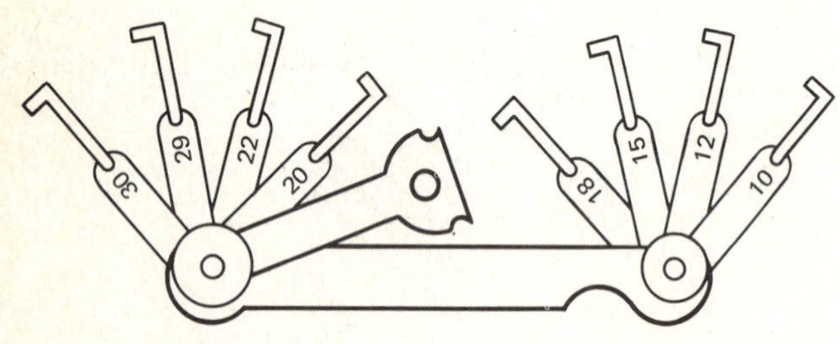

Fig. 3-6. Set of wire feeler gauges.

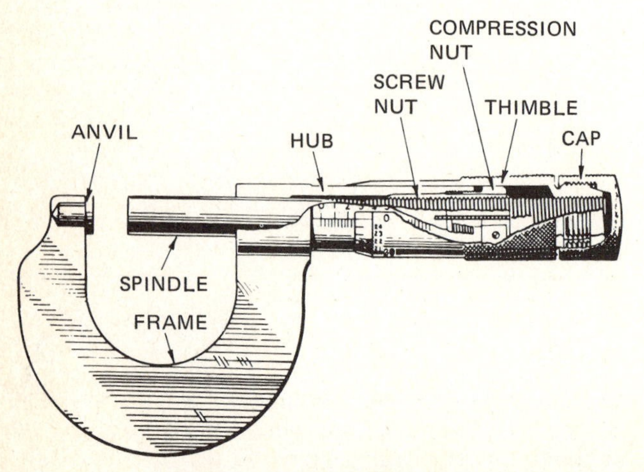

Fig. 3-7. Outside micrometer.

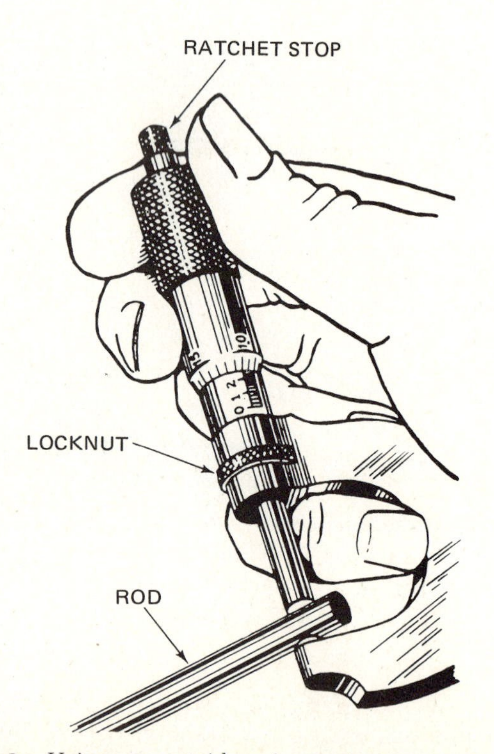

Fig. 3-8. Using an outside micrometer to measure the diameter of a rod.

larger. For instance, the micrometer shown in Fig. 3-15 is a 4-inch [101.6 mm] mike. It measures objects with 3- to 4-inch [76.2 to 101.6 mm] diameters.

⊘ 3-8 Reading the Mike After you have adjusted the mike, you have to read the mike to find out the diameter of the rod. To do this, you take two readings and add them together. The markings on the mike are in thousandths of an inch, that is, in decimals.

First look at the markings on the hub (see Fig. 3-9). Note that the hub is marked with numbers that run from 0 to 9. Each of these numbers indicates 1/10 inch, or 0.1 inch [2.54 mm]. The 2 that is exposed in Fig. 3-9 indicates 2/10 inch, or 0.2 inch [5.08 mm]. Note also that there are markings between the numbers. Each marking indicates 25/1,000 inch, or 0.025 inch [0.635 mm].

Note that there are marks on the end of the thimble. These marks indicate thousandths of an inch. There are 25 of these marks, one for each 0.001 inch. Every time the thimble is turned one revolution, it moves exactly 0.025 inch [0.635 mm]. When the thimble is turned four times, it moves 0.1 inch [2.54 mm].

Look at Fig. 3-9 again. The 2 and one of the markings next to the 2 are exposed on the hub. This means that the distance between the spindle and the anvil is 0.2 inch [5.08 mm] plus 0.025 inch [0.635 mm]. However, something must be added to this number to get the actual distance. Note that the thimble has been turned back almost one complete revolution from the 0.225 mark on the hub. In fact, it has been turned back 0.024 inch [0.61 mm]. So 0.024 inch [0.61 mm] must be added to 0.225 inch [5.715 mm] to get the actual setting, which is 0.249 inch [6.32 mm]. Other mike settings are shown in Fig. 3-10.

REMEMBER: Add the reading on the hub to the reading on the thimble to get the actual measurement.

⊘ 3-9 Decimal Equivalents The table of decimal equivalents (Fig. 3-11) will help you change the decimals you read on the mike into fractions. For example, if the reading were 0.25 inch, you would look at the table and see that this is 1/4 inch [6.35 mm]. A reading of 0.563 inch would be 9/16 inch [14.287 mm]. Many mikes carry a table of decimal equivalents either on the frame or on the thimble. The table can also be used to change fractions into decimals.

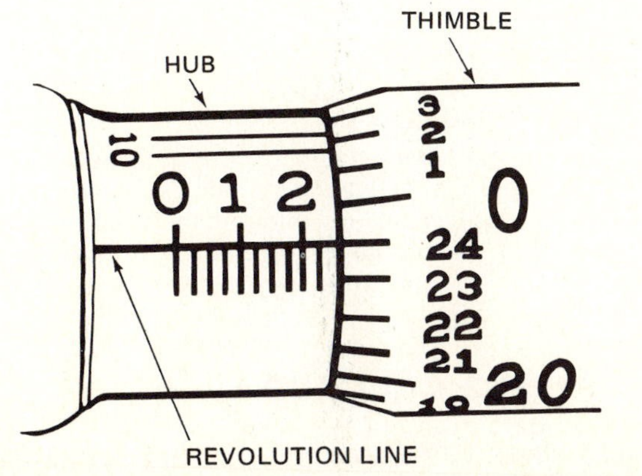

Fig. 3-9. Hub and thimble markings on a micrometer.

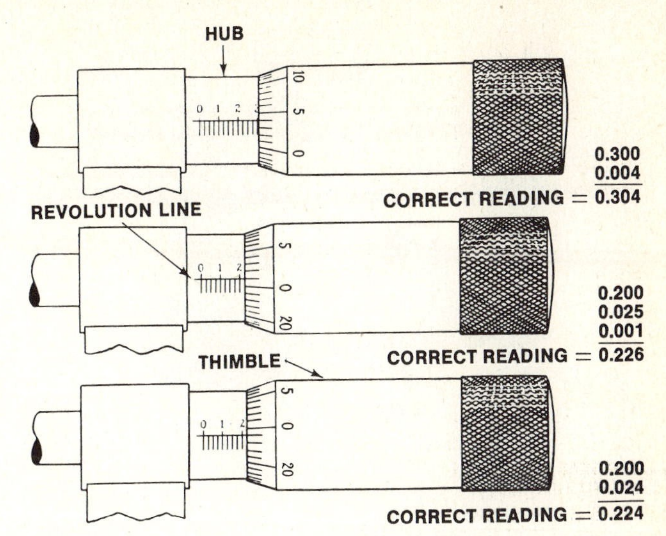

Fig. 3-10. Examples of reading micrometer settings.

1/64	.0156	17/64	.2656	33/64	.5156	49/64	.7656
1/32	.0312	9/32	.2812	17/32	.5312	25/32	.7812
3/64	.0468	19/64	.2969	35/64	.5469	51/64	.7969
1/16	.0625	5/16	.3125	9/16	.5625	13/16	.8125
5/64	.0781	21/64	.3281	37/64	.5781	53/64	.8281
3/32	.0937	11/32	.3437	19/32	.5937	27/32	.8437
7/64	.1094	23/64	.3594	39/64	.6094	55/64	.8594
1/8	.125	3/8	.375	5/8	.625	7/8	.875
9/64	.1406	25/64	.3906	41/64	.6406	57/64	.8906
5/32	.1562	13/32	.4062	21/32	.6562	29/32	.9062
11/64	.1719	27/64	.4219	43/64	.6719	59/64	.9219
3/16	.1875	7/16	.4375	11/16	.6875	15/16	.9375
13/64	.2031	29/64	.4531	45/64	.7031	61/64	.9531
7/32	.2187	15/32	.4687	23/32	.7187	31/32	.9687
15/64	.2344	31/64	.4844	47/64	.7344	63/64	.9843
1/4	.25	1/2	.5	3/4	.75	1	1.0

Fig. 3-11. Table of decimal equivalents.

NOTE: Metric micrometers are marked in millimeters and centimeters, and there is no need to change decimals to fractions.

⊘ 3-10 Inside Micrometers Inside micrometers (Fig. 3-12) are used to measure hole diameters, such as the diameter, or bore, of an engine cylinder (see Fig. 3-13). Extension rods, such as those above the inside micrometer in Fig. 3-12, can be attached so that the mike can be used to measure large diameters.

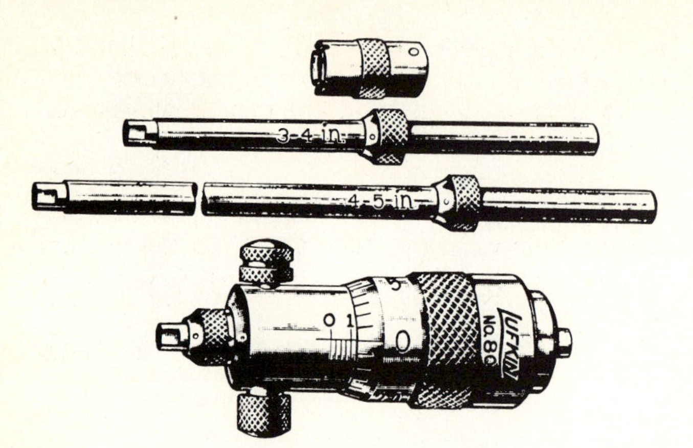

Fig. 3-12. Inside micrometer.

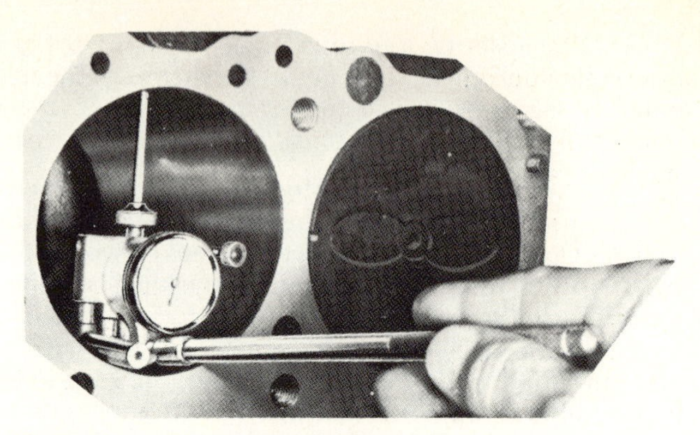

Fig. 3-14. Dial indicator being used to detect wear in an engine cylinder.

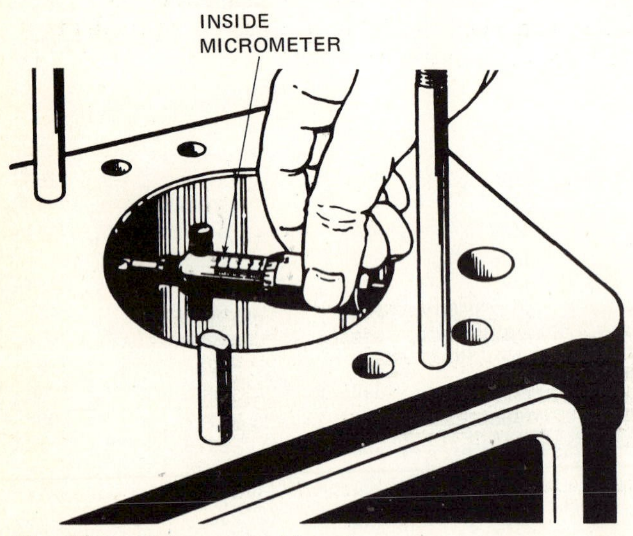

Fig. 3-13. Using an inside micrometer to measure the diameter, or bore, of a cylinder.

Here is how to use the dial indicator to find the actual diameter of an engine cylinder. First insert the indicator in the cylinder and note the position of the needle. Then remove the dial indicator and use a micrometer, as shown in Fig. 3-15. You must adjust the mike until the needle is in the same position as it was in the cylinder. Then read the setting of the mike and get the actual dimension in thousandths of an inch (or in millimeters).

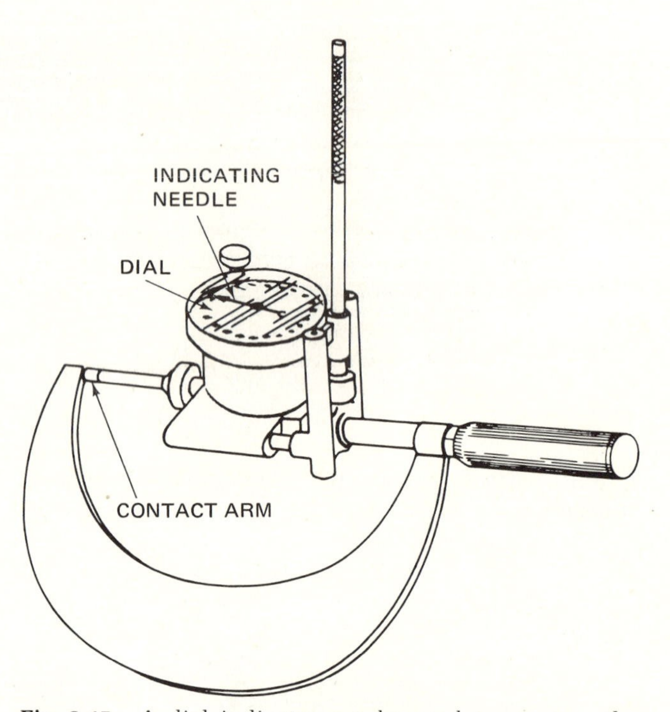

Fig. 3-15. A dial indicator can be used to measure the diameter, or bore, of an engine cylinder. Once the reading is taken, the dial is set to zero and then the reading is measured with a micrometer. (*Pontiac Motor Division of General Motors Corporation*)

⊘ 3-11 Micrometer Adjustments Many of the machines used in the automotive shop have micrometer adjustments. These machines have adjusting knobs or dials similar to the micrometer. The knobs or dials move the cutting or grinding edges of the machines in thousandths of an inch. The adjustments are read in the same way as the mike adjustments.

⊘ 3-12 Dial Indicators The dial indicator is a gauge that uses a dial face and a needle to register measurements (Fig. 3-14). The dial indicator has a movable contact arm. When the arm moves, the needle rotates on the dial face to measure the movement in thousandths of an inch. The dial indicator is used to measure end play in shafts or gears, movement of contact points, cylinder bores in engine blocks, and so on. Figure 3-14 shows a dial indicator used to measure the bore, or diameter, of an engine cylinder. As the dial indicator is moved up and down, any difference in the diameter will cause the needle to move. Differences in the cylinder diameter at various points indicate cylinder wear.

CHAPTER 3 CHECKUP

NOTE: Since the following is a chapter review test, you should review the chapter before taking the test.

Try the tests that follow to test your memory and understanding of what you have been studying about measurements. If any of the questions stump you, turn back into the chapter and reread the pages that will give you the information you need.

Completing the Sentences The sentences that follow are incomplete. After each sentence there are several words or phrases, only one of which will correctly complete the sentence. Write each sentence in your notebook, selecting the proper word or phrase to complete it correctly.

1. To read the mike, you add the exposed marking on the hub to the mark on the: (*a*) spindle, (*b*) end of the thimble aligned with the hub line, (*c*) anvil, (*d*) hub.
2. A reading of 0.125 inch is: (*a*) $\frac{1}{4}$ in, (*b*) $\frac{1}{8}$ in, (*c*) $\frac{1}{2}$ in, (*d*) $\frac{1}{16}$ in.
3. One millimeter is equal to: (*a*) 10 cm, (*b*) 0.039 in, (*c*) 0.39 in.
4. One inch is equal to: (*a*) 25.4 mm, (*b*) 25.4 cm, (*c*) 0.10 ft.

Testing Yourself Now check up on yourself by taking the following test. Turn back into the chapter and look up the information you need if any of the questions stump you.

1. What are the two basic kinds of blade feeler gauges?
2. Which kind of micrometer should you use to measure the diameter of a rod?
3. What are the two kinds of steel scales?
4. Why would you need the table of decimal equivalents when you are using a micrometer?
5. Do you need the table of decimal equivalents if you are using a metric mike?

chapter 4

FUNDAMENTAL SCIENCE

This chapter discusses the fundamental scientific principles that enter into the operation of the automobile engine. Heat and combustion are explained, and the meaning of energy, work, power, torque, and friction is discussed. All these terms are used to describe actions in the engine. As you read the chapter, you will come to understand what they mean and how they relate to engine operation.

⊘ 4-1 What Our Universe Is Made Of You may think it strange to start our discussion of automobile engines by talking about the universe and what it is made of. But by doing this we can very quickly explain some of the puzzling aspects of engine operation. For example, when 1 gallon [3.785 l (liters)] of gasoline is burned in the engine, more than 1 gallon [3.785 l] of water is produced. Why? You'll find the answer in this chapter. It is important to know about this because water in an engine can be very damaging to the engine. You will want to know not only how the water is formed but also how the engine manages to get rid of it. You will find out about this and many other interesting aspects of engine operation in the pages that follow.

⊘ 4-2 Atoms We can look around us and count thousands of different substances and materials, from wood to steel, glass to cloth, gasoline to water. We can see that the world is made of a tremendous variety of things. Yet, the amazing fact is that all these things are made up of about 100 types of basic "building blocks" called *atoms*. And these atoms, in turn, are made up of various quantities of only three basic particles.

In this modern world of the atom bomb, where "splitting the atom" is commonplace in scientific laboratories and nuclear power plants, we have all heard something about the atom. There are more than 100 varieties of atoms. Each variety has a special structure and a special name: iron, copper, hydrogen, sulfur, tin, oxygen, and so on. A piece of iron, for example, is made up of a tremendous number of one particular type of atom. A quantity of the gas oxygen is made up of a great number of another type of atom. Any substance made up entirely of only one type of atom is called an *element*. The Table of Elements that follows lists a number of the more common elements.

TABLE OF ELEMENTS

NAME	SYMBOL	ATOMIC NUMBER	APPROXIMATE ATOMIC WEIGHT	ELECTRON ARRANGEMENT
Aluminum	Al	13	27	.2)8)3
Calcium	Ca	20	40	.2)8)8)2
Carbon	C	6	12	.2)4
Chlorine	Cl	17	35.5	.2)8)7
Copper	Cu	29	63.6	.2)8)18)1
Hydrogen	H	1	1	.1
Iron	Fe	26	56	.2)8)14)2
Magnesium	Mg	12	24	.2)8)2
Mercury	Hg	80	200	.2)8)18)32)18)2
Nitrogen	N	7	14	.2)5
Oxygen	O	8	16	.2)6
Phosphorus	P	15	31	.2)8)5
Potassium	K	19	39	.2)8)8)1
Silver	Ag	47	108	.2)8)18)18)1
Sodium	Na	11	23	.2)8)1
Sulfur	S	16	32	.2)8)6
Zinc	Zn	30	65	.2)8)18)2

The atoms of the approximately 100 different elements can combine in many ways to form hundreds of thousands of different combinations, or *compounds*. This can be compared to the 26 letters of the alphabet, which can be combined in many ways to form the many thousands of words in our language. Thus, salt, water, wood, glass, and the very blood and bones of our bodies are made up of compounds produced by combining different atoms. Common table salt is made up of atoms of the elements sodium and chlorine. Water is made up of the elements hydrogen and oxygen.

⊘ 4-3 Size of Atoms Individual atoms are far too small for us to see, even with the most powerful microscope. There are billions upon billions of atoms in a single drop of water. However, even though no one has ever seen an atom, we have fairly definite knowledge of how atoms are constructed. But before we discuss their construction, let's first talk about the size of atoms. To get a rough idea of how small atoms are, let's talk about the simplest atom of all, the atom of the element hydrogen, a gas. We start with 1 cubic inch [3.387 cm^3 (cubic centimeters)] of hydrogen gas (at 32 degrees Fahrenheit [0°C (degrees Celsius, or centigrade)] and atmospheric pressure). This cube (Fig. 4-1) contains close to 880,000,000,000,000,000,000 (880 billion billion) atoms. Suppose we were able to expand this cube until it was large enough to contain the earth. That means each edge would measure 8,000 miles [12,872 km (kilometers)]. If the atoms were expanded in proportion, then each atom would measure about 10 inches [254 mm] in diameter on this tremendously enlarged scale.

NOTE: You might think that the atoms would be closely packed together since there are so many of them, but this is not so. The distance between atoms is considerably greater than the diameter of the atoms themselves.

⊘ 4-4 The Hydrogen Atom The hydrogen atom is the simplest of all atoms. It is made up of two particles (Fig. 4-2). One of these particles is at the center, or *nucleus,* of the atom; the other particle is whirling around the particle at the center at tremendous speed. The center, or nuclear, particle is called a *proton* (it has a tiny charge of *positive electricity*).

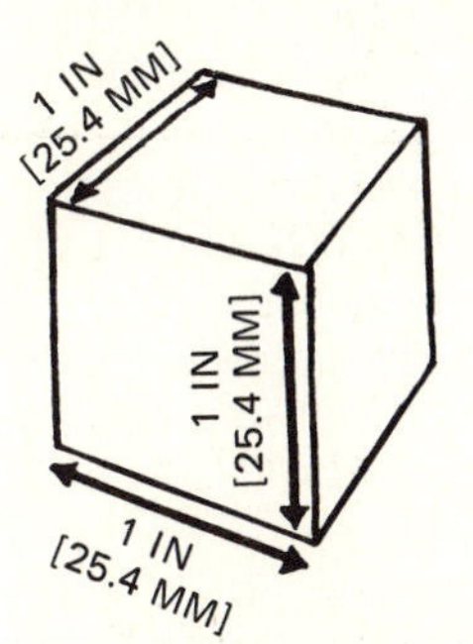

Fig. 4-1. One cubic inch [16.38 cubic centimeters] of hydrogen gas at atmospheric pressure and at 32 degrees Fahrenheit [0°C] contains about 880 billion billion atoms.

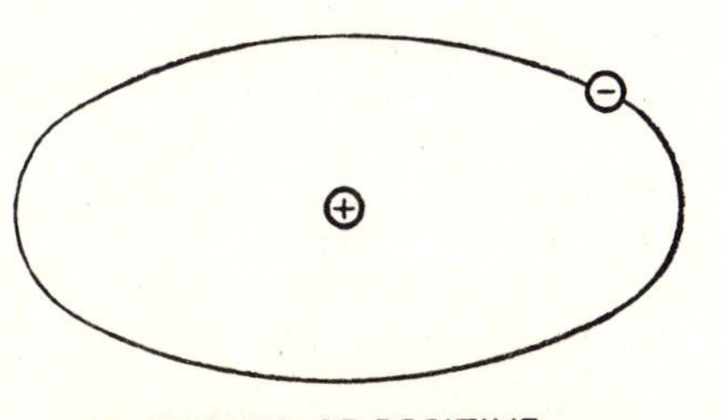

Fig. 4-2. The hydrogen atom consists of two particles: a proton, with a positive electric charge, and an electron, with a negative electric charge.

The particle that is whirling around the proton is called an *electron* (it has a tiny charge of *negative electricity*).

NOTE: Positive charge is indicated by a plus (+) sign. Negative charge is indicated by a minus (–) sign.

The electron is kept whirling in its path, or *orbit,* around the proton by a combination of forces. One force is the attraction that two particles of opposite electric charge have for each other. (Positive attracts negative, and negative attracts positive.) This attraction tends to pull the electron toward the proton. But balancing this is the tendency of the electron to move in a straight line and fly away from the proton. This is somewhat similar to the balancing of forces you have when you swing a ball attached to a rubber band in a circle (Fig. 4-3). As you swing the ball, the rubber band stretches because the ball tends to fly away from your hand. But the rubber band (the attracting force) keeps the ball moving in a circle around your hand.

⊘ 4-5 The Helium Atom The next element above hydrogen, as we go from the simplest to the more complicated atoms, is helium, another gas. The helium atom has 2 protons (+ charges) in its nucleus and 2 electrons (– charges) circling the nucleus (Fig. 4-4). In addition to the 2 protons, the helium nucleus

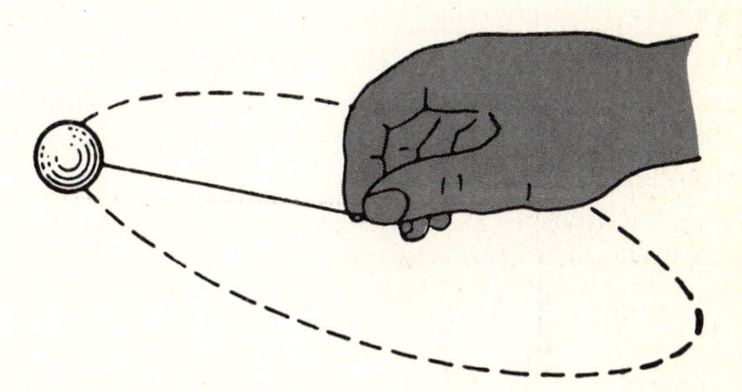

Fig. 4-3. The electron in a hydrogen atom circles the proton. This is like a ball on a rubber band swung in a circle around your hand.

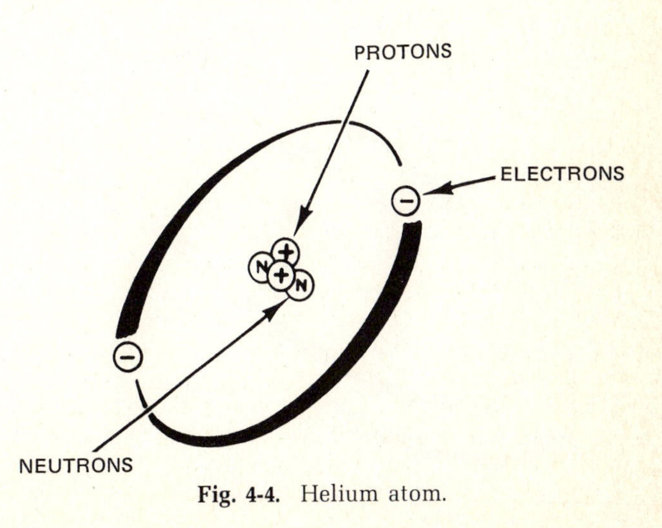

Fig. 4-4. Helium atom.

contains two other particles which are electrically neutral (have no charge) and are therefore called *neutrons*. The 2 neutrons seem to have the ability to hold the 2 positively charged protons together in the nucleus. If it were not for the neutrons, the 2 protons would fly apart. For just as two opposing electric charges (positive and negative) attract each other, so do similar electric charges repel each other. A positive charge repels a positive charge. A negative charge repels a negative charge.

⊘ **4-6 More Complex Atoms** The next element above helium in complexity is lithium, a very light metal. The lithium atom (Fig. 4-5) has a nucleus with 3 protons and 4 neutrons. Three electrons, one for each proton, circle the nucleus.

Then, in order of complexity, we have the following atoms: beryllium, with 4 protons, 5 neutrons, and 4 electrons; boron, with 5 protons, 5 neutrons, and 5 electrons; carbon, with 6, 6, and 6; nitrogen with 7, 7, and 7; oxygen, with 8, 8, and 8; and so on. Note that each atom normally contains the same number of electrons as protons. This makes the atom electrically neutral (since negative charges equal positive charges).

Chemical Reactions

⊘ **4-7 Molecules** Nearly everything around you, as well as your own body, is made up of combinations of different elements. It is rare that we see an element in its pure form. Whenever the atoms of different elements combine, *molecules* are formed. A molecule of common table salt is made up of two atoms, one atom of sodium and one atom of chlorine. A molecule of water is made up of one atom of oxygen and two atoms of hydrogen. These are simple molecules. However, many molecules are very complicated and contain many atoms. For example, the molecule of serum albumin (found in blood plasma) contains nearly 10,000 atoms.

The formation of molecules from atoms or from other molecules is called a *chemical reaction*. During a chemical reaction there is a sharing, or interchange, of electrons between the atoms involved as the atoms combine into molecules. The nuclei of the atoms remain unchanged.

⊘ **4-8 Combustion** Combustion, or fire, is a common chemical reaction in which atoms of the gas oxygen combine with atoms of other elements, such as hydrogen or carbon. One type of combustion process occurs in the automobile engine. In the engine, air and fuel (gasoline vapor) are mixed, compressed, and then ignited, or set on fire (Chap. 5, "Engine Fundamentals," explains this sequence in detail). The air (our atmosphere) contains oxygen; about 20 percent, or one-fifth, of the air is oxygen. Gasoline is made up essentially of hydrogen and carbon (thus, gasoline is called a *hydrocarbon*). Let's see what happens when the air-fuel mixture burns.

An oxygen atom has 8 protons and 8 neutrons in its nucleus and 8 electrons circling the nucleus in two separate paths, or orbits (Fig. 4-6). The inner orbit has 2 electrons, and the outer orbit has 6 electrons. The outer orbit can carry, or hold, 8 electrons; it will, in fact, accept 2 additional electrons if free electrons move close enough to the atom. The hydrogen atom has 1 electron, as already mentioned.

When gasoline burns in the engine, it splits up into hydrogen and carbon. Then these two elements combine with the oxygen in the air. For instance, let us look at what happens when the hydrogen combines with oxygen. During this action, 2 hydrogen atoms lose their electrons to 1 oxygen atom, as shown in Fig. 4-7. These 2 electrons enter the outer electron orbit of the oxygen atom, thereby "filling up," or "satisfying," this orbit. However, this gives the oxygen atom a negative electric charge. Also, the 2 hydrogen atoms that have lost their electrons have a positive electric charge. The result of all this is that the 2 hydrogen atoms are attracted to the 1 oxygen atom. The 3 atoms form a molecule which has the chemical symbol H_2O and the common name *water*.

At the same time, the carbon atoms, released when the gasoline splits into hydrogen and carbon, are also combining with oxygen. A carbon atom has

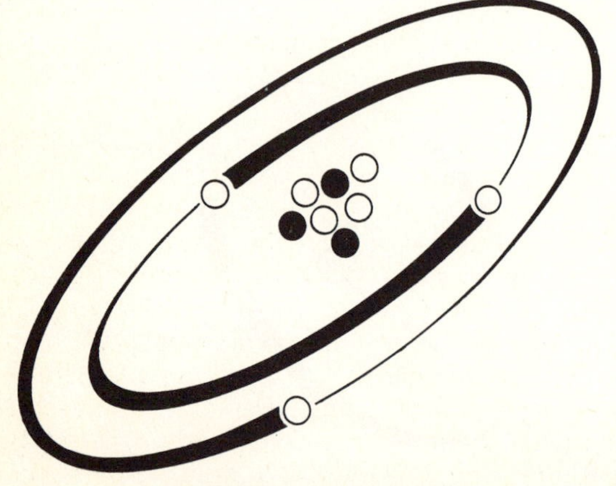

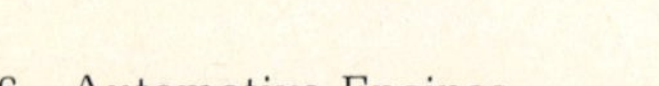

Fig. 4-5. Lithium atom.

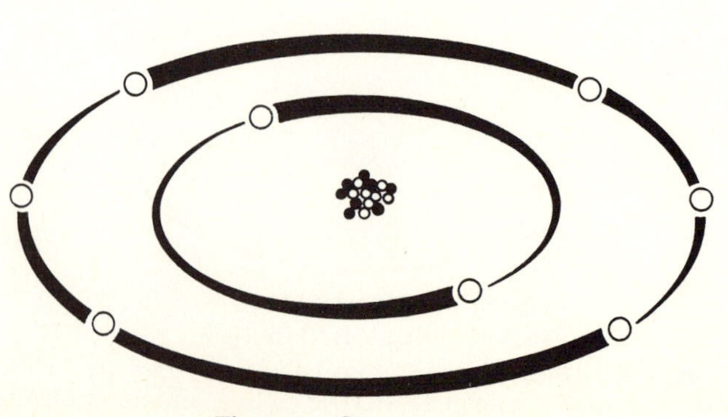

Fig. 4-6. Oxygen atom.

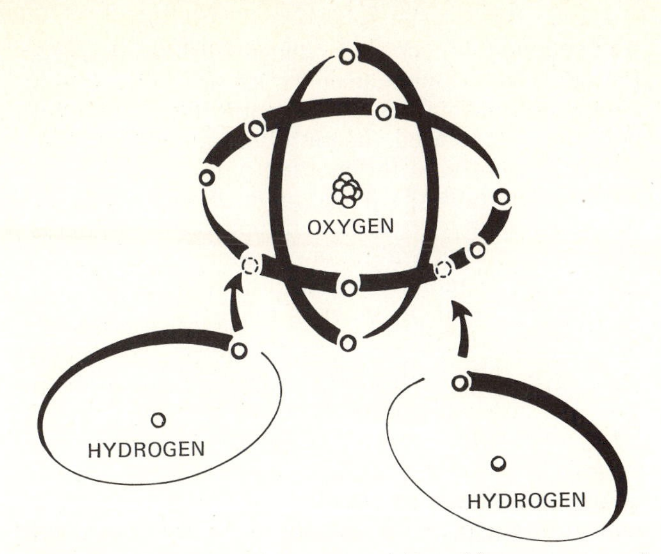

Fig. 4-7. One atom of oxygen uniting with two atoms of hydrogen to form one molecule of water (H_2O).

6 protons and 6 neutrons in its nucleus, with 6 electrons circling the nucleus in two orbits (Fig. 4-8). In the combustion process, the 4 electrons in the outer orbit are "snatched" by 2 oxygen atoms somewhat as shown in Fig. 4-9. This gives the oxygen atoms a negative electric charge. As a result, the carbon atom attaches to 2 oxygen atoms to form a molecule of carbon dioxide, or CO_2, gas.

Now let's see what has happened. In the combustion process, oxygen in the air unites with hydrogen and carbon in the gasoline to form water and carbon dioxide. Since the combustion is accompanied by high temperatures (temperatures may be above 4000 degrees Fahrenheit [2205°C]), the water is in the form of vapor, or steam. It therefore goes out the exhaust with the carbon dioxide. Interestingly enough, more than 1 gallon [3.785 l] of water is formed for every gallon [3.785 l] of gasoline that is burned (that is, it would measure more than 1 gallon [3.785 l] if it were cooled enough to return to liquid form). You see, when the hydrogen part of the gasoline burns (or unites with oxygen), it is actually picking up the oxygen from the air. That is why we obtain more than 1 gallon [3.785 l] of water (H_2O) when we burn 1 gallon [3.785 l] of gasoline.

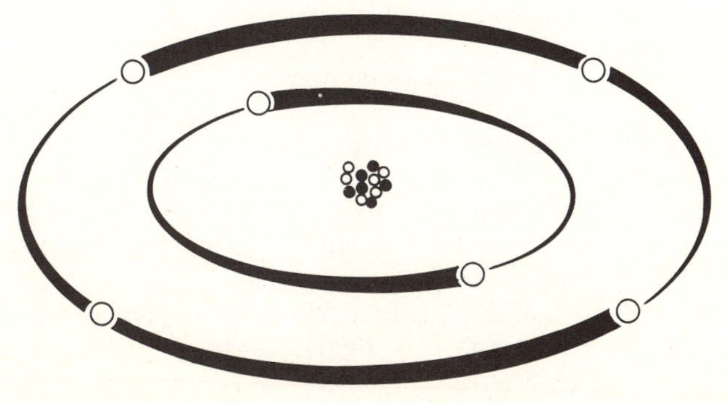

Fig. 4-8. Carbon atom.

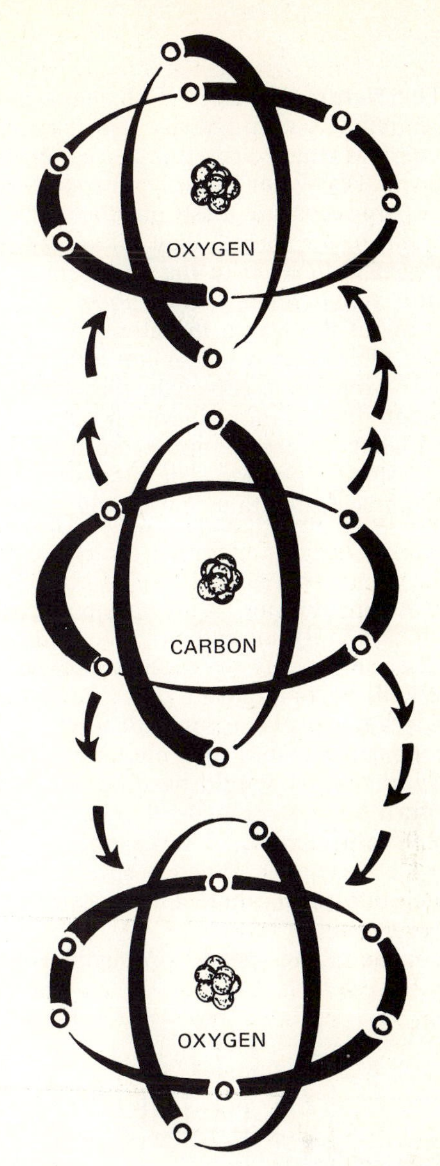

Fig. 4-9. Carbon dioxide (CO_2) molecule.

NOTE: When the engine is cold, some of the water condenses on cold engine parts. This water then works its way down into the crankcase, where it mixes with the engine oil to form sludge. You will find more on this subject in Chap. 17, ⊘ 17-3.

NOTE: With ideal (perfect) combustion, all the hydrogen and carbon in the gasoline would be converted into harmless water (H_2O) and into carbon dioxide (CO_2). However, in the engine, ideal combustion does not take place. Some hydrocarbons are left over; also, some carbon monoxide (CO) is produced instead of carbon dioxide (CO_2). These products of incomplete combustion contribute to atmospheric pollution, and they are the reason that modern cars are equipped with emission controls. These emission controls and how they work are described in later chapters.

Heat

⊘ **4-9 The Nature of Heat** If someone asked you to define "heat," you might say that it is something that keeps a person warm, that raises the temperature, that makes water boil or iron melt. It is true that heat produces all these effects, but scientists look at heat in a different way. They say that heat is simply an indication of *the rapid motion of the atoms and molecules of a substance.*

You might think that the atoms and molecules in a piece of iron or wood or any other solid substance are motionless. However, they are in motion even though they do move in rather restricted paths. But the higher the temperature, the more violently and the faster they move. The atoms in a piece of hot iron move faster than the atoms in a piece of cold iron.

⊘ **4-10 Change of State** It may seem strange to say that heat is simply an indication of the speed of atomic motion. Yet, as we discuss *change of state,* this idea will become clear.

If we were to place a pan of ice cubes over a fire, the ice cubes would soon melt, or turn to water. Presently, the water would become hot and would boil, or turn to vapor (Fig. 4-10). Almost every substance can exist in any of three states: as a *solid, liquid,* or *gas* (*vapor*). When a substance changes from one state to another, it is said to undergo a *change of state.*

A change in the speed of molecular motion, if great enough, will cause a change of state. For example, the water molecules in ice move relatively slowly and in fairly restricted paths. But as the temperature is increased, the molecules move faster and faster. Presently, at the melting point of ice (32 degrees Fahrenheit [0°C]), the molecules move so fast that they begin to break out of their restricted paths. The ice turns to water. As the molecular speed is increased still more (temperature continues to rise), the boiling point of water is reached (212 degrees Fahrenheit [100°C]). At this point, the molecules are moving so rapidly that great numbers of them actually fly out of the water in the form of steam. The water boils.

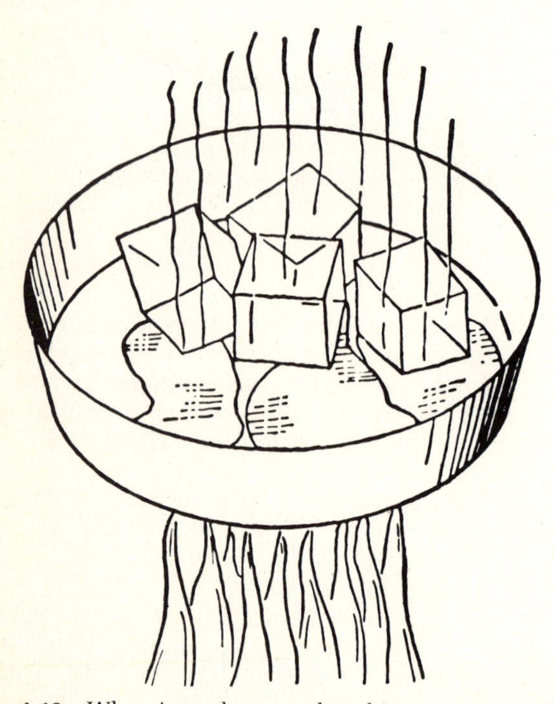

Fig. 4-10. When ice cubes are placed in a pan over a fire, they first melt, or turn into water. Then the water evaporates. Each of these changes, from a solid to a liquid to a vapor, is called a *change of state.*

⊘ **4-11 Producing Change of State** Let us take another look at the pan of ice cubes held over a fire (Fig. 4-10). We noted that combustion, or fire, is a chemical reaction in which atoms of oxygen combine with atoms of other elements, such as hydrogen and carbon. But how can this produce the increase in molecular speed, that is, heat? A simple explanation is that during the chemical reaction, the newly formed molecules (H_2O and CO_2, for example) are set into extremely rapid motion. They bombard the bottom of the pan from the outside. This bombardment knocks the molecules of metal in the pan into more rapid motion. These metal molecules, in turn, begin to bombard the ice molecules that are in contact with the pan on the inside. The water molecules are thus "hammered" into more rapid motion. As this continues, the ice melts. Then, as the bombardment continues, the water molecules are set into such rapid motion that they are knocked clear out of the water; the water boils.

⊘ **4-12 Light and Heat Radiations** The preceding is only a partial description of what takes place in a fire. In addition to the *swiftly* moving molecules that the fire produces, it produces *radiations.*

These radiations we see as light and *feel* as heat. They are produced by interesting actions that take place inside the atoms of fuel and oxygen as they combine to form molecules during combustion. Scientists do not fully understand these actions, but they believe that during combustion electrons jump between the orbits in the atoms. These jumps are accompanied by tiny flashes (or emissions) of radiant energy. These emissions we see as light and feel as heat.

⊘ **4-13 Expansion of Solids Due To Heat** When a piece of iron is heated, it expands. For example, a steel rod that measures exactly 10 feet [3.048 m] in length at 100 degrees Fahrenheit [37.8°C] will measure 10.07 feet [3.069 m] in length at 1000 degrees Fahrenheit [537.8°C] (Fig. 4-11). In other words, the steel rod will expand nearly 1 inch (0.84 inch [2.14 mm] to be more accurate) as it is heated from

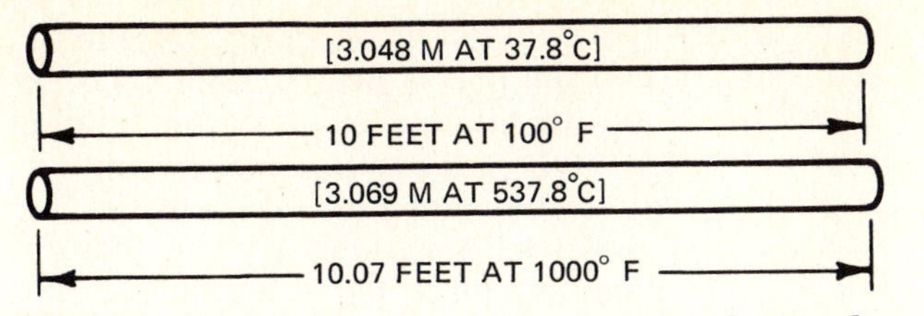

Fig. 4-11. A steel rod that measures 10 feet [3.048 m] at 100 degrees Fahrenheit [37.8°C] will measure 10.07 feet [3.069 m] at 1000 degrees Fahrenheit [537.8°C].

100 to 1000 degrees Fahrenheit [37.8 to 537.8°C]. The reason for this is that as the steel becomes hotter, the molecules move faster and faster. They "push" adjacent molecules away, so that the molecules, in effect, need more room. They spread out, causing expansion. As the temperature is increased still further, the steel will actually melt, or undergo a change of state.

Expansion of solids with heat is important in the engine. As a cold engine becomes hot, the engine parts expand. This expansion must be taken into account when the engine is designed. It is also important in engine servicing.

⊘ 4-14 Expansion of Liquids and Gases Due To Heat Not only solids such as iron, copper, and aluminum, but also liquids and gases expand when heated. If 1 cubic foot [0.028 m^3 (cubic meter)] of water at 39 degrees Fahrenheit [3.89°C] were heated to 100 degrees Fahrenheit [37.8°C], its volume would increase to 1.01 cubic feet [0.029 m^3]. If 1 cubic foot [0.028 m^3] of air at 32 degrees Fahrenheit [0°C] were heated to 100 degrees Fahrenheit [37.8°C], holding pressure constant, its volume would increase to 1.14 cubic feet [0.032 m^3]. These expansions result from more rapid molecular motion in which molecules move each other further apart so that they "spread out" and take up more room.

⊘ 4-15 Increase in Pressure with Temperature A different effect results if the volume is held constant while the cubic foot of air is heated from 32 to 100 degrees Fahrenheit [0 to 37.8°C]. If we start with a pressure of 15 psi (pounds per square inch) [1.055 kg/cm^2 (kilograms per square centimeter)], we would find that the pressure would increase to about 17 psi [1.195 kg/cm^2] at 100 degrees Fahrenheit [37.8°C]. This increase in pressure can be explained by the molecular theory of heat we have been discussing (⊘ 4-9 to 4-11).

Actually, gas or air pressure in a container is due to the unending bombardment of the sides of the container by the fast-moving molecules of gas or air (Fig. 4-12). Of course, a single molecule bumping against the sides of the container would have little effect. However, since there are billions upon billions of molecules bumping the walls, their combined "bumps" add up to a definite "push," or pres-

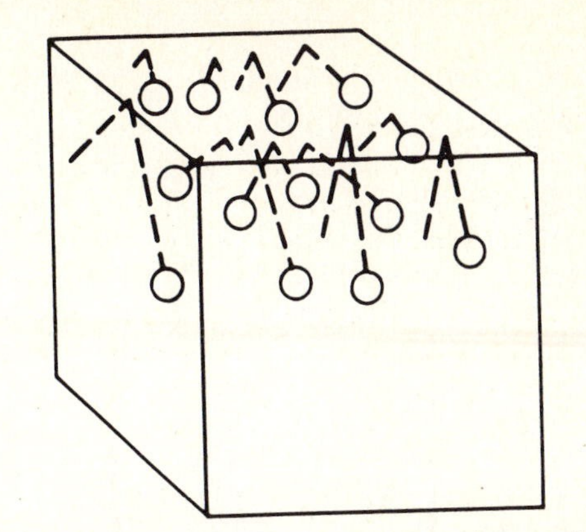

Fig. 4-12. Gas pressure in a container is the result of the ceaseless bombardment of the inner sides of the container by the fast-moving molecules of gas. For simplicity, this bombardment is shown on only one side of the container. Actually, the bombardment takes place against all inside surfaces. The molecules are shown greatly enlarged. Of course, there are almost countless billions of molecules in action—not just the few shown here.

sure. As the temperature is increased, the molecules and atoms of air move faster. They bump the walls of the container harder and more often, thus registering a stronger "push," or a higher pressure.

Another way to increase pressure in a container is to *compress* the gas in the container into a smaller volume. This is what happens in the engine cylinders. The mixture of air and gasoline vapor is squeezed to about a ninth or tenth of its original volume. The molecules are packed much closer together, and so they have much shorter distances to move before hitting the cylinder head or piston. The result—they hit the head or piston much more often, and the pressure is much greater.

But still greater pressure is achieved in the engine cylinder when the compressed air-fuel mixture is ignited. As this happens, the mixture burns (as has already been explained) and the temperature of the burning gas may go as high as 6000 degrees Fahrenheit [3330°C]. This means that the gas molecules are moving at very high speed. They hit the top of the piston so hard and so often that a push, or pressure, of 2 to 3 tons [1,800 to 2,700 kg (kilograms)] is registered on the pistonhead! This pressure, or push, is due entirely to the countless number of fast-moving molecules bombarding the piston.

⊘ 4-16 The Thermometer as an Application of Expansion The ordinary glass-stem thermometer (Fig. 4-13) is a familiar application of the expansion of liquids with increasing temperature. The liquid, usually mercury or a special form of alcohol, is largely contained in the glass bulb at the lower end of the thermometer. As temperature increases, the liquid expands. Since it now needs more room, part of it is forced up into the hollow glass stem. The higher the temperature, the farther the liquid is

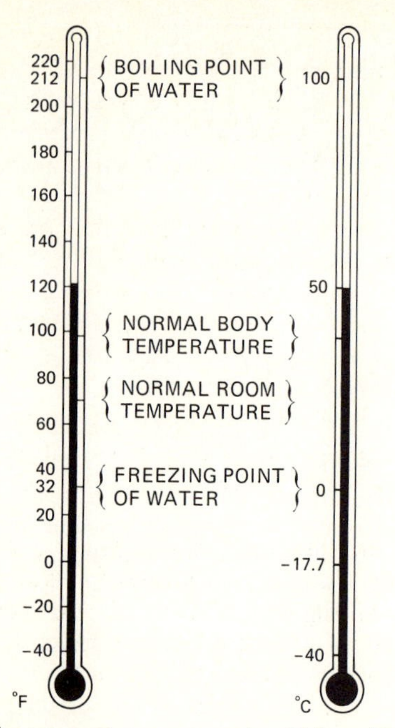

Fig. 4-13. Thermometers comparing Fahrenheit and Celsius (also called centigrade) readings.

forced up into the stem. The stem is marked off to indicate the temperature in degrees.

⊘ 4-17 The Thermostat as an Application of Expansion Different metals expand different amounts as they are heated. For instance, aluminum expands twice as fast as iron as their temperatures are increased. This difference in expansion rates is used to advantage in thermostats. These devices are temperature-sensitive and can be made to do various jobs as temperature changes. For instance, the thermostat in your home heating system turns the furnace on when the temperature goes down and turns it off when the temperature comes up to the shutoff point.

Various thermostats are used in the automobile to open and close electric circuits, control the engine cooling system, control the heating of the ingoing air-fuel mixture, and do various other jobs. One type of thermostat is shown in Fig. 4-14. It consists of a coil made of two strips of different metals, brass and steel, for example, welded or otherwise fastened to each other. When the coil is heated, one metal expands faster than the other. If the faster-expanding metal is on the inside of the strip, the coil will attempt to straighten out and unwind. Later in this book we shall discuss the various types of thermostats and the jobs they do in the automobile.

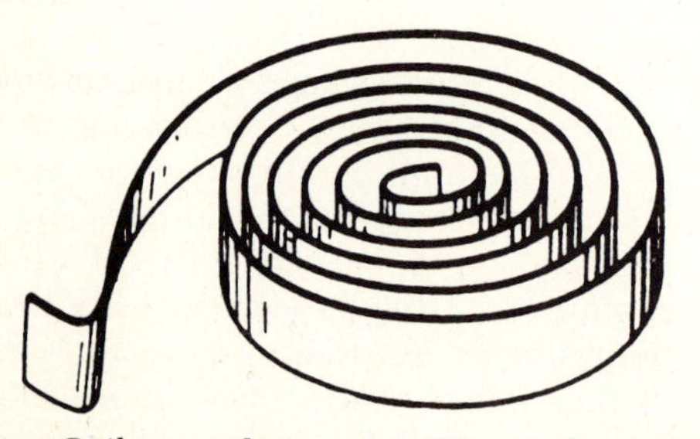

Fig. 4-14. Coil-type thermostat. The coil winds up or unwinds as the temperature goes up or down. This motion can be used to operate a control.

Check Your Progress

Progress Quiz 4-1 The following questions will help you find out how well you are understanding what you are reading. If you have any trouble answering any of the questions, you should reread the past few pages. Most students reread their lessons several times in order to make sure that they understand them. Do not be discouraged if you cannot answer all the questions at once. Just keep in mind that the material you have been reading is not quite so easy to remember as the plot of a story or a movie. So, if you run into trouble as you take the following quiz, just reread the past few pages and try the questions again. As you do this, you will begin to learn how to pick out the important facts you should remember. After you have done this a number of times in the next few sections of the book, you will find that it will become much easier to read and retain the essential facts. This will mean that you are becoming an expert student. And the expert student, the one who can remember the facts, is headed for success. So being an expert student is a big step toward being an expert—and successful—mechanic.

Completing the Sentences The sentences that follow are incomplete. After each sentence there are several words or phrases, only one of which will correctly complete the sentence. Write each sentence in your notebook, selecting the proper word or phrase to complete it correctly.

1. A substance made up entirely of only one type of atom is called: (*a*) a particle, (*b*) a molecule, (*c*) an element.
2. The hydrogen atom is made up of two particles which are called: (*a*) proton and neutron, (*b*) proton and electron, (*c*) proton and nucleus.
3. In the chemical reaction known as combustion, each oxygen atom acquires: (*a*) 1 electron, (*b*) 2 electrons, (*c*) 2 protons.
4. Gasoline is called a hydrocarbon because it is made up essentially of: (*a*) carbon and hydrogen, (*b*) carbon and oxygen, (*c*) hydrogen and oxygen.
5. When gasoline burns, two of the compounds that are formed are: (*a*) oxygen and hydrocarbon, (*b*) water and carbon dioxide, (*c*) water and oxygen.
6. One way of looking at heat is to say that with increasing temperature, the: (*a*) molecules move faster, (*b*) molecules move slower, (*c*) molecules vaporize.
7. Most substances can exist in any of three states:

(a) solid, gas, or vapor; (b) liquid, gas, or vapor; (c) solid, liquid, or gas.

8. As a piece of iron is heated, the more rapid movement of the molecules causes the iron to: (a) expand, (b) shrink, (c) increase in pressure.

9. If you heated a closed container of air, you would find that inside the container the: (a) volume would increase, (b) pressure would increase, (c) pressure would decrease.

10. The thermostat is a device that produces mechanical movement as the: (a) temperature changes, (b) pressure changes.

Physical Principles Related to Engine Operation

⊘ 4-18 What is Meant by "Physical Principles"? By "physical principles" we mean the rules, or "laws," that govern the different conditions that exist and the actions that take place in the world around us. When we release a stone from our hand, it drops to the ground. When a vacuum exists in an engine cylinder, air rushes in to "satisfy" the vacuum. When a car goes around a curve too fast, it skids. These actions occur because of "conditions" that exist in the physical world around us: these conditions are physical principles, or physical laws. Let us discuss those physical principles that help to explain how and why an engine operates.

⊘ 4-19 Gravity Gravity is the attractive force that exists between all objects. When we release a stone from our hand, it falls toward the earth because of the gravitational attraction, or pull, of the earth. The stone also exerts a gravitational attraction on the earth. However, because the stone is so much smaller, the stone moves and the earth does not. If the stone and the earth were the same size (had the same *mass*), then the two would move toward each other and would meet at a halfway point. When a car is driven up a hill, a considerable part of the power produced by the engine is used in overcoming gravity (that is, in raising the car against the gravitational attraction of the earth). Likewise, a car can coast down a hill with the engine turned off because of the gravitational attraction of the earth on the car.

We normally measure gravitational attraction in terms of weight. We put an object on a scale and note that it "weighs" 10 pounds [4.5 kg], for example. What we mean by this is that the object has sufficient mass for the earth to register that much pull on it. It is the gravitational attraction, or pull, of the earth that gives an object its weight.

⊘ 4-20 Atmospheric Pressure We do not usually think of the air as having any weight. But the air is an "object," and it has weight (has sufficient mass for the earth to register a gravitational attraction on it). At sea level and average temperature, 1 cubic foot [0.028 m^3] of air weighs about 0.08 pound [0.036 kg], or about $1\frac{1}{4}$ ounces. This does not seem to be very much; however, we must consider that the blanket of air, the *atmosphere*, surrounding the earth is many miles thick. This means that there are, in effect, many thousands of cubic feet of air piled on top of one another.

Actually, we find that the total weight, or pull of the earth, of this air amounts to about 14.7 psi [1.05 kg/cm^2] at sea level. This means that the pressure of the air, or *atmospheric pressure* is about 2,160 pounds [980 kg] on every square foot, which is more than 1 ton per square foot [919 kg per 0.093 m^2 (square meter)]. Since the average human body has a surface area of several square feet, it is sustaining a total atmospheric pressure of several tons!

When first reading this, a person may wonder why all this tremendous pressure that the human body is sustaining doesn't crush it. The reason is that the internal pressures of our bodies balance this external pressure of air. Fish have been found thousands of feet below sea level, where the pressures are more than 100,000 psi [7,030 kg/cm^2]. The fish can live because their internal pressures balance these tremendous external pressures.

Atmospheric pressure is not constant. It changes with weather. It also changes with altitude, that is, as you move upward from sea level into the mountains or into the sky. In ⊘ 4-21, on vacuum, we shall find out how variations in atmospheric pressures are measured and how these variations are used to predict changes in weather. For instance, decreasing atmospheric pressure may mean a storm is coming. Changes in weather and in atmospheric pressure are related because the air expands and becomes lighter as it is heated and contracts and becomes heavier as it is cooled. For example, 1 cubic foot of air at 0 degrees Fahrenheit [−17.8°C] weighs about 0.085 pound [0.035 kg]. One cubic foot of air at 100 degrees Fahrenheit [37.8°C] weighs only about 0.070 pound [0.0318 kg]. Thus, as the air is heated or cooled on sunny or cloudy days, it becomes lighter or heavier. Atmospheric pressure, therefore, decreases or increases.

Atmospheric pressure decreases with increasing altitude; that is, it decreases as you climb a mountain or fly upward in a plane. The reason for this is that the higher you go, the more of the total blanket of air, or atmosphere, is put below you where it cannot press down on you. For example, at 30,000 feet [9,144 m] above the earth's surface, the air pressure is less than 5 psi [0.3515 kg/cm^2]. At 100,000 feet [30,480 m], the air pressure is no more than 0.15 [0.0105 kg/cm^2] psi. A person could not live at this height without being enclosed in a sealed and pressurized airplane or space vehicle or without wearing a space suit which sustains the surrounding pressure and regulates the oxygen supply.

⊘ 4-21 Vacuum A vacuum is the absence of air or other matter. When astronauts blast off into space, they pass through the atmosphere and leave it behind. Out in space, there are practically no atoms of air. This is a vacuum.

But we do not need to leave the earth to find a vacuum. We can create a vacuum anywhere on earth with a long glass tube, closed at one end, plus a dish of mercury (a heavy metal that is liquid at normal temperatures). To produce a vacuum, we completely fill the tube with mercury and then close the end tightly. Next, we turn the tube upside down, put the end into the dish of mercury, and open this end. When we open the end, some of the mercury will run down out of the tube, leaving the upper end of the tube empty (Fig. 4-15). Since no air can enter, the upper part of the tube is actually a vacuum.

The device shown in Fig. 4-15 is called a *barometer;* it is used to measure atmospheric pressure. You might wonder why all the mercury does not run out of the tube when it is turned upside down since mercury is so heavy. The reason is that the atmospheric pressure won't let it run out. The air presses down on the surface of the mercury in the dish, and this pressure holds the mercury up in the tube (Fig. 4-16). This is somewhat like putting your hand, palm down, into soft mud. As you push down, the downward pressure causes some of the mud to squirt up between your fingers.

You can prove that it is air pressure that holds the mercury up in the tube by opening the top end of the tube. Now, air pressure will be admitted above the column of mercury in the tube. This air pressure will force the mercury in the tube down to the level of the mercury in the dish.

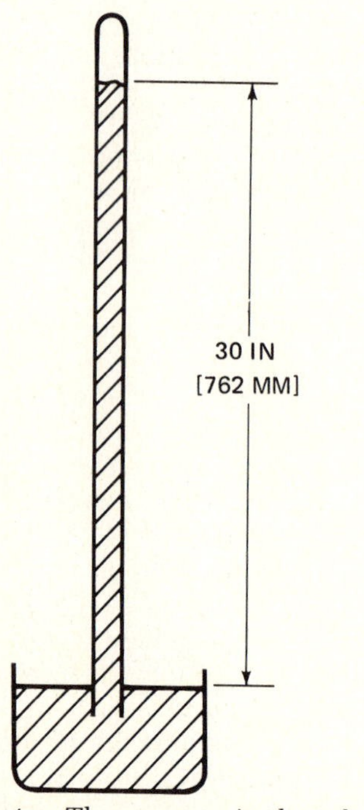

Fig. 4-15. Barometer. The mercury in the tube will stand at about 30 inches [762 mm] above the surface of the mercury in the dish at an atmospheric pressure of 15 psi [1.05 kg/cm²].

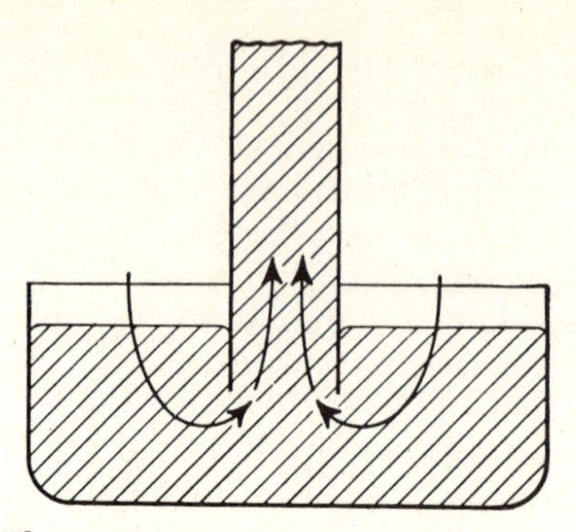

Fig. 4-16. The pressure of the air, acting on the surface of the mercury and through the mercury, holds the mercury up in the tube.

The barometer (Fig. 4-15) is a very useful device because it indicates air pressure. When air pressure goes up, the air pushes harder on the mercury and forces it higher up into the tube. But when air pressure goes down, there is a weaker push on the mercury, and so the mercury moves to a lower level in the tube. The barometer can foretell the coming of a storm. A storm is normally accompanied by a decrease in atmospheric pressure (brought on by the presence of heated and therefore lighter air). Thus, when the barometer "drops" (mercury goes down in the tube), the chances are a storm is coming.

There are, of course, many methods of producing a vacuum besides using a tube of mercury. Pumps of various types produce vacuum. The automobile engine is, in one sense, a vacuum pump. That is, it produces a partial vacuum in its cylinders; atmospheric pressure then pushes the air-fuel mixture into the cylinders. We shall discuss this in more detail later in the book.

⊘ 4-22 Vacuum Gauge Vacuum gauges are used in the automotive service industry to measure the amount of vacuum that engines produce, as explained in later chapters. The vacuum gauge is really a pressure gauge. It measures the pressure in an enclosed space and compares it with the atmospheric pressure. If the pressure in the enclosed space is lower than the atmospheric pressure, the vacuum gauge records so many "inches [millimeters] of mercury." Figure 4-17 shows a vacuum gauge that is used to check automotive engines. When the vacuum gauge is connected to the engine, it measures the amount of vacuum the engine is producing. (If the measurement is low or unsteady, then there is engine trouble.)

The vacuum gauge contains a bellows or diaphragm that is linked to an indicating needle on the dial face. When vacuum is applied, the bellows or diaphragm moves, and this causes the needle to move, indicating the amount of vacuum.

⊘ 4-23 Humidity Almost all air contains some water vapor (evaporated water). When the air is carrying a good deal of water vapor, it is said to have

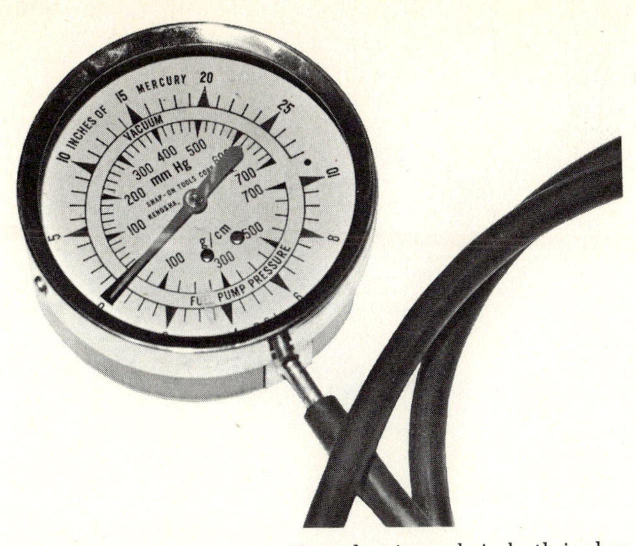

Fig. 4-17. Vacuum gauge. Note that it reads in both inches and millimeters of mercury (mmHg). (*Snap-on Tools Corporation*)

high humidity. When the air is carrying very little water vapor, it is said to have *low humidity*. Air over bodies of water (lakes, rivers, and oceans) has high humidity, while air over deserts has low humidity. Humidity is measured in terms of percentages. A reading of 0 percent humidity means the air contains no water vapor. A reading of 100 percent humidity means the air contains all the water vapor it can hold. A reading of 50 percent humidity means that the air contains just half as much water vapor as it is capable of holding.

Humidity has an effect on how well an automobile engine will run. An engine will not produce as much power if it is running in hot, moist (high-humidity) conditions as it will if it is running in cool, dry (low-humidity) conditions. The cool air is denser (the molecules are closer together), so that more air enters the engine; and this means more power. Excess moisture in the air enriches the mixture because the cylinders get less oxygen but the same amount of fuel. This richer mixture often causes an engine to run smoother on damp or rainy days.

⊘ 4-24 Atmospheric Factors Affecting Engine Combustion As you study this book, you will learn that changes in temperature, atmospheric pressure, and humidity affect combustion in the engine. They affect the way the fuel burns and the power output of the engine. In addition, they may affect the amount of unburned gasoline (hydrocarbon) and the amount of carbon monoxide (CO) that comes out through the tail pipe. Accurate testing of engines requires that all readings be corrected to take into account temperature, atmospheric pressure, and humidity. We will describe these factors and how they affect engine performance in more detail in later chapters.

⊘ 4-25 Work The engineer does not define the term "work" the way we do in everyday life. You may say you are "working" when you stand holding a package. You work a problem or work by sitting at a desk in an office. But to the engineer, work means changing the position of an object against an opposing force. In this sense, work means that an object must be moved by the application of a force (a push, a pull, or a lift). For example, when a weight is lifted from the ground, work is done on the weight; that is, the weight is lifted, or moved upward, against the force of gravity. In a similar way, when a coil spring is compressed, work is done on the spring (Fig. 4-18); a force (or push) is exerted through a certain distance.

Work is measured in terms of *distance* and *force*. For example, if a 5-pound weight is lifted 1 foot, then the work done on the weight will be 1 foot × 5 pounds, or 5 foot-pounds. If the 5-pound weight is lifted 2 feet, then the work done will be 10 foot-pounds. As another example, if you push down on a spring (Fig. 4-18) with an average force of 25 pounds and your hand moves a distance of 6 inches (or $\frac{1}{2}$ foot) in compressing the spring, then you have done 12.5 foot-pounds ($\frac{1}{2}$ foot × 25 pounds) of work on the spring.

In the metric system, work is measured in meter-kilograms. Thus our example would be: Lifting a 2.268-kilogram weight 0.305 meter requires 0.305 meter × 2.268 kilograms, or 0.692 meter-kilograms of work. Lifting the 2.268-kilogram weight 0.610 meter requires 1.383 meter-kilograms of work.

NOTE: Work can also be measured in other units, such as inch-pounds and inch-ounces [centimeter-grams], and so forth. However, in engineering, the foot-pound [meter-kilogram] is the most commonly used unit for measuring of work.

⊘ 4-26 Energy Energy is the capacity or ability to do work. When work is done on an object, *energy* (or ability to do work) is stored in that object. For

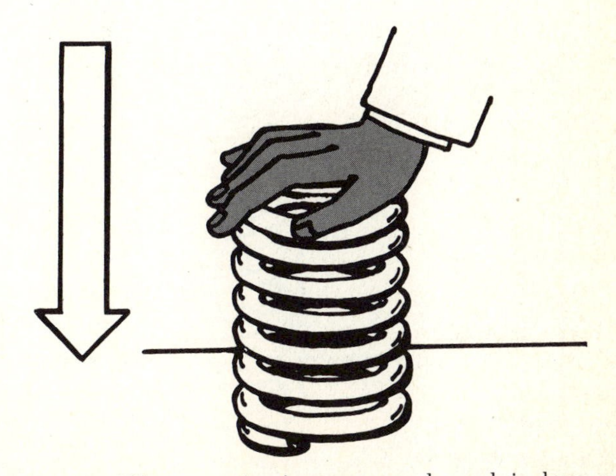

Fig. 4-18. When a spring is compressed, work is done on that spring and energy is stored in it.

example, when a spring is compressed (Fig. 4-18), work is done on the spring and energy is stored in it. If the spring is released (Fig. 4-19), the stored energy can do work on another object. For instance, the spring can lift a weight against the force of gravity. Energy is also stored in an object when the object is lifted from the earth. For example, lifting a 20-pound [9.072 kg] weight 4 feet [1.219 m] requires 80 foot-pounds [11.056 m-kg] of work. This amount of work is, in effect, stored in the weight. In other words if the weight is released and allowed to drop on a stake that is being driven into the ground, it will do 80 foot-pounds [11.056 m-kg] of work on the stake.

Energy is the measure of ability. Work is the application of this ability to produce motion.

⊘ 4-27 Power Work can be done slowly, or it can be done rapidly. You can lift a weight off the floor slowly, or you can lift it very quickly (if it is not too heavy). The speed with which work is done is measured in terms of *power*. A machine that does a great deal of work in a comparatively short time is called a *high-powered machine*. A machine that takes a much longer time to do the same amount of work is called a *low-powered machine*. Power, then, is the rate, or speed, at which work is done.

⊘ 4-28 Torque Torque is a turning or twisting effort. When you reel in a fish, you are applying torque to the reel crank. You apply torque to the steering wheel of an automobile to guide it around a curve. You apply torque to the top of a screw-top jar when you loosen the top (Fig. 4-20).

Torque should not be confused with power or with work. In an automobile, torque is the rotary or twisting force that the engine applies, through shafts and gears, to the car wheels. Power is something else—it is the rate at which the engine works. Work is the energy expended, or the product of force and distance. Both work and power imply motion. Torque does not; it is merely a turning effort and may or may not result in motion.

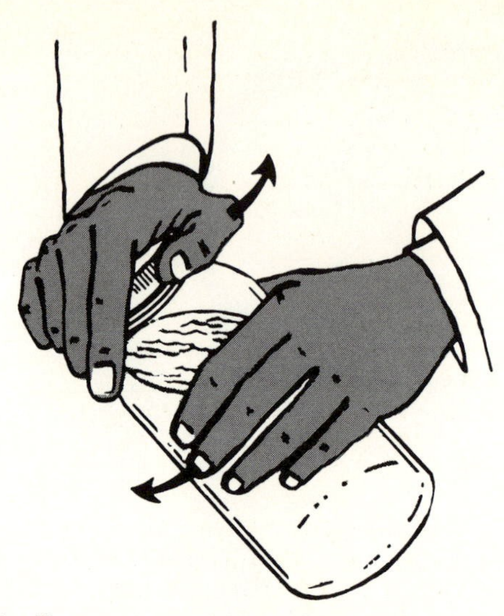

Fig. 4-20. Torque, or twisting effort, must be applied to loosen and remove the top from a screw-top jar.

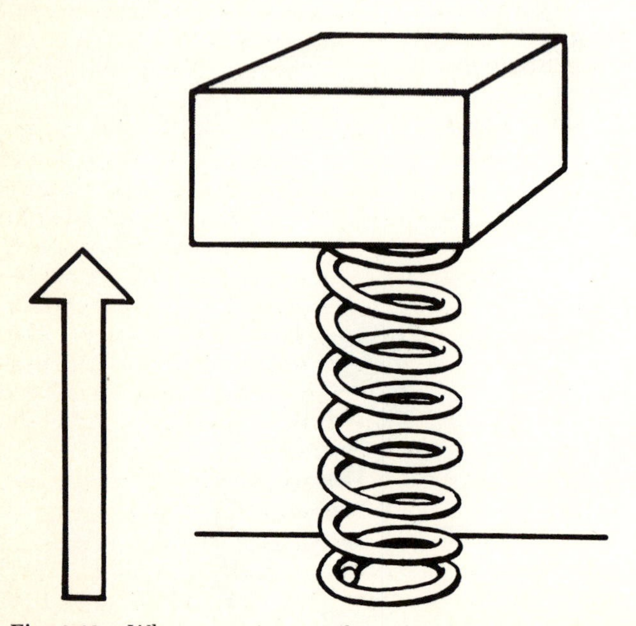

Fig. 4-19. When a spring is released, it can do work on another body. For example, it can lift a weight against the force of gravity.

Torque is measured in pound-feet (do not confuse this with foot-pounds of work). For instance, if you apply a 20-pound push to a windlass crank that is $1\frac{1}{2}$ feet long (from center to handle), you would be applying a torque of 30 pound-feet (Fig. 4-21). In the metric system, torque is measured in kilogram-meters. Thus, our example would be: If you apply a 9.072-kg kilogram push to a windlass crank that is 0.457 meter long (from center to handle), you would be applying a torque of 4.146 kilogram-meters.

⊘ 4-29 Horsepower Engine performance is measured in terms of power—*horsepower*. A horsepower is the power of one horse, or a measure of the speed at which a horse can work. It may seem strange that in this modern day we still compare the power of

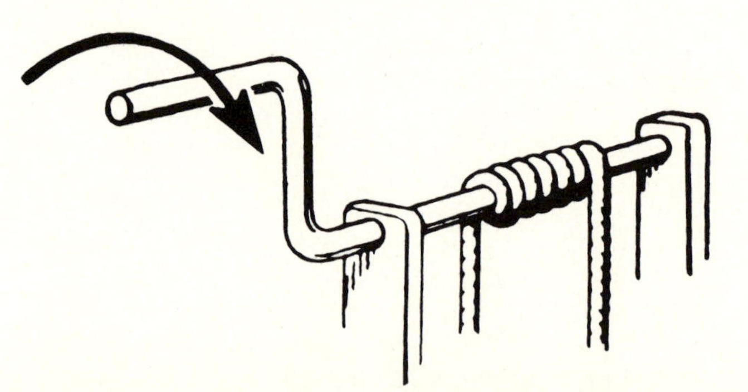

Fig. 4-21. Torque is measured in pound-feet [kg-m]. It is calculated by multiplying the push by the crank offset, or the distance of the push from the rotating shaft.

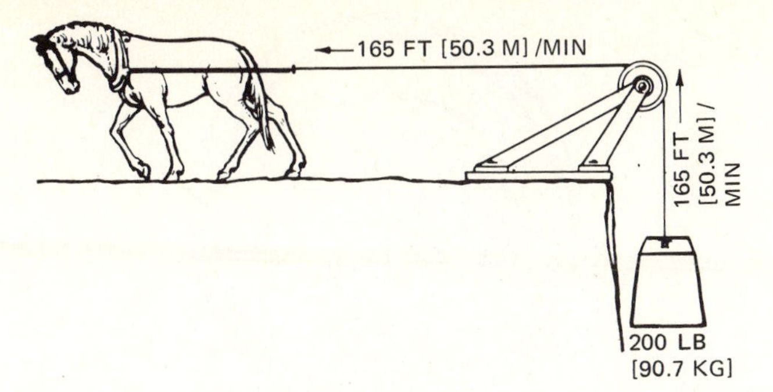

Fig. 4-22. One horse can do 33,000 foot-pounds of work a minute.

an engine with the speed at which a horse can work. However, years ago, when engines were first developed, some sort of measuring stick was needed to compare the power of different engines. Since the horse was then the most common source of power, it was natural that the power of engines should be measured in terms of the power of horses, that is, in horsepower. (If human beings had harnessed cats instead of horses to do their work for them, we would probably be referring to engines in terms of "catpower.")

It was found that an average horse can raise a 200-pound [90.718 kg] weight a distance of 165 feet [50.3 m] in 1 minute. Figure 4-22 shows a simplified version of the method by which this test was made. The horse walks 165 feet [50.3 m] in 1 minute, and the cable, running over the pulley, raises the 200-pound [90.7 kg] weight 165 feet [50.3 m] in that minute. The amount of work done is 33,000 foot-pounds [4,562.2 m-kg]. The length of time it takes to do this amount of work is 1 minute. The amount of power required is 1 hp [0.746 kW]. In other words, this is the amount of work one horse can do in 1 minute.

NOTE: In the metric system, the power produced by an engine is often measured in kilowatts, that is, in terms of the amount of electricity the engine could produce if it were driving an electric generator. Since 1 hp is equivalent to 0.746 kW, a 200-hp engine would be equivalent to a 149-kW engine.

If 400 pounds [181.436 kg] are raised 165 feet [50.292 m] in 1 minute, two horses are required (one for each 200-pound weight). The amount of work done is 66,000 foot-pounds (165 feet × 400 pounds) [9,127.8 m-kg (50.292 m × 181.436 kg)]. When this amount of work is done in 1 minute, then 2 hp [1.492 kW] are required. Similarly, if 33,000 foot-pounds [4,563.9 m-kg] of work is done in 2 minutes, then only ½ hp [0.373 kW] is required since work is being done at the rate of 16,500 foot-pounds per minute [2,281.95 m-kg/min]. The formula for horsepower is

$$\text{hp} = \frac{\text{ft-lb/min}}{33{,}000} = \frac{L \times W}{33{,}000 \times t}$$

where hp = horsepower
L = length through which W is forced, ft
W = push exerted through distance L, lb
t = time required to force W through L, min

PROBLEM: You have a heavy box loaded with sand and want to drag it 500 ft across a level lot in 2 min. You find that it requires an average pull of 2,000 lb to pull the box. How much horsepower would be required?

SOLUTION: Substituting in the formula for horsepower,

$$\text{hp} = \frac{L \times S}{33{,}000 \times t} = \frac{500 \times 2{,}000}{33{,}000 \times 2} = 15.15 \text{ hp}$$

PROBLEM: You are asked to select a gasoline engine capable of raising a 3,000-lb coal-mine elevator 220 ft in 1 min. Ignoring friction and all other power looses, what would be the minimum kilowatt rating of the engine that you could select? (Remember, 1 hp = 0.746 kW.)

SOLUTION: Use the formula

$$\text{hp} = \frac{L \times W}{33{,}000 \times t} = \frac{220 \times 3{,}000}{33{,}000 \times 1} = 20 \text{ hp}$$

Then, since 1 hp = 0.746 kW, 20 hp = 16.9 kW. An engine with a minimum kilowatt rating of 16.9 kW would be required.

⊘ 4-30 Inertia Inertia is a property of all material objects. It causes an object to resist any change of direction or speed of travel. A motionless object (zero speed) tends to remain at rest. Furthermore, it resists any attempt to make it move. But once an object is in motion, it resists any attempt to change its direction, hurry it up, or slow it down.

Consider the automobile. When it is standing still, its inertia must be overcome by applying force before it will move. To increase its speed, additional force must be applied. Also, when the automobile is in motion, its inertia causes it to resist any speed reduction. To decrease its speed, the brakes must be applied. In other words, the automobile has a certain energy content, stored in the form of speed. The brakes must absorb this energy (turning it into heat energy as the brake shoes press against the brake drums). In a similar manner, when the automobile goes around a curve, its inertia tries to keep it moving in a straight line. The friction of the tires on the road must overcome this tendency, or the inertia of the automobile will send it into a skid.

To say it in another way, the automobile may be thought of as a container into which energy is put. Application of power to the wheels overcomes car inertia; the automobile moves and energy is stored in it. The higher the speed, the more energy is stored. The torn and twisted wreckage of two cars that have collided at high speed is ample and tragic evidence of the amount of energy stored in cars when they are moving at high speed.

⊘ 4-31 Friction Friction is the resistance to motion between two objects that are in contact with each other. If you put this book on a tabletop and then push it across the table, you would find it took a certain amount of push to move it. Then, if you put another book on top of this book, you would find you had to push harder to push the two books together across the tabletop. Thus friction, or resistance to motion, increases with the *load*. The higher the load, the greater the resistance to motion (that is, the greater the friction).

In the automobile engine the bearings are heavily loaded. Some bearings sustain loads of more than 6,000 psi [421.8 kg/cm^2]. With such heavy loads, the friction between the bearing and journal surfaces would be tremendous if it were not for the lubricating oil. However, the oil, in effect, "floats" the rotating journal so that actual metal-to-metal contact is avoided. The friction, then, is between moving layers of oil rather than between dry metal parts. This friction is relatively low.

Friction is classified into three types: *dry*, *greasy*, and *viscous*.

1. *DRY FRICTION* This is the friction between two dry objects—a board being dragged across a floor, for instance. If the board and the floor are rough, the friction is relatively high. If the board and the floor are smooth, the friction is lower. You can think of dry friction as an interference to motion between two objects caused by surface irregularities that tend to catch on each other. Even objects that have been machined to a very smooth finish still have very small irregularities which offer resistance to relative motion.

2. *GREASY FRICTION* This is the friction between two objects thinly coated with oil or grease. It can be assumed that the thin film tends to fill in the low spots in the surfaces. This reduces the tendency for the surface irregularities to catch on each other. However, high spots will still catch and wear as the two surfaces move over each other.

In an automobile engine, greasy friction may occur in an engine on first starting. Most of the lubricating oil may have drained away from the bearing surfaces and from the cylinder walls and piston rings. When the engine is started, only the small amount of oil remaining on these surfaces protects them from undue wear. Of course, the lubricating system quickly supplies additional oil, but before this happens, greasy friction exists on the moving surfaces. The lubrication between the surfaces where greasy friction exists is not sufficient to prevent wear. That is the reason that automotive engineers say that initial starting and warm-up of the engine is hardest on the engine and wears it the most.

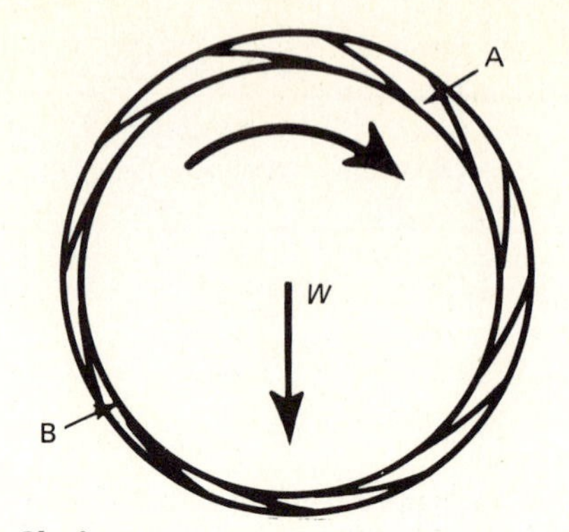

Fig. 4-23. Shaft rotation causes layers of clinging oil to be dragged around with it. The oil moves from the wide clearance A and is wedged into the narrow clearance B, thereby supporting the shaft weight *W* on an oil film.

3. *VISCOUS FRICTION* "Viscosity" refers to the tendency of liquids, such as oil, to resist flowing. A heavy oil is more viscous than a light oil and flows more slowly (has a higher viscosity, or higher resistance to flowing). *Viscous friction* is the friction, or resistance to relative motion, between adjacent layers of liquid.

In an engine bearing supplied with sufficient oil, layers of oil adhere to the bearing and journal surfaces. In effect, these clinging layers of oil are carried around by the rotating journal and wedge under the journal (Fig. 4-23). The wedging action lifts the journal so that the oil itself supports the weight, or load. Now, since the journal is supported (or "floats") on layers of oil, there is no metal-to-metal contact. However, the layers of oil must move over each other, and it does require some energy to make them so move. The resistance to motion between these oil layers is called *viscous friction*.

Check Your Progress

Progress Quiz 4-2 The past few pages discuss some important physical principles that will give you a better understanding of engine design, construction, and operation. Engine operation is based on these principles; when you know them, you know the "why" as well as the "what," and this will help you in your work in the shop or office. Here is your chance to find out how well you remember what you have been reading. If any of the following questions stump you, just reread these pages. Don't be discouraged. Remember that most good students have to reread their lessons several times before they understand the essential facts.

Completing the Sentences The sentences that follow are incomplete. After each sentence there are several words or phrases, only one of which will correctly complete the sentence. Write each sentence in your notebook, selecting the proper word or phrase to complete it correctly.

1. At sea level, atmospheric pressure is about: (*a*) 1.5 psi, (*b*) 15 psi, (*c*) 150 psi.
2. When air is heated, it: (*a*) expands and becomes heavier, (*b*) expands and becomes lighter, (*c*) contracts and becomes heavier.
3. When there are only a few atoms of air, widely scattered, the condition is called a: (*a*) vacuum, (*b*) pressure of air, (*c*) storm, (*d*) pump action.
4. Changing the position of an object against an opposing force is called: (*a*) energy, (*b*) power, (*c*) work, (*d*) torque.
5. The capacity, or ability, to do work is called: (*a*) power, (*b*) energy, (*c*) torque.
6. If you lifted a 10-lb weight a distance of 3 ft, you would be doing: (*a*) 3 ft-lb of work, (*b*) 30 ft-lb of work, (*c*) 300 ft-lb of work.
7. An engine that could raise 16,500 lb a distance of 100 ft in 1 min would have to produce about: (*a*) 5 hp, (*b*) 10 hp, (*c*) 25 hp, (*d*) 50 hp.
8. The characteristic of an object which causes it to resist any tendency to change its direction of travel is called: (*a*) energy, (*b*) power, (*c*) inertia.
9. Torque, which is twisting or turning effort, is measured in: (*a*) pound-feet, (*b*) foot-pounds, (*c*) foot-pounds per minute, (*d*) pound-feet per minute.
10. The three types of friction are: (*a*) dry, greasy, and liquid; (*b*) dry, greasy, and viscous; (*c*) dry, wet, and liquid.

CHAPTER 4 CHECKUP

NOTE: Since the following is a chapter review test, you should review the chapter before taking the test.

You have just completed Chap. 4 of this book and have taken an important step forward into a better future for you. The chapter you have just finished may not be as interesting as the later chapters, which deal directly with the engine. General principles are often harder to understand and remember than specific details. But these general principles are important to you. Once you do understand them, you will find that you can answer many puzzling questions about how and why engines operate. The following questions not only will give you a chance to check up on how well you understand and remember these principles but also will help you to remember them better. The act of writing the answers to the questions will fix the facts more firmly in your mind.

NOTE: Write your answers in your notebook. Then later you will find your notebook filled with valuable information which you can refer to quickly.

Physical Properties and Their Measurements Following are a list of several physical properties and a list of the units in which they are measured (not in order, however). In your notebook write the list of physical properties as it appears below. Then, next to each property write the units in which it is measured. For example, next to "temperature" you would write "degrees Fahrenheit and degrees Celsius."

PHYSICAL PROPERTIES
Torque
Horsepower
Work
Atmospheric pressure
Temperature

UNITS
Degrees Fahrenheit and degrees Celsius
Pound-feet and kilogram-meters
Pounds per square inch and kilograms per square centimeter
Foot-pounds per minute and meter-kilograms per minute
Foot-pounds and meter-kilograms

Completing the Sentences The sentences that follow are incomplete. After each sentence there are several words or phrases, only one of which will correctly complete the sentence. Write each sentence in your notebook, selecting the proper word or phrase to complete it correctly.

1. The smallest particle into which an element can be divided is called: (*a*) a molecule, (*b*) an atom, (*c*) a hydrocarbon.
2. The electron has a tiny charge of: (*a*) positive electricity, (*b*) neutral electricity, (*c*) negative electricity.
3. Atoms are held together in a molecule by: (*a*) electric charges, (*b*) centrifugal force, (*c*) chemical reaction, (*d*) combustion.
4. When you heat an object, you cause its: (*a*) speed to increase, (*b*) molecules to move faster, (*c*) molecules to vaporize.
5. You cannot produce a change of state unless you change: (*a*) molecular speed, (*b*) molecular composition, (*c*) chemical composition.
6. The reason the atmospheric pressure goes down when air temperature goes up is that: (*a*) heated air is heavier, (*b*) cold air is lighter, (*c*) heated air is lighter.
7. In the barometer, the mercury is held up in the tube by: (*a*) air pressure, (*b*) vacuum, (*c*) temperature.
8. The rate, or speed, at which work is done is called: (*a*) energy, (*b*) power, (*c*) torque.
9. A 100-hp [74.6 kW] engine, with a suitable mechanism (neglecting friction), can lift 100,000 lb [45,359 kg] a distance of 16 ft [4.877 m] in about: (*a*) 30 s, (*b*) 1 min, (*c*) 2 min, (*d*) 10 min.
10. If you were turning an 18-in [0.457 m] crank and found that a 20-lb [9.072 kg] push was required, you would be exerting on the shaft of the crank a torque of: (*a*) 8 lb-ft [1.106 kg-m], (*b*) 18 lb-ft [2.487 kg-m], (*c*) 30 lb-ft [4.146 kg-m], (*d*) 360 lb-ft [49.752 kg-m].

Problems Work out the following problems in your notebook.

1. You weigh 150 lb. You are carrying an object weighing 40 lb, and you walk up a flight of stairs that has a total measurement of 10 ft vertically and 12 ft horizontally. What is the total amount of work done?

2. You are pulling a cart that weighs 1,000 lb along a level road and find that you must exert a pull of 55 lb. After you have pulled it for 100 ft, how much work have you done on the cart?

3. How much horsepower would be required to pull the cart mentioned in the preceding problem 30.48 m in 10 s ($\frac{1}{6}$ min)? How many kilowatts?

4. You turn the crank of an eggbeater and find that you must exert a force of 2 lb [0.90 kg] on the crank handle. The handle is 3 in [76.2 mm] ($\frac{1}{4}$ ft) [0.0762 m] from the center of the shaft. What torque (in pound-feet and in kilogram-meters) do you exert?

Definitions In the following, you are asked to define certain terms. Write the definitions in your notebook. Writing the definitions will help you to remember them.

1. What is combustion?
2. What is change of state?
3. What is gravity?
4. What is vacuum?
5. Define work.
6. Define energy.
7. Define power.
8. What is a horsepower?
9. What is torque?
10. Define friction.

SUGGESTIONS FOR FURTHER STUDY

If you are interested in the basic principles discussed in this chapter, you might like to study them further. Almost any up-to-date high school physics book will give you much additional information on these principles. Your local library probably has several physics books that you will find of interest. Also, if you have a chance, you could talk over various points that might not be clear to you with your local high school science or physics teacher. Teachers are almost invariably fine people who are sincerely interested in helping you acquire more knowledge.

chapter 5

ENGINE FUNDAMENTALS

This chapter discusses fundamentals of engine construction and operation. In addition, it describes briefly the engine accessories that are necessary to the operation of the engine. These accessories include the fuel system, lubricating system, electric system, and cooling system. These accessories are, in fact, part of the engine itself. However, since this book deals primarily with the engine, the accessories are not described in full detail. They are discussed only to the extent that they enter into the actual operation of the engine. Other books in the McGraw-Hill Automotive Technology Series cover engine accessories in a comprehensive manner. Automotive Electrical Equipment *discusses the electric system in detail.* Automotive Fuel, Lubricating, and Cooling Systems *covers the accessories named in its title.*

⊘ 5-1 The Engine Cylinder The engine (Fig. 5-1) is the source of power that makes the wheels turn and the car move. It is usually referred to as an "internal-combustion engine." This is because the fuel (gasoline) is burned inside the engine—within the *engine cylinders*, or *combustion chambers*. The combustion, or burning, of the gasoline creates high pressure. The high pressure thus produced causes a shaft to turn, or rotate. This rotary motion is carried to the car wheels by the power train so that the wheels rotate and the car moves.

Most automotive engines have four, six, or eight cylinders. Since the actions are similar in all cylinders, let us examine one cylinder in detail. Figure 5-2 shows the construction of a single cylinder of an engine with the cylinder and piston sliced in half. Essentially, the cylinder is nothing more than a cylindrical air pocket that is closed at one end and open at the other. The piston fits snugly into the open end of the cylinder but is still loose enough to slide easily up and down inside the cylinder. The piston is made of aluminum or other suitable metal. There are grooves cut in the side of the piston, and piston rings are fitted into these grooves (Fig. 5-3). These rings fit tightly against the cylinder wall and provide such a good seal that very little air can escape between the piston and cylinder wall. Thus, when the piston is pushed up in the cylinder, the air in the cylinder is trapped above the piston and is compressed.

Fig. 5-1. Cutaway view of 365-hp V-8 engine. (*Ford Motor Company*)

Figure 5-4 shows this action very simply. In Fig. 5-4*a*, the piston is shown below the cylinder. The cylinder is drawn as though it were transparent so that the piston can be seen as it moves up into the cylinder (Fig. 5-4*b*). The air above the piston is compressed as it is pushed up into the cylinder. (Neither the piston rings nor the means of pushing the piston up into the cylinder are shown.) If we put some gasoline vapor into the compressed air and then applied a lighted match or spark to the air-vapor mixture, it is obvious what would happen. The gasoline vapor would burn. High pressure would be created, and the piston would be blown out of the cylinder, as shown in Fig. 5-4*c*. Actually, in a modified form, this is what happens in the cylinder. A mixture of air and gasoline vapor enters the cylinder, the piston is pushed up in the cylinder, the

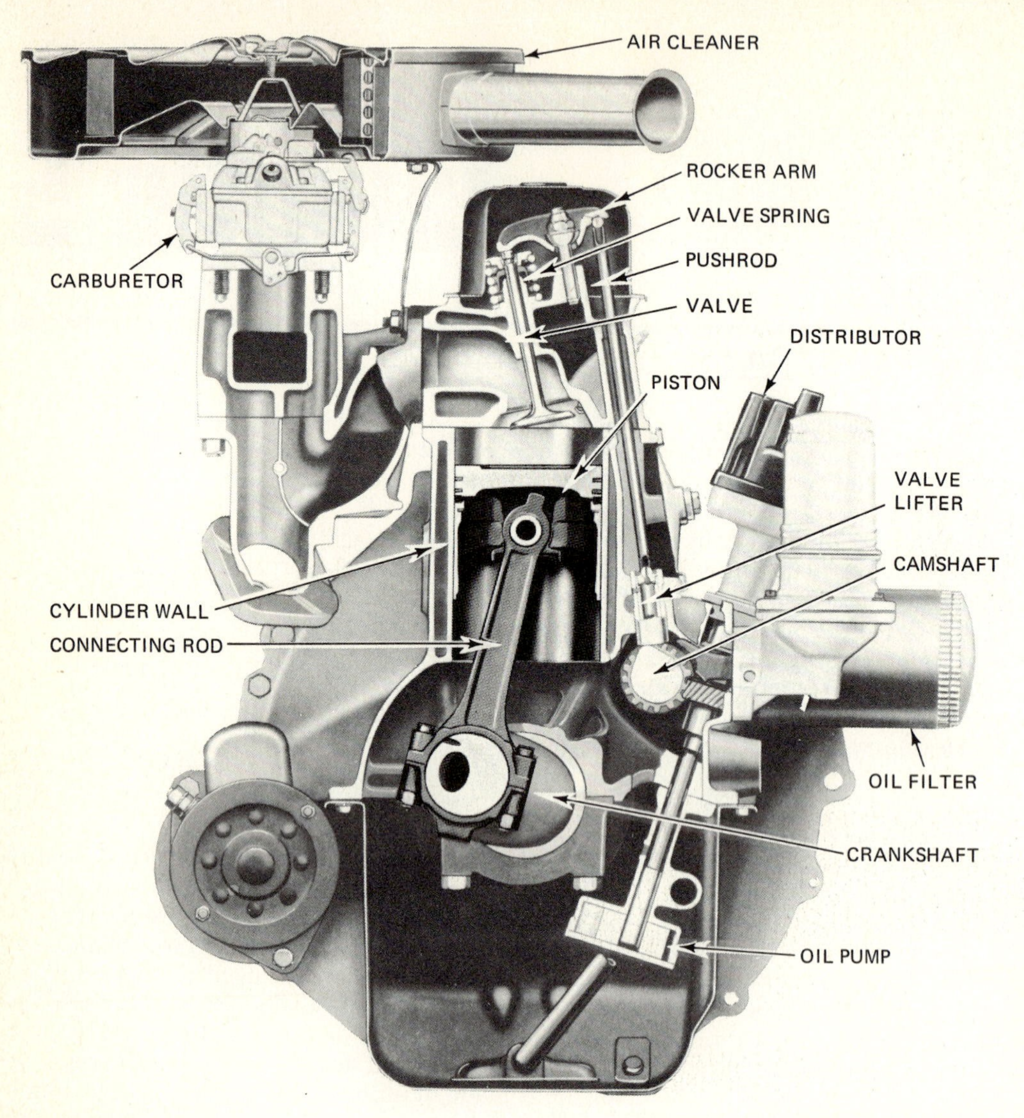

Fig. 5-2. Sectional view of a six-cylinder engine. The piston is near the top of its stroke. The piston and cylinder are shown cut in half. (*Ford Motor Company*)

mixture is ignited, and the piston is pushed down in the cylinder.

⊘ 5-2 Reciprocating to Rotary Motion The up-and-down movement of the piston is called *reciprocating motion*. The piston moves in a straight line. This straight-line motion must be changed to *rotary*, or turning, motion in order to make the car wheels rotate. Reciprocating motion is changed to *rotary motion* by a crank on the crankshaft and a connecting rod (Figs. 5-5 to 5-9).

The crank is an offset section of the crankshaft. It swings around in a circle as the shaft rotates (Fig. 5-6). The connecting rod connects the crankpin of the crank and the piston (Figs. 5-7 and 5-8). The crank end of the connecting rod is attached to the crankpin by fastening the rod-bearing cap to the connecting rod with the rod-cap bolts (Fig. 5-8). The cap and rod have bearings which permit the crankpin to rotate freely within the rod. The piston end of the rod is attached to the piston by the piston pin, or wrist pin. The piston pin is held in two bearings in the piston. A bearing in the piston-pin end of the connecting rod (or bearings in the piston) permits the rod to swing back and forth on the piston pin.

NOTE: The crank end of the connecting rod is sometimes called the rod "big end," and the piston end is sometimes called the rod "small end."

Now let us see what happens as the piston moves up and down in the cylinder (Fig. 5-9). The crankpin moves in a circle as the crankshaft rotates. As the piston starts moving down, the connecting rod tilts to one side so that the lower end can follow the circular path of the crankpin. If you follow the sequence of action shown in Fig. 5-10 (or steps numbered 1 to 8), you will note that the connecting rod tilts back and forth on the piston pin while the lower end moves in a circle along with the crankpin.

⊘ 5-3 The Valves There are two *openings*, or *ports*, in the enclosed end of the cylinder, one of which

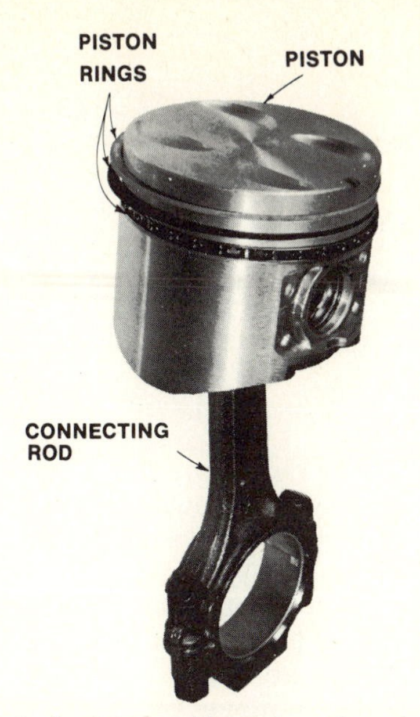

Fig. 5-3. Typical piston with piston rings in place and connecting rod attached. When the piston is installed in the engine cylinder, the rings are compressed into the grooves in the piston. (*Chrysler Corporation*)

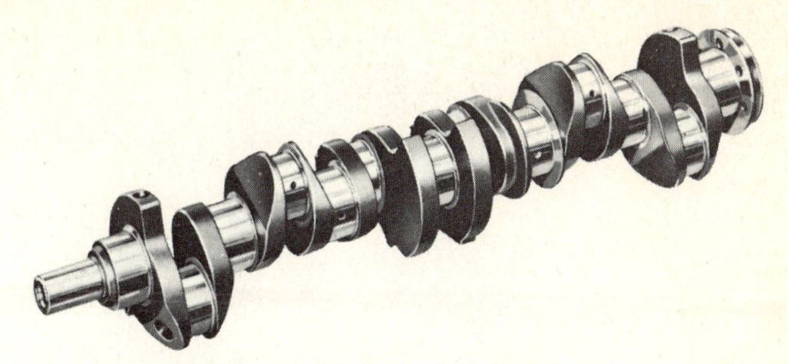

Fig. 5-5. Engine crankshaft. (*Ford Motor Company*)

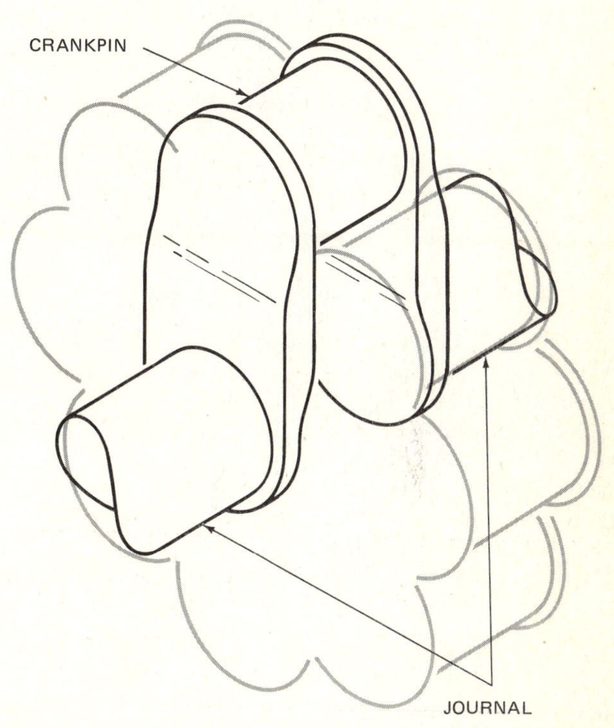

Fig. 5-6. As the crankshaft rotates, the crankpin swings in a circle around the shaft.

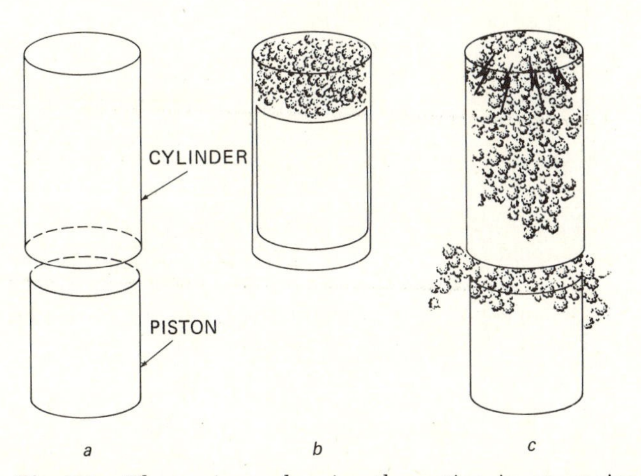

Fig. 5-4. Three views showing the action in an engine cylinder. (*a*) The piston is a second cylinder that fits snugly into the engine cylinder. (*b*) When the piston is pushed up into the engine cylinder, air is trapped and compressed. The cylinder is drawn as though it were transparent so that the piston can be seen. (*c*) The increase of pressure as the air-fuel mixture is ignited pushes the piston out of the cylinder.

is shown in Fig. 5-2. One of these ports permits the entrance of the mixture of air and gasoline vapor into the cylinder. The other port permits the burned gases to exhaust, or escape, from the cylinder after combustion.

The two ports have *valves* assembled into them, and these valves close off one or the other port, or both ports, during various stages of the actions taking place in the cylinder. The valves are accurately machined metal plugs that close the openings when the valves are *seated,* that is, when they have moved up into the openings. Figure 5-11 shows a valve and a valve seat of the type used in the engine illustrated in Fig. 5-2. This type of valve is called a *poppet valve.* The valve is shown open in Fig. 5-11, that is, pushed down off its seat. When closed, the valve moves up so that the outer edge of its head rests on the seat. In this position, the valve port is closed so that no air or gas can pass through it.

A spring on the valve stem (Fig. 5-2) tends to hold the valve on its seat. The lower end of the spring rests against a flat section of the cylinder head. The upper end rests against a flat washer, or spring retainer, which is attached to the valve stem by a retainer lock, or keeper. The spring is under compression, which means it tries to expand and thus tries to keep the valve seated.

The valve-opening mechanism includes a *valve*

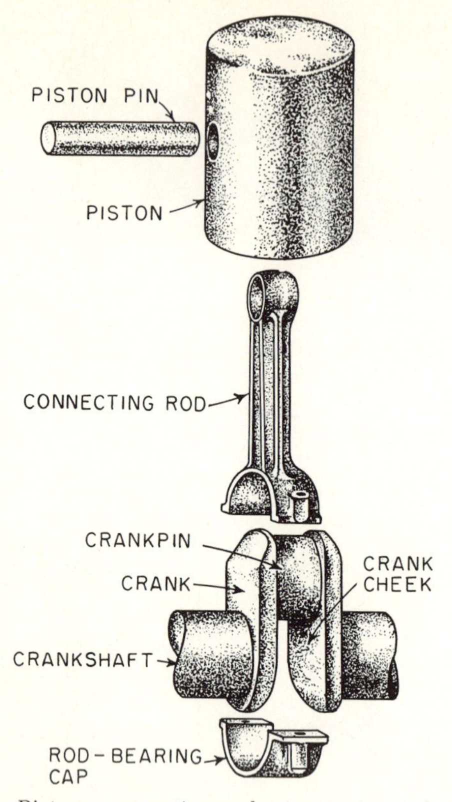

Fig. 5-7. Piston, connecting rod, piston pin, and crank of a crankshaft in disassembled view. The piston rings are not shown.

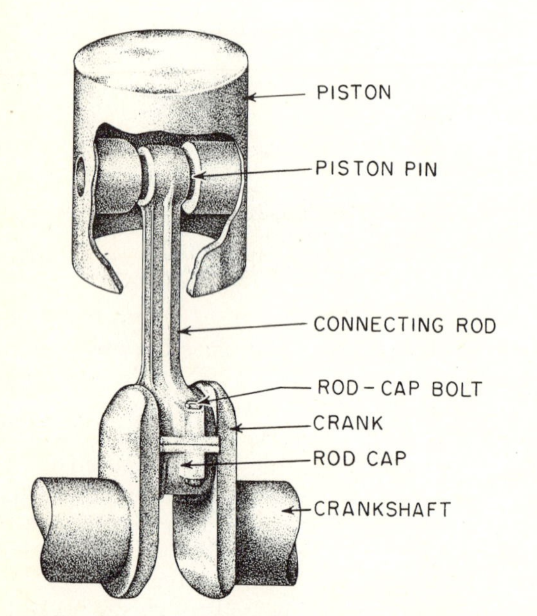

Fig. 5-8. Piston and connecting-rod assembly attached to a crankpin on a crankshaft. The piston rings are not shown. The piston is partly cut away so you can see how it is attached to the connecting rod.

lifter and a *cam* on a *camshaft* (Fig. 5-12). As the camshaft turns, the "bump," or lobe, on the cam comes around under the valve lifter, moving it upward. The valve lifter, in turn, pushes up on the pushrod. The pushrod then causes one end of the rocker arm to move up. The rocker arm pivots on its supporting shaft so that the valve end of the rocker arm is forced down, thus opening the valve. After the cam lobe moves out from under the valve lifter, the valve spring forces the valve up on its seat again. Figure 5-12 shows an overhead-valve mechanism. There are other types (see ⊘ 6-10).

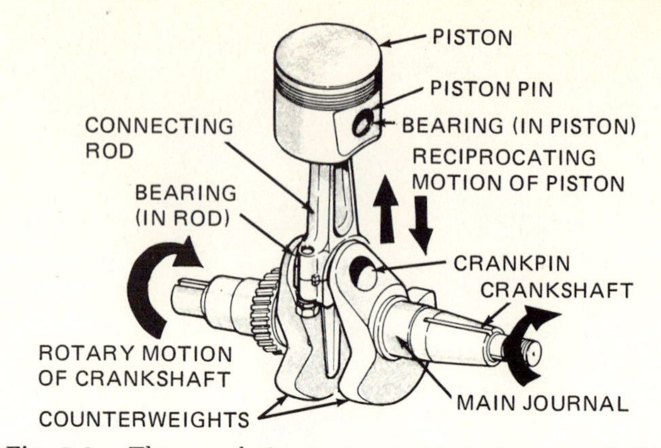

Fig. 5-9. The crankpin moves in a circle around the crankshaft while the piston moves up and down.

Figure 5-13 shows a typical camshaft. It has a cam for each valve in the engine, or two cams per cylinder. The camshaft is driven by gears, a toothed belt, or a chain, from the crankshaft, and it turns at one-half crankshaft speed. The cam lobes are so positioned on the camshaft as to cause the valves to open and close in the cylinders at the proper time with respect to the activities taking place in the cylinders. (See Figs. 5-14 and 5-15.)

NOTE: In all the engines we have discussed so far, the camshaft is located in the cylinder block. (See Figs. 5-1 and 5-2, for example.) However, in many engines now being produced, the camshaft is located on top of the cylinder head. This type of engine is called an *overhead-camshaft* (OHC) engine. We shall describe OHC engines in detail in Chap. 6.

⊘ 5-4 Engine Operation The activities taking place in the engine cylinder can be divided into four stages, or strokes. The term "stroke" refers to piston movement. The upper limit of piston movement (position 1 in Fig. 5-10) is called *top dead center* (TDC). The lower limit of piston movement is called *bottom dead center* (BDC). A stroke occurs when the piston moves from TDC to BDC or from BDC to TDC. In other words, the piston completes a stroke each time it changes direction of motion.

When the entire cycle of events in the cylinder requires four strokes (or two crankshaft revolutions), the engine is called a *four-stroke-cycle engine*, or a *four-cycle engine*. The term "Otto cycle" is also applied to this type of engine (after Friedrich Otto, a German scientist of the nineteenth century). The four piston strokes are *intake, compression, power*, and *exhaust*. (Two-stroke-cycle engines are also in use; in these, the entire cycle of events is completed in two strokes, or one crankshaft revolution.)

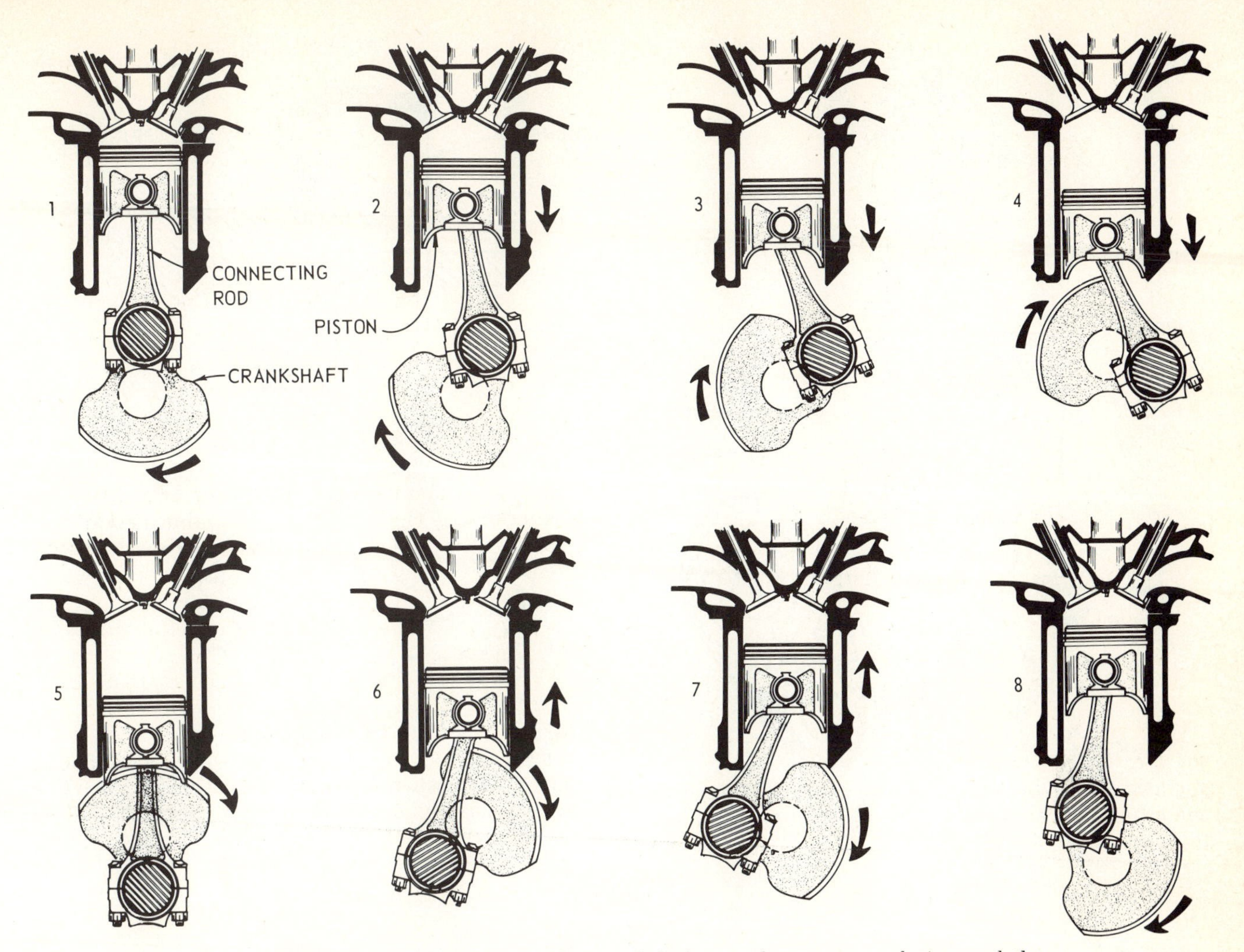

Fig. 5-10. Sequence of actions as the crankshaft completes one revolution and the piston moves from top to bottom to top again.

NOTE: For the sake of simplicity in the following discussion, the valves are considered to open and close at TDC and BDC. Actually, they are not timed to open and close at these points, as is explained in a later chapter. Also, the illustrations showing the four strokes (Figs. 5-16 to 5-19) are greatly simplified and show the intake and exhaust valves separated and placed on either side of the cylinder so that both can be seen.

1. *INTAKE* (*FIG. 5-16*) On the intake stroke, the intake valve has opened. The piston is moving down, and a mixture of air and vaporized gasoline is being "drawn" into the cylinder through the valve port. The mixture of air and vaporized gasoline is delivered to the cylinder by the fuel system and carburetor (discussed in ⊘ 5-8).

NOTE: Actually, the piston does not "draw" the air-fuel mixture into the cylinder. Instead, atmospheric pressure (or pressure of the air) outside the engine pushes air into the cylinder. This air passes through the carburetor, where it picks up a charge of gasoline vapor, and then through the intake mani-

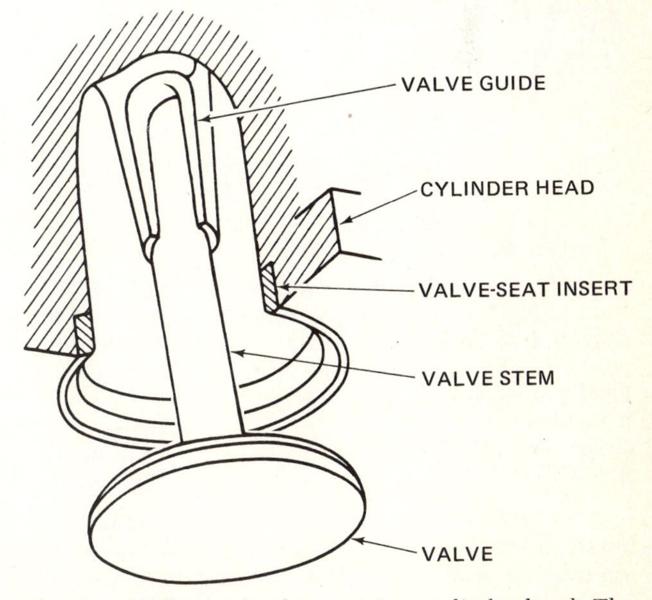

Fig. 5-11. Valve and valve seat in a cylinder head. The cylinder head and valve guide have been cut away so that the valve stem can be seen.

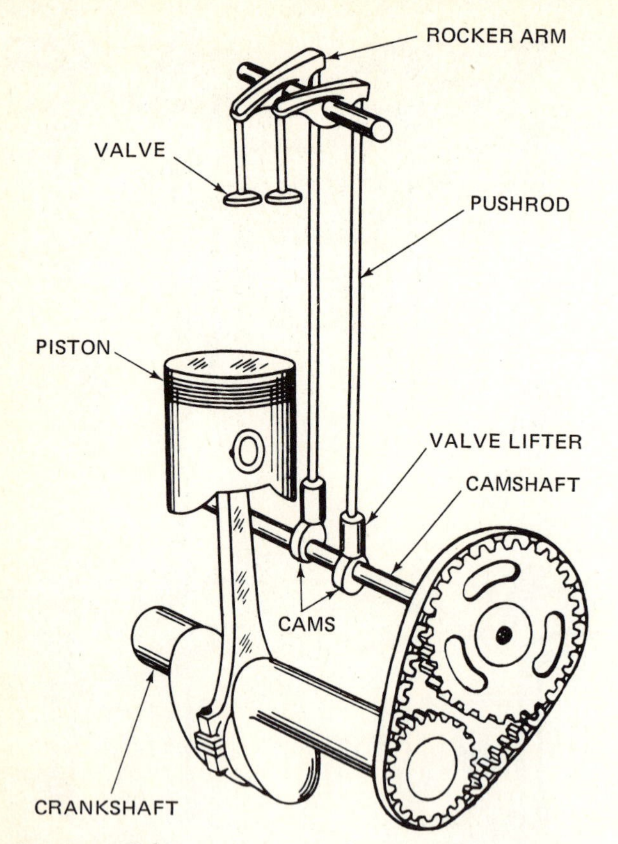

Fig. 5-12. Valve-operating mechanism for an I-head, or overhead-valve, engine. Only the essential moving parts for one cylinder are shown.

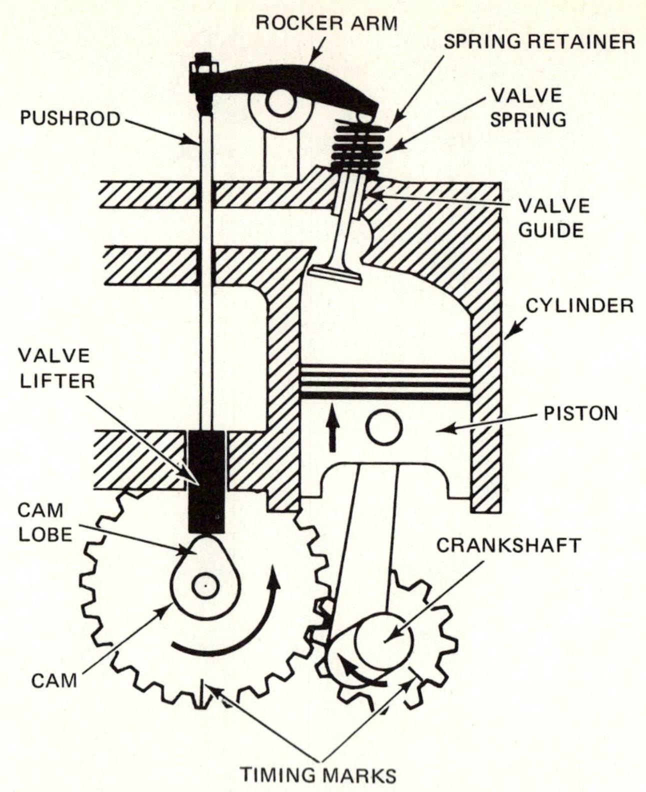

Fig. 5-14. Valve train on an engine using a pair of gears to drive the camshaft. The cam lobe has pushed the valve lifter and pushrod up, so the valve is opened.

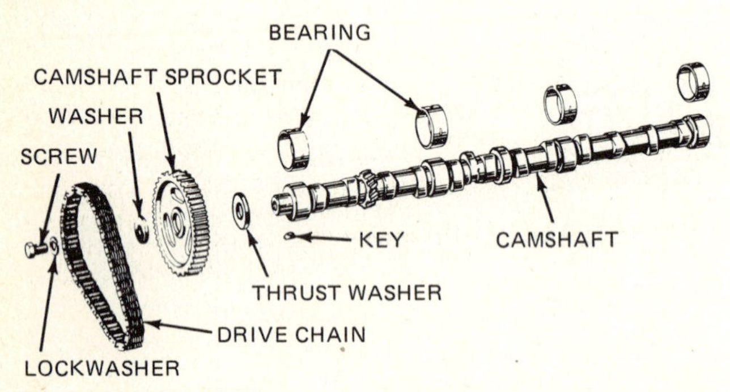

Fig. 5-13. Camshaft and related parts for a six-cylinder engine. (*Ford Motor Company*)

fold and intake-valve port. For a comprehensive discussion of vacuum and atmospheric pressure, see *Automotive Fuel, Lubricating, and Cooling Systems,* another book in the McGraw-Hill Automotive Technology Series.

2. COMPRESSION (FIG. 5-17) After the piston reaches BDC, or the lower limit of its travel, it begins to move upward. As this happens, the intake valve closes. The exhaust valve is also closed, and so the cylinder is sealed. As the piston moves upward (pushed now by the revolving crankshaft and connecting rod), the air-fuel mixture is compressed. By the time the piston reaches TDC, the mixture has

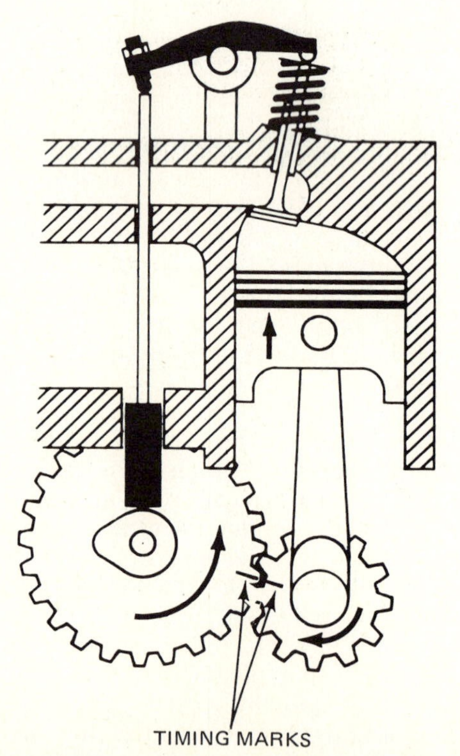

Fig. 5-15. Valve train of Fig. 5-14, showing how the valve closes when the cam lobe moves out from under the valve lifter.

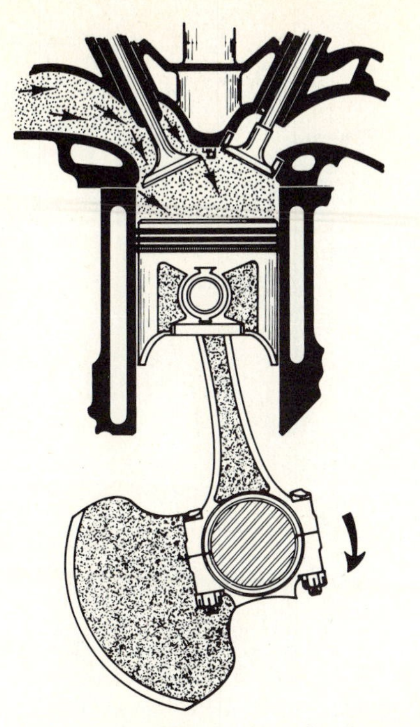

Fig. 5-16. Intake stroke. The intake valve (*upper left*) has opened. The piston is moving downward, drawing a mixture of air and gasoline vapor into the cylinder.

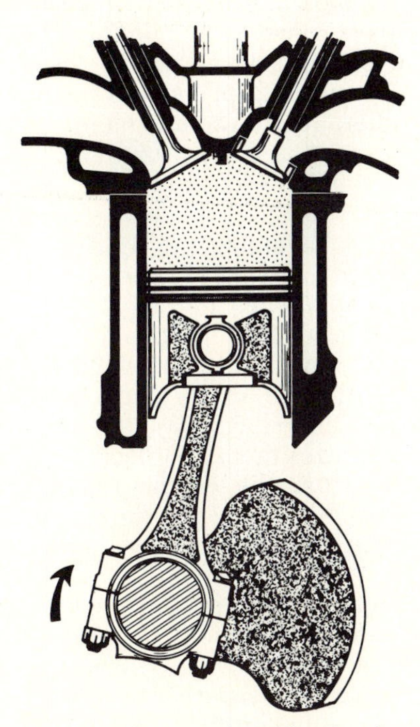

Fig. 5-17. Compression stroke. The intake valve has closed. The piston is moving upward, compressing the mixture.

been compressed to as little as one-seventh of its original volume or even less. This compression of the air-fuel mixture increases the pressure in the cylinder. Or, to say it another way, the molecules that compose the air-fuel mixture are pushed closer

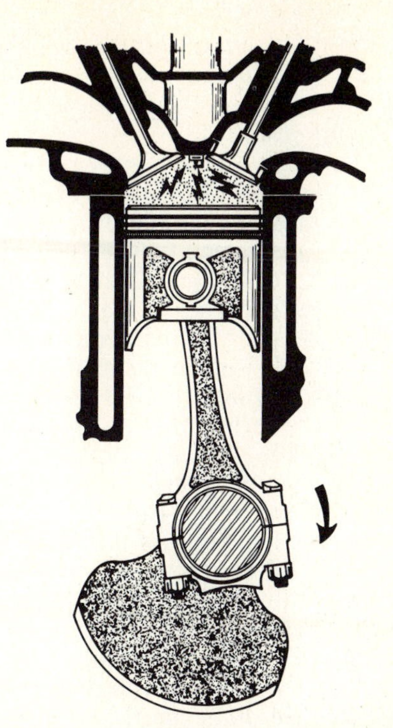

Fig. 5-18. Power stroke. The ignition system has delivered a spark to the spark plug that ignites the compressed mixture. As the mixture burns, high pressure is created which pushes the piston downward.

together, and therefore they bump into the cylinder walls and pistonhead more often. The increasing frequency of these bumps means a stronger "push" is registered on the walls and head; the pressure is increased. As they are pushed closer together, the moleules also collide with each other more frequently, which, in turn, makes them move faster. We know that more rapid motion and increased temperature mean the same thing. Therefore, when the air-fuel mixture is compressed, not only does the pressure in the cylinder increase, but also the temperature of the mixture increases.

3. *POWER* (*FIG. 5-18*) As the piston reaches TDC on the compression stroke, an electric spark is produced at the *spark plug*. The spark plug consists essentially of two wire electrodes that are electrically insulated from each other. The *ignition system* (part of the electric system discussed later in this chapter) delivers a high-voltage surge of electricity to the spark plug to produce the spark. The spark ignites, or sets fire to, the air-fuel mixture. The mixture begins to burn very rapidly, increasing the cylinder pressure to as much as 600 psi (pounds per square inch) [42.1848 kg/cm^2] or even more. This means that the hot gases are pushing against every square inch [6.452 cm^2] of the combustion chamber and pistonhead with a pressure of 600 psi [42.1848 kg/cm^2] or more. For example, a piston 3 inches [76.2 mm] in diameter with a head area of about 7 square inches [45.16 cm^2] would have a pressure on it of over 2 tons [1,814.36 kg]. This terrific push against the piston forces it downward, and a

power impulse is transmitted through the connecting rod to the crankpin on the crankshaft. The crankshaft is rotated as the piston is pushed down by the pressure above it.

Let us take a look at the activities we have just described from the molecular point of view. That is, let us see how the increased pressure can be explained by considering the air-fuel mixture as a vast number of molecules. We have already noted (in the previous paragraph) that compressing the mixture increases both its temperature and its pressure. The molecules move faster (higher temperature) and bump into the cylinder walls and into the pistonhead more often (higher pressure). When combustion takes place, the hydrocarbon molecules of gasoline are violently split apart into hydrogen and carbon atoms. The hydrogen and carbon atoms then unite with oxygen atoms in the air (see ⊘ 4-8, "Combustion"). All this activity sets the molecules into extremely rapid motion (still higher temperature, which may momentarily reach 6000 degrees Fahrenheit [3316°C]). The molecules begin to bombard the cylinder walls and pistonhead much harder and more often. In other words, the pressure is increased greatly.

It may be a little difficult, at first, to visualize a 2-ton [1,814.36 kg] push on the pistonhead as resulting from the bombardment of molecules that are far too small to be seen. But remember that there are billions upon billions of molecules in the combustion chamber, all moving at speeds of many miles [kilometers] a second. Their combined hammering on the pistonhead causes the high pressure that is registered.

4. *EXHAUST (FIG. 5-19)* As the piston reaches BDC again, the exhaust valve opens. Now, as the piston moves up on the exhaust stroke, it forces the burned gases out of the cylinder through the exhaust-valve port. Then, when the piston reaches TDC, the exhaust valve closes and the intake valve opens. Now, a fresh charge of air-fuel mixture will be drawn into the cylinder as the piston moves down again toward BDC. The above four strokes are continuously repeated during the operation of the engine.

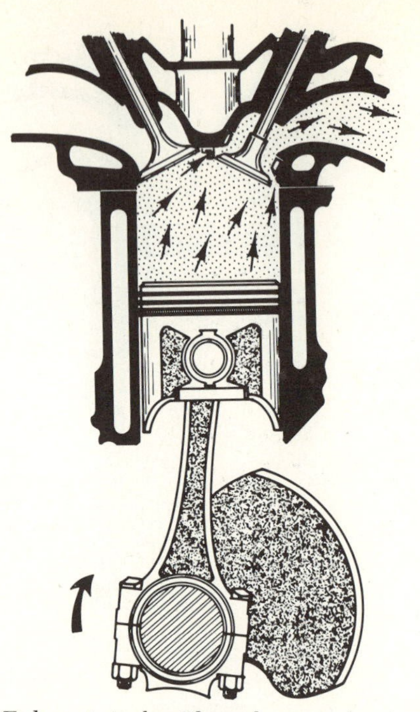

Fig. 5-19. Exhaust stroke. The exhaust valve (*upper right*) has opened. The piston is moving upward, forcing the burned gases from the cylinder.

⊘ 5-5 Multiple-cylinder Engines A single-cylinder engine provides only one power impulse every two crankshaft revolutions and is delivering power only one-fourth of the time. To provide for a more continuous flow of power, modern engines use four, six, eight, or more cylinders. The power impulses are so arranged as to follow one another or overlap (on six- and eight-cylinder engines). This gives a more nearly even flow of power from the engine.

⊘ 5-6 Flywheel Even though the power impulses in a multicylinder engine follow each other or overlap to provide a fairly even flow of power, additional leveling off of the power impulses is desirable. This would make the engine run still more smoothly. To achieve this, a *flywheel* is used (Fig. 5-20). The flywheel is a fairly heavy steel wheel attached to the rear end of the crankshaft.

To get a better idea of how the flywheel does its job, let us look at a single-cylinder engine. This engine delivers power only one-fourth of the time: during the power stroke. During the other three strokes it is absorbing power—to push out the exhaust gas, to produce a vacuum on the intake stroke, and to compress the air-fuel mixture. Thus, during the power stroke the engine tends to speed up. During the other strokes, it tends to slow down. Any rotating wheel, including the flywheel, resists any effort to change its speed of rotation [owing to inertia (see ⊘ 4-30)]. When the engine tends to speed up, the flywheel resists the speedup. When the engine tends to slow down, the flywheel resists the slowdown. Of course, in the single-cylinder engine, some speedup and slowdown times occur, but the flywheel minimizes them. In effect, the flywheel absorbs power from the engine during the power stroke (or speedup time) and then gives it back to the engine during the other three strokes (or slowdown time).

In the multicylinder engine, the flywheel acts in a similar manner to smooth out further the peaks and valleys of power flow from the engine. In addition, the flywheel forms part of the clutch, as already explained (⊘ 2-9). The flywheel also has teeth on its outer circumference that mesh with the electric-starting-motor drive pinion when the engine is being cranked to start it. This is explained in more detail in ⊘ 5-11.

Check Your Progress

Progress Quiz 5-1 The following quiz will help you check yourself on the progress you have been mak-

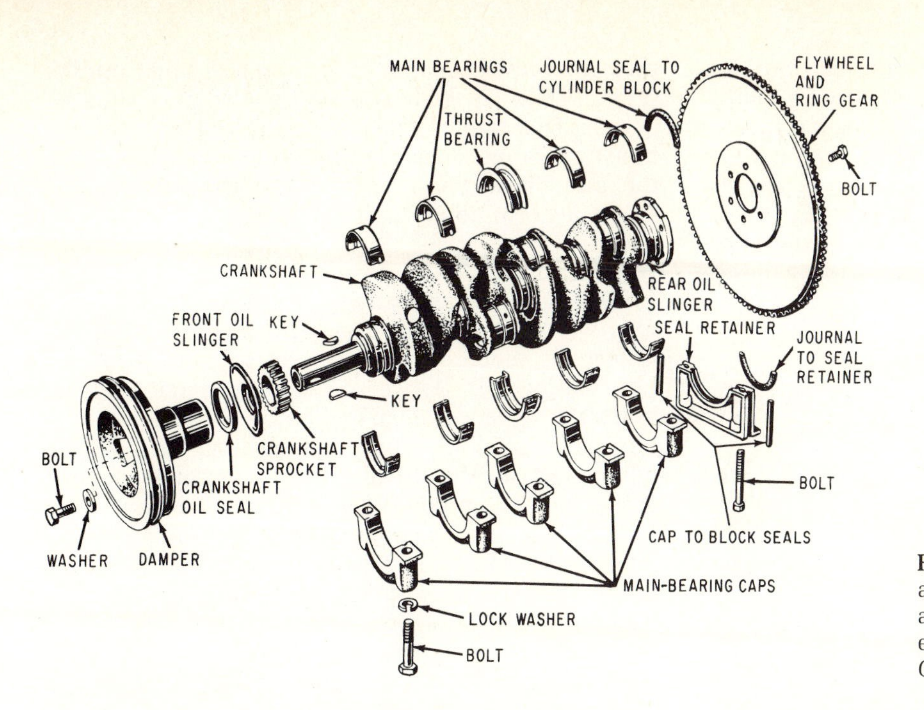

Fig. 5-20. Crankshaft and related parts used in an eight-cylinder V-type engine. (*Ford Motor Company*)

ing in understanding this chapter. If you have difficulty with the questions, reread the past few pages. Remember, most students reread their lessons several times, so don't be discouraged if you don't remember everything the first or second time. Notice that the questions usually refer to the most important facts, the most essential details, in the pages you have just read. The questions spotlight these important details and help you to remember them. These questions are put here for your own benefit to enable you to check up on yourself. If you are not sure about the answer to any question, reread the preceding pages of this chapter before attempting to give the answer.

Correcting Parts Lists The purpose of this exercise is to give you practice in spotting unrelated parts in a list. For example, in the list "cylinder, piston, rings, brake shoe, connecting rod," you would see that "brake shoe" does not belong because it is the only part named that does not belong in an engine.

In each of the following lists, you will find one item that does not belong. Write each list in your notebook, but *do not write* the item that does not belong.

1. The four piston strokes are intake, exhaust, reverse, compression, and power.
2. To operate, the engine requires a fuel system, lubricating system, cooling system, brake system, and electric system.
3. Engine parts include the piston, piston pin, connecting rod, crankshaft, clutch, and rod cap.
4. The valve system includes the camshaft, valve lifter, poppet valve, flywheel, valve spring, and valve guide.
5. Engine parts include the cylinder block, cylinder head, pistons, crankshaft, connecting rods, differential, camshaft, and valves.

Completing the Sentences The sentences that follow are incomplete. After each sentence there are several words or phrases, only one of which will correctly complete the sentence. Write each sentence in your notebook, selecting the proper word or phrase to complete it correctly.

1. The connecting rod is attached to the piston by the: (*a*) rod cap, (*b*) piston pin, (*c*) cap bolts, (*d*) big-end bearing.
2. In the standard engine each cylinder has: (*a*) one valve, (*b*) two valves, (*c*) three valves, (*d*) four valves.
3. The four strokes in an Otto-cycle engine are, in order: (*a*) intake, power, exhaust, and compression; (*b*) intake, exhaust, compression, and power; (*c*) intake, compression, power, and exhaust.
4. The two types of engine valves are: (*a*) intake and port, (*b*) intake and inlet, (*c*) intake and exhaust.
5. The valve is opened as the cam lobe on the cam raises the: (*a*) valve lifter, (*b*) bearing, (*c*) piston pin, (*d*) valve guide.
6. The rod-bearing cap attaches the connecting rod to the: (*a*) crankpin, (*b*) camshaft, (*c*) piston pin, (*d*) rod big end.
7. During combustion, the pressure in the cylinder may increase to as much as: (*a*) 60 psi [4.2185 kg/cm^2], (*b*) 600 psi [42.1848 kg/cm^2], (*c*) 6,000 psi [421.848 kg/cm^2].
8. During the power stroke, the intake and exhaust valves are, respectively: (*a*) closed and opened, (*b*) opened and closed, (*c*) closed and closed.
9. The camshaft has a separate cam for each: (*a*) engine valve, (*b*) piston, (*c*) cylinder, (*d*) crankpin.
10. The device for smoothing out the power impulses from the engine is called the: (*a*) crankshaft, (*b*) camshaft, (*c*) flywheel, (*d*) clutch.

⊘ 5-7 Engine Accessories The engine requires four separate mechanisms, or accessory systems, for its operation. These are the *fuel system, lubricating system, electric system,* and *cooling system.* In addition, an *exhaust system* is provided to carry away the burned gases exhausted from the engine cylinders. These various systems are actually essential parts of the complete engine. However, they are discussed separately since each system has a definite and individual job to do. Brief descriptions of these systems follow. In later chapters the systems are described in greater detail.

⊘ 5-8 Fuel System The gasoline-engine fuel system (Fig. 5-21) supplies the engine with a combustible mixture of fuel and air. It consists of a *fuel tank,* a *fuel pump,* a *fuel filter,* a *carburetor,* an *intake manifold,* and connecting *fuel lines* (metal tubes). The fuel pump pumps the liquid gasoline from the fuel tank to the carburetor. The carburetor mixes the gasoline with air and delivers the mixture to the intake manifold. From there, the mixture passes the intake-valve ports (when valves are open) and enters the engine cylinders.

The carburetor varies the proportion of fuel and air to suit different operating conditions. For example, the mixture must be rich, that is, have a high proportion of fuel, when the engine is cold. The reason for this is that fuel does not vaporize readily at low temperatures. Therefore, extra fuel must be added to the air-fuel mixture so that there will be enough vaporized fuel to form a combustible mixture. For cold starting, the ratio may be as low as 1 pound [0.454 kg] of gasoline for every 9 pounds [4.082 kg] of air. However, when the engine is warmed up and in operation at medium speeds, a relatively lean air-fuel mixture is required; the ratio may be about 1 pound [0.454 kg] of gasoline for every 15 pounds [6.804 kg] of air.

⊘ 5-9 Exhaust System An exhaust system is shown in Fig. 5-22. After the air-fuel mixture burns in the engine cylinder, the exhaust valve opens and the upward-moving piston forces the burned gases from the cylinder on the exhaust stroke. The gases pass into the exhaust manifold (Fig. 14-35) and from there

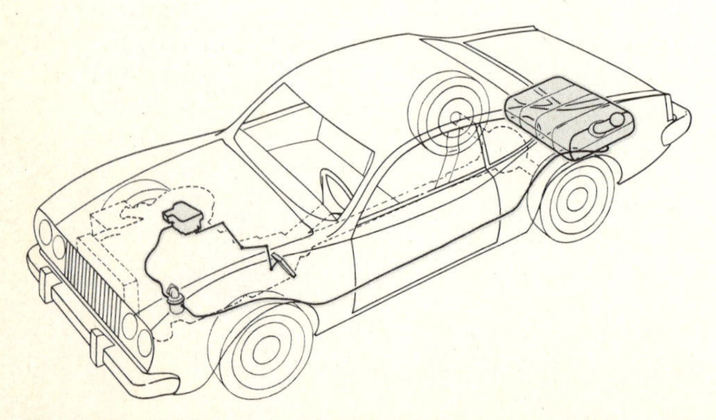

Fig. 5-21. Fuel system for a car with a V-8 engine. (*Ford Motor Company*)

through the exhaust pipe into the exhaust muffler (Fig. 13-25). The exhaust muffler provides a series of passages and chambers through which the exhaust gases must pass before being discharged into the air. These passages and chambers muffle the exhaust noise, thus quieting the engine.

⊘ 5-10 Lubricating System When two metal parts move over each other, they will wear away rapidly unless the metal surfaces in contact are lubricated. Lubricating oil between moving metal surfaces holds the surfaces apart. In effect, each surface slides on a film of oil. So long as the metal surfaces are held apart by the oil film, they cannot come into actual contact and therefore cannot wear each other. The action might be compared to pulling a boat up a river. As long as the water is deep enough to float the boat, there is no appreciable wear on either the bottom of the boat or the bed of the river. But if the water level should fall so that the protective films of water are removed from between the bottom of the boat and the river bed, there is a different action. Now, dragging the boat along the river bed will cause "wear" on both the bottom of the boat and the river bed. In a like manner, depriving moving metal parts of lubricating oil will permit them to slide on each other and wear rapidly.

The lubricating system used in the automotive engine is designed to supply oil to all moving parts. In a sense, these parts "float" on a film of oil which prevents actual metal-to-metal contact. In the engine, oil is supplied to crankshaft and camshaft journals rotating in their supporting bearings. Also, oil is supplied to the cylinder walls so that the pistons and rings will slide easily and smoothly without undue piston, ring, or wall wear. In a like manner, other moving engine parts are supplied with oil.

Automotive-engine lubricating systems are of the *pressure-feed* type. Figure 5-23 shows a typical system. A pump forces oil onto the various moving metal parts in the engine. Holes are drilled through the cylinder block and crankshaft, and oil is forced through these holes. The oil flows onto the bearing surfaces of the crankshaft. This oil coating allows the crankshaft to rotate freely in its bearings. Also, the crankpins can rotate in the connecting-rod big-end bearings. Oil thrown off these bearings covers the cylinder walls so that the walls, pistons, and piston rings are lubricated. Some oil gets to the piston-pin bearings so that they are lubricated. Other oil passages carry oil to the cylinder head so that the rocker arms and valve stems are lubricated. Oil flows through holes in the cylinder block to lubricate the camshaft bearings. The oil drains off all these parts and returns to the oil pan at the bottom of the engine. From there, it is picked up again by the oil pump and recirculated through the engine.

In addition to providing lubrication, the oil also carries away some of the heat from the moving engine parts. The oil picks up heat, becomes hotter, and then, when it returns to the oil pan, gives up heat

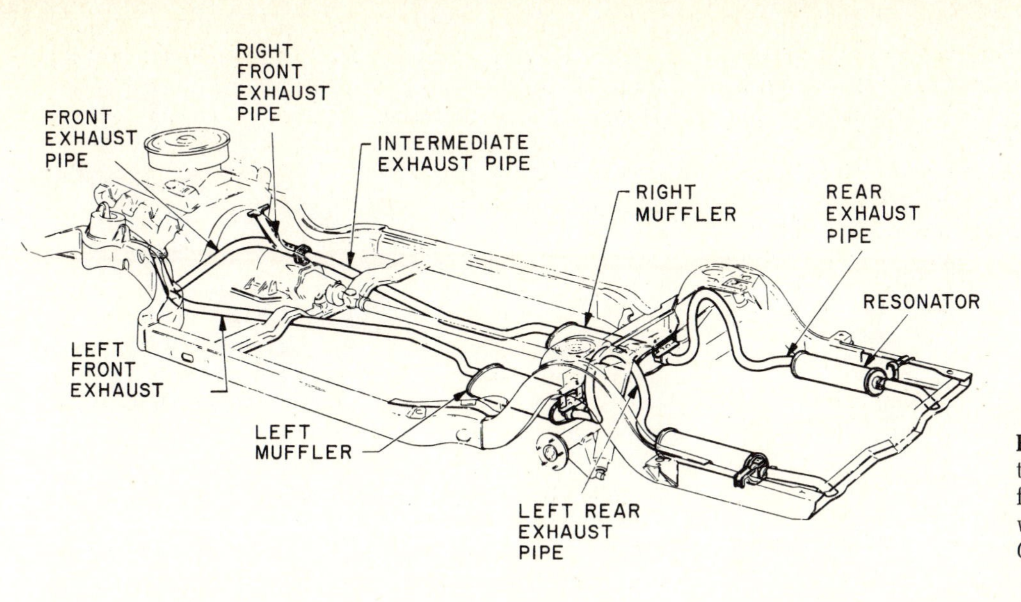

Fig. 5-22. Exhaust system in place on a car frame. (*Buick Motor Division of General Motors Corporation*)

and cools off. The oil pan gives up heat to the passing air circulating around and under it.

NOTE: Further details of engine lubricating systems are to be found in "Engine Lubricating Systems," Chap. 17.

⊘ 5-11 Electric System The *electric system* (Fig. 5-24) includes the *storage battery, starting motor, alternator, regulator,* and *ignition system,* as well as the wires and switches for connecting these various units. Lights, radio, heater, indicating gauges, and other electrically operated devices, although they are a part of the electric system, are usually considered accessory items since they are not absolutely essential to the operation of the car.

What follows is merely an introduction to the electrical units in the system. For details, refer to *Automotive Electrical Equipment,* another book in the McGraw-Hill Automotive Technology Series.

1. *STORAGE BATTERY* The storage battery is an *electrochemical* device, which means that its operation depends upon both electrical and chemical actions. The battery is a source of electric current when the engine is being cranked with the starting motor. It also supplies current when the alternator is not able to carry the electric load. When current is withdrawn from the battery, chemical actions take place to produce the flow of current. The chemicals in the battery are, in a sense, used up by this action. Thus, after a certain amount of current has been withdrawn for a certain length of time, the battery becomes "discharged." To "recharge" the battery, current from some external source, such as an alternator or battery charger, must be forced through it in the reverse, or charging, direction.

2. *STARTING MOTOR* The starting motor is a special direct-current electric motor that starts the engine by rotating the crankshaft when the starting-motor switch is closed. Closing the switch connects the motor to the battery. There is a special gearing arrangement between the starting-motor drive pinion and the engine flywheel that causes the flywheel and crankshaft to turn when the starting motor operates.

3. *ALTERNATOR* The alternator converts mechanical energy from the automobile engine into a flow of electric current. This current restores the battery to a charged condition when it has become run down, or discharged. It also operates electrical devices, such as the ignition system, lights, and radio. The alternator is usually mounted on one side of the engine and is driven by the engine fan belt.

4. *REGULATOR* Under some conditions, the alternator could produce too much current, and this would damage the various connected electrical devices. To prevent such damage, an alternator output regulator is used. The regulator controls the amount of current the alternator produces, allowing the alternator to produce a high current when the battery is in a discharged condition and electrical devices are turned on. When the battery becomes charged and electric units are turned off, the regulator reduces the current produced to the amount needed to meet the operating requirements of the system. In Fig. 5-24, the regulator is shown as a separate unit. However, in many late-model alternators the regulator is built into the alternator itself.

5. *IGNITION SYSTEM* The ignition system (Fig. 5-25, p. 52) provides the high-voltage surges, or electric sparks, that ignite the compressed air-fuel mixture in the engine cylinders. After the fuel system has delivered the air and gasoline-vapor mixture to the engine cylinder and the mixture has been compressed by the piston compression stroke, it must be ignited. The ignition system does this job by producing sparks at the spark-plug gap in the engine cylinders. The sparks set the compressed mixture on fire so that it burns and creates the high pressure that drives the piston down on the power stroke. Chapter 15 describes the operation of the ignition system in detail.

6. *WIRING AND SWITCHES* The wiring (Fig. 5-24) connects the various electric units and switches and

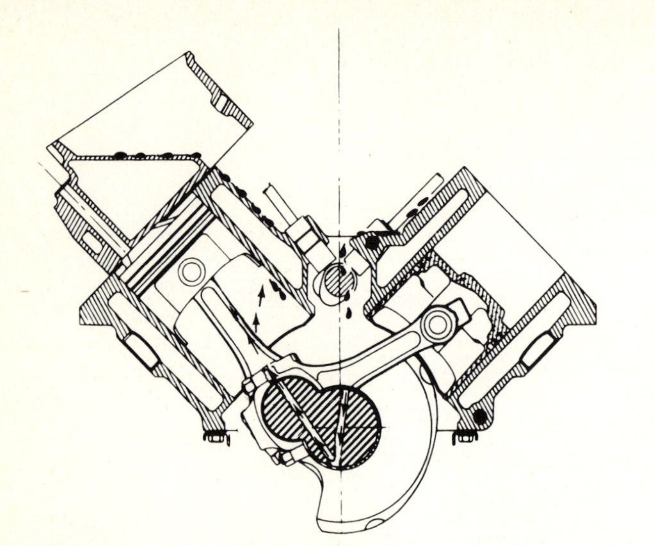

CYLINDER WALLS ARE OILED BY OIL THROWN OFF PRESSURE-FED CONNECTING-ROD BEARINGS

CYLINDER-WALL AND CAMSHAFT-LOBE OILING

OIL-FILTER BYPASS VALVE

OIL FILTER AND DISTRIBUTOR OILING

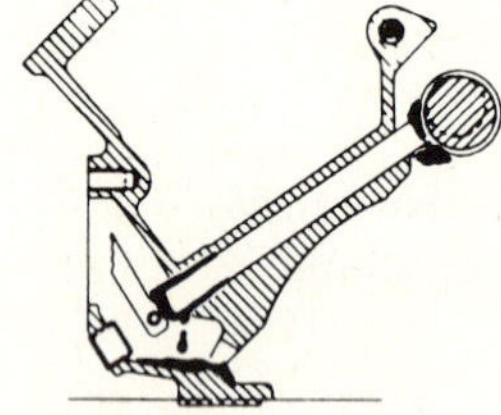

FUEL-PUMP PUSHROD OILING

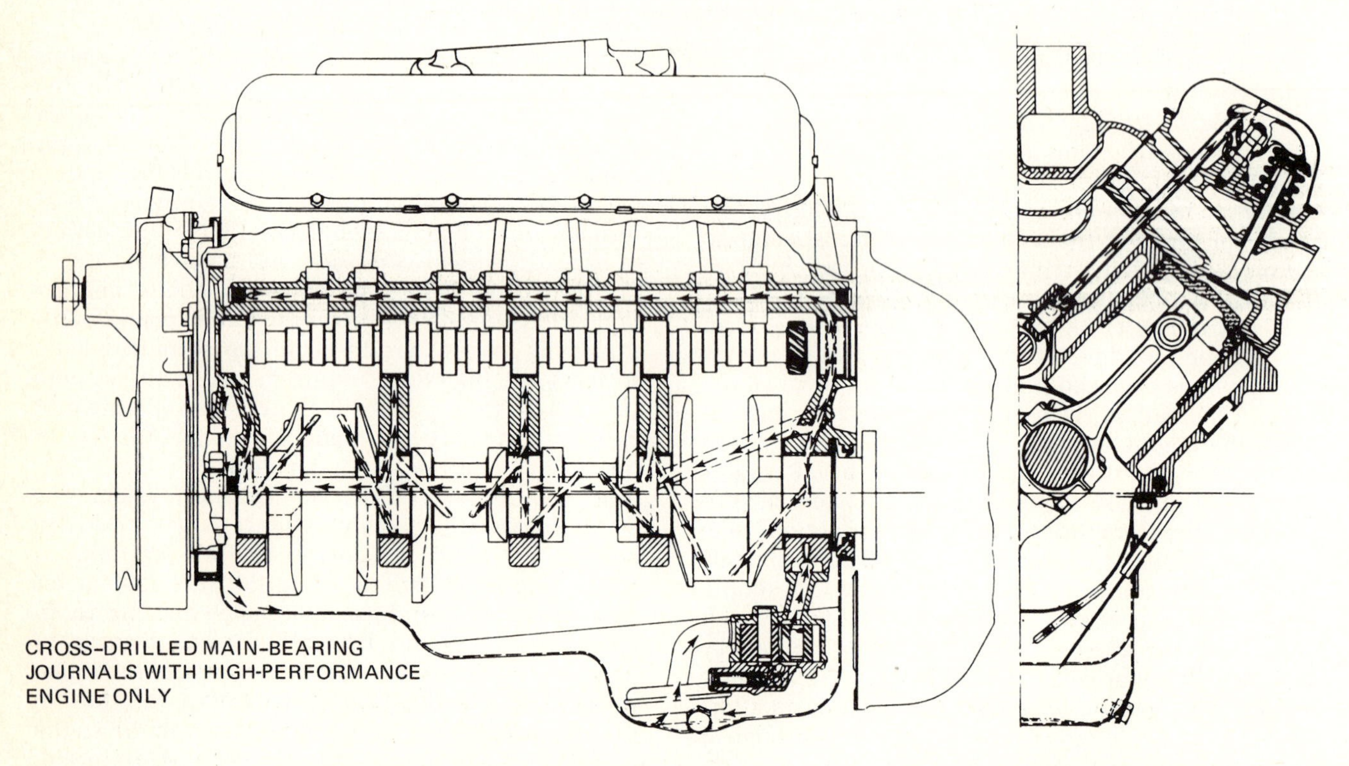

CRANKCASE AND CRANKSHAFT OILING

VALVE-MECHANISM OILING

Fig. 5-23. Lubricating system of a V-8 overhead-valve engine. Arrows show the flow of oil to the moving parts in the engine. (*Chevrolet Motor Division of General Motors Corporation*)

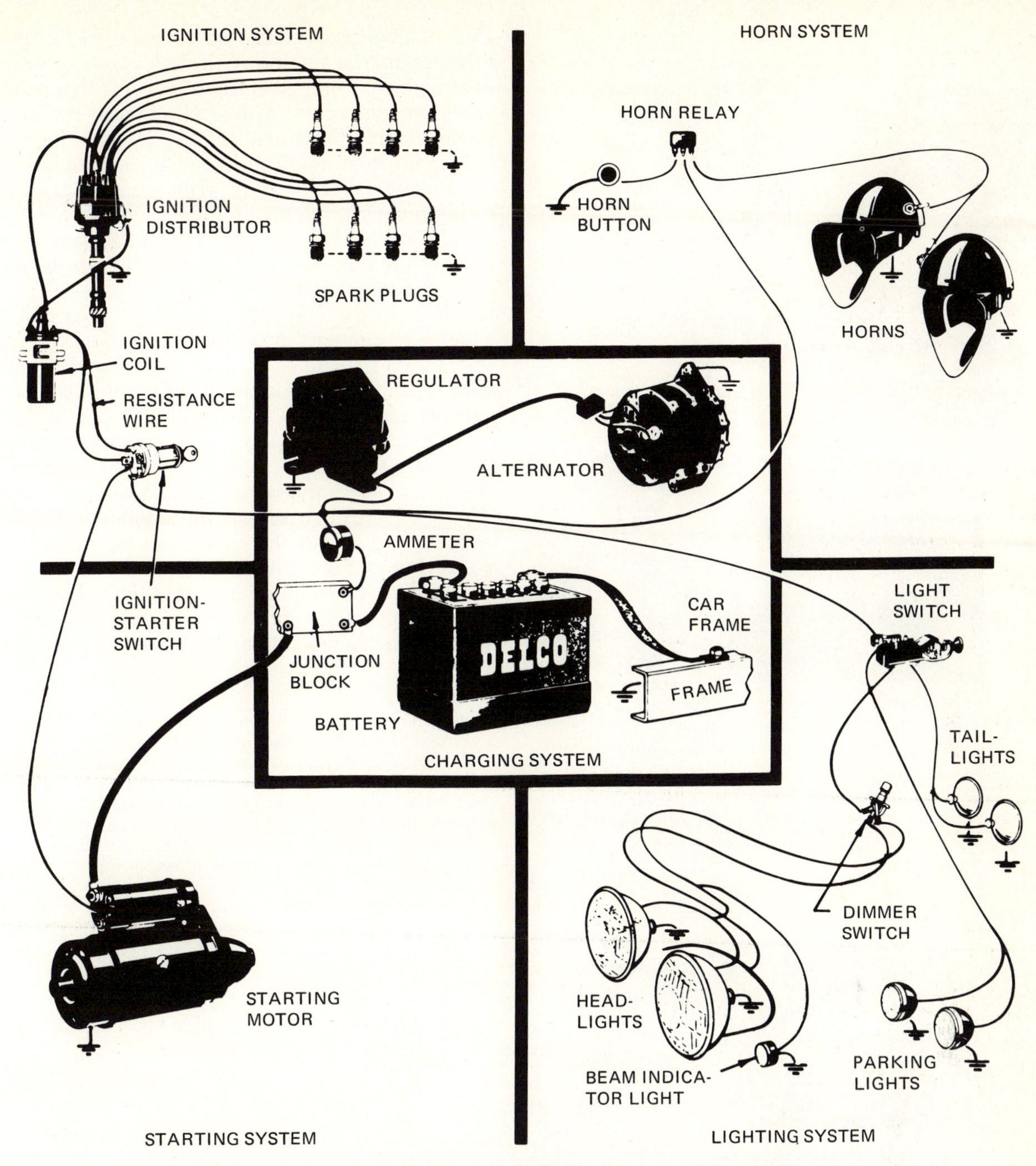

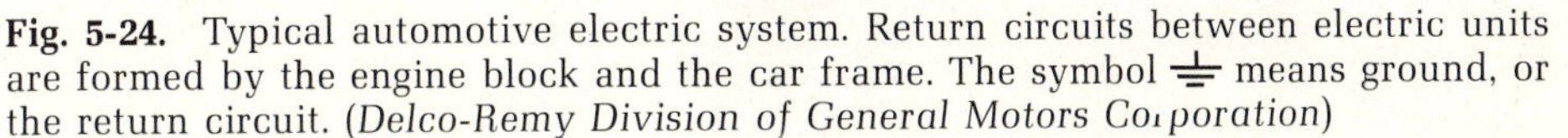

Fig. 5-24. Typical automotive electric system. Return circuits between electric units are formed by the engine block and the car frame. The symbol ⏚ means ground, or the return circuit. (*Delco-Remy Division of General Motors Corporation*)

serves as a path through which the electric current can flow from one unit to another. Switches placed in these circuits are forms of valves that can close or open the circuits to permit or prevent the flow of current. The wires are made up of conducting materials, such as copper, that freely conduct current between the electric units. Some materials, such as rubber and glass, are nonconductors, or insulators; they do not allow current to flow through them. Such substances are used to cover and insulate the wires so that the current will be kept within the proper circuits and paths (and will not *short-circuit*).

Note that the automotive wiring system shown in Fig. 5-24 is a one-wire system. The electric units are normally connected to each other by one wire. The return circuit is through the car frame and the engine block. This return circuit is also called the *ground;* all the electric units are connected to it.

7. LIGHTS, HEATER, RADIO, AND INDICATING DEVICES The lights and heater add to the flexibility, comfort, and convenience of the car, and the indicating gauges and lights keep the driver informed as to the engine temperature, oil pressure, amount of fuel in the tank, and battery charging rate.

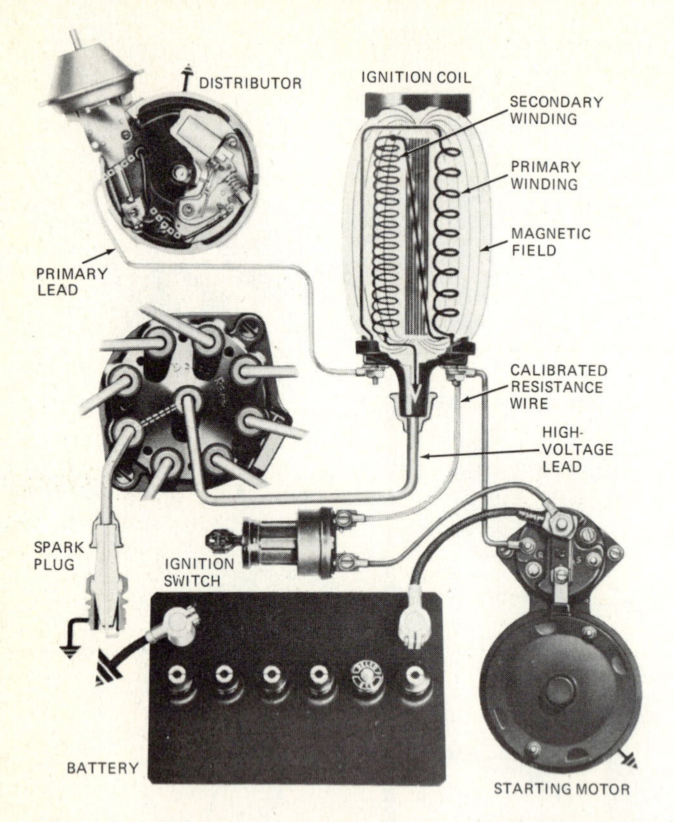

Fig. 5-25. Typical ignition system. It consists of the battery (source of power), ignition switch, ignition coil (shown schematically with magnetic lines of force indicated), distributor (shown in top view with its cap removed and placed below it), spark plugs (one shown in sectional view), and wiring. (*Delco-Remy Division of General Motors Corporation*)

⊘ 5-12 Cooling System The burning of the air-fuel mixture in the engine cylinders produces a great deal of heat. Temperatures of several thousand degrees are generated in the cylinders as the mixture burns. Some of this heat is carried out of the engine by the lubricating oil. Some of it escapes in the hot exhaust gases. Some of it is absorbed by the evaporating fuel entering the cylinders in the air-fuel mixture.[1] The *cooling system* (Fig. 5-26) carries away most of the remaining excess heat. This prevents the engine from becoming too hot. Excessive engine temperatures would damage or ruin the engine. At excessive temperatures the lubricating oil would lose its effectiveness, and so wear would increase very rapidly. High temperatures would also damage various engine parts.

Circulating coolant is the cooling medium in the cooling system. The combustion chambers of the engine are surrounded by *pockets,* or *water jackets* (Fig. 5-26), through which the coolant can flow. A water pump causes the coolant to be pumped from the bottom of the radiator, through the water jackets, and back to the radiator again. As the coolant passes through the water jackets, it absorbs heat and becomes hot. Then, as it enters the top of the radiator,

[1]You will recall that when a liquid *evaporates* (undergoes a change of state), it absorbs heat (as explained in ⊘ 4-10 and 4-11). This action utilizes some of the heat developed in the engine and provides a certain amount of cooling as the fuel evaporates. In aircraft piston engines this effect is of great importance, and under certain circumstances requiring maximum power, the pilot may enrich the mixture in order to keep engine temperature down.

Fig. 5-26. Cutaway view of a six-cylinder overhead-camshaft engine showing the cooling system. Arrows indicate the direction of water flow through the system. Only the top and the bottom of the radiator are shown. (*Pontiac Motor Division of General Motors Corporation*)

it starts to cool off. The radiator has numerous coolant passages through which the coolant flows. Around these small coolant passages are numerous air passages. The engine fan pulls air through these air passages, thus removing heat from the radiator. Thus, as the coolant passes down through the radiator, it is cooled. The cooler coolant is then pumped back through the water jackets in the engine. The pump keeps the coolant in continual circulation so that the coolant continues to transfer heat from the engine to the air passing through the radiator.

Check Your Progress

Progress Quiz 5-2 Once again the following questions allow you to check up on yourself. You have made a good start in reading the book, and if you have taken the previous "Progress Quizzes," you know how well you have been absorbing the information you have been reading. Naturally, you are not going to be able to remember everything, but you probably do not have too much difficulty recalling the essential facts. These essential facts are referred to in the following questions. Answering these questions will help you fix the facts firmly in your mind. If you find questions that stump you, turn back in the chapter to the pages that will help you answer them.

Correcting Parts Lists The purpose of this exercise is to help you to spot unrelated parts in a list. For example, in the list "distributor, coil, spark plugs, fuel pump, ignition switch," you would see that "fuel pump" does not belong since it is the only part that does not belong to the ignition system.

In each of the following lists, you will find one item that does not belong. Write each list in your notebook, but *do not write* the item that does not belong.

1. Fuel-system parts include the carburetor, fuel pump, oil pump, fuel line, and fuel tank.
2. The electric system includes the storage battery, starting motor, alternator, carburetor motor, regulator, ignition distributor, and coil.
3. The ignition system includes the ignition switch, distributor, spark plugs, coil, spark pump, and wiring.
4. The cooling system includes the water pump, water jackets, radiator, coolant, and float bowl.

Completing the Sentences The sentences that follow are incomplete. After each sentence there are several words or phrases only one of which will correctly complete the sentence. Write each sentence in your notebook, selecting the proper word or phrase to complete it correctly.

1. The fuel system contains a fuel tank, fuel line, fuel: (*a*) valve, and muffler; (*b*) pump, and carburetor; (*c*) pressure gauge, and valve; (*d*) vacuum gauge, and carburetor.
2. When the carburetor is delivering a rich mixture, it means that the: (*a*) mixture has more gasoline than air, (*b*) proportion of gasoline is increased, (*c*) proportion of air is increased (*d*) throttle opening is increased.
3. A mixture of 15 lb [6.804 kg] of air and 1 lb [0.454 kg] of gasoline is considered to be a: (*a*) relatively rich mixture, (*b*) relatively lean mixture, (*c*) relatively noncombustible mixture, (*d*) good mixture for acceleration.
4. The ignition system supplies a high-voltage surge to the spark plug in the cylinder toward the end of the: (*a*) intake stroke, (*b*) compression stroke, (*c*) power stroke, (*d*) exhaust stroke.
5. In the cooling system, the water pump circulates coolant between the engine water jacket and the: (*a*) radiator, (*b*) water heater, (*c*) water tank.

CHAPTER 5 CHECKUP

NOTE: Since the following is a chapter review test, you should review the chapter before taking the test.

Once again you will want to test your knowledge of the subjects covered in the chapter you have just completed. The questions that follow have two purposes: One is to test your knowledge; the other is to help you to review the chapter and fix the facts firmly in your mind. It may be that you will not be able to answer all the questions offhand. If this happens, turn back into the chapter and reread the pages that will give you the answer. For instance, under "Listing Parts" you are asked to list the parts in the valve mechanism. If you cannot remember them all, turn back to the illustration of the valve mechanism in the chapter and refer to it while writing your list. The act of writing the names of the parts will help you to remember them.

NOTE: Write your answers in your notebook. Then later, when you finish *Automotive Engines*, you will find your notebook filled with valuable information to which you can quickly refer.

Completing the Sentences The sentences that follow are incomplete. After each sentence there are several words or phrases, only one of which will correctly complete the sentence. Write each sentence in your notebook, selecting the proper word or phrase to complete it correctly.

1. The parts that must be added to the piston to ensure a good seal with the cylinder wall are the: (*a*) piston pins, (*b*) connecting rods, (*c*) piston rings, (*d*) gaskets.
2. The part that tends to keep the valve closed is called the: (*a*) guide, (*b*) lifter, (*c*) spring, (*d*) cam, (*e*) retainer.
3. The crankshaft has at one end a: (*a*) piston, (*b*) connecting rod, (*c*) camshaft, (*d*) flywheel, (*e*) cylinder head.

4. For each crankshaft revolution, the camshaft revolves: (*a*) one-half turn, (*b*) one turn, (*c*) two turns.
5. To bring gasoline from the fuel tank to the carburetor, the fuel system includes: (*a*) an accelerator pump, (*b*) a vacuum pump, (*c*) a fuel pump, (*d*) a float.
6. The term "enriching the mixture" means: (*a*) adding more air, (*b*) adding greater air speed, (*c*) adding greater vacuum, (*d*) adding more gasoline.

Listing Parts In the following, you are asked to list parts that make up the various automotive components discussed in this chapter. Write these lists in your notebook.

1. List the four piston strokes in the four-cycle engine.
2. List the parts that move up and down in the engine cylinder.
3. List the parts in the valve mechanism.
4. List the parts through and past which the air moves as it passes from outside the engine into the engine cylinder.
5. List the most essential, or basic, parts of the carburetor.
6. List the major components of the fuel system.
7. List the major parts of the fuel pump.
8. List the major components of the electric system.
9. List the main parts in the ignition system.
10. List the parts through which the water moves in the cooling system.

Purpose and Operation of Components In the following, you are asked to write the purpose and operation of certain components of the automobile discussed in this chapter. If you have any difficulty in writing your explanations, turn back in the chapter and reread the pages that will give you the answer. Then write down your explanation. Don't copy; try to tell it in your own words. This is a good way to fix the explanation firmly in your mind. Write the answers in your notebook.

1. Briefly describe the actions that take place during the four piston strokes.
2. What is the purpose of the connecting rod?
3. List the parts in the valve mechanism.
4. What is the purpose of the flywheel?
5. Describe the action of the lubricating system.
6. Describe the action of the ignition system.
7. Describe the action of the cooling system.

SUGGESTIONS FOR FURTHER STUDY

If you would like to study the engine electric, cooling, lubricating, and fuel systems further, there are several things you can do. First, you can read *Automotive Fuel, Lubricating, and Cooling Systems* and *Automotive Electrical Equipment,* which are other books in this McGraw-Hill Automotive Technology Series. Also, you can inspect your own and your friends' cars, as well as cars in the school automotive shop. You can go to a friendly service garage where repair work on these parts is done. By watching what goes on the the ordinary work of the day you will learn much about these automotive components. Perhaps you can borrow shop repair manuals from your school automotive-shop library or from car-dealer service shops. You may also be able to buy some of these from the car manufacturers. In addition, your school automotive shop may have cutaway and working models on exhibit that are used as teaching aids. Studying these will help you understand the construction and operation of the actual units.

chapter 6

ENGINE TYPES

This chapter describes various ways in which engines are classified. All automotive engines are of the internal-combustion type; that is, combustion of the fuel takes place inside the engine (as opposed to engines in which combustion of the fuel takes place outside the engine, for example, steam engines). Engines are classified according to (1) number of cylinders, (2) arrangement of cylinders, (3) arrangement of valves, (4) type of cooling system, (5) number of cycles (two or four), (6) type of fuel, (7) type of cycle (Otto or diesel).

There are also a number of automotive power plants that cannot be classified in the above manner because of their unorthodox design. These include the gas turbine, Wankel engine, and Stirling engine. They are discussed separately at the end of this chapter.

⊘ 6-1 Number and Arrangement of Cylinders American passenger-car engines have four, six, or eight cylinders. Imported cars offer a greater variety, using engines of two, three, four, five, six, eight, or twelve cylinders. Engines with two, four, six, and eight cylinders are described and illustrated in following sections. Cylinders can be arranged in several ways: in a row (in-line); in two rows, or banks, set at an angle (V-type); in two rows opposing each other (flat or pancake); or like spokes on a wheel (radial-airplane type). Figure 6-1 shows various cylinder arrangements.

⊘ 6-2 Two-cylinder Engines The cylinders of two-cylinder engines can be arranged in three ways: in-line, opposed, and V. Figure 6-2 is a cutaway view of an in-line two-cylinder engine used in a small German automobile. Figure 6-3 is a cutaway view of an opposed, or flat, two-cylinder engine. This engine is used in the DAF of Holland car and is air-cooled; that is, air circulating around the cylinders removes excess heat from the engine (see ⊘ 6-11). In the opposed two-cylinder engine, the

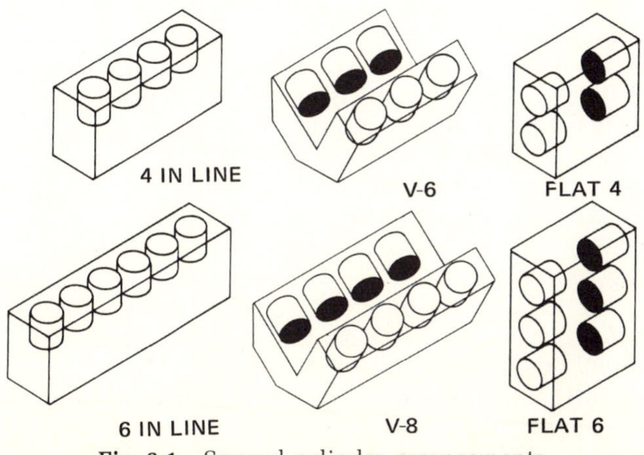

Fig. 6-1. Several cylinder arrangements.

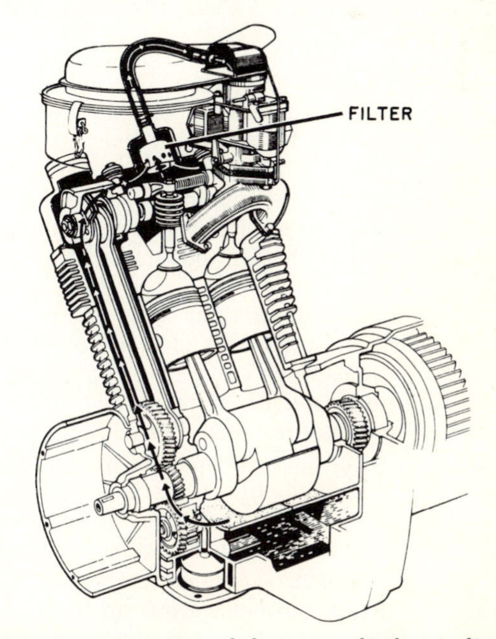

Fig. 6-2. Cutaway view of the two-cylinder, in-line, air-cooled NSU Prinz engine. The arrows (*left*) show the action of the crankcase ventilator. (*NSU of Germany*)

Fig. 6-3. Cutaway view of a two-cylinder, air-cooled engine with cylinders opposing each other. (*DAF of Holland*)

crankshaft and camshaft are positioned between the two cylinders. The valves are located in the cylinder head (I-head arrangement—see ⊘ 6-10).

⊘ 6-3 Four-cylinder Engines The cylinders of a four-cylinder engine can be arranged in any of three ways: in-line, V, or opposed. In the V-type engine, the cylinders are arranged in two banks, or rows, of two cylinders each and the rows are set at an angle to each other. In the opposed-type engine, the cylinders are arranged in two banks, or rows, of two cylinders each and the rows are set opposite each other.

1. *IN-LINE ENGINES* Figure 6-4 is a cutaway view of a four-cylinder in-line engine. The cylinders are arranged in one row, or line. A very similar engine is shown in Fig. 6-5. In this *slant-four engine,* the cylinders are slanted to one side to permit a lower hood line. In a sense, this engine is just one-half of a V-8 engine. There are relatively few slant-four engines produced in the United States, but it is an interesting design.

Another slant-four engine, with an overhead-camshaft (OHC) and integrated clutch, transmission, and differential, is shown in Fig. 6-6. This engine is especially interesting because of its overhead camshaft. Other engines previously described have the camshaft located in the lower part of the engine, the valves being operated by lifters, pushrods, and rocker arms (see Figs. 5-1, 6-4, 6-5, etc.). However, with the camshaft located in the cylinder head, as shown in Fig. 6-6, there is no need for pushrods, nor, in some engines, for rocker arms or lifters. This design has certain advantages, as explained later.

2. *V-4 ENGINES* The V-4 engine has two rows, or banks, of two cylinders each, and the rows are set at an angle, or a V, to each other. The crankshaft has only two cranks, with connecting rods from opposing cylinders in the two banks being attached to the same crankpin. Each crankpin has two connecting rods attached to it. Figure 6-7 is a phantom view of a V-4 engine with the internal moving parts emphasized. This type of engine is rather difficult to balance with counterweights on the crankshaft (see ⊘ 7-15), and in the engine illustrated balance is achieved by using a balance shaft that turns in a direction opposite to the crankshaft (Fig. 6-8).

3. *OPPOSED FOUR-CYLINDER ENGINES* Figure 6-9 shows the opposed, or flat, four-cylinder engine used by Volkswagen. The two rows, or banks, of cylinders oppose each other. The flat design, sometimes called a *pancake engine,* requires very little headroom, so that the engine compartment can be very compact. The Volkswagen engine is air-cooled and is mounted at the rear of the car (Fig. 6-10).

⊘ 6-4 Six-cylinder Engines As with four-cylinder engines, six-cylinder engines may be in-line, V, or opposed. Most six-cylinder engines are in-line, although there are some V-6 and flat-six engines.

1. *IN-LINE ENGINES* Figure 6-11 shows a six-cylinder in-line engine that is partly cut away so that the internal construction can be seen. The valves are overhead; this is an I-head, or overhead-valve, engine, and the crankshaft is supported by seven main bearings. Thus, there is a bearing on each side of every crank for additional support and rigidity.

Figure 6-12 shows a slant-six overhead-valve engine. This engine is similar to other six-cylinder in-line engines except that the cylinders are slanted to one side, similar to the four-cylinder engine shown in Fig. 6-5, so that the hood line can be lowered. This engine is also interesting because it has been supplied with either a cast-iron or a die-cast aluminum cylinder block. This feature will be discussed in ⊘ 7-6, where cast-iron blocks are compared with aluminum blocks.

Figure 6-13 is a partial cutaway view of a six-cylinder in-line OHC engine, with the camshaft driven by a neoprene belt reinforced with fiber-glass cords. The belt has a facing of woven nylon fabric on the tooth side. The teeth on the belt fit teeth molded into the outer diameters of the drive and driven pulleys. This is very similar to the metal chain-and-sprocket arrangement used in chain-driven camshafts. However, the neoprene belt is quieter and does not require lubrication.

2. *V-6 ENGINES* Several V-6 engines have been built. This design uses 2 three-cylinder rows, or banks, that are set at an angle, or V, to each other. The crankshaft has only three cranks, with connecting rods from opposing cylinders in the two banks being attached to the same crankpin. Each crankpin has two connecting rods attached to it. Figure 6-14 illustrates one version of this design.

3. *FLAT-SIX ENGINES* Figure 6-15 shows the flat-six engine used in the Chevrolet Corvair. It is air-cooled and mounted at the rear of the vehicle.

Fig. 6-4. Partial cutaway view of a four-cylinder, in-line, overhead-valve engine. (*Chevrolet Motor Division of General Motors Corporation*)

⊘ 6-5 Eight-cylinder Engines At one time the eight-cylinder in-line engine was widely used, but it has been replaced by the V-8 engine (Fig. 6-16). In the V-8, the cylinders are arranged in two rows, or banks, of four cylinders each, with the two rows set at an angle to each other. In effect, this engine is much like 2 four-cylinder in-line engines mounted on the same crankcase and working to a single crankshaft. The crankshaft in the V-8 has four cranks, with connecting rods from opposing cylinders in the two rows being attached to a single crankpin. Thus, two rods are attached to each crankpin, and two pistons work to each crankpin. The crankshaft is usually supported on five bearings.

The V-8 engine shown in Fig. 6-16 has overhead valves operated by valve lifters, pushrods, and rocker arms from a single camshaft located between the two cylinder banks. Some high-performance engines have overhead camshafts, with the camshafts located in the cylinder heads. One version of this design has a single overhead camshaft in each cylinder head. Another version has two overhead camshafts in each cylinder head—one for the intake valves, the other for the exhaust valves. So this version has a total of four camshafts in the engine. An engine of this type is shown in Fig. 6-17. Advantages of the overhead camshaft are discussed later.

⊘ 6-6 Twelve- and Sixteen-cylinder Engines These engines have been used in automobiles, buses, trucks, and industrial installations. The cylinders are usually arranged in two banks (V type or pancake type), three banks (W type), or four banks (X type). The pancake engine is similar to the V-type except that the two banks are arranged in a plane, but opposing; the cylinders work to the same crankshaft. At present, the only automobiles being made with 12-cylinder engines are the Jaguar, Maserati, and Ferrari.

Fig. 6-5. Sectional view from the end of a four-cylinder, in-line, overhead-valve engine. The cylinders are slanted to one side to permit a lower hood line. (*Pontiac Motor Division of General Motors Corporation*)

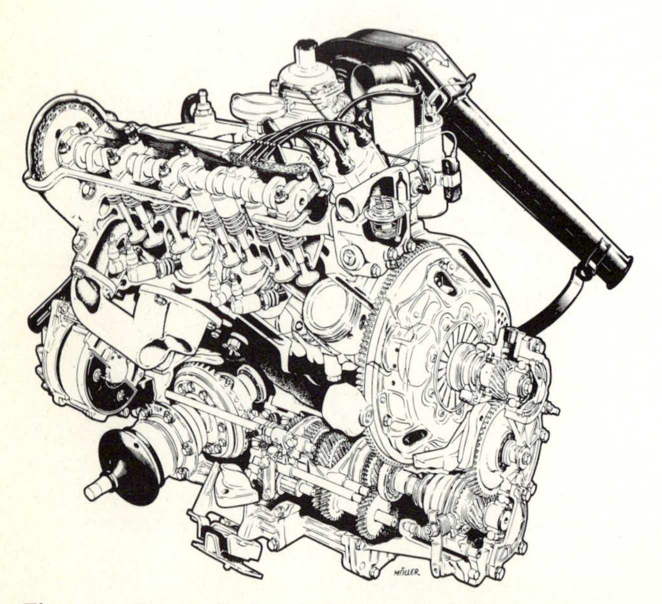

Fig. 6-6. Four-cylinder, in-line, OHC engine with integrated clutch and transmission. (*Saab Car Division of Saab-Scania*)

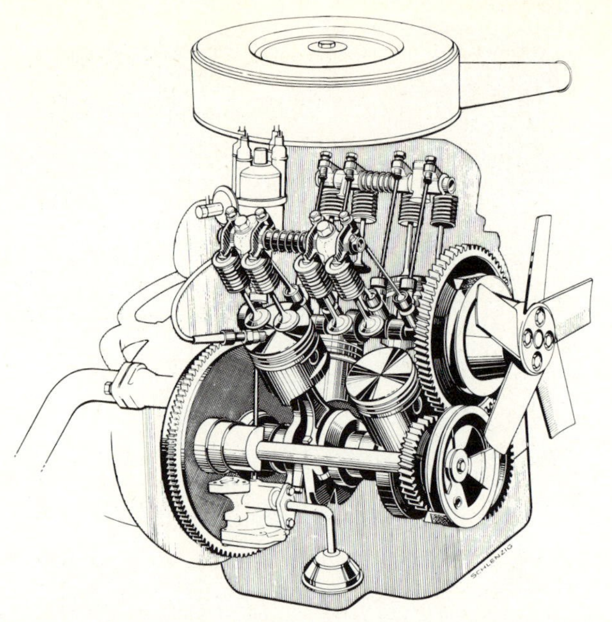

Fig. 6-7. Phantom view of a V-4 engine showing the major moving parts. (*Ford Motor Company of Germany*)

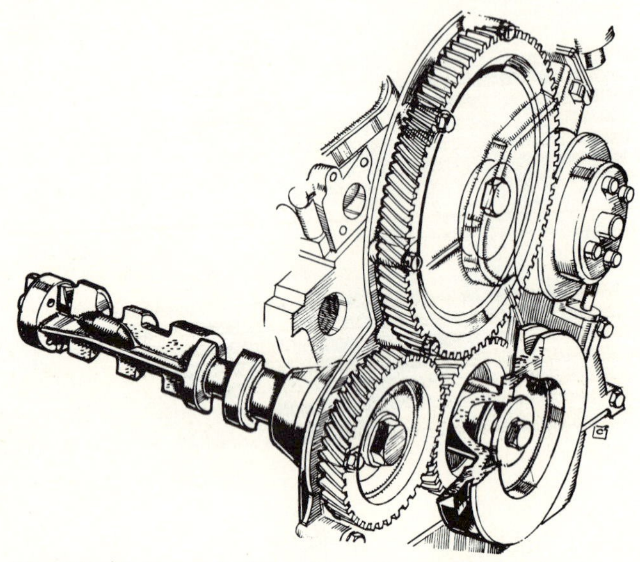

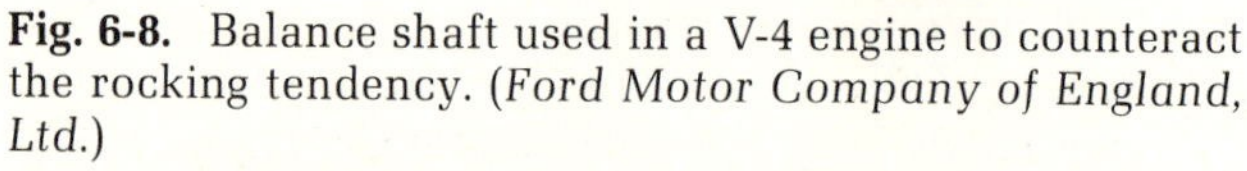

Fig. 6-8. Balance shaft used in a V-4 engine to counteract the rocking tendency. (*Ford Motor Company of England, Ltd.*)

Fig. 6-9. Flat four-cylinder engine with two banks of two cylinders each, opposing each other. This is an air-cooled engine. (*Volkswagen*)

Fig. 6-10. Mounting arrangement for the flat four-cylinder engine at the rear of the automobile. (*Volkswagen*)

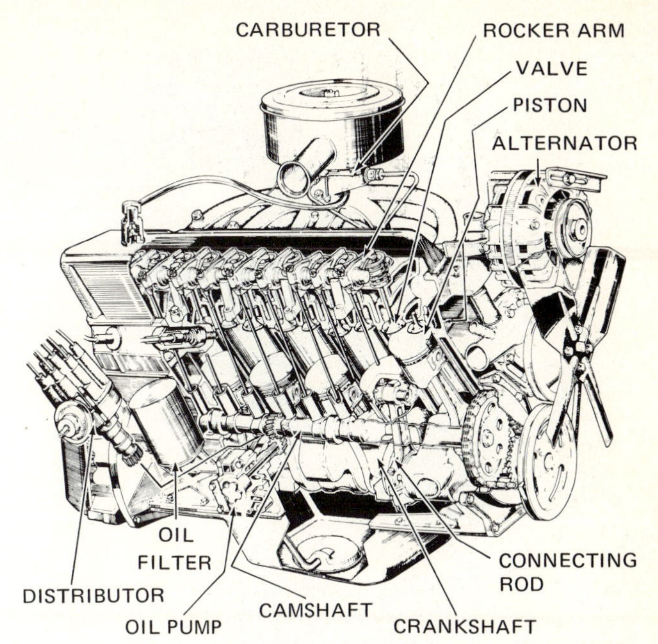

Fig. 6-12. Slant-six, in-line, overhead-valve engine, cut away to show the internal parts. Cylinders are slanted to permit a lower hood line. (*Chrysler Corporation*)

Fig. 6-11. Six-cylinder, in-line engine with overhead valves, partly cut away to show the internal construction. (*Ford Motor Company*)

Fig. 6-13. Partial cutaway view of a six-cylinder engine with the overhead camshaft driven by a toothed neoprene belt. (*Pontiac Motor Division of General Motors Corporation*)

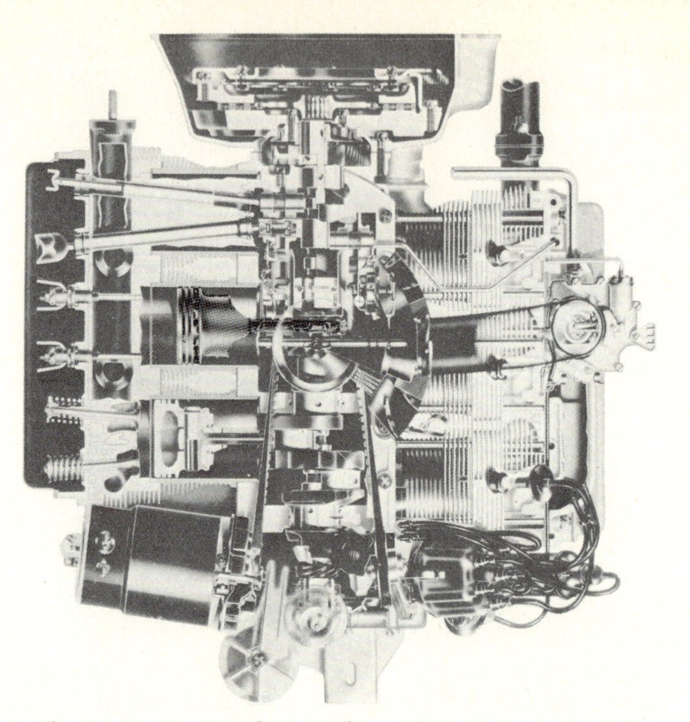

Fig. 6-15. Sectional view from the top of a flat six-cylinder, overhead-valve, air-cooled engine, sometimes referred to as a *pancake engine*. (*Chevrolet Motor Division of General Motors Corporation*)

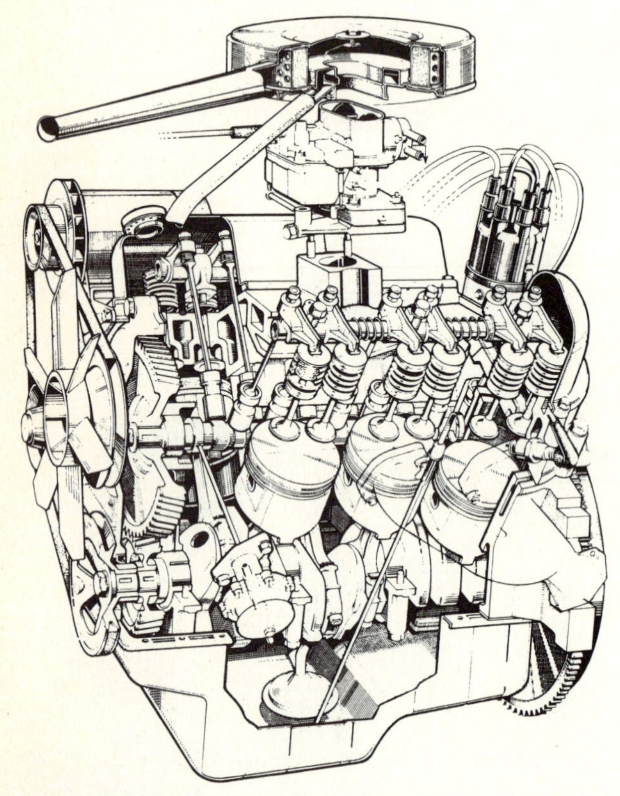

Fig. 6-14. Cutaway view of a V-6, overhead-valve engine. (*Ford Motor Company of Germany*)

Fig. 6-16. Cutaway view of 365-hp V-8 engine. (*Ford Motor Company*)

⊘ 6-7 Radial Engines The radial engine has the cylinders radiating from a common center, like the spokes of a wheel. All connecting rods work to a common crankpin; the crankshaft has only one crankpin. The radial engine is air-cooled (⊘ 6-11) and is used mainly for aircraft applications. Another design is the multiple-bank radial, which is essentially two or more radial engines, one mounted back of another, using a multiple-crankpin crankshaft.

⊘ 6-8 Eight-cylinder In-line Engines Compared with V-8 Engines Although eight-cylinder in-line engines were once widely used in automobiles, they have been superseded by V-8 engines. Engineers

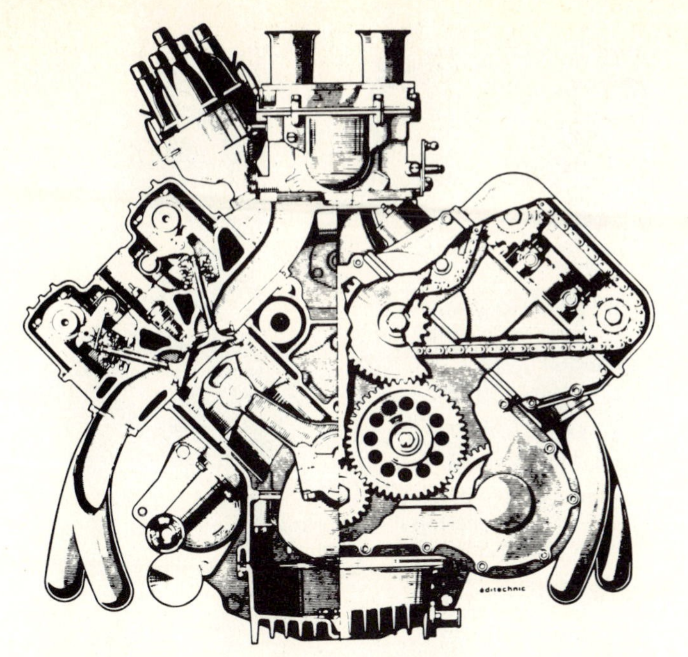

Fig. 6-17. Sectional view from the end of a V-8 engine with four overhead camshafts. Right bank has been cut away to show the camshaft drive arrangement. Left bank has been cut away to show the internal construction of the engine and location of the valves and other components. (*Renault*)

mention several advantages of the V-8 engine: It is a shorter, lighter, and more rigid engine. Also, the arrangement permits the use of intake manifolding that ensures relatively even distribution of the air-fuel mixture to all cylinders (since all cylinders are relatively close together). This contrasts with the in-line engine, where the end cylinders could be fuel-starved while the center cylinders received adequate fuel. (Some eight-cylinder in-line engines used two carburetors for better fuel distribution.)

The more rigid V-8 engine permits higher running speeds and combustion pressures (higher power outputs) with less difficulty from flexing, or bending, of the cylinder block and crankshaft. Flexing throws the engine out of line, increases frictional losses and wear, and may also set up internal vibrations.

The shorter engine makes possible more passenger space on the same wheel base or a shorter-wheel-base car. In addition, the V-8 engine permits a lower hood line and thus a lower vehicle profile. This is because the carburetor and other parts can be nested between the two rows of cylinders so that they do not take up headroom above the cylinders.

⊘ 6-9 Firing Order The firing order, or order in which the cylinders deliver their power strokes, is selected as part of the engine design. The best design provides a well-distributed pattern along the crankshaft. A design that permits two cylinders at the same end of the crankshaft to fire one after another is avoided as far as possible.

In-line engine cylinders are numbered from front to rear. Several different cylinder-numbering systems are used for V-6 and V-8 engines. For example, Ford V-8 engine cylinders are numbered:

Front of Car

LEFT BANK	RIGHT BANK
⑤	①
⑥	②
⑦	③
⑧	④

General Motors numbers the cylinders of its V-8 engines (except for some Buicks) in this way:

Front of Car

LEFT BANK	RIGHT BANK
①	②
③	④
⑤	⑥
⑦	⑧

The Chrysler Corporation's V-8 engines are numbered in the same way as the General Motors V-8 engines, above. Some Buick engines have the cylinders numbered:

Front of Car

LEFT BANK	RIGHT BANK
②	①
④	③
⑥	⑤
⑧	⑦

The Volkswagen flat-four engine has the cylinders numbered:

Front of Car

LEFT BANK	RIGHT BANK
③	①
④	②

The Corvair flat-six engine has the cylinders numbered:

Front of Car

LEFT BANK	RIGHT BANK
⑥	⑤
④	③
②	①

The General Motors V-6 engine has the cylinders numbered:

Front of Car

LEFT BANK	RIGHT BANK
①	②
③	④
⑤	⑥

As you can see, these different cylinder-numbering systems can be confusing. Therefore, when you are working on an engine, you should always check the shop manual or other specific reference work when you want to know how the cylinders are numbered.

Now that we have looked at cylinder-numbering systems, let us see about firing orders of various engines. The two firing orders used in four-cylinder in-line engines are 1-3-4-2 or 1-2-4-3.

Two firing orders are possible in six-cylinder in-line engines: 1-5-3-6-2-4 or 1-4-2-6-3-5. All modern six-cylinder in-line engines use 1-5-3-6-2-4.

The V-6 engine, mentioned previously, uses this firing order: 1-6-5-4-3-2. Note how this scatters the power strokes along the crankshaft.

General Motors V-8 engines use different firing orders. For example, most V-8 engines use this firing order: 1-8-4-3-6-5-7-2. Late-model Cadillac engines use this firing order: 1-5-6-3-4-2-7-8. Chrysler Corporation V-8 engines follow this firing order: 1-8-4-3-6-5-7-2. Ford V-8 engines use two firing orders: 1-5-4-2-6-3-7-8 and 1-3-7-2-6-5-4-8.

When you work on an engine, you must know the proper firing order, especially when you are doing ignition work. Therefore, you must look up the firing order in the shop manual.

NOTE: On many engines, the firing order is cast into the intake manifold.

⊘ 6-10 Arrangement of Valves The intake and exhaust valves in the engine can be arranged in various positions in the cylinder head or block (Fig. 6-18). The I-head, or overhead-valve, design is by far the most common.

1. *L-HEAD ENGINE* In the L-head engine, the combustion chamber and cylinder form an inverted L (Fig.6-18). The intake and exhaust valves are located side by side, and all valves for the engine are arranged in one line (except for V-8 L-head engines, in which they are arranged in two lines). This design permits the use of a single camshaft to operate all valves. Since the valve mechanisms are in the block, removal of the cylinder head for major overhaul of the engine is relatively easy.

However, the L-head engine, though rugged and dependable, cannot be adapted to higher compression ratios. One reason is that the valves require a certain minimum space to move up into when they open. This space plus the minimum clearance required above the top of the piston determines the minimum possible clearance volume, or the volume

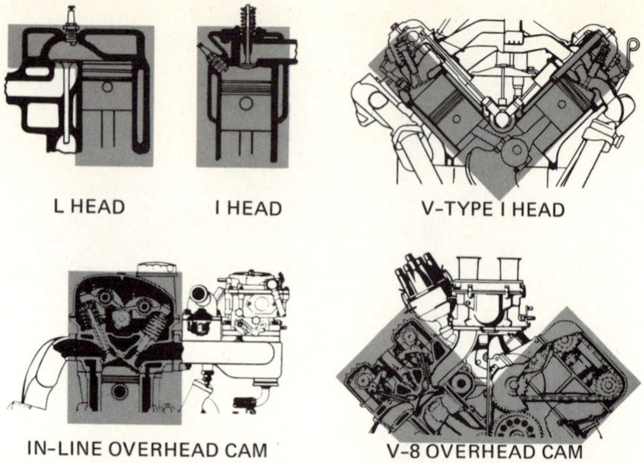

Fig. 6-18. Valve arrangements. Compare these line drawings with the sectional and cutaway views of various engines shown elsewhere in the book.

with the piston at top dead center (TDC). Since the clearance volume cannot be decreased below this minimum, there is a limit as to how much the compression ratio of the engine can be increased. [Recall that the compression ratio is the ratio between cylinder volume at bottom dead center (BDC) and clearance volume, or the volume at TDC.] The I-head (overhead-valve) engine is better suited to higher compression ratios, as explained in following paragraphs.

2. *I-HEAD ENGINE* In the I-head, or overhead-valve, engine, the valves are carried in the cylinder head (Fig. 6-19). In in-line engines, the valves are arranged in a single row, as shown in Figs. 6-4 and 6-12. In V-8 engines, the valves can be arranged in a single row in each bank (Fig. 6-16) or in a double row in each bank (Fig. 6-19). Regardless of arrangement, a single camshaft actuates all valves, with the

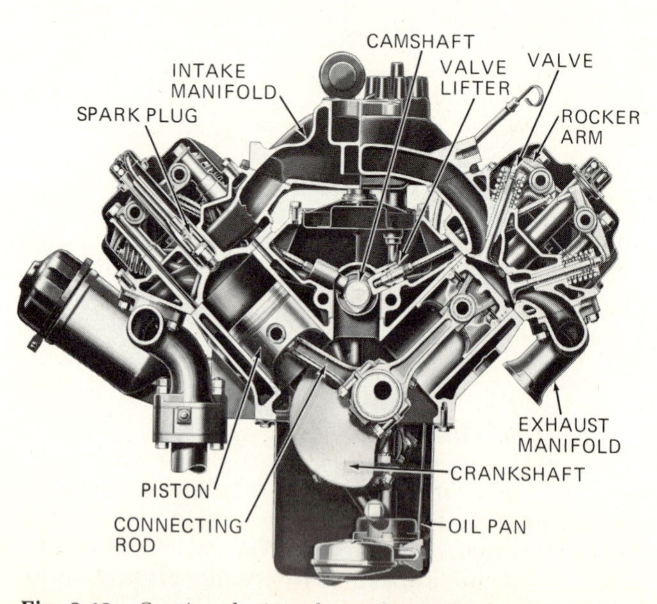

Fig. 6-19. Sectional view from the end of a V-8 engine with overhead valves. (*Chrysler Corporation*)

valve lifters, pushrods, and rocker arms carrying the motion from the cams to the valves (see Fig. 5-12).

The overhead-valve arrangement has come into widespread use since it is more adaptable to the higher-compression-ratio engines. In an engine with overhead valves, it is practical to reduce the clearance volume a proportionally greater amount than in an L-head engine. If you study the illustrations of the various I-head engines in this book, you will see that the method of grouping the valves directly above the piston permits a small clearance volume. In some I-head engines, there are pockets in the pistonheads into which the valves can move when the valves are open with the piston at TDC. In some engines, the clearances between the piston and valves are only a few thousandths of an inch.

3. OVERHEAD-CAMSHAFT ENGINES (FIGS. 5-26 and 6-22) As already noted, the I-head engine, in its most common version, uses pushrods and rocker arms to operate the valves. This design is often referred to as a *pushrod engine*. The pushrods and rocker arms impose some inertia that tends to affect valve action. That is, the pushrods and rocker arms flex (bend) slightly before they move to open the valve. This slows valve action somewhat and can also cause irregular valve action. At low speeds, this does not matter. But as speed increases, the flexing also increases. This causes increasing irregular valve action, which tends to limit top engine speed. However, in overhead-camshaft (OHC) engines, the cams work directly on the rocker arms or valve lifters. This results in quicker valve response so that higher engine speeds are possible.

The single-overhead-camshaft engine (one camshaft in each cylinder head) is called a *SOHC engine*. The double-overhead-camshaft engine is called a *DOHC engine*. Figures 6-6, 6-13, and 6-17 show various OHC engines. The overhead camshaft (or camshafts) is driven by a metal chain and sprockets or by a neoprene belt. OHC engines are discussed in more detail in later chapters.

4. V-TYPE AND PANCAKE ENGINES V-type and pancake engines may use either L or I heads. Most engines, however, use the I-head arrangement for the reasons given in previous paragraphs. All the newer high-output engines use I heads.

⊘ 6-11 Type of Cooling System Engines are classified as *liquid-cooled* or *air-cooled*. Most present-day American automotive engines are liquid-cooled. The Corvair, Volkswagen, and some other automotive engines are air-cooled (see Figs. 6-9 and 6-15). Also, the small one- and two-cylinder engines used on power lawn mowers and other garden equipment are air-cooled (Figs. 6-20 and 6-21). In air-cooled engines, the cylinder barrels are usually separate and are equipped with metal fins which give a large radiating surface. This permits engine heat to radiate from the cylinders. Many air-cooled engines are also equipped with metal shrouds which direct the air flow around the cylinders for improved cooling.

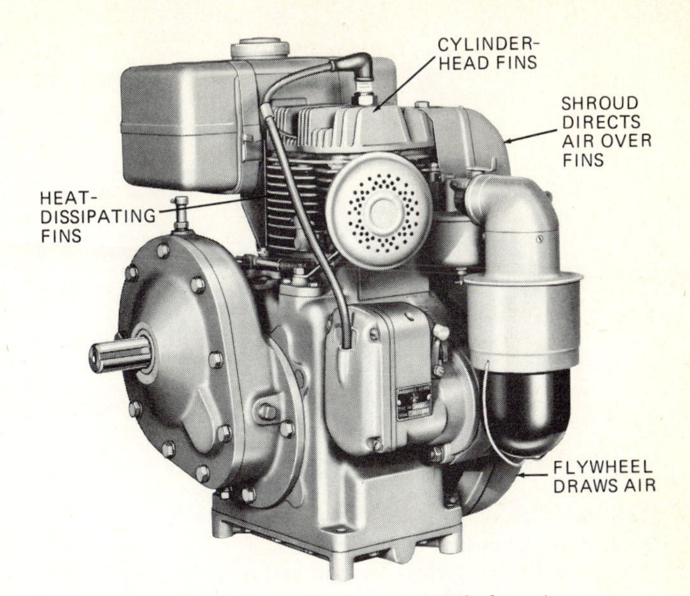

Fig. 6-20. One-cylinder, air-cooled engine.

Liquid-cooled engines ordinarily use water with an antifreeze compound added to serve as the cooling medium. The mixture is called the *coolant*. The engines have water jackets surrounding the cylinders and the combustion chambers in the cylinder block and head (Figs. 5-26 and 6-22). The coolant can circulate freely through these jackets. The coolant enters from the bottom of the engine radiator and circulates through the engine, where it absorbs heat and thus becomes quite hot. It then exits from the engine water jackets and pours into the radiator upper tank. From there, the coolant passes down through the radiator to the lower tank.

The radiator has two sets of passages: coolant passages (from top to bottom) and air passages (from front to back). The air passing through (pulled by the engine fan and the forward motion of the car) picks up heat from the hot coolant passing down through the radiator. The result is that the coolant entering the lower tank is cool and ready for another trip through the engine. This constant circulation is kept going by the water pump, which is mounted on the front end of the engine and is driven by the fan belt. The fan is usually mounted on the water-pump pulley, and so both turn together. Chapter 16 contains detailed descriptions of the two basic types of cooling systems, air-cooled and liquid-cooled.

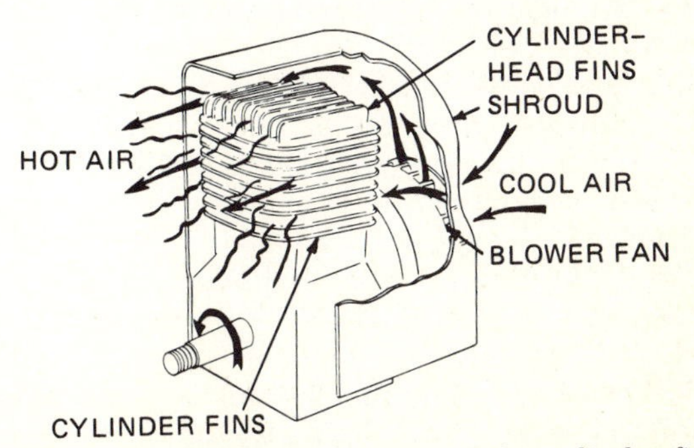

Fig. 6-21. Circulation of air around the cylinder fins.

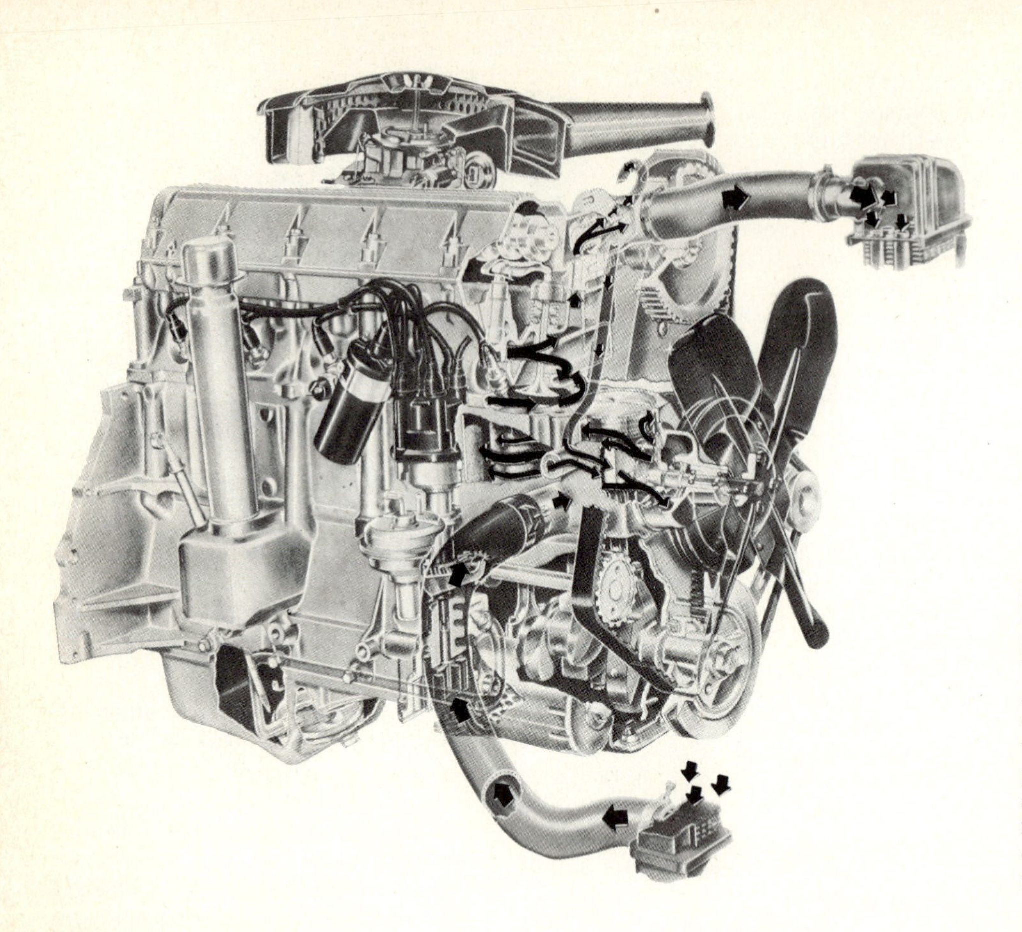

Fig. 6-22. Six-cylinder, in-line, overhead-camshaft engine partly cut away to show the engine cooling system. Arrows indicate the direction of water flow in the engine water jackets. Only small parts of the radiator are shown at the lower and upper right. (*Pontiac Motor Division of General Motors Corporation*)

Check Your Progress

Progress Quiz 6-1 Here is your opportunity to check up on yourself to find out how well you are remembering the material covered in the past few pages. If some of the questions stump you, go back and reread the pages that will give you the answer.

Completing the Sentences The sentences that follow are incomplete. After each sentence there are several words or phrases, only one of which will correctly complete the sentence. Write each sentence in your notebook, selecting the proper word or phrase to complete it correctly.

1. The smallest number of cylinders in use in American automobile engines is: (*a*) two, (*b*) four, (*c*) six, (*d*) eight.
2. The V-8 engine crankshaft is usually supported on: (*a*) two bearings, (*b*) three bearings, (*c*) four bearings, (*d*) five bearings.
3. The most common type of V engine is the: (*a*) V-4, (*b*) V-6, (*c*) V-8, (*d*) V-12.
4. The crankpins on the six-cylinder crankshaft are arranged in pairs: (*a*) 90° apart, (*b*) 120° apart, (*c*) 180° apart.
5. One firing order for six-cylinder engines is: (*a*) 1-3-2-6-5-4, (*b*) 1-4-3-2-6-5, (*c*) 1-5-3-6-2-4, (*d*) 1-2-3-5-4-6.
6. The crankshaft on the V-8 engine has: (*a*) two crankpins, (*b*) three crankpins, (*c*) four crankpins, (*d*) six crankpins.
7. The engine in which the cylinders are arranged like the spokes of a wheel is: (*a*) a spoke engine, (*b*) a radial engine, (*c*) an in-line engine, (*d*) a V-type engine.
8. Widely used valve arrangements are: (*a*) L, I, F, E; (*b*) L, I, and OHC; (*c*) SOHC, DOHC, and T.
9. The type of engine using valve lifters, pushrods, and rocker arms is a: (*a*) V type, (*b*) I-head type, (*c*) L-head type, (*d*) T-head type.
10. In regard to cooling, engines are classified as: (*a*) air-cooled and liquid-cooled, (*b*) water-cooled and liquid-cooled, (*c*) air-cooled and oil-cooled.

⊘ 6-12 Number of Cycles Engines can be classified as either *two-stroke-cycle* or *four-stroke-cycle.* In the four-stroke-cycle engine (usually called a *four-cycle engine*), already discussed in ⊘ 5-4, the complete cycle of events requires four piston strokes: intake, compression, power, and exhaust. In the two-stroke-cycle engine (*two-cycle engine*), the intake and compression strokes, as well as the power and exhaust strokes, are in a sense combined. This permits the engine to produce a power stroke every two piston strokes, or every crankshaft rotation.

In the two-cycle engine, the piston acts as a valve, clearing valve ports in the cylinder wall as it nears BDC. A fresh air-fuel mixture enters through the intake port, and the burned gases exit through the exhaust port. The complete cycle of operation is as follows. As the piston nears TDC, ignition takes place (Fig. 6-23). The high combustion pressures

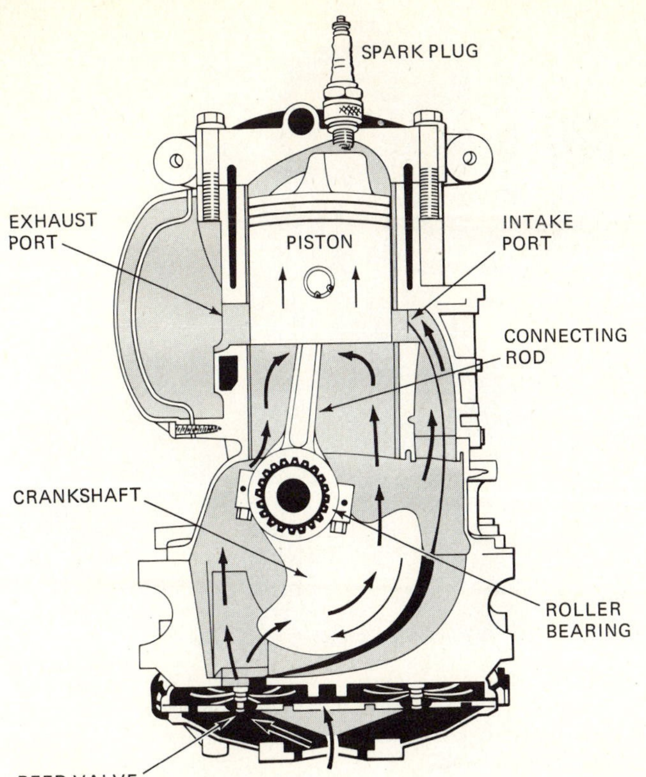

Fig. 6-23. Sectional view of a two-cycle engine with the piston nearing TDC. Ignition of the compressed air-fuel mixture occurs approximately at this point. (*Johnson Motors*)

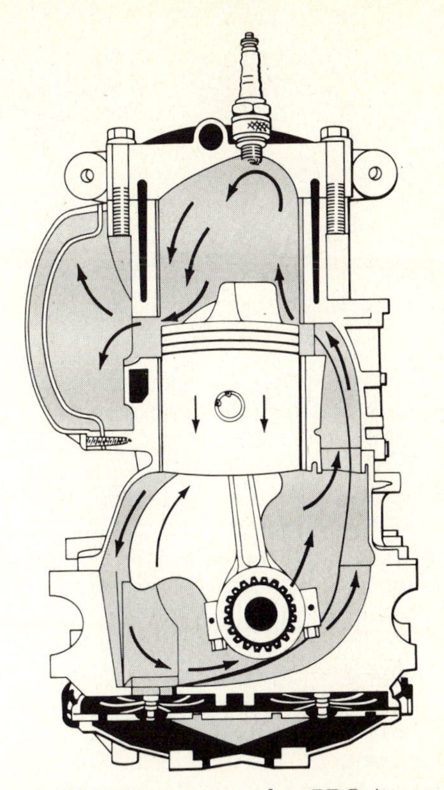

Fig. 6-24. As the piston approaches BDC, it uncovers the intake and exhaust ports. Burned gases stream out through the exhaust port, and a fresh charge of air-fuel mixture enters the cylinders, as shown by the arrows. (*Johnson Motors*)

drive the piston down, and the thrust through the connecting rod turns the crankshaft. As the piston nears BDC, it passes the intake and exhaust ports in the cylinder wall (Fig. 6-24). Burned gases, still under some pressure, begin to stream out through the exhaust port. At the same time, the intake port, now cleared by the piston, begins to deliver air-fuel mixture, under pressure, to the cylinder. The top of the piston is shaped to give the incoming mixture an upward movement. This helps to sweep the burned gases ahead and out through the exhaust port.

After the piston has passed through BDC and starts up again, it passes both ports, thus sealing them off (Fig. 6-25). Now the fresh air-fuel mixture above the piston is compressed and ignited. The same series of events takes place again and continues as long as the engine runs.

We mentioned that the air-fuel mixture is delivered to the cylinder under pressure. In most engines, this pressure is put on the mixture in the crankcase. The crankcase is sealed except for a leaf, or reed, valve at the bottom (Fig. 6-26). The reed valve is a flexible, flat metal plate that rests snugly against the floor of the crankcase. There are holes under the reed valve that connect to the engine carburetor. When the piston moves up, a partial vacuum is produced in the sealed crankcase. Atmospheric pressure lifts the reed valve off the holes, and air-fuel mixture enters the crankcase (Fig. 6-23). After

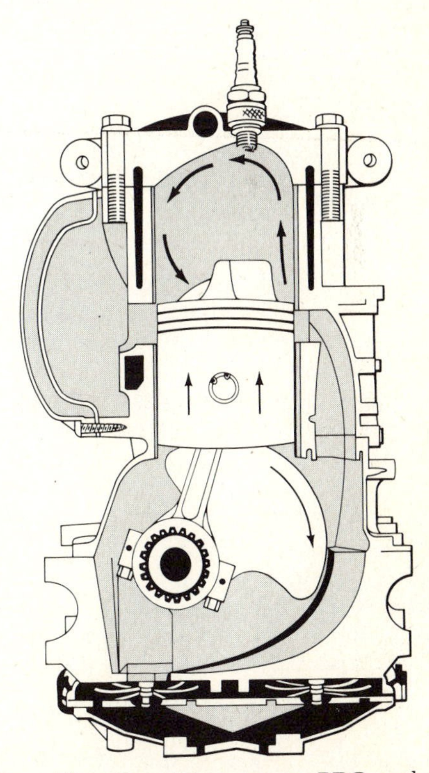

Fig. 6-25. After the piston passes BDC and moves up again, it covers the intake and exhaust ports. Further upward movement of the piston traps and compresses the air-fuel mixture. (*Johnson Motors*)

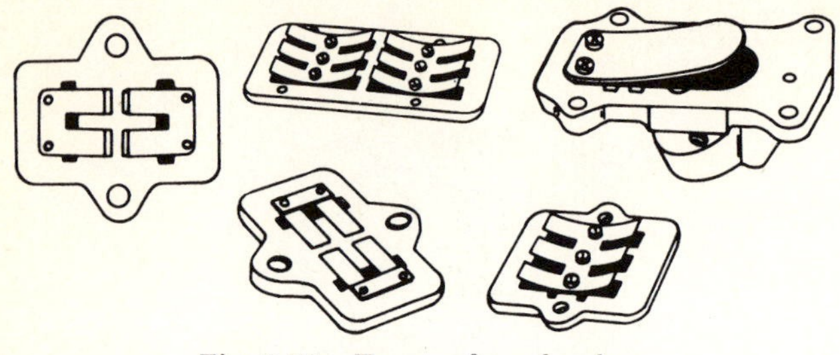

Fig. 6-26. Types of reed valves.

the piston passes TDC and starts down again, pressure begins to build up in the crankcase. This pressure closes the reed valve so that further downward movement of the piston compresses the trapped air-fuel mixture in the crankcase. The pressure which is built up on the air-fuel mixture then causes it to flow up through the intake port into the engine cylinder when the piston moves down far enough to clear the intake port (Fig. 6-24).

Instead of using a reed valve in the crankcase, some engines have a third, or transfer, port in the cylinder (Fig. 6-27). In this type of engine, the intake port is cleared by the piston as it approaches TDC. When this happens, the air-fuel mixture pours into the crankcase, filling the partial vacuum left by the upward movement of the piston. Then, as the piston moves down, the intake port is cut off by the piston. The air-fuel mixture in the crankcase is compressed, and the other actions then take place, as already described.

Another type of two-cycle engine uses valves in the cylinder head to exhaust the burned gases (Fig. 6-28). As the piston moves down past the intake ports (notice there is a ring of them around the cylinder), the exhaust valve opens. The engine shown in Fig. 6-28 is a diesel engine, and only air enters through the intake ports. (Diesel engines are discussed in ⊘ 6-15.)

⊘ 6-13 Comparison of Two-cycle and Four-cycle Engines The four-cycle engine requires four piston strokes to complete the cycle of events: intake, compression, power, and exhaust. The two-cycle engine does the job in two piston strokes. In the four-cycle engine, only one of every four piston strokes is a power stroke. Thus there is one power stroke for every two crankshaft revolutions. In the two-cycle engine, every other piston stroke is a power stroke. Thus there is one power stroke for every crankshaft revolution. Figure 6-29 compares the operating cycles of the two types of engine.

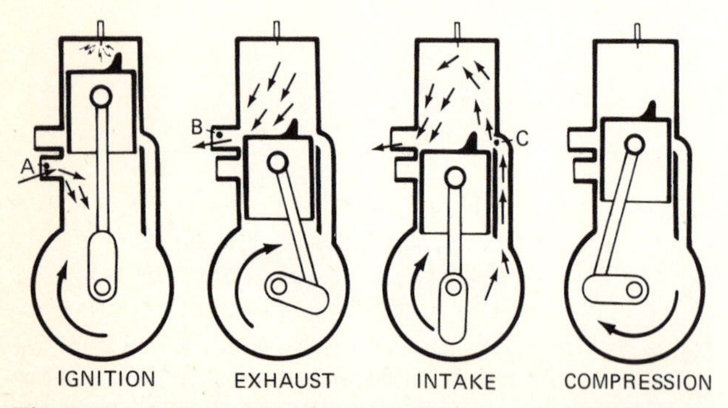

Fig. 6-27. Actions in a three-port, two-cycle engine. Note the intake port (A), exhaust port (B), and transfer port (C).

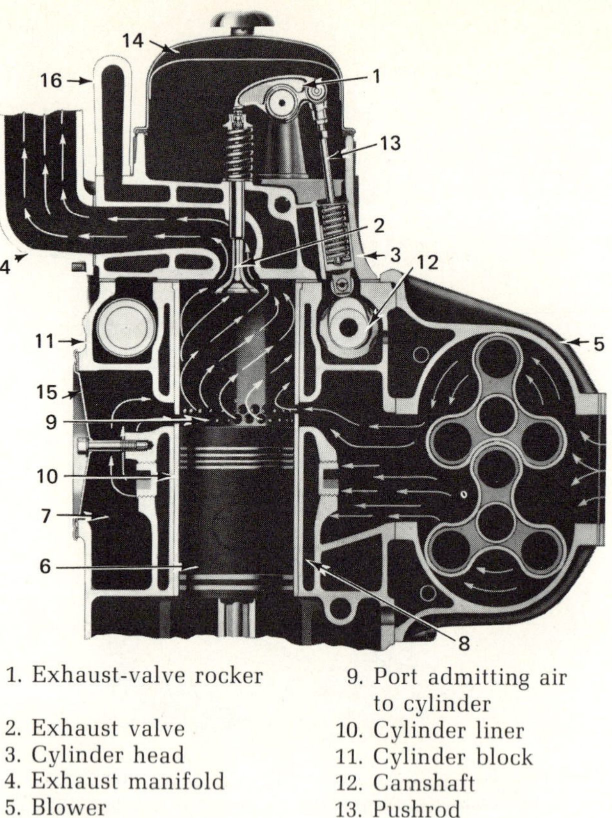

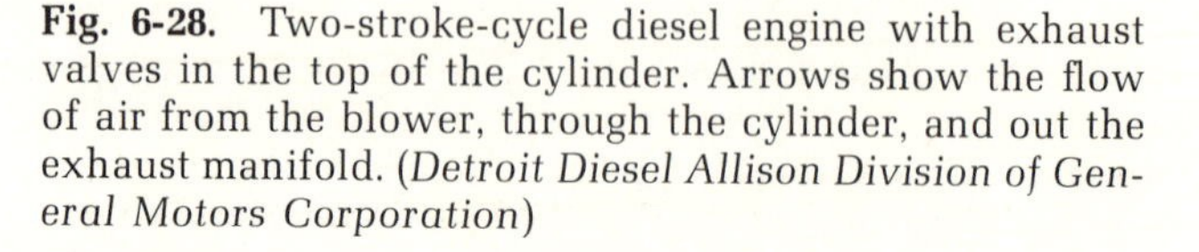

1. Exhaust-valve rocker
2. Exhaust valve
3. Cylinder head
4. Exhaust manifold
5. Blower
6. Piston
7. Air box
8. Cooling-water passage
9. Port admitting air to cylinder
10. Cylinder liner
11. Cylinder block
12. Camshaft
13. Pushrod
14. Rocker cover
15. Hand-hole cover
16. Water manifold

Fig. 6-28. Two-stroke-cycle diesel engine with exhaust valves in the top of the cylinder. Arrows show the flow of air from the blower, through the cylinder, and out the exhaust manifold. (*Detroit Diesel Allison Division of General Motors Corporation*)

You might think that because the two-cycle engine produces twice as many power strokes as the four-cycle engine, the two-cycle engine is twice as powerful. This is not so. In the two-cycle engine, when the intake and exhaust ports have been cleared by the piston, there is always some mixing of the fresh mixture and burned gases. Not all the burned gases can get out of the cylinder. The burned gases left in the cylinder reduce the amount of fresh air-fuel mixture that can enter. In addition, only part of the piston stroke is devoted to getting the air-fuel mixture into the cylinder. This further reduces the amount of mixture entering the cylinder.

In the four-cycle engine, however, nearly all of the burned gases are forced out of the cylinder by the piston on the exhaust stroke. More air-fuel mixture can then enter because a complete piston stroke is devoted to getting the air-fuel mixture into the

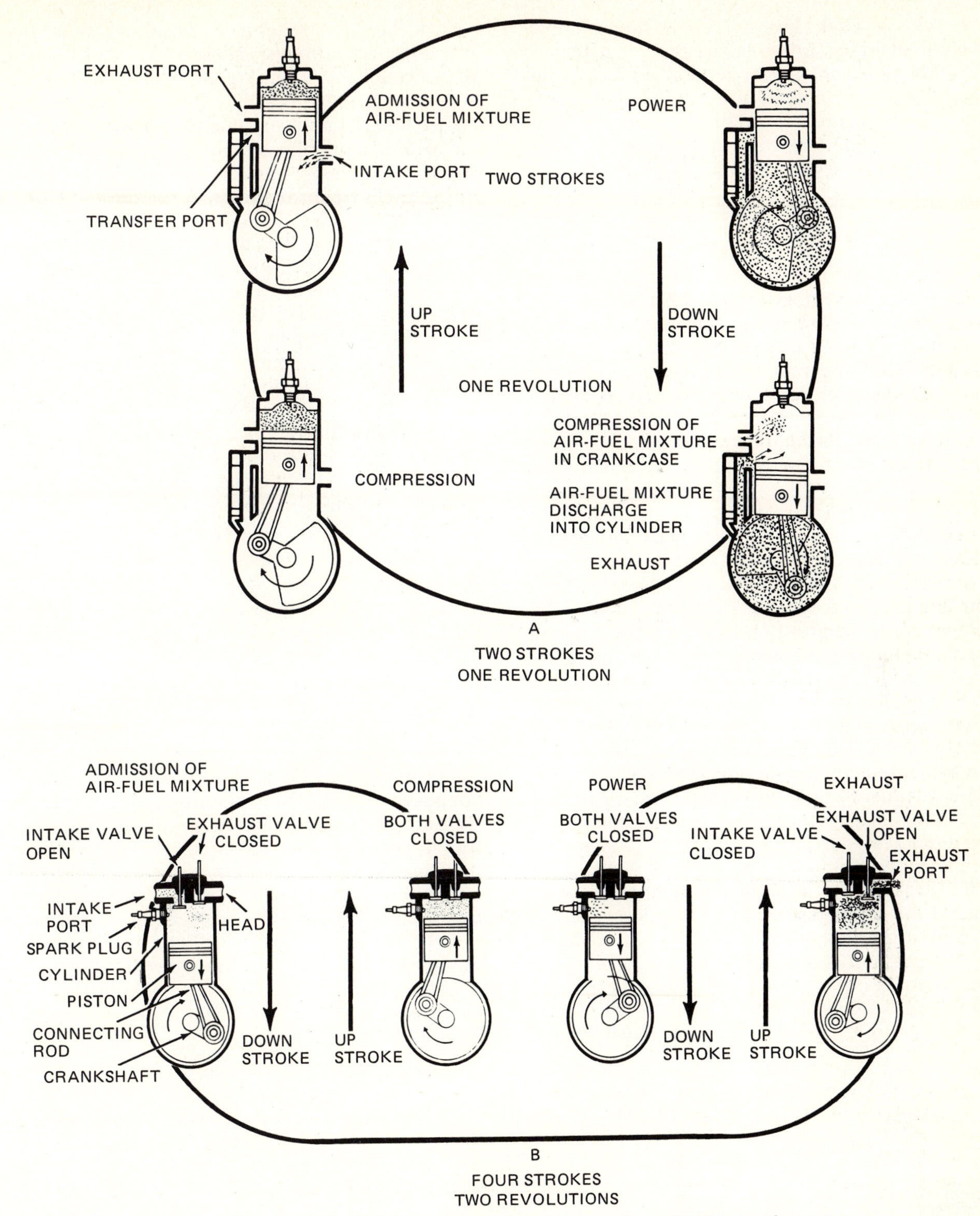

Fig. 6-29. Comparison of the operation of a two-cycle and a four-cycle engine.

cylinder. The result is a more powerful piston stroke. Therefore, the power stroke of the four-cycle engine is stronger than the power stroke of the two-cycle engine.

Even though the power stroke of the two-cycle engine is comparatively weak, it is still powerful enough for many jobs. Two-cycle engines are widely used as power plants for lawn mowers, motor boats, snow removers, motor scooters, power saws, and other such equipment. These engines are usually air-cooled. Because the two-cycle engines have no valve trains or liquid cooling systems, they are relatively lightweight, simple in construction, and easy to service. These are desirable characteristics for engines used to run equipment that must be handled and moved around.

⊘ 6-14 Type of Fuel Internal-combustion engines can be classified according to the type of fuel they use. Automotive engines, in general, use gasoline. Some bus and truck engines use liquified petroleum gas (LPG); these are essentially gasoline engines adapted for LPG. Diesel engines use diesel fuel oil. (Chapter 13 describes these fuels in detail.)

⊘ 6-15 Diesel Engines The diesel engine operates on a different principle than the Otto-cycle engine since the fuel is not mixed with the air entering the cylinder during the intake stroke. Air alone is compressed during the compression stroke, and the diesel fuel oil is injected or sprayed into the cylinder at the end of the compression stroke. In diesel engines the compression ratios used are as high as 21:1 and provide pressures of about 500 psi [351.5 kg/cm^2] at the end of the compression stroke. When air is rapidly compressed to this pressure, it is heated to a temperature as high as 1000 degrees Fahrenheit [540°C]. This temperature is high enough to ignite spontaneously the fuel oil that is injected or sprayed into the cylinder at this instant. The combustion of the oil can be controlled by the speed with which the oil is introduced into the cylinder. Thus in the diesel engine the combustion is not a rapid burning of the fuel already present in the cylinder, as in the gasoline engine, but a slower burning that produces an even increase of pressure. This allows a more complete utilization of the energy in the fuel.

The four-stroke-cycle diesel engine requires four piston strokes for each power stroke, as in the gasoline engine. These are intake, compression, power, and exhaust. On the intake stroke the piston moves downward and pulls air into the cylinder past the intake valve. The intake valve closes as the compression stroke starts, and the air is compressed. At the end of the compression stroke the diesel fuel oil is sprayed or injected into the combustion chamber, where it burns and creates high pressure. The piston is pushed down during the power stroke, at the end of which the exhaust valve opens to allow the burned gases to escape during the exhaust stroke.

In the two-stroke-cycle diesel engine a blower, or rotary-type pump, is used to create an initial pressure on the incoming air (Fig. 6-30). The piston serves as a valve or valves, and on its downward stroke the piston clears the ports through which the air enters and exhaust gases escape. The engine shown in Fig. 6-30 has an exhaust valve in the top of the cylinder through which the burned gases are forced when the valve opens and the piston clears the intake ports (Fig. 6-28). As the piston moves upward, it passes the intake ports and the exhaust valve closes. The air thus trapped in the cylinder is highly compressed; the fuel oil is sprayed into the cylinder, and the power stroke takes place.

The fuel oil used in diesel engines does not burn rapidly unless it is finely atomized and thoroughly mixed with the compressed air. To ensure adequate mixing, particularly in smaller engines, combustion chambers of various shapes are used. These special shapes produce turbulence, or whirling, of the compressed air, which improves the mixing of the fuel with the air during the combustion process.

⊘ 6-16 Diesel-Engine Applications Diesel engines have been made in a great variety of sizes and outputs, from a few to 5,000 hp [3,730 kW (kilowatts)]. They are used in automobiles, trucks, buses, farm and construction machinery, ships, electric power plants (up to about 5,000 kW), and other mobile and stationary applications.

Automobiles manufactured in the United States do not have diesel engines (trucks, buses, and other commercial equipment may have them). However, in Europe, automobiles with diesel engines are rather common. For example, Mercedes-Benz has produced more than half a million diesel-powered automobiles since 1950. Many of these have been imported to the United States. Figure 6-31 is a sectional view of the Mercedes-Benz four-cylinder inline diesel engine for automobiles. This engine has a 121 cubic inch [1,983 cm^3] displacement and a compression ratio of 21:1. At 4,200 rpm, it peaks at 60 hp [44.76 kW].

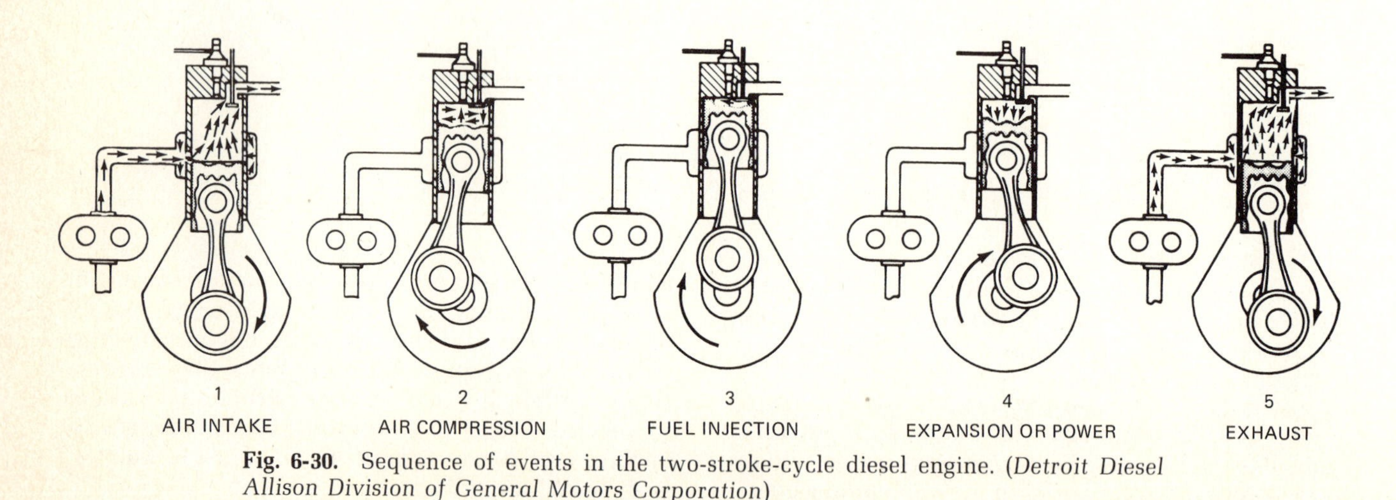

Fig. 6-30. Sequence of events in the two-stroke-cycle diesel engine. (*Detroit Diesel Allison Division of General Motors Corporation*)

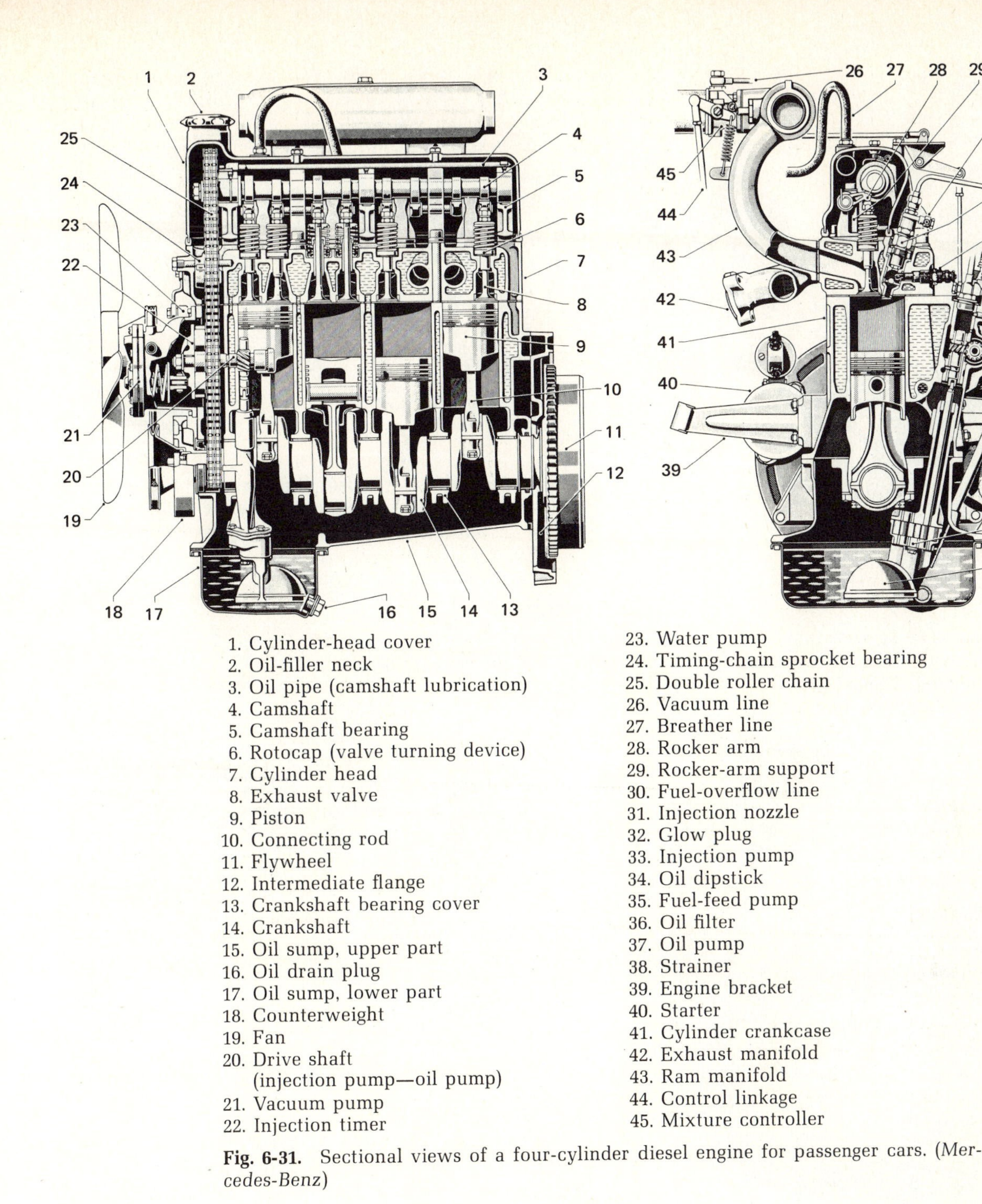

1. Cylinder-head cover
2. Oil-filler neck
3. Oil pipe (camshaft lubrication)
4. Camshaft
5. Camshaft bearing
6. Rotocap (valve turning device)
7. Cylinder head
8. Exhaust valve
9. Piston
10. Connecting rod
11. Flywheel
12. Intermediate flange
13. Crankshaft bearing cover
14. Crankshaft
15. Oil sump, upper part
16. Oil drain plug
17. Oil sump, lower part
18. Counterweight
19. Fan
20. Drive shaft (injection pump—oil pump)
21. Vacuum pump
22. Injection timer
23. Water pump
24. Timing-chain sprocket bearing
25. Double roller chain
26. Vacuum line
27. Breather line
28. Rocker arm
29. Rocker-arm support
30. Fuel-overflow line
31. Injection nozzle
32. Glow plug
33. Injection pump
34. Oil dipstick
35. Fuel-feed pump
36. Oil filter
37. Oil pump
38. Strainer
39. Engine bracket
40. Starter
41. Cylinder crankcase
42. Exhaust manifold
43. Ram manifold
44. Control linkage
45. Mixture controller

Fig. 6-31. Sectional views of a four-cylinder diesel engine for passenger cars. (*Mercedes-Benz*)

For heavy-duty applications, some diesel engines have four valves per cylinder—two intake valves and two exhaust valves (Fig. 6-32). The additional valves improve engine breathing and thus increase engine output, particularly at high speeds. Figure 6-33 shows a diesel engine which has four valves per cylinder and overhead camshafts.

Another diesel-engine application is shown in Fig. 6-34, which is a three-cylinder four-cycle diesel engine for marine use. This engine has wet cylinder liners. These are replaceable sleeves that are in contact with the cooling water.

⊘ 6-17 Gas Turbines The *gas turbine,* now making its appearance as an automotive truck-type power plant, consists essentially of two sections—a *gasifier section* and a *power section.* Figure 6-35 is a simplified sectional view of a turbine, and Fig. 6-36 is a cutaway view of an actual unit. The compressor in the gasifier has a rotor with a series of blades around its outer edge. As it rotates, air between the blades is carried around and thrown out by centrifugal force. This action supplies the burner with air at relatively high pressure. Fuel is sprayed into the compressed air. (The fuel can be gasoline, kerosene,

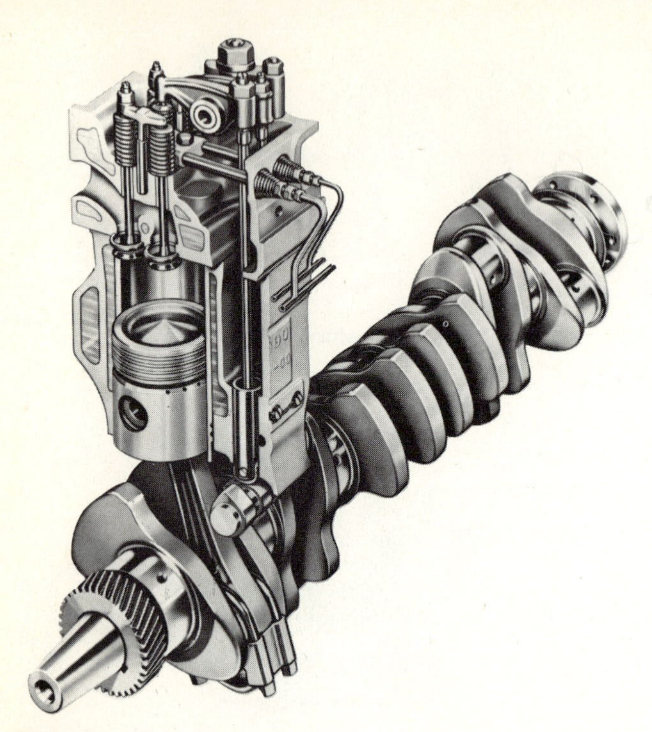

Fig. 6-32. Cutaway view of one cylinder of a four-cycle diesel engine using two intake and two exhaust valves in each cylinder. (*Cummins Engine Company, Incorporated*)

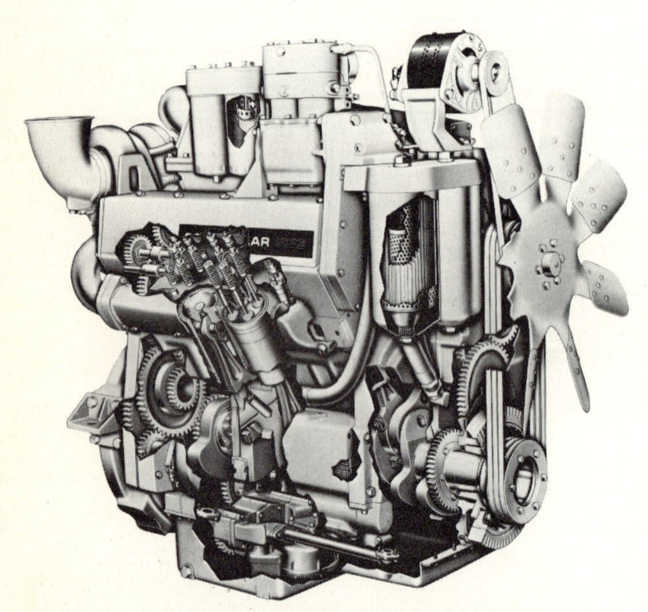

Fig. 6-33. Partial cutaway view of a V-8 diesel engine using overhead camshafts. (*Caterpillar Tractor Company*)

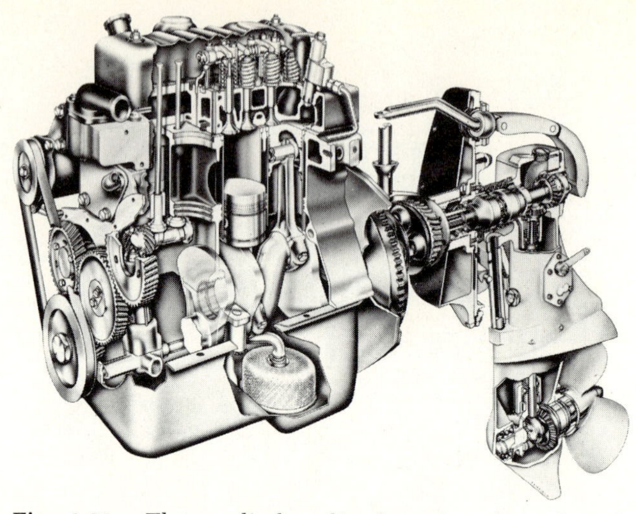

Fig. 6-34. Three-cylinder diesel engine for inboard-outboard motorboat. The propeller drive is to the right. (*Perkins Engines, Limited*)

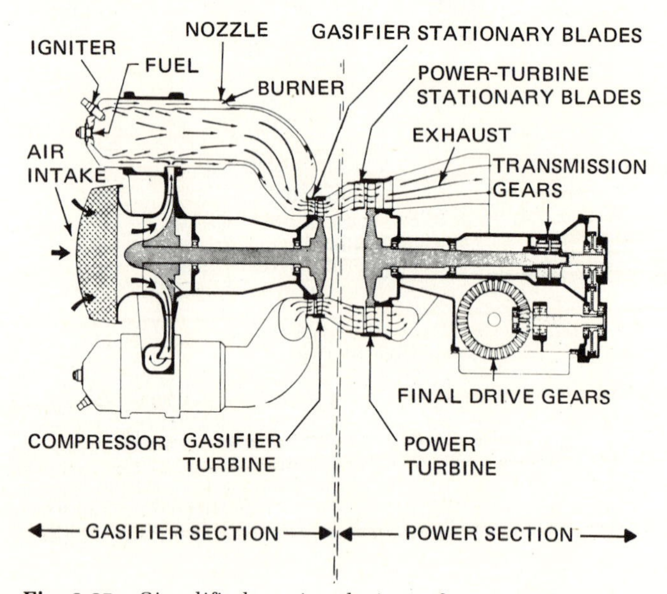

Fig. 6-35. Simplified sectional view of a gas turbine.

or oil.) As the fuel burns, the pressure is increased. The high-pressure, high-temperature gas then passes through the gasifier-nozzle diaphragm. A series of stationary blades directs this high-pressure gas against a series of curved blades on the outer edge of the gasifier-turbine rotor. The resulting high pressure against the curved blades causes the gasifier-turbine rotor to spin at high speed. And since the gasifier-turbine rotor and the compressor rotor are mounted on one shaft, the compressor rotor also spins at high speed. This action continues to supply the burner with an ample amount of compressed air as long as fuel is supplied to the burner.

After the high-pressure, high-temperature gas leaves the gasifier section, it enters the power turbine. Here it strikes another series of stationary curved blades and is directed against a series of curved blades on the outer edge of the power-turbine rotor. The resulting high pressure against these rotor blades spins the rotor at high speed. In some models, the turbine may turn up to 50,000 rpm. However, this high speed is reduced by a series of transmission gears before the power is applied to the wheels of the vehicle.

⊘ 6-18 Wankel Engines The *Wankel engine* uses a three-lobe rotor that rotates eccentrically in an

Fig. 6-36. Cutaway view of a gas turbine. (*Caterpillar Tractor Company*)

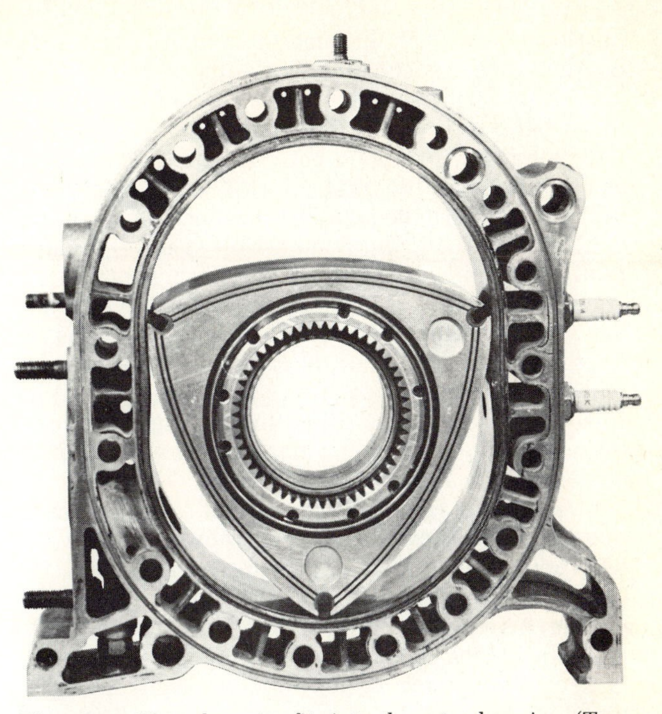

Fig. 6-38. How the rotor fits into the rotor housing. (*Toyo Kogyo Company, Limited*)

oval chamber (Figs. 6-37 and 6-38). The rotor is mounted on the crankshaft through external and internal gears. The four cycles of intake, compression, power, and exhaust are going on simultaneously around the rotor when the engine is running. Figure 6-39 gives you an idea of how the Wankel engine works. The rotor lobes A, B, and C seal tightly against the side of the oval chamber. The rotor has oval recesses in its three faces between the lobes (shown by dashed lines in Fig. 6-39). These recesses correspond to the combustion chambers in the piston engine.

Let us follow the rotor as it goes through the four cycles. At I (upper left), lobe A has passed the intake port and air-fuel mixture (1) is ready to enter. At II (upper right), lobe A has moved on around so that the space between lobes A and C and the chamber wall (2) increases and air-fuel mixture enters. At III (lower right), the intake continues as the com-

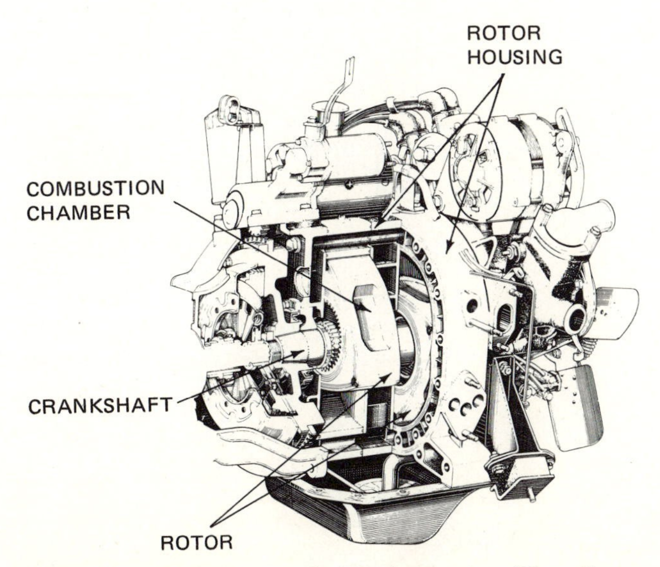

Fig. 6-37. Cutaway view of a Wankel engine. (*Toyo Kogyo Company, Limited*)

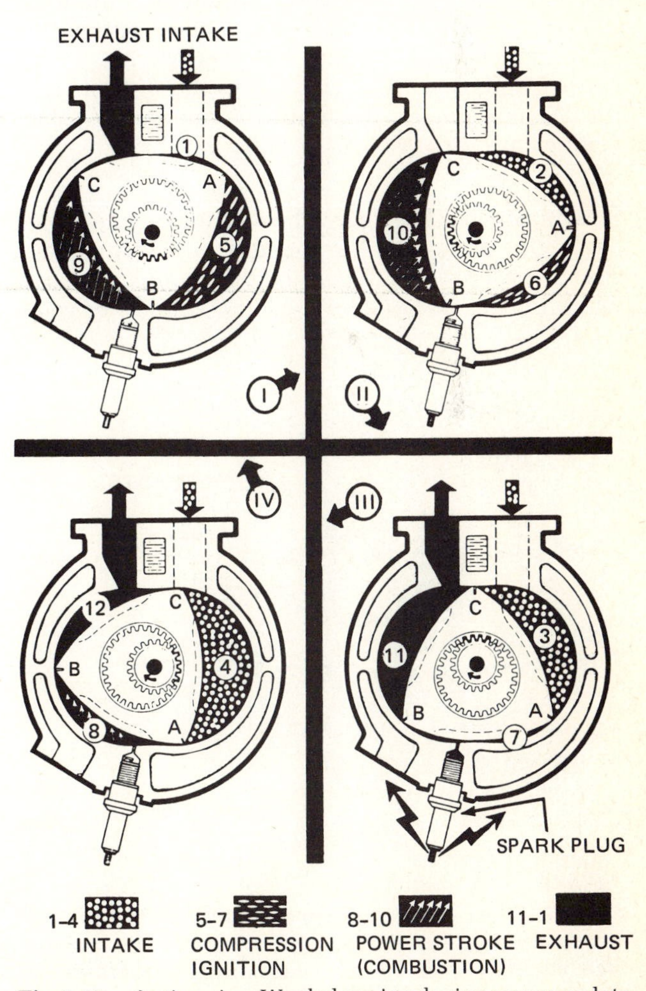

Fig. 6-39. Actions in a Wankel engine during one complete revolution of the rotor.

bustion space (3) continues to increase. The combustion space reaches its maximum at IV (4).

To see what happens to the air-fuel mixture, let us go back to I (upper left) again. Here, an air-fuel charge has been trapped between lobes A and B (5) as lobe A passes the intake port. Further rotation of the rotor reduces the combustion space (6), as shown at II. Then, at III, the combustion space (7) is at a minimum, and so the mixture reaches maximum compression. At this instant, the spark plug fires and ignites the mixture. Now the power cycle begins. At IV, the expanding gases (8) force the rotor around, and this action continues in I (9) and II (10). With further rotation of the rotor, the leading lobe clears the exhaust port and the burned gases are forced out as shown at III (11) and IV (12).

As will be seen in Chap. 10, the Wankel engine has three power cycles for each rotor revolution and delivers power almost continuously. In a way, the Wankel engine is equivalent to a two-cylinder piston engine. A two-rotor Wankel engine would be equivalent to a four-cylinder piston engine. Figure 6-40 is a cutaway view of a two-rotor Wankel engine developed by NSU of Germany. Toyo Kogyo of Japan is a leading manufacturer of Wankel engines and Wankel-powered passenger cars. This company has manufactured several hundred thousand Wankel-powered Mazda cars. Other automotive manufacturers, (including General Motors in the United States) are studying the possibility of using the Wankel engine for their automobiles. (Wankel engines are discussed more fully in Chap. 10.)

⊘ 6-19 Stirling Engines The *Stirling engine* makes use of the fact that the pressure in a container of gas increases when the gas is heated and decreases when the gas is cooled. A specific amount of gas is sealed into the engine and alternately heated and cooled. When it is heated, its increased pressure pushes a power piston down. When the gas is cooled, its decreased pressure, in effect, pulls the power piston up.

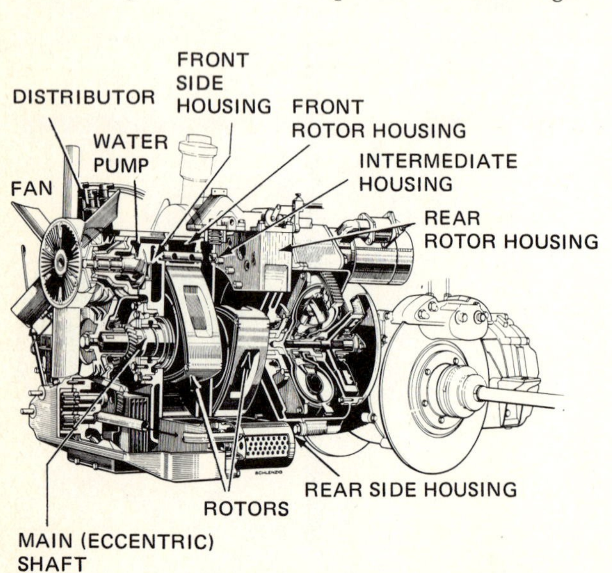

Fig. 6-40. Cutaway view of a two-rotor Wankel engine with attached torque converter and transmission. This engine is the same as the single-rotor engine, except that there are two rotors on a single shaft. Each rotor turns in its own chamber. (*NSU of Germany*)

Figure 6-41 shows the Stirling engine. The upper, or displacer, piston does not produce power but simply causes the air to move between the hotter and cooler sections of the engine. The power piston moves up and down between the power chamber and the buffer space. The two pistons are linked by a rhombic drive to a pair of synchronizing gears which, in turn, drive the output shaft. The name "rhombic" comes from the fact that the four links form a rhombus, which is a geometric figure of four parallel but unequal sides.

Let us follow the engine through a complete cycle of events and see how it operates. We shall start with the pistons in the position shown in Fig. 6-41. The working gas has been heated, and therefore its pressure has increased. This increase in pressure, applied to the head of the power piston, forces the piston to move down. (Note that the working gas expands during this action.)

The power piston, in moving down, pushes downward through its two connecting links against the two gears. The gears therefore rotate and turn the engine output shaft. Meanwhile, the rotation of the gears causes the two links to the displacer piston to pull the displacer piston down. This increases the hot space above the displacer piston so that some

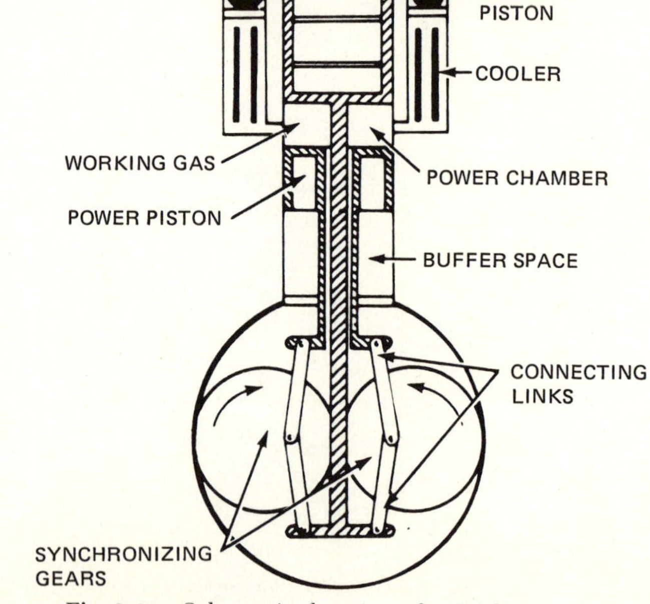

Fig. 6-41. Schematic drawing of a Stirling engine.

of the working gas can flow from the power chamber toward this space. As it moves through the cooler on the way to this upper space, the gas is cooled and its pressure drops.

Notice that there is a buffer space below the power piston. This buffer space contains a specific amount of gas. When the power piston moves down, this gas is compressed and its pressure increases. Now, after the power piston has reached BDC, the pressure of this buffer gas becomes higher than the pressure of the working gas. Remember that the working gas has been cooled, and so its pressure has dropped. Now, with the buffer gas at a higher pressure than the working gas, the power piston is pushed up. This motion is transmitted through the two links to the gears, rotating them on around to the position shown in Fig. 6-41.

Meanwhile, the working gas that has flowed into the hot space becomes heated. The heating effect results from the continuous combustion of a fuel such as kerosene or oil. The heating of the working gas causes its pressure to increase, and once again the power piston is forced down by the increasing pressure.

The action, therefore, results from the repeated heating and cooling of the working gas. When the working gas is heated, it drives the power piston down. When it is cooled, the buffer-gas pressure forces the piston up. You might think that this heating and cooling would take considerable time and result in an awkward, slow, and inefficient engine. However, recent experimental engines, using helium as the working gas at an average pressure of 1,500 psi [105.462 kg/cm^2], have operated at 3,000 rpm with an efficiency of 30 percent, which is as good as or better than most automotive engines.

Some engineers have proposed the Stirling engine for such small-engine applications as lawn mowers. They are very easy to start, silent in operation, and simple in basic design. Other engineers believe that they will appear first for large stationary applications, such as remote electric power plants or pumping stations. One possibility is that they will be used to power space satellites and space stations. In space, no source of heat other than the sun would be needed. With a large reflector to gather heat from the sun and a large radiator to radiate the heat away, a Stirling engine could operate without any fuel at all.

⊘ 6-20 Experimental Automotive Stirling Engines Figure 6-42 is a cutaway view of a four-cylinder experimental Stirling engine proposed for automotive applications. The design shown in Fig. 6-42 will produce 180 hp [134.28 kW] and fits into an automobile as shown in Fig. 6-43. Note that this design does not use the rhombic drive. Instead, it uses a *swish plate*. A swish plate, also called a *wobble plate*, is a metal disk that is set at an angle to the shaft on which it is mounted. The swish plate changes the reciprocating motion of the pistons to rotary motion.

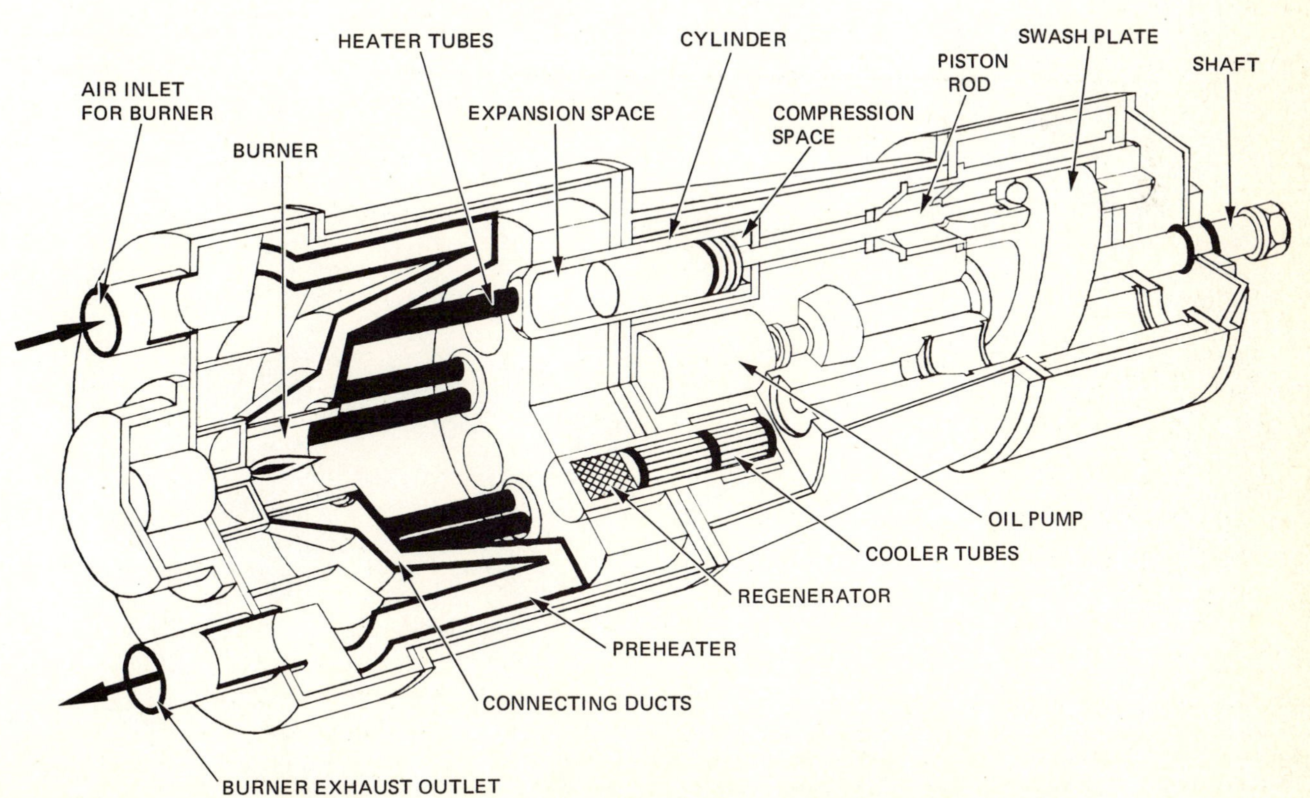

Fig. 6-42. Cutaway view of a four-cylinder, 180-hp Stirling engine.

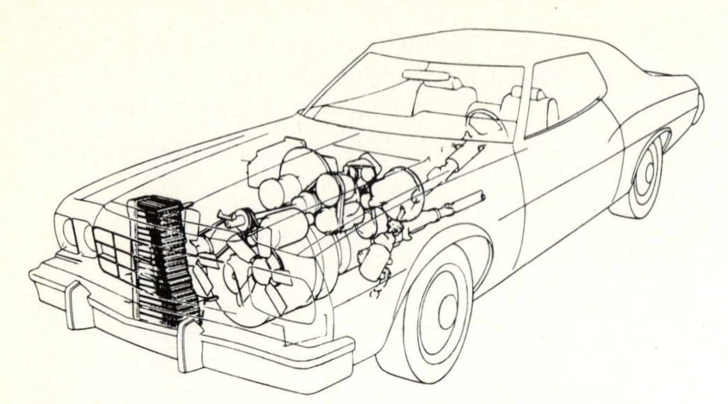

Fig. 6-43. How the Stirling engine shown in Fig. 6-42 would fit into an automobile.

⊘ 6-21 Other Possible Power Plants In recent years several alternative power plants have been proposed to the internal-combustion engine for running automobiles. These include steam engines, electric batteries, and spinning flywheels.

1. *STEAM ENGINES* As mentioned previously, the steam engine is an external-combustion engine; that is, the combustion takes place outside the engine. Water or other liquid is heated in a boiler where it turns into steam, or vapor. Figure 6-44 shows the design of one experimental car using a steam engine. The engine burns kerosene, and the heat produced evaporates the liquid. The liquid used is *Freon*, the same liquid used in refrigerators. It evaporates at a much lower temperature than water. The Freon vapor, at high pressure and temperature, enters the main drive engine where it spins rotors. The rotary motion is then carried through the transmission and differential to the car wheels. As the Freon vapor leaves the engine, it is carried up to the condenser which covers the top of the car. There, the Freon gives up heat and returns to liquid. It then flows back down to the burners where it is turned into high-pressure vapor again.

2. *ELECTRIC BATTERIES* Electric batteries can be used to supply the energy needed to run a car. The lead-acid storage battery (used in internal-combustion engines to start the engine) is the battery used in most electric cars. However, these batteries are very heavy. And they do not provide enough electricity to run the car more than a few miles. Thus after about 50 miles [80.467 km], the batteries need to be recharged. Despite these drawbacks there are

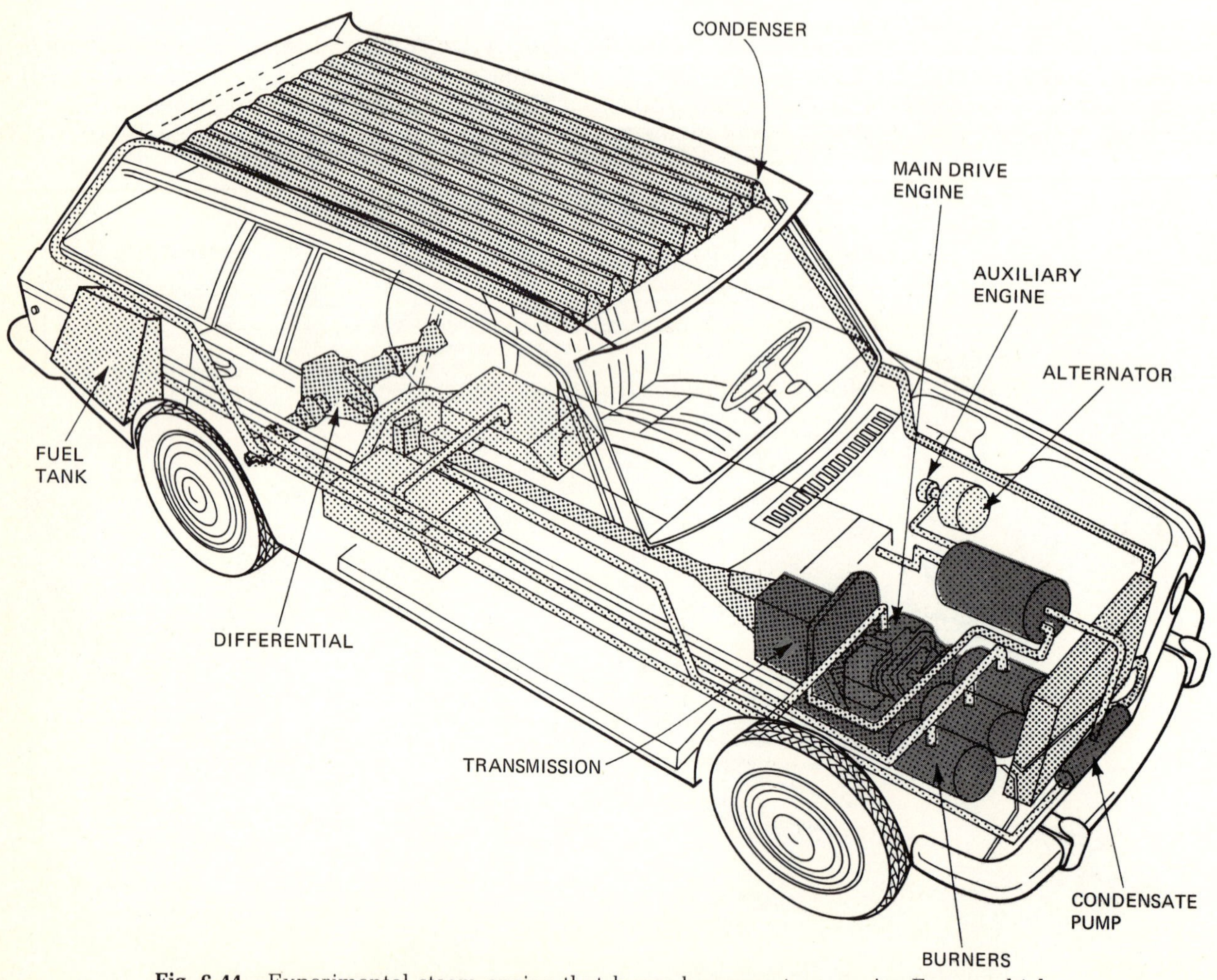

Fig. 6-44. Experimental steam engine that burns kerosene to vaporize Freon, which drives the steam engine.

still a few electric vehicles in operation (in addition to electric golf carts). For example, there are an estimated 50,000 electric vans in use, mostly in short-mileage delivery service.

General Motors has developed an experimental electric car (Fig. 6-45) which uses silver-zinc batteries. These batteries hold much more energy per pound than the heavy lead-acid batteries. Even though this experimental car shows better mileage per battery charge, there are still serious drawbacks. The silver-zinc batteries are very expensive. Also, any electric car requires electric energy to charge the batteries. And the electric-power resources in the United States are already strained to the limit. A large number of new electric power plants would have to be built if electric cars became common.

3. FLYWHEELS The car powered by a large, heavy flywheel is a rather "far-out" idea. The flywheel could be located in the back, as shown in Fig. 6-46. The flywheel would be charged by using an electric motor at a charging station to spin it at high speed. Then the energy of the flywheel would be used to run the car. One study shows that an ordinary car equipped with a driving flywheel could travel 100 miles [160.93 km] between charges. It could accelerate to 60 miles per hour [96.561 km/h (kilometers per hour)] in 15 seconds and reach a top speed of 70 miles per hour [112.65 km/h].

The flywheel power plant was actually tried in a bus system in Europe. The buses were driven by 3,300-pound [1,496.85 kg] flywheels. It worked but was considered impractical because the flywheel had to be charged about every ½ mile [0.805 km]. Nonetheless, experimental and design work continues. Recent design improvements on flywheels have considerably increased the theoretical energy that could be stored per pound of flywheel.

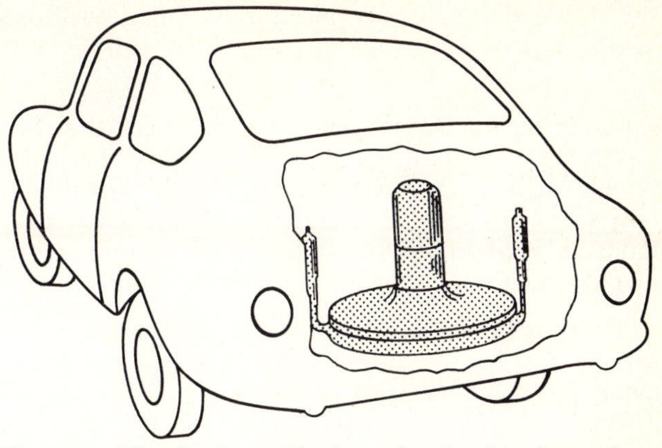

Fig. 6-46. Flywheel car. The large flywheel is shown here located at the back of the car.

Check Your Progress

Progress Quiz 6-2 Here is your chance to find out how well you have understood the material covered in the last half of this chapter. If you are not sure about some of the answers to the following ques-

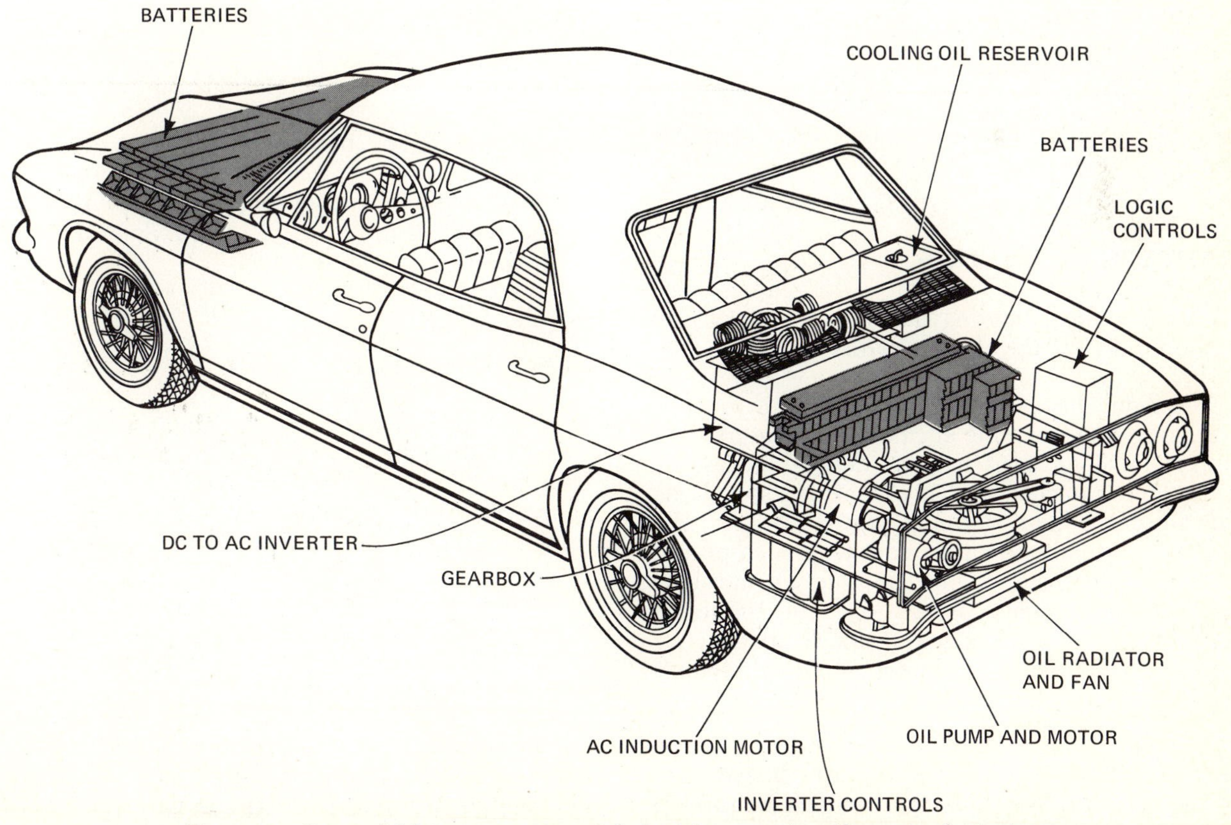

Fig. 6-45. General Motors experimental electric car using silver-zinc batteries.

tions, reread the past few pages and try the questions again.

Completing the Sentences The sentences that follow are incomplete. After each sentence there are several words or phrases, only one of which will correctly complete the sentence. Write each sentence in your notebook, selecting the proper word or phrase to complete the sentence correctly.

1. There are two classifications of engine by cycle: (*a*) one-cycle and two-cycle, (*b*) two-cycle and three-cycle, (*c*) two-cycle and four-cycle.
2. The two-cycle engine produces a power stroke: (*a*) every crankshaft revolution, (*b*) every two crankshaft revolutions, (*c*) every four crankshaft revolutions.
3. The two-cycle engine has valve ports in the: (*a*) pistons, (*b*) cylinder walls, (*c*) piston rings.
4. For fuel the diesel engine uses: (*a*) LPG, (*b*) gasoline, (*c*) fuel oil.
5. In the diesel engine, fuel is injected into the cylinder at the end of the: (*a*) intake, (*b*) compression, (*c*) power, (*d*) exhaust stroke.
6. In the diesel engine rapid compression of the air to about 500 psi [35.154 kg/cm^2] will produce an air temperature of about: (*a*) 100°F [37.8°C], (*b*) 1000°F [537.8°C], (*c*) 2000°F [1093.3°C], (*d*) 5000°F [2760°C].
7. Compression ratios in the diesel engine are as high as: (*a*) 5:1, (*b*) 10:1, (*c*) 15:1.
8. The two major parts of the gas turbine are the: (*a*) gasifier section and combustion section, (*b*) gasifier section and power section, (*c*) power section and turbine section.
9. In comparing the Stirling and the Wankel engines, it will be noted that the Stirling engine is a sealed unit, whereas the Wankel has: (*a*) a three-lobed rotor, (*b*) two pistons per cylinder, (*c*) a rhombic drive.

CHAPTER 6 CHECKUP

NOTE: Since the following is a chapter review test, you should review the chapter before taking the test.

You have been making real progress in your studies of the automobile engine and in the past six chapters have covered a great deal of necessary background information. This information will be of considerable help to you as you move on to the later parts of the book which deal with such practical things as details of engine construction and servicing and maintenance procedures. The following questions give you a chance to check up on your success in remembering and understanding the facts discussed in this chapter.

Picking Out the Right Answer Several answers are given for each of the following questions. Read each question carefully and decide which answer is the correct one. Then write the question, with the right answer, in your notebook.

1. How many cylinders are used in American automobile engines? (*a*) 4, 8, and 12; (*b*) 4, 6, and 8; (*c*) 6, 8, and 12.
2. What are the two most common valve arrangements in automotive engines? (*a*) OHC and I-head, (*b*) L-head and OHC, (*c*) I-head and F-head.
3. Which type of valve arrangement requires the use of rocker arms? (*a*) I-head, (*b*) T-head, (*c*) L-head.
4. Large American-made automobiles are powered by what type of engine? (*a*) L-head V-8, (*b*) I-head V-8, (*c*) four-cylinder L-head.
5. What are three ways to classify automobile engines? (*a*) by valve number, cylinder arrangement, and valve arrangement; (*b*) by cylinder ports, cylinder arrangement, and valve arrangement; (*c*) by number of cylinders, cylinder arrangement, and valve arrangement.

Lists In the following, you are asked to write certain lists and also to make a couple of simple drawings. Write and draw in your notebook. As you write or draw, you are helping yourself to remember since the action of writing or drawing helps to fix the information in your mind. More than that, you are putting the information in your notebook where you can quickly refer to it.

1. List the various cylinder arrangements given in the book.
2. List the various valve arrangements given in the book.
3. List the differences between the two-cycle and the four-cycle engines.
4. List three types of internal-combustion-engine fuel.
5. List differences between Otto-cycle and diesel engines.
6. Make simple drawings (copy them from the book if you wish) of the L-head, I-head, and OHC valve arrangements.

SUGGESTIONS FOR FURTHER STUDY

If you would like to learn more about diesel engines, you can probably find books about them in your local public or school library. In addition, your high school science or automotive mechanics teacher may have books or manuals on diesel engines. If a diesel-engine manufacturer has a branch or a dealership in your area, you can probably find out a good deal about diesel engines from the people who work there. You might also ask a local truck or bus operator who has a fleet of diesel-engine vehicles. You may also be able to obtain diesel-engine manuals for a modest sum by writing to the engine manufacturer. In writing, you should list specific engine numbers and explain why you need the manuals.

chapter 7

PISTON-ENGINE CONSTRUCTION: CYLINDER BLOCKS, HEADS, CRANKSHAFTS, AND BEARINGS

In this chapter we consider the construction of automotive piston engines in greater detail, with emphasis on the engine cylinder blocks, cylinder heads, crankshafts, and bearings. Chapters 8 and 9, which describe pistons, piston rings, and valve mechanisms, complete the analysis of construction details of automotive piston engines. Chapter 10 describes the construction and operation of Wankel engines. These chapters supply the background information that will permit you to move on to the chapters on trouble diagnosis and servicing later in the book.

⊘ 7-1 Engine Construction Thus far the engine has been considered from the operational point of view. We have seen how the mixture of fuel and air is compounded by the carburetor, drawn into the cylinders, compressed, ignited, and burned. We have noted that this combustion process creates a high pressure that forces the piston down, rotates the crankshaft, and thus moves the vehicle. The air-fuel mixture is admitted to the cylinder by the opening of the intake valve; the burned gases are exhausted through the exhaust-valve port when the exhaust valve opens. Let us now look at the engine from the constructional point of view and examine its various component parts.

⊘ 7-2 Engine Cylinder Block The *cylinder block* of a liquid-cooled engine is the basic framework, or foundation, of the engine. Everything else is put inside the block or fastened onto it. (In air-cooled engines, the cylinders are often separate parts, as shown in Fig. 7-10). Figure 7-1 shows three views of a cylinder block for a V-8 engine. The parts that are installed in or fastened to the block are described in ⊘ 7-4.

The cylinder block is cast in one piece from gray iron or iron alloyed (mixed) with other metals, such as nickel or chromium. Some cylinders are cast from aluminum. The engine shown in Fig. 6-12, for example, has been supplied with either a die-cast aluminum or cast-iron cylinder block. Aluminum cylinder blocks are described in detail in ⊘ 7-6.

The iron cylinder block is cast by pouring molten iron into a mold that is usually made of sand. The various openings in the casting are formed by putting sand cores into the mold. Figure 7-2 shows the lower half of a mold for 2 six-cylinder engine blocks with the cores in place. Note the cores which form the water jacket and crankcase in Fig. 7-2. After the upper half is put into place, molten iron is poured into the mold. The molten iron fills all the spaces between the mold and the cores. After the casting has cooled and hardened, it is removed from the mold. The cores are broken up and shaken out of the casting.

If you examine Fig. 7-1 carefully, you will find several arrows pointing to "water-jacket plugs." These plugs are in the openings through which the cores have been removed to leave the water jackets. You will recall that the water jackets are spaces surrounding the cylinders through which coolant (water and antifreeze) flows. The water jackets are formed by the inner shells of the cylinders and the outer shell of the cylinder block. In Fig. 7-3 you can see the water jackets in one bank of a V-6 engine.

⊘ 7-3 Machining the Block After the cores are removed, the cylinder block is machined. Examine the three views of the block in Fig. 7-1 again. Note the many finished surfaces, passages, bores, and tapped holes in the block. The machining operation includes:

1. Drilling holes for attachment of various parts
2. Machining the cylinders
3. Boring the camshaft-bearing holes
4. Smoothing the surfaces to which parts are attached

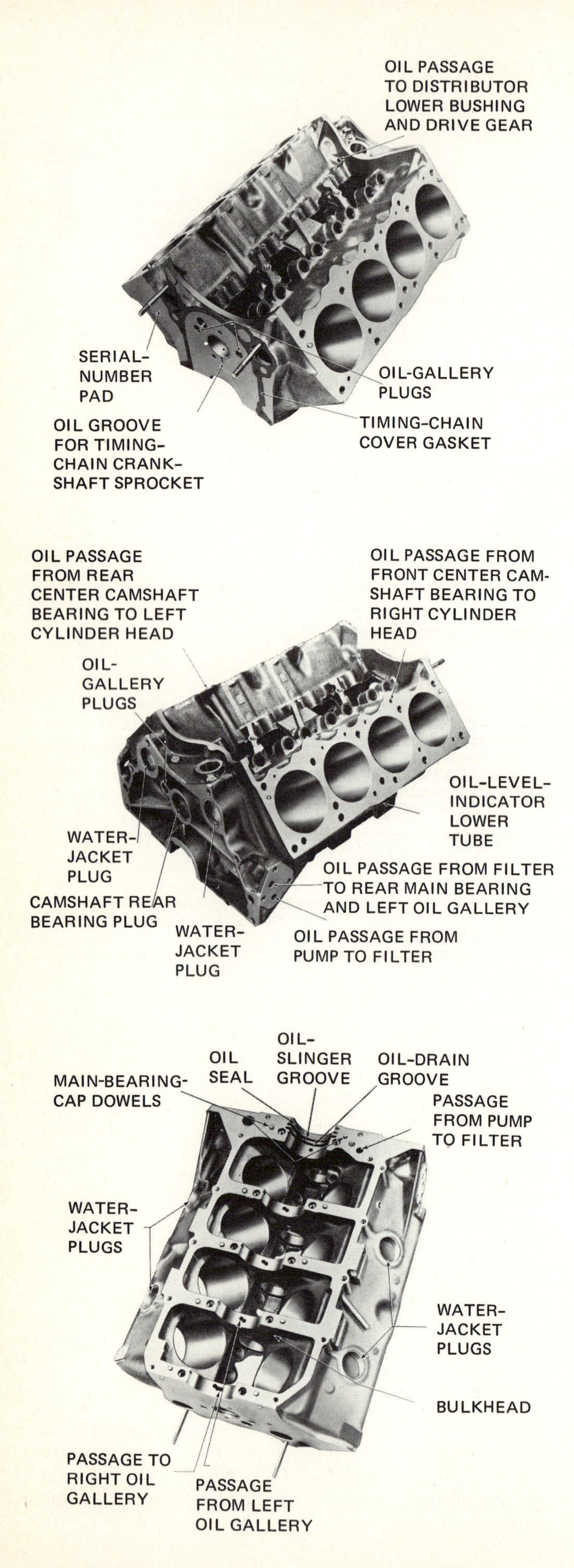

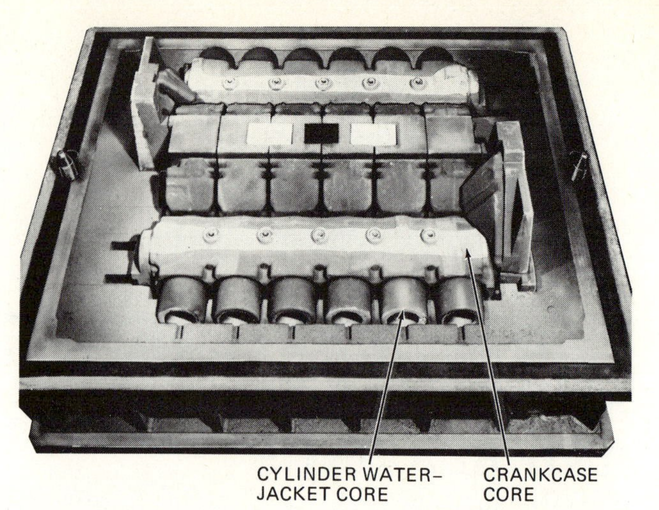

Fig. 7-2. Cores for 2 six-cylinder engine blocks in place in the lower half of the mold. (*Ford Motor Company*)

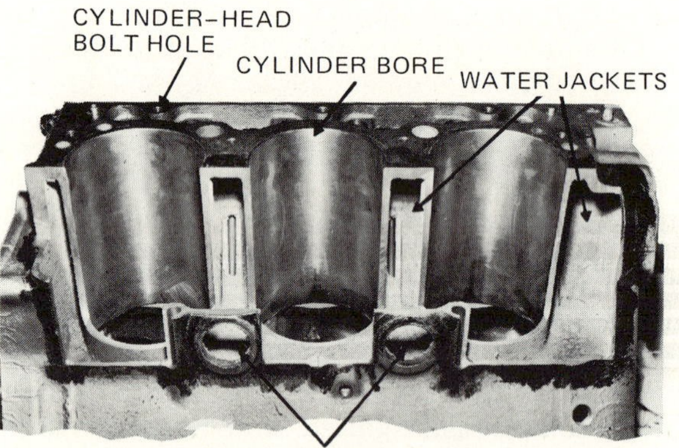

Fig. 7-3. One bank of a V-6 engine block partly cut away so the internal construction can be seen. (*General Motors Corporation*)

5. Drilling oil passages
6. Boring the valve-lifter bores
7. Cleaning out water passages
8. Boring or machining the bores for the main bearings

⊘ 7-4 Parts Attached to Block Figure 7-4 shows the parts that are attached to and mounted on the cylinder block for a V-8 engine. The cylinder heads are mounted at the top, along with the cylinder-head covers, intake and exhaust manifolds, carburetor, and other related parts. The oil pan is attached to the bottom of the cylinder block. In addition, the water pump and timing-chain cover (not shown in Fig. 7-4) are attached at the front of the block. Figure

Fig. 7-1. Three views of a cylinder block from a V-8, overhead-valve, liquid-cooled engine. Note locations of the coolant and oil passages and plugs. (*Pontiac Motor Division of General Motors Corporation*)

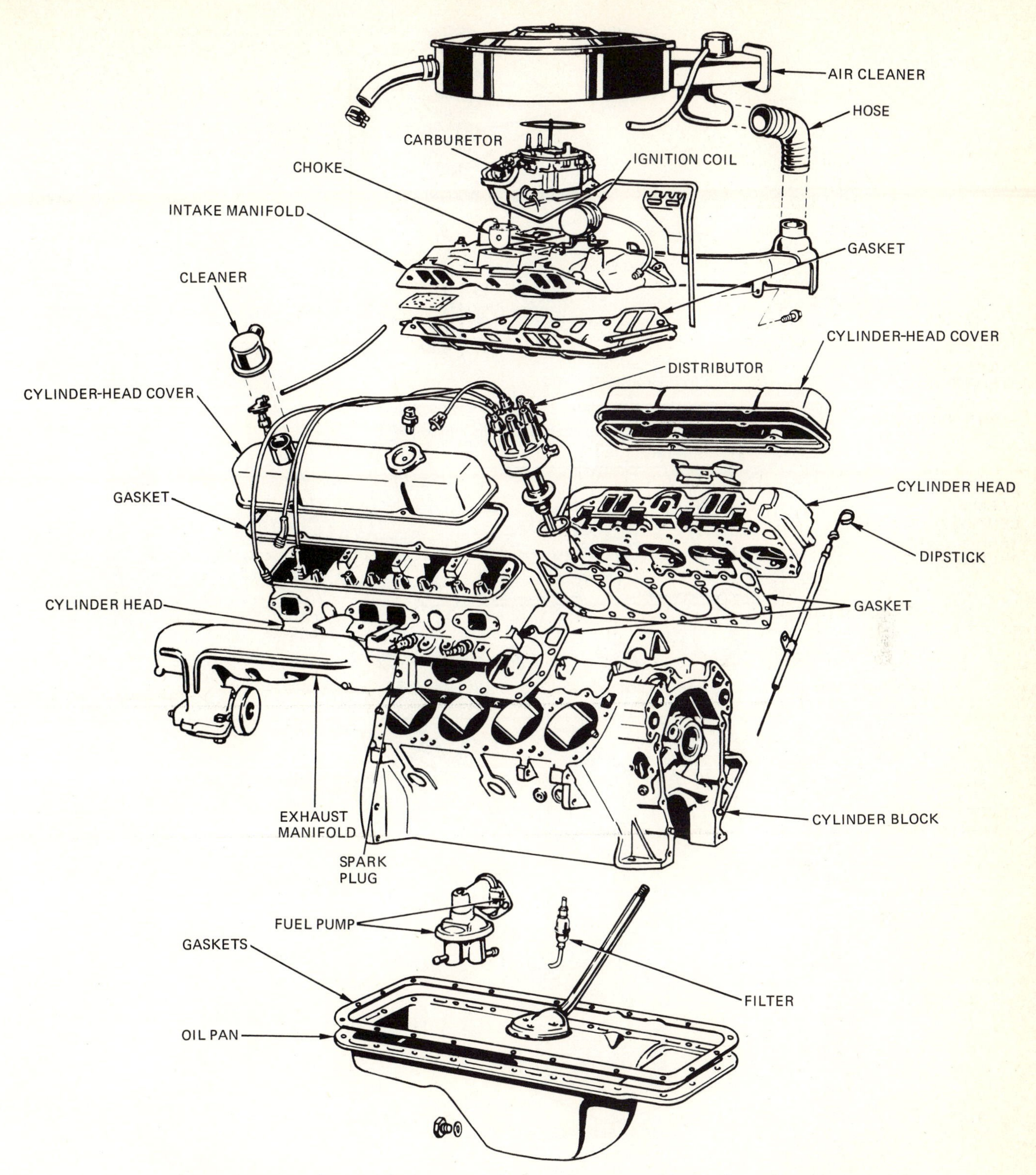

Fig. 7-4. External parts that are attached to the engine block when the engine is assembled. (*Chrysler Corporation*)

6-16 shows how all these parts go together in the assembled engine. The purpose of all these parts will be explained later.

Figure 7-5 shows parts that are installed inside the cylinder block of a six-cylinder engine. For the sake of simplicity, only one piston and connecting rod are shown, although the engine actually uses six pistons and rods. Figure 6-11 shows how all the parts go together in the assembled engine. The crankshaft is hung from the bottom of the cylinder block by

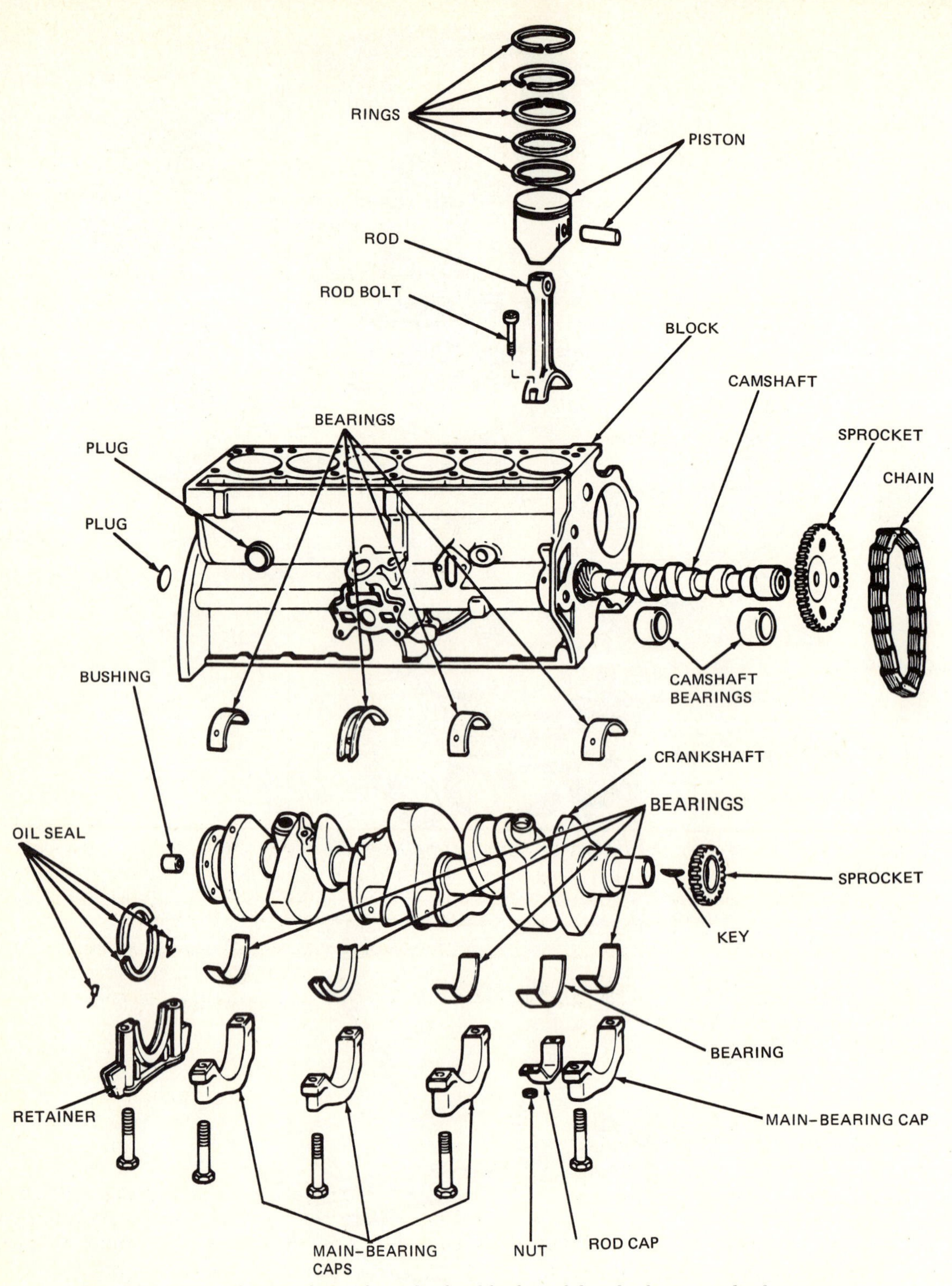

Fig. 7-5. Internal parts from the cylinder block and head of a six-cylinder engine. (*Chrysler Corporation*)

bearings and caps. Figure 7-6 is a picture of a crankshaft and related parts for a six-cylinder engine.

The crankshaft bearings, called the *main bearings*, or *mains*, are of the split type. The upper bearing halves fit into half-round sections of the cylinder block. These sections are webs, or bulkheads (see bottom view in Fig. 7-1), that are part of the block casting. The lower bearing halves fit into bearing caps which are attached to the cylinder block by cap screws to support the crankshaft.

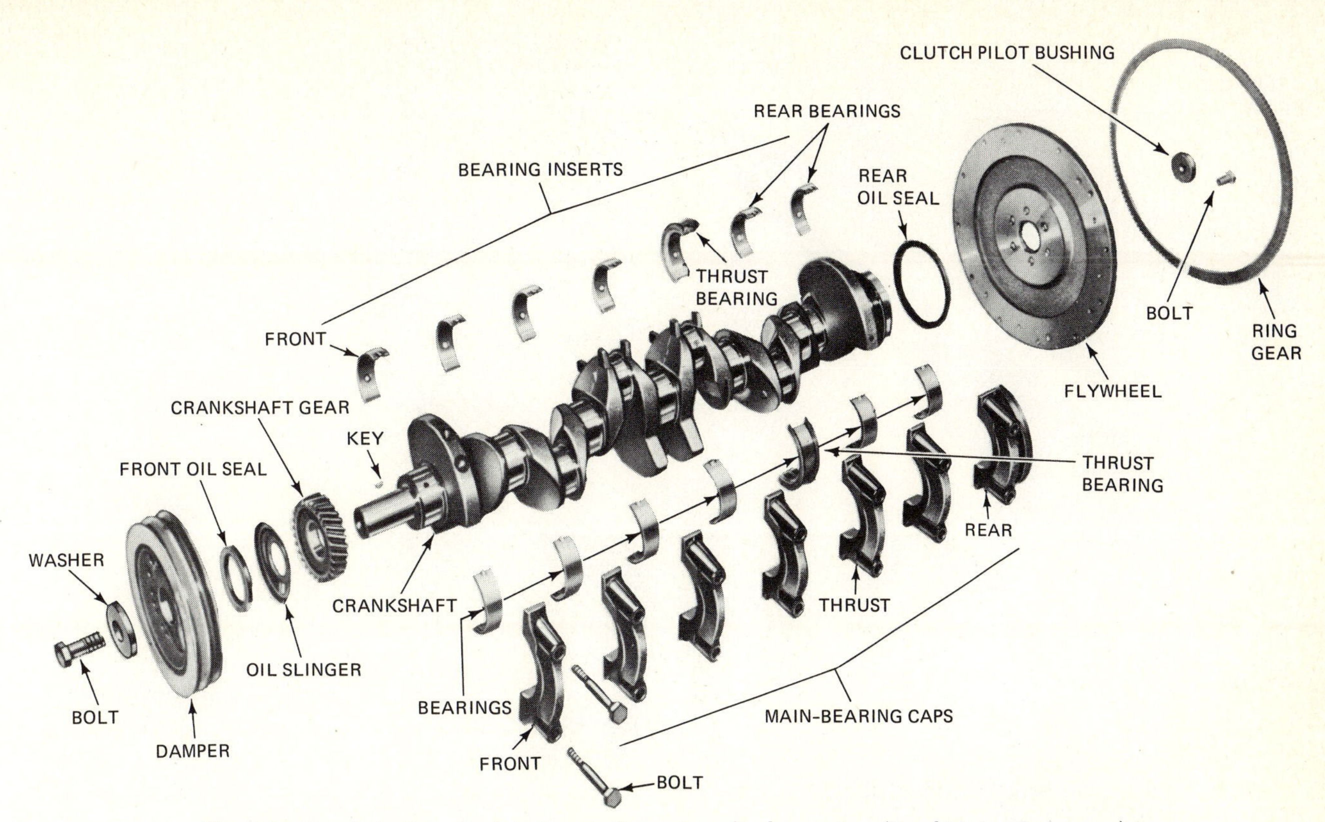

Fig. 7-6. Crankshaft and related parts for a six-cylinder engine. (*Ford Motor Company*)

NOTE: In order to establish the proper fit of the bearing caps to the cylinder block, and thus the proper fit of the main bearings, this is the procedure that is followed during original engine assembly: The caps are bolted into place. Then the bores in the cylinder block and cap are bored to take the bearings. This means that each cap fits in one particular spot on the block. Therefore, when caps are removed for any service job, they must be put back onto the block in the same place and in the same position as they were originally.

The flywheel is attached to the rear of the crankshaft. The crankshaft gear or sprocket and pulley with damper are attached to the front of the crankshaft. We shall discuss all these parts later.

After the crankshaft is installed, the pistons, with their piston rings and connecting rods, are installed in the block. The connecting rods are fastened to the crankpins on the crankshaft by bearings and caps with rod bolts and nuts.

NOTE: The parts shown in Fig. 7-5, when assembled into the cylinder block, make an assembly called a *short block*. This is a service item that can be purchased. Sometimes a service problem can be solved by buying a short block. The rest of the old engine, including head and manifolds, are then installed on the new short block. More will be said about this in the chapters on engine service.

⊘ 7-5 Oil Pan The *oil pan* (Fig. 7-7) is attached to the bottom of the cylinder block, and together they form the *crankcase*. They enclose, or *encase*, the cranks on the crankshaft. The oil pan holds from about 4 to 9 quarts [3.79 to 8.52 l] of oil, depending on the engine design. Note that gaskets (explained later) are used to seal the joint between the cylinder block and the oil pan. The gaskets prevent loss of oil from the oil pan.

When the engine is running, the oil pump sends oil from the oil pan up to the moving engine parts. The engine lubricating system is described in Chap. 17.

⊘ 7-6 Aluminum Cylinder Block Some engines have aluminum cylinder blocks. Aluminum is a relatively light metal, weighing much less than cast iron. Aluminum also conducts heat more rapidly than cast iron. Lightness and heat conductivity are the two main reasons that aluminum has been used for cylinder blocks. However, aluminum is too soft to use as cylinder-wall material. It would wear very rapidly. Therefore, aluminum cylinder blocks must have cast-iron cylinder liners (with one exception, the Vega engine). Cylinder liners are sleeves that are either cast into the cylinder block or installed later.

In the cast-in type, the cylinder liners are installed in the mold and the aluminum is poured

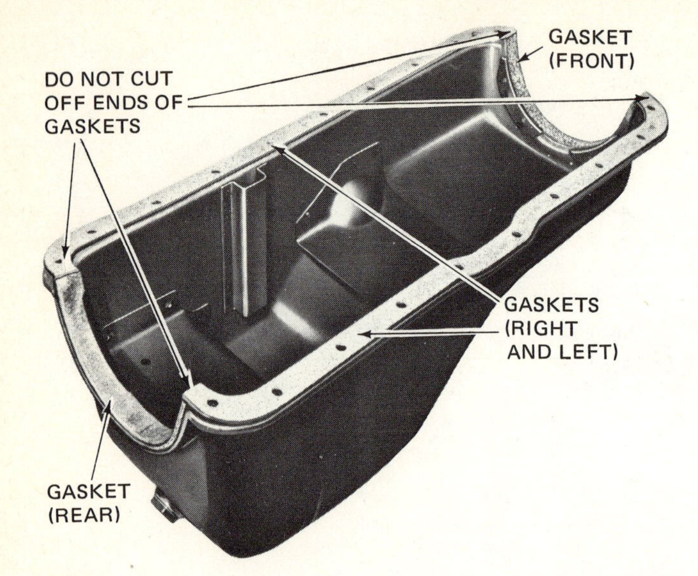

Fig. 7-7. Oil pan with gaskets in place, ready for pan replacement. (*Chrysler Corporation*)

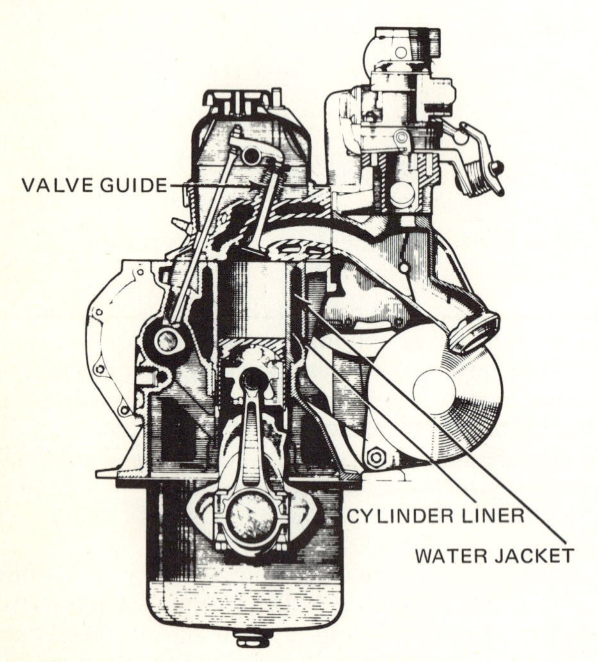

Fig. 7-8. Sectional view of an engine that uses wet cylinder liners. (*Renault*)

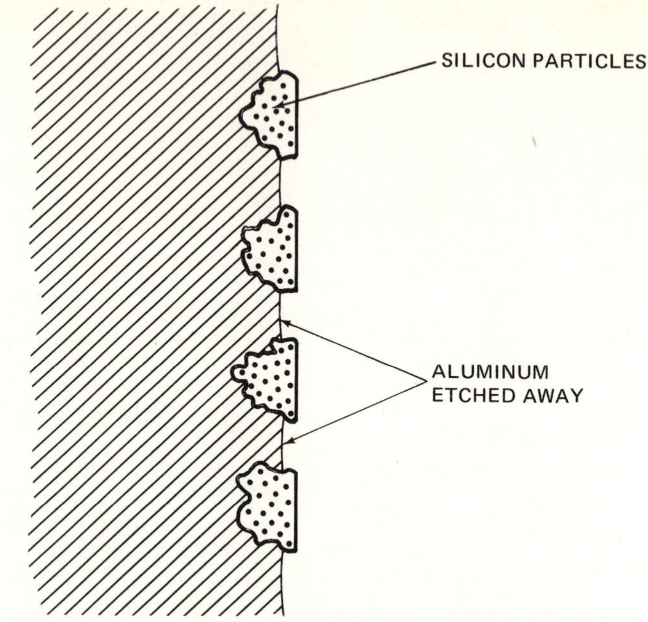

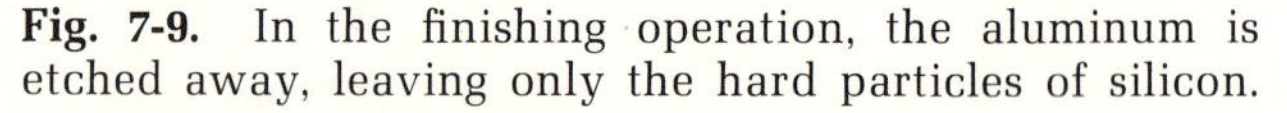

Fig. 7-9. In the finishing operation, the aluminum is etched away, leaving only the hard particles of silicon.

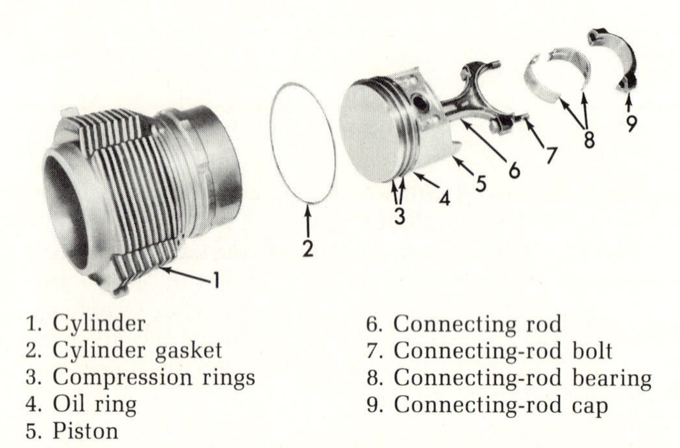

1. Cylinder
2. Cylinder gasket
3. Compression rings
4. Oil ring
5. Piston
6. Connecting rod
7. Connecting-rod bolt
8. Connecting-rod bearing
9. Connecting-rod cap

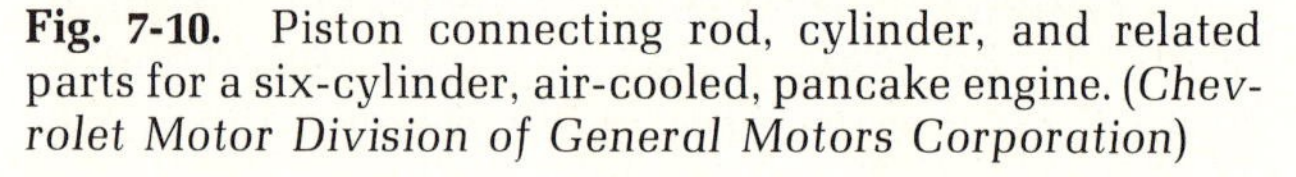

Fig. 7-10. Piston connecting rod, cylinder, and related parts for a six-cylinder, air-cooled, pancake engine. (*Chevrolet Motor Division of General Motors Corporation*)

around them. They thus become a permanent part of the cylinder block.

Two kinds of liners, dry and wet, can be installed later. The dry type of cylinder liner is pressed into the cylinder block with high pressure. This type touches the cylinder block along the liner's full length. The wet type of liner touches the cylinder block only at the top and bottom. The rest of the liner is in direct contact with the coolant. Figure 7-8 is a sectional view of an engine using wet cylinder liners. These liners are removable and can be replaced if they become worn or damaged.

⊘ 7-7 Vega Engine The *Vega engine* has a cast-aluminum cylinder block which does not use cylinder liners. Instead, the aluminum used is loaded with silicon particles. Silicon is an extremely hard material. After the cylinder block is cast, the cylinders are honed and then treated with a chemical that etches (or eats away) the surface aluminum. This leaves only the silicon particles exposed, and so the pistons and rings slide on the silicon (Fig. 7-9).

⊘ 7-8 Air-cooled Engine Figure 7-10 shows a cylinder for an *air-cooled engine,* along with the piston

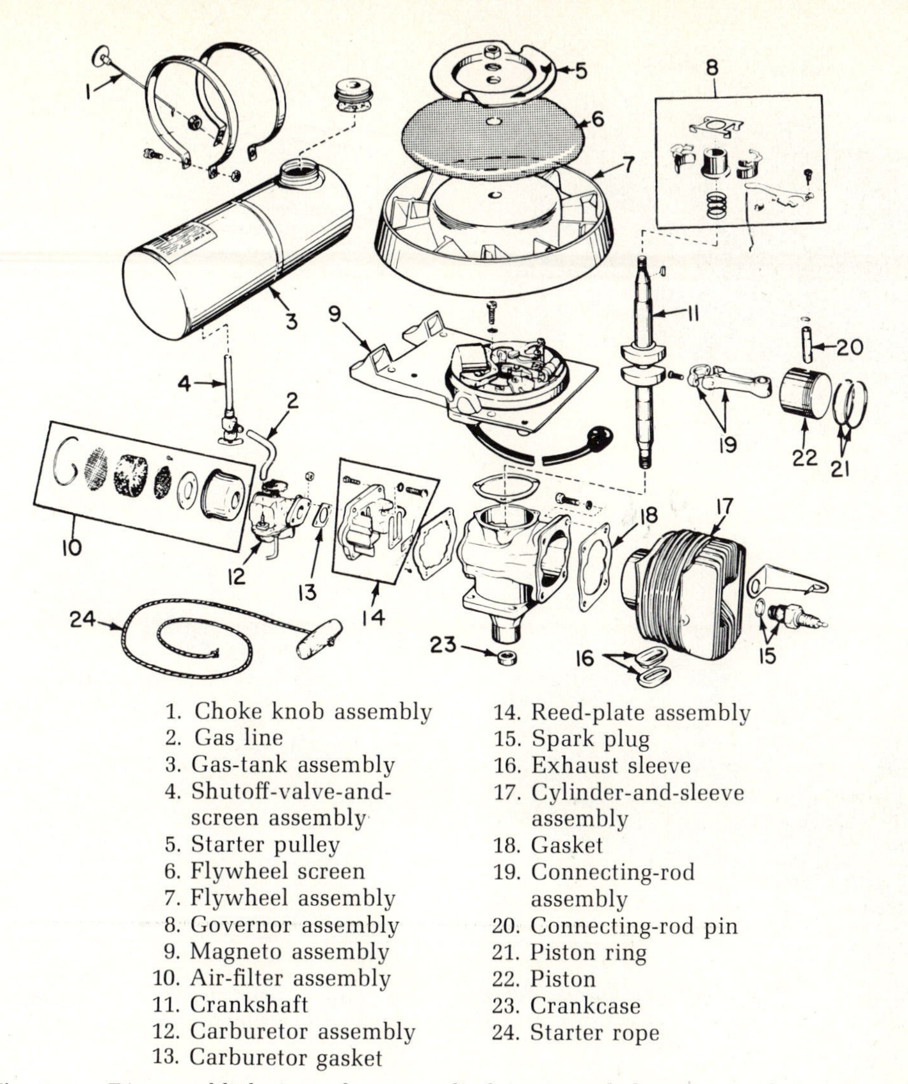

1. Choke knob assembly
2. Gas line
3. Gas-tank assembly
4. Shutoff-valve-and-screen assembly
5. Starter pulley
6. Flywheel screen
7. Flywheel assembly
8. Governor assembly
9. Magneto assembly
10. Air-filter assembly
11. Crankshaft
12. Carburetor assembly
13. Carburetor gasket
14. Reed-plate assembly
15. Spark plug
16. Exhaust sleeve
17. Cylinder-and-sleeve assembly
18. Gasket
19. Connecting-rod assembly
20. Connecting-rod pin
21. Piston ring
22. Piston
23. Crankcase
24. Starter rope

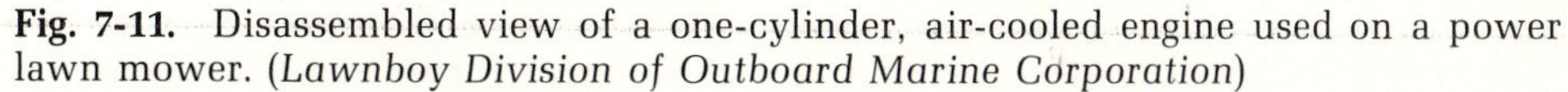

Fig. 7-11. Disassembled view of a one-cylinder, air-cooled engine used on a power lawn mower. (*Lawnboy Division of Outboard Marine Corporation*)

and other related parts. Note that in the air-cooled engine, the cylinders are separated and have fins. The fins are integral metal rings which aid in radiating heat from the cylinder. In the completed air-cooled engine, the cylinders are installed on the crankcase. Then the cylinder head is installed on top of the cylinders. Figures 6-9 and 6-15 show air-cooled engines. Figure 7-11 is a disassembled view of a one-cylinder, two-cycle air-cooled engine used on a power mower. Figures 6-20 and 6-21 show a similar engine and the shroud that directs air around the cylinder to help cool the engine.

⊘ 7-9 Cylinder Head Figure 7-12 shows two views of a cylinder head for a V-8 engine. Two such cylinder heads are used on a V-8 engine, one for each bank. (Figures 6-16 and 6-19 show V-8 engines.)

The cylinder head is cast in one piece from iron, iron mixed (alloyed) with other metals, or aluminum alloy. Most cylinder heads are made of cast iron. The cylinder head includes water jackets and passages from the valve ports to the openings in the manifolds. Many machining operations are required to convert the rough casting into a finished cylinder head, including:

1. Drilling tapped holes for the spark plugs, for bolts and studs to attach the manifolds, and for other parts
2. Finishing coolant and oil-passage holes
3. Finishing mounting surfaces
4. Drilling valve-guide holes
5. Finishing valve seats

There are three general types of cylinder heads: *L head*, *I head*, and *overhead camshaft*. L-head engines are described in ⊘ 6-10. The only places you will see L heads are in antique cars and some small engines, such as those used in power lawn mowers. The I-head engine, with the valves in the cylinder head, is the usual arrangement in automotive engines.

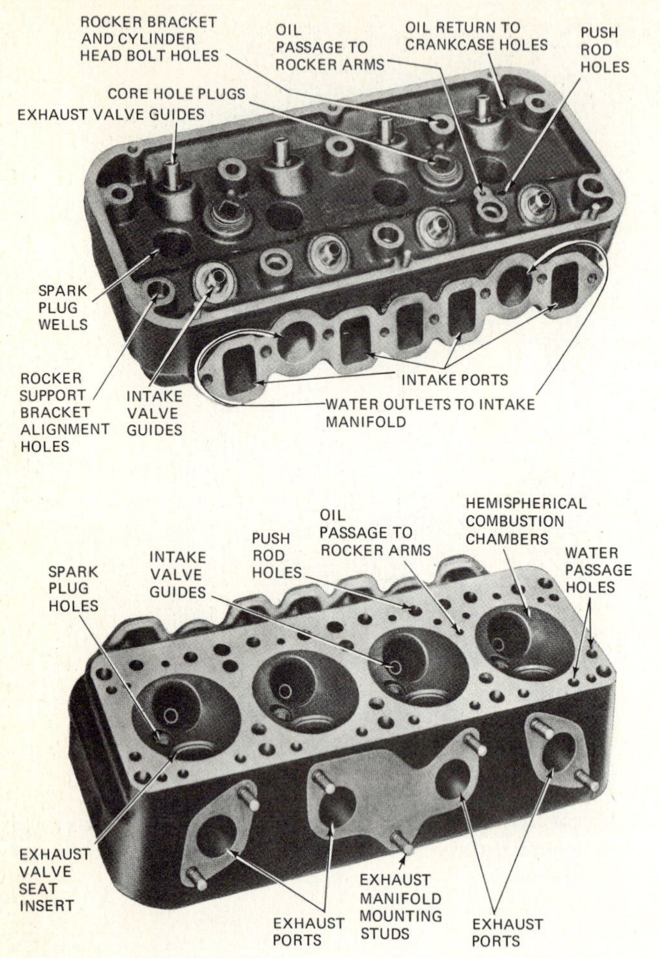

Fig. 7-12. Top and bottom views of a cylinder head from a V-8 overhead-valve engine. (*Chrysler Corporation*)

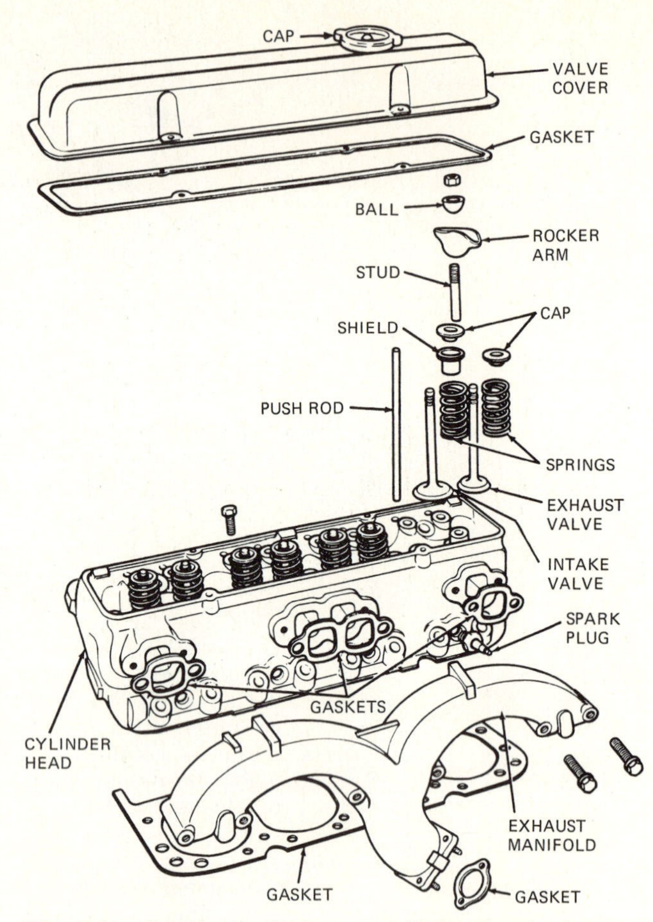

Fig. 7-13. Parts attached to a cylinder head of a V-8 engine. (*Chevrolet Motor Division of General Motors Corporation*)

Figures 6-11, 6-16, and 6-19 show I-head engines, and there are many other pictures of I heads in this book. The cylinder head shown in Fig. 7-12 is for an I-head engine. Figure 7-13 shows the essential parts that are installed in an I-head cylinder head. For the sake of simplicity, the valves for only one cylinder are shown, along with their related rocker arms, pushrods, and valves lifters.

The cylinder head for overhead-camshaft engines has an arrangement which supports the camshaft, as shown in Fig. 7-14. Many arrangements are used. Overhead-camshaft engines are discussed in more detail in Chap. 9.

Air-cooled engines use a different type of cylinder head, as shown in Fig. 7-15. These cylinder heads do not have water jackets. Instead, they have fins for cooling. Figure 7-16 shows a cylinder head for an air-cooled engine with the valves and related parts in place.

⊘ 7-10 Combustion Chamber The cylinder head forms the top of the *combustion chamber;* the upper part of the cylinder walls and pistonhead form the remainder of the chamber (Fig. 7-17). The shape of the combustion chamber plays a very important part in engine performance. There are two general shapes: wedge and hemispheric (Fig. 7-18). Figures 6-5 and 6-19 show engines with the two types of combustion chambers. In the wedge-shaped chamber, the flame has a relatively long distance to travel from its starting point—the spark plug—to the farthest point. In the hemispheric chamber, the spark plug is centrally located and the flame has a shorter distance to travel. The different shapes thus have different effects on the combustion process in the chamber. These effects will be discussed in later sections dealing with engine knock and smog.

⊘ 7-11 Gaskets The joint between the cylinder block and cylinder head must be tight. It must be able to withstand the pressure and heat developed in the combustion chambers. The block and head cannot be machined flat enough to provide the necessary seal. Therefore, *gaskets* are used to secure a good seal (Fig. 7-19). Head gaskets are made of thin sheets of soft metal or asbestos and metal (Fig. 7-20). All cylinder, coolant, valve, and head-bolt openings are cut out. When the gasket is installed, tightening of the head bolts squeezes the soft metal so that the joint is sealed. As a rule, gaskets can be used only

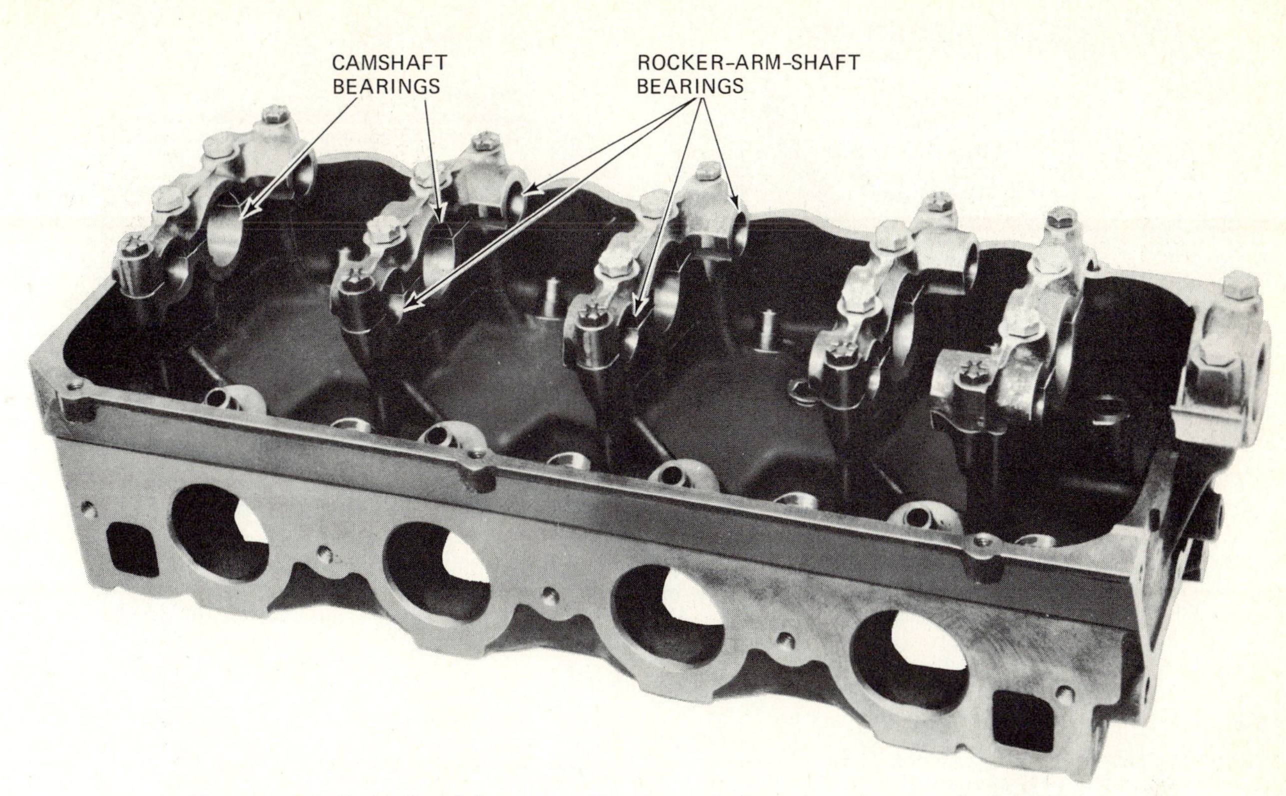

Fig. 7-14. Cylinder head for an overhead-camshaft engine, showing the bearings to support the camshaft and rocker-arm-shaft bearings. (*Ford Motor Company*)

once. If removed and then reinstalled, they cannot be further compressed to provide an effective seal. Gaskets are also used to seal joints between other parts, such as between the oil pan, manifolds, or water pump and the head or block.

⊘ 7-12 Exhaust Manifold The exhaust manifold (Fig. 7-21) is essentially a series of interconnected tubes for carrying the burned gases away from the engine cylinders. On I-head in-line engines the exhaust manifold is bolted to the side of the cylinder head. On V-8 engines there are two exhaust manifolds, one for each bank of cylinders. The exhaust manifolds are bolted to the outsides of the two banks. On some cars, they are interconnected by a crossover pipe and they exhaust through a common muffler and tail pipe. On others, each manifold is connected to a separate exhaust pipe, muffler, and tail pipe (Fig. 7-22). The exhaust manifold normally is tied in closely with the intake manifold. The purpose of this is to provide a certain amount of heat transfer from the exhaust manifold to the intake manifold during engine warm-up. This heat transfer improves vaporization of the fuel and provides better initial engine performance just after the engine is started (see ⊘ 14-25).

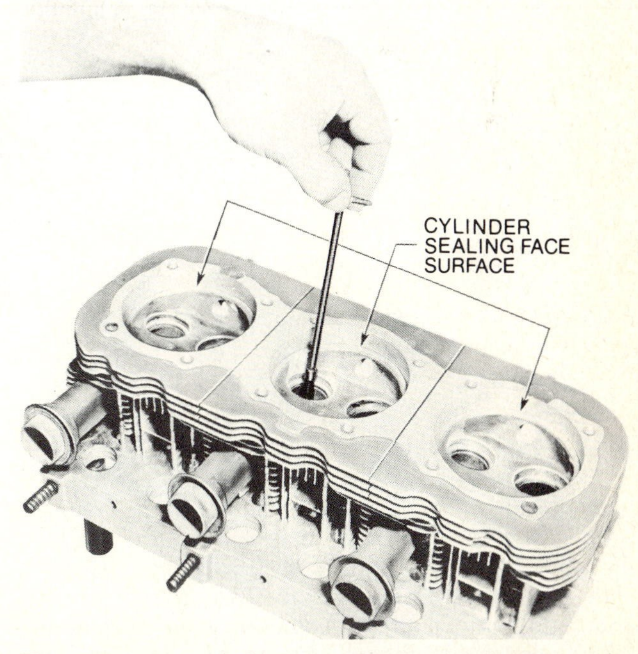

Fig. 7-15. One of the two cylinder heads for a six-cylinder, air-cooled, pancake engine. The tool is being used to clean the valve-guide bores. (*Chevrolet Motor Division of General Motors Corporation*)

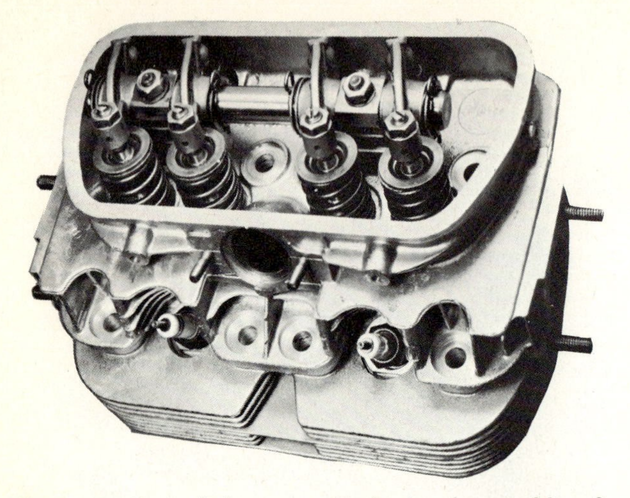

Fig. 7-16. One of the two cylinder-head assemblies for a four-cylinder, air-cooled, pancake engine. (*Volkswagen*)

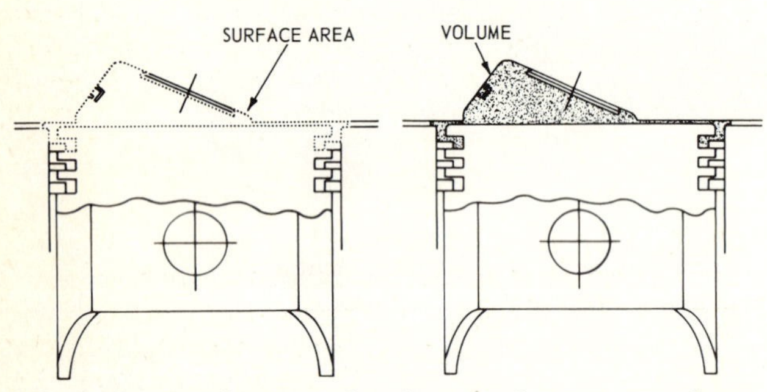

Fig. 7-17. Combustion chamber. Surface area is shown in dotted line.

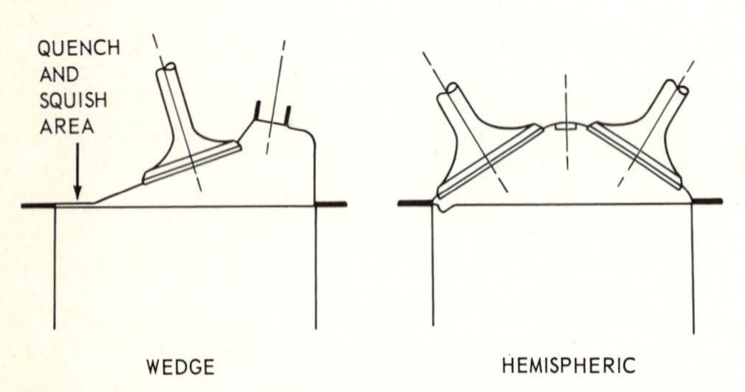

Fig. 7-18. Wedge and hemispheric combustion chambers. (*General Motors Corporation*)

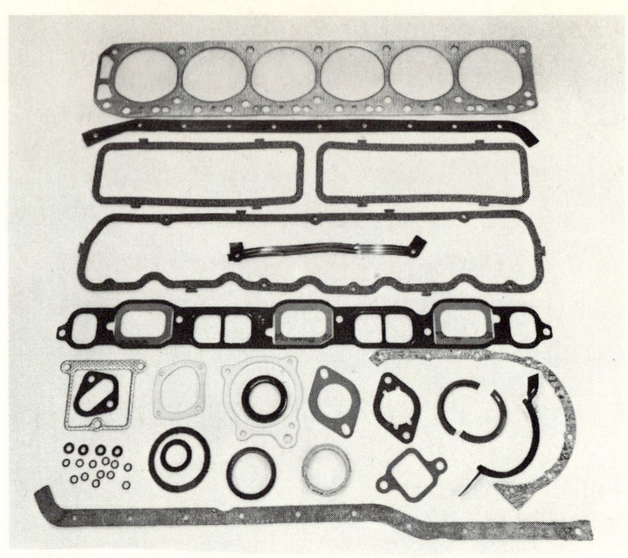

Fig. 7-19. Engine overhaul gasket set for a six-cylinder engine, showing the gaskets and seals used in the engine. (*McCord Replacement Products Division of McCord Corporation*)

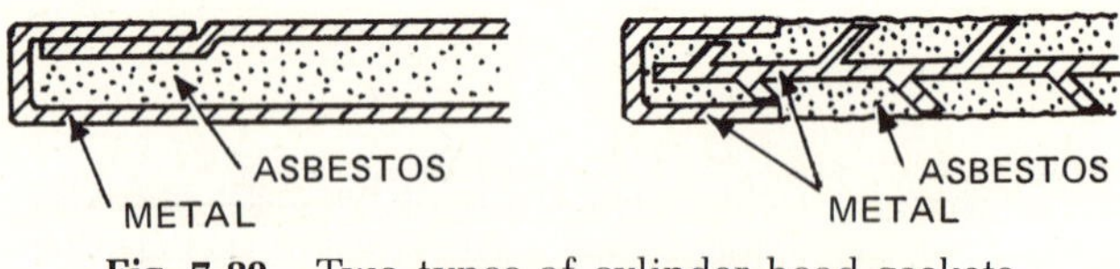

Fig. 7-20. Two types of cylinder-head gaskets.

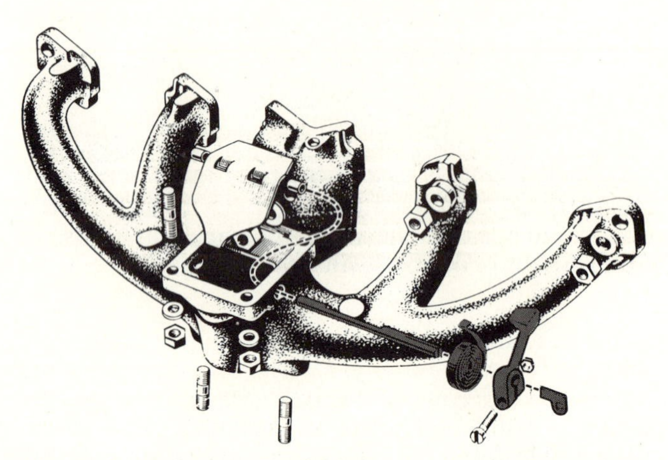

Fig. 7-21. Exhaust manifold for a six-cylinder, in-line engine with heat-control valve and parts disassembled.

Dual Exhaust System

The dual exhaust system used on one V-8 engine is shown in Fig. 7-22. Each exhaust manifold exhausts into a separate exhaust pipe, which, in turn, exhausts into its own muffler, resonator, and tail pipe. The resonators further reduce exhaust noises. The resonators are, in effect, secondary mufflers. The use of two separate exhaust systems, one for each bank of cylinders, improves the "breathing" ability of the engine, allowing it to exhaust more freely. This tends to reduce the amount of exhaust gas left in the cylinder at the end of the exhaust stroke and thus improves engine performance.

⊘ 7-13 Intake Manifold Essentially, the intake manifold (Figs. 7-23 to 7-25) is a series of tubes for carrying the air-fuel mixture from the carburetor to the engine intake-valve ports. The carburetor is normally mounted in a central position on the intake manifold. The intake manifold is attached to the side of the cylinder head on I-head in-line engines. On

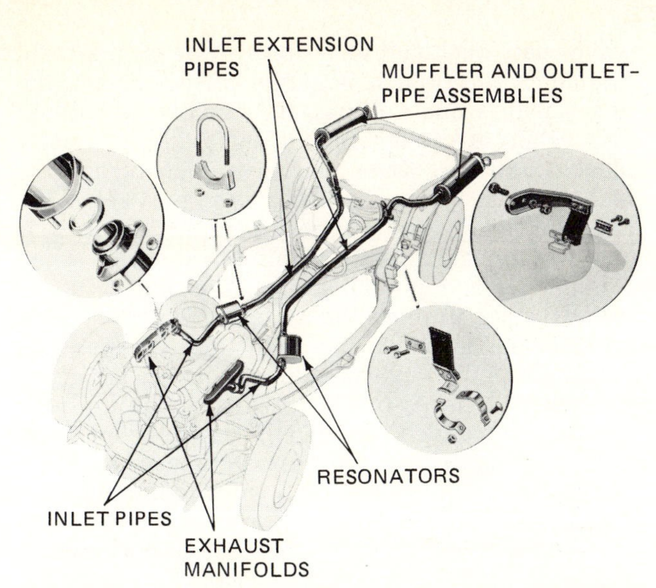

Fig. 7-22. Dual exhaust system for a V-8 engine. Each bank of cylinders has its own exhaust system. The circles show details of assembly and attachment. (*Ford Motor Company*)

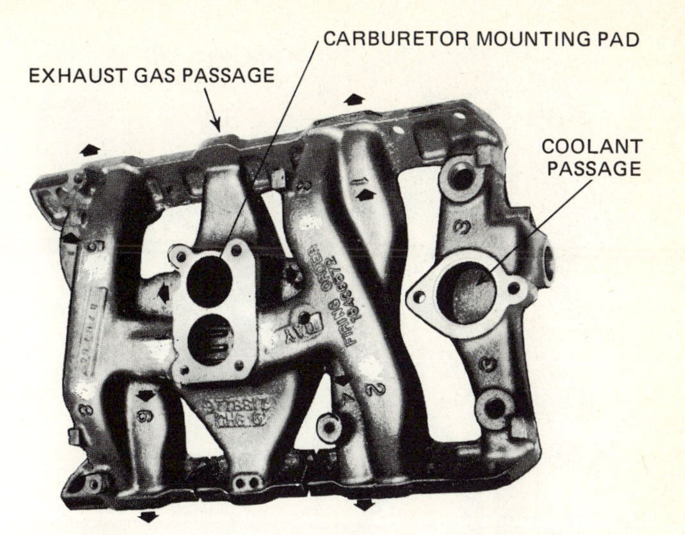

Fig. 7-24. Intake manifold for an I-head V-8 engine. The arrows show the flow of air-fuel mixture from the two barrels of the carburetor to the eight cylinders in the engine. The central passage connects the two exhaust manifolds. Exhaust gas flows through this passage during engine warm-up. (*Pontiac Motor Division of General Motors Corporation*)

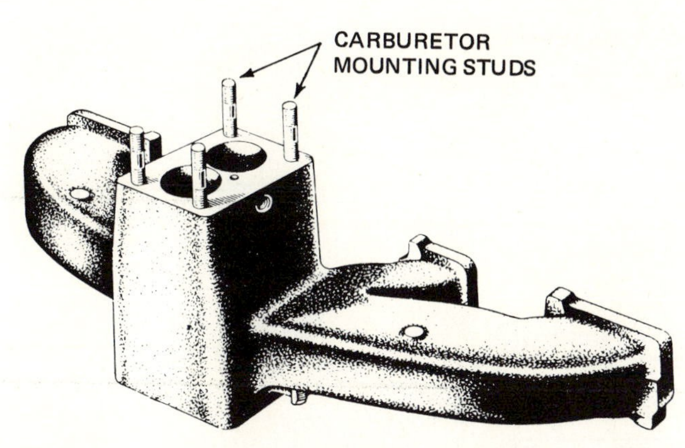

Fig. 7-23. Intake manifold for an in-line, six-cylinder engine.

I-head V-8 engines the intake manifold is situated between the two banks of cylinders and is attached to the insides of the two cylinder heads. Figure 7-12 shows the intake ports on an I head from a V-8 engine over which the intake manifold is attached. A gasket is used between the intake manifold and the mating surface of the block or head.

Figure 7-24 shows an intake manifold for an I-head V-8 engine. A two-barrel carburetor mounts on the intake manifold, each barrel supplying four of the eight cylinders with air-fuel mixture. A two-barrel carburetor is, in effect, two separate carburetors assembled together. Each barrel has its own venturi, throttle valve, and fuel nozzles. (See Chap. 14, "Automotive Carburetors.") The arrows in Fig. 7-24 indicate the pattern of air-fuel distribution from the two carburetor barrels to the eight cylinders of the engine. Note that each carburetor barrel supplies four cylinders. This arrangement permits a very even distribution of air-fuel mixture. Each cylinder receives its "share," and no cylinder is starved.

Many V-8 engines use four-barrel carburetors, often called "quad" (for the number 4) carburetors. In these carburetors, two of the barrels make up the primary side, and the other two the secondary side. For most operating conditions, the primary side takes care of the engine requirements. But for acceleration and high-power operation, the secondary side supplies additional air-fuel mixture (see Chap. 14). Figure 7-25 shows an intake manifold for a four-barrel carburetor. It has four openings in the carburetor mounting pad to match the four barrels of the carburetor. The primary and secondary openings at the top are interconnected to cylinders 2, 3, 5, and 8. The primary and secondary openings at the bottom are interconnected to cylinders 1, 4, 6, and 7.

⊘ 7-14 Tuned Intake and Exhaust Systems

For high volumetric efficiency or better engine breathing, valves are made as large as possible and the passages in the intake and exhaust manifolds are arranged so as to allow maximum air flow. For example, Fig. 7-26 shows how the intake port was altered on one high-performance engine to improve the intake of air. The hump was added to throw the air upward toward the roof of the intake port, and this new design showed a decided improvement in volumetric efficiency.

Another method of improving volumetric efficiency in high-performance engines is to *tune the intake and exhaust systems*. "Tuning," in this case, means designing the proper length and size of intake manifold between the carburetor and intake ports. Tuning a manifold is something like tuning a pipe

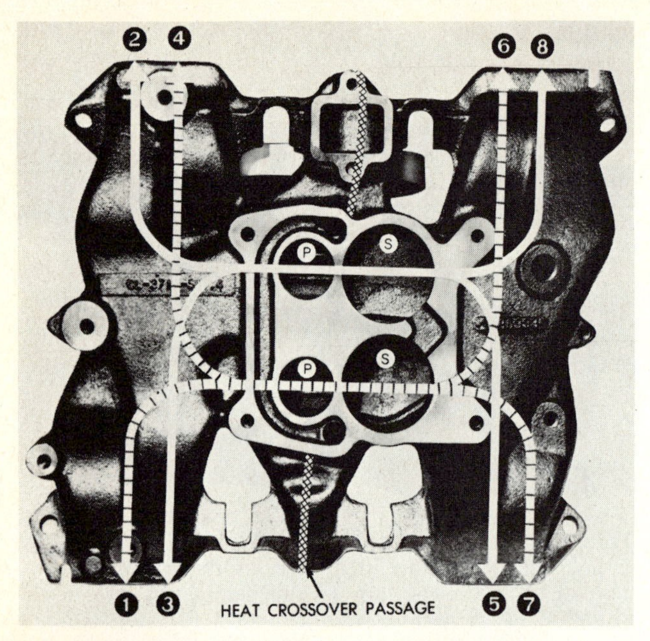

Fig. 7-25. Intake manifold for a V-8 engine using a four-barrel carburetor. "P" means primary barrel; "S" means secondary barrel. (*Cadillac Motor Car Division of General Motors Corporation*)

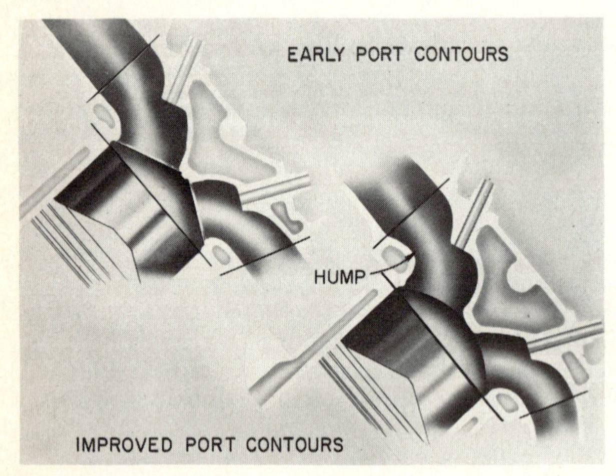

Fig. 7-26. Early and improved intake-port contours. Adding the hump improves volumetric efficiency. (*Ford Motor Company*)

organ. In the pipes of a pipe organ, air is set into vibration. As the air in a pipe vibrates, high-pressure waves pass rapidly up and down the pipe, and this action produces the sound. (Sound is nothing more than high-pressure waves passing through air.) In a tuned intake manifold, the incoming air-fuel mixture is made to vibrate, not to produce sound but to ram more of the air-fuel mixture into the cylinder. Under ideal conditions, a high-pressure wave in the mixture will reach an intake-valve port just as the valve opens.

The high-pressure waves in the mixture are initiated by the sound of the intake valve closing. The waves pass back and forth in the tube, or manifold branch. If the branch is of the correct length, the waves will resonate, or pass back and forth, without loss of any great amount of energy. Then, under ideal conditions, when the valve opens again, the high-pressure wave will hit the valve port at just the right instant to produce the ramming effect. It is obvious, however, that a manifold cannot be tuned for effective action at all speeds. The velocity of the sound waves through the mixture will not vary a great deal. However, the time intervals between valve closing and valve opening will vary greatly at different speeds. As a rule, the intake manifold will be tuned to be in phase when the engine is operating near or at top speed. This is the time when volumetric efficiency begins to drop off and the ramming effect is most needed.

One design using two carburetors is shown in Fig. 7-27. Chrysler, the developer of this configuration, termed the process "ram-charging" or "ram induction" because of the ramming effect. The design shown used intake branches 3 feet [0.914 m] long. Many later designs use considerably shorter branches.

Exhaust systems are also tuned in many engines for the same reason. With a tuned system, the high-pressure waves forming in the exhaust gas give it added momentum which, in effect, hurries it along and ensures a more nearly perfect exhausting of the burned gases from the cylinder. This of course improves engine performance.

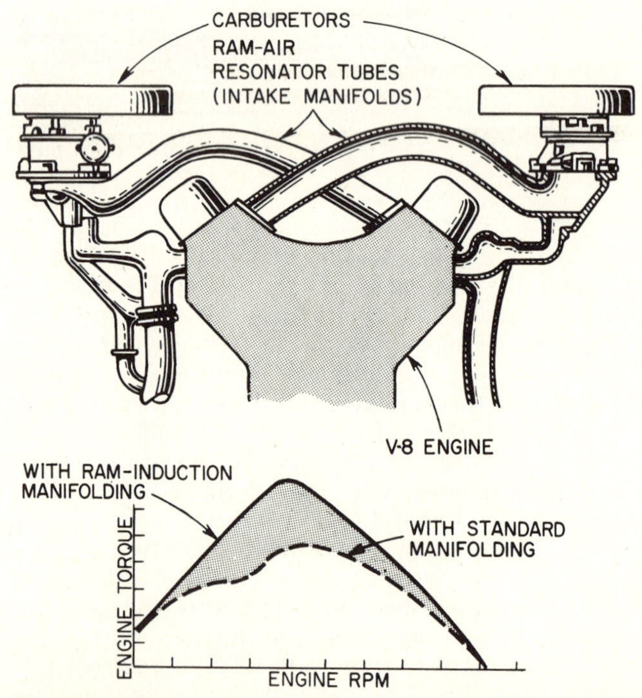

Fig. 7-27. Tuned or ram-induction system using long intake-manifold passages (or resonator tubes). Note the improvement in engine torque with ram induction, as shown by the graph. (*Chrysler Corporation*)

Check Your Progress

Progress Quiz 7-1 Once more you have the chance to stop and find out how well you have been absorbing the material you have been reading. If anything in the quiz stumps you, turn back into the chapter and reread the pages to find the answer.

Correcting Parts Lists The purpose of this exercise is to help you spot unrelated parts in a list. For example, in the list "head, oil pump, water pump, main-bearing caps, differential housing, clutch housing," you would see that "differential housing" does not belong because it is the only part that is not attached to the cylinder block.

In each of the following lists, you will find one item that does not belong. Write each list in your notebook, but *do not write* the item that does not belong.

1. Parts in the engine exhaust system include the exhaust manifold, muffler, oil pan, heat-control valve, thermostat.
2. Parts attached to the cylinder block include the clutch housing, cylinder head, oil pan, water pump, timing cover, drive shaft.
3. Openings in the cylinder block include the water jackets, cylinder bores, stud holes, coolant passages, connecting rods.
4. Parts of the engine lubricating system include the oil pan, pump, intake manifold, strainer, oil lines.
5. Parts driven off the camshaft include the oil pump, water pump, fuel pump, ignition distributor.

Completing the Sentences The sentences that follow are incomplete. After each sentence there are several words or phrases, only one of which will correctly complete the sentence. Write each sentence in your notebook, selecting the proper word or phrase to complete the sentence correctly.

1. The oil pan, crankshaft, water pump, and cylinder head are attached to the: (*a*) car frame, (*b*) engine cylinder block, (*c*) manifolds.
2. The fastening device that is threaded at both ends is called a: (*a*) bolt, (*b*) stud, (*c*) screw, (*d*) lag.
3. The flat pieces that are put between the engine block and those parts attached to the engine block are called: (*a*) studs, (*b*) fasteners, (*c*) gaskets, (*d*) flatteners.
4. The water pump is attached to the block at the: (*a*) front, (*b*) side, (*c*) top, (*d*) back.
5. The most complicated head, considering the number of parts attached to it, is the: (*a*) I head, (*b*) L head, (*c*) T head, (*d*) V head.
6. The part that carries the burned gases from the engine cylinders is called the: (*a*) gas manifold, (*b*) intake manifold, (*c*) exhaust manifold, (*d*) manifold control.
7. The V-8 engine normally has: (*a*) a single exhaust manifold, (*b*) two exhaust manifolds, (*c*) four exhaust manifolds.
8. The device located in the exhaust system that reduces exhaust noise is called the: (*a*) muffler, (*b*) exhaust manifold, (*c*) tail pipe.
9. The V-8 engine normally has: (*a*) a single intake manifold, (*b*) two intake manifolds, (*c*) four intake manifolds.
10. In the ideal tuned intake system, a high-pressure wave of air-fuel mixture reaches the intake port just as the: (*a*) intake valve opens, (*b*) intake valve closes, (*c*) mixture fires.

⊘ **7-15 Crankshaft** The *crankshaft* is a one-piece casting or forging of heat-treated alloy steel of considerable mechanical strength (Figs. 5-20 and 7-28). The crankshaft, it will be remembered, takes the downward thrust of the pistons during the power stroke. Pressure exerted by the pistons through the connecting rods against the crankpins on the crankshaft causes the shaft to rotate. This rotary motion is transmitted through the clutch and the power train to the car wheels.

The problems of static and dynamic balance and torsional vibration must be considered in the design of a crankshaft. The cranks on the crankshaft, being offset from the center line of the shaft, naturally introduce an out-of-balance condition. In addition, the lower ends of the connecting rods are moving in circles along with the crank journals. These reciprocating and rotating masses introduce

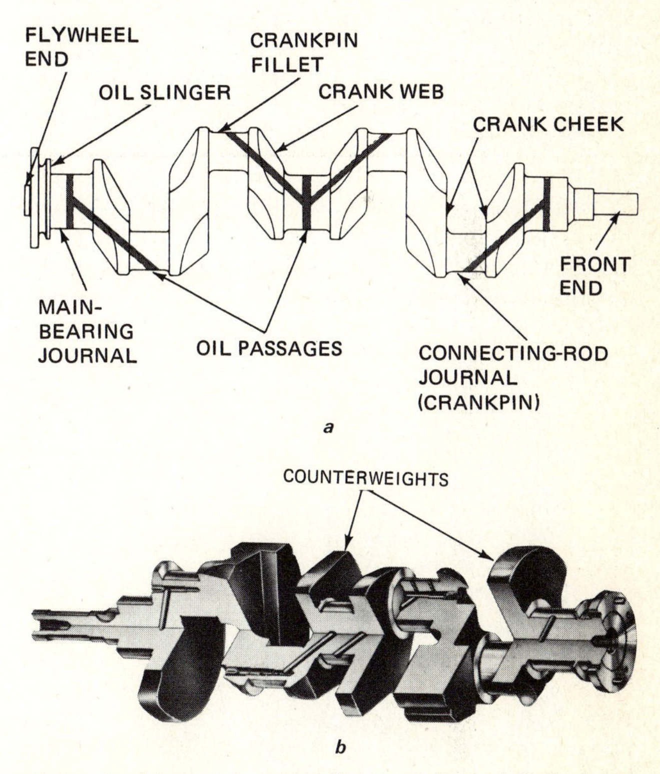

Fig. 7-28. (*a*) Line drawing of a typical crankshaft showing the names of the parts; (*b*) cutaway view of a crankshaft for a V-8 engine. (Note the oil passages drilled to the crankpin for lubricating rod bearings.) (*Johnson Bronze Company; Ford Motor Company*)

various forces that could set up serious vibrations during shaft rotation if it were not for the counterweights that tend to counterbalance these forces. Figure 7-29 shows the resultant force R in a V-8 engine caused by the rotation of the crank and the motion of the two pistons and connecting rods attached to the crankpin of the crank. It is this resultant force that the counterweights must balance.

Crankshafts generally have drilled oil passages (Fig. 7-28) through which oil can flow from the main to the connecting-rod bearings (see ⊘ 7-21). An improved method of drilling oil passages in the crankshaft is illustrated in Fig. 7-30: Both the crankshaft and the crank journals are drilled straight through, with two holes drilled in each crank journal (one for each connecting-rod bearing). Diagonal holes are then drilled to connect between the crankshaft and crank journals. The large holes drilled in the crank journals lighten the crankshaft and also serve as sludge traps. Cup plugs are used to seal these traps.

In the assembled engine the front end of the crankshaft carries three devices. One of these is a gear, or sprocket, that drives the camshaft (Figs. 9-5 to 9-7). The camshaft is driven at one-half the speed of the crankshaft. A second device is the vibration damper (see ⊘ 7-17), which combats torsional vibration in the crankshaft. As a part of the vibration damper, there is a pulley with one or more grooves. V belts fit these grooves and drive the engine fan and water pump as well as the alternator. There is an additional groove in the pulley on cars equipped with power steering. This additional groove is fitted with a V belt that drives the power-steering hydraulic pump. There may also be another groove for belt-driving the compressor in cars equipped with air conditioning.

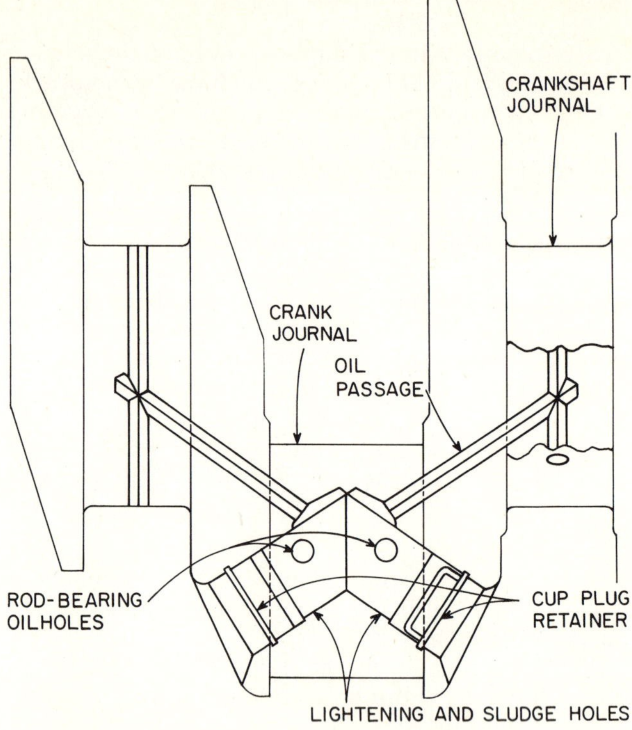

Fig. 7-30. V-8 engine crankshaft drilling for oil passages and sludge traps. (*Ford Motor Company*)

⊘ 7-16 Flywheel The flow of power from the engine is not smooth. Although the power strokes from the various cylinders may overlap to some extent in engines with six or more cylinders, there are periods when more power is being delivered to the crankshaft than at other times (Fig. 7-31). When more power is being delivered, the crankshaft tends to speed up; with less power, it tends to slow down. This would produce a roughly running engine if it were not for the flywheel.

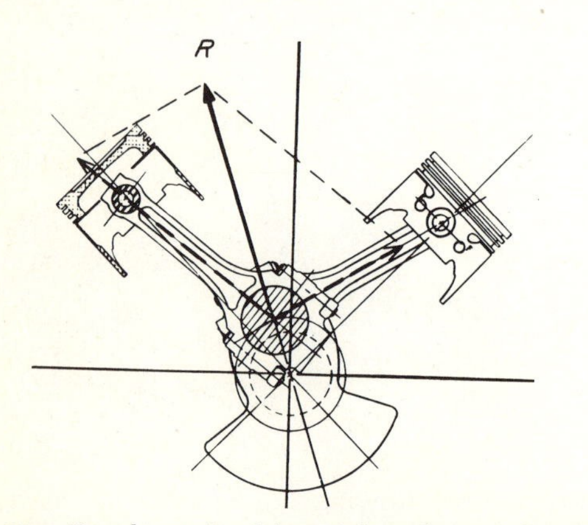

Fig. 7-29. Resultant R of inertial and centrifugal forces from the rotating and reciprocating masses of crankshaft and connecting-rod-and-piston assemblies. (*Oldsmobile Division of General Motors Corporation*)

The flywheel, a comparatively heavy wheel bolted to the rear end of the crankshaft (Fig. 7-6), tends to resist any change of speed because of its inertia. Inertia, it will be recalled (⊘ 4-30), is the property that causes a body to resist any attempt to change its speed or direction of motion. The flywheel absorbs power during the intervals that the engine attempts to speed up. During the intervals when less power is being produced by the engine, the flywheel resists the engine's attempt to slow down by giving up part of its energy of rotation. In addition to this function, the flywheel has gear teeth around its outer rim that mesh with the starting-motor drive pinion when the starting motor is operated to crank the engine for starting. The rear face of the flywheel also serves as the driving member of the engine clutch.

⊘ 7-17 Vibration Damper The transmission of the power impulses to the crankshaft tends to set up *torsional vibration* in the crankshaft. When a piston and rod move down on the power stroke, they suddenly impose a load of as much as 2 tons [1,814.36 kg] on the crankpin. This heavy load tends to twist the crankshaft; the shaft is actually twisted slightly. When the end of the power stroke is reached, the push against the crank is relieved, and

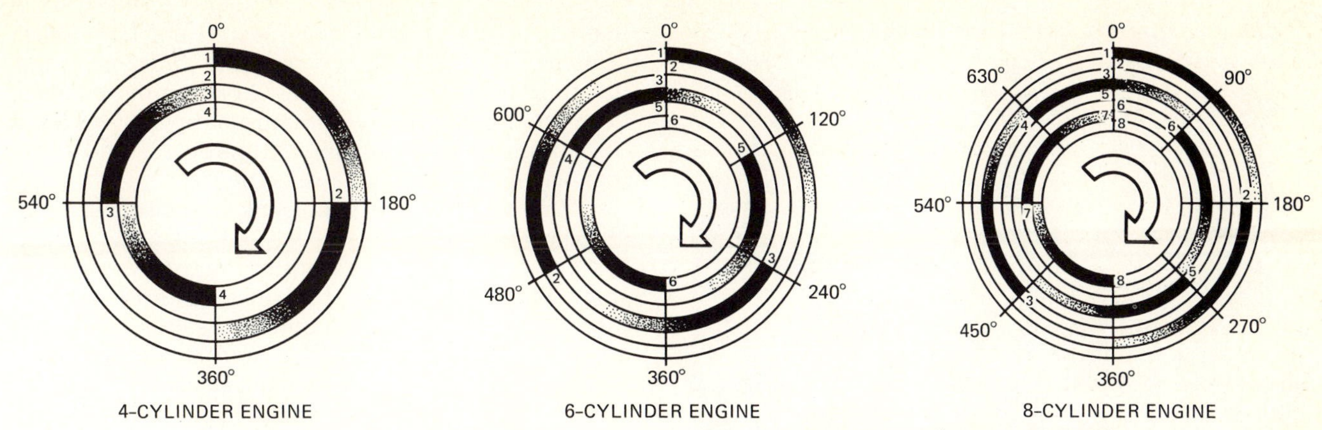

Fig. 7-31. Power impulses in four-, six-, and eight-cylinder engines during two crankshaft revolutions. The complete circle represents two crankshaft revolutions, or 720 degrees. Less power is delivered toward the end of the power stroke. This is indicated by the lightening of the shaded areas that show power impulses. Note the power overlap on six- and eight-cylinder engines.

so the shaft, having been twisted, attempts to return to its original shape. The crankshaft acts as a spring, however, and overrides this position, going beyond the original shaft and twisting slightly in the opposite direction. It then returns, overriding in the other direction. This sets up an oscillating motion, or torsional vibration, within the crankshaft which is repeated every power stroke. If this torsional vibration were not controlled, succeeding power impulses would continue to add to the original oscillation of the shaft until, at certain speeds, the shaft might be broken by the excessive twisting it would undergo. To control this torsional vibration, devices variously called *vibration dampers, torsional balancers,* or *crankshaft-torque impulse neutralizers* are used. They are usually mounted to the front end of the crankshaft (Fig. 7-6) and include the fan-belt pulley.

The typical vibration damper consists of two parts: a small inertia ring, or damper flywheel, and the fan-belt pulley. These parts are bonded to each other by a rubber insert that is approximately $\frac{1}{4}$ inch [6.35 mm] thick (Fig. 7-32). The pulley is mounted on the front end of the crankshaft. As the crankshaft tends to speed up or slow down, the damper flywheel imposes a dragging effect because of its inertia. This effect, which slightly flexes the rubber insert, tends to hold the pulley and crankshaft to a constant speed. The action tends to check the twist-untwist, or torsional vibration, of the crankshaft.

⊘ 7-18 Engine Bearings In the engine, there must be relative motion between the piston and the connecting rod, between the connecting rod and the crankpin on the crankshaft, and between the crankshaft and the supporting bearings in the cylinder blocks. At all these points (as well as at other places in the engine) *bearings* must be installed (Figs. 7-33 and 7-34). The bearings are called *sleeve bearings* because they are shaped like sleeves that fit around rotating journals or shafts. (The part of a shaft that rotates in a bearing is called a *journal*).

Connecting-rod and crankshaft (or main) bearings are of the split, or half, type; that is, the bearing is split into two halves. Figure 7-35 illustrates a typical sleeve-type bearing half. With main bearings, the upper half is assembled into the counterbore in the cylinder block and the lower half is held in place in the bearing cap (Fig. 7-33). Figure 8-1 shows the bearings (4 and 7), in disassembled and assembled

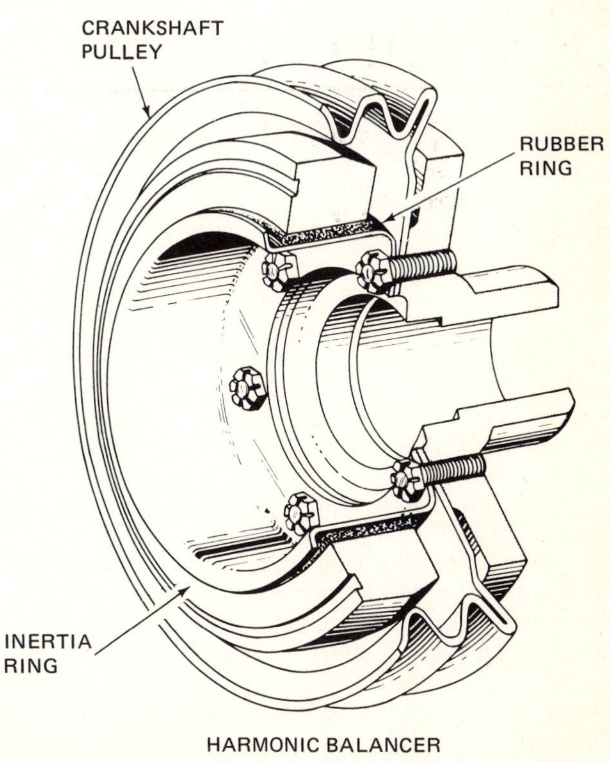

Fig. 7-32. Partial cutaway view of a torsional vibration damper. (*Pontiac Motor Division of General Motors Corporation*)

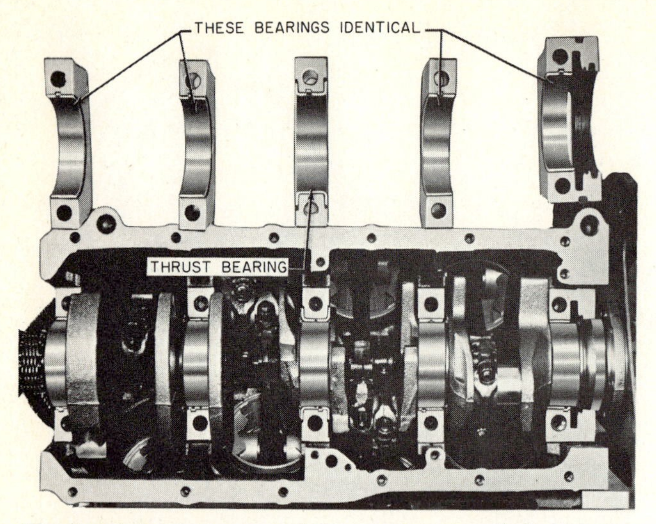

Fig. 7-33. Arrangement of the main, or crankshaft, bearings in a V-8 engine. The engine is shown from the bottom. The bearing caps with bearings have been removed and placed above the engine. The center bearing is an end-thrust bearing. (*Buick Motor Division of General Motors Corporation*)

views, used in a connecting rod. The big-end bearing is the split type, but the piston-pin bearing, where used, is the full-round, or bushing, type.

In some high-performance engines, the intermediate main-bearing caps are tied in to the sides of the cylinder block with two additional bolts, as shown in Fig. 7-36. This provides the additional bearing-support strength required for the high performance.

Figure 7-37 shows a sleeve-type bearing half with the various parts named. Note that it has a back to which a lining has been applied. The back, usually made of steel or bronze, gives the bearing rigidity and strength. The lining consists of two or more thin layers of relatively soft material (the bearing material) only a few thousandths of an inch thick (Fig. 7-38). Several metals have been used: lead, tin, copper, aluminum, and antimony, for example. The rotating journal is supported by this thin layer of bearing material. One reason for using a soft material is that when wear does take place, the bearing rather than the more expensive engine part will wear. When wear has gone beyond a certain point, the bearing rather than the more costly engine part can be replaced.

We noted in the previous paragraph that the bearing material "supports" the load of the rotating journal. Actually, the bearing surfaces are flooded with lubricating oil from the engine lubricating system so that the journals are, in effect, "floated" on films of oil (⊘ 5-10). It can be said, however, that the bearings support the films of oil, which, in turn, support the journals. So the bearings do, of course, "carry the load."

The main bearings in most engines do not have the oil-distributing grooves shown in Fig. 7-37. They may or may not have annular, or ring, grooves. The main bearings on many engines do not have these grooves. On other engines, only the upper halves of the main bearings have them. On still other engines, both the upper and lower main-bearing halves have the annular grooves. Connecting-rod big-end bearings usually do not have oil grooves.

V-8 engines have five crankshaft bearings, one at each end and one on each side of every crank. Six-cylinder engines have either four or seven crankshaft bearings. In the four-bearing arrangement, there is a bearing on each end of the crankshaft and a bearing between cranks 2 and 3 and between cranks 4 and 5. In the seven-bearing arrangement (in almost universal use today), there is a bearing on each end of the crankshaft and a bearing on each side of every crank. The engine shown in Fig. 7-6 has seven crankshaft bearings. With seven bearings, there is less stress and vibration in the crankshaft and cylinder block; the engine runs more smoothly.

⊘ 7-19 Main Thrust Bearing One of the main bearings in the engine is a *thrust bearing,* designed to prevent excessive end play of the crankshaft. Figure 7-33 shows an end thrust bearing in place in the bearing cap of a V-8 engine. Figure 7-39 shows a crankshaft thrust bearing. Thrust bearings have thrust faces on their two sides. Flanges on the crankshaft ride against these thrust faces. This holds the crankshaft in position so that it does not have excessive endwise movement. There is, of course, some clearance between the thrust faces of the bearing and the flanges on the crankshaft. These clearances permit oil to flow between the surfaces to provide adequate lubrication. In V-8 engines, the intermediate or center main bearing is usually the thrust bearing (Fig. 7-33). In in-line engines, the rear main bearing is usually the thrust bearing (Fig. 7-34).

⊘ 7-20 Press-fit Bushings A typical example of a *press-fit bushing* is the bushing used in the piston end (small end) of the connecting rod on some engines (Fig. 8-1, No. 7). The outside diameter of the bushing is slightly larger than the inside diameter of the hole into which it is to be installed. Thus, the bushing must be forced, or pressed, into place. The bushing is then reamed or honed to size. The chapters in this book covering engine service explain how these operations are performed.

⊘ 7-21 Engine-Bearing Lubrication The engine bearings are flooded with oil by the lubricating system. For example, the bearing shown in Fig. 7-37 has an oilhole that aligns with an oilhole in the engine block. Oil feeds through this hole constantly, keeping the annular groove and the distributing groove in the bearing filled with oil. Oil constantly feeds from these grooves onto the bearing surfaces. The oil works its way outward to the edges of the bearing. As it reaches the outer edges, it is thrown off and falls back into the oil pan. Thus oil is constantly

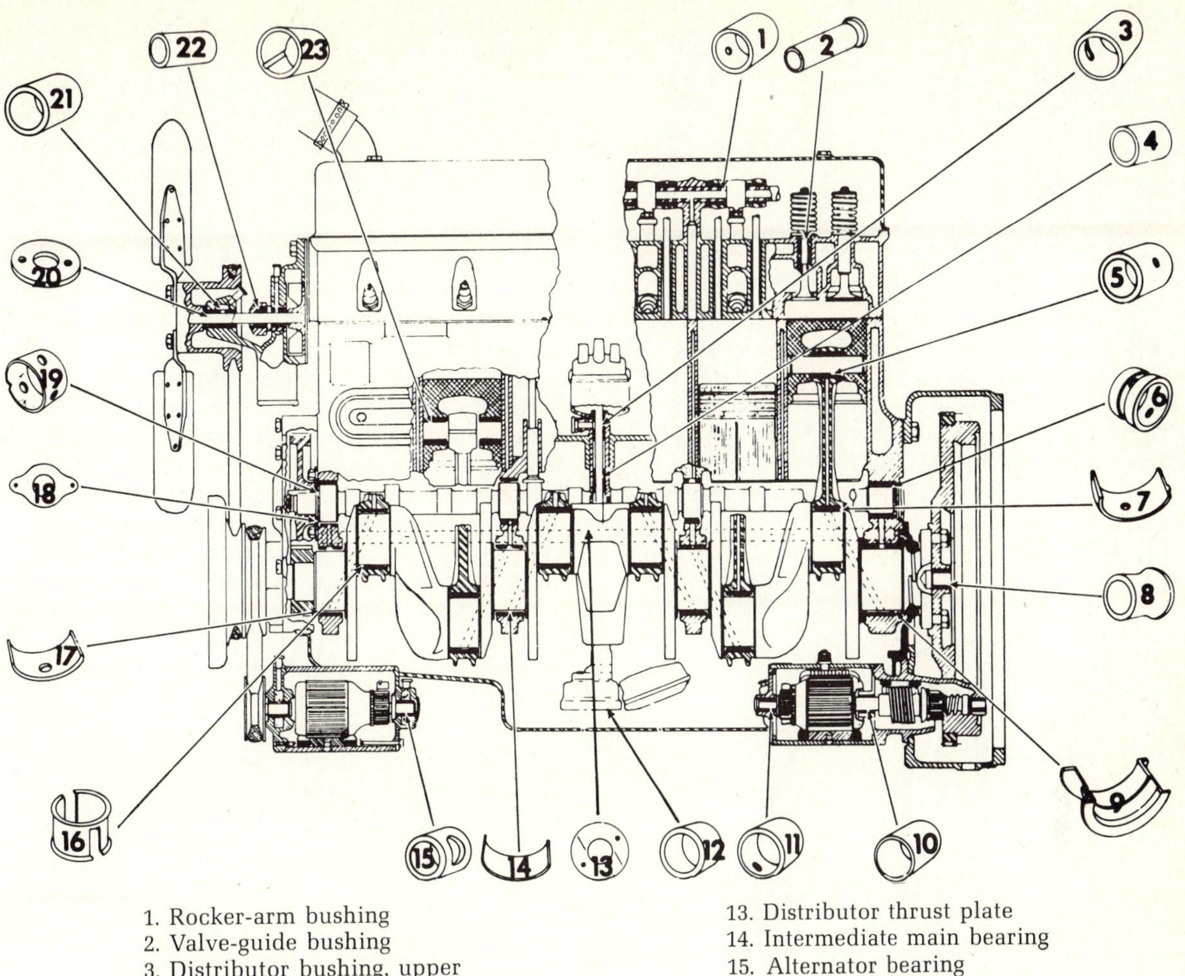

1. Rocker-arm bushing
2. Valve-guide bushing
3. Distributor bushing, upper
4. Distributor bushing, lower
5. Piston-pin bushing
6. Camshaft bushing
7. Connecting-rod bearing
8. Clutch pilot bushing
9. Flanged main bearing
10. Starting-motor bushing, drive end
11. Starting-motor bushing, commutator end
12. Oil-pump bushing
13. Distributor thrust plate
14. Intermediate main bearing
15. Alternator bearing
16. Connecting-rod bearing, floating type
17. Front main bearing
18. Camshaft thrust plate
19. Camshaft bushing
20. Fan thrust plate
21. Water-pump bushing, front
22. Water-pump bushing, rear
23. Piston-pin bushing

Fig. 7-34. Various bearings and bushings used in a typical engine. (*Johnson Bronze Company*)

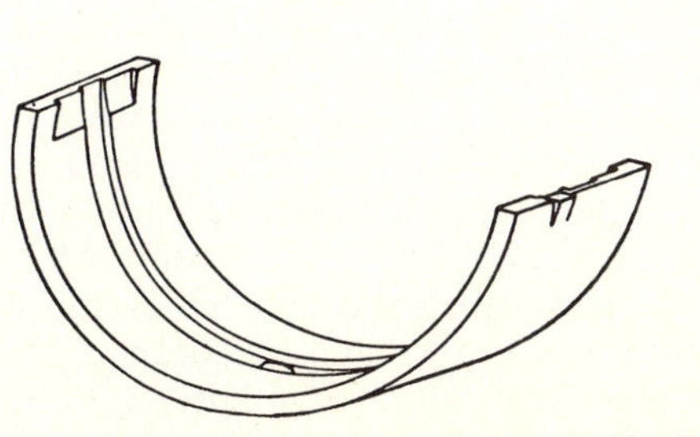

Fig. 7-35. Typical sleeve-type bearing half. (*Federal-Mogul Corporation*)

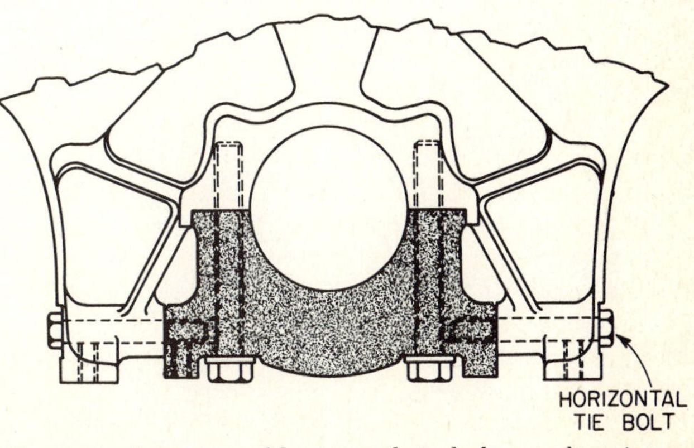

Fig. 7-36. Location of horizontal tie bolts used on intermediate main-bearing caps. (*Chrysler Corporation*)

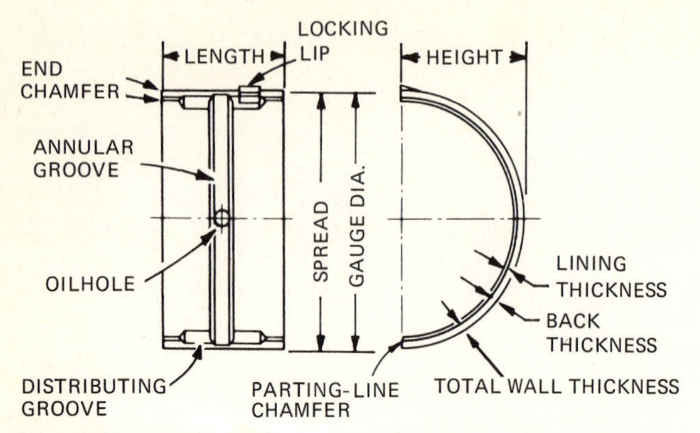

Fig. 7-37. Typical sleeve-type bearing half with parts named. (Many bearings do not have annular and distributing grooves.) (*Federal-Mogul Corporation*)

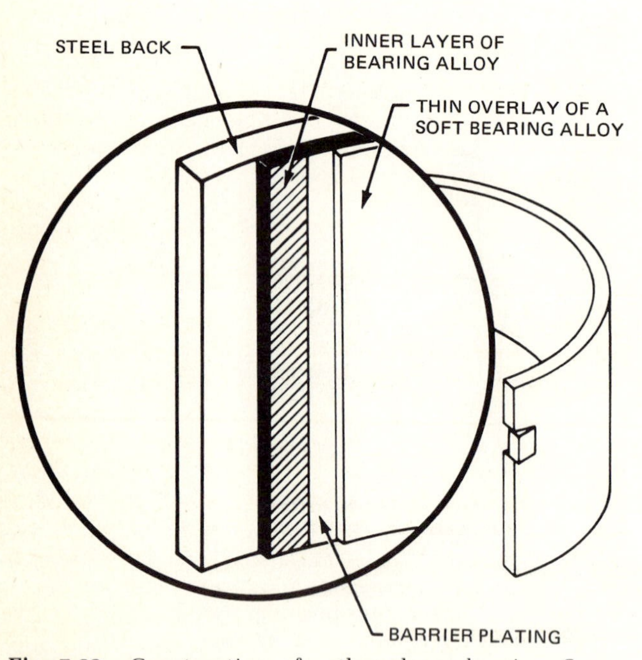

Fig. 7-38. Construction of a three-layer bearing. Some bearings have three layers, as shown; others have two layers. (*Federal-Mogul Corporation*)

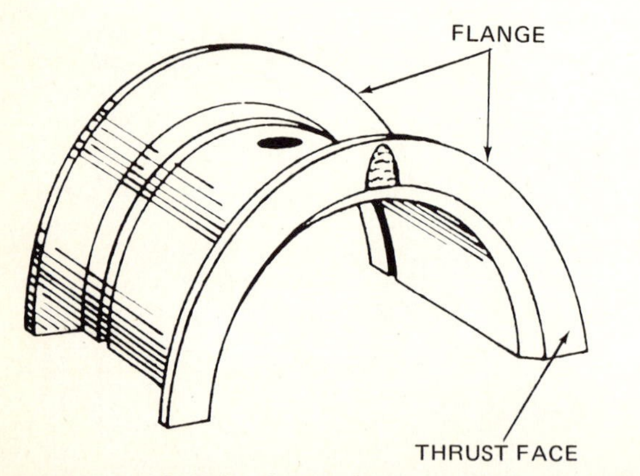

Fig. 7-39. Crankshaft thrust bearing. (*Federal-Mogul Corporation*)

circulating across the faces of the bearings in the engine.

One function of the oil, of course, is to provide lubrication; that is, it keeps the bearing and rotating journal separated by a film of oil so that there is no actual metal-to-metal contact. In addition, the oil helps to cool the bearing. The oil is relatively cool as it comes from the oil pan. As it spreads across the bearing and passes off the bearing edges, it warms up, thus removing heat from the bearing. This keeps the bearing at lower operating temperatures. A third function of the oil is to act as a flushing medium. It tends to flush out particles of dirt or grit that may have worked into the bearing (and on other engine parts). These particles either settle to the bottom of the oil pan or are removed from the oil by the oil filter, as explained in Chap. 17, which discusses engine lubricating systems.

In addition to these functions, the "throw-off" of oil from the engine bearings helps lubricate other engine parts. For instance, the cylinder walls, pistons, and rings are lubricated by the oil that is thrown onto the walls by the rotating crankshaft and connecting rods.

It is important to note that in order to function properly, the oil must circulate *through* the bearing; it must flow. In order to permit this, the shaft-journal diameter is made somewhat smaller than the bearing diameter. The difference in diameters is called the *oil clearance* (Fig. 7-40). It is obvious that the greater this clearance, the faster the oil will flow through the bearing. Proper clearance varies somewhat with different engines, but 0.0015 inch [0.037 mm] would be a typical clearance. As the clearance becomes greater (from bearing wear, for instance), the amount of oil flowing through and being thrown off increases. With a 0.003-inch [0.076 mm] clearance (only twice 0.0015 inch [0.037 mm]), the oil throw-off increases as much as five times. A 0.006-inch [0.152 mm] clearance allows 25 times as much oil to flow through and be thrown off.

As bearings wear and oil flow-through and throw-off increase, more and more oil is thrown onto the cylinder walls. The piston and rings cannot handle these excessive amounts of oil. Part of the oil works its way up into the combustion chambers, where it is burned, forming carbon on the piston, rings, and valves. This causes loss of power, increased engine wear, and other troubles (see Chap. 22, "Diagnosing Engine Troubles"). It is also true that excessive oil clearance in some bearings may cause other bearings to be oil-starved, so that they fail from lack of oil. The reason for this is that the oil pump can put out only so much oil. If the oil clearances of bearings are very large, most of this oil will pass through the oil clearances of the nearest bearings and there won't be enough oil left for the more distant bearings. An engine with this trouble usually has low oil pressure. The oil clearances are so great that the oil pump cannot build up normal pressures.

On the other hand, if oil clearances are not sufficiently great, the lubricating oil films in the

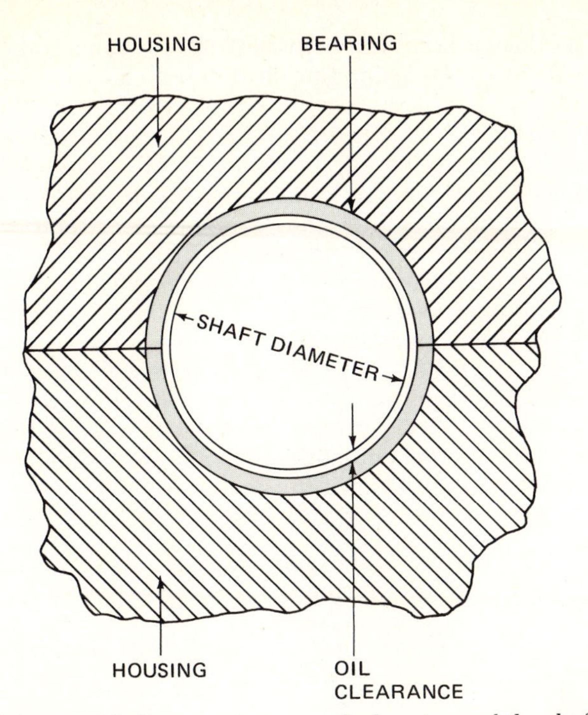

Fig. 7-40. Oil clearance between the bearing and the shaft journal.

bearing may not be thick enough to prevent metal-to-metal contact between the bearing and the shaft journal. Extremely rapid wear and early bearing failure will result. Furthermore, not enough oil will flow through, and therefore oil throw-off will be insufficient to provide adequate lubrication of such other engine parts as the cylinder walls, pistons, and rings.

⊘ 7-22 Engine-Bearing Types Early engines and some later-model heavy-duty engines used a "poured" bearing. This bearing was prepared by fitting a shaft-sized jig, or mold, into the counterbore where the bearing was to be and then pouring molten bearing material into the space between the jig and counterbore. After the metal cooled, the jig was removed and the metal was then scraped or machined down to the proper size to fit the shaft to be installed. This was a laborious process.

Today bearing installation is much simpler. In the first place, the bearings are supplied as replaceable shells (Fig. 7-35) consisting of a hard back coated with a layer of the bearing material. Secondly, in many engines these bearings are so precisely made that they can be replaced without any machining or fitting of the bearing, counterbore, or shaft journal (provided that the journal is not unduly worn). These bearings are called *precision-insert*, or *precision-type, bearings.* On many engines using these bearings, main bearings can be replaced without removing the crankshaft. The old bearing is simply slipped out and the new bearing slipped in by use of a special roll-out tool (Figs. 25-13 and 25-14). The tool is inserted into the oilhole in the crankshaft journal. Then, when the crankshaft is rotated, the tool forces the bearing shell out of its position between the shaft and cylinder block.

Some engines use *semifitted bearings*. These are approximately the correct size but do have a few thousandths of an inch of extra bearing material. This extra material is bored out after the bearings are installed to establish proper fit and alignment with the shaft journals. The machining compensates for any slight irregularity in the alignment of the counterbores in the cylinder block and bearing caps. Figure 7-41 illustrates a semifitted crankshaft bearing of the thrust type. This illustration shows the "semifitted" principle, which requires removal of bearing stock for proper fit. It also shows the thrust faces found in the crankshaft thrust bearing. One of the main (or crankshaft) bearings in the engine is always a thrust bearing. The purpose of the thrust faces has already been explained (⊘ 7-19).

⊘ 7-23 Bearing Requirements Bearings must be able to withstand the varying loads imposed on them without being damaged or wearing with excessive rapidity. But bearings must have other desirable characteristics. Some of these characteristics are listed and described below (not necessarily in order of importance).

1. *LOAD-CARRYING CAPACITY* Modern engines are lighter and more compact, yet more powerful, than the engines of a few years ago. We have already noted that higher compression ratios and consequent higher combustion pressures have made it possible to step up horsepower output without increasing engine weight. These higher combustion pressures and horsepowers have brought with them greater loading of the engine bearings. For example, only a few years ago connecting-rod bearings on many automobile engines sustained loads of 1,600 to 1,800 psi [112.48 to 126.54 kg/cm^2]. Today, connecting-rod bearing loads of more than 6,000 psi [421.82 kg/cm^2] are not uncommon.
2. *FATIGUE RESISTANCE* When a piece of metal

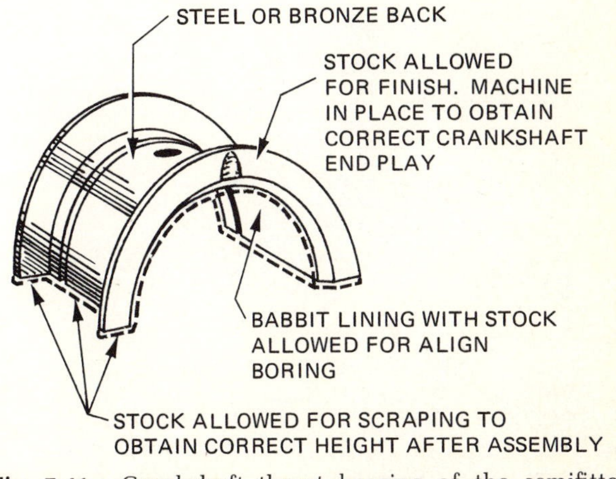

Fig. 7-41. Crankshaft thrust bearing of the semifitted type. (*Federal-Mogul Corporation*)

is repeatedly subjected to stress so that it flexes or bends (even slightly), it may harden and ultimately crack or break. An example of this is what happens when you repeatedly bend a piece of wire or sheet metal. Ultimately it will break. As far as bearings are concerned, they are subject to repeated and varying loads that tend to flex them. The bearing material must be able to stand this stress without any undue tendency to crack and break down.

3. *EMBEDDABILITY* "Embeddability" is the ability of a bearing to permit foreign particles to embed in it. Despite air cleaners and oil filters and screens, particles of dirt and dust do get into the engine, and some of them find their way to the engine bearings. A bearing protects itself by letting particles actually embed in the bearing-lining material. If the bearing material were too hard to allow this, the particles would simply lie on the surface of the bearing. They would soon scratch the shaft journal turning in the bearing and also gouge out the bearing. This, in turn, would cause overheating and rapid bearing wear, and so the bearing and the journal would soon fail. There is a limit to the number of particles a bearing can embed, however. If too many particles are embedded, the bearing will become overloaded with them and fail. Also, if particles are too large, they will not completely embed; they will scratch the shaft journal or gouge out grooves in the bearing. This also could lead to bearing failure.

4. *CONFORMABILITY* "Conformability" is associated with embeddability. It refers to the ability of the bearing material to conform to variations in shaft alignment or journal shape. For example, suppose a bearing is installed under a shaft that is slightly bent (or which bends as it is loaded). This causes a certain area of the bearing to be heavily loaded while other areas are very lightly loaded. If the bearing material has high conformability, it will "flow" slightly away from the heavily loaded area to the lightly loaded area. In effect, this redistributes the bearing material so that the bearing is more uniformly loaded. A similar action takes place when foreign particles are embedded in the bearing. As they embed, they displace material, thus producing a local high spot (Fig. 7-42). However, the material flows away from the high spot, thus tending to prevent heavy local loading that could cause bearing failure.

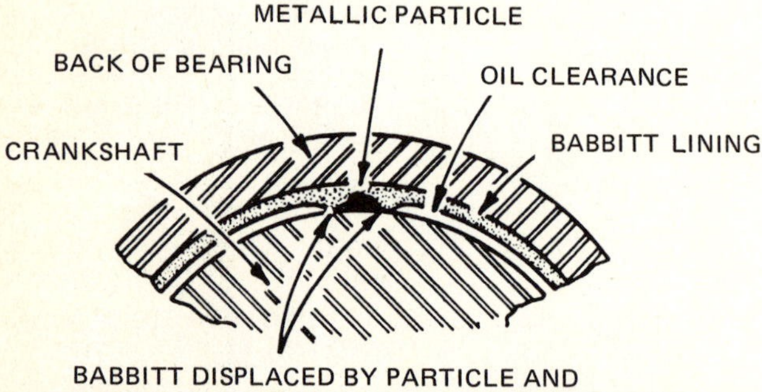

Fig. 7-42. Effect of a metallic particle embedded in the bearing material (the babbitt lining). (*Federal-Mogul Corporation*)

5. *CORROSION RESISTANCE* Certain acids appear in the oil as by-products of the combustion process and engine operation. Manufacturers of lubricating oil add certain compounds to their oils to combat the acids and prevent them from corroding engine bearings and other parts. These acids would otherwise attack certain types of bearing materials and cause them to fail rapidly.

6. *WEAR RATE* The bearing material must be sufficiently hard and tough so that it will not wear too quickly. At the same time, it must not be so hard as to have poor embeddability and conformability or cause undue wear of the shaft journal it supports.

⊘ 7-24 Bearing Materials As has already been mentioned, the bearing back is usually steel or bronze. Steel is the most common backing material in use today; precision-insert bearings are steel-backed. The bearing material applied to the back is a mixture of several metals. For example, one bearing material is made of lead, tin, antimony, and copper. Another bearing material is made of lead, tin, mercury, calcium, and aluminum. Still other combinations include copper, antimony, and tin; silver, copper, and cadmium; copper, lead, and silver; aluminum, tin, and copper. Figure 7-43 shows schematically how different types of bearings are manufactured.

As can be seen, many possibilities exist. It is up to the bearing designer to compound a bearing material that will stand up under the specific stresses and conditions to which the bearing will be subjected in operation. What is satisfactory for one installation might not be good for another. Corrosion, for example, might be a factor in some types of engine operation. Under some operating conditions, acid appears in the engine oil, and this can corrode some types of bearing material. Another example is changes in bearing loading under different operating conditions (see Figs. 7-46 and 7-47, which show bearing loading during two engine speeds). All these factors must be considered by the bearing designer as in selecting bearing material for an engine.

⊘ 7-25 Bearing Loading We have already noted that the pressure in the engine cylinder varies from below atmospheric to several hundred pounds per square inch (see Fig. 9-50) [100 psi = 7.031 kg/cm^2]. These varying pressures are transmitted through the piston and connecting rod to the crankpin, which imposes a varying load on the bearings. But there are other forces at work that also impose loads on the bearings.

For example, let us analyze the forces that act on the connecting-rod, or big-end, bearing. In addition to the pressure loads, there are centrifugal loads and inertia loads. Centrifugal loads result from the

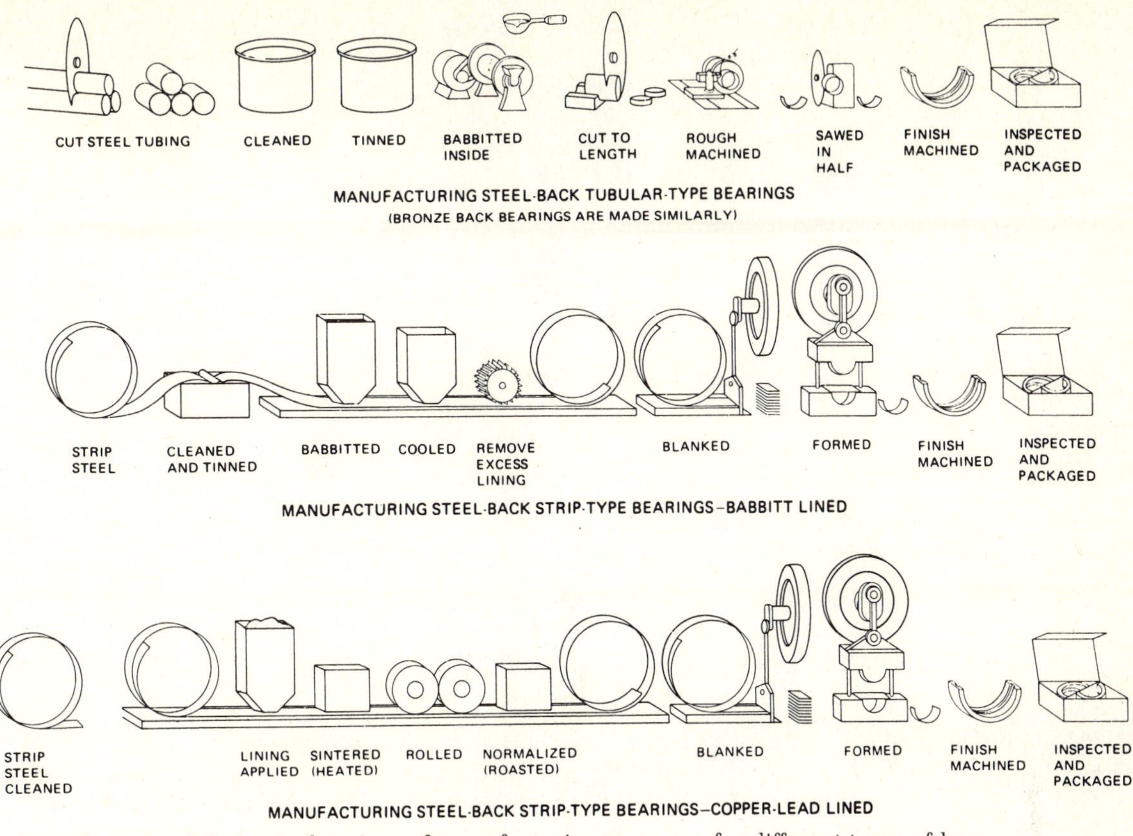

Fig. 7-43. Schematic drawings of manufacturing processes for different types of bearings. (*Federal-Mogul Corporation*)

centrifugal force on the rod big end that attempts to throw it outward from the center line of the crankshaft (Fig. 7-44). This force imposes a load on that part of the connecting-rod bearing which is toward the center of the crankshaft. The centrifugal load is constant for any particular speed and increases as the speed increases. It circles the bearing uniformly and can be quite large. For example, let us assume that the crankpin is offset 2 inches [50.8 mm] and that the connecting rod weighs 2 pounds [0.907 kg], of which 1 pound [0.454 kg] (at the big end) is effective in producing centrifugal force. With these assumptions, we can calculate that the centrifugal force acting on the connecting-rod bearing at 4,000 engine rpm would be more than 900 pounds [408.231 kg].[1] In other words, this is the bearing load produced by centrifugal force. This load is always on that section of the bearing which is toward the center of the crankshaft.

Now let's consider *inertia loads*. Inertia, you will recall (⊘ 4-30), is that characteristic of all material objects that causes them to resist any change of speed or direction of travel. The piston and upper part of the connecting rod are constantly changing speed and direction of travel. At the end of each stroke [at TDC (top dead center) and BDC (bottom dead center)] the piston is brought to a complete stop. Then it is accelerated to the high speed it attains in the middle of each stroke. At 4,000 engine rpm, for example, the piston will accelerate from a "standing start" to about 88 feet per second [26.822 m/s (meters per second)] (a mile a minute [1.609 km/min] in 0.00375 second). Then, in the next 0.00375 second, it must slow down and stop again. Even though a piston may weigh only 1 pound [0.454 kg] (cast-iron pistons weigh a little more), it takes a considerable force to stop the piston, start it again, accelerate it to high speed, and then stop it once more. This force, remember, is imposed on the connecting-rod big-end bearing and produces what are known as *inertia loads*.

Inertia loads vary greatly from a minimum toward the middle of the stroke to a maximum at around TDC and BDC when the maximum change of speed is taking place (or when the piston is brought to a stop and reversed in direction). For example, at 4,000 rpm, a 1-pound [0.454 kg] piston moving in a 4-inch [101.6 mm] stroke will impose a maximum inertia load of more than 700 pounds [317.513 kg] on the connecting-rod bearing. There is an additional load due to the inertia effect of the connecting rod, which is also brought to a stop and

[1] Centrifugal force $= \frac{W}{g}\left(\frac{2\pi N}{60}\right)^2 r = \frac{1}{32.16}\left(\frac{2\pi 4{,}000}{60}\right)^2 \frac{1}{6} = 914$ lb [414.584 kg]

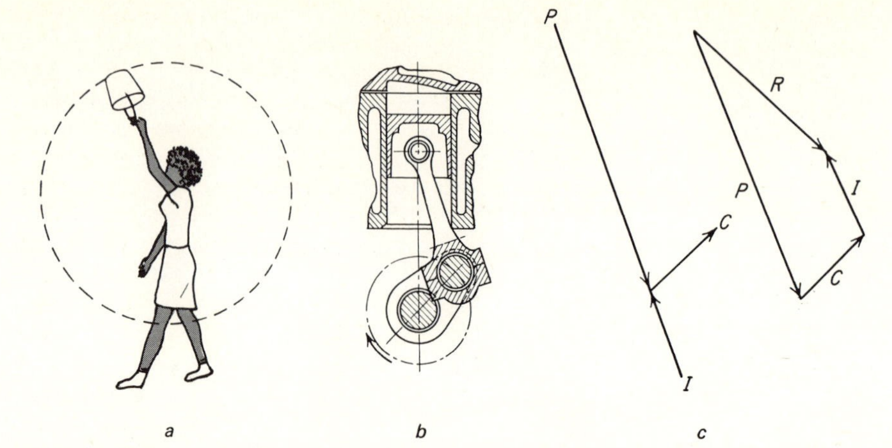

Fig. 7-44. The woman swinging a pail of water in (*a*) illustrates one of the three forces acting on the connecting-rod bearing (*b*). This is *centrifugal force*, which keeps the water in the pail by pushing it outward, just as centrifugal force pushes the connecting rod outward away from the center of rotation. This imposes a centrifugal load on the bearing in the direction shown in (*c*) by arrow *C*. Arrow *P* represents combustion pressure acting along the connecting rod on the bearing. Arrow *I* is the inertia load, which is acting in the opposite direction to *P* in the piston and rod position shown. *R* is the resultant load. (*Federal-Mogul Corporation*)

then moved in the opposite direction. Remember that this maximum load occurs as the piston passes through TDC (or BDC) or during the time that the piston is brought to a stop and then started moving in the opposite direction. Inertia loads increase with engine speed, just as centrifugal loads do.

⊘ 7-26 The Effective Bearing Load

The three different loads on the connecting-rod bearing sometimes add up and sometimes oppose each other, but generally they work at various angles to each other. You might compare this to two people pushing at angles on a box, as shown in Fig. 7-45. As they exert forces *A* and *B*, respectively, they work partly against each other, but part of their effort does add up to produce resultant force *C*. If they exerted different forces (as at forces *D* and *E*), then the resulting force would be to the right, as shown (resultant force *F*). In the drawings, note that the lengths of the arrows are proportional to the forces exerted.

In a similar manner, as the bearing loads work at different angles to each other, they produce a resultant loading force that is, in effect, the same as a single loading force. For example, Fig. 7-44*c* shows the power, inertia, and centrifugal loads on the bearing at one certain engine speed and piston position on the power stroke. Note that the pressure load opposes the inertia load. The inertia load partly cancels out the pressure load since it is in the opposite direction to the pressure load. The centrifugal load is an an angle.

Figure 7-44*c* shows how to combine the three loads to find out the resultant load *R*. The arrows are proportional in length to the forces exerted, and they point in the directions in which the forces are exerted. After the three force arrows are drawn, the resultant arrow *R* is drawn in. Its length represents the strength of the resultant force, and its direction indicates the direction in which the force is exerted.

⊘ 7-27 The Bearing-Load Graph

Figure 7-46 is a graph of the resultant bearing loads (from pressure, centrifugal, and inertia loads) imposed on the connecting-rod bearing in a certain engine equipped with a $3\frac{1}{8}$-inch [79.375 mm] bore and a $4\frac{3}{8}$-inch [111.125 mm] stroke operating at 3,700 rpm. The small figures represent the degrees of crankshaft rotation; 0 to 180 degrees of crankshaft rotation represents the power stroke, 180 to 360 degrees the exhaust stroke, 360 to 540 degrees the intake stroke, and 540 to 720 degrees (or 0 degrees) the compression stroke. The distance from the intersection of the two straight lines to any point on the curved line represents the amount of bearing load imposed, and the direction in which the measurement is taken indicates the direction of loading. For example, at 90

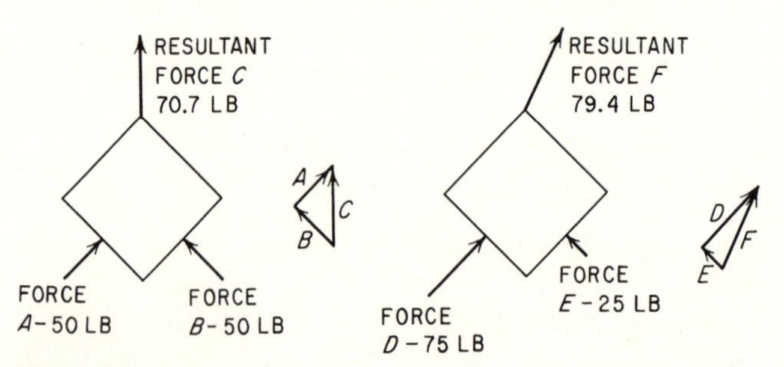

Fig. 7-45. Forces at angles to each other partly cancel each other out. *Left*, the two 50-pound forces give a resultant force of 70.7 pounds. *Right*, the 25- and 75-pound forces give a resultant force of 79.4 pounds. Lengths of arrows are proportional to the loads. Thus, arrow *D* (75 pounds) is three times as long as arrow *E* (25 pounds).

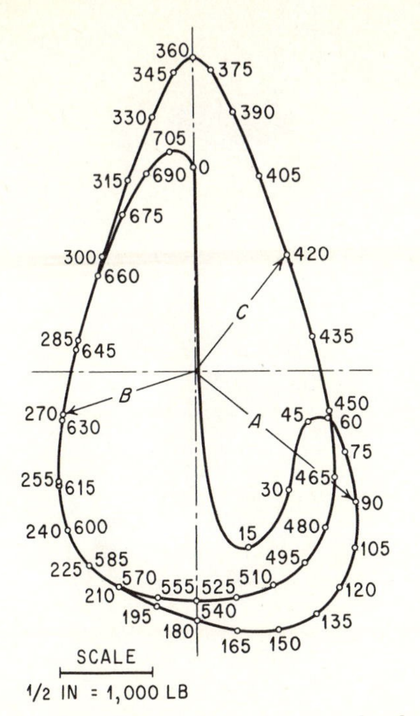

Fig. 7-46. Graph of the resultant force, or loads, imposed on the connecting-rod bearings in a $3\frac{1}{8} \times 4\frac{3}{8}$ inch [79.375 × 111.125 mm] engine operating at 3,700 rpm. The figures represent degrees of crankshaft rotation: 0 to 180 degrees is the power stroke, 180 to 360 degrees is the exhaust stroke, 360 to 540 degrees is the intake stroke, and 540 to 720 degrees (or 0 degrees) is the compression stroke. (*Federal-Mogul Corporation*)

degrees (or halfway through the power stroke), the bearing load is shown by arrow *A*, which scales at 2,100 pounds [952.54 kg] and is pointing down to the lower right. Then, at 270 degrees, which is halfway through the exhaust stroke, the bearing load is as shown on arrow *B*. This is almost horizontal to the left, and scales about 1,470 pounds [666.78 kg]. Arrow *C* indicates the bearing load at 420 degrees, or at a point 60 degrees after the intake stroke has begun. This scales at 1,480 pounds [671.32 kg] and is toward the upper right.

The graph in Fig. 7-46 provides a great deal of interesting information. For instance, under the running conditions at which the graph was compiled, at the beginning of the power stroke there is an almost zero bearing loading for a few degrees of crankshaft rotation (because of pressure load canceling the centrifugal and inertia loads). Moreover, in a matter of 15 degrees of crankshaft rotation, the load changes from about 2,000 pounds [907.18 kg] on the upper part of the bearing (at 0 degrees) to about 1,800 pounds [816.46 kg] on the lower part of the bearing (at 15 degrees). Note also that the load decreases after this because of pressure drop and also swings up toward the right. Then, at about 45 degrees, the load again increases and swings downward. You might find it useful to compare this graph of connecting-rod-bearing loading with the graph of cylinder pressures (Fig. 9-50).

Remember that the graph (Fig. 7-46) shows only one set of conditions at one engine speed. The bearing loadings (and graph, of course) change with changed operating conditions. For example, Fig. 7-47 compares the bearing loads shown in Fig. 7-46 with bearing loads obtained at 1,000 rpm (with a wide-open throttle) on the same engine. Note that under the latter conditions, an entirely different set of bearing loads is imposed, with the maximum coming early in the power stroke and amounting to 3,780 pounds [1,714.57 kg]. Note also how small the inertia and centrifugal loads are at 1,000 rpm (upper part of graph). As is obvious from the graph, this sort of operation is very hard on the bearings; it subjects them to periodic heavy shock loads. Bearings will have a relatively short life under such conditions.

We should also note that we have been discussing only connecting-rod bearings in the past few pages. However, other engine bearings (main, crankshaft, piston pin, and so forth) also have varying loads imposed on them; somewhat similar bearing-load analyses could be made on these other bearings.

⊘ 7-28 Engine Vibration Mountings

The crankshaft counterweights largely balance the inertial and centrifugal loads imposed by the crankpin-piston-rod combination. The torsional-vibration damper largely controls torsional vibration of the crankshaft. But there are other forces at work in and around the engine that produce vibrations of various frequencies and intensities. A basic one is, of course, the periodic combustion pressures and their action

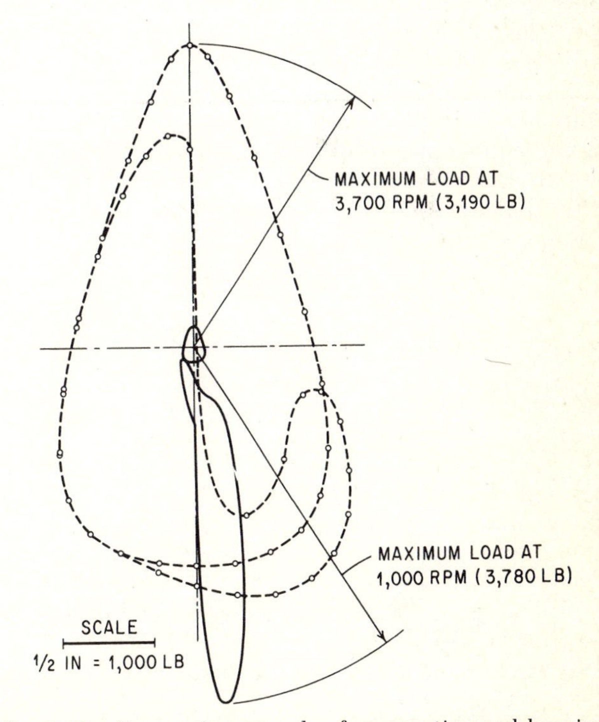

Fig. 7-47. Comparison graph of connecting-rod-bearing loads at 3,700 rpm (shown in the dotted line) with loads at 1,000 rpm at wide-open throttle (shown in the solid line). These graphs were compiled on the same engine. (*Federal-Mogul Corporation*)

in applying torque to the crankshaft. This torque is carried through the crankshaft to the transmission and produces a countertorque in the engine. That is, as the crankshaft turns in one direction, the engine attempts to move in several directions.

If the engine were mounted rigidly on the car frame, the driver and passengers would be subjected to various unpleasant engine noises and vibrations. Also, various car components, such as the radiator, relays, and instruments, would be subjected to this vibration, and it would shorten their service life. In addition, twisting of the car frame (which occurs during normal operation) would place great strain on the engine block and could possibly break engine mounting lugs. To prevent all this, the engine is mounted on the frame through flexible mountings that interpose a collar or pad or rubber or similar material between the engine and frame. Figures 7-48 and 7-49 show the mounting arrangements for the front and rear of an engine.

Check Your Progress

Progress Quiz 7-2 Once again here is your chance to find out how well you have understood and remembered the material you have been reading. This repeated checking up keeps you on your mental toes and also helps you review the material you have read. If any of the questions seem hard to answer, reread the pages that will give you the answers.

Correcting Parts Lists The purpose of this exercise is to help you spot unrelated parts in a list. For example, in the list "connecting rod, piston, piston pin, main bearing, bearing cap," you would see that "main bearing" does not belong since it is not a part of the connecting-rod-and-piston assembly.

In each of the following lists, you will find one item that does not belong. Write each list in your notebook, but *do not write* the item that does not belong.

1. Parts directly assembled on the crankshaft include the flywheel, vibration damper, piston, timing gear or sprocket.
2. Bearings assembled on the crankshaft include the main bearings, piston-pin bearings, connecting-rod bearings.
3. Desirable sleeve-bearing characteristics include embeddability, fatigue resistance, conformability, convertibility, corrosion resistance.
4. Materials for sleeve bearings mentioned in the text include lead, tin, copper, antimony, carbon.
5. The three loads acting on the connecting-rod bearing include pressure loads, depression loads, centrifugal loads, inertia loads.

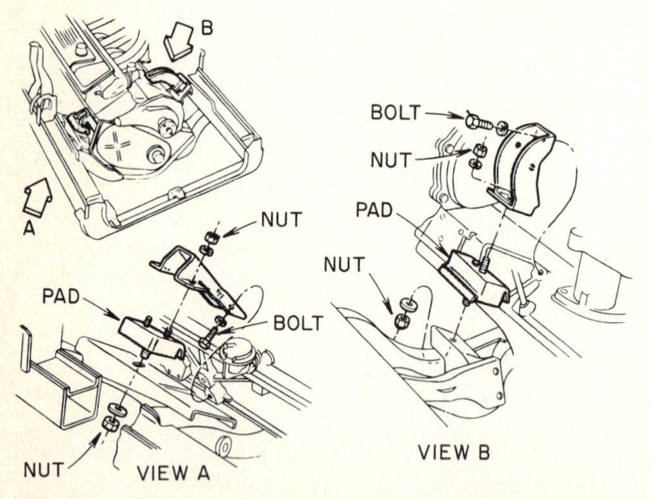

Fig. 7-48. Front-engine flexible mounts used to support the engine on the frame. (*Chrysler Corporation*)

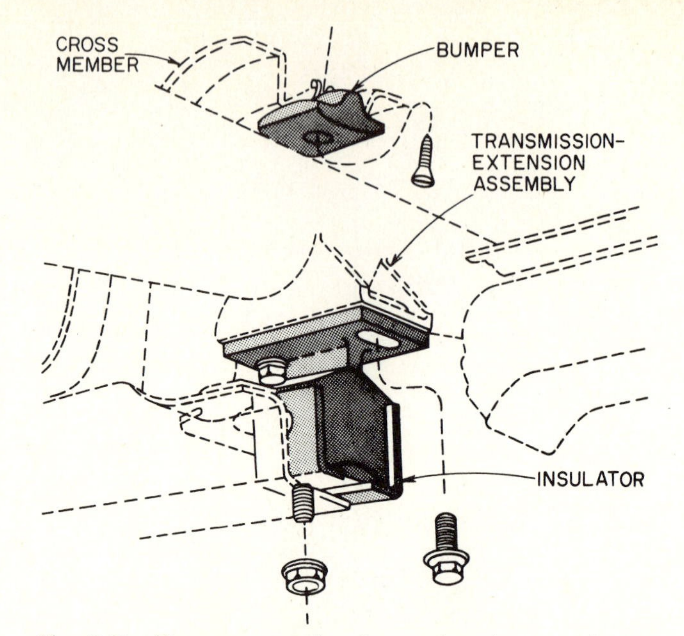

Fig. 7-49. Rear support for the engine showing the insulator arrangement. (*Chrysler Corporation*)

Completing the Sentences The sentences that follow are incomplete. After each sentence there are several words or phrases, only one of which will correctly complete the sentence. Write each sentence in your notebook, selecting the proper word or phrase to complete it correctly.

1. There is an overlapping of power impulses on: (*a*) four- and six-cylinder engines, (*b*) four- and eight-cylinder engines, (*c*) six- and eight-cylinder engines.
2. Twisting and untwisting of the crankshaft is called: (*a*) torsional balance, (*b*) torsional vibration, (*c*) power impulses.
3. The difference in diameters between a shaft journal and its supporting sleeve bearing is called: (*a*) bearing slack, (*b*) oil clearance, (*c*) distributing groove.
4. A typical oil clearance would be: (*a*) 0.00015 in [0.0038 mm], (*b*) 0.0015 in [0.038 mm], (*c*) 0.015 in [0.381 mm], (*d*) 0.15 in [3.81 mm].

5. The bearing that prevents excessive endwise movement of the crankshaft is called the: (*a*) push bearing, (*b*) thrust bearing, (*c*) face bearing.
6. The characteristic of the bearing that allows foreign particles to embed in it is called: (*a*) absorption, (*b*) embeddability, (*c*) conformability, (*d*) fatigability.
7. The rotating effect of the connecting rod on the connecting-rod bearing produces the: (*a*) inertia load, (*b*) pressure load, (*c*) centrifugal load.
8. When several different forces act at angles on an object, the combining of these forces produces a: (*a*) resultant force, (*b*) remaining force, (*c*) canceling force.

CHAPTER 7 CHECKUP

NOTE: Since the following is a chapter review test, you should review the chapter before taking the test.

You have made real progress in your study of the automotive engine and are rapidly approaching the point where you will be able to take the information you are learning into the automotive shop and begin to do actual repair work on engines. The following checkup will help you check yourself on how well you are retaining the essential information that you will need in your shopwork. Turn back into the chapter and do some rereading if you have any doubts about the answers to any of the following questions.

Where you are asked to write something down, write it in your notebook. By now your notebook should have a good deal of valuable information in it. If you have not been keeping a notebook, *now* is a good time to start one.

Completing the Sentences The sentences that follow are incomplete. After each sentence there are several words or phrases, only one of which will correctly complete the sentence. Write each sentence in your notebook, selecting the proper word or phrase to complete it correctly.

1. The crankshaft is suspended from the lower part of the: (*a*) crankcase, (*b*) cylinder block, (*c*) cylinder head, (*d*) oil pan.
2. The ignition distributor and fuel pump are driven from the: (*a*) flywheel, (*b*) crankshaft, (*c*) camshaft, (*d*) fan belt.
3. The sealing devices placed between the cylinder block and the parts attached to it are called: (*a*) gaskets, (*b*) studs, (*c*) bolts, (*d*) cylinder heads.
4. The manifold heat-control valve is assembled into the: (*a*) intake manifold, (*b*) exhaust manifold, (*c*) crankcase, (*d*) cylinder head, (*e*) tail pipe.
5. The muffler is located in the: (*a*) exhaust manifold, (*b*) intake manifold, (*c*) exhaust line, (*d*) tail pipe.
6. Attached to the rear end of the crankshaft is the: (*a*) vibration damper, (*b*) flywheel, (*c*) drive pulley, (*d*) timing gear.
7. If oil clearance in a bearing is excessive, then there will probably be excessive: (*a*) oil pressure, (*b*) oil throw-off, (*c*) oil compression.
8. Important bearing characteristics include: (*a*) embeddability, compression, and fatigue; (*b*) embeddability, conformability, and fatigue resistance.
9. Three loads imposed on the connecting-rod bearing are due to: (*a*) centrifugal force, inertia, and combustion pressure; (*b*) atmospheric pressure, inertia, and torsional vibration; (*c*) inertia, engine speed, and centrifugal force.
10. Two of the three connecting-rod-bearing loads that increase as engine speeds increase are: (*a*) pressure and inertia loads, (*b*) pressure and centrifugal loads, (*c*) centrifugal and inertia loads, (*d*) torsional and pressure loads.

Definitions and Lists In the following you are asked to write the function or define the purpose of different engine parts or to make lists. Write these in your notebook. The act of writing these items does two things: It tests your knowledge, and it also helps fix the information firmly in your mind. Turn back into the chapter if you are not sure of an answer, and reread the pages that will give you the information.

1. What are the three functions of the flywheel?
2. List 10 parts attached to or installed in the cylinder block.
3. List 5 parts attached to an I head.
4. What is the purpose of the manifold heat-control valve?
5. What is the purpose of the intake manifold?
6. List five desirable characteristics of a sleeve bearing.
7. What is the purpose of the vibration damper?
8. What are the three types of load imposed on the connecting-rod bearing?

SUGGESTIONS FOR FURTHER STUDY

To study engine construction further, go to a friendly service shop where engine repair work is done and watch different engines being torn down and serviced. If possible, handle various engine parts and examine them closely. Note how pistons, connecting rods, crankshafts, and other parts are made and assembled into the engine. Examine as many cylinder blocks as you can, and notice the many machined surfaces on the top, sides, end, and bottom for attachment of different parts.

You will also be able to handle and study various engine parts in your own school automotive shop. Your school shop may also have cutaway and working models on display, and a study of these will help you understand engine construction.

chapter 8

PISTON-ENGINE CONSTRUCTION: PISTONS AND PISTON RINGS

Chapter 7 described cylinder blocks, heads, crankshafts, and bearings. This chapter continues the description of piston-engine construction and discusses pistons and piston rings. Chapter 9 will conclude the discussion of piston-engine construction by describing valves and valve trains.

⊘ 8-1 Connecting Rod The *connecting rod* (Fig. 8-1) is attached at one end to a crankpin on the crankshaft and at the other end to a piston (through a piston pin, or wrist pin). The connecting rod must combine great strength and rigidity with light weight. It must be strong enough to maintain rigidity when carrying the thrust of the piston during the power stroke. At the same time it must be as light as possible so that the centrifugal and inertia loads on the bearings will be no greater than necessary (see ⊘ 7-25).

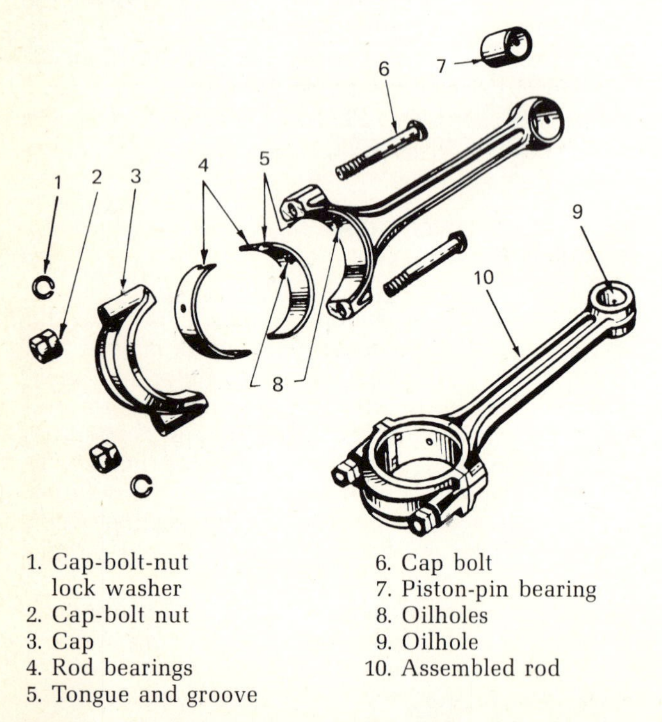

1. Cap-bolt-nut lock washer
2. Cap-bolt nut
3. Cap
4. Rod bearings
5. Tongue and groove
6. Cap bolt
7. Piston-pin bearing
8. Oilholes
9. Oilhole
10. Assembled rod

Fig. 8-1. Connecting rod with bearings and bearing cap in disassembled view (*top*) and assembled view (*bottom*). (*Chrysler-Plymouth Division of Chrysler Corporation*)

Connecting rods are forged from high-strength alloy steel and carefully balanced so that all rods in an engine will be of the same weight. If they were of different weights, engine balance would be thrown off and engine roughness would result. In the original factory assembly, rods and caps are individually matched to each other and usually carry identifying numbers so that they will not be mixed up if the engine is disassembled. They must never be mixed during any service job. Each cap must be replaced on the rod from which it was removed. Placing the wrong cap on a rod could result in poor bearing fit and bearing failure.

The crankpin end of the connecting rod (rod big end) is provided with a split-type sleeve bearing, as shown in Fig. 8-1. It is attached to the crankpin by means of the bearing cap, rod bolts, and nuts.

The piston end of the connecting rod (rod small end) is attached to the piston by means of a *piston pin* (also called a *wrist pin*). Bosses are provided in the piston (Fig. 8-2) with holes into which the piston pin is assembled. The pin also goes through the connecting rod. If the pin is to swing back and forth in the rod, a sleeve bearing, or bushing, is provided for it. Five methods of attaching the piston and connecting rod with the piston pin are used (Fig. 8-3). One method locks the piston pin in the piston by a lock bolt (Fig. 8-4). On this type of installation, the connecting rod has a bushing that provides a bearing surface between the rod and the pin so that the rod can rock back and forth on the pin.

Another method provides a press fit of the piston pin in the connecting rod. The press fit is tight enough to prevent the piston pin from moving out of position. This is a commonly used design. A variation of this design (Fig. 8-5), used in some imported cars, locks the connecting rod to the piston pin by means of a clamp screw. When this method is used

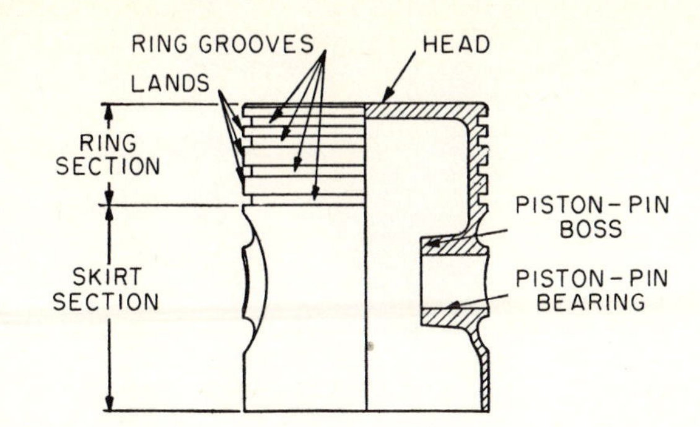

Fig. 8-2. Piston in partial sectional view, with its parts named.

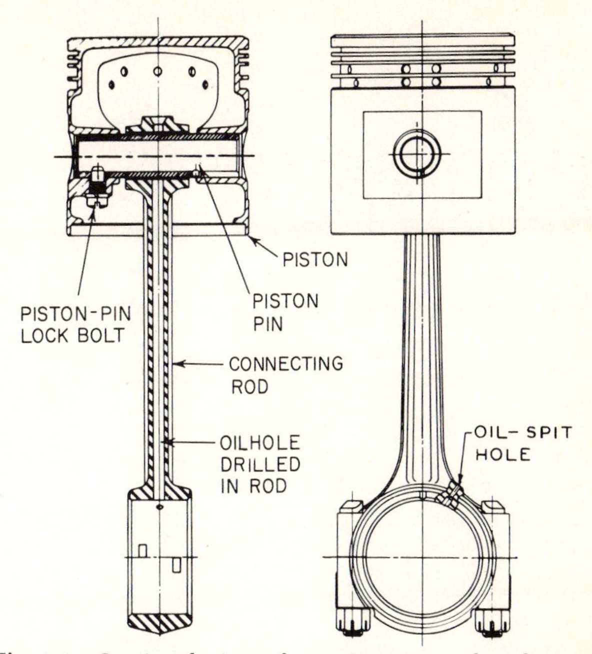

Fig. 8-4. Sectional view of a connecting rod and piston showing the oilhole, which lubricates the piston pin, and the oil-spit hole, which lubricates the cylinder wall. Note the lock bolt, which locks the piston pin to the piston. (*Oldsmobile Division of General Motors Corporation*)

with a cast-iron piston, the piston pin rocks back and forth in the bearing surfaces of the piston as the rod tilts one way and the other. When this method is used with an aluminum piston, the bore of the piston-pin boss is often used as the bearing surface for the pin.

A different method of attaching the piston and connecting rod with the piston pin has a bushing in the connecting rod, the pin not being locked to either the rod or piston. This method is called *free floating*. In this design the pin is prevented from moving out and scoring the cylinder walls by means of *snap rings* (also called *lock rings*) in the piston bosses (Figs. 8-6 and 8-7).

Various methods of providing lubrication of the piston-pin bushings have been used. Where the connecting rod has a bushing in which the pin moves (Figs. 8-4 and 8-7) some designs have an oil-passage hole drilled the entire length of the connecting rod from the crankpin-journal bearing to the piston-pin bushing (Fig. 8-1). Oil feeds through this passage from the crankpin oilhole to the piston-pin-bushing hole to provide lubrication. On the designs where the piston moves in the piston bearing surfaces (Figs. 8-5 and 8-7), lubrication is from oil scraped from the cylinder wall by the oil-control rings. The piston has grooves, holes, or slots through which oil can feed from the oil-control-ring groove to the piston-pin bearing surfaces. Some engines depend upon oil splash and throw-off from the crankpins for lubrication of the piston-pin bearing surfaces.

Many connecting rods have an oil-spit hole drilled on one side, as shown in Fig. 8-4. As the crankpin rotates within the rod bearing, an oil-passage hole in the crankpin *indexes*, or *lines up*, with this oilhole. Oil feeds through so that for a moment oil spits, or streams, from the hole in the connecting rod. This oil is thrown against the cylinder wall to provide additional cylinder-wall lubrication. On in-line engines, the holes are arranged to

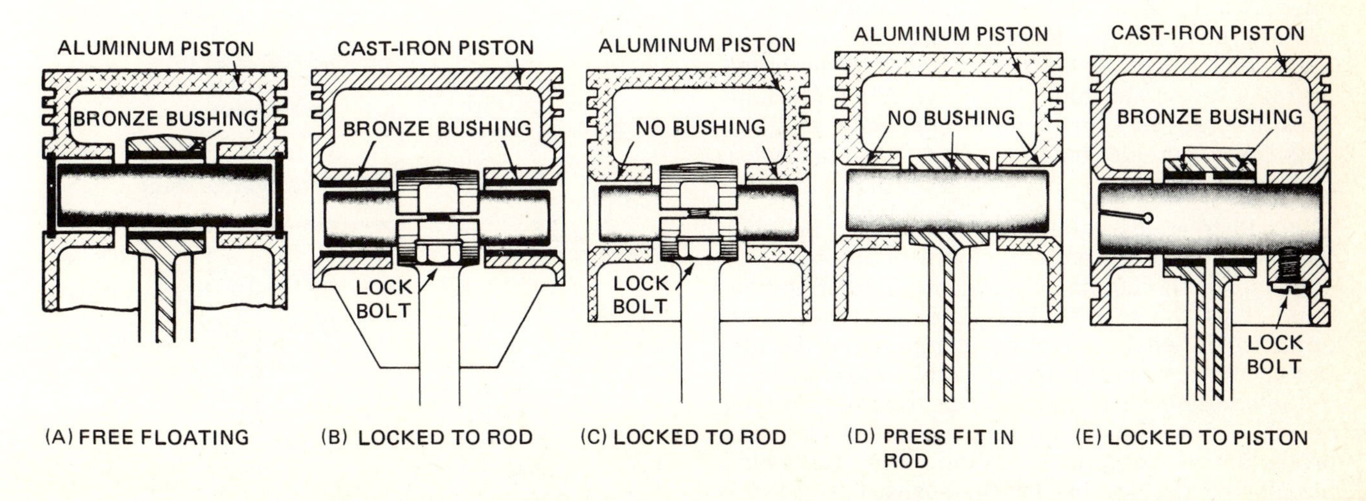

Fig. 8-3. Five piston-pin arrangements. (*Sunnen Products Company*)

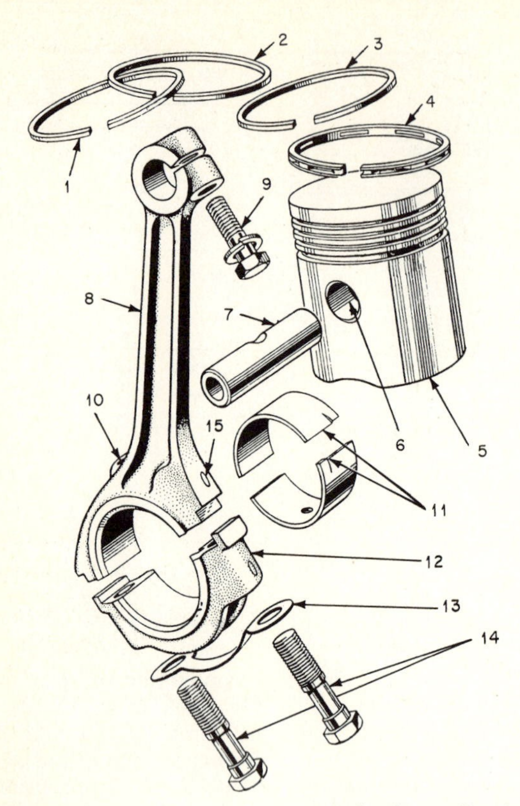

1–3. Piston rings
4. Piston oil ring
5. Piston
6. Piston-pin lubricating hole
7. Piston pin
8. Connecting rod
9. Clamping screw and washer
10. Cylinder-wall oil-spit hole
11. Connecting-rod bearing
12. Connecting-rod cap
13. Lock washer
14. Cap bolts
15. Connecting-rod and cap marking

Fig. 8-5. Connecting-rod-and-piston assembly in disassembled view. Note the clamping screw, which clamps the rod on the piston pin. (*MG Car Company, Limited*)

index just as the piston approaches TDC (top dead center). Thus a large area of the cylinder wall is covered with oil.

On many V-8 engines, cylinder walls and piston pins are lubricated by oil jets from opposing connecting rods. That is, each rod has a groove or hole that indexes with an oil-passage hole in the crank journal every crankshaft revolution. When this happens, a jet of oil spurts into the opposing cylinder in the other cylinder bank (see Fig. 5-23, upper right).

⊘ 8-2 Pistons Essentially, the piston is a cylinder that is open at the bottom, closed at the top, and attached to the connecting rod at an intermediate point (Figs. 8-2 to 8-7). The piston moves up and down in the engine cylinder, compressing the air-fuel mixture, transmitting the combustion pressure to the crankpin through the connecting rod, forcing out the burned gases on the exhaust stroke, and producing a vacuum in the cylinder that "draws in" the air-fuel mixture on the intake stroke.

The piston may seem to be a fairly simple part,

Fig. 8-6. Piston and connecting-rod assembly. This type has lock rings to hold the piston pin in position in the piston and connecting rod. (*Chrysler Corporation*)

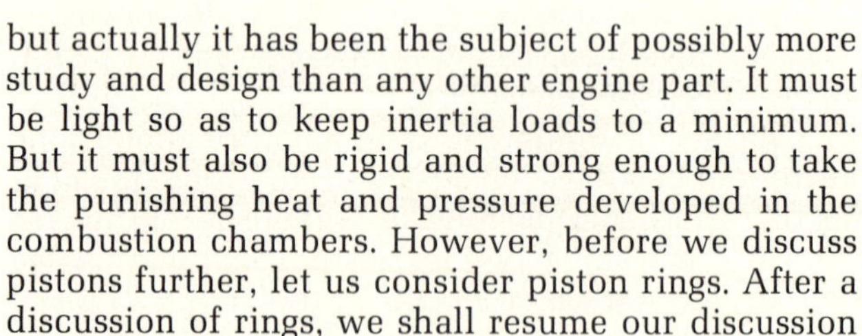

but actually it has been the subject of possibly more study and design than any other engine part. It must be light so as to keep inertia loads to a minimum. But it must also be rigid and strong enough to take the punishing heat and pressure developed in the combustion chambers. However, before we discuss pistons further, let us consider piston rings. After a discussion of rings, we shall resume our discussion of pistons (⊘ 8-13).

⊘ 8-3 Piston Rings A good seal must be maintained between the piston and the cylinder walls so that *blow-by* is prevented. "Blow-by" describes the escape of unburned air-fuel mixture and burned gases from the combustion chamber, past the piston rings, and into the crankcase. In other words, these gases

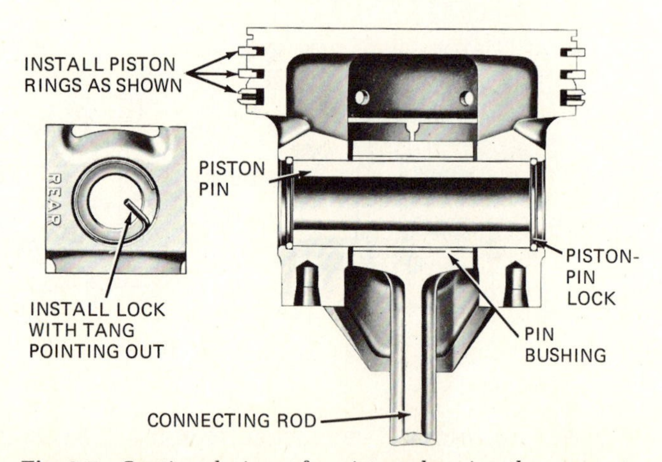

Fig. 8-7. Sectional view of a piston showing the connecting rod and piston pin in place. (*Cadillac Motor Car Division of General Motors Corporation*)

"blow by" the piston rings. It would be difficult to machine a piston to fit the cylinder accurately enough to prevent excessive blow-by. Even if this were possible, the differences in expansion between the cylinder and piston during engine operation, caused by variations in temperature, would change the fit so that it would be either too loose to too tight.

Piston rings, assembled into grooves in the piston as shown in Figs. 8-6 and 8-7, are used to provide a good seal, and so blow-by is kept at a minimum. Also, they seal in the compressed air-fuel mixture; they maintain good compression. Piston rings also perform a second function: They scrape oil off the cylinder walls on the piston downstrokes (power and intake). This prevents too much oil from working up past the piston and getting into the combustion chamber. (Oil in the combustion chamber burns, leaving a residue of carbon that fouls spark plugs, valves, and piston rings.) Piston rings perform a third function: They help cool the piston by transmitting a considerable amount of heat from the piston to the cylinder walls. The cylinder walls are cooled by the water circulating in the water jackets, and this cooling effect is thus carried through the rings to the piston. This helps guard against excessive and possibly damaging piston temperatures.

Piston rings are of two types, according to their purpose: compression rings and oil-control rings. There are usually three or more rings on a piston. The upper rings are the compression rings; they seal in compression and prevent blow-by. The lower rings are the oil-control rings; they control the oil on the cylinder walls by scraping off excess oil and returning it to the crankcase, leaving only enough oil on the cylinder walls to provide piston and ring lubrication.

Figure 8-8 shows a compression ring (top) and an oil-control ring (bottom); the various parts are named. Figures 8-6 and 8-7 show pistons with two compression rings and one oil-control ring installed on them. Note that the rings have a joint (they are split) so that they can be expanded and slipped over the pistonhead and into the recessed grooves cut in the piston. Rings for automotive engines usually have butt joints, but in some heavy-duty engines, the joints may be angled or they may be of the lap or sealed type, among others.

The reason that two compression rings may be used on a piston instead of only one is that more than one ring is required to do the job. For example, the top compression ring may seal in most of the combustion pressure, but the second ring is required to complete the job. Even with two rings, some slight amount of blow-by may be experienced. However, two rings keep it down to an unimportant minimum (when rings, cylinder wall, and other engine parts are in good condition). Most automobile engines have two compression rings, but some heavy-duty engines have more.

The compression ring is larger in diameter outside the cylinder than it is inside the cylinder. When it is placed in the cylinder, the ring is compressed so that the joint is nearly closed. Compressing the ring into the cylinder in this way places an initial tension on it; the ring presses tightly against the cylinder wall.

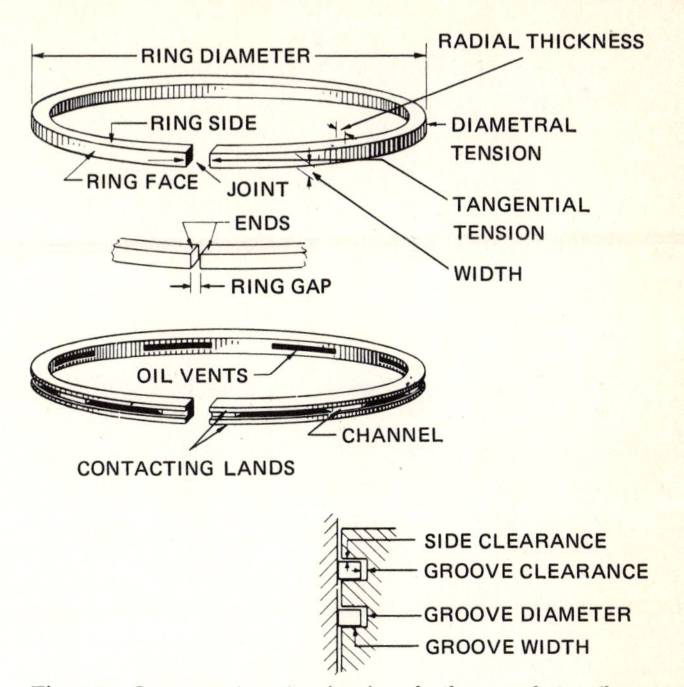

Fig. 8-8. Compression ring (*top*) and oil-control ring (*bottom*), with various parts named. (*Sealed Power Company*)

Most automobile engines use one oil-control ring. (However, two oil-control rings are used on many truck and diesel engines to provide adequate oil control.) The lower compression ring removes most of the oil from the cylinder wall, and the oil-control ring removes the remainder of the excess oil. Actually, a thin film of oil is left on the cylinder wall. If all the oil were removed, actual metal-to-metal contact between rings and wall would result, which would cause very rapid ring and wall failure. A ring expander (⊘ 8-12) may be used under the one-piece oil-control ring, as shown in Fig. 8-22. Most engines use the three-piece type of oil-control ring (see ⊘ 8-7).

⊘ 8-4 Compression Rings Compression rings are made of cast iron. This material has excellent wearing qualities and also provides adequate initial ring tension, or pressure, on the cylinder walls.

1. *RING SHAPES* The shape of compression rings has undergone many changes in the past 30 years. Some of the different shapes are shown in Fig. 8-9. The *chamfered* (beveled) and *scraper* types are of special interest since they are widely used for top and second compression rings in automotive engines today. Figure 8-10 shows a chamfered, or beveled, compression ring. Cutting off the inner edge unbalances the internal tension forces in the ring. When the ring is compressed (joint is closed) as it is installed in the cylinder, these unbalanced tensions in the ring cause it to twist slightly. A similar

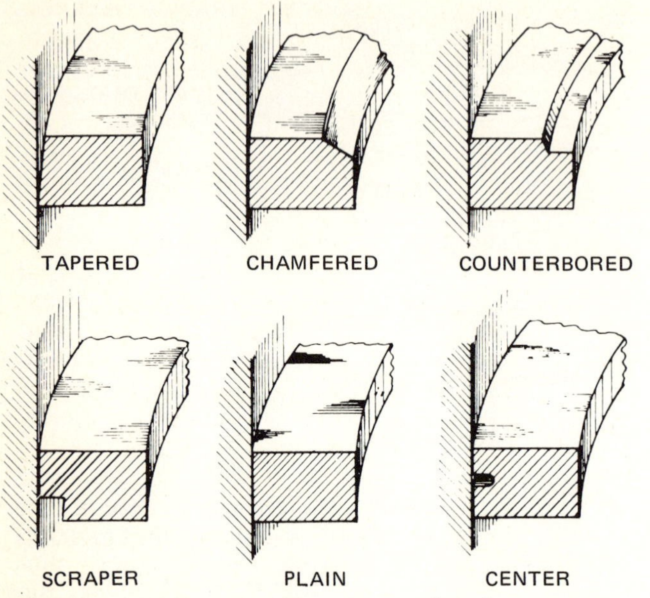

Fig. 8-9. Compression-ring shapes. (*Muskegon Piston Ring Company*)

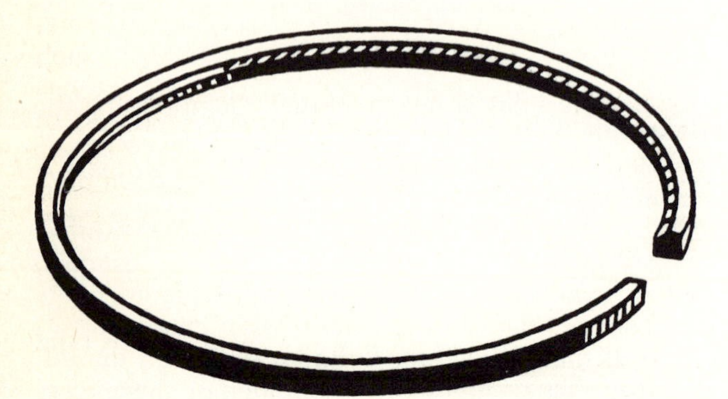

Fig. 8-10. Compression ring with chamfer (bevel).

twisting takes place with the scraper type of compression ring.

These rings are effective in controlling the oil that gets past the oil-control rings; on the intake stroke, they scrape all but a very small amount of the remaining oil from the cylinder wall (Fig. 8-11). Then, on the exhaust and compression strokes, when the rings are moving upward, they tend to "skate" over the film of oil on the cylinder wall so that they have less tendency to carry oil up into the combustion-chamber area. At the same time, wear is minimized on the upstrokes.

On the power stroke, combustion pressures press down on the top and on the back of the rings. This overcomes the internal tensions in the rings, causing them to untwist and straighten and thus present full-face contact with the cylinder walls for effective sealing (Fig. 8-12).

NOTE: Instead of being chamfered, many compression rings for heavy-duty applications have a *counterbore* (Fig. 8-9). If you check automotive-parts dealers, you will usually find that they list both the chamfered and the counterbored rings for most engines. The counterbored rings are more expensive, but they do a better job of controlling oil in the heavy-duty engine.

2. HEADLAND RING The headland ring has a modified L-shaped cross section (Fig. 8-13). An advantage claimed for this ring is that it reacts more swiftly to the buildup in combustion pressures. As combustion starts, the increasing pressures act quickly upon the upper lip of the ring, forcing the ring outward and into a good sealing contact with the cylinder wall. With other rings, the gas pressure must travel halfway around the ring before it gets behind the ring to force it out. This quicker response of the headland ring is designed to reduce blow-by. In addition, because the ring is located only $\frac{1}{16}$ inch [1.587 mm] below the top of the piston (rather than the $\frac{3}{8}$ to $\frac{1}{2}$ inch [9.525 to 12.70 mm] in other designs), there is almost no gas between the piston and cylinder wall above the top ring. In other designs, the

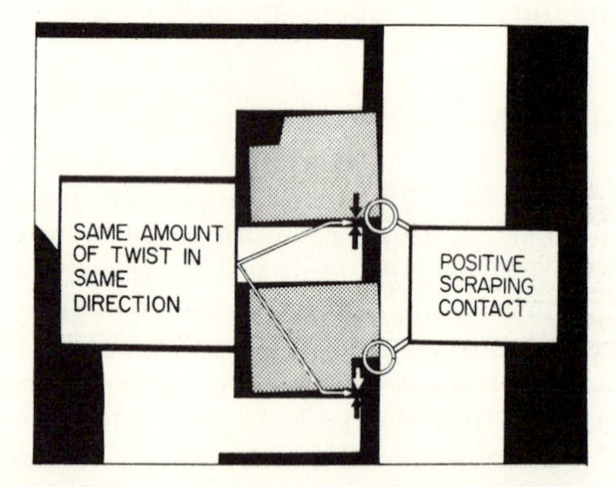

Fig. 8-11. Action of counterbored and scraper compression rings during the intake stroke. Internal forces in the rings tend to twist them so that positive scraping contact is established between the rings and the cylinder wall. This helps to remove any excessive oil that has worked past the oil-control rings. (*Service Parts Division of Dana Corporation*)

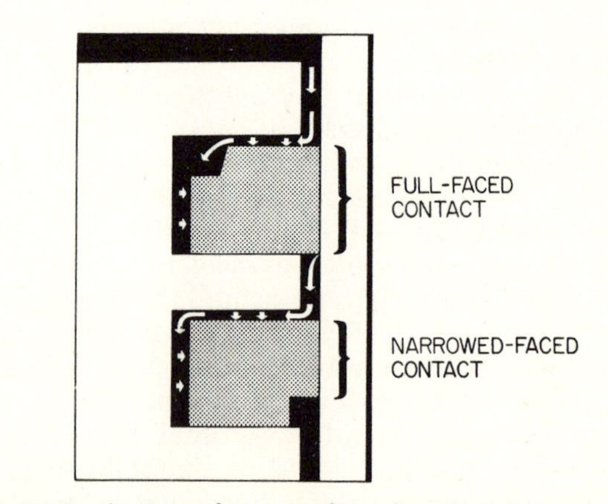

Fig. 8-12. Action of counterbored and scraper compression rings during a power stroke. The combustion pressure presses the rings against the cylinder wall with full-faced contact. This forms a good seal. (*Service Parts Division of Dana Corporation*)

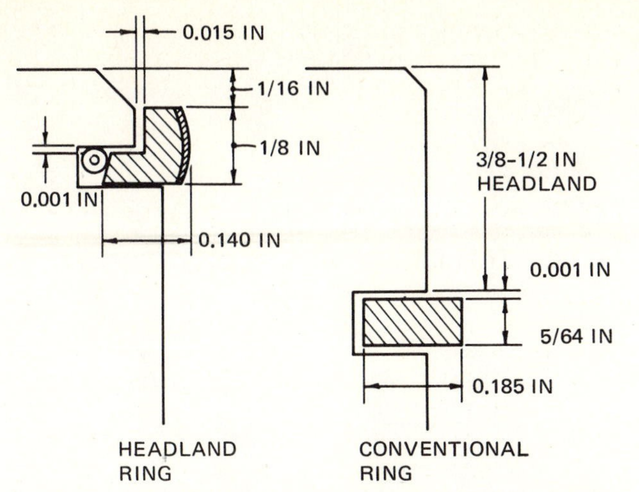

Fig. 8-13. Headland-piston-ring installation compared with that of a conventional ring. Note that the headland ring is located nearer the top of the piston. This arrangement eliminates the space between the piston and cylinder wall found in the conventional design. (*Sealed Power Corporation*)

air-fuel mixture between the piston and cylinder wall has little chance to burn because it is cooled below the combustion point by the adjacent relatively cool metal surfaces. Therefore, this unburned fuel exits with the exhaust gases. This not only reduces engine efficiency, but also contributes to the formation of smog (see Chaps. 18 and 19).

3. RING COATINGS Various coatings have been used on compression rings as an aid to effective *wear-in* and also to prevent rapid wear (thus prolonging ring and cylinder-wall life). These coatings include phosphate, graphite, iron oxide, tin, molybdenum, and chromium. By "wear-in" we mean this: When new, the rings and cylinder wall have certain irregularities and do not fit, or mate, adequately. However, after a time, these irregularities are worn away so that a much better fit, or mating, between the rings and wall comes about. Relatively soft substances such as phosphate, graphite, and iron oxide, which wear rapidly, help this wear-in. In addition, they have oil-absorbing properties which permit them to "soak up" a certain amount of oil for improved lubrication.

Another factor to consider in regard to coating materials for compression rings is the type of wear the rings will encounter. At one time, the most common type of ring wear was caused by abrasives that entered the engine through the air intake. These abrasives, or fine dust particles, would circulate with the engine oil, deposit on the cylinder walls, and cause abrasive wear of the rings. The best substance to resist abrasive wear is chromium because of its hardness. Thus, where abrasive wear is a problem, compression rings are chromium-plated. Today most engines use chromium-plated compression rings. Because chromium is so hard, however, you might be led to believe that a chromium-plated ring would cause rapid cylinder-wall wear. But in the manufacture of these rings, the chromium plate is lapped to a very smooth finish. For example, one manufacturer specifies that surface irregularities must be no greater than 0.0001 inch (one ten-thousandth inch [0.00254 mm]) in the finished ring. With this extreme smoothness, wear-producing high spots are at a minimum. Cylinder wear is therefore low.

In recent years, abrasive wear has become of less importance because of the greatly improved air-filtering systems used at the air intake of carburetors. However, a relatively new type of wear, called *scuff wear,* has gained importance. Modern engines must produce high outputs, and this means higher cylinder-wall, piston, and ring temperatures. These higher temperatures increase the possibility that the rubbing metal surfaces will reach the melting point in spots. Local hot spots may develop that are so hot that the metal will momentarily melt. Small-area welds then occur. These are only momentary welds. Movement of the rings breaks the welds, but scuffed places, or depressions, are left in the rings. Rings and walls are scratched, and failure soon occurs.

To combat scuff wear, high-temperature ring-coating materials are required. Molybdenum is such a material. The iron from which compression rings are made melts at about 2250 degrees Fahrenheit [1233°C]. Chromium melts at about 3450 degrees Fahrenheit [1898°C]. But molybdenum will not melt until it reaches a temperature of about 4800 degrees Fahrenheit [2648°C]. You can see, therefore, that compression rings coated with molybdenum will work at higher temperatures without danger of scuff wear. Thus, where scuff wear is a problem, this type of ring can be used.

NOTE: Recent developments in plasma thermal-application techniques (where metals are turned into gas at high vacuums and temperatures) have made it possible to apply new types of coatings to compression rings. These include carbides of tungsten, titanium, and tantalum, which, even though more expensive, appear to be highly resistant to both abrasive and scuff wear. These coatings will probably begin to appear in high-performance engines in the near future.

⊘ 8-5 Why Two Compression Rings? Two compression rings are used to reduce the pressure drop across each ring. At the start of the power stroke, pressures in the combustion chamber may reach 1,000 psi [70.31 kg/cm^2]. Pressure in the crankcase is about atmospheric. Thus, the pressure differential could be around 1,000 psi [70.31 kg/cm^2]. It would be difficult for a single compression ring to hold this much pressure effectively. However, with the second ring, the pressure is, in effect, divided between the two rings. Not only does this reduce blow-by, or loss of pressure past the upper ring, but it also reduces the load on the upper ring so that it does not press quite as hard on the cylinder wall. Ring friction and cylinder and ring wear are thus reduced.

⊘ 8-6 Oil-control Rings The oil-control rings prevent too much oil from working up into the combustion chamber. As we mentioned in our discussion of engine bearings, the oil throw-off from the bearings lubricates the cylinder walls, rings, and pistons. In addition, some connecting rods have an oil-spit hole which spits oil onto the cylinder wall to provide additional lubrication. Actually, under most circumstances, there is far more oil thrown on the cylinder walls than is needed for lubrication. Most of it must be scraped off and returned to the crankcase. However, this scraped-off oil does serve a purpose. It carries away with it particles of carbon that have formed, as well as dust or dirt particles that have come into the engine with the air-fuel mixture. Most of these particles are then removed from the oil by the oil filter. Smaller particles may continue to circulate in the oil, but generally they are so small that they represent no threat to the engine. Thus, the oil circulates as well as lubricates, and this circulation of the oil helps keep the cylinder walls, pistons, and rings clean.

The oil circulation also has some cooling effect. The oil picks up heat from the cylinder walls and also from the piston and rings as it passes around and through the rings after being scraped from the walls. The oil then drops back into the oil pan, where it is cooled and then recirculated in the engine by the oil pump. Tests have shown that in some engines the circulating oil removes about 20 percent of the excess heat from the engine, with the engine cooling system removing the other 80 percent.

NOTE: In addition to lubricating, cleaning, and cooling, the oil does a fourth job: It helps to create a seal between the rings and the cylinder wall and between the rings and the piston. Because of its ability to cling to metal surfaces, the oil improves the sealing effect of the rings. You can test this effect yourself on an old engine with tapered walls and worn rings. Such an engine does not hold compression and is subject to excessive blow-by. If you remove the spark plugs, pour in a little heavy oil, and then test the compression with a gauge, you will find the compression has improved. The heavy oil temporarily improves the seal between the rings and the cylinder wall and between the rings and the piston. (This is a test that is actually performed on engines that have lost compression to determine whether or not cylinder and ring wear is the cause. It is discussed more fully in ⊘ 20-5.)

⊘ 8-7 Types of Oil-control Rings There are three general types of oil-control rings: the one-piece, slotted, cast-iron type (no longer widely used), the one-piece, pressed-steel type, and the dual-rail and expander type. Figure 8-14 shows several oil-control rings of the one-piece, cast-iron type. Note that all these rings have holes or slots between the upper and lower bearing surfaces. These openings give the oil that is scraped from the cylinder walls somewhere to go. The oil passes through these openings and through holes drilled in the back of the oil-ring grooves in the piston. From there, it returns to the oil pan. Most rings of this type are slotted (or channeled) to provide the most room for the oil and thus permit maximum oil circulation.

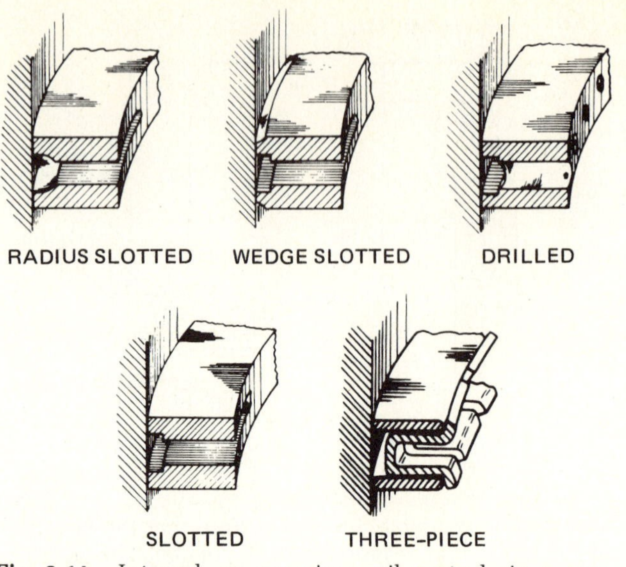

Fig. 8-14. Integral, or one-piece, oil-control rings compared with a three-piece ring. (*Muskegon Piston Ring Company*)

Some rings of the cast-iron, one-piece type are installed with expander rings, either of the hump or of the coil-spring type. The expander spring increases the pressure of the ring on the cylinder wall and thus improves the oil-scraping effect.

In recent years, the one-piece, cast-iron oil-control ring has been replaced by the dual-rail and expander type and the one-piece, pressed-steel type (Figs. 8-15 to 8-19). One reason is that the cast-iron ring lost effectiveness as compression ratios and speeds increased. As you can see in Fig. 8-17, oil can pass around behind the ring when the ring is not resting on the upper side of the groove. Ring inertia, especially at higher speeds, prevents the ring from accelerating as rapidly as the piston so that the oil path above the ring is open longer. Also, the higher compression increases the vacuum in the cylinder on the intake stroke. This causes more oil to flow up past the ring.

The dual-rail and expander type and the one piece, pressed-steel type of oil-control rings (Figs. 8-15 to 8-19) are able to maintain contact with the sides of the ring groove under nearly all conditions. On the dual-rail and expander type, the expander not only forces the two rails outward and into contact with the cylinder wall. It also forces the two rails apart (Fig. 8-18) so that they rest against the two sides of the groove. This closes off the oil path

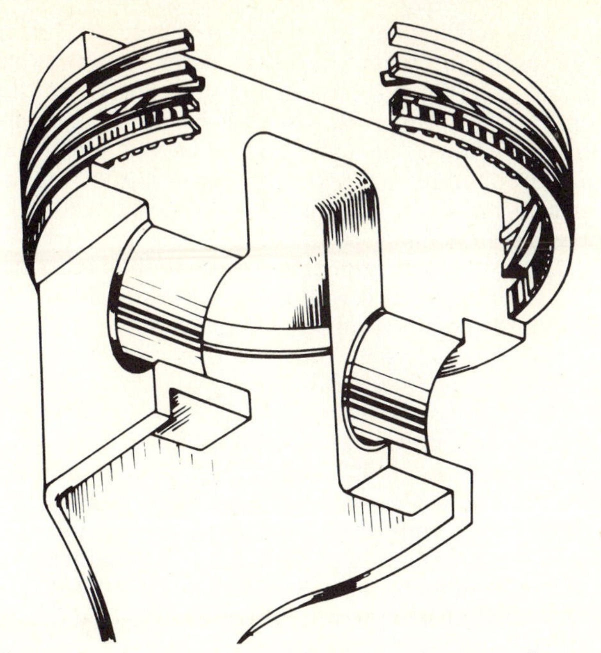

Fig. 8-15. Cutaway view of rings and piston showing construction. The second compression ring has an inner tension ring. The oil ring consists of an expander spacer and two rails. (*TRW Inc.*)

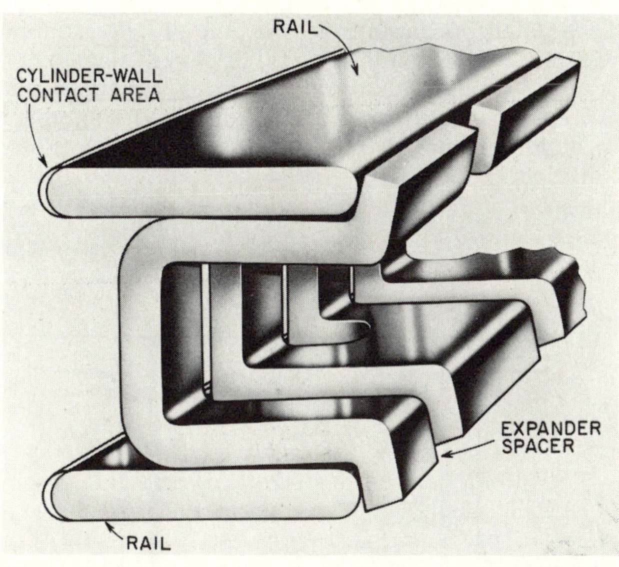

Fig. 8-16. Close-up view of an oil-control ring with two rails and an expander spacer. (*Service Parts Division of Dana Corporation*)

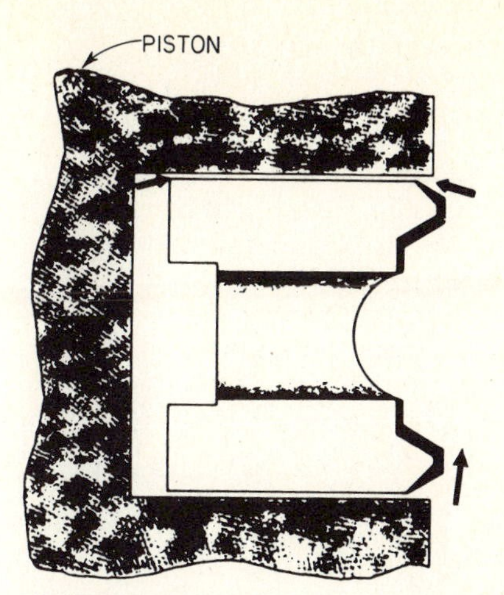

Fig. 8-17. Oil paths (indicated by arrows) around and past a one-piece oil-control ring. (*Service Parts Division of Dana Corporation*)

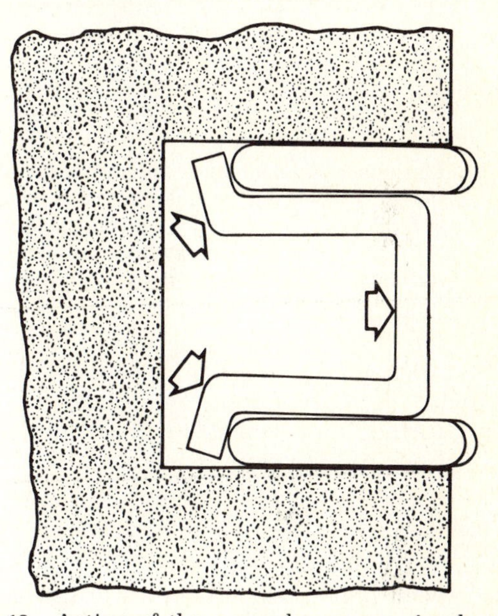

Fig. 8-18. Action of the expander spacer. As shown by the arrows, it forces the rails out against the cylinder wall and up and down against the sides of the ring grooves. (*Service Parts Division of Dana Corporation*)

around the back and up over the ring. The one-piece, pressed-steel ring works in a similar manner.

Oil-control rings operate very efficiently to prevent too much oil from working past the rings and up into the combustion chamber, where it would burn. Consider this fact: If the engine burned as much as one drop of oil on every power stroke, then the engine would burn 1 quart [0.946 l] of oil every 2 miles [3.219 km]. But many engines in good condition will operate thousands of miles [kilometers] without requiring the addition of any great amount of oil. Therefore, it is obvious that the oil-control rings do a very good job.

⊘ 8-8 Why Only One Oil-control Ring? On most early engines using a full-skirt piston, each piston had two compression rings and two oil-control rings. Figures 8-4 and 8-5 show full-skirt pistons of the type using four rings. However, as engine height was reduced by the requirements of the lower hood lines, it was necessary for the piston stroke and height to be reduced. Thus, semislipper and slipper pistons came into use (see Figs. 8-24 and 8-25). These pistons use only three rings and have the skirts cut away so they will not interfere with the counterweights on the crankshaft (see Fig. 8-26). It was possible to

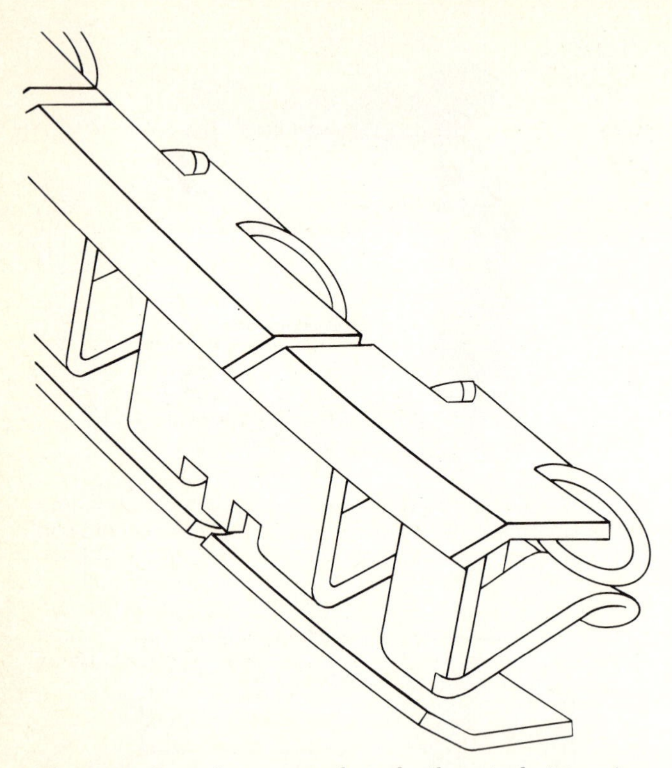

Fig. 8-19. One-piece, pressed-steel, oil-control piston ring. The segmental construction of this ring, with its three-way spring effect, provides pressure against the upper and lower sides of the ring groove, as well as against the cylinder wall. (*Muskegon Piston Ring Company*)

eliminate one oil-control ring because of the improvements in ring and piston design. Someday, perhaps, further improvements will permit the elimination of one of the compression rings, so that pistons with only two rings will come into use.

⊘ 8-9 Effect of Engine Speed on Oil Control As engine speed increases, oil-control rings have a harder time controlling the oil. There are several reasons for this. For one thing, the engine parts and engine oil become hotter at high speeds, which means the oil becomes thinner and so can pass the rings more easily. In addition, at high speeds more oil is pumped, more oil is thrown off the bearings, and more oil gets onto the cylinder walls. All this means that the oil-control rings have a harder job to do. And they have less time to do it. As a consequence, the oil-control rings are less effective at high speeds; more oil works past them and gets into the combustion chamber, where it is burned. This increases oil consumption considerably. An engine uses two or three times as much oil at high speeds as at low speeds. Much, but not all, of this is caused by the reduced effectiveness of the oil-control rings at high speed (see ⊘ 17-6 for further discussion of the effect of speed on oil consumption).

⊘ 8-10 Effect of Engine Wear on Oil Control As engine miles [kilometers] pile up, parts wear. The cylinder walls wear unevenly, or taper, with more wear at the top where the combustion pressures are the greatest (Fig. 8-20). Remember that the higher the combustion pressures, the harder the rings are pressed against the cylinder walls (see Fig. 8-12). Uneven cylinder-wall wear makes it harder for the rings to maintain good contact with the wall. The rings have to change in size, expanding as they move up into the worn area. The result is that the rings do a poorer job of scraping off oil. More oil works up into the combustion chambers where it is burned.

We have all seen old cars that emit clouds of blue smoke as they go down the highway. The engines of those cars probably have worn cylinder walls, which allow oil to get up into the combustion chambers where it is burned. The burning oil gives the exhaust gas a bluish color. Such engines are called *oil pumpers;* they "pump" oil up into the combustion chambers. They are also sometimes called *smoggers* because the exhaust pollutes the air and contributes to the formation of smog (see Chaps. 18 and 19).

There are also other causes for burning oil in the combustion chambers. We shall discuss these when we look at engine troubleshooting and service later in the book.

⊘ 8-11 Oil Consumption The amount of oil that an engine uses depends on engine speed and engine wear, as we have mentioned in previous paragraphs. All engines burn some oil. A new engine operated at moderate speed would probably not require any additional oil between oil changes. But an old engine that is operated at high speed could require 1 quart [0.946 l] of oil every few hundred miles.

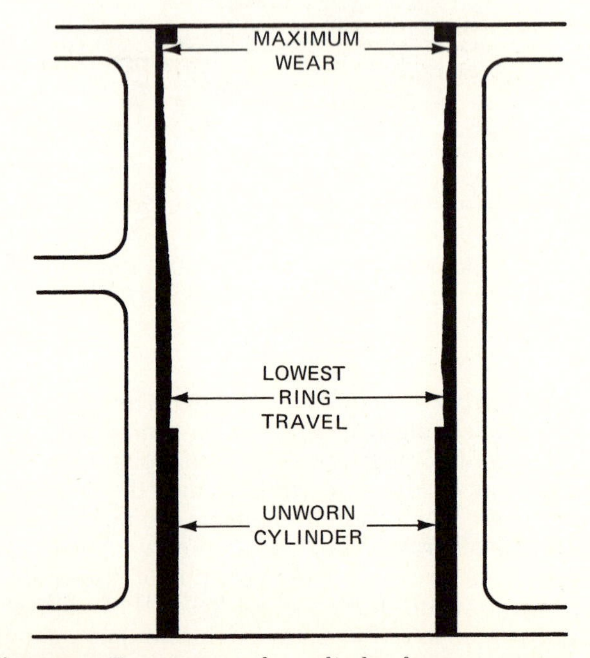

Fig. 8-20. Taper wear of a cylinder from movement of the rings on the wall.

⊘ 8-12 Replacement Rings As cylinder walls wear, power is lost and oil is burned in the combustion chambers. There comes a time when the engine is losing so much power and burning so much oil that repair is required. This means the engine must be torn down so that the cylinder bores can be checked to determine the amount of taper wear. If the taper wear is not too great, satisfactory repair can often be made by installing a set of special replacement rings. Figure 8-21 illustrates such a set of replacement rings installed on a piston which uses four rings. A similar set of rings is shown in disassembled view in Fig. 8-22.

The lower of the two compression rings (2 in Fig. 8-21) has a ring expander inside it. The ring expander is a steel spring in the shape of a wavy or humped ring (Fig. 8-23) that adds tension, or cylinder-wall pressure, to the compression ring. With the ring expander, the compression ring can be made somewhat thinner (from back to front); the ring expander more than makes up for any loss of tension from the reduced thickness. Figure 8-23 shows the shape and location of a ring expander placed inside a compression ring. This combination offers the advantage of high flexibility with high tension (which gives high wall pressure). In a tapered or out-of-round bore, the compression ring must expand and contract—it must change shape—as it moves up and down in the cylinder. This means that the ring must have good flexibility as well as relatively high tension. The combination of ring and ring expander gives the compression ring a better chance to conform to the changing shape of the bore as the ring moves up and down in the cylinder.

Note that the upper oil-control ring (3 in Fig. 8-21) is of the slotted, or channel, type. A ring expander is used under it. The bottom oil-control ring (4 in Fig. 8-21) is made up of three parts: upper and lower rails with an expander spring between them.

Check Your Progress

Progress Quiz 8-1 Here is your chance to check up on yourself once more and find out how well you have been absorbing the material you have been reading. If any of the questions seem hard, just review the past few pages to find the answers.

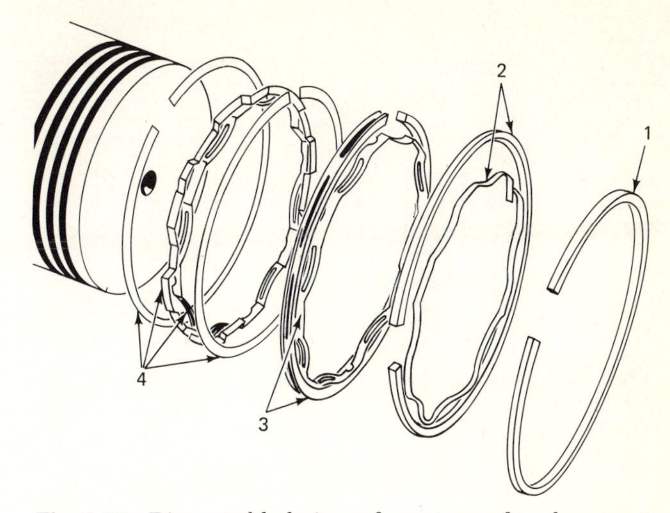

Fig. 8-22. Disassembled view of one type of replacement piston rings: (1) top compression ring; (2) second compression ring, which includes an expander ring; (3) oil-control ring, which includes an expander ring; (4) lower oil-control ring. The last has a three-part construction; it consists of upper and lower rails with an expanding spring. (*Grant Piston Rings*)

Completing the Sentences The sentences that follow are incomplete. After each sentence there are several words or phrases, only one of which will correctly complete the sentence. Write each sentence in your notebook, selecting the proper word or phrases to complete it correctly.

1. In the piston-and-connecting-rod assembly the piston pin may be: (*a*) free or locked to piston or rod, (*b*) free or locked to crankpin or rod, (*c*) free or locked to head or camshaft.
2. The escape of burned gases from the combustion chamber past the pistons and into the crankcase is called: (*a*) gas loss, (*b*) blow-by, (*c*) bypass, (*d*) passed gas.

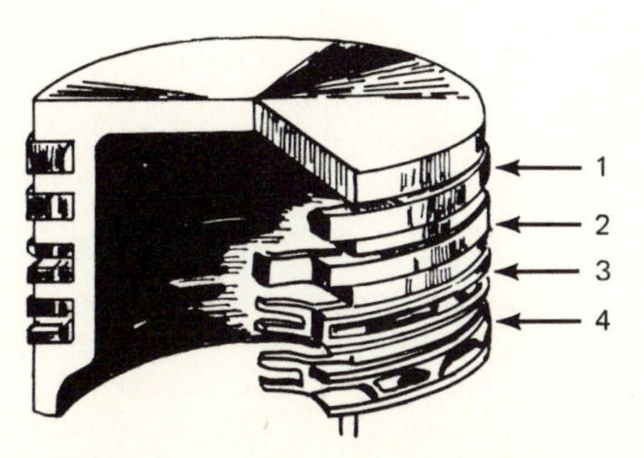

Fig. 8-21. Piston rings designed for installation in a cylinder with a tapered bore. Rings 1 and 2 are compression rings. Rings 3 and 4 are oil-control rings.

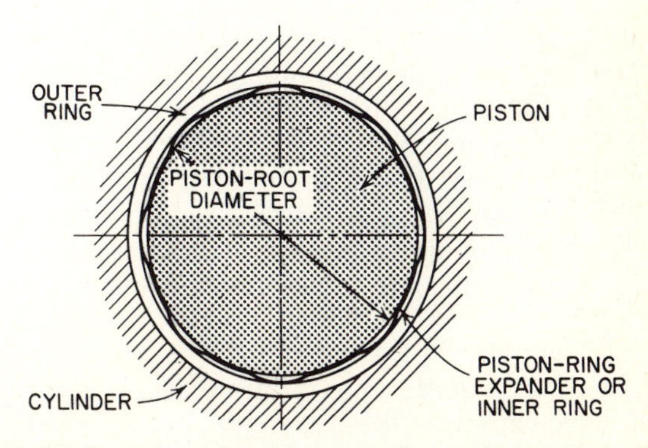

Fig. 8-23. Location of a ring expander under a piston ring (top sectional view). (*Muskegon Piston Ring Company*)

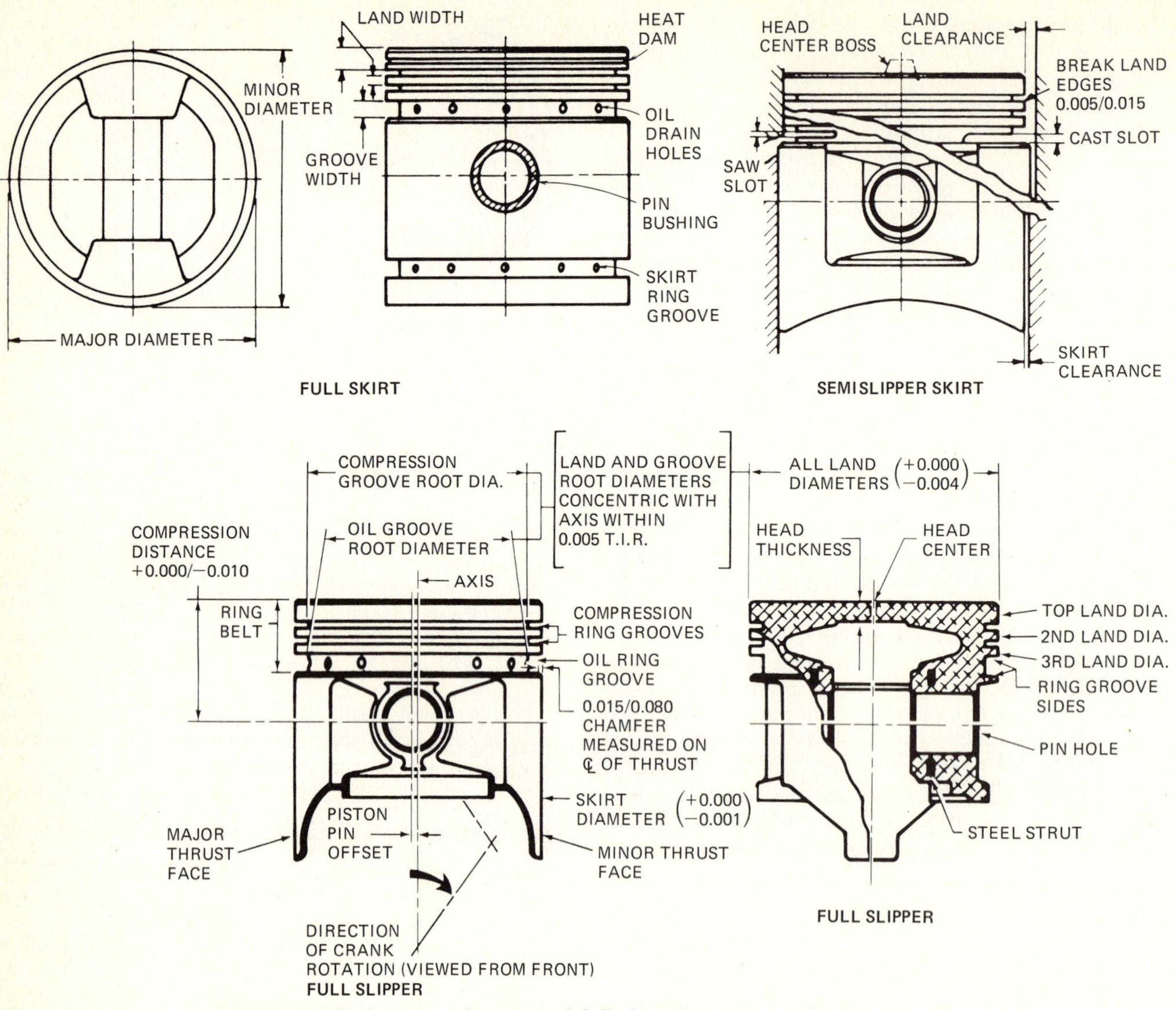

Fig. 8-24. Full-skirt, semislipper, and full-slipper pistons with parts named.

3. Piston rings do three jobs. They: (*a*) cool, seal, and control oil, (*b*) seal, prevent piston slap, and hold compression, (*c*) seal, compress, and control.
4. The beveled, or grooved, compression ring holds its twisted position on all piston strokes except the: (*a*) intake stroke, (*b*) compression stroke, (*c*) power stroke, (*d*) exhaust stroke.
5. The ring expander combined with the compression ring offers the advantage of: (*a*) high flexibility with high tension, (*b*) high rigidity with high tension, (*c*) high oil control with high compression.
6. The purpose of the oil-spit hole in the connecting rod is to: (*a*) spit out excess oil, (*b*) lubricate the main bearings, (*c*) lubricate the cylinder walls.
7. In cross section, the headland ring is: (*a*) square, (*b*) rectangular, (*c*) L-shaped, (*d*) oval.
8. The three general types of oil-control rings are the one-piece cast-iron, the one-piece pressed-steel, and the: (*a*) headland, (*b*) dual-rail and expander, (*c*) counterbored.

⊘ 8-13 Pistons The piston undergoes great stresses during engine operation. With each power stroke, a pressure of several tons is suddenly applied to the pistonhead. At highway speeds, this occurs 30 to 40 times a second. Combustion temperatures of thousands of degrees occur with each power stroke. The piston must be able to take these pressure and heat stresses millions of times during its life without undue wear or other damage.

The piston must be strong and rigid. But it must be as light as possible to reduce inertia loads on the bearings as well as inertia losses. The power used up in stopping and starting the piston at the ends of the piston strokes increases with piston weight.

In modern automotive engines pistons vary in weight from about 10 to about 30 ounces [0.284 to 0.852 kg]. All pistons in an engine must be of the same weight. Otherwise, an unbalanced condition and noticeable vibration would result. To simplify the service job when oversize replacement pistons

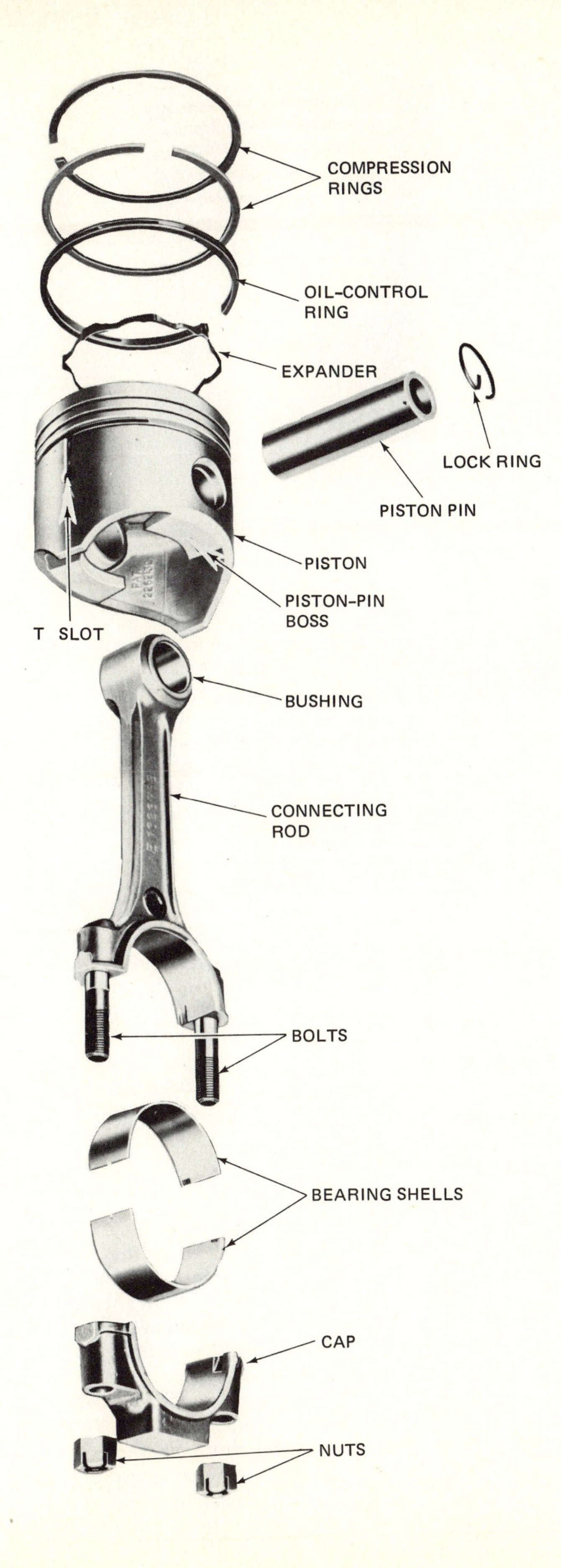

Fig. 8-25. Slipper-type piston and connecting-rod assembly disassembled so that the internal construction of the piston can be seen. Note that the piston skirt is short under the piston-pin bosses. This allows clearance between the piston and crankshaft counterweights when the piston is at BDC. The arrangement permits lower and more compact engine construction. (*Chrysler Corporation*)

are required (for instance, after a cylinder-rebore job), oversize pistons are made to weigh the same as the standard pistons. Thus, only those pistons in cylinders requiring service need to be replaced.

⊘ 8-14 Piston Designs Pistons have been made in a number of sizes and shapes. Pistons for the older-style, long-stroke, small-bore engines were generally of the full-skirt type (Fig. 8-24). Then, as lower hood lines and oversquare engines[1] became popular, the semislipper and full-slipper pistons came into use (Fig. 8-24). On these pistons, the number of piston rings is reduced to three: two compression and one oil-control ring, as already noted (⊘ 8-8). One reason for using the slipper piston is that on the short-stroke oversquare engine, the piston skirt must be cut away to make room for the counterweights on the crankshaft (Figs. 8-25 and 8-26). Also, the slipper piston, being shorter and having its skirt cut away, is lighter. This reduces the inertia loads on the bearings and also makes for a more responsive engine.

Another way to reduce the weight of the piston is to make it of a light metal. Thus, today most automotive-engine pistons are made of aluminum, which is less than half as heavy as iron. (Iron pistons were common in earlier engines.) Aluminum expands more rapidly than iron with increasing tem-

[1]Oversquare engines are engines in which the bore is greater than the stroke, as explained in ⊘ 11-1.

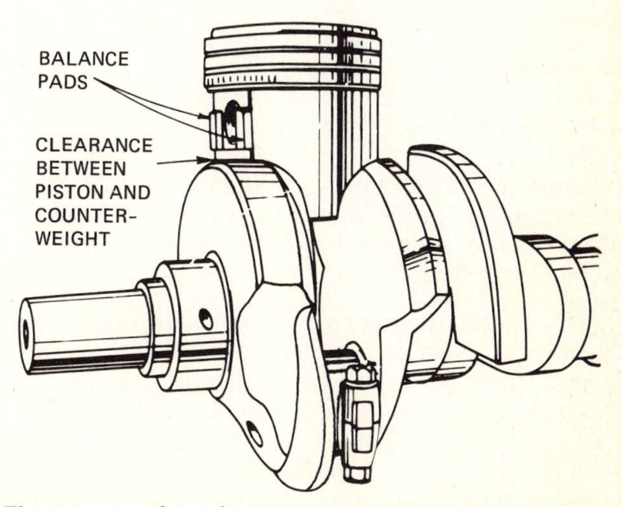

Fig. 8-26. Modern slipper piston and connecting rod assembled to the crankshaft. Note the small amount of clearance between the piston and counterweights on the crankshaft. (*Chevrolet Motor Division of General Motors Corporation*)

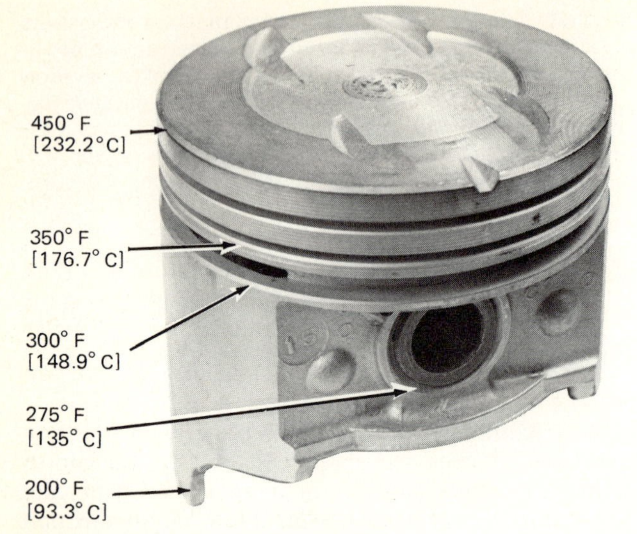

Fig. 8-27. Typical operating temperatures of various parts of a piston. (*Ford Motor Company*)

Fig. 8-29. Full-skirt piston with horizontal and vertical slots cut in the skirt. The horizontal slot reduces the path for heat travel. The vertical slot allows for expansion without an increase in piston diameter.

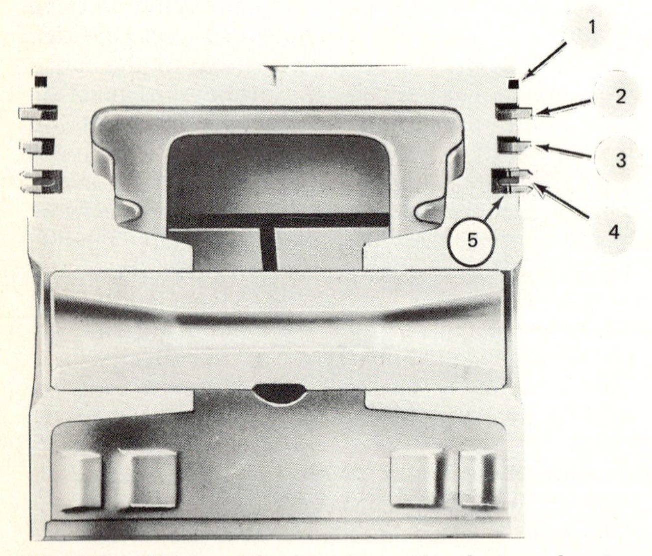

Fig. 8-28. Piston with rings in sectional view showing heat dam and ring shapes: (1) heat dam, (2) upper compression ring, (3) lower compression ring, (4) oil ring, (5) ring expander.

perature, however; and, since the cylinder block is of iron, special provisions must be made to maintain proper piston clearance at operating temperatures.

⊘ 8-15 Piston Clearance *Piston clearance* is the distance between the piston and the cylinder wall. Proper clearance varies with different engines, but it is generally in the neighborhood of 0.001 to 0.004 inch [0.0254 to 0.102 mm]. In operation, this clearance is filled with oil so that the piston and rings move on oil films.

If the clearance is too small, there will be loss of power from excessive friction, excessive wear, and possible seizure of the piston in the cylinder. This last condition would cause complete engine failure. On the other hand, if the clearance is excessive, *piston slap* will result. Piston slap is caused by the sudden tilting of the piston in the cylinder as the piston starts down on the power stroke. The piston shifts from one to the other side of the cylinder with sufficient force to produce a distinct noise. Usually, piston slap is a problem only in older engines with worn cylinder walls and worn or collapsed piston skirts; any of these conditions will produce excessive clearance.

Pistons run many degrees hotter than the adjacent cylinder walls, and therefore expand more than the walls. This expansion must be controlled in order to avoid loss of adequate piston clearance. Such a loss could lead to serious engine trouble. (The problem is more acute with aluminum pistons because aluminum expands more rapidly than iron with increasing temperature.) One way to control this expansion is to reduce the flow of heat from the pistonhead to the piston skirt. Another way to control it is to utilize one of several methods of forcing the piston skirt to change shape slightly as it heats up.

⊘ 8-16 Heat Control in Pistons The head of the piston should run hot, and the skirt comparatively cool. If the pistonhead runs too cool, thermal efficiency will be reduced (because of the heat loss through the piston) and the chilling effect of the cold metal on the combustion process will increase the percentage of unburned gasoline escaping in the exhaust gas. However, if the pistonhead runs too hot, surface ignition could result (⊘ 12-9), with its related engine roughness and short piston life.

The skirt of the piston should run comparatively cool so as to avoid excessive expansion due to high temperatures. This is especially true with the full-skirt piston.

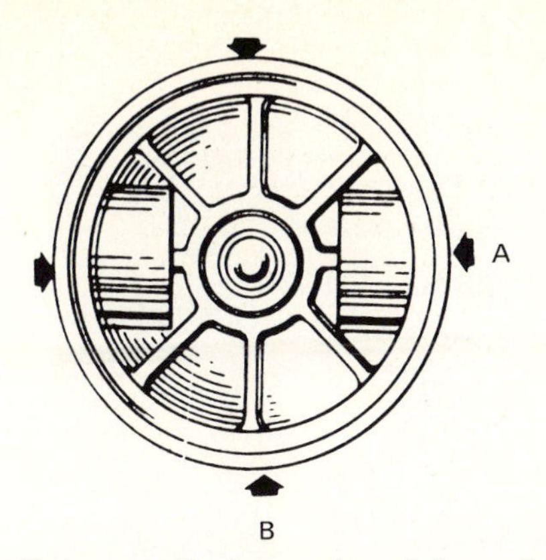

Fig. 8-30. Cam-ground piston viewed from the bottom. When the piston is cold, its diameter at A (the piston-pin holes) may be 0.002 to 0.003 inch [0.051 to 0.076 mm] smaller than at B. (*Chrysler Corporation*)

Figure 8-27 shows the operating-temperature pattern of a full-skirt piston. Various methods of reducing heat flow to the piston skirt have been used. One is to use a heat dam just below the pistonhead. A heat dam is a groove cut circumferentially in the piston between the top ring groove and the pistonhead—almost like an extra ring groove (Fig. 8-28). This groove reduces the heat path from the pistonhead to the skirt. Therefore the skirt runs cooler and does not expand so much.

In the older full-skirt piston, sometimes horizontal slots were cut in the skirt just below the lower oil-control ring groove (Fig. 8-29). These slots reduced the heat path to the skirt. Also, some skirts had a vertical slot (Fig. 8-29) which allowed the metal to expand without causing an appreciable increase in the skirt diameter.

⊘ 8-17 Expansion Control in Pistons Expansion control of piston skirts utilizes several methods of forcing the skirt to change shape slightly as it warms up. One method is to cam-grind the piston. Another method is to cast into the piston a belt, band, or strut. A third method is to use pistons with barrel-shaped and tapered skirts.

1. *CAM-GROUND PISTONS* Cam-grinding of pistons is one method of expansion control. The pistons are called *cam-ground* because they are finish-ground on a machine that uses a cam to move the piston toward and away from the grinding wheel as the piston rotates. This gives the piston an elliptical shape (Figs. 8-30 and 8-31). When a cam-ground piston is cold, it is elliptical, with the long axis (B in Fig. 8-30) perpendicular to the piston-pin bosses. The area of normal clearance of the cold piston is small and is at the thrust faces of the piston (Fig. 8-32). The clearance at other areas on the piston is excessive. This condition changes, however, as the piston warms up.

With increasing heat, the head of the piston expands equally in all directions. The relatively stiff piston-pin bosses are more effective in transmitting the outward forces than the thrust-face parts of the skirt. As the distance across the piston-pin bosses increases, the thrust faces tend to be drawn inward. This is called the *hoop-stretching effect* because it is much like the action of a hoop that is pulled out in one direction. This action causes the diameter of the hoop to be decreased in the direction perpendicular to the pull. The total effect is that the piston assumes a more nearly round shape as it reaches operating temperature, so that the area of normal clearance is increased (Fig. 8-32).

2. *BELTS, BANDS, AND STRUTS* Another method of controlling piston expansion is the use of belts, bands, or struts made of carbon steel, which has an expansion rate nearly equal to that of cast iron. Figure 8-33 shows a belted piston, partly cut away so that the belt can be seen. The belt is a ring, round in cross section, which is cast into the top of the skirt just below the horizontal slot. When the pistonhead expands with temperature increase, the band retards expansion along the thrust-face axis.

Figure 8-34 shows a banded piston, which has a band, stamped from sheet steel and rectangular in cross section, cast into the top of the piston skirt. The band has a relatively large radial depth at the piston-pin-boss locations and a relatively small radial depth at the thrust faces. The band stiffens the boss areas so that they move out as the piston heats, thus producing a modified hoop-stretching effect.

Another device is the strut, a stamped steel plate cast into the piston-pin-boss sides of the skirt, as shown in Fig. 8-35. As temperature increases, the action is similar to that in belted or banded pistons.

3. *BARREL-SHAPED AND TAPERED SKIRTS* Some pistons for high-performance engines have skirts that are barrel-shaped and tapered. The piston shown in Fig. 8-36 is barrel-shaped above the center line of the piston pins and tapered below this line. The combination shape allows the piston to have the close clearance necessary to prevent piston slap and at the same time adequate clearance in the critical expansion areas to prevent excessive wear and possible piston seizure.

⊘ 8-18 Pistonhead Shape The simplest pistonhead is the flat head (Fig. 8-28). However, the demand for higher-compression engines has made it necessary to reduce the clearance volume. A limiting factor here is that the valves must have room to open without striking the pistonhead. A solution is to provide notches in the pistonhead for adequate valve clearance when the piston is at top dead center (TDC). Also, some pistons have a trough, or are dished, to improve the turbulence or swirling of the air-fuel mixture (Fig. 8-37). Such turbulence improves the combustion process.

⊘ 8-19 Piston-Pin Offset In many engines, the piston pin is offset from the center line of the piston toward the major thrust face. This is the face that bears most heavily against the cylinder wall during the power stroke (Fig. 8-38). If the pin is centered, the minor thrust face will remain in contact with the cylinder wall until the end of the compression

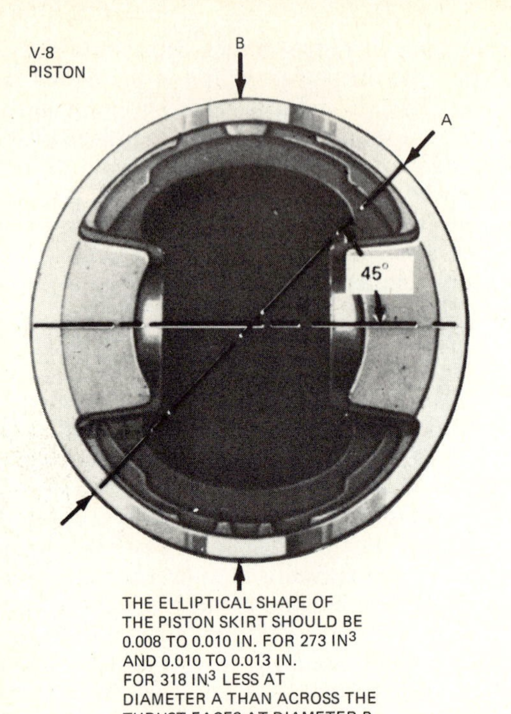

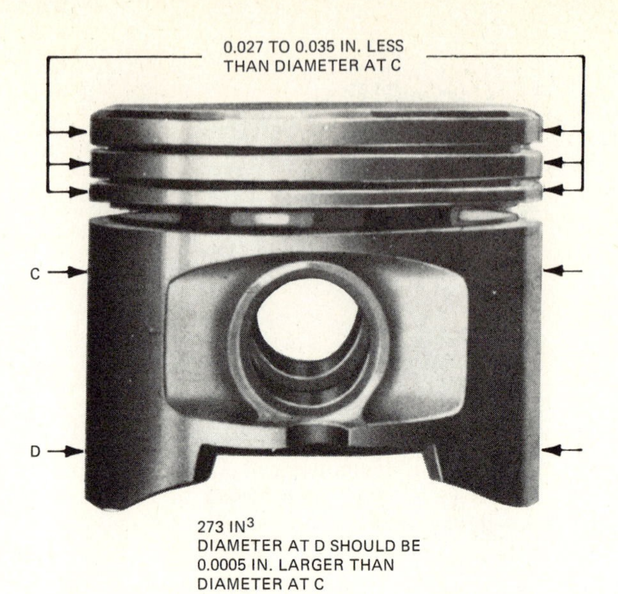

Fig. 8-31. Cam-ground piston. Piston skirt on the 273 cubic inch engine piston is also tapered (diameter at D larger than at C). (*Chrysler Corporation*)

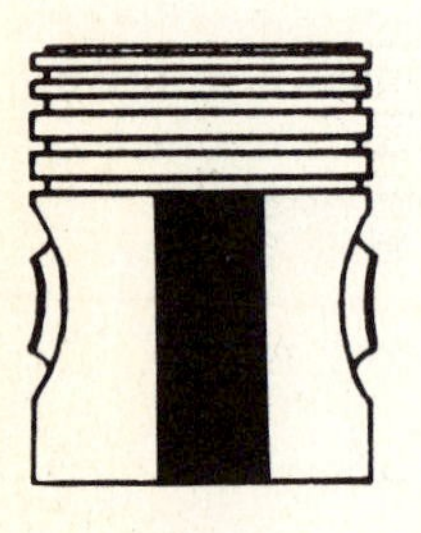

Fig. 8-32. As the cam-ground piston warms up, the expansion of the skirt distorts the piston from an elliptical to a round shape. This increases the area of normal clearance between the piston and the cylinder wall.

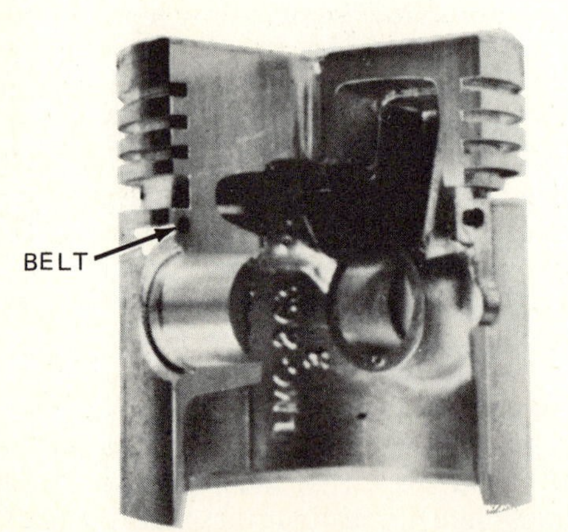

Fig. 8-33. Piston with cast-in belt to provide expansion control. (*TRW Inc.*)

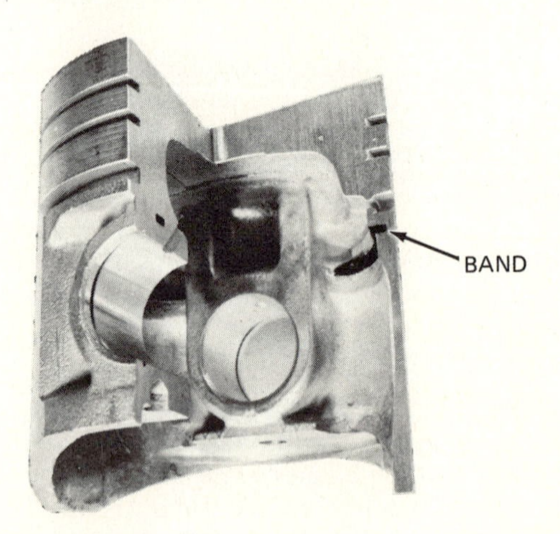

Fig. 8-34. Piston with cast-in band to provide expansion control. (*TRW Inc.*)

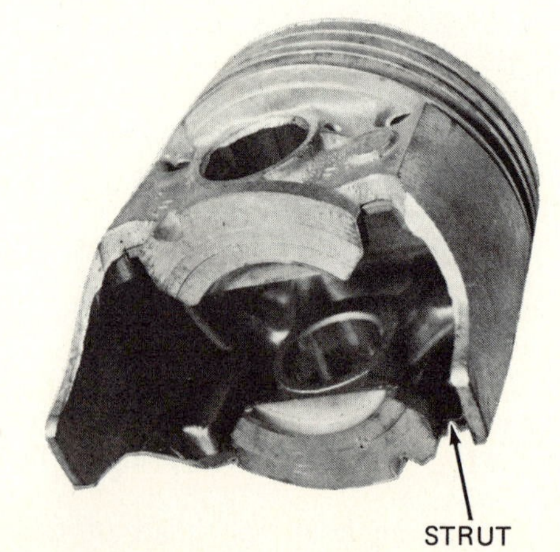

Fig. 8-35. Piston with cast-in strut to provide expansion control. (*TRW Inc.*)

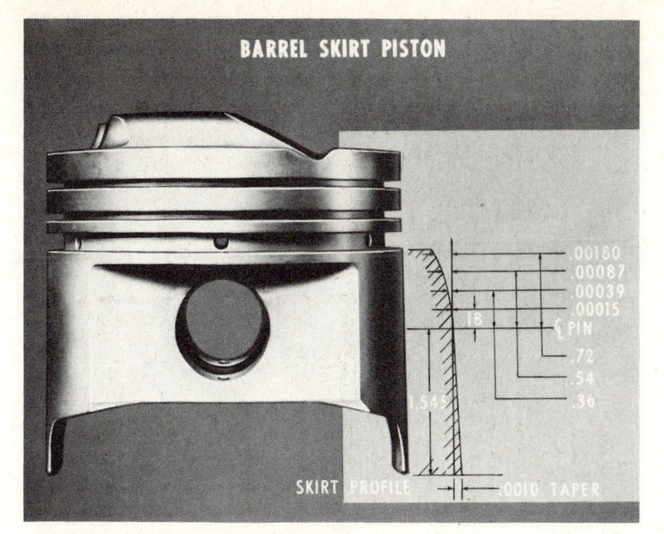

Fig. 8-36. Piston with barrel skirt and taper. (*Chevrolet Motor Division of General Motors Corporation*)

stroke. Then, as the power stroke starts, the rod angle will change from left to right (Fig. 8-38). This causes a sudden shift of the side thrust on the piston from the minor thrust face to the major thrust face. If there is any appreciable clearance, the shift of the piston from one side to the other will produce noticeable piston slap.

However, if the piston pin is offset (Fig. 8-39), the compression pressure will cause the piston to tilt as the piston nears TDC, as shown, so that the lower end of the major thrust face will make contact first with the cylinder wall. Then, after the piston passes TDC and the reversal of side thrust occurs, full major-thrust-face contact will be made with less tendency for piston-slap to occur.

The tilting action occurs because there is more combustion pressure on the right-hand part of the piston [which measures $R + O$ (piston radius plus offset)] than on the left-hand side of the piston (which measures $R - O$).

⊘ 8-20 Ring-Groove Fortification The compression rings are in periodic motion relative to the ring grooves. At the beginning of the power stroke, as the piston goes up over TDC and starts down again, the top compression ring is forced down hard against the lower side of the ring groove. The increasing combustion pressures produce this action. Then, during the intake stroke, the vacuum in the cylinder causes the top compression ring to move up and into contact with the upper side of the ring groove. Thus, the compression ring is repeatedly striking the upper and lower sides of the ring groove.

In some high-performance engines, the impact of the compression ring striking the lower side of the ring groove may approach 2,000 G.[2] The repeated impacts can cause rapid ring-groove wear. It is the top ring groove that suffers the most because it receives the greater part of the combustion pressures. To combat this wear, pistons for some high-performance engines have top-ring-groove fortification. The fortification consists of a ring of cast iron or nickel-iron alloy which is cast into the piston (Fig. 8-40). For cast pistons in medium-duty engines, the inserts are stamped from steel sheets. However, inserts cannot be used in forged pistons. Instead, if ring fortification is required, it is accomplished by spraying the groove area with molten metal having the proper wear resistance. Then the groove is machined to the proper dimensions.

⊘ 8-21 High-Performance Pistons Aluminum pistons can be either cast or forged. Cast pistons are made by pouring molten aluminum into molds. Forged pistons are made from slugs of aluminum, which, when subjected to the high forging pressures, flow, or extrude, into dies to form the pistons. Both types of pistons must be heat-treated.

[2] One G is the force acting on a body due to the gravitational pull of the earth. This force, acting on a falling body, increases its velocity by 32.16 feet per second [9.760 m/s] each second it falls.

Fig. 8-37. Pistons for modern internal-combustion engines.

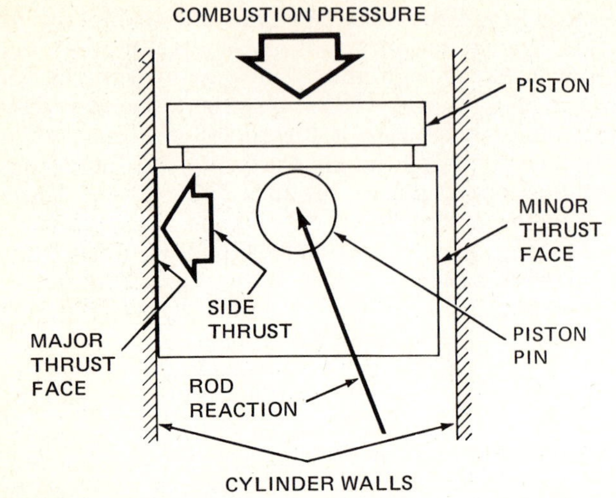

COMBUSTION PRESSURE
PISTON
MINOR THRUST FACE
SIDE THRUST
PISTON PIN
MAJOR THRUST FACE
ROD REACTION
CYLINDER WALLS

Fig. 8-38. As combustion pressure is applied to the pistonhead and the connecting-rod angle changes from left to right, side thrust on the piston causes it to shift abruptly toward the major thrust face.

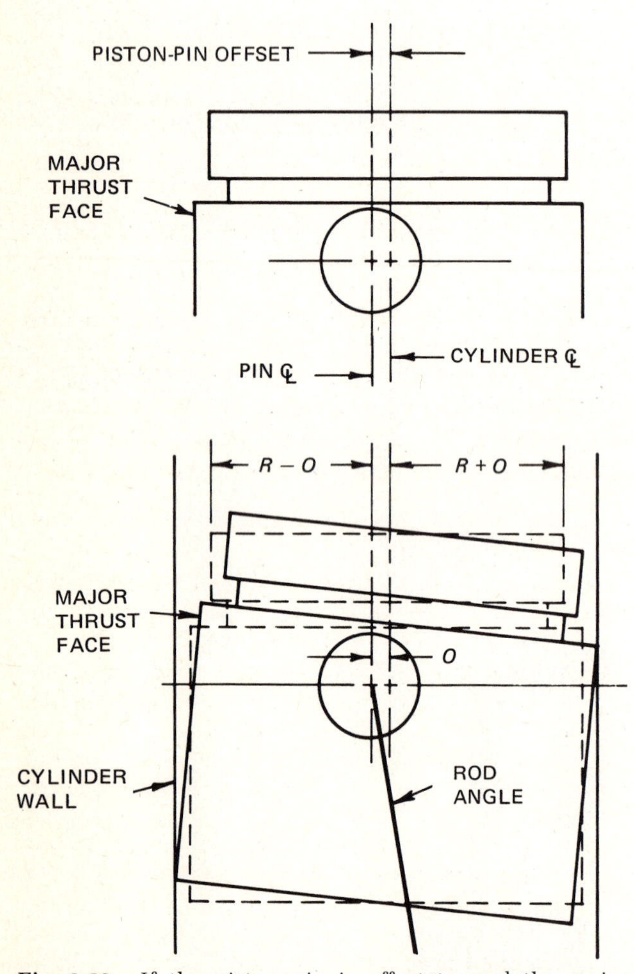

Fig. 8-39. If the piston pin is offset toward the major thrust face, combustion pressure will cause the piston to tilt to the right, as shown, to reduce piston slap. *R* is the radius of the piston; *O* is the offset of the piston pin. (*Bohn Aluminum and Brass Company, Division of Universal American Corporation*)

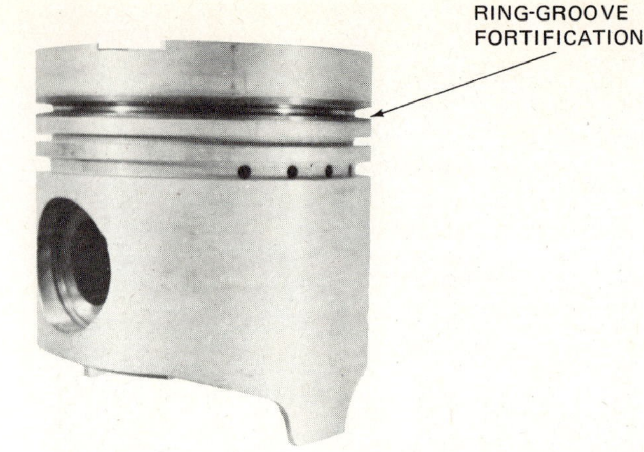

Fig. 8-40. Piston with top-ring-groove cast-in fortification. (*TRW Inc.*)

The forged piston is denser than the cast piston and forms a better heat path to allow the heat to move away from the pistonhead, as shown in Fig. 8-41. The flags indicate the temperatures at comparable points in the two pistons with both operating under the same conditions.

The forged piston also has a grain flow that improves its wearing ability (Fig. 8-42). Notice that the flow is vertical in the skirt, that is, in the direction that the piston moves in the cylinder.

The forged piston is also lighter than the cast piston, even though it is just as strong. Thus, it has lower inertial forces, which results in less bearing wear and a more responsive engine. Taking all these factors together, it can be seen that the forged piston, even though it is more expensive, is the preferred piston for high-performance engines.

Certain high-performance pistons also have special skirt configurations for added strength. The oval skirt (Fig. 8-43) and the undulated skirt (Fig. 8-44) are designed to provide high strength. They are designed for use in high-performance automobiles but are strong enough for competition engines. The outboard piston-pin-boss piston (Fig. 8-45) is designed for maximum strength and for use in competition engines. Note how the piston-pin bosses are located outside the walls of the piston and also that the thrust faces of this piston are relatively small.

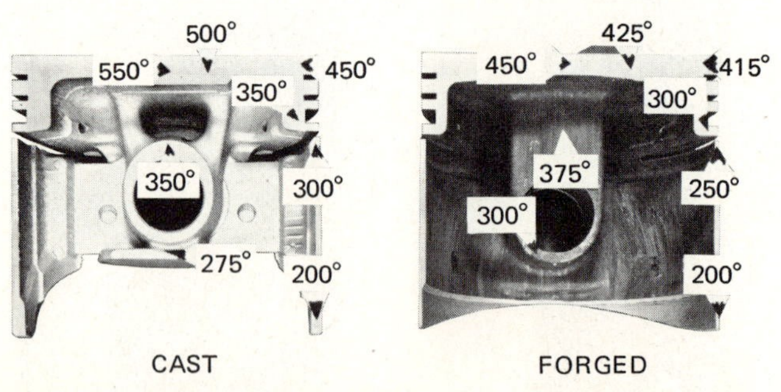

Fig. 8-41. Cast and forged pistons, cut in half to show the operating temperatures (in degrees Fahrenheit) at comparable points in the two pistons. (*TRW Inc.*)

Fig. 8-42. Forged aluminum piston, cut in half and etched to bring out the grain flow. The etched lines show the directions in which the metal flowed during the forging process. (*TRW Inc.*)

Fig. 8-43. Piston with oval skirt. (*TRW Inc.*)

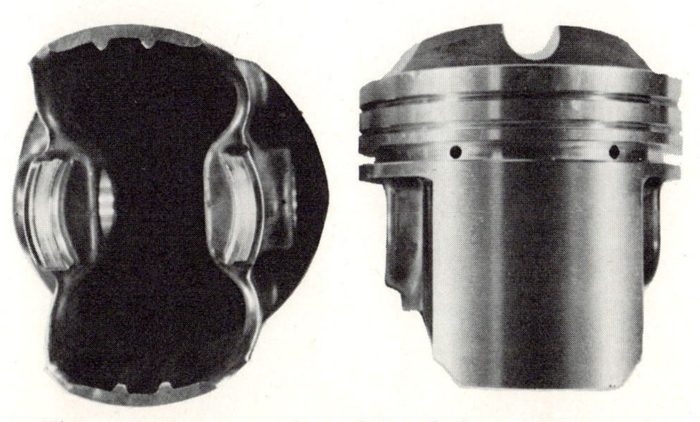

Fig. 8-44. Piston with undulated skirt. (*TRW Inc.*)

Fig. 8-45. Piston with outboard piston-pin bosses. (*TRW Inc.*)

Check Your Progress

Progress Quiz 8-2 Once again, here is a chance to check up on yourself and find out how well you are remembering what you have been studying in this chapter. If any of the questions give you trouble, simply reread the preceding pages and try again.

Completing the Sentences The sentences that follow are incomplete. After each sentence there are several words or phrases, only one of which will correctly complete the sentence. Write each sentence in your notebook, selecting the proper word or phrase to complete it correctly.

1. Piston slap results from excessive clearance between the: (*a*) piston and piston pin, (*b*) piston and piston rod, (*c*) piston and rings, (*d*) piston and cylinder wall.
2. If you compared the oversize replacement pistons with the standard pistons, you would find that the oversize pistons: (*a*) weigh less, (*b*) weigh the same, (*c*) weigh more.
3. Oversquare engines use pistons of the: (*a*) semi- or full-slipper type, (*b*) full-skirt type, (*c*) counterweight type.
4. Aluminum pistons are either cast or: (*a*) ground, (*b*) forged, (*c*) pressed.
5. Of the two types of pistons mentioned in the previous question, the lighter one is the: (*a*) cast type, (*b*) cam-ground type, (*c*) forged type.
6. Piston clearance is the distance between the piston and the: (*a*) cylinder head, (*b*) counterweights, (*c*) cylinder wall, (*d*) connecting rod.
7. In an operating engine, the hottest part of the piston is the: (*a*) skirt, (*b*) ring grooves, (*c*) pin bosses, (*d*) head.
8. Pistons which are slightly elliptical when cold but round when at operating temperature are called: (*a*) cam-ground pistons, (*b*) slipper pistons, (*c*) round-ground pistons.
9. One method of controlling piston expansion is to cast into the piston: (*a*) expanders or bars; (*b*) springs or plates; (*c*) struts, bands, or belts.
10. The ring groove most in need of fortification is the: (*a*) oil-ring groove, (*b*) top compression-ring groove, (*c*) lower compression-ring groove.

CHAPTER 8 CHECKUP

NOTE: Since the following is a chapter review test, you should review the chapter before taking the test.

You have now finished two of the three chapters on engine construction. After you have finished the next chapter, on valves and valve trains, you will be ready to study the chapters on the fuel, lubricating, and cooling systems. Then you will study the trouble diagnosis and repair of engines. You are making good progress. Now check up on yourself and find out how well you are remembering what you have been studying.

Completing the Sentences The sentences that follow are incomplete. After each sentence there are several words or phrases, only one of which will correctly complete the sentence. Write each sentence in your notebook, selecting the proper word or phrase to complete it correctly.

1. Three methods of installing the piston pin are: (*a*) pin locked to rod, locked to crankpin, or free; (*b*) pin locked to rod, locked to piston, or free; (*c*) pin locked to journal, locked to piston, or free.
2. Three jobs the piston rings do are: (*a*) seal, attach piston pin, control oil; (*b*) hold intake, compression, seal; (*c*) seal, control oil, cool.
3. With reference to the number of piston rings, automotive engines usually use: (*a*) two or three rings, (*b*) three or four rings, (*c*) four or five rings.
4. The oil that circulates on the cylinder walls and pistons and through the oil-control rings can be said to do four jobs. It: (*a*) lubricates, seals, cleans, and cools; (*b*) lubricates, cools, oils, and cleans; (*c*) lubricates, circulates, oils, and cleans.
5. Under normal operating conditions, the difference in the temperature between the pistonhead and lower part of the piston skirt may be between: (*a*) 20 and 30°F [−6.7 and 1.1°C], (*b*) 200 and 300°F [93.8 and 146.7°C], (*c*) 2000 and 3000°F [1093.3 and 1648.9°C].
6. Clearance between the cylinder wall and piston of a new engine is somewhere around: (*a*) 0.0001 to 0.0004 in [0.00254 to 0.0102 mm], (*b*) 0.001 to 0.004 in [0.0254 to 0.102 mm], (*c*) 0.01 to 0.04 in [0.254 to 1.02 mm].
7. As the cam-ground piston warms up, the hoop-stretching effect causes it to: (*a*) elongate, (*b*) shift the pin bosses inward, (*c*) assume a more nearly round shape.
8. The purpose of offsetting the piston pin is to: (*a*) prevent piston slap, (*b*) improve piston clearance, (*c*) improve piston-pin lubrication.

Definitions and Lists In the following you are asked to write the function of or describe certain engine parts, or make lists. Write these in your notebook. The act of writing does two things: It tests your knowledge, and it also helps to fix the information in your mind. Turn back into the chapter if you are not sure of an answer, and reread the pages that will give you the information you need.

1. List the various parts that make up the connecting-rod assembly.
2. List and describe the three general types of oil-control rings.
3. Explain the four jobs that the lubricating oil does on the cylinder walls, rings, and pistons.
4. Explain the three functions of the piston rings.
5. What is the hoop-stretching effect, and how does it work?
6. Explain the advantages of the headland ring.
7. Describe the various methods of lubricating the piston-pin bushings.
8. Why are two compression rings used?
9. Why is only one oil-control ring used on many late-model engines?
10. What is blow-by?
11. Explain how belts, bands, or struts cast into the piston help control the expansion of the piston.
12. What is meant by heat control in pistons?
13. What are the advantages of forged aluminum pistons?

SUGGESTIONS FOR FURTHER STUDY

To study pistons, piston rings, and connecting rods further, go to a friendly service shop where engine repair work is done, so that you can see the various types of pistons and rings. Handle these parts if you can. You will be able to handle and examine these parts in your own school automotive shop, of course. Notice particularly the various kinds of defects in the parts, and try to find out what caused these defects. Study the manufacturers' shop manuals to learn more about the various engine parts. Be sure to write the important facts in your notebook.

chapter 9

PISTON-ENGINE CONSTRUCTION: VALVES AND VALVE TRAINS

Chapters 7 and 8 described cylinder blocks, heads, crankshafts, bearings, pistons, and piston rings. This chapter concludes the description of piston-engine construction by discussing valves and valve trains. The information in Chaps. 7 to 9 on piston-engine construction will enable you to move ahead with confidence into the trouble-diagnosis and servicing sections in the latter part of the book.

⊘ 9-1 Valves and Valve Trains The *valve train* is made up of the different parts in the engine that carry motion from the crankshaft to the valves and cause them to open and close. The three basic arrangements—L head, I head, and overhead camshaft (OHC)—are discussed later. (Probably the only places you will see L-head engines are in antique cars and small machines such as power mowers.)

Regardless of design, the valves must open and close at the proper times as the engine runs in order to admit fresh charges of air-fuel mixture into the cylinders and allow the burned mixture to exit from the cylinders. A valve opens and closes as many as 2,000 times a mile [1.609 km]. This means that it has opened and closed up to 200 million times in an engine that has gone 100,000 miles [160,900 km]. And with each cycle of opening and closing, the exhaust valve takes temperatures of up to 1400 degrees Fahrenheit [761°C], seating and sealing every time it closes despite the high temperatures and pressures.

⊘ 9-2 Cams and Camshafts A *cam* is a device that can change rotary motion into linear, or straight-line, motion. The cam has one or more high spots, or lobes; a follower riding on the cam will move away from or toward the camshaft as the cam rotates (see Fig. 9-1).

In the engine, cams are used to control the opening and closing of the intake and exhaust valves. Figure 9-2 shows the valve mechanisms, called the *valve train*, in an I-head engine. The cams are formed as integral sections of the camshaft. Each cam has one lobe. There are two cams for each cylinder, one each for the intake and exhaust valves. Thus, a six-cylinder engine would have 12 cams, and an eight-cylinder engine 16 cams (Fig. 9-3). In addition, the camshaft has another cam (or an eccentric) to operate the fuel pump and also a gear designed to drive the ignition distributor and the oil pump (see Fig. 9-4).

The camshaft is driven from the crankshaft by sprockets and a chain or toothed belt (Figs. 6-13, 9-5, and 9-6) or by two gears (Fig. 9-7). The camshaft sprocket or gear is twice as large as the crankshaft sprocket or gear. This gives a gear ratio of 1:2. The camshaft turns at half the speed of the crankshaft. Thus, every two revolutions of the crankshaft produces one camshaft revolution and one opening and closing of the two valves in each cylinder. In other words, the intake and exhaust valves open and close once for every two crankshaft revolutions. The piston makes four strokes for each two crankshaft revolutions. During one of these four strokes, the intake stroke, the intake valve is open. During another of these four strokes, the exhaust stroke, the exhaust valve is open. Thus, you can see why the 1:2 gear ratio between the camshaft and the crankshaft is

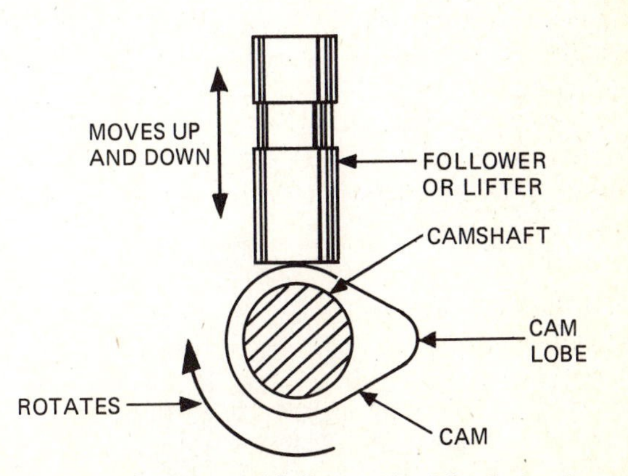

Fig. 9-1. Simple cam and follower (or lifter). As the cam revolves, the follower follows the cam surface by moving up and down.

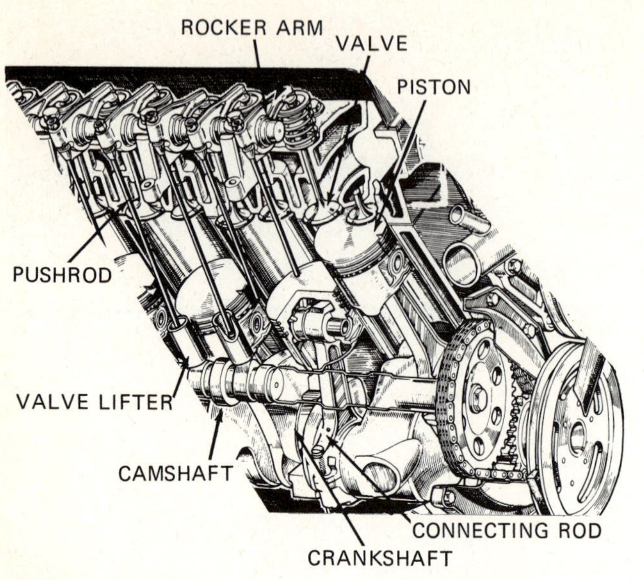

Fig. 9-2. Valve-train parts in an engine. (*Chrysler Corporation*)

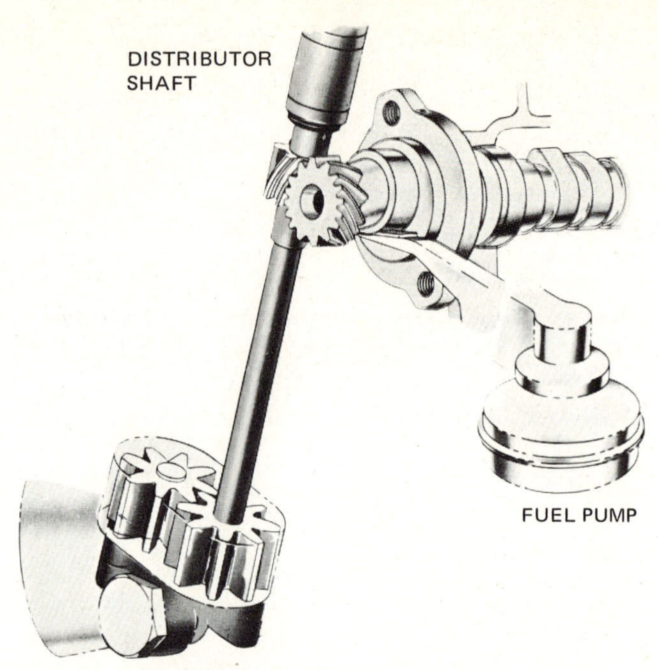

Fig. 9-4. Oil-pump, distributor, and fuel-pump drives. The gear on the camshaft drives the oil pump and distributor. The eccentric on the camshaft drives the fuel pump. (*Buick Motor Division of General Motors Corporation*)

necessary. To summarize the actions, the valves are opened by the positive mechanical action of the cam lobes. The valves are closed by means of valve-spring pressure.

The gears are called *timing gears*. The chain is called the *timing chain*. The toothed belt is called the *timing belt*. The reason for this is that the gears, chain or toothed belt, and sprockets "time" the valve action. One difference between driving the camshaft by a chain or toothed belt and sprockets and driving it by gears is that when a chain or toothed belt is used, the camshaft and crankshaft both turn in the same direction. When gears are used, the camshaft rotates in the opposite direction from the crankshaft.

In in-line engines the camshaft is mounted in bearings in the lower part of the cylinder block (Figs. 6-4 and 6-12). In V-type engines, the camshaft is located directly above the crankshaft and between the two banks of cylinders (Figs. 6-16 and 6-19). In engines with overhead camshafts, the camshaft is mounted in bearings on top of the cylinder head, as explained in ⊘ 9-13.

⊘ 9-3 Valves Each cylinder has two valves: an intake valve and an exhaust valve. Some high-performance engines have four valves per cylinder—two intake valves and two exhaust valves (⊘ 9-13). The intake valve is usually larger than the exhaust

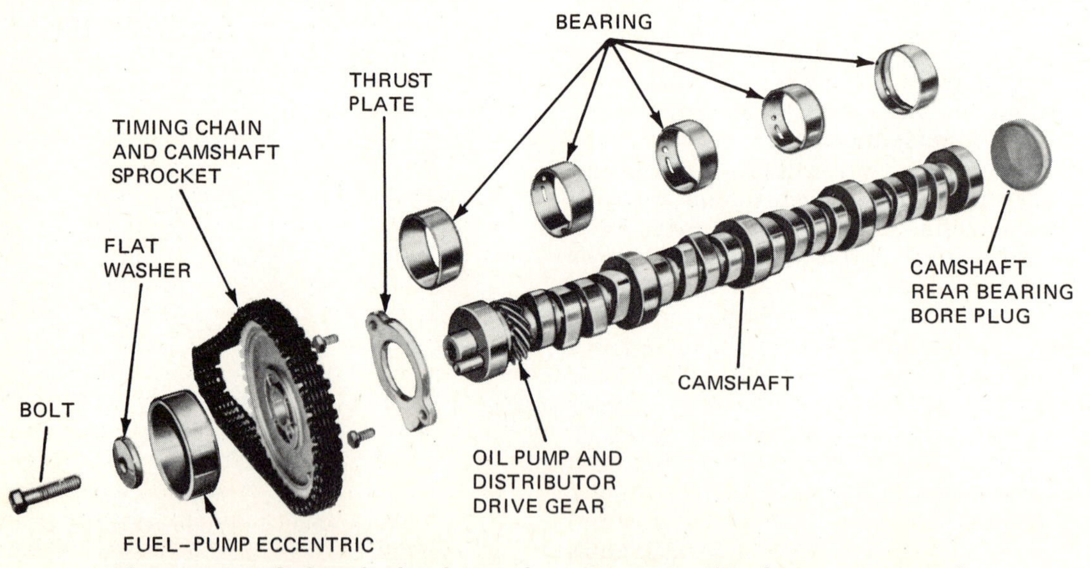

Fig. 9-3. Camshaft and related parts for a V-8 engine. (*Ford Motor Company*)

Fig. 9-5. Crankshaft and camshaft sprockets with chain drive for a six-cylinder engine, showing timing marks on sprockets. Note that the larger of the two sprockets is on the camshaft so that it turns at one-half crankshaft speed. (*Chrysler Corporation*)

Fig. 9-6. Crankshaft and camshaft sprockets with chain drive for a V-8 engine. (*Chrysler Corporation*)

valve for the following reason: When the intake valve is opened, the only driving force moving air-fuel mixture into the cylinder is atmospheric pressure. When the exhaust valve is opened, the piston is moving up and there is a high pressure driving the exhaust gases out. Thus, the intake port (and thus the intake valve) must be larger to allow enough air-fuel mixture to enter.

Various types of valves have been used in the past, including the sliding-sleeve and rotary types. But the valve that is in general use today is the mushroom, or poppet, valve (Figs. 9-8 and 9-9). When the poppet valve is closed, it is held on the valve seat by the valve spring. The fit between the valve and seat is tight, and so a good seal is formed.

Modern engines using low- or no-lead gasoline work the valves harder. The reason for removing the lead from gasoline is that emission controls can get

Fig. 9-7. Crankshaft and camshaft gears for a six-cylinder engine. Note the timing marks on the gears. (*Buick Motor Division of General Motors Corporation*)

fouled up from lead in gasoline. (We shall discuss this in detail in later chapters.) Meantime, the fact that there is little or no lead in gasoline creates a wear problem for valve faces and valve seats. Lead in gasoline forms a fine coating on the valve faces and seats. This coating acts as a lubricant. Without the lead coating, the valve faces and seats lack lubricant and can wear rather rapidly. For this reason, in many engines the valve faces have special coatings which reduce wear significantly. For severe service, the valve faces may be made of stellite, a very hard metal (Fig. 9-19). In addition, many valves and valve seats are ground to give what is called an *interference angle* to ensure improved seating and sealing. This is described in ⊘ 9-8.

⊘ 9-4 Valve-Head Shape The shape of the *valve head* is of great importance. Some engineers believe that the head should be rigid and unyielding. Others believe that the head should be flexible (Fig. 9-10). The more flexible the head, the better it will conform to seat irregularities. If the valve face and valve seat do not mate perfectly, there will be leakage between the two when the valve is closed. Leakage can become a very serious problem, especially with exhaust valves. If there is leakage past the exhaust valve, the hot combustion gases leaking through will soon burn both the valve face and the valve seat. Burned valves must be replaced, and burned valve seats must be refinished. Thus, a flexible valve head, in conforming to any seat irregularities, is less likely to burn as a result of leakage. However, a flexible head, having less metal to carry away heat, will run hotter. Figure 9-11 gives the head and upper-stem temperatures for the standard (B) and the elastic (D) valve heads shown in Fig. 9-10. The higher temperatures are undesirable because they reduce the

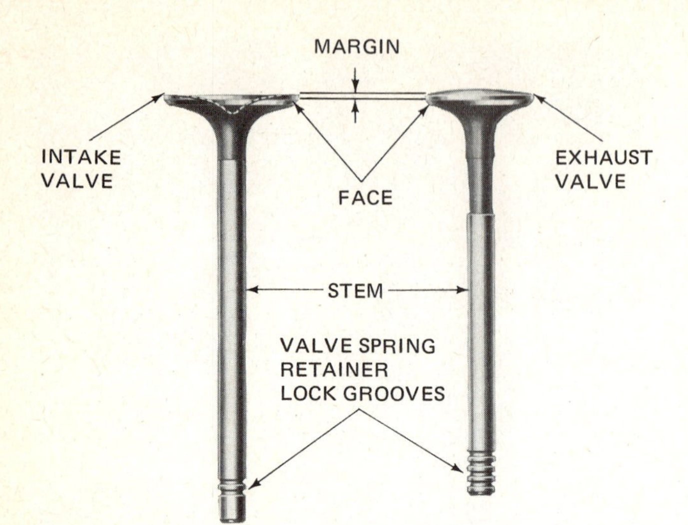

Fig. 9-8. Typical poppet valves. (*Chrysler Corporation*)

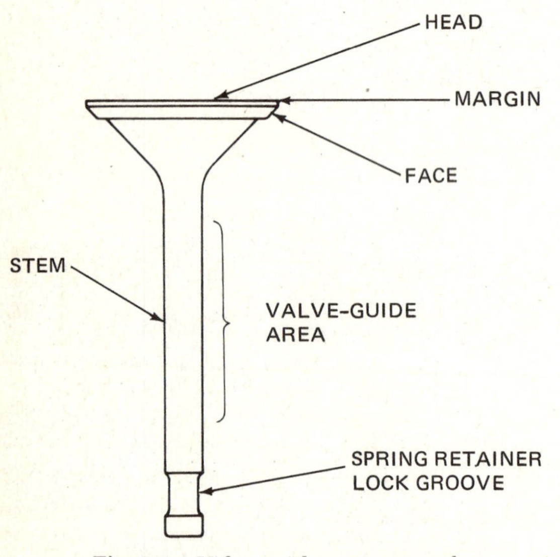

Fig. 9-9. Valve with parts named.

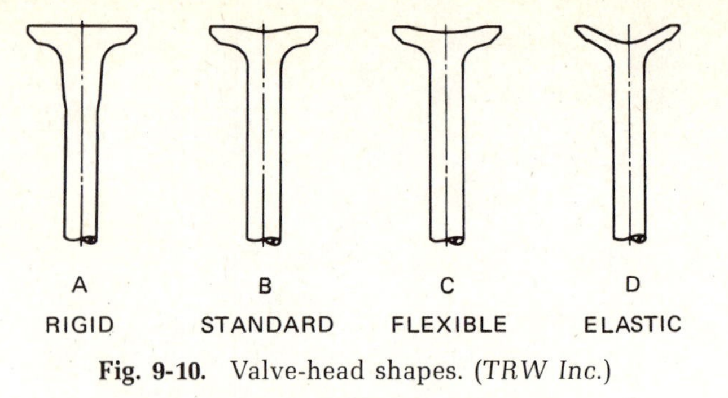

Fig. 9-10. Valve-head shapes. (*TRW Inc.*)

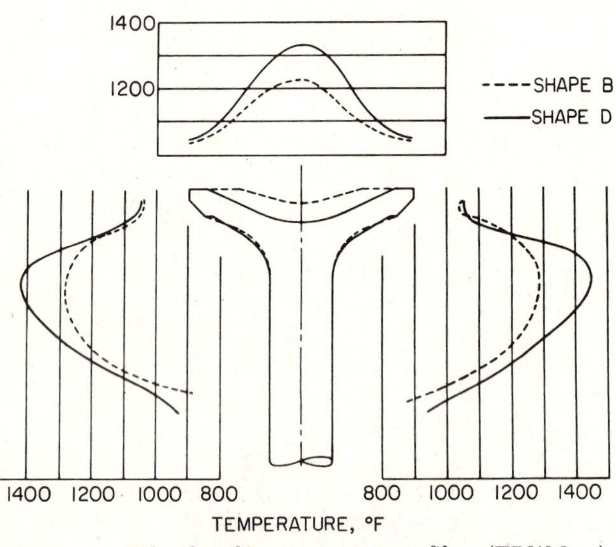

Fig. 9-11. Valve-head temperature profiles. (*TRW Inc.*)

strength of the metal and can lead to valve distortion and burning or breakage.

The shape of the valve under the head is also important, especially with the intake valve. As the air-fuel mixture approaches the opened intake valve, the curved area influences the motion of the mixture. The curves are designed to fit the curves of the intake manifold and valve port so that the mixture will flow smoothly into the cylinder. The wrong valve-head shape can produce turbulence in the mixture and thus lowered volumetric efficiency (that is, less mixture enters the cylinder).

⊘ 9-5 High-Performance Valves and Valve Cooling

When temperature problems exist because of high-performance demands, exhaust valves are given facings of special metals. The facings are applied as rings which are welded to the valve head and then finished to the proper angles. Tungsten-cobalt is one metal combination that is used. These metals can withstand the higher temperatures successfully. Note that only the valve face is made of these metals. They are expensive, and, in addition, they may not have the correct properties to give normal wear in the valve-guide and valve-tip area.

1. *HIGH-PERFORMANCE VALVES* High-performance valves may also have stems and tips of special alloys that are welded on. These special alloys are selected to withstand the stem wear in the valve-guide area and the scrubbing wear on the valve tip that is caused by the motion of the rocker-arm tip as it slides across the valve tip.

Some valves have hollow stems (Fig. 9-12). This makes a lighter-weight valve that has less inertia. It therefore follows the valve-train action more quickly, which makes for a more responsive engine.

2. *VALVE COOLING* The intake and exhaust valves are essentially similar. However, the intake valve is required to pass only the relatively cool air-fuel mixture, while the exhaust valve must pass the very hot burned gases. The exhaust valve may actually become red hot. Temperatures above 1200 degrees Fahrenheit [650°C] are common. Figure 9-11 shows a typical temperature pattern of exhaust valves. Note that the area near the valve face is somewhat cooler than the area that is closer to the center. The valve stem is also relatively cool. The

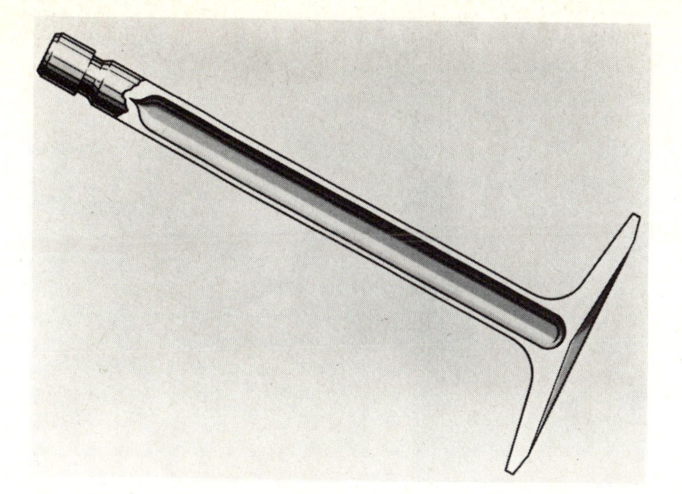

Fig. 9-12. Intake valve with hollow stem. (The valve has been cut away to show the hollow stem.) (*Ford Motor Company*)

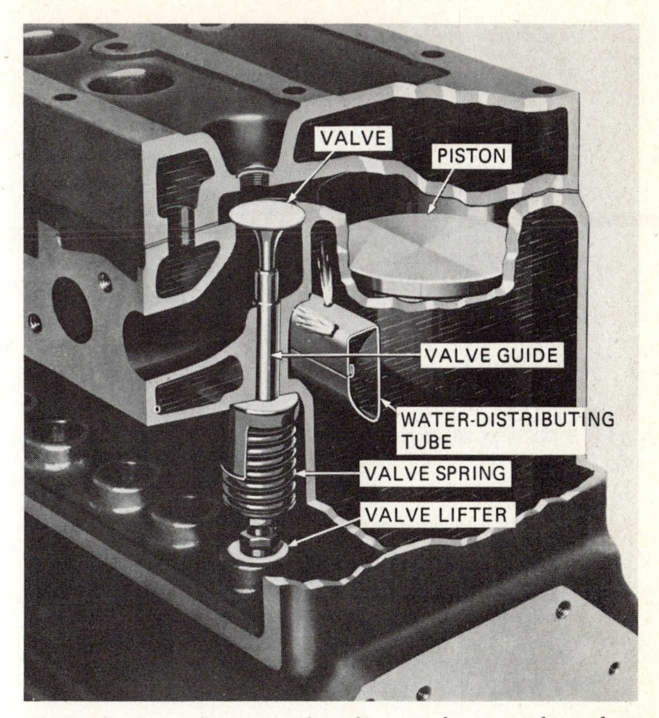

Fig. 9-13. Use of a water-distributor tube to cool a valve. The location of a valve in an L-head engine is also shown.

reason that the valve face is not as hot as the inner part of the valve head is that when the valve is closed, it is cooled to some extent by the valve seat; that is, heat passes from the valve to the seat. The valve stem is also cooled since it passes heat to the valve guide. However, it is estimated that about 75 percent of the heat is lost from the valve through the valve seat while only about 25 percent is lost through the valve guide. Therefore, the valve seat plays the most important role in keeping the valve from overheating.

It is obvious that in order to avoid excessive valve temperatures (which cause rapid valve failure) the valve must seat properly and the valve seat and guide must be adequately cooled. To accomplish this, water-distributing tubes have been used in the cylinder-block water jacket in L-head engines (Fig. 9-13), and water nozzles have been used in the cylinder in some I-head engines (Fig. 9-14). These devices provide for additional coolant circulation and cooling around the critical areas.

Another method of ensuring good coolant circulation around potential hot spots in the cylinder head is to use deflectors which are cast into the cylinder head. Figure 9-15 shows the location of the deflectors in a late-model-engine cylinder head.

The importance of proper valve seating is also emphasized in Fig. 9-11. If the valve face and the valve seat do not mate properly, or if they are rough or worn, then full-face contact will not be attained. This will reduce the area of contact through which heat transfer (and valve cooling) can take place. At the same time, uneven contact may mean that hot exhaust gases will leak between the valve face and its seat in some spots. These spots will naturally run hotter. Actually, a poor seat may cause the valve to run several hundred degrees hotter than normal, with local hot spots running at even higher temperatures. Naturally, these higher temperatures greatly shorten valve life.

Another factor of importance in valve and

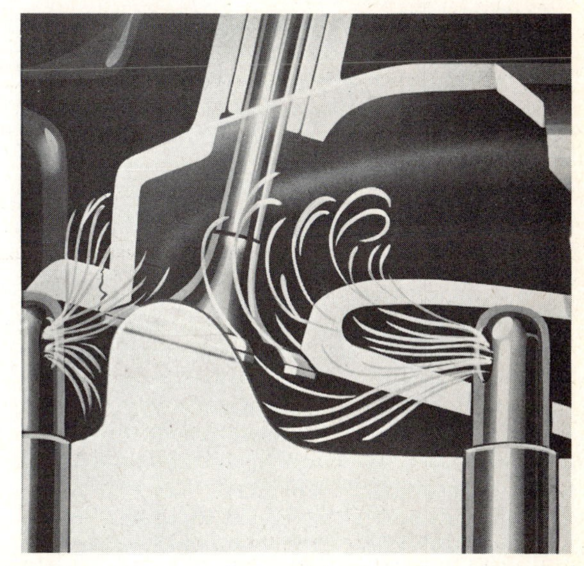

Fig. 9-14. Water nozzles used in the cylinder head of an overhead-valve engine to aid in valve-seat cooling. (*Chevrolet Motor Division of General Motors Corporation*)

valve-seat cooling is the location of the valves in the cylinder head. In some engines, the valve ports are "siamesed," that is, located so that two valves can use the same manifold passages. (The word "siamese" comes from the Siamese twins who were born physically joined together.) Figure 9-16 shows how valve ports are siamesed in some engines. Note (at the top) that the intake-valve ports in cylinders 1 and 2 are connected to a common passage from the intake manifold. Also, the exhaust-valve ports in cylinders 2 and 3 are siamesed, or connected, to a

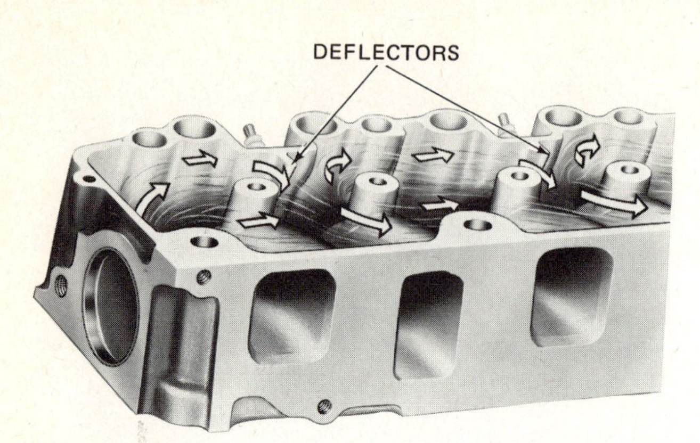

Fig. 9-15. Water circulation in the cylinder head of a late-model six-cylinder, I-head engine. Arrows show the direction of water flow. Note the effect of the deflectors. (*Ford Motor Company*)

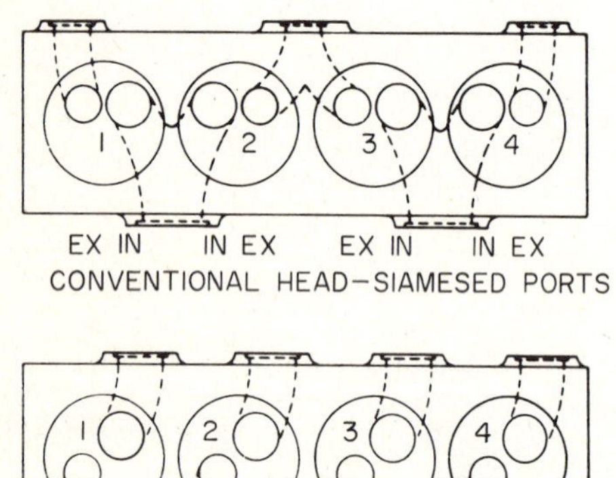

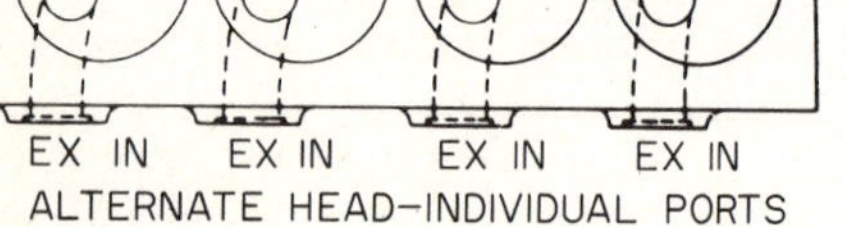

Fig. 9-16. Two basic valve-seat arrangements: siamesed ports and individual ports. (*TRW Inc.*)

common passage to the exhaust manifold. This arrangement, with two exhaust-valve ports side by side, produces a greater valve and valve-seat cooling problem because it reduces the amount of space between the two exhaust valves through which the coolant may pass. However, in the arrangement shown at the bottom of Fig. 9-16 the valve ports are spaced so that the exhaust ports are as far away from each other as possible. In this arrangement, adequate cooling is possible because coolant passages of sufficient size can be located around each exhaust-valve port.

⊘ 9-6 Sodium-cooled Valves To assist in valve cooling, many of the higher-output heavy-duty engines use sodium-cooled exhaust valves. This type of valve (Fig. 9-17) has a hollow stem which is partly filled with metallic sodium. Sodium melts at 208 degrees Fahrenheit [97.8°C]. Thus at operating temperatures the sodium is liquid. As the valve moves up and down, opening and closing, the sodium is thrown upward into the hotter part of the valve where it absorbs heat. As the sodium drops down

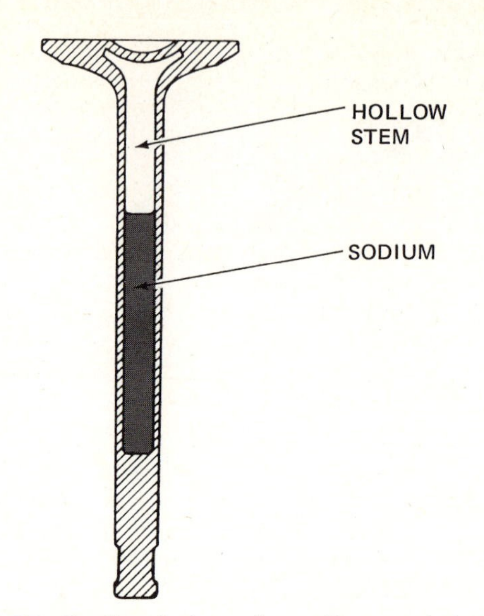

Fig. 9-17. Sectional view of a sodium-cooled valve.

into the cooler stem again, it gives up its heat to the stem. This circulation of liquid sodium in the valve stem cools the valve head. Therefore the valve runs cooler. Actually, a sodium-cooled valve will run as much as 200 degrees Fahrenheit [93.8°C] cooler than a solid-stem valve of similar design. Other factors being equal, this means longer valve life.

CAUTION: Sodium is a highly reactive element. If a piece of sodium is dropped into water, it will burst into flame with almost explosive violence. If it gets on the skin, it will cause deep and serious burns. Of course, as long as it is safely sealed in the valve stem, there is no danger. But if the hollow stem of a sodium-cooled valve is cracked or broken, then a potentially dangerous situation results. Old or damaged sodium valves should be disposed of safely. Some manufacturers recommend burying them deep underground.

⊘ 9-7 Valve Seats The exhaust-valve seat is subjected to the extremely high temperatures of the burned gases. For this reason, in some engines the exhaust-valve seats are made of special heat-resistant steel-alloy insert rings (Fig. 9-18). These rings hold up better than the block or head materials. Also, when the rings become worn so much that they cannot be finished with a valve-seat grinder, they can be replaced.

However, there are drawbacks to the use of these valve-seat insert rings. First, they complicate engine manufacture and service. Secondly, the interface between the insert and the cylinder head interferes with the flow of heat, and so the valve seat runs hotter. This means that the valve itself will run hotter and will wear more rapidly, which may make it necessary to use more expensive high-performance valves. Thus, if possible, engine designers prefer to use the *integral* type of valve seat, which is part of the cylinder head itself. For severe service, valve seats may be hardened by a special electric-induction process (Fig. 9-19). For some heavy-duty engines, however, valve-seat inserts are required.

⊘ 9-8 Interference Angles The use of an interference angle in the grinding of the valves and valve seats has become common practice. As shown in Fig. 9-19, the interference angle is achieved by grinding the valve at an angle $\frac{1}{2}$ to 2 degrees flatter than the seat angle. This produces greater pressure at the outer edge of the valve seat. The valve edge thus tends to cut through any deposits that have formed so that a good seal is produced. (An interference angle is not always recommended where the valve is faced with stellite and the seat is induction-hardened.)

Today most cars have either valve-seat inserts or interference angles. Different manufacturers use different interference angles. For example, General Motors recommends 46-degree seats and 45-degree valves. Ford recommends seats at 45 degrees and valves at 44 degrees. Chrysler recommends 45 degrees for both intake valves and seats, but 45 degrees for exhaust-valve seats and 43 degrees for exhaust-valve faces. American Motors Corporation recommends 30-degree intake-valve seats and 29 degrees for intake valves, but 44.5 degrees for exhaust-valve seats and 44 degrees for exhaust-valve faces.

⊘ 9-9 L-Head Valve Train The L-head engine uses a relatively simple valve train, or valve mechanism: The valves are located near, and in a straight line with, the camshaft. Figure 9-20 shows this valve mechanism with the various parts named. The valve spring is compressed between the cylinder block at one end and a spring retainer at the other. The spring retainer is attached to the end of the valve stem with a retainer lock (Fig. 9-21).

The valve rides in a valve guide assembled into the cylinder block (Fig. 9-20). The valve *lifter* (or *tappet,* as it is sometimes called) rides on the cam and moves up and down as the cam lobe passes under it. This motion is carried to the valve, causing it to open. Valve-spring pressure then recloses the valve as the cam lobe moves out from under the valve lifter.

⊘ 9-10 I-Head Valve Train In the I-head, or overhead, valve train (Figs. 6-11 and 9-2) two parts are installed in addition to those used in the L-head valve train. These parts are the pushrod and the rocker arm. There are several types of rocker arms. One type is shown in Fig. 9-22. The pushrods extend through openings in the cylinder head and block to the valve lifters above the cams on the camshaft. As the pushrod is moved upward by the lifter, it causes the rocker arm to pivot, or rock, on its mounting shaft. The valve end of the rocker arm then bears down on the valve stem, pushing it down so that the valve opens. Figure 9-23 shows how rocker arms

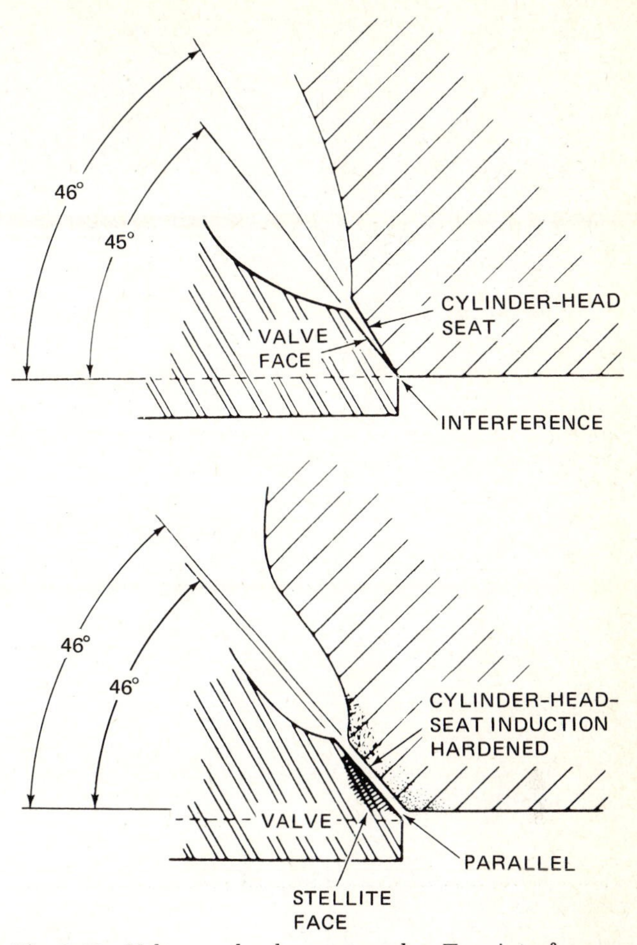

Fig. 9-19. Valves and valve-seat angles. Top, interference angle. Bottom, for very severe service, the valve seat can be hardened by a special electric-induction process. This process induces local heating by induction coils. The valve shown here is stellite-faced. Stellite is extremely resistant to heat and wear. Note that the faces of the valve seat and valve are parallel. (*Chevrolet Motor Division of General Motors Corporation*)

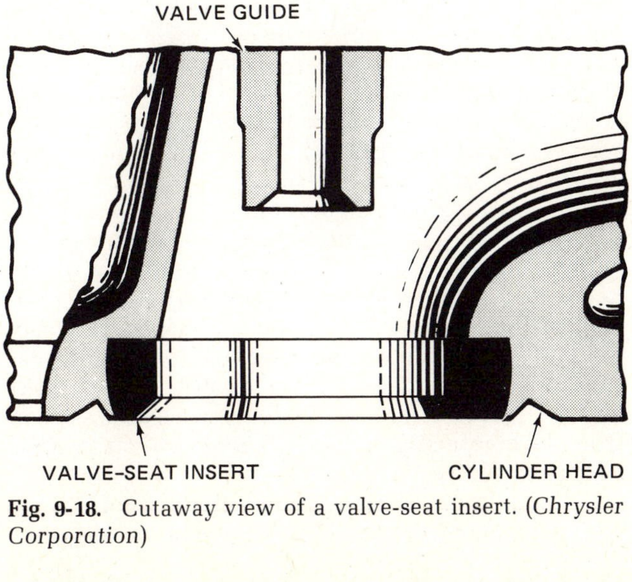

VALVE GUIDE

VALVE-SEAT INSERT CYLINDER HEAD

Fig. 9-18. Cutaway view of a valve-seat insert. (*Chrysler Corporation*)

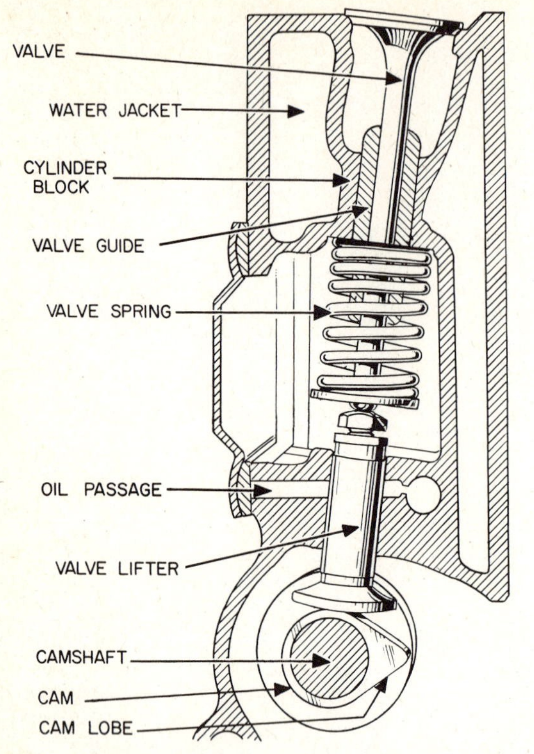

Fig. 9-20. Valve mechanism used in an L-head engine. The valve is raised off its seat with every camshaft rotation.

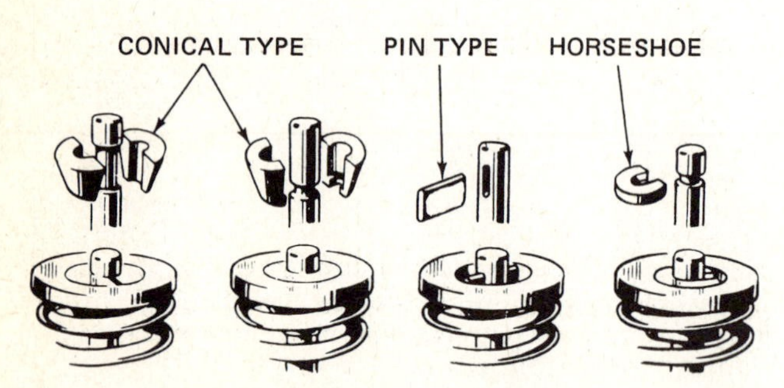

Fig. 9-21. Types of valve-spring retainer locks, or keepers.

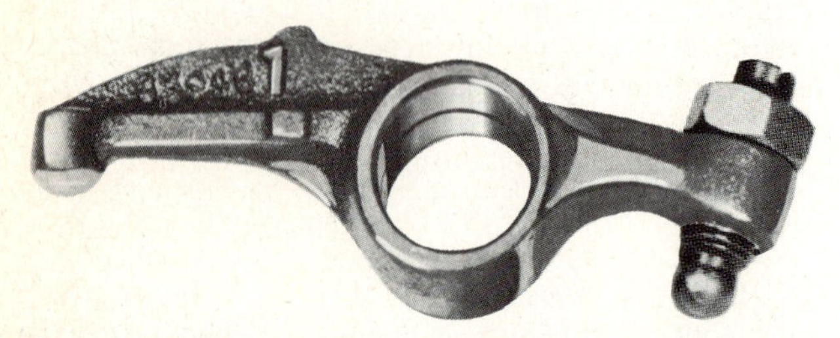

Fig. 9-22. One type of rocker arm used in overhead-valve engines. (*Chevrolet Motor Division of General Motors Corporation*)

of the type shown in Fig. 9-22 are assembled to a shaft mounted on the cylinder head. In this design, proper clearance between the valve stem and rocker arm is maintained by an adjustment screw and a locknut assembled into the rocker arm. The lower end of the adjustment screw is ball-shaped, and it rests in a socket in the upper end of the pushrod. Measurement of the clearance is made between the valve stem and rocker arm, but adjustment is made on the pushrod side of the rocker arm. Other types of rocker arms are discussed in ⊘ 9-12.

In some I-head engines, the valves are carried in replaceable valve guides. In others, the guides are integral, or part of the head; that is, they are holes bored in the head and are not separate from it. On these, if the guides become worn, they can be reamed out to a larger size and valves with oversize stems can be installed.

⊘ 9-11 Comparison of L- and I-Head Engines We have already mentioned (⊘ 6-13) that the I-head engine has replaced the L-head engine for automotive use. There are a number of reasons for this development. One is that the I-head engine can be built to have considerably high compression ratios. Figure 9-24 shows why. The L-head engine must have a certain minimum volume into which the valves can move. There must be sufficient space around the valves when they are open to permit free passage of gas. This volume plus the necessary minimum volume above the piston form the total clearance volume of the cylinder. This clearance volume cannot be reduced below a certain minimum, and this minimum fixes the ultimate compression ratio of the engine. Also, the L-head engine has the disadvantage of having a much longer flame-travel path, which increases the possibility of knocking. Then, too, in the L-head engine the greater metal-surface area exposed to the combustion process makes it a dirtier engine insofar as exhaust gas is concerned. That is, the cooler metal surfaces cool the adjacent air-fuel mixture below the combustion point. With large areas of metal surface, more air-fuel mixture fails to burn and therefore adds to the smog-producing compounds that are exhausted from the engine. (See Chap. 12 for more details.)

The I-head engine can be designed to have a much smaller clearance volume inasmuch as the valves and pistons are both working into the same general area. Many pistons have relief notches cut into their heads or are dished to give the valves space to move into when the piston is near or at top dead center (TDC). (See Fig. 8-37.) With this smaller clearance volume, the compression ratio can be higher. The advantages of higher compression ratios are discussed in ⊘ 11-5.

It is practical to make the valves larger in the I-head engine. Also, the incoming air-fuel mixture and the outgoing exhaust gases have shorter paths to travel. All this adds up to higher volumetric efficiency; that is, the engine can breathe more easily (⊘ 11-7). Furthermore, with the smaller area of relatively cool metal surface exposed to the burning air-fuel mixture, the I-head engine loses less heat from the combustion process. This improves thermal efficiency (⊘ 11-17).

An advantage of the L-head engine is, of course, its relative simplicity. This engine does not need

Fig. 9-23. Rocker-arm assembly for a six-cylinder, overhead-valve engine. (*Chevrolet Motor Division of General Motors Corporation*)

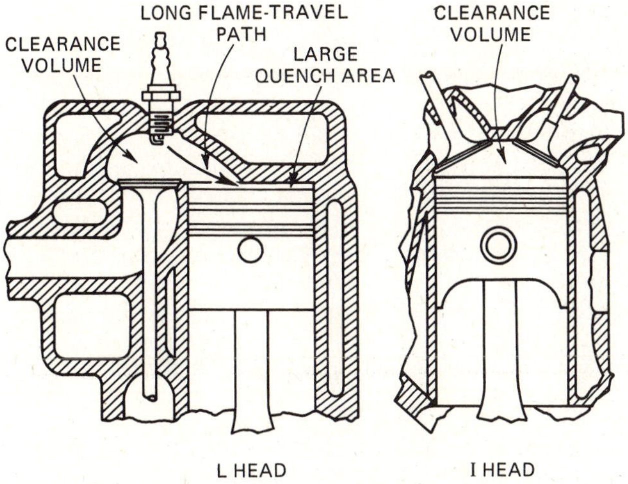

Fig. 9-24. Comparison of L- and I-head combustion chambers and valve arrangements.

pushrods, rocker arms, and rocker-arm supports. Also, the cylinder head is much simpler. However, the advantages of the I-head engine outweigh these greatly, and so the I-head engine has come into universal use for automotive vehicles.

Check Your Progress

Progress Quiz 9-1 Here is your chance to check up on yourself once more and find out how well you have been absorbing the material you have been reading. If any of the questions seem hard, just review the past few pages.

Completing the Sentences The sentences below are incomplete. After each sentence there are several words or phrases, only one of which will correctly complete the sentence. Write each sentence in your notebook, selecting the proper word or phrases to complete it correctly.

1. The three basic valve train arrangements are: (*a*) L head, I head, and overhead camshaft (OHC); (*b*) flat head, round head and overhead; (*c*) A head, B head, and C head.
2. Cams are used to control the: (*a*) opening and closing of the valves, (*b*) speed of the engine, (*c*) engine oil consumption, (*d*) connecting-rod bearing clearances.
3. The camshaft turns at: (*a*) twice the speed of the crankshaft, (*b*) the same speed as the crankshaft, (*c*) half the speed of the crankshaft.
4. The valves are closed by: (*a*) mechanical action of the cam lobes, (*b*) valve-spring pressure, (*c*) the pushrods.
5. In keeping the valve from overheating, the most important part of the valve is the: (*a*) tip, (*b*) stem, (*c*) face, (*d*) spring.
6. About 75 percent of the heat is lost from the valve through the: (*a*) valve guide, (*b*) valve seat, (*c*) valve stem, (*d*) valve spring.
7. To assist in valve cooling, some valves may have hollow stems filled with: (*a*) water, (*b*) sodium, (*c*) carbon monoxide, (*d*) mineral oil.
8. Some engines use valve-seat: (*a*) clearances, (*b*) lubrication, (*c*) knurling, (*d*) inserts.
9. When the valve face angle is ground $\frac{1}{2}$ to 2 degrees flatter than the seat angle, this is called: (*a*) competition angle, (*b*) sloppy workmanship, (*c*) interference angle, (*d*) bearing clearance.
10. The I-head valve train requires two parts in addition to those the L-head valve train requires. They are: (*a*) valve lifter and pushrod, (*b*) rocker arm and valve lifter, (*c*) pushrod and rocker arm, (*d*) rocker arm and valve spring.

⊘ 9-12 Rocker Arms There are two general types of rocker arms: the cast or forged type (Fig. 9-22) and the steel stamping type (Figs. 9-25 and 9-26). Rocker arms are supported on the cylinder head by either shafts, studs, or T supports.

1. *BALL-PIVOT ROCKER ARMS* The ball-pivot type of rocker arm is shown in Fig. 9-25; and the rocker arm is shown in sectional view in Fig. 9-26. This rocker arm is a heavy steel stamping, shaped as shown. The pushrod end is formed into a socket in which the end of the pushrod rides. The rocker

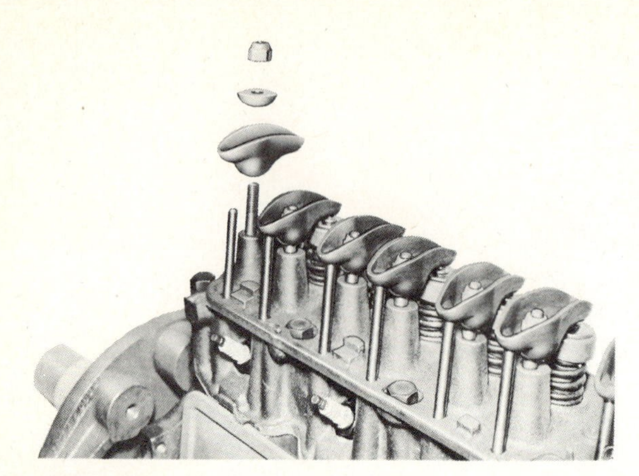

Fig. 9-25. Rocker arm using a ball pivot with associated parts. (*Chevrolet Motor Division of General Motors Corporation*)

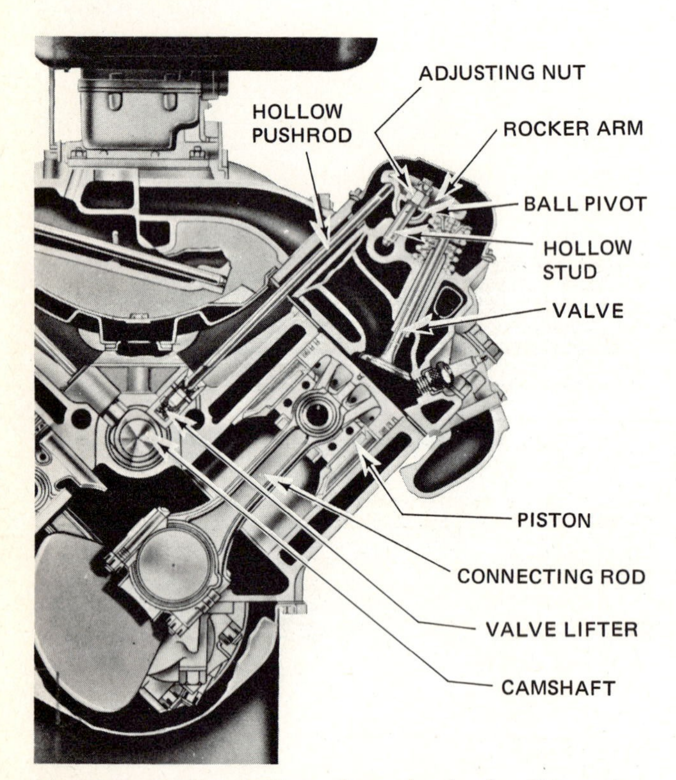

Fig. 9-26. Sectional view of one bank of a V-8 engine using ball-pivoted valve rocker arms. (*Pontiac Motor Division of General Motors Corporation*)

arm is supported by a ball pivot mounted on a stud. The stud is hollow and opens into an oil gallery in the head. Oil feeds through the stud to the ball pivot for lubrication. Also, the pushrod is hollow and feeds oil from the valve lifter to the contact area between the pushrod and the rocker arm. Adequate lubrication of the moving parts of the valve train is thus ensured. The valve clearance on this design is measured between the rocker arm and the valve stem, as with other designs. However, on this design, adjustment is made by turning the adjustment nut above the ball pivot up or down on the mounting stud. This adjustment raises or lowers the rocker arm to increase or decrease valve clearance.

NOTE: Most engines now use hydraulic valve lifters (⊘ 9-21). In many of these engines, there is no provision for adjusting valve lash. The lifters hydraulically take up any clearance when the valves are closed. On this type of engine, some manufacturers specify that if valves and valve seats are ground, then the valve stem should be shortened by grinding off the tip. Other manufacturers supply longer or shorter pushrods to take care of changes in valve-stem height. Either method restores the proper linkage adjustments to the valve train (see ⊘ 23-16 to 23-19). Some engines using hydraulic valve lifters do have provisions for adjustment as explained on following pages.

2. T-SHAPED ROCKER-ARM SUPPORTS The T-shaped design is shown in Fig. 9-27. The rocker arm is a heavy steel stamping. Each T-shaped rocker-arm support carries two rocker arms, as shown. The rocker-arm supports are positioned on the cylinder head by the rocker-arm retainer. Each retainer positions two rocker-arm supports. Clips on the rocker-arm retainer hold the rocker arms on their supports and prevent them from slipping off.

3. ROCKER ARMS FOR OHC ENGINES Overhead-camshaft (OHC) engines were discussed briefly in ⊘ 6-10. A more complete discussion follows.

⊘ 9-13 Overhead-Camshaft Engines The overhead-camshaft (OHC) engine may have one camshaft per cylinder bank [single overhead camshaft (SOHC)] or two camshafts per cylinder bank [double overhead camshaft (DOHC)]. Figure 6-6 shows a four-cylinder SOHC engine. Figure 6-17 shows a V-8 DOHC engine. The camshaft may be driven by sprockets and chain (Figs. 6-6, 6-17, and 9-31) or by a neoprene belt (Figs. 6-13, 9-28, and 9-32). The belt is reinforced with fiber-glass cords and has a facing of woven nylon fabric on the toothed side. The teeth in the belt fit the teeth on the drive pulley. Several arrangements are used to carry the cam action to the valve stems. In some OHC engines, cam action is carried directly to the valve stem through a cap called the *valve tappet* (Figs. 9-28 and 9-32). This cap fits over the valve stem and spring. In other OHC engines, the cam action is carried through a rocker arm (Figs. 9-30 and 9-35).

Now let us look at the Vega four-cylinder SOHC engine shown in Figs. 9-28 and 9-29. The cam-lobe motion is carried directly to the valve stem through the valve tappet. Because the pushrod and rocker arm are not used, there is no problem with valve-train flexing as there is in engines using rocker arms and pushrods (see ⊘ 6-10, item 3). Therefore, the engine is more flexible because the valves respond more quickly. Higher engine speeds are possible because there are fewer parts to move. This engine has an aluminum cylinder block with silicon cylin-

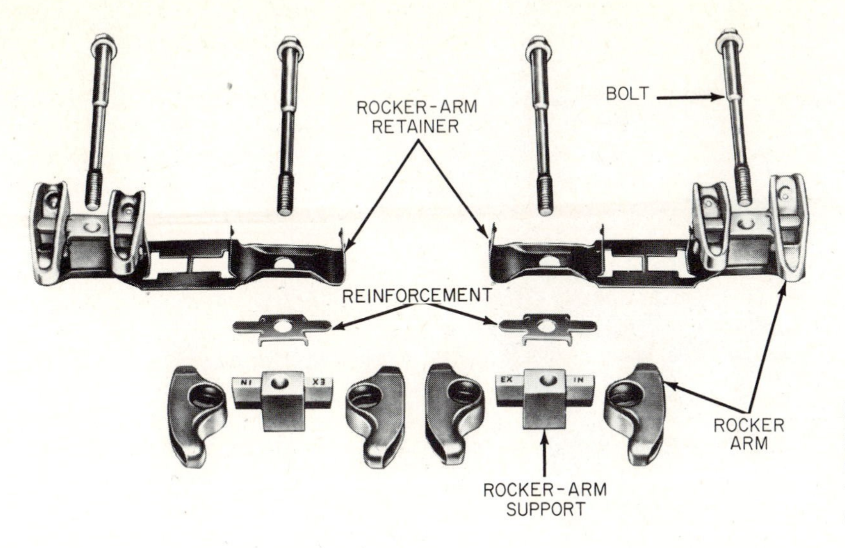

Fig. 9-27. Rocker arms, supports, and retainers. (*Cadillac Motor Car Division of General Motors Corporation*)

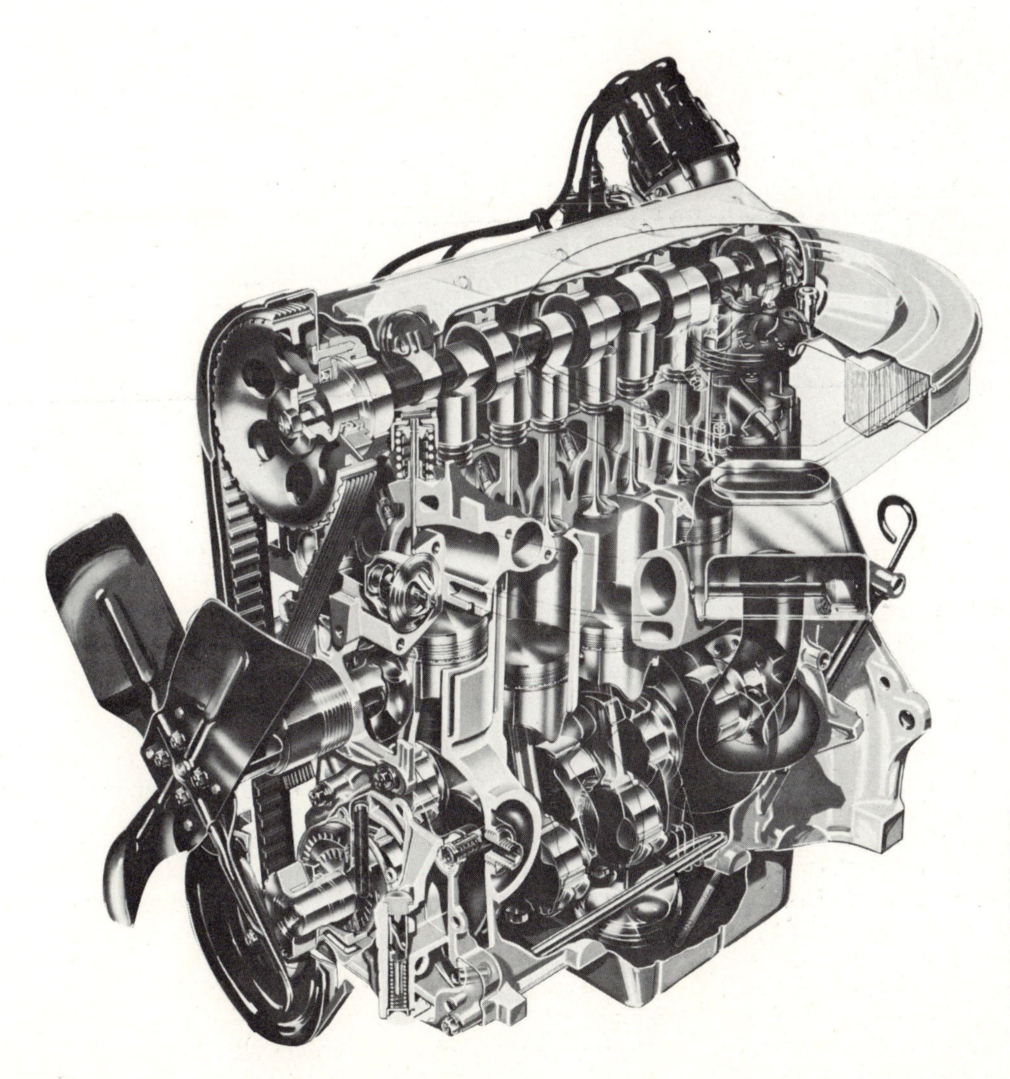

Fig. 9-28. Cutaway view of the Vega four-cylinder engine. This is an overhead-valve engine with the camshaft in the cylinder head. (*Chevrolet Motor Division of General Motors Corporation*)

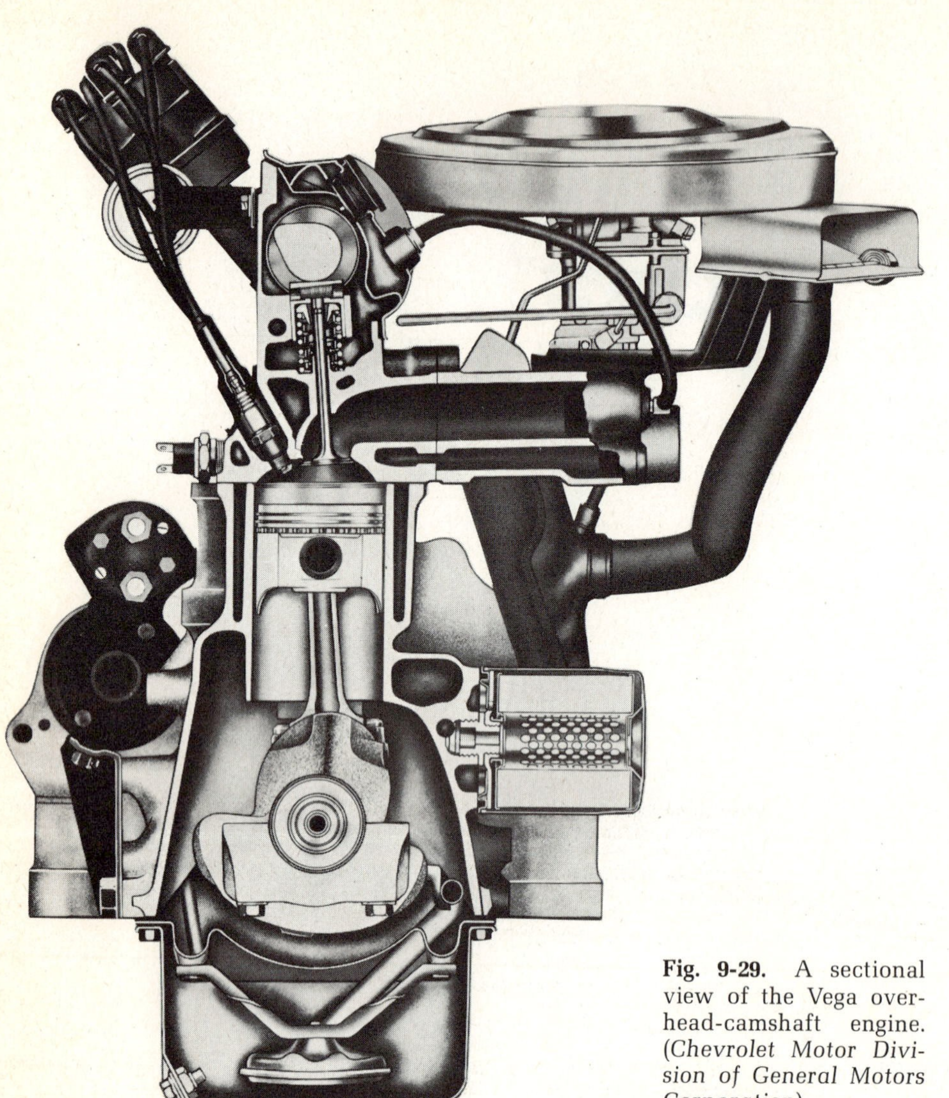

Fig. 9-29. A sectional view of the Vega overhead-camshaft engine. *(Chevrolet Motor Division of General Motors Corporation)*

der-bore wear surfaces on which the pistons and rings slide (see ⊘ 7-7).

The Opel SOHC engine shown in Figs. 9-30 and 9-31 uses ball-pivot rocker arms and a chain and sprockets to drive the camshaft. Note, in Fig. 9-31, the damper block and chain tensioner. Their purpose is to maintain proper tension on the chain and prevent it from whipping. Whipping would create uneven valve timing and also noise.

Figure 7-14 shows the cylinder head for a SOHC engine. Note the bearings that hold the camshaft and rocker-arm shaft.

Figures 9-32 and 9-33 show the high-performance Cosworth Vega DOHC engine. This engine uses a toothed neophrene belt and toothed pulleys to drive the camshafts. Note that this engine has four valves per cylinder: two intake valves and two exhaust valves. Figure 9-34 shows top and bottom views of the cylinder head for this engine with the valves in place. The four valves allow more air-fuel mixture to enter and the exhaust gases to leave more easily. In other words, the four valves per cylinder increase the volumetric efficiency of the engine (see ⊘ 11-7). Note that the combustion-chamber surfaces of the cylinder head have been polished to improve combustion and heat transfer. This engine has no rocker arms and uses the same type of valve tappets as the SOHC Vega engine shown in Figs. 9-28 and 9-29. The cylinder block of this engine is die-cast aluminum with silicon cylinder-bore wear surfaces on which the pistons and rings slide (see ⊘ 7-7).

Figure 6-17 shows, in end sectional view, a V-8 engine using four overhead camshafts, two in each cylinder head. This engine has a rather complicated camshaft-drive arrangement. The crankshaft gear drives a second gear, which then drives a third gear. The third gear is mounted on the same shaft as a chain sprocket, which drives a chain. The chain then drives the two camshafts in one of the cylinder heads. The arrangement is duplicated for the other bank. This engine uses rocker arms. One end of the rocker arm is pivoted on a shaft. The center part has a roller which rolls on the cam as it rotates. The other end of the rocker arm rests on the end of the valve stem.

A somewhat similar rocker-arm arrangement is

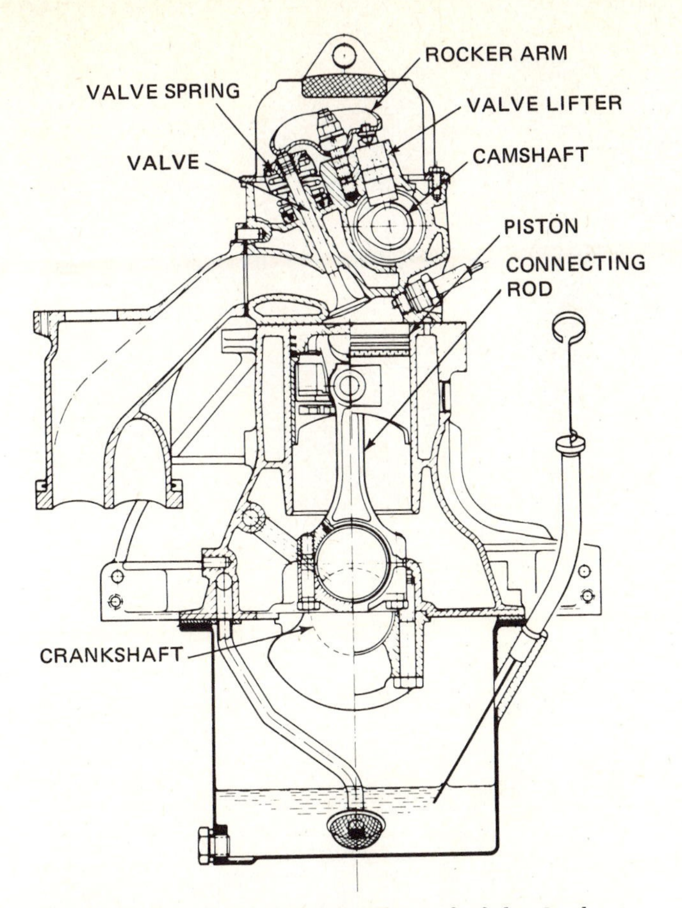

Fig. 9-30. Sectional view from the end of the Opel overhead-camshaft engine. (*Buick Motor Division of General Motors Corporation*)

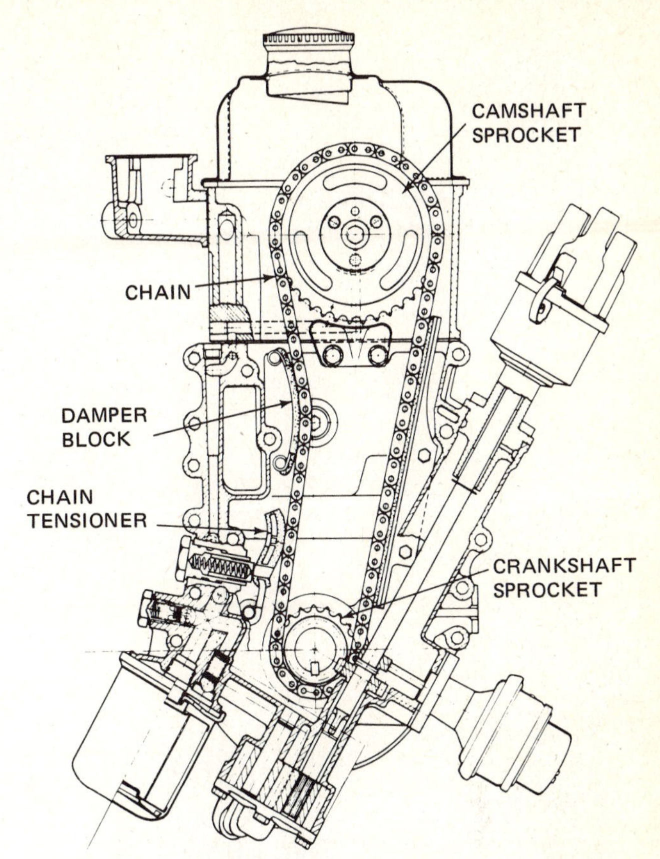

Fig. 9-31. Camshaft drive arrangement for the Opel overhead-camshaft engine. (*Buick Motor Division of General Motors Corporation*)

shown in Fig. 9-35. This is the valve train for the engine illustrated in Fig. 6-13. The rocker arm floats. It is attached at one end to the automatic *valve-lash adjuster* (which corresponds in some ways to the hydraulic valve lifter to be discussed in ⊘ 9-21). The other end of the rocker arm rests on the tip of the valve stem. Midway, a hump on the rocker arm rests against a cam on the camshaft.

The automatic valve-lash adjuster (Fig. 9-35) has a hollow cylinder in which a small piston fits. Oil under pressure from the engine lubricating system flows into the hollow cylinder, lifting the piston. One end of the rocker arm is attached to the piston by a clip. The upward movement of the piston raises the rocker arm so that it moves up into contact with the cam. This action reduces the valve lash to zero. Now, when the cam lobe comes around against the rocker arm, the rocker arm is forced down. It pivots on the piston in the valve-lash adjuster, and thus the valve-stem end of the rocker arm moves down, forcing the valve to move down and open.

As soon as the cam lobe passes the rocker arm, the valve spring moves the valve upward to the closed position, forcing the rocker arm upward. Any oil that may have leaked out of the valve-lash adjuster is now replenished by the engine lubricating system, and the piston pushes upward to bring the rocker arm into contact with the cam. The rocker arm is thus put into position for the next valve-opening cycle.

One advantage of the overhead-camshaft engine is that with the elimination of the pushrod and, in some engines, the valve lifter, the inertia of the valve train is greatly reduced and there is less deflection in the system. With decreased inertia, there is less tendency for valve *toss* or *float* to occur. When valve toss or float occurs, the valve momentarily floats free from the direct influence of the cam contour. This could occur at high speed, for example, if the inertia of the fast-moving valve lifter, pushrod, and rocker arm momentarily overcame the valve spring. Clearances would appear between the valve-train parts. Then, as the spring again took command, the parts would come together with a strong pounding effect. Valve float can seriously damage valve-train parts and cause the valve to strike the top of the piston.

Figure 9-36 illustrates valve toss or float. The dashed line shows the theoretical valve lift, the valve movement that would occur if the valve followed the cam contour exactly. The solid line shows what can happen under some high-speed conditions. Note that the actual valve movement starts late (from 1 to 2). This is because the pushrod and rocker arm flex, or bend. Then, at about the halfway point, the pushrod and rocker arm react like springs that

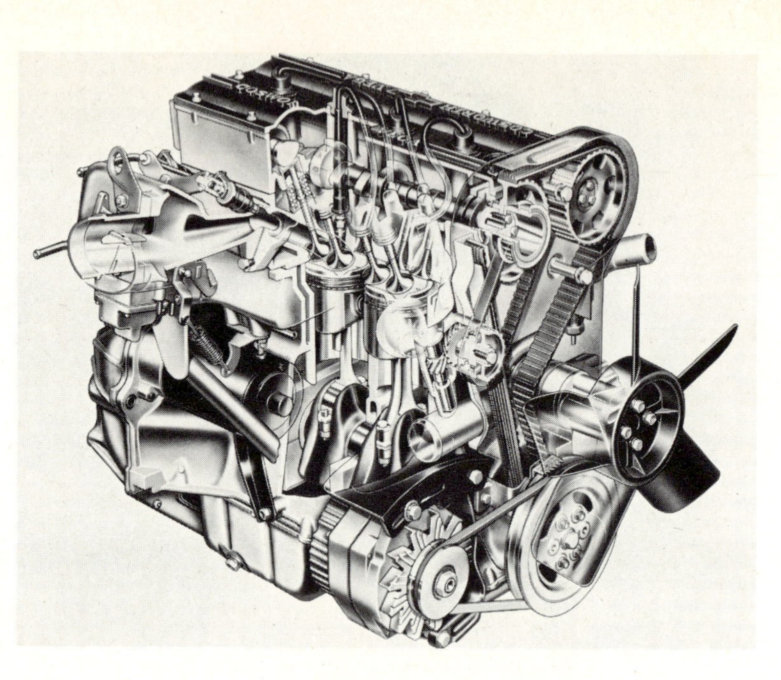

Fig. 9-32. Phantom view of the Cosworth Vega overhead-camshaft engine. (*Chevrolet Motor Division of General Motors Corporation*)

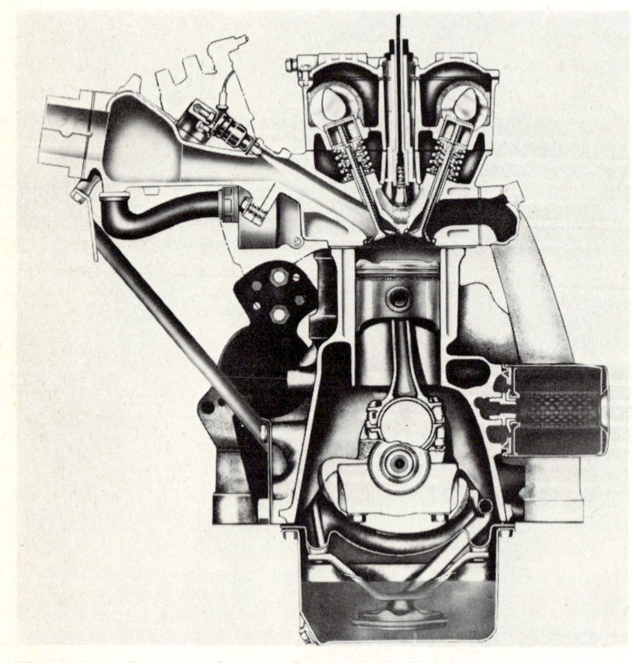

Fig. 9-33. Sectional view from end of the Cosworth Vega engine. (*Chevrolet Motor Division of General Motors Corporation*)

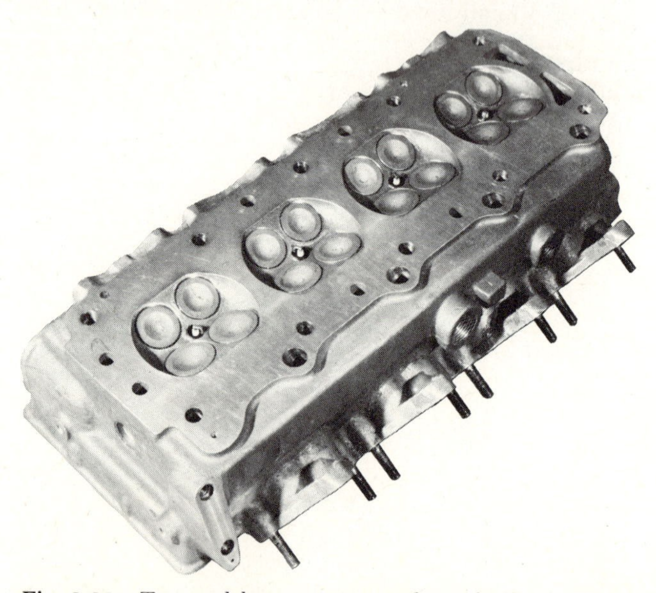

Fig. 9-34. Top and bottom views of a cylinder head for a four-cylinder, high-performance engine—the Cosworth Vega engine. Note that it has four valves for each cylinder: two intake valves and two exhaust valves. (*Chevrolet Motor Division of General Motors Corporation*)

have been compressed (from 2 to 3). That is, they unflex, or straighten, and toss the valve, causing it to float off the contour of the cam lobe. The valve moves freely between 2 and 4. The amount of float is shown in Fig. 9-36. Then as the valve starts to close, it generally follows the cam contour down to 5, where the cam begins to slow the rate of valve closing. But at this point, the inertia of the moving valve train causes the pushrod and rocker arm to flex again, letting the valve down hard onto the valve seat (at 6). The pushrod once more acts as a spring, and as it straightens it lifts the valve off the

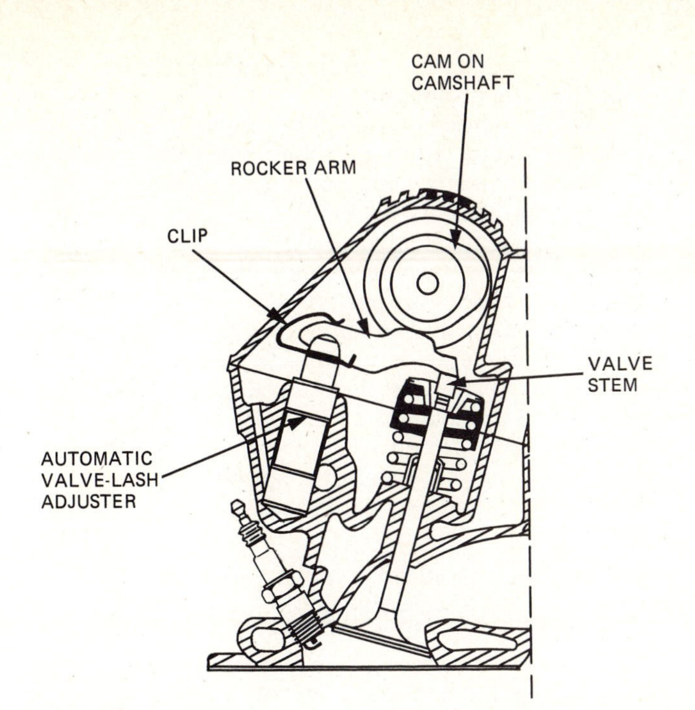

Fig. 9-35. Sectional view of the valve mechanism in an SOHC engine using a toothed belt to drive the camshaft. (*Pontiac Motor Division of General Motors Corporation*)

seat. In other words, the valve bounces (between 6 and 7 and between 7 and 8).

OHC engines have fewer valve-train parts, as explained previously. There is less inertia in the valve train and no pushrod to flex. This means that higher speeds can be achieved with lower valve-spring pressures for better performance.

Another advantage of the OHC engine is that the cam contours can be made sharper to provide more rapid opening and closing of the valves. This means that the engine can breathe better and volumetric efficiency is higher (see ⊘ 11-7). Thus, engine performance is improved, especially at high speeds.

NOTE: Although the full names for these engines are *single-overhead-camshaft* (SOHC) *engine* and *double-overhead-camshaft* (DOHC) *engine*, these names are usually shortened to *overhead-cam engine*. The I-head engine that uses pushrods is often called a *pushrod engine* to distinguish it from the overhead-cam engine.

⊘ 9-14 Valve Oil Seals On overhead-valve engines, there is always considerable oil on top of the cylinder head. Oil is needed here to lubricate the pushrods, rocker arms, and valve stems. This means that oil will try to work down the valve guides past the valve stems and enter the combustion chamber. If this happens, there will be rapid carbon buildup. Valves will not work properly. The compression ratio will be increased by the carbon buildup to the point where knocking will occur. Spark plugs will foul up due to the carbon. All this means that the engine will perform poorly.

The newer, higher-compression engines are more susceptible to this problem because of the higher vacuum they develop during the intake stroke. A higher vacuum during intake means a greater pressure differential pushing oil past the valve stem and down into the cylinder. Oil can also seep past the exhaust valve into the combustion chamber. For these reasons, many late-model engines have special provisions to prevent oil seepage past the valve stems into the combustion chamber. Figure 9-37 shows one type of seal and shield. The seal, which is a rubberlike ring, fits between the spring-retainer skirt and an undercut in the valve stem. The shield covers the top two turns of the spring. The seal prevents oil from seeping down the valve stem past the locks and retainer. The shield prevents undue amounts of oil from being thrown through the spring onto the valve stem. A variation of this design is shown in Fig. 9-38. Here, the shield is on the inside of the valve springs rather than on

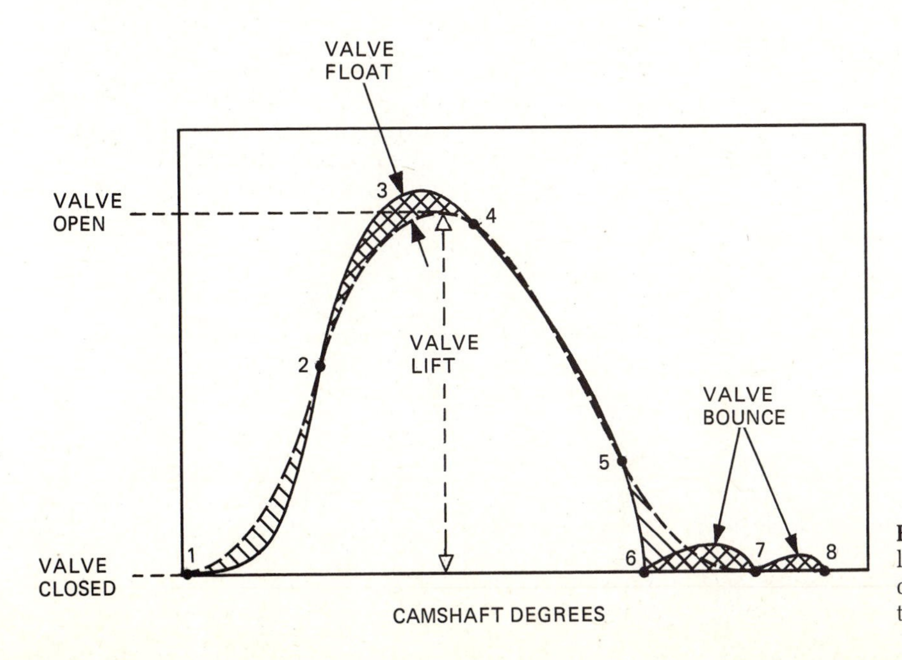

Fig. 9-36. Theoretical (dashed line) and actual (solid line) curve of valve lift in an I-head valve train.

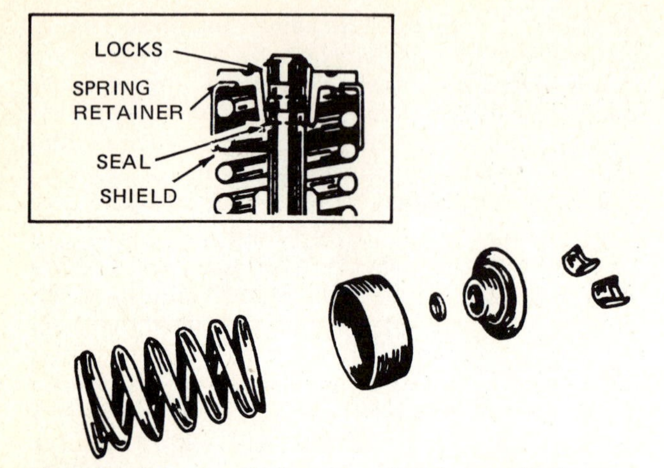

Fig. 9-37. Disassembled and sectional assembled views of the valve-and-spring assembly with oil seal and shield. (*Chevrolet Motor Division of General Motors Corporation*)

the outside. The purpose of the shield remains the same, however: to protect the valve stem from excessive oil.

Another type of seal to prevent oil seepage past the valve stem is shown in Fig. 9-39. This type of oil seal fits down against the cylinder head and around the valve stem.

⊘ 9-15 Valve Springs Valve springs are called upon to handle a very difficult job. They must have sufficient tension to close the valves quickly, even at high speed, and at the same time their tension must be low enough to prevent undue wear of valves, rocker arms, pushrods, lifters, and cams. Springs must maintain this tension under greatly varying speeds and temperatures. Also, they must have a natural frequency of vibration well above the top speed at which they will operate.

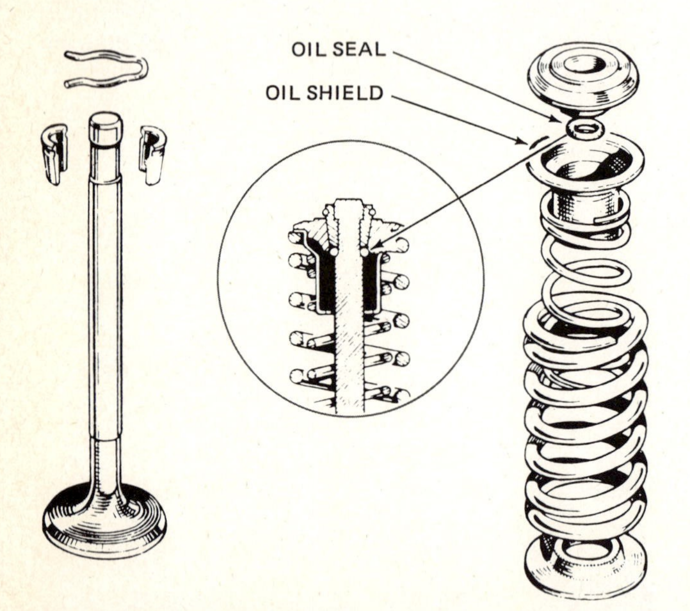

Fig. 9-38. Disassembled and sectional assembled views of the valve-and-spring assembly with oil seal and shield. (*MG Car Company, Limited*)

In some engines, two springs per valve are used to achieve good operating characteristics. In other engines, the spring coils are closer together at one end than at the other (or valve-stem) end. Still other engines use cone-shaped springs. In these, each coil has a different natural frequency of vibration, which puts spring resonance periods at very high frequencies. If the natural frequency of vibration were too low, it would coincide with some specific operating speed. This means that the natural frequency would be reinforced by the operating cycle, with the result that the spring would set up very powerful vibrations. This could render the spring ineffectual or even cause damage to valve-train parts.

⊘ 9-16 Valve Rotation If the exhaust valve were to rotate a small amount each time it opened, many valve troubles would be minimized. For example, one common cause of valve burning is the depositing of combustion products on the valve faces. Such deposits prevent normal valve seating, face-to-seat contact, and heat transfer. Soon, as the deposits collect, valves begin to overheat and burn. The poor seating also permits exhaust-gas leakage, which accelerates the burning process. Another cause of valve trouble is the valve sticking open. This condition usually results from accumulations of decomposed, or carbonized, oil (brought about by the high temperatures) on the valve stem. Ultimately, these deposits work into the clearance between the valve stem and valve guide; the valve sticks, or "hangs up," in the guide and does not close. Then, with poor or no seating, the valve soon overheats and burns.

Whenever a valve burns or does not seat properly, loss of compression and loss of combustion pressure will result. This means that the engine cylinder where the offending valve is located will be "weak." The cylinder will not deliver its share of the power.

However, if the valve is rotated as it opens, there will be less chance of valve-stem accumulations causing the valve to stick. In addition, valve rotation results in more uniform valve-head temperatures. Some parts of the valve seat may be hotter than others; actual hot spots may develop. If the same part of the valve face continues to seat on a hot spot, that part of the valve face will reach a higher temperature and tend to wear, or burn, faster. But if the valve is rotated, no one part of the valve face will be subjected to this higher temperature. Valve-head temperature will be more uniform.

⊘ 9-17 Types of Valve Rotators In the usual engine design, the rocker arm is slightly offset from the centerline of the valve (Fig. 9-40). This means that every time the valve is opened, there will be a push on the stem that is slightly off-center. This off-center push tends to make the valve rotate slightly as it is

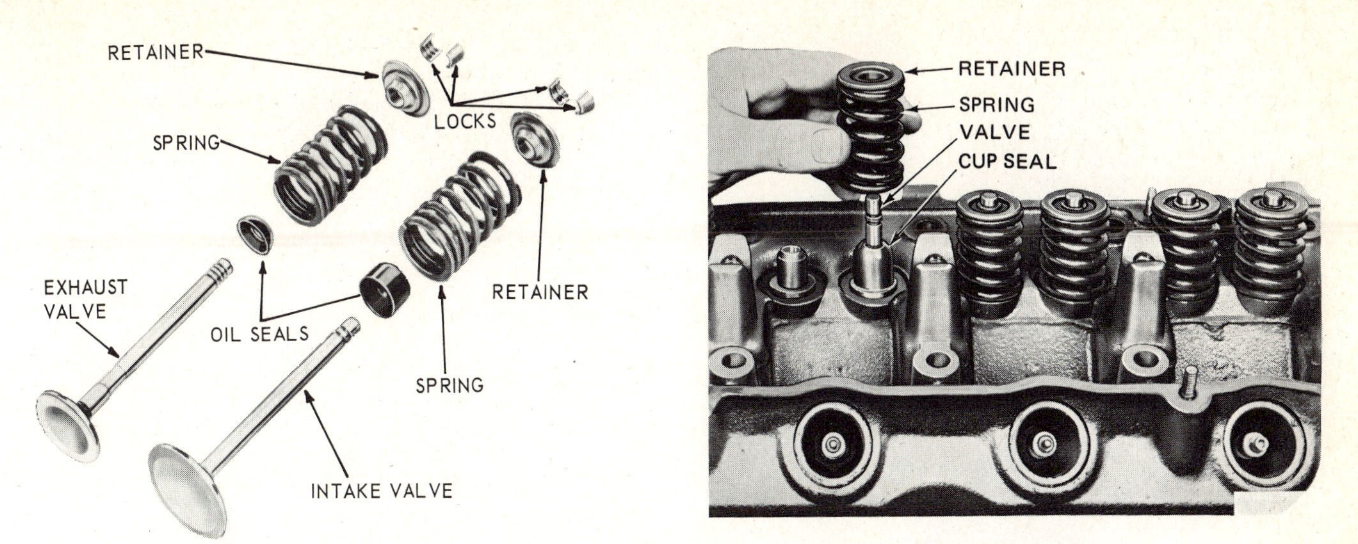

Fig. 9-39. *Left,* disassembled view of valves, springs, and seals, and related parts; *right,* parts assembled in head. (*Chrysler Corporation*)

pushed open. In the design shown in Fig. 9-40, the contact point can be adjusted by adding or removing spacers from one or the other side of the rocker arm.

There are also special valve-rotating mechanisms that are part of the valve train. The *free-type* valve rotator relieves all spring pressure on the valve so that it is rotated by engine vibration. The *positive-type* valve rotator actually rotates the valve stem each time the valve opens.

⊘ 9-18 Free-type Valve Rotator Figure 9-41 illustrates details of the free-type valve rotator. In this design, the spring-retainer lock has been replaced by two parts: a washer-type lock (split) and a tip cup. As the valve lifter moves up, the rocker arm presses down against the tip cup, and the tip cup then carries the motion to the lock and valve retainer. The valve retainer is pushed down, thereby taking up the valve-spring pressure. Then the bottom of the tip cup moves down against the end of the valve stem so that the valve is opened. Note that the spring pressure is taken off the valve stem; the valve is free. Since it is free, it can rotate. Engine vibration causes the valve to rotate.

⊘ 9-19 Positive-type Valve Rotator The positive-type valve rotator is shown in Figs. 9-42 to 9-44. This design contains a device that applies a rotating force on the valve each time it is opened, thereby ensuring positive rotation. Figure 9-43 shows the details of the rotator, and Fig. 9-44 shows it installed on a valve in an I-head engine. A seating collar (A in Fig. 9-43) is spun over the outer lip of the spring retainer (B). The valve spring rests on the seating collar. The collar encloses a flexible washer (C) placed below a series of spring-loaded balls (D). In Fig. 9-43, the middle view shows how the balls and springs are positioned in grooves in the spring retainer. The tops of the grooves (or races) are inclined as shown in the top view (E), which is a section (X-X) cut from the middle view. When the lifter is raised, the rocker arm lifts the valve and applies an increased pressure (as the valve spring is compressed) on the seating collar. This action flattens the flexible washer (C) so that the washer applies the spring load on the balls (D). As the balls receive this load, they roll up the inclined races, causing the retainer to turn a few degrees. This turning motion is applied through the retainer lock to the valve stem, and therefore the valve turns. When the valve closes, the spring pressure is reduced so that the balls return to their original positions, ready for the next valve motion.

The positive-type valve rotator can be installed on valves in either the L- or I-head engine. It can be installed at the tip end of the valve, as shown in Fig. 9-44, or at the valve-guide end between the

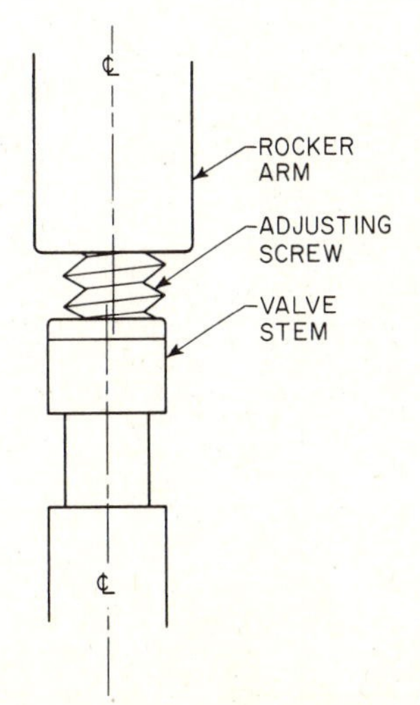

Fig. 9-40. In the Volkswagen engine, the center line of the rocker-arm adjusting screw is offset from the center line of the valve stem, thus promoting valve rotation.

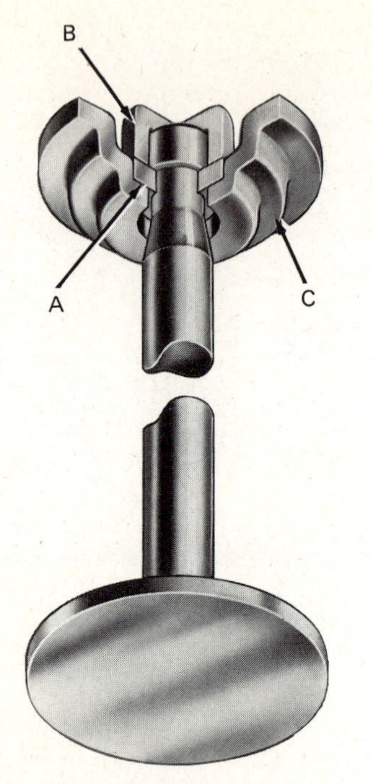

Fig. 9-41. Construction of a "free-type" valve rotator: (A) spring-retainer lock, (B) tip cup, (C) spring retainer. (*TRW Inc.*)

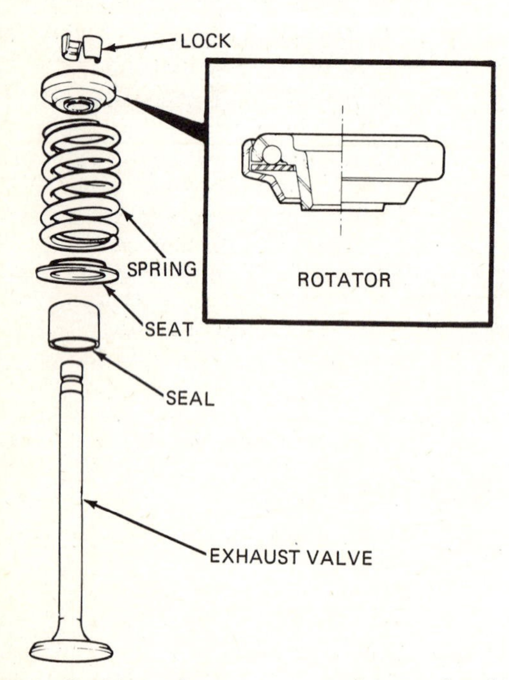

Fig. 9-42. Positive valve rotator on exhaust valve. (*American Motors Corporation*)

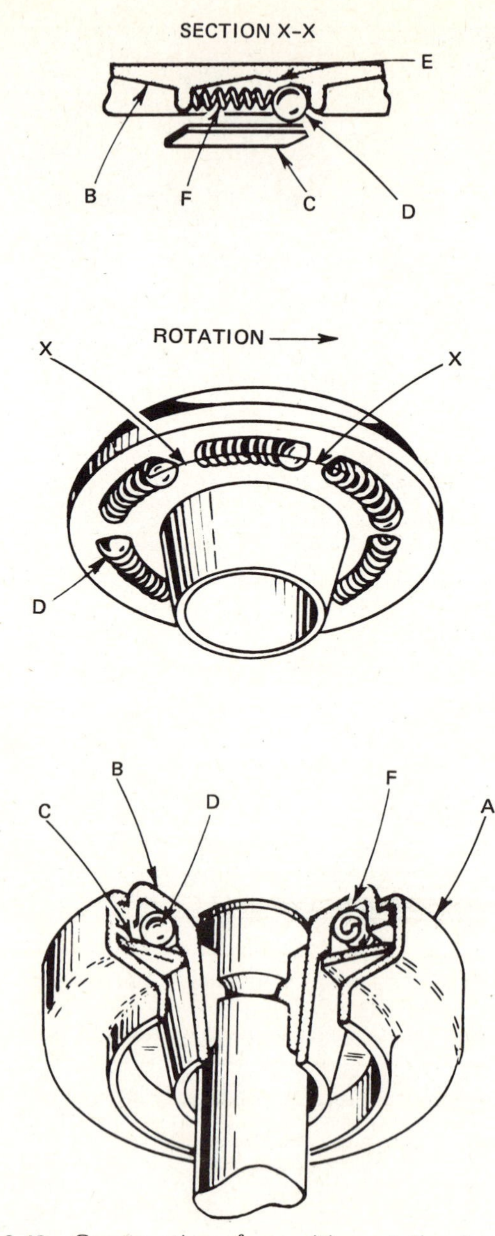

Fig. 9-43. Construction of a positive-rotation-type valve rotator: (A) seating collar, (B) spring retainer, (C) flexible washer, (D) balls, (E) inclined race, (F) ball return spring. (*TRW Inc.*)

valve spring and the cylinder block, or head. In the latter installation, the turning motion would be carried from the rotator through the spring, spring retainer, and lock to the valve stem.

⊘ 9-20 Valve Lifters There are two types of *valve lifters* (also called *valve tappets*): the *solid*, or *mechanical*, type and the *hydraulic* type (discussed in ⊘ 9-21). The mechanical valve lifter is essentially a round cylinder which is placed between the cam on the camshaft and the pushrod. The valve lifter is rotated in much the same way as the valve is rotated by the rocker arm; that is, it is slightly offset from the center of the cam. This rotation of the valve lifter prevents sludge formation from accumulating in the lifter bore in the cylinder block. At the same time, it rotates the pushrod, keeping the pushrod bearing surfaces between the lifter and rocker arm clean. A rough or worn cam lobe can often be pinpointed by noting which pushrod is not turning or is turning slower than the other pushrods. The rotation of the pushrods distributes the wear from the cam over the face of the lifter.

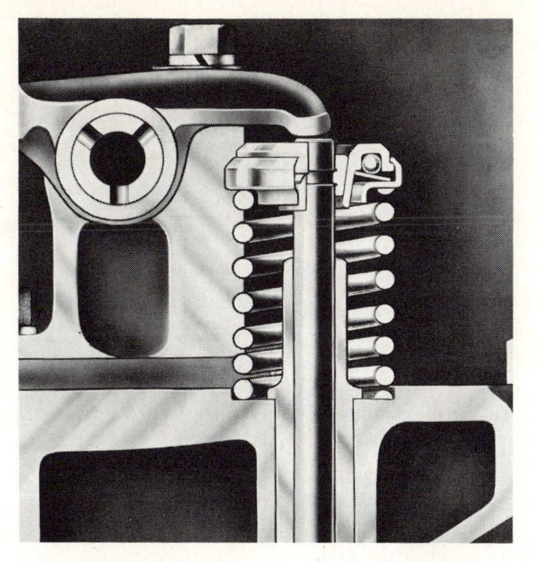

Fig. 9-44. Installation of the positive-rotation-type valve rotator on the valve in an I-head, or overhead-valve, engine. (*TRW Inc.*)

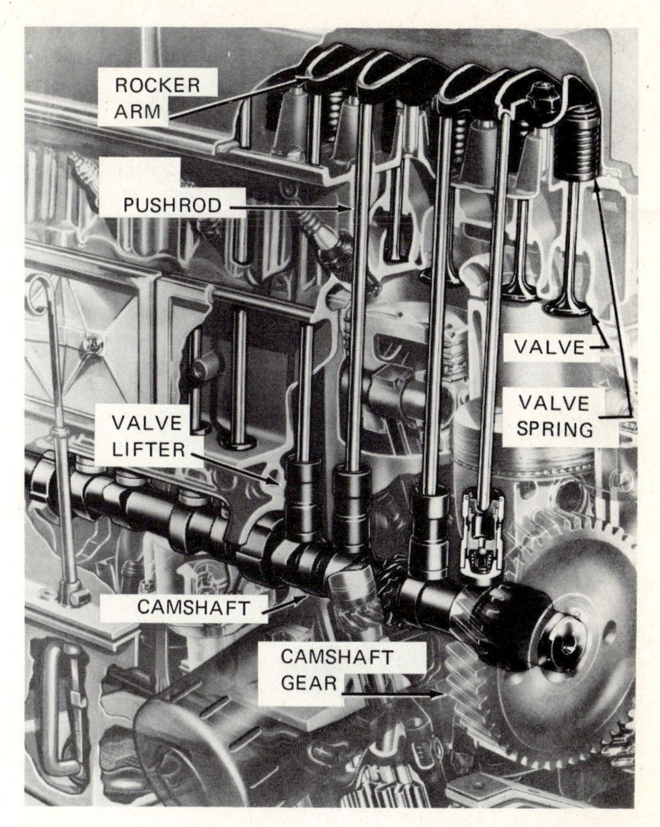

Fig. 9-45. Cutaway view of an I-head engine showing the location of the hydraulic valve lifter in the valve train. (*Pontiac Motor Division of General Motors Corporation*)

⊘ 9-21 Hydraulic Valve Lifters Many engines now use *hydraulic valve lifters*. This type of lifter, which provides zero valve clearance, is very quiet in operation. There is no "click" (or *tappet noise*, as it is called) as the adjustment screw on the valve lifter meets the valve stem (or pushrod), as there may be with the mechanical valve lifter, particularly when clearances are high. On the hydraulic valve lifter there is no adjustment screw (on most engines) and no clearance. As a rule, this type of valve lifter requires no adjustment in normal service. Variations due to wear of parts or temperature changes are taken care of hydraulically.

Figure 9-45 shows a hydraulic valve lifter installed on an overhead-valve engine. Figure 9-46 shows the operation of this type of valve lifter, and Figs. 9-47 and 9-48 show details of its construction. Oil is fed into the valve lifter under engine-oil-pump pressure from an oil gallery that runs the length of the engine (a V-8 engine would have two oil galleries, one for each bank).

When the valve is closed, oil from the engine oil pump is forced into the hydraulic valve lifter through the oilholes in the lifter body and plunger (see Fig. 9-46 for an illustration of the following action). As the oil enters the plunger, it acts on the disk valve in the bottom of the plunger, forcing it open. Oil now passes the disk valve and enters the space under the plunger. The plunger is therefore forced upward until it comes into contact with the valve pushrod (or valve stem on an L-head engine). This action takes up any clearance in the system.

Now, when the cam lobe moves around under the lifter body, the lifter is raised. Since there is no clearance, there is no tappet noise. As the lifter is raised, the sudden increase in oil pressure in the body chamber under the lifter causes the disk valve to close. The oil is therefore trapped in the chamber, and the lifter acts as a simple one-piece lifter: It moves up as an assembly and causes the valve to open. Then, when the lobe moves out from under the lifter, the valve spring forces the valve to close and the lifter to move down. The pressure on the oil in the chamber under the plunger is reduced, and the disk valve opens. Oil from the engine oil pump is again forced past the disk valve to replace whatever oil may have leaked from the chamber. Slight amounts of oil may leak past the disk valve and between the plunger and the lifter body. As this oil is replaced, the plunger moves up as much as is necessary to bring the pushrod seat into contact with the pushrod. This eliminates any clearance in the valve train.

NOTE: Some hydraulic valve lifters use a disk valve; others use a ball-check valve (Figs. 9-47 and 9-48). The action is the same for both designs.

⊘ 9-22 Valve Timing In previous discussions of engine and valve action it was assumed that the intake and exhaust valves opened and closed at top dead center (TDC) and bottom dead center (BDC). Actually (as can be seen from Fig. 9-49) the valves are not timed to open or close at these points in the engine cycle of operation. For example, in the valve-timing diagram illustrated in Fig. 9-49, the exhaust valve starts to open at 47 degrees before BDC on the power stroke. It remains open through the

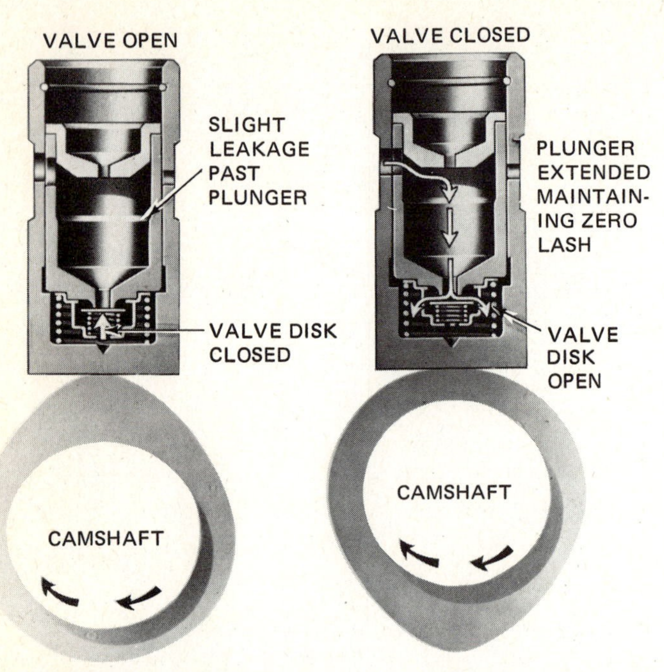

Fig. 9-46. Hydraulic valve lifter with valve open and closed. (*Ford Motor Company*)

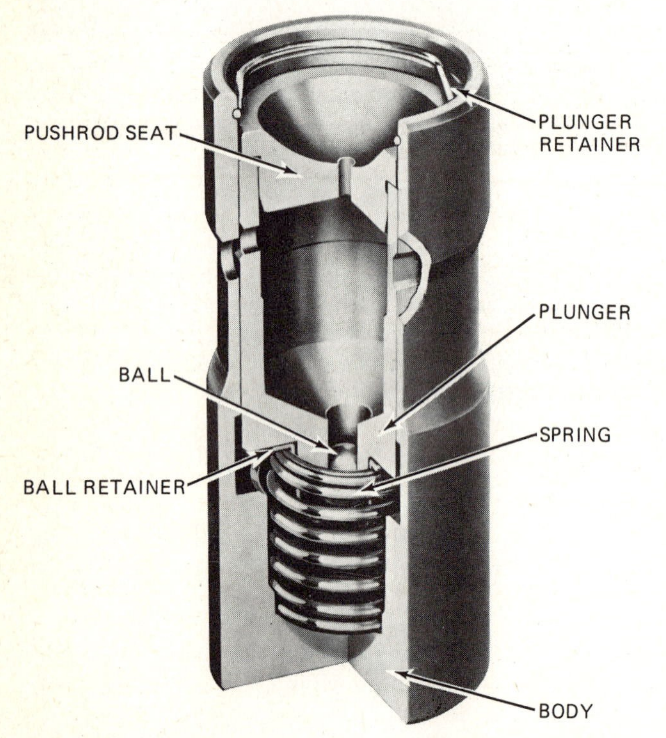

Fig. 9-47. Cutaway view of a hydraulic valve lifter. (*Chevrolet Motor Division of General Motors Corporation*)

remainder of the power stroke and through the entire exhaust stroke. The valve does not close until 21 degrees after TDC on the intake stroke. This additional exhaust-valve opening gives more time for the exhaust gases to leave the cylinder. When the exhaust valve starts to open (47 degrees before BDC), the combustion pressure has dropped considerably;

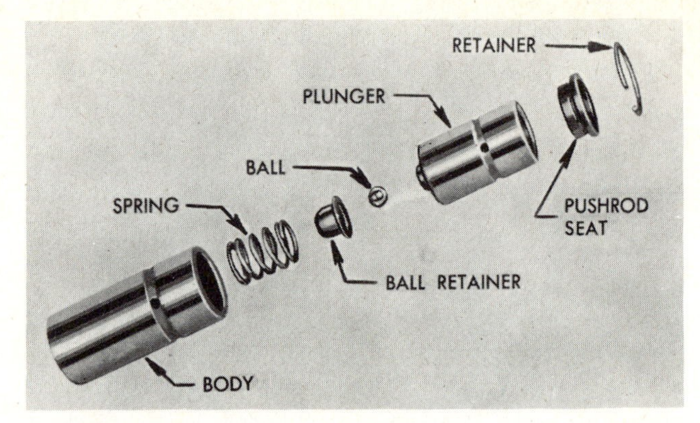

Fig. 9-48. Disassembled view of the hydraulic valve lifter shown in Fig. 9-47. (*Buick Motor Division of General Motors Corporation*)

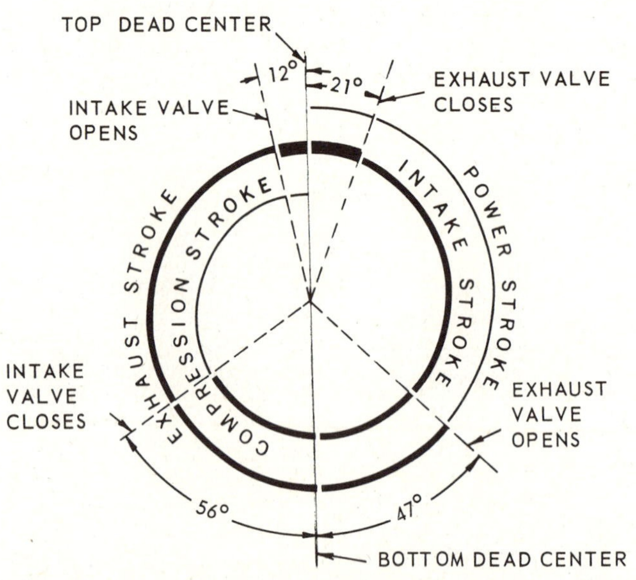

Fig. 9-49. Intake- and exhaust-valve timing. The complete cycle of events is shown as a 720-degree spiral, which represents two complete crankshaft revolutions. Timing of valves differs for different engines.

most of the available power of the burning air-fuel mixture has already been transmitted to the downward-moving piston. Opening the exhaust valve this early gives the exhaust gases additional time to leave the cylinder.

NOTE: To get an idea of how much the combustion pressure has dropped at 47 degrees before BDC on the power stroke, refer to Fig. 9-50, which shows the pressures in an engine cylinder during the four piston strokes. In the curve shown, the pressure has dropped from a peak combustion pressure of almost 700 psi [49.216 kg/cm^2] to approximately 100 psi [7.031 kg/cm^2] at 47 degrees before BDC on the power stroke.

The intake valve starts to open at 12 degrees before TDC on the exhaust stroke, thus giving a 33-degree overlap during which both valves are at

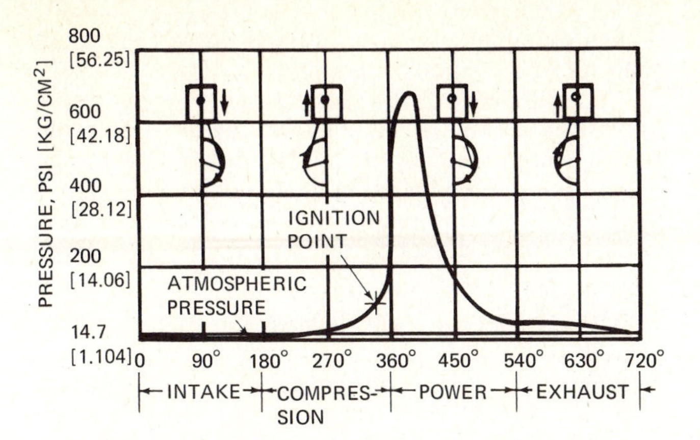

Fig. 9-50. Pressures in an engine cylinder during the four piston strokes. The four strokes require two crankshaft revolutions (360 degrees each), or a total of 720 degrees of rotation. This curve is for a particular engine operating at one definite speed and throttle opening. Changing the speed and throttle opening would change the curve, particularly the power curve.

least partly open. The intake valve then stays open until 56 degrees past BDC on the compression stroke. This gives the air-fuel mixture additional time to enter the cylinder. (The delivery of adequate amounts of air-fuel mixture to the cylinders is a critical item in engine operation.) Actually, the cylinder is never quite "filled up" when the intake valve closes. Thus there is no loss of compression resulting from the fact that the intake valve stays open well past BDC on the compression stroke. At BDC, as the compression stroke starts, pressure in the cylinder is below atmospheric (see Fig. 9-50). Pressure does not reach atmospheric until the piston is well past BDC. (See ⊘ 11-7 for a discussion of volumetric efficiency.)

NOTE: Figure 9-49 shows valve timing for one engine, but different engines have different degrees of valve timing. In some engines, valves open and close earlier or later than shown in Fig. 9-49 and remain open for different degrees of cam rotation.

Timing of the valves is obtained by the relationship between the gears or sprockets on the camshaft and crankshaft, as well as by the contours of the cam lobes. Changes in the relationship between the driving gear or sprocket and the driven gear or sprocket result in changes in the timing at which the valves open and close. For example, suppose the engine backfires, causing the timing chain to be thrown back as the camshaft is moving forward. This could cause the chain to "jump time." The chain slips a tooth, and so the camshaft jumps ahead that one tooth. As a result, the valves would open and close earlier. Let us say that the valve action is moved ahead by 15 degrees. Then, as in the example in Fig. 9-49, the exhaust valve would open at 62 degrees before BDC on the power stroke and close at 6 degrees after TDC on the exhaust stroke. The intake-valve actions would be moved ahead by a like amount. This change in valve timing would seriously reduce engine performance and cause engine overheating. To make installation easier, the gears or sprockets are marked so that they can be properly aligned (see Figs. 9-5 to 9-7).

⊘ 9-23 Cams for Mechanical and Hydraulic Valve Lifters Cam contours differ for hydraulic and mechanical valve lifters. Figure 9-51 shows a cam for a mechanical valve lifter. The base circle (A) is that part of the cam on which the lifter rides when the valve is closed. With the mechanical lifter, there must be some clearance (or valve lash) in the valve train to allow for dimensional changes as the engine warms up and the valve-train parts expand. Therefore, as the first part of the opening ramp comes around under the valve lifter, it raises the lifter enough to take up this clearance (B). Following that, the valve-train parts move as a unit, causing the valve to begin opening. The valve continues to open as the opening ramp (C) passes under the valve lifter. The shape of the opening ramp determines the acceleration, deceleration, and total amount of valve opening.

The valve reaches the fully opened position as the nose of the cam (D) comes under the lifter. The width of the nose (or number of degrees of rotation that the nose is under the lifter) determines the time that the valve remains wide open. After the nose passes from under the valve lifter, the closing ramp (E) lets the valve spring move the valve-train parts toward the closed position. The acceleration and deceleration of the valve-train parts are determined by the shape of the closing ramp as well as by the spring characteristics. Finally, after the valve reaches the closed position, the clearance reappears

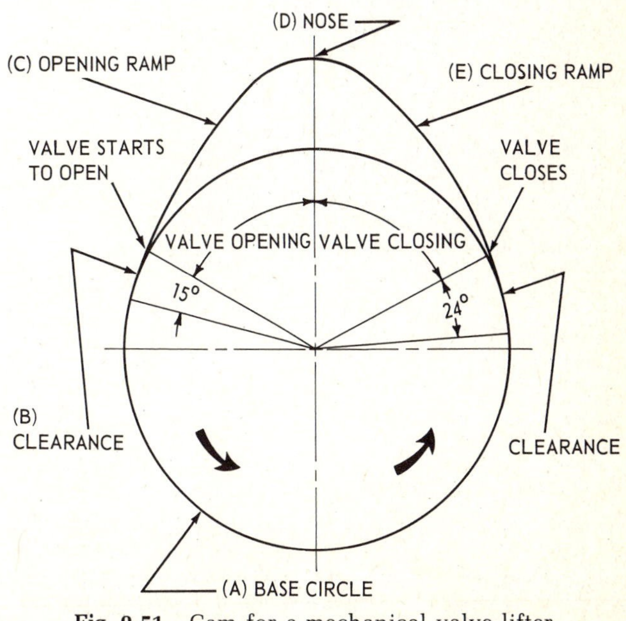

Fig. 9-51. Cam for a mechanical valve lifter.

in the valve train as the base circle comes around under the lifter once again.

⊘ 9-24 Comparison of Cams for Mechanical and Hydraulic Lifters There is no need for clearance, or lash, in the valve train using a hydraulic valve lifter. Thus, the cam contour for this valve lifter is different from that used for the mechanical lifter. Figure 9-52 compares the opening ramps for the mechanical and hydraulic valve lifters. With the mechanical lifter, as the opening ramp of the cam comes up under the lifter, the first movement of the lifter takes up the clearance in the valve train, as already noted. Then, some additional motion is required to take up the initial valve-train deflection. Thus as the valve-train parts (especially the pushrod) are loaded, they are compressed slightly. (This loading is caused by the opposition of the valve spring and inertia of the parts.) Then, the valve begins to move.

The opening ramp of the cam for the mechanical valve lifter is called a *constant-velocity ramp*. First there is a short acceleration period at the junction of the ramp with the base circle. Then the ramp imparts a constant velocity to the valve lifter for the rest of the initial phase during which the valve-train clearance and deflection are taken up. After this, as the valve begins to open, acceleration is again imparted to the valve lifter and train. This ramp contour avoids undue impact loading on the valve-train parts during the initial phase.

The opening ramp of the cam for the hydraulic valve lifter starts out differently and is called an *accelerated ramp* (bottom, Fig. 9-52). This ramp immediately introduces acceleration, which continues throughout the initial phase (including the take-up of the hydraulic-valve-lifter lag and valve-train deflection). Thus, the valve-train parts have already received some acceleration before the valve starts to move, which allows the valve to open more quickly. The take-up of the valve-lifter lag provides a cushioning effect which permits this arrangement without undue shock loading.

The last part of the closing ramp for hydraulic valve lifters must be higher than the opening ramp by an amount equal to the leakdown rate of the valve lifter during the open-valve phase. This allows the valve lifter to come down onto the base circle without undue deceleration.

⊘ 9-25 High-Performance Cams For high-performance engines, such as those used in racing and dragstrip cars, the cams are contoured to give a longer duration of valve opening and thus greater overlap. Figure 9-53 shows the effect on valve action of a high-performance cam for "road-and-drag" racing. Note that the intake valve opens 30 degrees before TDC and that the exhaust valve does not close until 27 degrees after TDC, giving an overlap of 57 degrees. The exhaust valve opens at 68 degrees before BDC, and the intake valve does not close until 68 degrees after BDC. The long duration of valve opening gives the engine a much better chance to breathe at high speeds. However, low-speed and idling characteristics become very poor because of the great overlap. Thus, to eliminate rough low-end operation, the idle speed must be increased considerably. Of course, to achieve this high performance, economy must be sacrificed. An engine with this type of camshaft will not give very good fuel economy.

Figure 9-54 shows the valve timing for a high-performance competition engine. This timing is even more radical than that shown in Fig. 9-53. The intake and exhaust valves are both open 312 degrees.

Another method of improving engine performance at the higher end is to use a higher-lift cam, that is, a cam that lifts the valve farther and thus opens it wider. As an example, a typical standard cam for a modern engine provides a total valve lift, or valve opening, of slightly less than 0.400 inch [10.16 mm]. A high-performance cam for the same engine provides a valve lift of about 0.450 inch

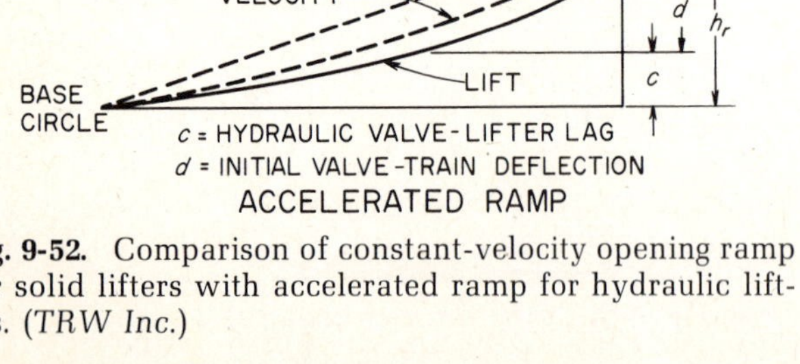

Fig. 9-52. Comparison of constant-velocity opening ramp for solid lifters with accelerated ramp for hydraulic lifters. (*TRW Inc.*)

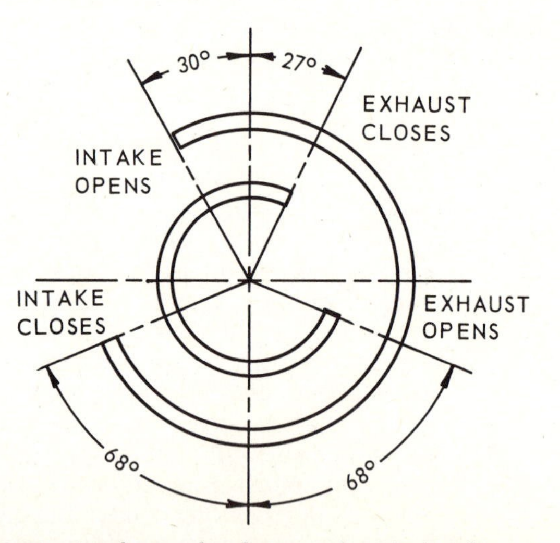

Fig. 9-53. Intake- and exhaust-valve timing for cams of a "road-and-drag" high-performance camshaft.

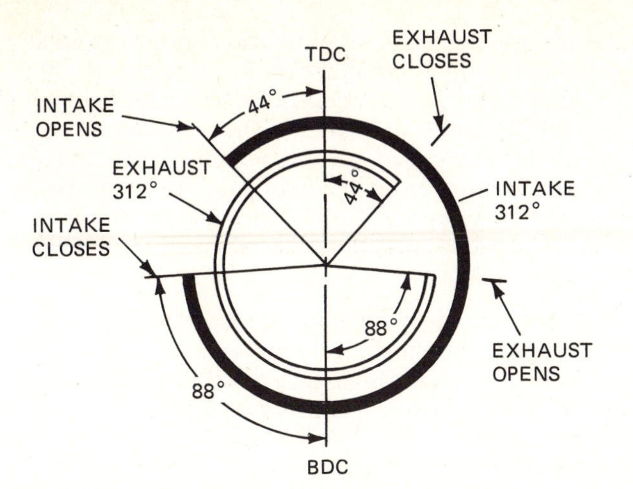

Fig. 9-54. Valve timing for a V-8 competition engine. (*Chrysler Corporation*)

[11.43 mm]. The difference of 0.050 inch [1.27 mm] allows a greater volume of gas to pass in a given time and thus improves engine breathing and high-speed performance. However, there is a limit to the amount of valve lift that cams can provide because of the close clearances between valves, when open, and the pistons at TDC. If the valve lift is too great, the piston could strike the valve and cause severe damage. Also, with close clearances, any carbon buildup on the pistonhead could result in the same situation.

NOTE: There is a point beyond which any additional valve lift fails to provide any additional engine power. The purpose of the valve is to open and shut the port into the combustion chamber. Once the valve opening is greater than the port area, no further valve lift can cause the port to carry additional air-fuel mixture.

Actually, according to automotive engineers, there are four general classifications of racing camshafts: road, road and drag, super road and drag, and track and drag. Each has special cam contours designed for the specific type of operation to which the engine will be subjected. Also, there are many modifications and special cam grinds developed by racing-engine specialists.

CHAPTER 9 CHECKUP

NOTE: Since the following is a chapter review test, you should review the chapter before taking the test.

You have now completed the part of the book that is designed to give you the theoretical background on engines that you need for automotive-engine maintenance and repair work. You have been making excellent progress and by now should have a good practical knowledge of engine components, their construction, and their operation. Again, a checkup has been included to permit you to check yourself on how well you have been absorbing the material. If you are not sure of the answers to any of the questions, turn back into the chapter and reread the pages that will give you the answers. Also, when you are asked to list parts, as, for instance, those in the hydraulic valve lifter, you may wish to refer to the text and illustrations pertaining to the lifter. Write your answers in your notebook.

NOTE: It is still not too late to start a notebook in case you have not been keeping one. A notebook becomes increasingly important as you move into the shopwork part of this book since you will want to write down and keep important details of making engine repairs and operating repair machinery.

Correcting Parts Lists The purpose of this exercise is to help you spot unrelated parts in a list. For example, in the list "valve, spring, lifter, retainer, spark plug," you would see that "spark plug" does not belong since it is the only part that is not part of the valve mechanism.

In each of the following lists, you will find one item that does not belong. Write each list in your notebook, but *do not write* the item that does not belong.

1. Parts that operate off the camshaft include the valves, fuel pump, water pump, oil pump, distributor.
2. Parts of the poppet valve include the head, margin, skirt, face, stem.
3. Parts in the L-head valve train include the valve, spring, retainer, crankshaft, lifter, lock, adjustment screw.
4. Parts in the I-head valve train include the lifter, pushrod, rocker arm, connecting rod, valve, spring, adjustment screw.
5. Parts in the hydraulic valve lifter include the pushrod seat, plunger, ball, ball retainer, piston ring, body, spring.

Completing the Sentences The sentences that follow are incomplete. After each sentence there are several words or phrases, only one of which will correctly complete the sentence. Write each sentence in your notebook, selecting the proper word or phrase to complete it correctly.

1. In operation it is not uncommon for the exhaust-valve-head temperatures to reach: (*a*) 1000°F [537.8°C], (*b*) 1500°F [815.6°C], (*c*) 2000°F [1093.3°C], (*d*) 2500°F [1371.1°C].
2. In normal operation the part of the exhaust valve that gets the hottest is the: (*a*) face, (*b*) middle of stem, (*c*) edge of margin, (*d*) center of head.
3. Cooling of the exhaust valve is assisted by two parts, the: (*a*) guide and lifter, (*b*) guide and spring, (*c*) guide and seat, (*d*) guide and cam.
4. The sodium-cooled valve, in comparison with solid-stem valves operating under similar conditions, will run as much as: (*a*) 200°F [93.8°C] cooler,

(*b*) 400°F [204.4°C] cooler, (*c*) 100°F [37.8°C] hotter, (*d*) 200°F [93.8°C] hotter.
5. The conical type of valve-spring-retainer lock fits between the: (*a*) valve stem and spring, (*b*) valve stem and spring retainer, (*c*) valve lifter and stem.
6. The I-head valve train includes two items not found in the L-head valve train. These are the: (*a*) pushrod and lifter, (*b*) spring retainer and lock, (*c*) pushrod and rocker arm, (*d*) rocker arm and valve lifter.
7. In the I-head valve train the adjustment screw is located in the: (*a*) rocker arm, (*b*) pushrod, (*c*) valve lifter.
8. Valve rotation improves valve life since it tends to prevent the accumulation of deposits on the: (*a*) valve stem and head, (*b*) valve margin and face, (*c*) valve stem and face.
9. In the free-valve type of valve rotator, valve-spring pressure on the valve stem is relieved as the tip cup moves down against the: (*a*) valve head, (*b*) spring-retainer lock, (*c*) valve spring, (*d*) roller balls.
10. As the hydraulic valve lifter moves up, opening the valve, the ball-check valve in the plunger is: (*a*) opening, (*b*) closing, (*c*) open, (*d*) closed.
11. When the valve is seated, it is in contact with two stationary parts, the: (*a*) seat and guide, (*b*) seat and retainer, (*c*) seat and retainer lock.
12. One advantage of using an exhaust-valve-seat insert is that the ring: (*a*) is more easily machined, (*b*) wears in more quickly, (*c*) withstands high exhaust-gas temperatures better.
13. One of the important reasons for the use of a valve guide, rather than a hole drilled in the block (or head), is that the guide: (*a*) wears in more quickly, (*b*) can be replaced when it wears, (*c*) is more easily machined.
14. In the cam, the distance between the base circle and the nose is called the: (*a*) flank, (*b*) lobe, (*c*) nose, (*d*) lift.
15. The opening ramp on the cam for the hydraulic lifter is: (*a*) a constant-velocity ramp, (*b*) a constant-acceleration ramp, (*c*) an accelerated ramp.

Definitions and Lists In the following you are asked to write the function or define the purpose of different engine parts, or to make lists. Write these in your notebook. The act of writing these items does two things: It tests your knowledge, and it also helps fix the information more firmly in your mind. Turn back into the chapter if you are not sure of an answer, and reread the pages that will give you the information you need.

1. List the parts in the L-head valve train.
2. List the parts in the I-head valve train.
3. What are three advantages of rotating the valves?
4. List the parts in a hydraulic valve lifter. Describe its operation.
5. What determines the amount that the valve will move as it opens?
6. What determines the number of degrees of cam rotation that the valve will stay open?
7. Explain the difference between cams for mechanical and hydraulic lifters.

SUGGESTIONS FOR FURTHER STUDY

To study valves and valve mechanisms further, go to a friendly service shop where engine repair work is done, so that you can see various types. Handle these parts if you can, and notice how they are constructed and put together. You will be able to handle and inspect these parts in your own school automotive shop, of course. Your school shop may have cutaway and working models that will help you improve your understanding of how valve mechanisms operate. You will also find factory shop manuals of interest. These manuals are published by engine and accessory manufacturers for the benefit of automotive mechanics. They contain much valuable information on the construction and servicing of automotive engines. A careful study of such manuals will be of value to you.

chapter 10

ROTARY ENGINES: TURBINE AND WANKEL ENGINES

Most of our discussion of engines so far has been centered on the piston engine. The piston engine is called a reciprocating engine *because the pistons move up and down in the cylinders. In this chapter we describe two types of* rotary engines: *the gas-turbine and the Wankel, or rotary-combustion, engine. Development work has been going on with both of these engines for a number of years. Now, they are beginning to appear in automotive vehicles on our highways.*

⊘ 10-1 Turbine The gas-turbine engine burns fuel at a steady rate. The resulting high-pressure gas is directed at the blades of a turbine rotor. When gas at high pressure hits the blades, it forces the rotor to spin (Fig. 10-1). The principle of the turbine is simple, but the actual machinery required to produce power is complex.

Figure 10-2 is an outside view of a turbine engine. Figure 10-3 is a cutaway view of the engine. Figure 10-4 is a simplified schematic diagram showing the various parts of the turbine engine. Study Figs. 10-3 and 10-4 as we describe the operation of the turbine. Note that the turbine has two basic sections: a gasifier section and a power section. The gasifier section brings in and compresses the air, mixes it with fuel, and burns the mixture in the combustor. The resulting high-pressure gas flows to the power-turbine rotor and spins the rotor. This rotary motion is carried through reduction gears to the wheels of the vehicle.

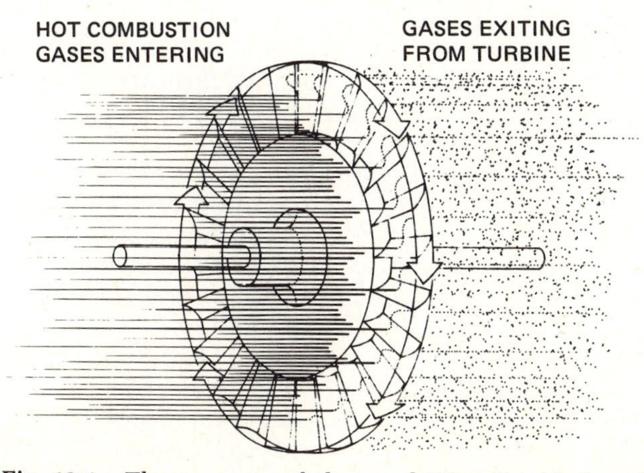

Fig. 10-1. The pressure of the gas hitting the blades of the turbine rotor forces the rotor to spin.

⊘ 10-2 Turbine Regeneration System After the gas passes through the power-turbine rotor, it is still hot. In other words, there is still a lot of energy in the gas. A regeneration system uses this heat, which would otherwise be wasted, to make the turbine more efficient. In the regeneration system of one

Fig. 10-2. Turbine engine. (*Ford Motor Company*)

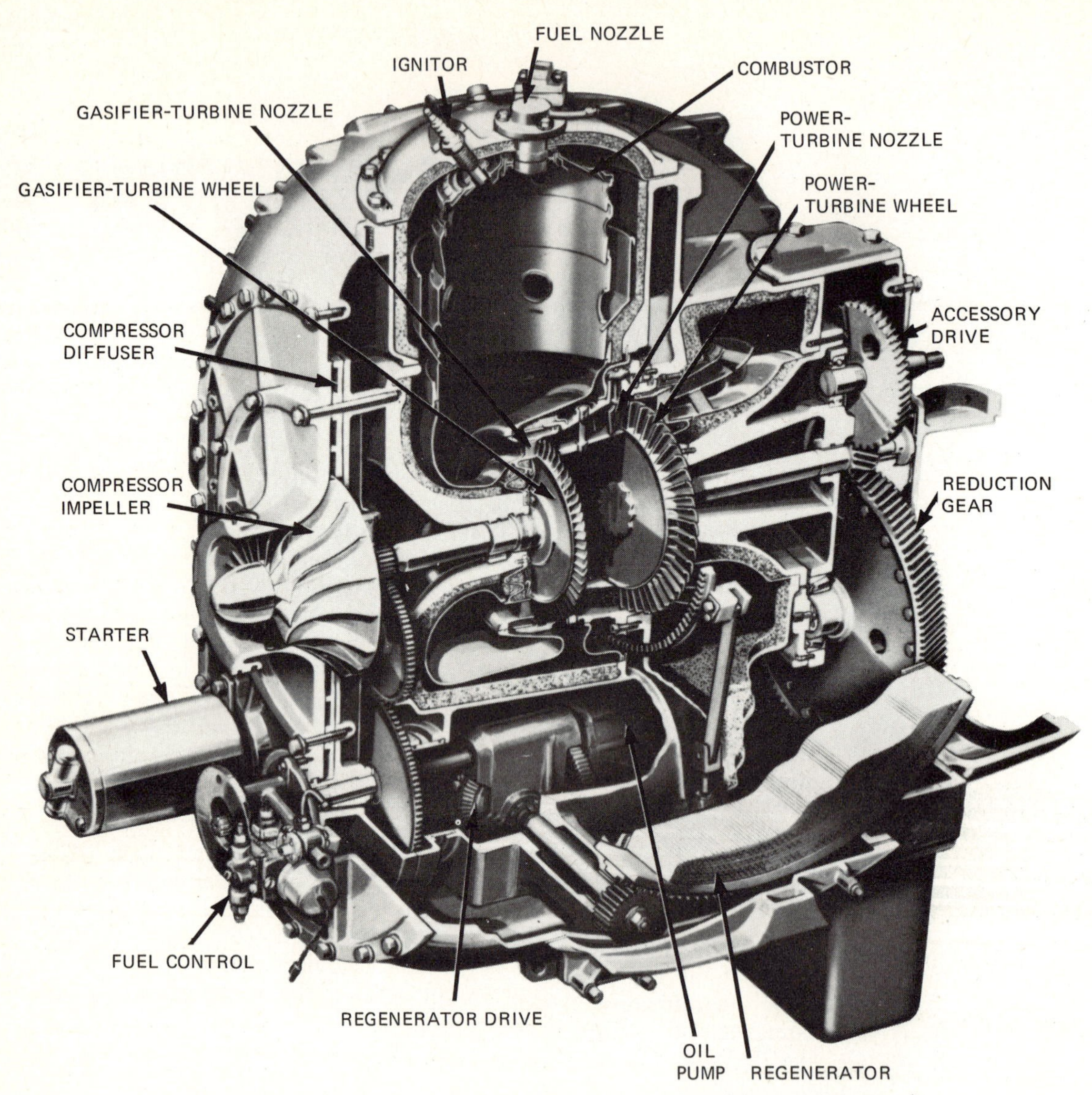

Fig. 10-3. Cutaway view of a turbine engine. (*Ford Motor Company*)

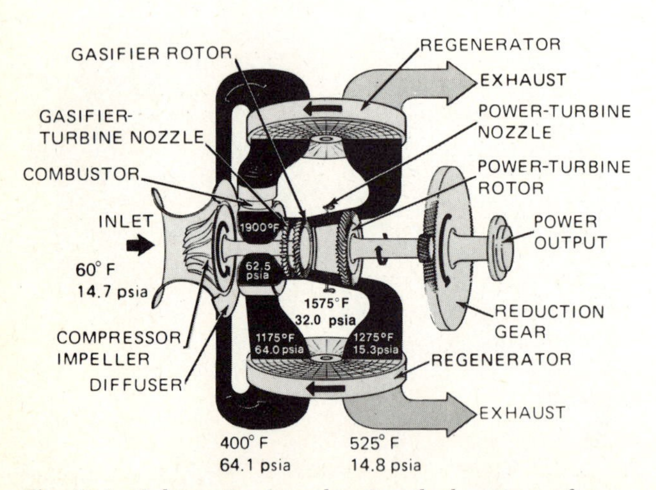

Fig. 10-4. Schematic view showing the locations of components in a turbine engine. (*Ford Motor Company*)

turbine model, two porous ceramic wheels are used. These wheels, shown at the top and bottom in Fig. 10-4, are 28 inches [711.2 mm] in diameter and about 2 inches [50.8 mm] thick. The driving arrangement for the regenerator wheels is shown in Fig. 10-5. (The turbine is shown spinning at 37,300 rpm, but the speed is reduced to 15 rpm through a series of gears.)

The regenerator wheels rotate at slow speed. The hot gas from the power-turbine rotor must pass through the porous wheel before it can leave the turbine. As the hot gas passes through the porous wheel, much of its heat is given up. For example, in Fig. 10-4 note that the temperature of the gas is 1275 degrees Fahrenheit [690.6°C] as it enters the porous wheel of the regenerator. Its temperature is 525 degrees Fahrenheit [273.9°C] as it leaves the wheel. The ceramic wheel has picked up this heat. As the wheel rotates, it carries this heat to the inlet area, where air from the compressor enters the regenerator.

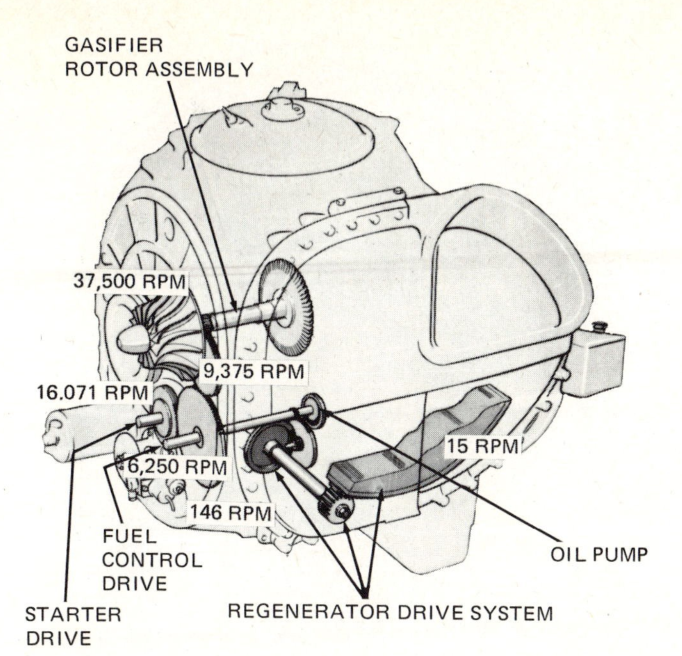

Fig. 10-5. Driving arrangement for the regenerator wheels. (*Ford Motor Company*)

This air is relatively cool, about 400 degrees Fahrenheit [204.4°C]. As the air passes through the hot porous wheel, it is heated to 1175 degrees Fahrenheit [635°C]. Thus the air going into the combustor is already hot. The air is heated further by burning fuel with it, which raises the pressure also. Thus, there is a stronger push on the turbine blades. The efficiency of the turbine is increased greatly by using air that is already hot.

NOTE: The turbine described here is the Ford turbine. Other companies have different designs. However, all turbines are essentially alike in that air is mixed with fuel and the mixture is burned in a combustor. The hot gases that result then flow through a turbine, which is spun at a high speed. Note also that the speeds and temperatures shown in Figs. 10-4 and 10-5 are for one particular operating condition. The speeds and temperatures will vary as different loads are put on the turbine.

⊘ 10-3 History of the Wankel Engine The Wankel engine has created quite a stir in recent years. However, it is not really a "new" engine, as some people believe. Felix Wankel, the German engineer who developed it, got his first engine going in 1957, after years of experimentation. Since then, the Wankel engine has been improved, and several automobile manufacturers are making them or are at least planning to make them. The Toyo Kogyo Company, a Japanese firm, has produced several hundred thousand Wankel engines for its Mazda cars. General Motors is considering the development of a small car powered by a Wankel engine. Some experts say that the Wankel is the engine of tomorrow. Whether or not the experts are right, you, as an automotive mechanic, should know what the Wankel engine is all about.

⊘ 10-4 Construction of the Wankel Engine We shall describe Toyo Kogyo's Wankel engine in detail. This engine is used in their Mazda automobiles. However, regardless of who makes the Wankel engine, the construction and operation of the engine will be similar.

Toyo Kogyo calls its Wankel the *Mazda rotary*. It is also known as a rotary-combustion (RC) engine because of the way it operates; that is, the combustion chambers rotate.

Figure 10-6 is a cutaway view of a two-rotor Wankel engine, with an attached torque converter and transmission. Since both rotors work the same way, we shall focus on the actions of a single rotor.

The rotor is enclosed on its two sides by flat housings. The rotor housing, in which the rotor turns, is shaped somewhat like a very fat figure 8 (Fig. 10-7). This shape is called an *epitrochoid curve*. The rotor has three lobes, or apexes (Fig. 10-8), and turns in the rotor housing. The rotor is sealed on its two sides to the two flat side housings. Figure 10-9 shows how the rotor fits into the rotor housing. The seals on the tips of each lobe are in contact with the inner face of the rotor housing. The rotor rotates in an eccentric pattern in the housing, as shown in Fig. 10-10. As the rotor turns, each of the three rotor lobes follows the epitrochoid curve. In other words, each of the seals on the lobes fits tightly against the inner face of the housing, providing a tight seal. This produces three separate chambers that are sealed from one another. These chambers increase and decrease in volume, as shown in Fig. 10-10. This action compares with the increase and decrease in volume in the cylinder of a reciprocating engine as the piston moves down and up.

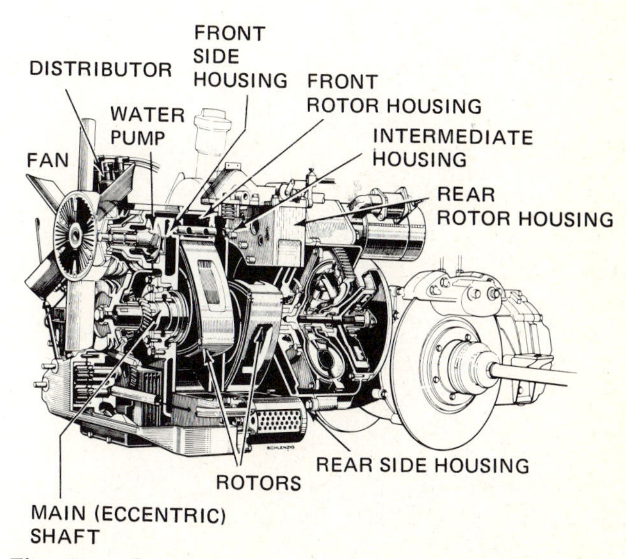

Fig. 10-6. Cutaway view of a two-rotor Wankel engine with attached torque converter and transmission. (*NSU of Germany*)

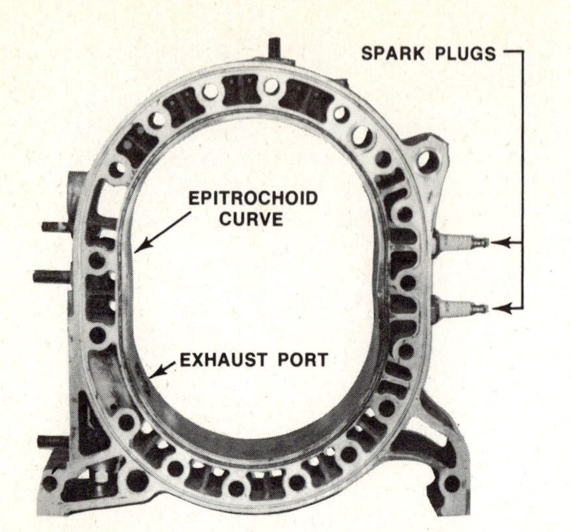

Fig. 10-7. Rotor housing. (*Toyo Kogyo Company, Limited*)

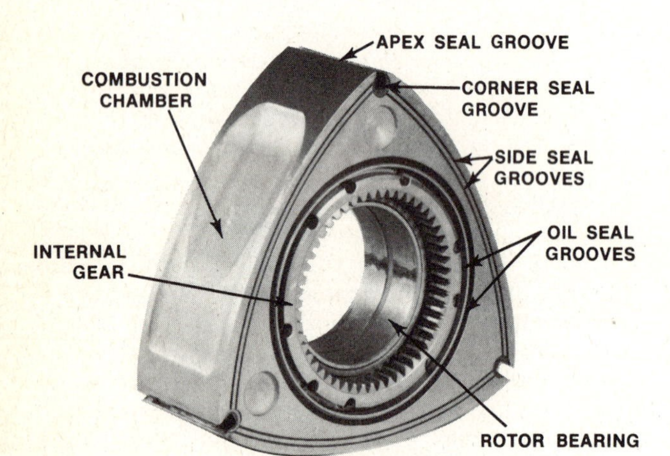

Fig. 10-8. Rotor for a Mazda Wankel engine. (*Toyo Kogyo Company, Limited*)

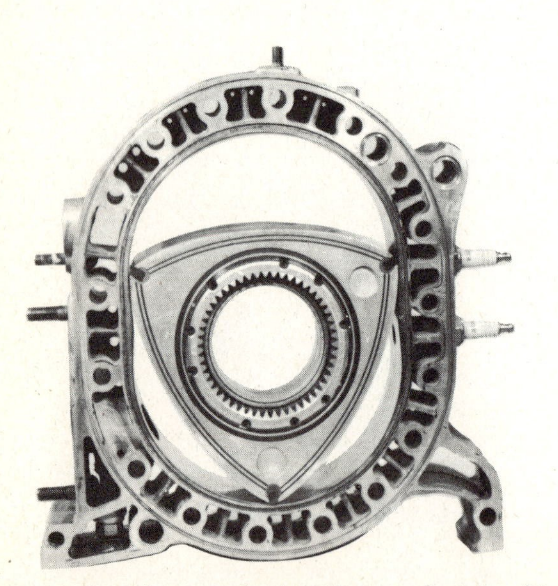

Fig. 10-9. How the rotor fits into the rotor housing. (*Toyo Kogyo Company, Limited*)

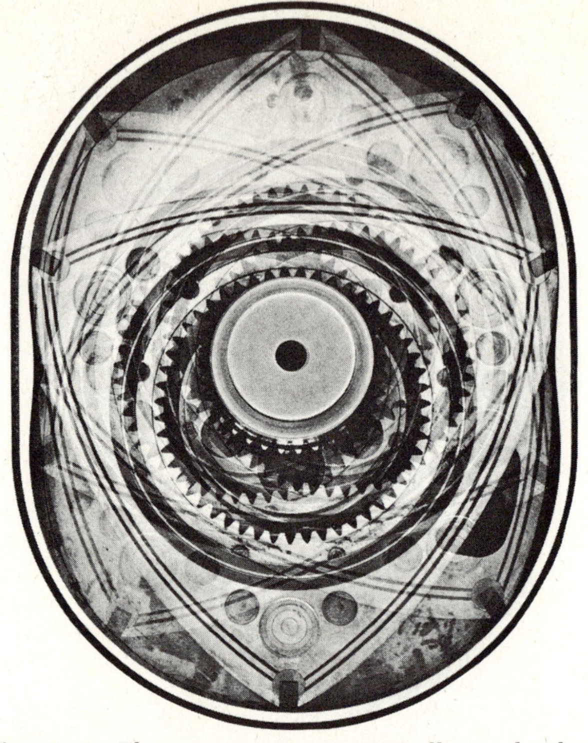

Fig. 10-10. The rotor rotates eccentrically, so the three apexes are always in sliding contact with the inner face of the rotor housing. (*Toyo Kogyo Company, Limited*)

The two sides of the rotor are enclosed and sealed by two flat-faced side housings. The front housing encloses one side of the rotor, and the rear housing encloses the other side. Figures 10-11 and 10-12 show the front and rear housings. Seals in the sides of the rotor provide a good seal between the rotor and the housing.

If the engine has two rotors, an intermediate housing is provided between the rotors, as shown in Fig. 10-6. Figure 10-13 shows the intermediate housing. Figure 10-14 shows how the side housings stack up in a one- and two-rotor Wankel engine. Notice the holes around the outer edges of all these housings (Figs. 10-11 to 10-13). These holes are for coolant circulation and are part of the engine cooling system. (We shall explain the cooling system in ⊘ 10-10.)

⊘ 10-5 Wankel-Engine Operation Figure 10-15 shows the complete series of actions that occur during one revolution of the rotor. There are four stages in the action cycle. Let us start at A (upper center) in Fig. 10-15. Here, the rotor has moved around so that one of the rotor lobes has cleared the intake port, as shown at 1. As the rotor continues to rotate (clockwise) in Fig. 10-15, the space between the rotor and housing increases, as shown at 2 in B. This produces a partial vacuum, which causes air-fuel mixture to enter, as shown by the small arrow under 2. This is the same action as in the piston engine when the piston moves down on the intake stroke.

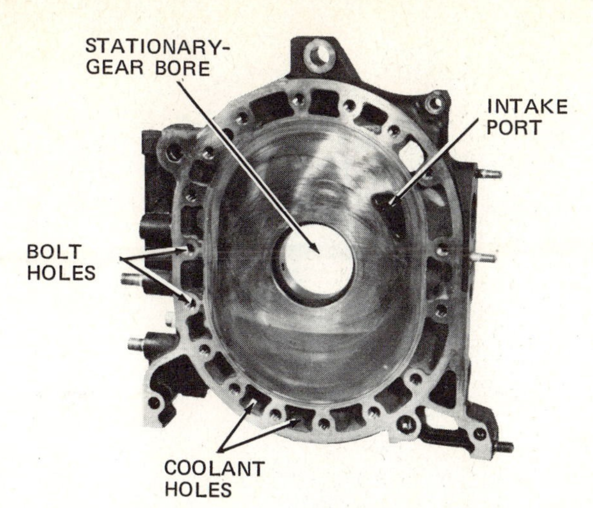

Fig. 10-11. Front side housing. Note the large holes around the outer edge for coolant flow. Note also the intake port to the right of the stationary gear. The small round holes are bolt holes. (*Toyo Kogyo Company, Limited*)

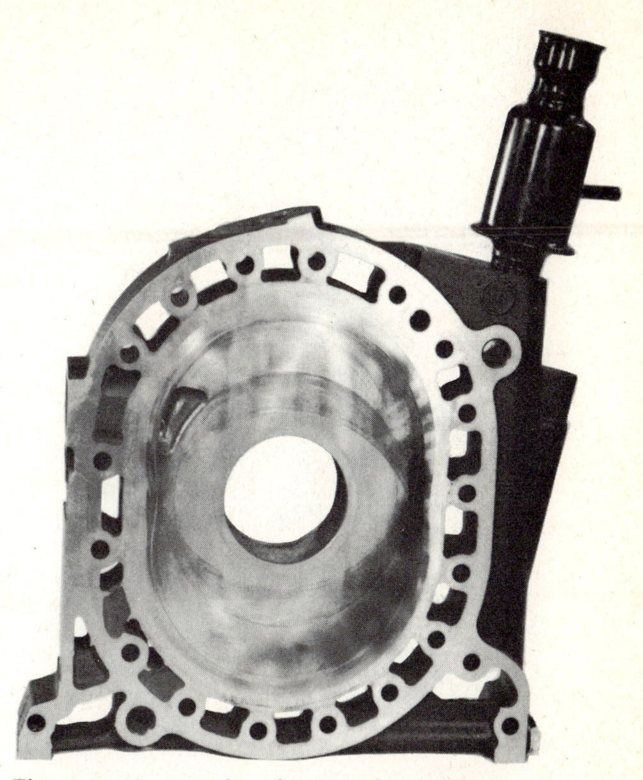
Fig. 10-13. Intermediate housing showing the coolant and bolt holes. (*Toyo Kogyo Company, Limited*)

Fig. 10-12. Rear side housing. Note the coolant holes and the bolt holes. (*Toyo Kogyo Company, Limited*)

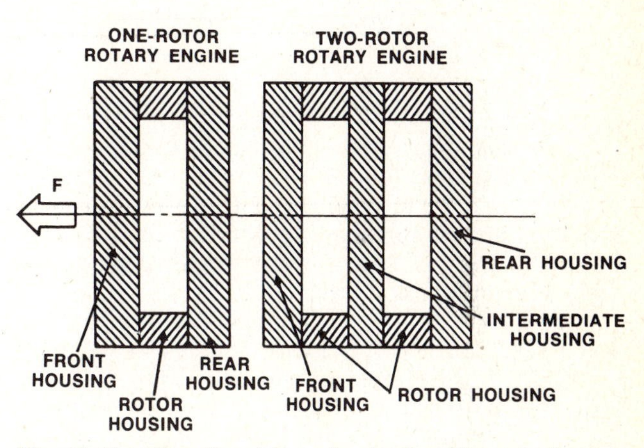

Fig. 10-14. How the side and rotor housings stack up in a one-rotor and a two-rotor Wankel engine. (*Toyo Kogyo Company, Limited*)

As the rotor moves farther around, the space between it and the housing continues to increase. (See 3 in C and 4 in D.) When the rotor reaches the point shown in D, the trailing lobe passes the intake port. Now, the air-fuel mixture is sealed between the two lobes of the rotor, as shown at 5 in E.

Now let us follow the air-fuel mixture as the rotor continues to turn. Look at F. Here, the mixture (6 in F) is starting to be compressed. The compression continues through 7 in A. At 8 in B, the mixture is nearing maximum compression. This is the same action as the piston approaching TDC on the compression stroke.

Next, combustion takes place. The combustion chamber formed by the housings and rotor is long and narrow. At 9 in C, the spark plugs fire and the compressed air-fuel mixture is ignited. Now, the hot gases push against the rotor and turn it farther around, as shown at 10 in D. The hot gases continue to expand, as shown at 11 in E and 12 in F. This is the same as the power stroke in the piston engine.

Note that the engine uses two spark plugs. The two plugs ensure more complete burning of the air-fuel mixture because the combustion is started at two points in the mixture. Exhaust emissions are thereby reduced. However, other Wankel engines operate satisfactorily with a single spark plug.

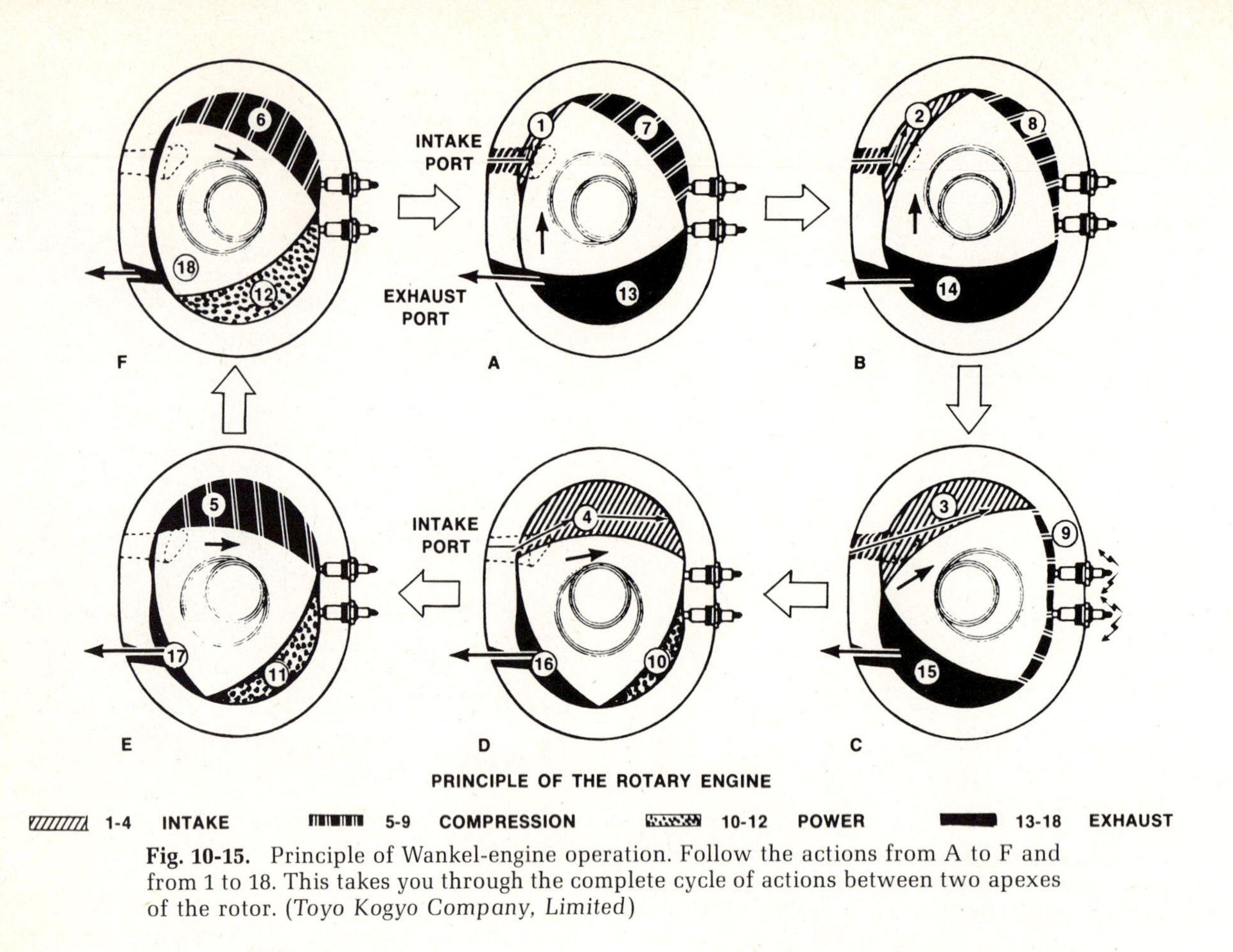

Fig. 10-15. Principle of Wankel-engine operation. Follow the actions from A to F and from 1 to 18. This takes you through the complete cycle of actions between two apexes of the rotor. (*Toyo Kogyo Company, Limited*)

At 13 in A, the leading lobe of the rotor has cleared the exhaust port in the housing. Now, the burned gases begin to exhaust from the space between the rotor lobes. This exhaust continues through 14 in B, 15 in C, 16 in D, 17 in E, and 18 in F. By that time, the leading rotor lobe has cleared the intake port, as shown at 1 in A. Now the whole chain of events takes place again.

We have looked at the actions taking place between one pair of rotor lobes. But note that there are three lobes and three chambers between the lobes. Thus three sets of actions occur at the same time in the engine. In other words, there are three power thrusts for every revolution of each rotor.

⊘ 10-6 Getting the Power to the Crankshaft Now let's see how the rotor transmits power to the crankshaft, or, more accurately, to the *eccentric shaft.* Figure 10-16 shows the eccentric shaft of a two-rotor engine. The shaft is supported by two main bearings. It has two rotor journals, which are offset, as shown. The offset rotor journals on the eccentric shaft do the same job as the crankpins on the crankshaft of a piston engine. Figure 10-17 shows how a rotor fits on one of the eccentric rotor journals; it also shows how the push on the rotor makes the eccentric shaft rotate. The arrows show the high-pressure gas pushing against the rotor. Most of the push occurs below the center of the eccentric shaft, which makes the shaft turn at the same time the rotor revolves.

A stationary gear meshes with the internal gear of the rotor to control the rotation of the rotor in the proper manner. You can see the internal gear in the rotor in Fig. 10-8. The stationary gear is shown in Fig. 10-18. The stationary gear is installed in the side housing and contains the bearing that supports one end of the eccentric shaft. Figure 10-19 shows how the stationary gear keeps the rotor moving in the proper path. The gear ratio between the rotor gear and the stationary gear is 2:3.

⊘ 10-7 The 1:3 Ratio You might assume that the rotor and eccentric shaft rotate together at the same speed. They do not. The eccentric shaft rotates three times while the rotor turns only once. This is a complicated movement, and you will have to think about it for a moment. Figure 10-20 will help you figure it out.

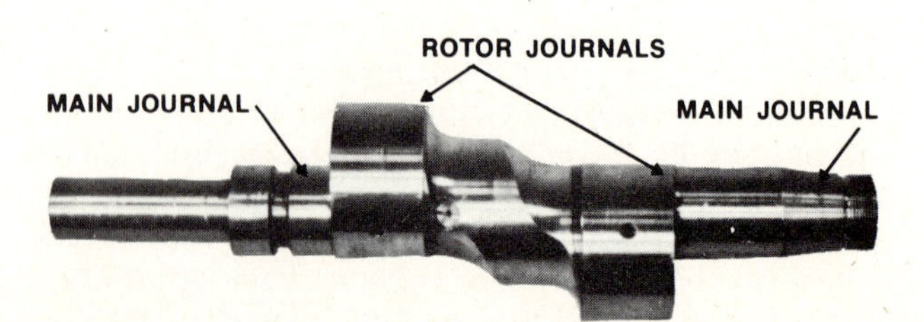

Fig. 10-16. Eccentric shaft. (*Toyo Kogyo Company, Limited*)

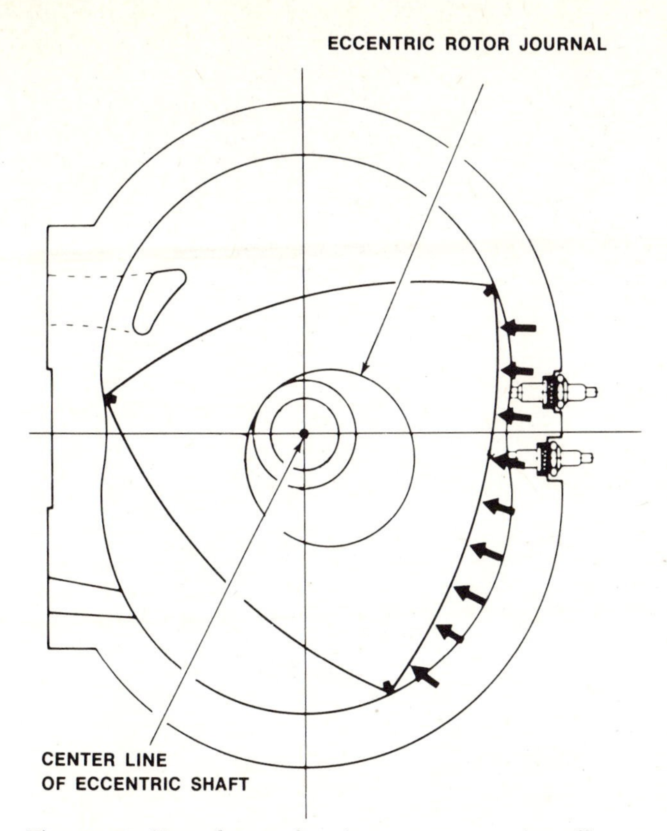

Fig. 10-17. How the combustion pressure, acting off center on the eccentric of the eccentric shaft, forces the shaft to rotate. (*Toyo Kogyo Company, Limited*)

Fig. 10-18. Stationary gear which is mounted in the side housing. (*Toyo Kogyo Company, Limited*)

Let's follow the rotor during one power thrust. Start at A in the upper-left corner of Fig. 10-20. The lobe marked "X" is at the lower right. The center line of maximum crank (journal) eccentricity is pointing to the right. Note that in B the rotor has turned only 30 degrees; that is, the rotor lobe X has advanced only 30 degrees. But, at the same time, the center line and the eccentric shaft have advanced 90 degrees. Now notice how the centerline and the eccentric shaft continue to move ahead of lobe X as it moves through C, D, E, and F. In F, lobe X has turned only 90 degrees from its position at A. But

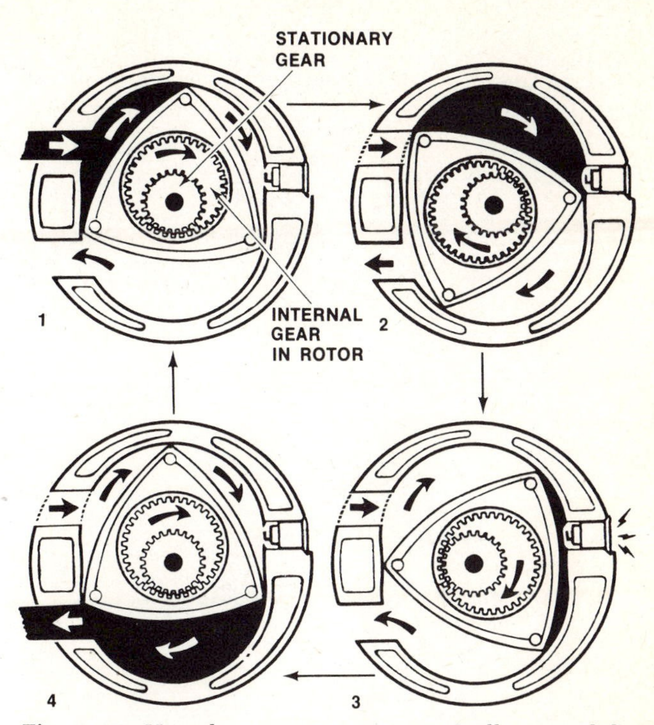

Fig. 10-19. How the rotor rotates eccentrically around the stationary gear. The rotor follows an orbit that keeps all three apex seals in sliding contact with the rotor housing. (*Toyo Kogyo Company, Limited*)

the centerline and the eccentric shaft have turned 270 degrees. When the lobe turns 30 degrees more, or to the position X′ in A, it will have turned 120 degrees, or one-third of a complete revolution of 360 degrees. By the time the rotor makes one complete revolution, the eccentric shaft will have turned three times; that is, there is a 1:3 ratio.

⊘ 10-8 Ignition System Figure 10-21 is a wiring diagram of an ignition system for a two-rotor engine. Note that there are two separate ignition distributors and two ignition coils. The leading distributor and coil are connected to the leading spark plugs in the two rotor housings. The trailing distributor and coil are connected to the trailing spark plugs in the two housings. The leading plugs are the plugs that are ahead of the trailing plugs in the direction of rotor rotation. The two-plug arrangement shown in Fig. 10-21 is used in the ignition system of the Toyo Kogyo Mazda Wankel engine. Other Wankel engines may use only one plug in each rotor housing.

NOTE: Do not be confused by the fact that there are, in effect, two ignition systems. Each ignition system handles two plugs, one plug in each rotor housing.

⊘ 10-9 Fuel System Figure 10-22 shows the location of the intake manifold and carburetor on a two-rotor rotary-combustion (RC) engine. The carburetor is a four-barrel unit, which is similar to the

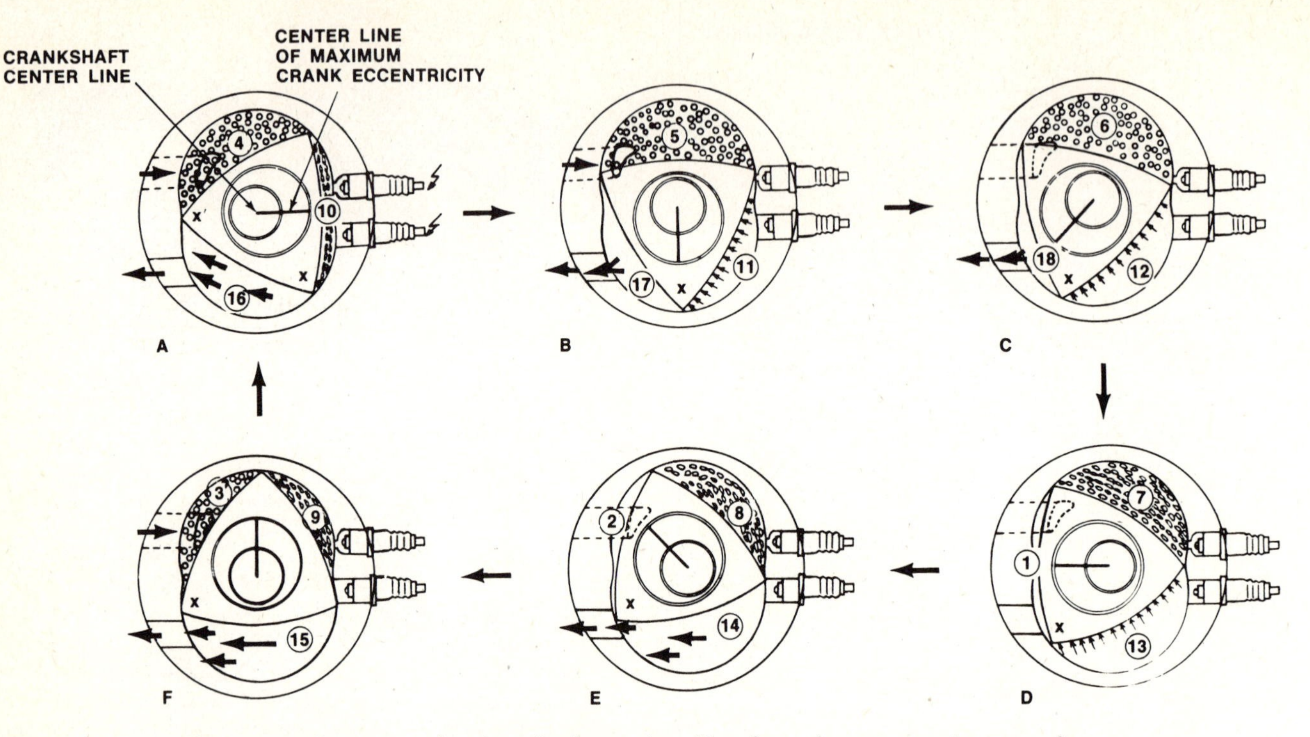

Fig. 10-20. Sequence of actions in the engine. Note how the rotor, as it moves from A to F and back to A again, causes the eccentric shaft to rotate one full revolution, or 360 degrees, even though the rotor has turned only 120 degrees. (*Toyo Kogyo Company, Limited*)

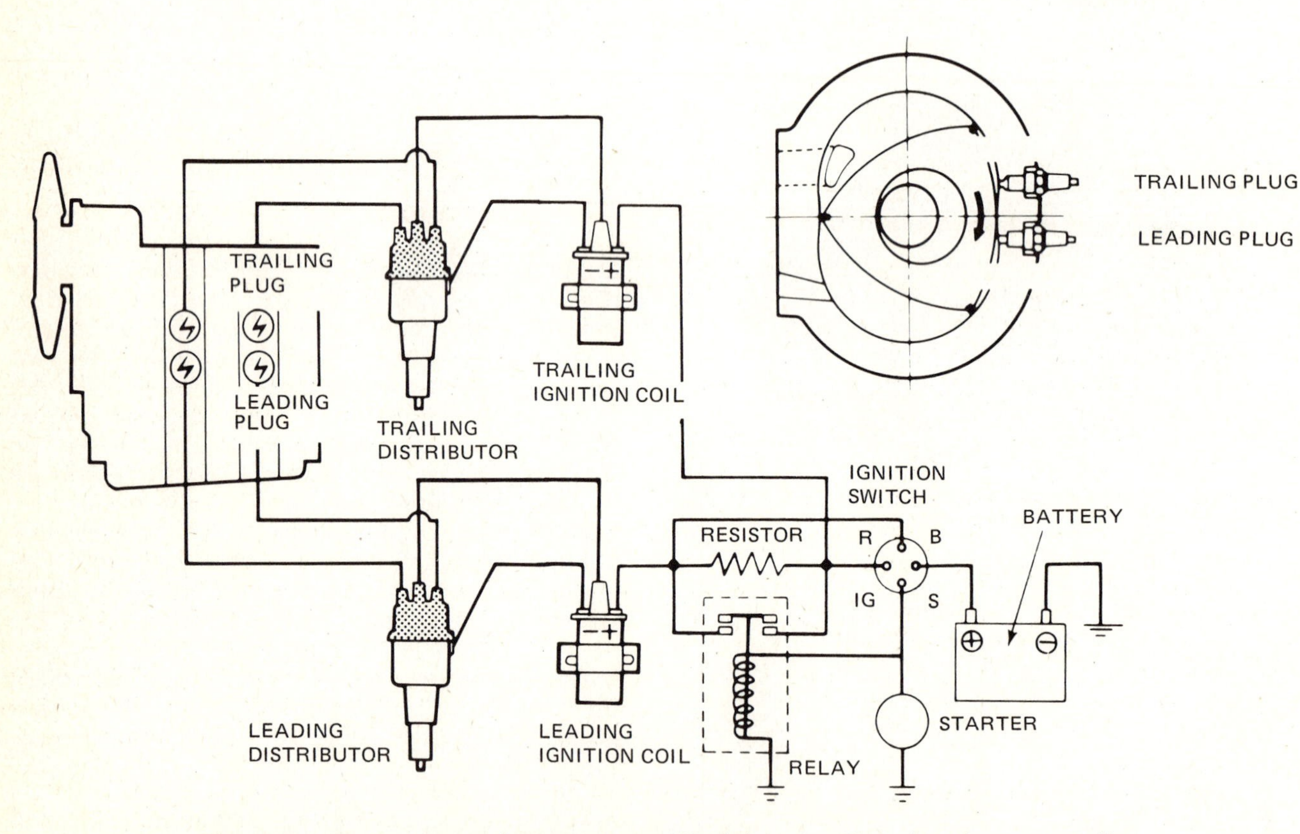

Fig. 10-21. Ignition system for a two-rotor Wankel engine. Note that there are really two ignition systems: one that fires the trailing plugs and one that fires the leading plugs. (*Toyo Kogyo Company, Limited*)

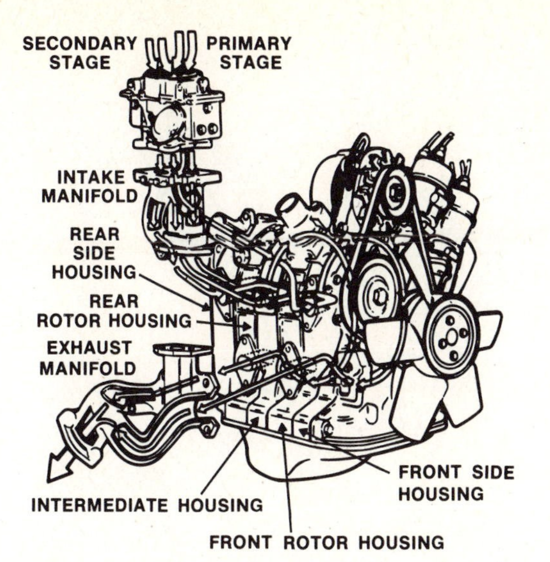

Fig. 10-22. Location of the manifold and carburetor. (*Toyo Kogyo Company, Limited*)

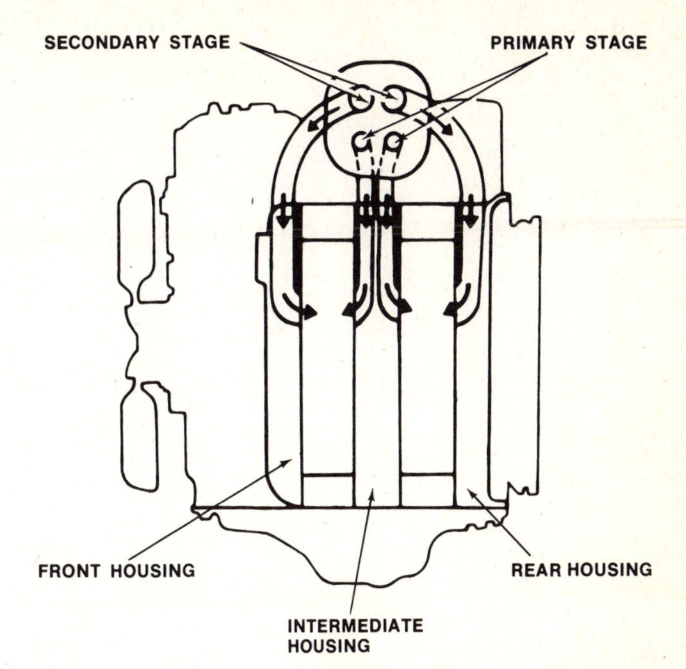

Fig. 10-23. Schematic view of the carburetor. Note how the two carburetor stages supply air-fuel mixture to the intake ports in the three side housings. (*Toyo Kogyo Company, Limited*)

four-barrel carburetor used on a reciprocating engine. These carburetors are discussed in Chap. 14.

The arrows in Fig. 10-22 indicate the paths the air and the air-fuel mixture take on their way to the engine. Note that the primary stage of the carburetor feeds the two rotors through intake ports in the intermediate housing. (You can see one of these ports in Fig. 10-13.) The primary stage provides the air-fuel mixture for all operating conditions through intermediate speed. For acceleration and full-power operation, the secondary stage comes into operation. In Fig. 10-22 note that the secondary stage feeds the air-fuel mixture through intake ports in the outer side housings. Figure 10-23 is a simplified schematic diagram showing how the two stages of the carburetor work.

⊘ 10-10 Cooling System Figure 10-24 shows the cooling system for the Toyo Kogyo Mazda Wankel engine. The housings are shown separated so that the flow of coolant (indicated by arrows) can be seen. The water pump in the Mazda Wankel engine is similar to the water pump used in the reciprocating engine. The water pump picks up the coolant from the bottom of the radiator and sends the coolant through the upper set of water jackets in the housings. The coolant flows through the upper water jackets of the front side housing, front rotor housing, intermediate housing, rear rotor housing, and rear side housing. In the rear side housing, the coolant reverses direction and returns through the lower water jackets of all the housings on its way back to the radiator.

Figure 10-25 is an end-sectional view of the intermediate housing, showing the upper and lower water jackets.

Just as a thermostat is used in the cooling system of the reciprocating engine, a thermostat is used in the cooling system of the Wankel engine, as shown in Figure 10-26. When the engine is cold, the thermostat valves are in the upper position. This blocks coolant circulation to the radiator. Coolant therefore continues to circulate between the pump and the engine through the bypass hole until the engine warms up. Then, the thermostat opens. This shuts off the bypass hole and opens the passage to the radiator. Now, the cooling system operates in the hot-engine mode.

⊘ 10-11 Lubricating System Figure 10-27 shows the lubricating system of the Wankel engine. The oil pump is located at the bottom of the engine (lower left) and is driven by a chain from the eccentric shaft. The arrows show the flow of oil from the oil pump to the engine. Part of the oil flows through the oil cooler, which keeps the oil temperature down. The rest of the oil flows through the bottom oil passage in the engine.

The oil flowing directly from the oil pump and the oil flowing from the oil cooler meet at the end of the engine and then flow to the oil filter. From the oil filter, the oil takes two paths. Following one path, the oil goes through the hollow eccentric shaft and lubricates the bearings and gears. This oil then drops down from the bearings and gears and returns to the oil pan at the bottom of the engine.

Following the other path, the rest of the oil flows through a hollow tubular dowel to the distributors and metering oil pump. The metering oil pump sends the proper amount of oil to the carburetor, where the oil mixes with the air-fuel mixture entering the engine. In the engine the oil lubricates the seals on the rotor. This arrangement is much like

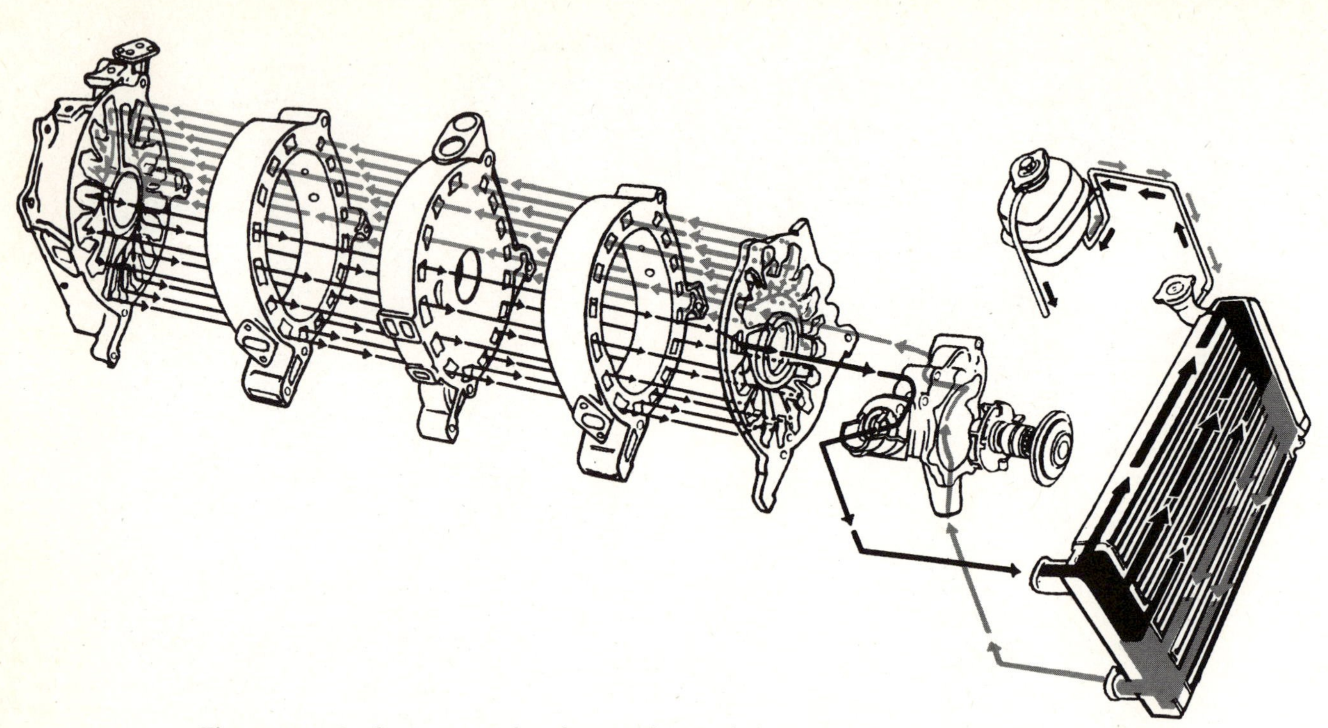

Fig. 10-24. Cooling system for the Mazda Wankel engine. (*Toyo Kogyo Company, Limited*)

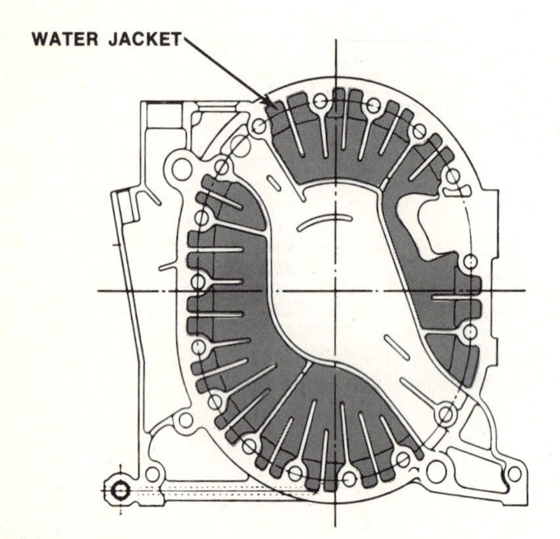

Fig. 10-25. Sectional view from the side of the intermediate housing showing the water jackets. (*Toyo Kogyo Company, Limited*)

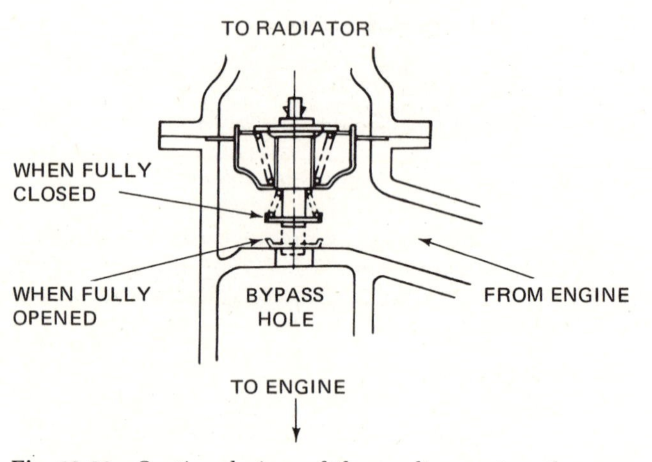

Fig. 10-26. Sectional view of the cooling-system thermostat showing the two extreme positions for cold- and hot-engine performance. (*Toyo Kogyo Company, Limited*)

the one in the two-cycle reciprocating engine. In the two-cycle engine, the oil is mixed with gasoline to provide lubrication of the bearings, piston, and rings.

The metering oil pump (Fig. 10-28) sends oil to the carburetor float bowl according to throttle opening. The metering oil pump is driven by the distributor drive gear installed at the front of the eccentric shaft. As the metering oil pump is driven, the plunger moves back and forth. The longer the stroke, the more oil is sent to the carburetor. The length of the stroke is determined by the linkage of the metering oil pump to the throttle. The more the throttle is opened, the longer the stroke and the more oil is delivered to the float bowl in the carburetor. When the oil reaches the carburetor float bowl, it mixes with the gasoline before entering the fuel nozzles in the carburetor barrels.

⊘ 10-12 Wankel-Engine Serviceability As you can tell from studying this chapter, the Wankel engine operates on the same four-cycle arrangement as the piston engine: intake, compression, power, and exhaust. The Wankel engine also has the same ignition, lubricating, fuel, and cooling systems as the piston engine. Many of the components in these systems are alike in both the Wankel and the piston engine,

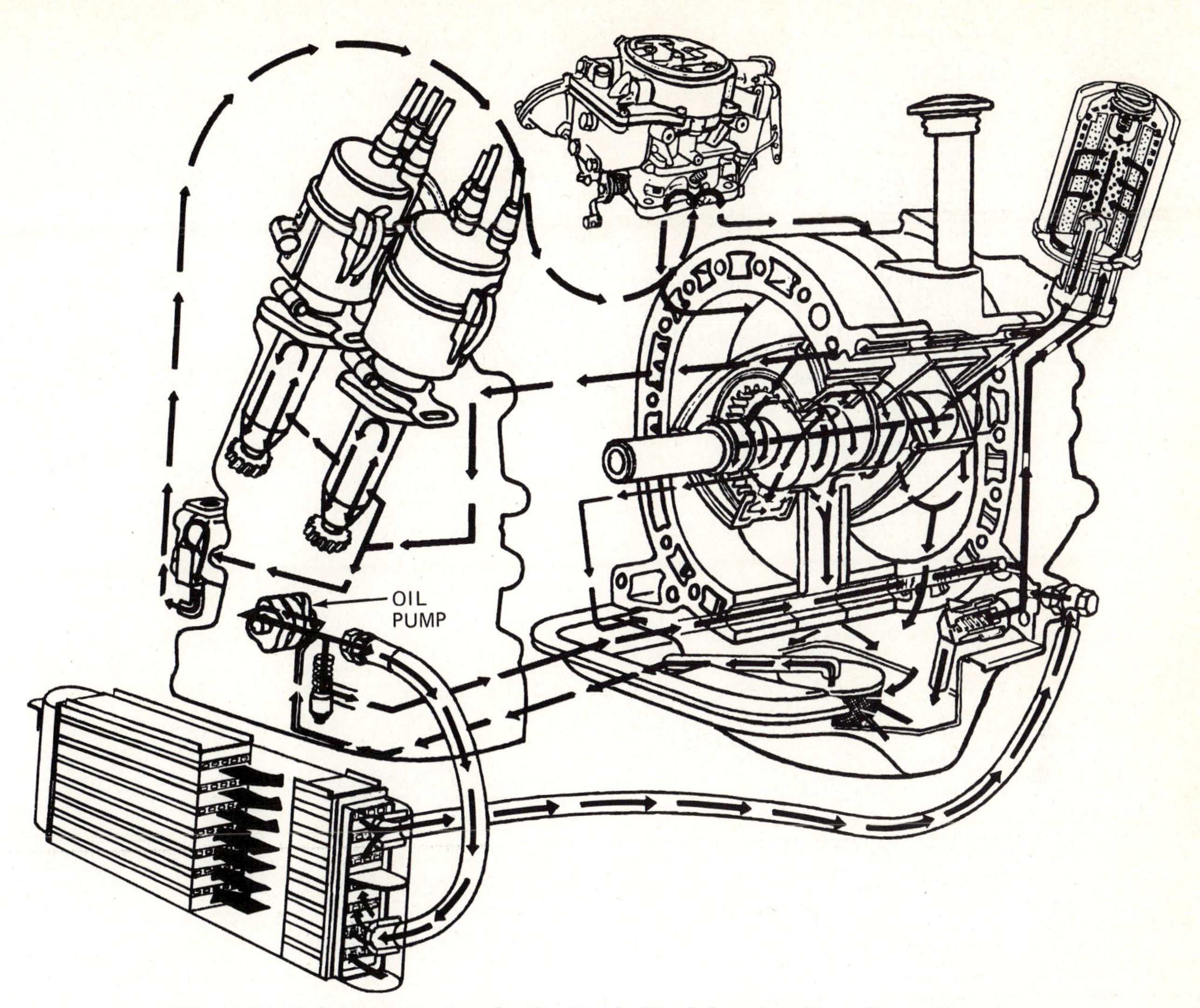

Fig. 10-27. Lubricating system for the Mazda Wankel engine. (*Toyo Kogyo Company, Limited*)

and they are serviced in the same way. However, the Wankel engine has fewer parts than the piston engine, and therefore fewer service operations are required (see Chap. 26). In addition, the Mazda Wankel is built according to the metric system, and so the mechanic needs metric wrenches and sockets.

CHAPTER 10 CHECKUP

NOTE: Since the following is a chapter review test, you should review the chapter before taking the test.

Again, this checkup will give you a chance to find out how well you understand and remember what you have just been studying on rotary engines. If you are not sure of the answers to any of the questions, turn back into the chapter and reread the pages that will give you the information.

Completing the Sentences The sentences that follow are incomplete. After each sentence there are several words or phrases, only one of which will correctly complete the sentence. Write each sentence in your notebook, selecting the proper word or phrase to complete it correctly.

1. Two types of rotary engines are the: (*a*) turbine and piston, (*b*) steam and Wankel, (*c*) turbine and Wankel.
2. The two basic sections of the gas turbine are the (*a*) gasifier section and power section, (*b*) gasifier section and regeneration section, (*c*) regeneration section and power section.
3. Each rotor in a Wankel engine has: (*a*) three lobes, (*b*) two lobes, (*c*) one lobe.
4. The rotor in the Wankel engine rotates: (*a*) in a circle, (*b*) eccentrically, (*c*) in an oval.
5. The two sides of the rotor in the Wankel engine are enclosed and sealed by: (*a*) the front side housing, (*b*) the rear side housing, (*c*) both the front side housing and the rear side housing.
6. The holes around the outer edge of the side housings in a Wankel engine are for: (*a*) gasoline, (*b*) coolant circulation, (*c*) air-fuel mixture.

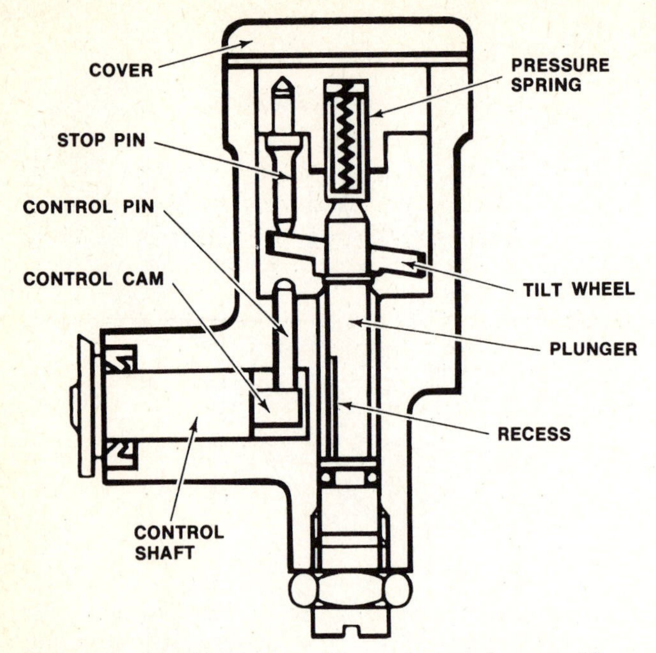

Fig. 10-28. Sectional view of the metering oil pump. (*Toyo Kogyo Company, Limited*)

7. The crankshaft in the piston engine is called the: (*a*) rotary shaft in the Wankel, (*b*) eccentric shaft in the Wankel, (*c*) camshaft in the Wankel.
8. In the Wankel engine, the eccentric shaft rotates three times while the rotor is rotating: (*a*) one time, (*b*) two times, (*c*) three times.
9. Two separate ignition systems are used on the Mazda Wankel engine, one system to fire the: (*a*) trailing plugs and the other to fire the leading plugs, (*b*) two plugs in the rear rotor housing, and the other to fire the two plugs in the front rotor housing, (*c*) neither (*a*) nor (*b*).
10. The eccentric shaft in the Wankel engine is supported by: (*a*) two main bearings, (*b*) three main bearings, (*c*) four main bearings.

Definitions and Lists In the following, you are asked to write the function or define the purpose of different engine parts. Write the answers in your notebook. The act of writing the facts will help you to remember them. Also, you will add to your notebook basic facts that you can refer to any time you want to refresh your memory. If you are not sure of an answer, reread the pages that will give you the information you need.

1. What is the basic principle of the turbine?
2. What are the two basic sections of the turbine?
3. What is the basic principle of the regenerator?
4. How many lobes, or apexes, does the Wankel rotor have?
5. How many chambers are formed by the rotor and the housing?
6. What does RC stand for?
7. How many housings does the two-rotor Wankel engine have?
8. What are the names of the housings in a two-rotor Wankel engine?
9. How many power thrusts are there for each complete rotor revolution?
10. How many times does the eccentric shaft turn as the rotor makes one complete revolution?
11. How many spark plugs does a Mazda two-rotor Wankel engine use?
12. Do all Wankel engines use the same number of spark plugs per rotor?
13. How many barrels does the Mazda two-rotor Wankel engine carburetor have?
14. In the two-rotor Wankel engine, where are the intake ports that are fed by the primary stage of the carburetor?
15. In the two-rotor Wankel engine, where are the intake ports that are fed by the secondary stage of the carburetor?

SUGGESTIONS FOR FURTHER STUDY

On a sheet of notebook paper, make two columns listing the parts in the Wankel engine and comparable parts in the piston engine. Head one column "Wankel Engine" and head the other column "Piston Engine." Start with "rotor" under "Wankel Engine" and "piston" under "Piston Engine." Then add "eccentric shaft" and "crankshaft"; next "rotor seals" and "piston rings." See the idea? You want to compare parts between the two engines. Of course, lots of the parts will not compare. For example, the Wankel has a metering oil pump and the piston engine does not. The piston engine has valves, and the Wankel does not. But the exercise of making these lists will help you understand the similarities and differences between the two engines.

Whenever you run across articles in magazines or newspapers about Wankel engines, cut them out and file them in your notebook. Also, if you are able to borrow the manufacturers' service manuals on Wankel engines, study them to learn more details. Make notes of important facts, and file the notes in your notebook. In this way, you will build up a file of important facts about Wankel engines that will be of value to you.

chapter 11

ENGINE-PERFORMANCE MEASUREMENTS

This chapter describes various ways in which engines and engine performance can be measured by cylinder diameter, length of piston stroke, horsepower, efficiency, and so on.

⊘ 11-1 Bore and Stroke The size of an engine cylinder is given by the *bore*, or diameter, and the *stroke*, or distance the piston travels from bottom dead center (BDC) to top dead center (TDC) (Fig. 11-1). Note that the bore is always mentioned first, as "a 4- by $3\frac{1}{2}$-inch [101.6 by 88.9 mm] cylinder." This means that the diameter, or bore, of the cylinder is 4 inches [101.6 mm] and the stroke is $3\frac{1}{2}$ inches [88.9 mm]. These measurements are used to figure piston displacement.

Before 1955, most engines were built with a relatively small bore and a long stroke, as for instance a 3- by 4-inch [76.2 by 101.6 mm] engine. More recently, however, engines are being designed with a large bore and a short stroke. For example, one popular Ford V-8 engine has a bore of 4 inches [101.6 mm] and a stroke of 3.50 inches [88.9 mm]. Such engines are called *oversquare engines*. (A *square* engine would have a bore and stroke of equal measurements.)

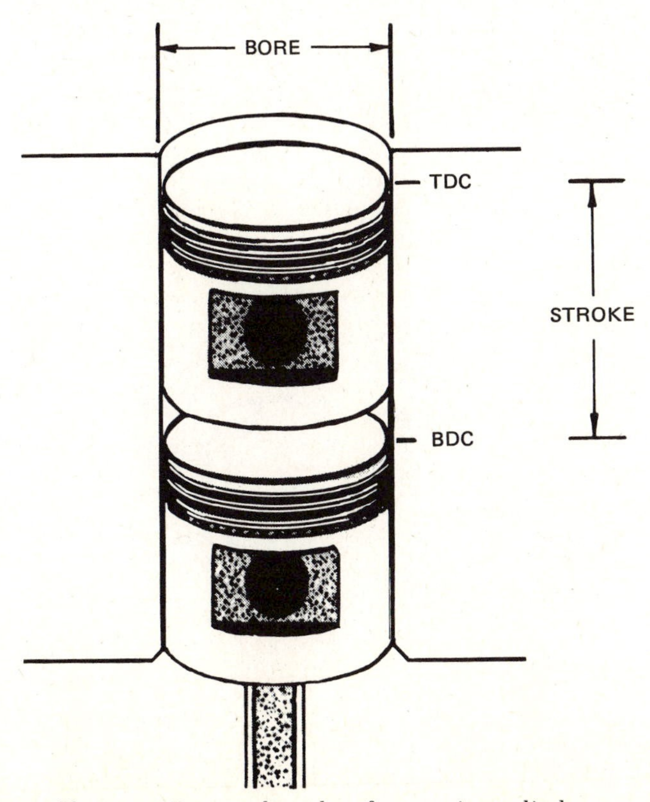

Fig. 11-1. Bore and stroke of an engine cylinder.

NOTE: In the metric system, bore and stroke are given in millimeters. Thus, a 4- by 3.5-inch cylinder would be a 101.6- by 88.9-mm cylinder.

There are several reasons for the popularity of the oversquare engine. With the shorter piston stroke, there is less friction loss because the piston does not move as far in the cylinder. Also, the shorter stroke reduces inertia and centrifugal loads on the engine bearings (see ⊘ 7-25 and 7-26). In addition, the shorter stroke permits a reduction of the engine height and thus a lower hood line.

Despite the advantages of the shorter-stroke oversquare engine, recent emphasis on reducing atmospheric pollutants in the exhaust has forced automobile manufacturers to lengthen the stroke. For example, one late-model Ford six-cylinder engine has a stroke of 3.91 inches [99.3 mm] compared with a 3.13-inch [79.5 mm] stroke for earlier years. This longer stroke, in effect, provides more burning time for better combustion. (In Chap. 18, we discuss air pollution and engine exhaust gases in detail.)

⊘ 11-2 Piston Displacement *Piston displacement* is the volume that the piston displaces, or "sweeps out," as it moves from BDC to TDC. You can picture this volume as a cylinder that is the diameter of the engine cylinder, the top and bottom being the piston-head at the TDC and BDC positions. To calculate piston displacement, you use the bore D and the length of stroke L. Thus, piston displacement of a 4- by $3\frac{1}{2}$-inch [101.6 by 88.9 mm] cylinder is the volume of a cylinder 4 inches [101.6 mm] in diameter and $3\frac{1}{2}$ inches [88.9 mm] long, or

$$\frac{\pi \times D^2 \times L}{4} = \frac{3.1416 \times 4^2 \times 3\frac{1}{2}}{4}$$

$$= \frac{3.1416 \times 16 \times 3\frac{1}{2}}{4}$$

$$= 43.98 \text{ in}^3$$

If the engine has eight cylinders, then the total displacement in the engine is 43.98 × 8, or 351.84 cubic inches.

NOTE: In the metric system, displacement is given in cubic centimeters. Thus, a 200 cubic inch displacement would be 3,280 cm^3 (cubic centimeters). And, since 1,000 cm^3 equals 1 liter (l), 3,280 cm^3 equals 3.28 l.

Displacement limitations are used in competitive racing. The greater the displacement, the more horsepower the engine can produce, other factors being equal. Therefore, by setting displacement limitations, a premium is placed on driving skill and mechanical ingenuity rather than engine size. Without the limitation, the size of the engines would get larger and larger until engine size would be the dominant factor. Thus, at the Indianapolis 500 (the Indy 500, as it is called) the maximum allowable displacement for a recent race was set at 305.1 cubic inches [5 l] for nonsupercharged engines.

In many races, the displacement is given in terms of liters. Thus, the Indy 500 specification (305.1 cubic inches) is 5 liters. (One liter is 61.02 cubic inches, and so 305.1 divided by 61.02 is 5.)

⊘ 11-3 Wankel-Engine Single-Chamber Capacity

The Wankel engine does not have pistons, and so of course you cannot figure piston displacement on the Wankel (see ⊘ 11-2). Instead, the equivalent measurement on the Wankel engine is called *single-chamber capacity*. This is the displacement the rotor produces as it rotates and the volume in one combustion chamber goes from maximum to minimum (Fig. 11-2). For example, suppose the volume is reduced 490 cm^3 [29.9 cubic inches] as it goes from maximum to minimum. This is the capacity of one chamber of the rotor. To figure total chamber capacity, you would multiply by the number of combustion chambers (three in a single-rotor engine, six in a two-rotor engine, etc.).

⊘ 11-4 Compression Ratio

The compression ratio of an engine is a measurement of how much the air-fuel mixtures are compressed in the engine cylinders. It is calculated by dividing the air volume in one cylinder with the piston at BDC by the air volume with the piston at TDC (Fig. 11-3).

NOTE: The air volume with the piston at TDC is called the *clearance volume* as it is the clearance that remains above the piston when it is at TDC.

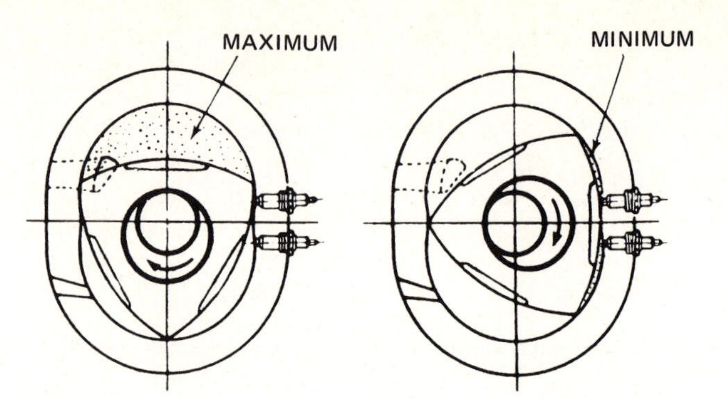

Fig. 11-2. Single-chamber capacity of a Wankel engine.

For example, the engine of one popular car has a cylinder volume of 42.35 cubic inches [649 cm^3] at BDC (*A* in Fig. 11-3). It has a clearance volume of 4.45 cubic inches [73 cm^3] (*B* in Fig. 11-3). The compression ratio is therefore 42.35 divided by 4.45 [649 ÷ 73], or 9.5/1 (that is, 9.5 : 1). In other words, during the compression stroke, the air-fuel mixture is compressed from a volume of 42.35 cubic inches [649 cm^3] to 4.45 cubic inches [73 cm^3], or to 1/9.5 of its original volume.

⊘ 11-5 Effect of Increasing the Compression Ratio

In recent years, engineers have designed engines with higher and higher compression ratios. Increasing the compression ratio of an engine offers several advantages. The power and economy of the engine also increase without a comparable increase in engine size or weight. In effect, an engine with a higher compression ratio "squeezes" the air-fuel mixture harder (compresses it more). When this happens, the air-fuel mixture gives up more power on the power stroke. Here's why: A higher compression ratio means a higher initial pressure at the end of the compression stroke. This, in turn, means that when the power stroke starts, higher combustion pressures are reached and a greater push is registered on the piston. The burning gases also expand to a greater

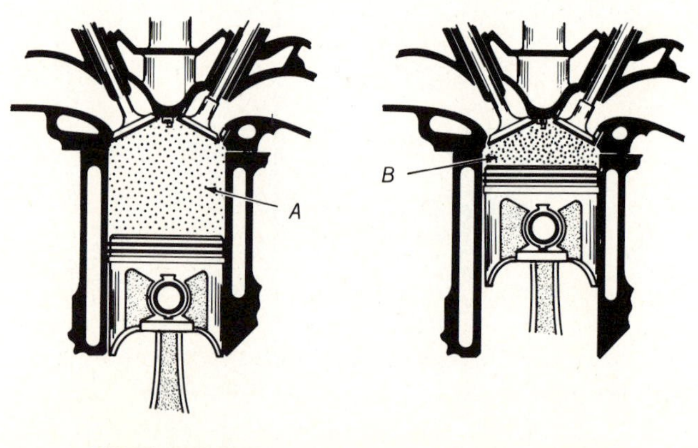

Fig. 11-3. Compression ratio is the volume in the cylinder with the piston at BDC divided by the volume with the piston at TDC, or *A* divided by *B*.

volume during the power stroke. The result is that more push is exerted on the piston for a larger part of the power stroke. Thus more power is obtained from each power stroke.

Because of these advantages, compression ratios of engines have increased year after year. In 1955, the average compression ratio of automobile engines made in the United States was less than 8:1. In 1969, it was nearly 10:1. In 1970 some high-performance engines had compression ratios of 11:1.

Increasing the compression ratio does introduce special problems, however. First, as the compression ratio goes up, the problem of detonation, or *knocking*, in the engine becomes more acute. A detailed discussion of knocking is given in ⊘ 12-6 to 12-10. For now, we'll summarize the problem. Any particular fuel will stand a certain amount of "squeezing," or compression, without causing knocking. But if it is "squeezed" further as a result of a higher compression ratio, knocking will occur. This knock, or ping, robs the engine of power and, if severe, may even cause engine parts to break. Thus, as compression ratios of engines have been increased, gasoline companies have had to supply new types of fuel that will operate in these higher-compression engines without knocking (see ⊘ 12-10 to 12-12).

A second factor to be considered in increasing the compression ratio of an engine is the effect of carbon accumulations in the cylinders. Carbon buildup results from incomplete fuel combustion. Many engines accumulate carbon, especially in city driving, where the engine works much of the time at part throttle. When carbon accumulates in an engine, it has the effect of increasing the compression ratio. The carbon reduces clearance volume, and the air-fuel mixture is therefore squeezed into a smaller volume. For example, 1 eight-cylinder engine with a compression ratio of 8.25:1 has a clearance volume of only about 5½ cubic inches [90 cm^3]. Suppose that a tablespoonful of carbon (about 1 cubic inch [16 cm^3]) is accumulated. This reduces the clearance volume to about 4½ cubic inches [74 cm^3] and thereby increases the effective compression ratio to about 10:1! The cylinder in which this carbon accumulation takes place would "knock its head off." Thus, with the higher-compression engines, more attention must be paid to engine maintenance. (Further discussions on this matter follow in later chapters of this book.)

We should also like to mention that compression ratios are being reduced to some extent as one result of the effort to reduce smog caused by engine exhaust. The reason for this is complicated, and we shall explain it in detail in Chap. 18. For the moment, let us say that lead-free gasoline is necessary in the antismog effort and this means lowering the octane, or antiknock, rating of gasoline.

⊘ 11-6 Delivery of Air-Fuel Mixture When the piston moves down on the intake stroke, a vacuum is produced in the cylinder. Atmospheric pressure (pressure of tne air) then pushes air into the cylinder. This air must first pass through the carburetor (where it picks up a charge of fuel), intake manifold, and intake-valve port. You might think that atmospheric pressure (15 psi [1.055 kg/cm^2] at sea level) would be great enough to push air through these passages and into the cylinder almost instantly. However, it does take an appreciable amount of time for the air to move through any restricting passage. (Anyone who has deflated a tire knows how long it takes for the air to escape after the valve core has been removed.)

Thus, as the air moves through the carburetor, intake manifold, and intake-valve port, it must pass through narrow passages and go around turns that, in effect, "hold it back." It takes time for the air-fuel mixture to get into the cylinder on the intake stroke. But even during engine idle, there isn't much time available. During idle (say at 550 rpm) the entire intake stroke takes less than 0.1 second. At high speed, this time is reduced to less than 0.01 second.

It is therefore obvious that the vacuum in the cylinder (caused by the downward-moving piston) is not going to be completely "satisfied." The air-fuel mixture is going to be shut off by the closing of the intake valve before the cylinder is "filled up." The higher the engine speed, the less air-fuel mixture will be taken in during the intake stroke. This is one reason that an engine will not increase in speed and in power output indefinitely. There is a certain speed at which an engine will produce maximum power; above this speed, power drops off.

⊘ 11-7 Volumetric Efficiency The amount of air-fuel mixture taken in by the engine on the intake stroke is a measure of the engine's *volumetric efficiency*. If the air-fuel mixture were drawn into the cylinder very slowly, it would be possible to "fill up" the cylinder. However, as mentioned in the previous section, the air-fuel mixture must pass very rapidly through a series of narrow passages and bends in the carburetor and intake manifold. In addition, the mixture is subjected to heat (from the engine and exhaust manifold), and it therefore increases in temperature. We know that when air is heated, it expands. The two conditions, rapid movement and heating, reduce the amount of air-fuel mixture that can enter the cylinder on the intake stroke. A full charge of air-fuel mixture cannot enter because the time is too short.

Volumetric efficiency is the ratio of the amount of air-fuel mixture that actually enters the cylinder to the amount that could enter under ideal conditons. For example, a certain engine has an air volume (*A* in Fig. 11-3) of 47 cubic inches [770 cm^3] per cylinder (piston at BDC). If the cylinder were allowed to completely "fill up," it would take in 0.034 ounce [0.964 gram] of air. However, suppose that the engine were running at a fairly high speed so that only 0.027 ounce [0.765 g] of air could enter. This means that volumetric efficiency would be only about 80 percent (0.027 is 80 percent of 0.034). Actually, 80 percent is a good volumetric efficiency for an engine running at fairly high speed. Volumetric efficiency of some engines drops to as low as 50 percent at high speeds. This is another way of saying that the cylinders are only "half-filled" at high speeds.

This is one reason why engine speed and output cannot continue to increase without limit. In effect, at higher speeds the engine has an increasingly hard time "breathing," or drawing in air. At a certain speed, the engine begins to "starve" for air, and it is at this point that the engine begins to "weaken" so that its power output drops off.

To improve volumetric efficiency, intake valves can be made larger and the number of valves per cylinder can be increased (Fig. 9-34). Also, valve lift can be increased; that is, the cam lobes on the cam can be made larger so that the valves open wider. However, when this is done, there is danger of the pistonhead striking the valve head. Unless engine design takes this into account, serious engine damage could result. This is discussed more fully in ⊘ 8-18 and 9-23.

Volumetric efficiency can also be improved by making the intake-manifold passages larger, wider, and as straight and short as possible. Also, the smoothness on the inside surfaces of the intake manifolds is important. Rough surfaces tend to restrict the flow of air-fuel mixture.

Finally, volumetric efficiency can be improved by using carburetors with extra air passages (or barrels), which open at high speed to improve engine breathing. (Carburetors are discussed in detail in Chap. 14.) All these improvements help the engine produce more power at higher speeds.

⊘ 11-8 Engine Power Output Power is the rate at which work is done. The rate at which an engine can do work is measured in horsepower (see ⊘ 4-29). An engine that can deliver 33,000 foot-pounds [4,560.6 m-kg] of work in 1 minute is a 1-hp [0.746 kW] engine. An engine that can deliver 6,600,000 foot-pounds [912,120 m-kg] of work in 1 minute is a 200-hp [149.2 kW] engine. The power that the engine actually delivers is called *brake horsepower* (bhp). Automotive engines are usually rated in terms of bhp, although the word "brake" is not always used. Thus, an engine that is rated as developing 160 bhp [119.36 kW] may be called a *160-hp* [119.36 kW] *engine*.

⊘ 11-9 Determining Brake Horsepower The term "brake horsepower" came from the fact that a *Prony brake* was one of the first devices used to measure engine horsepower output. The Prony brake (Fig. 11-4) contains a large brake drum around which a brake is clamped. A brake arm is attached to the brake at one end and rests on a scale at the other. The brake has a device (*A* in Fig. 11-4) that can be tightened so that the brake will exert a greater

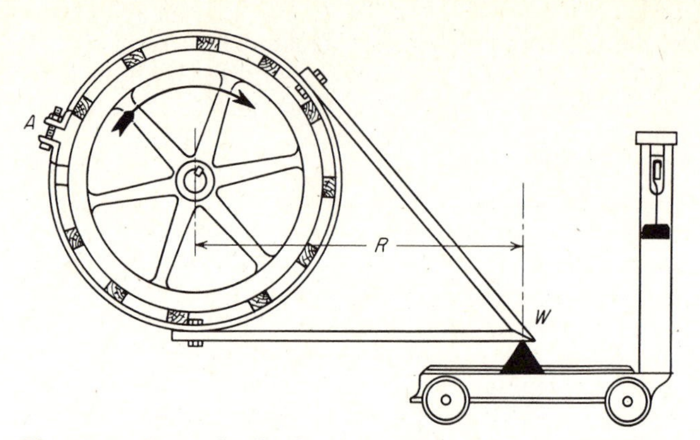

Fig. 11-4. Prony brake for determining the power output of an engine. *A* is the adjustment for tightening the brake on the brake drum. *R* is the length of the arm (from the center of the shaft to the end supported on scales). *W* is the weight, or force, exerted on the scales.

braking effect on the drum. The brake drum is driven by the engine. At the same time, there is a greater push exerted through the brake and brake arm on the scales.

The Prony-brake test is made by running the engine (and brake drum) at a steady speed and gradually tightening the brake on the brake drum. As this is done, the load on the engine is increased, and the throttle must be opened wider in order to maintain engine speed. Loading the engine in this way also increases the load, or weight, on the scales (*W* in Fig. 11-4). To find the maximum power that the engine can develop at any one speed, the load is increased gradually (the throttle being opened wider and wider at the same time to maintain speed) until the throttle is wide open. Further loading would then cause the engine speed (and power output) to drop off.

The maximum load on the scale is then used in the following formula to determine what horsepower the engine was developing:

$$\text{bhp} = \frac{2\pi RNW}{33{,}000} = \frac{RNW}{5{,}252}$$

where R = length of arm (from center of drum)
N = engine speed, rpm
W = load on scale, lb

For example, let us suppose the arm is 3 feet [0.914 m] long, the load on the scale is 100 pounds [45.359 kg], and the engine speed is 1,000 rpm. Substituting in the formula gives

$$\text{bhp} = \frac{3 \times 1{,}000 \times 100}{5{,}252} = 57.12 \text{ bhp}$$

There is a second formula for horsepower which is used more often today, especially by automotive technicians working with engines and chassis dynamometers (⊘ 11-10). This formula is

$$\text{bhp} = \frac{\text{Torque} \times \text{rpm}}{5{,}252}$$

The reason it is more commonly used is that with modern dynamometers, engine performance is measured in rpm, torque, and horsepower. You can see that this formula is easier to work with.

⊘ 11-10 Dynamometer Rating of Engine Horsepower Instead of the Prony brake, engine testers now use a dynamometer (Fig. 11-5). This device includes a mechanism which can load the engine by varying amounts. Thus, the dynamometer can measure the amount of horsepower the engine develops under different operating conditions. One of the loading mechanisms is an electric generator. The amount of electric current the generator produces is therefore a measure of the amount of horsepower the engine is developing.

Another type of dynamometer makes use of a *water brake*. The water brake contains a rotating device with numerous blades, or vanes. When water is put into the device, the rotating member has a restriction placed on it. The more water added, the more restriction is placed on the member and the more power is needed to drive it. As can be seen, the device can apply a varying load on the engine as water is added to or removed from the rotating mechanism.

Some dynamometers have meters that read torque and rpm as well as horsepower. Many dynamometers test detached engines, as shown in Fig. 11-5. Other dynamometers, called *chassis dynamometers*, test the engine while it is in the car (Fig. 11-6). On these, the rear wheels of the car are placed on rollers. The engine is started, and the transmission is placed in gear. The rollers are then driven by the engine. The rollers, in turn, are connected to the dynamometer so that engine output can be measured. The use of the chassis dynamometer is becoming more common in the automotive servicing field because it gives a very quick report on engine condition (by measuring engine output at various speeds and loads). The chassis dynamometer is also valuable in testing and adjusting automatic transmissions because checks and adjustments can be made in the shop (no road test is necessary).

Check Your Progress

Progress Quiz 11-1 Once again you have the chance to check up on yourself and find out how well you are remembering and understanding what you are reading. The material you have just covered is different from the material in previous chapters because it describes some of the ways in which engine performance is measured. You will appreciate that this is valuable information. Anyone interested in automotive engines should know how engine output is measured, what "compression ratio" means, and so on. You can find out how well you have remembered this material by answering the following questions.

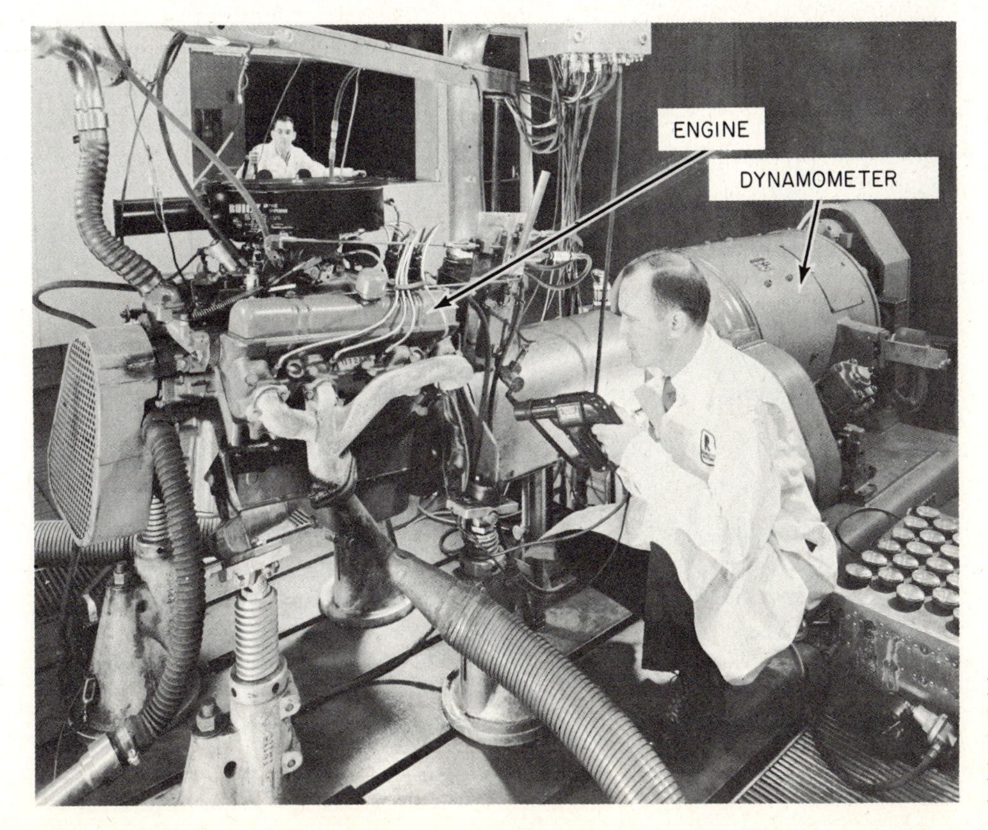

Fig. 11-5. Dynamometer used for testing engines and measuring their power output. (*General Motors Corporation*)

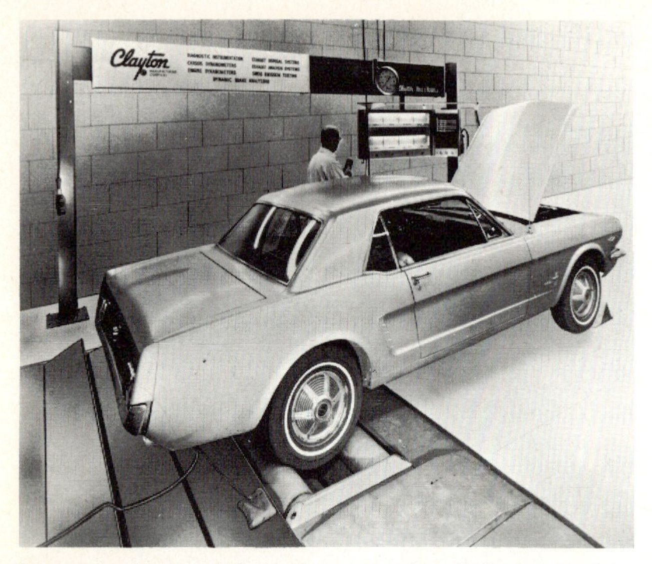

Fig. 11-6. Automobile in place on a chassis dynamometer. The rear wheels drive the dynamometer rollers, which are flush with the floor. Instruments on the test panel measure car speed, engine power output, engine vacuum, ignition-system operation, and so on. (*Clayton Manufacturing Company*)

Completing the Sentences The sentences that follow are incomplete. After each sentence there are several words or phrases, only one of which will correctly complete the sentence. Write each sentence in your notebook, selecting the proper word or phrase to complete it correctly.

1. The size of an engine cylinder is referred to in terms of its: (*a*) bore and length, (*b*) bore and stroke, (*c*) diameter and bore.
2. Piston displacement is calculated from the: (*a*) bore and stroke, (*b*) piston length and diameter, (*c*) cylinder diameter and length.
3. The air volume above the piston with the piston at TDC is called the: (*a*) compression ratio, (*b*) clearance volume, (*c*) piston displacement, (*d*) bore.
4. The air volume in the cylinder with the piston at BDC divided by the clearance volume is called the: (*a*) piston displacement, (*b*) cylinder ratio, (*c*) compression ratio.
5. Two special problems to be considered in the higher-compression engines are: (*a*) air-fuel ratio and speed, (*b*) ping and detonation, (*c*) detonation and carbon accumulations.
6. As carbon accumulates in a cylinder, it causes an increase in the: (*a*) clearance volume, (*b*) effective compression ratio, (*c*) piston displacement.
7. The amount of air-fuel mixture taken in by the engine on the intake stroke is a measure of the engine's: (*a*) clearance volume, (*b*) compression ratio, (*c*) volumetric efficiency.
8. A volumetric efficiency of 80 percent for an engine running at fairly high speed is: (*a*) good, (*b*) fair, (*c*) poor.
9. The more easily an engine can breathe, the higher is its: (*a*) volumetric efficiency, (*b*) piston displacement, (*c*) compression ratio.
10. The Prony brake is used to determine: (*a*) fhp, (*b*) bhp, (*c*) hp.

⊘ 11-11 Dynamometer Test Results A dynamometer test of an engine, even under controlled laboratory conditions, usually gives test results that do not agree with the advertised horsepower rating of the engine. For example, an engine advertised as being a "300-hp [223.8 kW] engine" might show only about 260-hp [193.96 kW] output maximum on the chassis dynamometer and possibly about 270-hp [201.42 kW] maximum on the engine dynamometer. One reason for this is that the advertised ratings are measured without such accessories as alternators, an air cleaner, and an exhaust system. Each of these cuts down the engine output by 2 to 3 hp [1.49 to 2.24 kW]. Also, on the chassis dynamometer there is a friction loss of several horsepower [kilowatts] in the transmission, universal joints, and rear axles.

In addition, the advertised ratings have been corrected for both atmospheric pressure and temperature. Other factors being equal, the power output of an engine will increase with increased air pressure and decrease with increased temperature (within certain limits of course). Increased air pressure increases the amount of air-fuel mixture forced into the engine cylinders. But increased temperature, which causes the air to expand, reduces the amount of air-fuel mixture entering the cylinders. Thus, in order to have a system that will give the same corrected ratings regardless of where the engine is tested, engine testers have adopted the following code: After an engine is tested on the dynamometer, the test results are adjusted for actual air temperature and pressure. For example, the test results would be adjusted upward if the actual air temperature were too high or pressure were too low (either of these causes a power loss). The test results would be adjusted downward if the actual air temperature were too low or pressure were too high. Actually, a single formula is used to make the corrections. This formula corrects the test results to a standard which corresponds to dry air at 60 degrees Fahrenheit [15.6°C] and at 15 psi [1.0546 kg/cm^2].

⊘ 11-12 Indicated Horsepower Another method of evaluating engines is by *indicated horsepower* (ihp). Indicated horsepower is based on the power actually developed inside the engine cylinders by the combustion process. A special indicating device, the oscilloscope, is required to determine ihp. This device measures the pressure (by electronic means) in the cylinder continuously throughout the four piston strokes (intake, compression, power, and exhaust). It makes a graph of these pressures, relating them to the piston position in the cylinder during the four piston strokes. Such a graph, taken from a typical engine test, is given in Fig. 9-50. The four small drawings in Fig. 9-50 show the crank, rod, and piston positions, as well as the directions of motion, during the four strokes. Note that the pressure in the cylinder is about atmospheric at the beginning of the intake stroke. Then it falls a little below atmospheric as air-fuel-mixture delivery to the cylinder lags slightly behind piston movement (that is, volumetric efficiency is less than 100 percent). When the compression stroke begins (both valves are closed), the pressure starts to increase as the piston moves upward in the cylinder. As it reaches a value of somewhat over 100 psi [7.031 kg/cm^2] and a little before the piston reaches TDC, ignition takes place (the graph shows an ignition-spark advance of about 20 degrees). Now, as the air-fuel mixture burns, pressure increases very rapidly, reaching a peak of around 580 psi [40.779 kg/cm^2] at about 25 degrees past TDC on the power stroke. Pressure begins to fall off fairly rapidly as the power stroke continues, but there is still a pressure of about 50 psi [3.515 kg/cm^2] at the end of the power stroke. When the exhaust stroke begins, the pressure falls off as the piston moves up, forcing the burned gases out of the cylinder. At the end of the exhaust stroke, the pressure has fallen to around atmospheric. Another intake stroke then begins.

From a graph such as the one shown in Fig. 9-50, the *mean effective pressure* (mep) in the cylinder can be determined. The mep is the average pressure during the power stroke minus the average pressures during the other three strokes. The mep is, in effect, the pressure that actually forces the piston down during the power stroke. From the mep and other engine data, the following formula is used to calculate ihp:

$$\text{ihp} = \frac{PLANK}{33{,}000}$$

where P = mean effective pressure, psi
L = length of stroke, ft
A = area of cylinder section, in
N = number of power strokes per minute (or rpm/2)
K = number of cylinders

In operation, some of the power developed in the engine is used up in overcoming friction in the engine. Thus ihp (the power developed in the engine) is always greater than bhp (the power delivered by the engine). Friction losses in an engine can be determined by subtracting bhp from ihp. ⊘ 11-18 explains how bhp and ihp can be used to determine engine efficiency.

⊘ 11-13 Friction Horsepower Friction losses occur in an engine and are sometimes quite high, despite the presence of adequate lubrication. There are numerous surfaces in the engine that are moving against each other (Fig. 11-7). Journals are turning in bearings. Pistons and piston rings are sliding up and down in cylinders. One of the crankshaft bearings has end-thrust flanges to contain the end movement of the crankshaft.

The friction losses in an engine are sometimes referred to in terms of *friction horsepower* (fhp). This expression means the amount of horsepower required to overcome friction losses in the engine. The fhp of an engine can be determined by driving the engine with an electric motor (the dynamo in an electric dynamometer can be used for this purpose). The engine is turned off (after running it to bring it up to operating temperature) and driven with no fuel in the carburetor and the throttle wide open. Under these conditions the amount of power required to drive the engine at different speeds is determined. At low speeds an engine uses up a relatively small amount of power in overcoming friction. However, as engine speed increases, the fhp increases rapidly. The graph in Fig. 11-8 shows the fhp increase in a typical engine. Note that at about 1,000 rpm, the fhp is only about 4 hp [2.984 kW]. However, at 2,000 rpm, the fhp is nearly 10 hp [7.46 kW]. At 3,000 rpm, it has increased to about 21 hp [15.666 kW], and at 4,000 rpm it is about 40 hp [29.84 kW].

One of the major causes of friction loss (or fhp) in the engine is *piston-ring friction*. Under some conditions, the friction of the rings on the cylinder walls accounts for 75 percent of all friction loss. For example, Fig. 11-8 shows a fhp of 40 hp [29.84 kW] at 4,000 rpm. It is possible that under certain conditions 75 percent of this total, or 30 hp [22.38 kW], is being used up by the piston rings in overcoming friction. Understanding this fact makes us realize more fully than ever the difficult job that the piston rings must do in the engine. It also points up one advantage of the short-stroke oversquare engine: With a short stroke, the piston rings do not have as far to slide on the cylinder walls; thus ring friction is reduced and friction losses in the engine are lowered.

⊘ 11-14 Relating bhp, ihp, and fhp We have already mentioned that bhp is the power delivered by the engine, ihp is the power developed in the engine cylinders by the combustion process, and fhp is the horsepower required by the engine to overcome friction losses. The relationship among these three is

$$\text{bhp} = \text{ihp} - \text{fhp}$$

In other words, the horsepower delivered by the engine is equal to the horsepower developed in the engine minus the horsepower losses resulting from friction.

⊘ 11-15 Torque *Torque* can be defined as a turning effort (see ⊘ 4-28). An example of torque can be seen in early-model cars in which the engine must be cranked by hand (Fig. 11-9). As the hand crank is turned, torque is applied to the engine crankshaft. In like manner, when the piston is moving down on the power stroke, it applies torque to the crankshaft

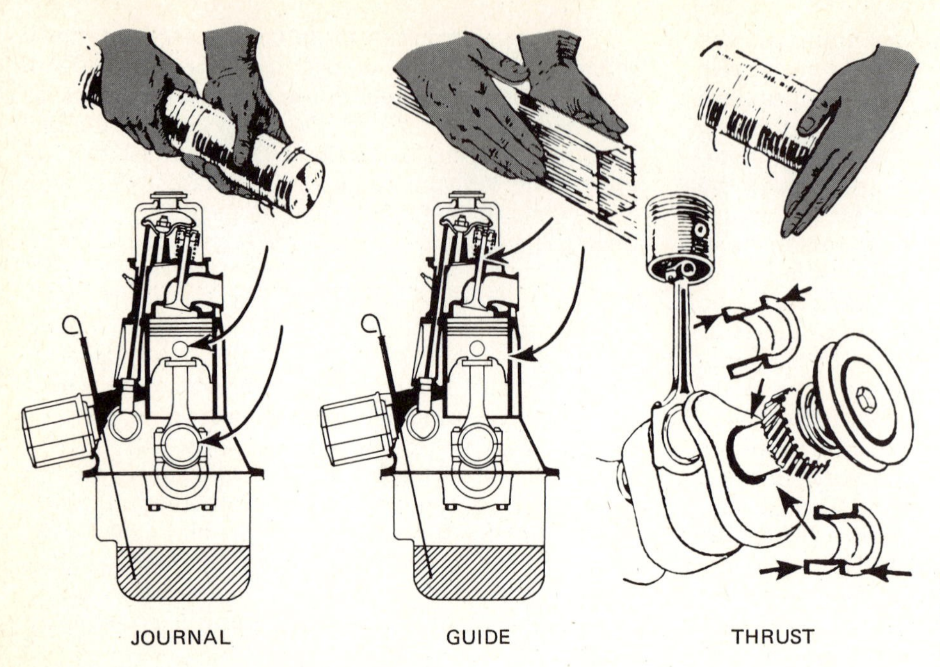

Fig. 11-7. Three types of friction-bearing surfaces in an automobile engine.

through the connecting rod and crank on the crankshaft (Fig. 11-9).

The harder the push on the piston, the greater the torque applied. Thus, the higher the combustion pressures in the engine, the greater the amount of torque developed.

Torque should not be confused with engine power. Torque is the rotary or twisting effort that the engine applies through the crankshaft. Power is the rate at which the engine works. Of course, the two are related, but it must be remembered that power takes into consideration engine speed, or rpm, while torque does not. Torque is merely the force times the distance the force acts from the shaft center, or pound-feet [kilogram-meters]. Torque can be applied without any motion at all; however, this is not a normal condition in engine operation. The dynamometer (⊘ 11-10) can be used to measure engine torque.

EXAMPLE: An engine that is tested is found to provide a torque of 440 pound-feet [60.8 kg-m] at 2,700 rpm. This can be converted into horsepower (since the rpm is known) by use of the formula presented in ⊘ 11-9:

$$\text{bhp} = \frac{\text{torque} \times \text{rpm}}{5{,}252} = \frac{440 \times 2{,}700}{5{,}252} = 226.18 \text{ bhp}$$

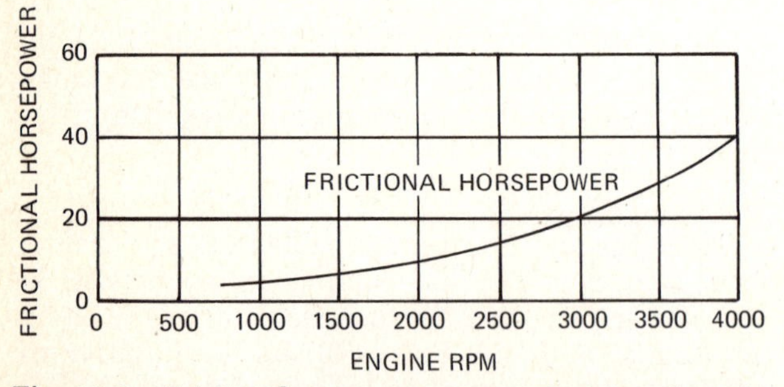

Fig. 11-8. Friction-horsepower curve showing the relationship between frictional horsepower (fhp) and engine speed (rpm).

⊘ 11-16 Torque Compared with Brake Horsepower

The torque that an engine can develop changes with engine speed. An engine develops more torque at intermediate speeds (open throttle) than at high speeds. This is because, at the lower speeds (open throttle), there is more time for air-fuel mixture to enter the cylinders. In other words, at intermediate speeds (open throttle), volumetric efficiency is high. This means that there will be a greater amount of

Fig. 11-9. Torque applied to the hand crank of the old-time car rotates the engine crankshaft. Likewise, the piston and connecting rod impart torque to the crank of the crankshaft, causing the crankshaft to rotate.

air-fuel mixture to burn during the power stroke and higher combustion pressures will develop. This means, in turn, that greater torque will be applied to the crankshaft.

At high speeds, there is less time for air-fuel mixture to enter the cylinders (the volumetric efficiency is low). Thus there is less air-fuel mixture to burn. Combustion pressures will not go so high, and torque will not be so great. Also, at high speeds the piston is moving so much faster that it tends to "keep step" with the increasing pressure as combustion starts, and so less thrust is exerted on the piston. Figure 11-10 shows, graphically, how torque changes as engine speed changes. Note that at low speed (400 rpm) the engine can develop considerable torque (about 170 pound-feet [23.49 kg-m]). Torque increases with engine speed until, in the range of 1,500 to 2,000 rpm, torque is at a maximum of around 215 pound-feet [29.713 kg-m]. Further increase of engine speed, however, results in a falling off of torque. This is due to the reduction in volumetric efficiency (the amount of air-fuel mixture that is able to enter the engine cylinders).

Horsepower also increases with engine speed (Fig. 11-11). It is obvious that this must be so because horsepower depends on rpm as well as on torque. Thus, as long as torque and rpm increase, horsepower output will also increase. Figure 11-11 gives the bhp curve of the same engine for which the torque curve is given in Fig. 11-10. Note that at low speed (400 rpm) the bhp is small (only about 14 bhp [10.444 kW]). But bhp increases steadily with increasing speed until a maximum of about 110 bhp [82.06 kW] is reached at around 3,500 rpm. Then, bhp falls off rapidly. This tapering off results from the rapid decrease of engine torque in the higher-speed ranges, as well as the rapid increase of fhp. The relationships can be readily seen in Fig. 11-12, where the torque, bhp, and fhp curves are shown together. Note that the bhp continues to increase for some

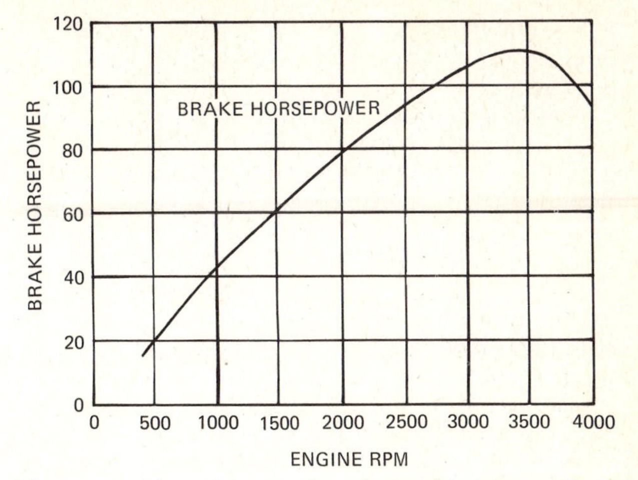

Fig. 11-11. Curve showing the relationship between brake horsepower (bhp) and engine speed (rpm).

time even after the torque starts to drop. This is the effect of increasing engine speed. Soon, however, fhp uses up so much horsepower that bhp starts dropping off. Aiding in this bhp-curve drop is the torque-curve drop.

NOTE: The curves shown in Figs. 11-8 and 11-10 to 11-12 are for one particular engine only. Different engines have different torque, bhp, and fhp curves. Peaks may occur at higher or lower speeds, and the torque-bhp-fhp-speed relationships may not be as indicated.

In recent years, engineers have come to rely more and more on torque rather than horsepower as a measure of engine performance. It is torque produced by the engine and delivered to the car wheels that gives the car "zip" during passing, for instance. The torque curve for late-model oversquare engines is flatter than that for engines of older designs. It is higher at low speeds and does not drop so much at high speeds. The reason is that the bore and valves are larger, giving better volu-

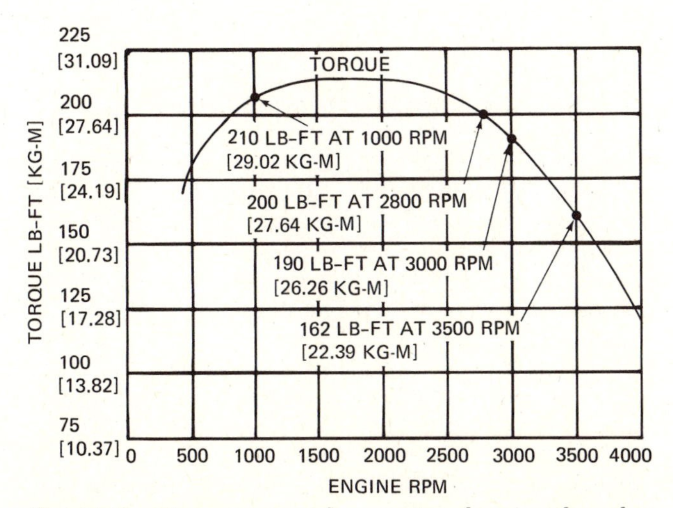

Fig. 11-10. Torque curve of an engine showing the relationship between torque (lb-ft [kg-m]) and engine speed (rpm).

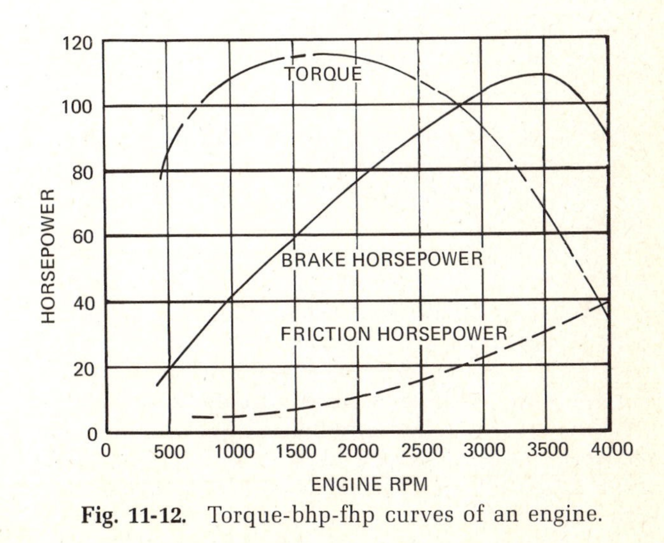

Fig. 11-12. Torque-bhp-fhp curves of an engine.

metric efficiency. At the same time, the shorter stroke reduces engine power losses from piston and ring friction.

There is another advantage to be found in the larger-bore, higher-displacement oversquare engines: their improved torque at low speeds. That is, these engines produce greater horsepower at the lower speeds (because horsepower depends on torque as well as on speed). With improved horsepower at intermediate speeds, it is possible to use a higher gear ratio in the differential so that a lower engine speed will produce the same car speed. A typical engine made in the 1930s, for example, had to turn 50 rpm for each mile per hour of car speed. This meant that to drive a car at 40 miles per hour [64.374 km/h] the engine would have to turn 2,000 rpm. A late-model engine, on the other hand, turns only about 35 rpm for each mile per hour of car speed; at 40 miles per hour [64.374 km/h] it would have to turn only 1,400 rpm. You can see how both friction horsepower (friction losses) and engine wear would be substantially reduced.

⊘ 11-17 Engine Efficiency "Efficiency" means the relationship between the effort exerted and the results obtained. As applied to engines, efficiency is the relationship between the power delivered (bhp) and the power that could be obtained if the engine operated without any power loss. Engine efficiency can be computed in two ways: as mechanical efficiency and as thermal efficiency.

1. MECHANICAL EFFICIENCY Mechanical efficiency is the relationship between bhp and ihp:

$$\text{Mechanical efficiency} = \frac{\text{bhp}}{\text{ihp}}$$

EXAMPLE: At a certain speed, the bhp of an engine is found to be 116 and its ihp is 135. The mechanical efficiency would be

$$\text{Mechanical efficiency} = \frac{\text{bhp}}{\text{ihp}}$$

$$= \frac{116}{135} = 0.86 \qquad \text{(or 86\%)}$$

That is, 86 percent of the power developed in the cylinders is delivered by the engine (the remaining 14 percent, or 19 hp [14.17 kW], being consumed as fhp).

2. THERMAL EFFICIENCY "Thermal" means of, or related to, heat. The thermal efficiency of an engine is the relationship between the power output and the energy in the fuel burned to produce this power output. We have already noted that a considerable part of the heat produced by the combustion process is carried away by the cooling water and lubricating oil. In addition, since the exhaust gases are still hot when they leave the engine, they carry away a considerable part of the heat produced by combustion. All these losses are heat, or thermal, losses that reduce the thermal efficiency of the engine; they do not add to the power output of the engine. The remainder of the heat, in causing the gases to expand, creates the high pressures that force the pistons down so that power is developed by the engine. Because there is a great deal of heat lost during engine operation, thermal efficiency of the gasoline engine may be as low as 20 percent and is seldom above 25 percent. Practical limitations prevent higher thermal efficiencies.

Improvements in engine design in recent years have increased engine thermal efficiency. The larger cylinder bores reduce heat losses somewhat. The heat has a greater distance to travel from the center of the combustion chamber to the cooler cylinder walls. Increasing the compression ratio also increases thermal efficiency because the combustion gases expand into a larger volume, thereby becoming cooler. Thus, more of the heat energy is utilized in the engine. Also, the engine cylinder wall, piston-head, and cylinder-head surfaces can be operated at higher temperatures. This reduces the heat losses because the temperature differential between the combustion gases and surrounding surfaces is reduced. Thermal efficiency is therefore increased.

⊘ 11-18 Overall Efficiency Gasoline enters the engine with a certain energy content, or a certain ability to do work. At every step in the process, from the burning of the air-fuel mixture in the cylinders to the rotation of the car wheels, energy is lost. Figure 11-13 illustrates these energy losses as determined for one engine in a car during a test run. Note that about 35 percent of the heat energy is carried away by the cooling water and lubricating oil. Another 35 percent is lost in the exhaust gases, which are still hot after leaving the cylinders. Friction (in the engine and power train) accounts for another 15 percent of heat energy loss. The amount of energy remaining, which may be as little as 15 percent of the energy contained in the gasoline, is actually responsible for propelling the car. This power is used to overcome rolling and air resistance and to accelerate the car.

1. ROLLING RESISTANCE Rolling resistance results from irregularities in the road over which the wheels ride. It also results from the flexing of the tires as they rotate.

NOTE: Rolling resistance is decreased with radial tires.

2. AIR RESISTANCE Air resistance is the resistance of the air to the passage of the car body through it. As the car speed increases, so also does the air resistance. At 90 miles per hour [144.8 km/h], for example, tests have shown that as much as 75 percent of the power the engine develops is used up in overcoming air resistance. Streamlining the car

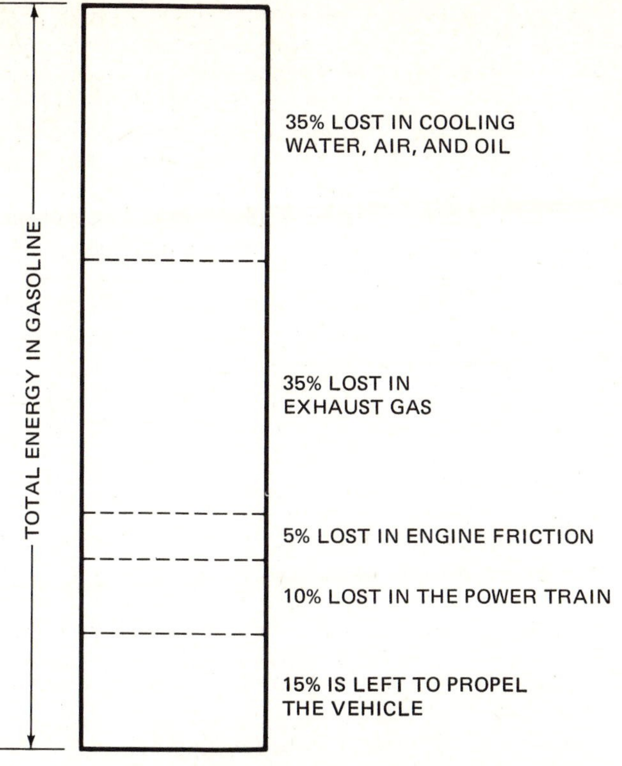

Fig. 11-13. Energy loss from the cylinder to the wheels.

Fig. 11-14. Streamlining the car means shaping it so that the air flows smoothly over and around it, rather than forming eddies that hold the car back and use up power.

body reduces power loss from air resistance (Fig. 11-14).

3. *ACCELERATION* Power is required to increase car speed. In effect, the power applied to accelerate the car overcomes the inertia of the car; energy in the form of car speed is built up, or stored, in the car.

Check Your Progress

Progress Quiz 11-2 Here is your chance to quiz yourself again to find out how well you are doing in your studies. In the past few pages, you have covered more ground on engine-performance measurements. Information on this subject is very valuable to any persons who have anything to do with engines because it gives them a better understanding of engine performance, design, and construction. Check your knowledge by answering the following questions. If you have any trouble, reread the past few pages and take the quiz again. Remember, the quiz is designed to help you remember the important facts about engines.

Completing the Sentences The sentences that follow are incomplete. After each sentence there are several words or phrases, only one of which will correctly complete the sentence. Write each sentence in your notebook, selecting the proper word or phrase to complete it correctly.

1. In comparison with the advertised horsepower rating of an engine, the horsepower as determined by a chassis dynamometer will be: (*a*) higher, (*b*) lower, (*c*) the same.
2. The power actually developed inside the engine cylinders is called: (*a*) ihp, (*b*) bhp, (*c*) fhp.
3. The average pressure during the power stroke minus the average pressures during the intake, compression, and exhaust strokes is called the: (*a*) ihp, (*b*) bhp, (*c*) compression ratio, (*d*) mep.
4. The power used in overcoming friction in the engine is called: (*a*) ihp, (*b*) fhp, (*c*) bhp.
5. One of the major causes of fhp in an engine is: (*a*) low volumetric efficiency, (*b*) piston-ring friction, (*c*) excessive mechanical efficiency.
6. Ihp minus fhp equals: (*a*) bhp, (*b*) mep, (*c*) SAE hp.
7. Engine torque is highest at: (*a*) low speed, (*b*) intermediate speed, (*c*) high speed.
8. One reason why torque drops off at high speed is that at high speed: (*a*) volumetric efficiency is lower, (*b*) the engine breathes better, (*c*) the fuel mixture is richer.
9. Bhp divided by ihp is: (*a*) mechanical efficiency, (*b*) thermal efficiency, (*c*) volumetric efficiency.
10. The percentage of the energy in the gasoline burned in the engine which is actually utilized in propelling the car may be as little as: (*a*) 15 percent, (*b*) 25 percent, (*c*) 35 percent, (*d*) 70 percent.

CHAPTER 11 CHECKUP

NOTE: Since the following is a chapter review test, you should review the chapter before taking the test.

You are making good progress in your study of the automobile engine. The chapter you have just completed gives you a good background on the various ways in which engines and engine performance are measured. When someone talks to you about piston displacement, compression ratio, or brake horsepower, you will know what is meant. The following questions will help you make sure that you remember the important points covered in the chapter. Review the chapter before taking the test. If the questions seem hard to answer, review the chapter again. Remember, good students may review their lessons several times in order to make sure that the essential facts are fixed firmly in their minds.

Completing the Sentences The sentences that follow are incomplete. After each sentence there are several words or phrases, only one of which will correctly complete the sentence. Write each sentence in your notebook, selecting the proper word or phrase to complete it correctly.

1. Knowing the bore and stroke, you can calculate the: (*a*) compression ratio, (*b*) piston displacement, (*c*) volumetric efficiency.
2. Knowing the clearance volume and the air volume in the cylinder with the piston at BDC, you can figure the: (*a*) compression ratio, (*b*) volumetric efficiency, (*c*) bhp, (*d*) ihp.
3. Knowing the weight of air an engine cylinder could take in and the weight of air actually taken in under running conditions, you can figure: (*a*) compression ratio, (*b*) piston displacement, (*c*) volumetric efficiency.
4. Knowing the speed at which an engine is running and the torque it is developing, you can figure: (*a*) bhp, (*b*) ihp, (*c*) fhp.
5. In the advertised horsepower rating of an engine, the rating has been corrected for: (*a*) speed and pressure, (*b*) air pressure and temperature, (*c*) fhp, (*d*) ihp.
6. If you know the engine speed, bore, stroke, number of cylinders, and mep in the cylinders, you can calculate: (*a*) bhp, (*b*) ihp, (*c*) fhp.
7. Knowing the ihp and fhp of an engine, you can calculate: (*a*) SAE, (*b*) hp, (*c*) bhp, (*d*) rpm, (*e*) compression ratio.
8. Streamlining reduces: (*a*) air resistance, (*b*) fhp, (*c*) rolling resistance.
9. To figure the mechanical efficiency of an engine under test, you need to know: (*a*) mep and ihp, (*b*) bhp and ihp, (*c*) mep and rpm.
10. The ratio between the power output of an engine and the energy in the fuel burned to produce that power is called: (*a*) thermal efficiency, (*b*) mechanical efficiency, (*c*) volumetric efficiency.

Problems Work out the following problems in your notebook. Refer back to the formulas in the chapter if you are not sure of them.

1. What is the piston displacement in a 3- by 4-in cylinder?
2. What is the compression ratio of an engine that has a clearance volume of 5.3 in^3 [86.851 cm^3] (piston at TDC) and an air volume of 45.05 in^3 [738.237 cm^3] with the piston at BDC?
3. An engine has a compression ratio of 8:1. It has a clearance volume of 5 in^3 [81.935 cm^3] and an air volume of 40 in^3 [655.483 cm^3] with the piston at BDC. After some months of adverse service, accumulations of carbon amount to an average of 1 in^3 [16.387 cm^3] per cylinder. What is the average effective compression ratio?
4. An engine has an air volume (piston at BDC) of 47 in^3 [770.192 cm^3] per cylinder. This means that at atmospheric pressure the cylinder could hold 0.034 oz [0.9639 g] of air. In an actual running test at 3,000 rpm, the engine is found to take in only 0.024 oz [0.6804 g] per cylinder on each intake stroke. What is the volumetric efficiency under these conditions?
5. In a Prony-brake test of an engine, where the brake arm is 4 ft long, the load on the scale was found to be 200 lb at 1,250 rpm. What bhp is the engine developing?
6. Determine the ihp of an eight-cylinder 3- by 4-in engine which shows a mep of 360 psi when running at 2,000 rpm.
7. Under certain test-operating conditions an engine has a rating of 31 fhp while developing 136 bhp. What is its ihp under these conditions?
8. What is the mechanical efficiency of the engine in Prob. 7 under the test conditions?
9. What torque is the engine described in Prob. 5 producing?

SUGGESTIONS FOR FURTHER STUDY

There are a number of engineering books that supply additional information on engine measurements as well as details of how engine tests are run. If you are interested in learning more about these matters, go to your local library and have your librarian help you find the books containing this additional information. It is also possible that your local high school automotive mechanics teacher or physics teacher may be able to help you. The operating instructions for the dynamometer contain a review of basic engine measurements as well as a description of how to make them.

chapter 12

AUTOMOTIVE-ENGINE FUELS

This chapter discusses the various kinds of fuel used in automotive engines: gasoline, diesel-engine fuel oil, and LPG (liquefied petroleum gas).

⊘ 12-1 Gasoline Gasoline is a hydrocarbon (HC) made up largely of hydrogen and carbon compounds. These two elements unite readily with oxygen, a common element which makes up about 20 percent of the air. When hydrogen unites with oxygen, water (H_2O) is formed. When carbon unites with oxygen, carbon monoxide (CO) and carbon dioxide (CO_2) are formed. If all the gasoline in an engine burned completely, all that would come out would be H_2O and CO_2. However, perfect combustion is not achieved in the engine, and so some CO and HC are present in the exhaust gases. These two compounds, plus a third compound, nitrogen oxides (NO_x), are the pollutants emitted from automobiles. Chapters 18 and 19 explain how control of these emissions is achieved.

NOTE: Gasoline is often referred to as "gas," which can cause some confusion. The sort of gas you burn in a gas stove or use to heat a house is actually a *gas* that is delivered through gas lines or pipes. So there is *gas* and there is "gas," or gasoline. Don't get confused.

⊘ 12-2 Source of Gasoline Gasoline is made from crude oil, or petroleum. No one knows exactly how crude oil was originally formed. It is found in "pools" or reservoirs in the ground. When a well is drilled down into a reservoir, the underground pressure (or pressure artificially applied from above ground) forces the oil up and out of the well. The crude oil is then put through a refining process, which produces gasoline, lubricating oil, greases, fuel oil, and many other products.

In the refining process, several compounds, called *additives*, are put into the gasoline. These additives give the gasoline the properties a good gasoline should have. Additives are described in ⊘ 12-15.

⊘ 12-3 Volatility of Gasoline Actually, gasoline is not a simple substance. It is a mixture, or blend, of a number of different hydrocarbons, each having its own characteristics. Aside from its combustibility, one of the important properties of gasoline is *volatility*.

Volatility refers to the ease with which a liquid vaporizes. The volatility of a simple compound like water is found by increasing its temperature until it boils, or vaporizes. A liquid that vaporizes at a low temperature has a high volatility; it is highly volatile. If its boiling point is high, its volatility is low. A heavy oil with a boiling point of 600 degrees Fahrenheit [315.5°C] has a very low volatility. Water has a relatively high volatility; it boils at 212 degrees Fahrenheit [100°C] at atmospheric pressure.

Gasoline is blended from different hydrocarbon compounds, each having a different volatility. The proportions of high- and low-volatility hydrocarbons in gasoline must be correct for the following engine-operating requirements.

1. *EASY STARTING* For easy starting with a cold engine, gasoline must be highly volatile so that it will vaporize readily at a low temperature. Thus, a percentage of the gasoline must have highly volatile hydrocarbons. (This percentage must be higher in the cold regions than in warm climates.)

2. *FREEDOM FROM VAPOR LOCK* If the gasoline is too volatile, engine heat will cause it to vaporize in the fuel pump. This can cause vapor lock. Vapor lock prevents normal fuel delivery to the carburetor and would probably produce stalling of the engine. Thus, the percentage of gasoline with highly volatile hydrocarbons must be kept low to prevent vapor lock. The use of a vapor-return line to return vaporized fuel from the fuel pump to the fuel tank is discussed in ⊘ 13-7.

3. *QUICK WARM-UP* The speed with which the engine warms up depends, in part, on how much gasoline vaporizes immediately after the engine starts. Volatility for this purpose does not have to be quite so high as for easy starting, but, all the same, it must be fairly high.

4. *SMOOTH ACCELERATION* When the throttle is opened for acceleration, there is a sudden increase

in the amount of air passing through the throttle valve. At the same time, the accelerator pump delivers extra gasoline. If this gasoline does not vaporize quickly, there will be a momentary time during which the air-fuel mixture will be too lean. This causes the engine to hesitate, or stumble. Immediately after, as the gasoline begins to evaporate, the mixture will become temporarily too rich. Here, again, there is poor combustion and a tendency for the engine to hesitate. Enough of the gasoline must be sufficiently volatile to ensure adequate vaporization for smooth acceleration.

5. *GOOD MILEAGE OR FUEL ECONOMY* For good fuel economy, or maximum miles per gallon [kilometers per liter], gasoline must have a high heat content, or energy, and low volatility. Gasoline with high overall volatility tends to reduce economy. It may produce an overrich mixture under many operating conditions. On the other hand, gasoline with volatility that is too low increases starting difficulty. Such gasoline reduces speed of warm-up and does not give quite as good acceleration. Thus, only a limited percentage of the hydrocarbons in gasoline can be of low volatility.

6. *FREEDOM FROM CRANKCASE DILUTION* Crankcase dilution results when part of the gasoline is not vaporized when it enters the engine cylinders. The liquid gasoline does not burn. It runs down the cylinder walls and enters the oil pan, where it dilutes the oil. This process washes lubricating oil from the cylinder walls (thus increasing the wear of walls, rings, and pistons). Also, diluted oil is less able to lubricate other engine parts, such as the bearings. To avoid damage from crankcase dilution, the gasoline must be volatile enough so that little of it enters the cylinders in liquid form.

7. *THE VOLATILITY BLEND* As you can see, no one volatility of gasoline will satisfy all engine-operating requirements. Gasoline must be of high volatility for easy starting and good acceleration. But it must also be of low volatility to give good fuel economy and combat vapor lock. Thus, gasoline is blended from various amounts of different hydrocarbons having different volatilities. Such a blend satisfies the various operating requirements.

⊘ 12-4 Antiknock Value During normal combustion in the engine cylinder, the pressure increases evenly. Under some conditions, the last part of the compressed air-fuel mixture explodes, or detonates. This produces a sudden and sharp pressure increase. This detonation may cause a rapping, or knocking, noise that sounds almost as though the pistonhead had been struck by a hard hammer. Actually, the sudden pressure increase does impose a sudden heavy load on the piston that is almost like the blow of a hammer. This sudden load can be very damaging to the engine. It can wear moving parts rapidly and even cause parts to break. Also, some of the energy in the gasoline is wasted. The sudden pressure increase does not permit best utilization of the fuel energy.

Some types of gasoline cause much more knocking than others. Because knocking is such an undesirable characteristic, gasoline producers improve their fuels to reduce knocking tendencies. Certain chemicals have been found to reduce knocking when added to the gasoline. The actual rating of the "antiknock" tendencies of a gasoline is given in terms of octane number, or octane rating.

⊘ 12-5 Heat of Compression To understand why knocking occurs, remember what happens to air or any other gas when it is compressed. We noted (⊘ 6-15) that the diesel engine compresses air to less than one-fifteenth of its original volume. This compression increases the air temperature to about 1000 degrees Fahrenheit [537.8°C]. This temperature rise is called *heat of compression*. Let us see how heat of compression affects knocking.

⊘ 12-6 Cause of Knocking, or Detonation Normally the spark at the spark plug starts the air-fuel mixture burning in the combustion chamber. A wall of flame spreads out in all directions from the spark (moving outward almost like a rubber balloon being blown up). The flame travels rapidly outward through the compressed air-fuel mixture until all the charge is burned. The speed with which the flame travels is called the *rate of flame propagation*. The movement of the flame wall during normal combustion is shown in Fig. 12-1, top. During combustion, the pressure increases to several hundred pounds per square inch. It may exceed 1,000 psi [70.3 kg/cm^2] in modern high-compression engines.

Under certain conditions, the last part of the compressed air-fuel mixture, or *end gas*, will explode before the flame front reaches it (Fig. 12-1, bottom). Remember that the *end gas* (the unburned mixture) is subjected to increasing pressure as the flame progresses through the combustion chamber. This increases the temperature of the end gas (because of heat of compression and also radiated heat from the combustion process). If the temperature goes high enough, this end gas will explode before the flame front arrives. The effect is almost the same as if the pistonhead had been struck by a heavy hammer. In fact, it sounds as though this had happened. The sudden shock load due to detonation of the end gas increases wear on bearings and may actually break engine parts if it is severe enough.

⊘ 12-7 Compression Ratio versus Knocking As compression ratios of engines have increased, so has the tendency for engines to knock. Here is the reason. With a higher compression ratio, the air-fuel mixture is more highly compressed at top dead center (TDC). *It is thus at a higher initial temperature.* With higher initial pressure and temperature, the temperature at which detonation occurs is reached sooner. Thus, high-compression engines have a

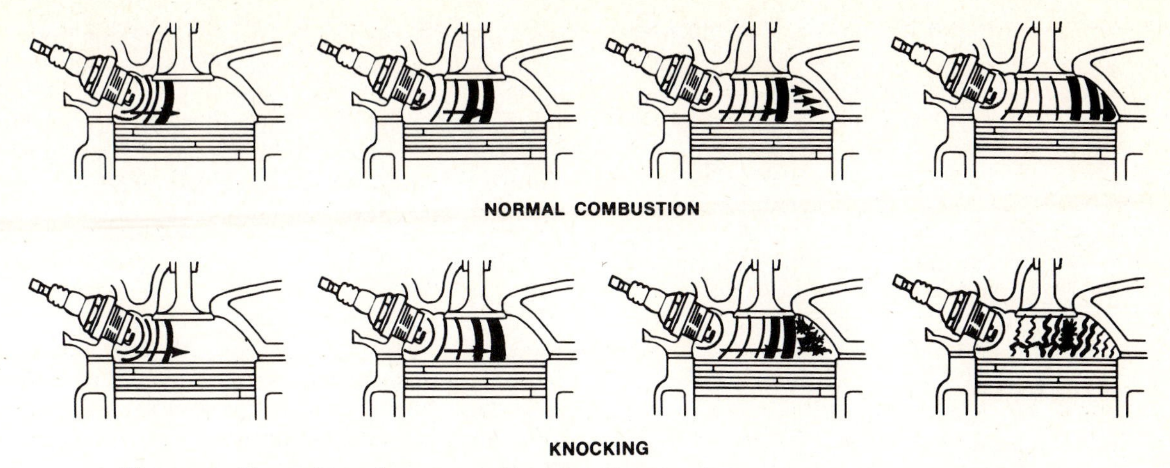

Fig. 12-1. Normal combustion without knocking is shown in the top row. The fuel charge burns smoothly from beginning to end, providing an even, powerful thrust to the piston. Knocking is shown in the bottom row. The last part of the fuel explodes, or burns, almost all at once, to produce detonation, or knocking. (*General Motors Corporation*)

greater tendency to knock. However, special slower-burning fuels have been developed for higher-compression engines. These special fuels have a greater resistance to being set off suddenly by heat of compression. They are less likely to explode suddenly. For ignition they depend only on the wall of flame traveling through the air-fuel mixture.

⊘ 12-8 Measuring Antiknock Values There are several methods of measuring the antiknock value of fuels. The rating is made in terms of *octane number*, or *octane rating*. A high-octane fuel is highly resistant to knock. A low-octane fuel knocks rather easily. For example, one fuel called *isooctane* is very resistant to knocking. It is given an octane number of 100. Another fuel, called *heptane*, knocks very easily. It is given a rating of 0. A mixture of half isooctane and half heptane (by volume) would have an octane number of 50. A mixture of 90 percent isooctane and 10 percent heptane would have an octane number of 90.

Actually, isooctane and heptane are reference fuels, used only to test and rate unknown fuels. The test is made approximately as follows: The fuel to be tested is used in an engine under various conditions and compression ratios. Its tolerance to knocking is noted. Then the two reference fuels, isooctane and heptane, are mixed in varying proportions and used to run the engine under identical conditions. For example, suppose that a mixture of 88 percent isooctane and 12 percent heptane produces the same knocking characteristics as the fuel being tested. Then the fuel being tested is considered to have the same octane number as the reference fuel: 88.

One test method, called the *modified borderline procedure*, rates fuels at various speeds. This test is made by running the engine at various speeds on a dynamometer. Then the amount of ignition spark advance the fuel can tolerate at each speed without knocking is determined. If the spark is advanced too much at any particular speed, knocking will occur. Thus, the test results give a curve that shows the knock characteristics of the fuel being tested at every engine speed (Fig. 12-2). Any spark advance above the curve causes knock.

We should note, however, that some fuels knock at high speeds and others knock at low speeds. For example, refer to Fig. 12-3, which shows the curves of two fuels A and B. Curve C is the amount of spark advance the distributor provides on the engine used in the test. (See ⊘ 15-20 to 15-24 for a discussion of spark-advance mechanisms.) If, at any particular speed, the distributor advances the spark more than the fuel can tolerate, the fuel will knock. Thus, at low speed, fuel A will knock since the spark advance is more than the fuel can tolerate. That is, curve C is above curve A at low speed. On the other hand, fuel A will not knock at high speed since the spark advance is not up to the amount the fuel can tolerate at high speed. But fuel B shows a different picture. It will not knock at low speed but will knock at high speed with the spark-advance curve shown. These curves, which apply only to fuels A and B, show that different fuels act differently at different speeds and in different engines.

⊘ 12-9 Types of Abnormal Combustion Two of the more common types of abnormal combustion are detonation and preignition. Let's define these terms.

DETONATION: A secondary explosion that occurs after the spark at the spark-plug gap
PREIGNITION: Ignition of the air-fuel mixture prior to the occurrence of the spark at the spark-plug gap

Thus far, we have discussed the type of knocking that results from detonation, or sudden explosion, of the last part of the fuel charge. This type of knocking is usually regular in character. It is most noticeable when the engine is accelerated or is under heavy load, as when climbing a hill. Under

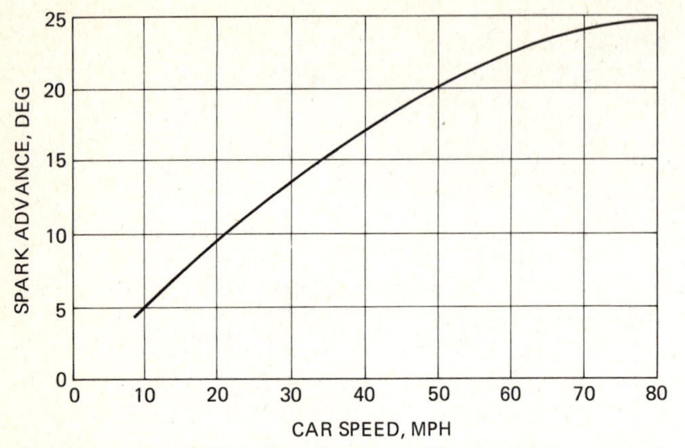

Fig. 12-2. Borderline knock curve. The fuel being tested will knock if the ignition spark is advanced to any value above the curve at any speed.

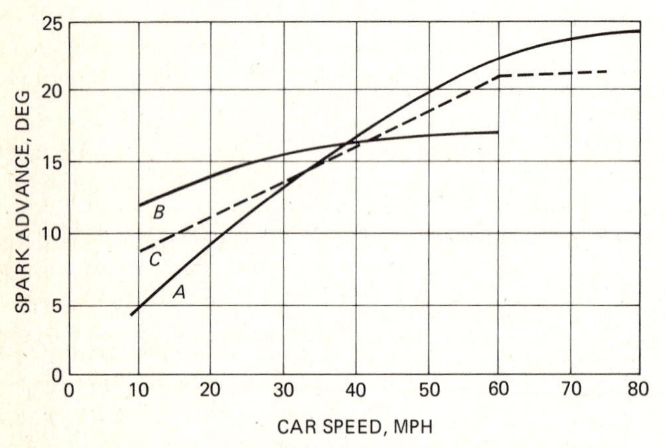

Fig. 12-3. Comparison of borderline knock curves of two fuels *A* and *B*. Curve *C* is the spark advance actually provided by the ignition distributor on the engine.

these conditions, the throttle valve is nearly or fully wide-open. The engine is taking in a full air-fuel mixture on every intake stroke. This means the compression pressures reached are at the maximum. Detonation pressures are more likely to be reached after the air-fuel mixture is ignited.

There are other types of abnormal combustion, however, including *surface ignition, preignition*, and *rumble*. Surface ignition can start from hot spots in the combustion chamber. That is, surface ignition can start from a hot exhaust valve or spark plug or from combustion-chamber deposits. In some cases the deposits may break loose so that particles float free and become hot enough to produce ignition. Surface ignition can occur before (preignition) or after the spark occurs at the spark plug. Also, it can cause engine rumble and rough operation or mild to severe knocking. In some cases, the hot spots act as substitutes for the spark plugs so that the engine continues to run even after the ignition switch is turned off. This condition can cause serious engine damage.

Surface ignition, preignition, and rumble are usually considered service problems. They result from inadequate servicing of the engine, installation of the wrong spark plugs (which run too hot), and use of incorrect fuels and lubricating oils for the engine and type of operation. With incorrect fuel or oil, engine deposits may occur. These engine deposits not only can cause surface ignition and rumble, but also they may increase the compression ratio so that the engine is more likely to knock.

⊘ 12-10 Chemical Control of Knocking Several chemicals added to gasoline tend to prevent detonation of the last part of the end gas during combustion. One theory is that the chemicals increase the reaction time of the fuel, that is, the time required for the end gas to explode. This increased time gives the flame front time to reach the end gas. The result is that the end gas enters into the normal combustion process instead of exploding. One of the compounds that is most successful in preventing knocking is tetraethyl lead, commonly called *ethyl* or *tel*. A small amount added to gasoline raises the octane number of the gasoline.

Special scavengers are also added. These prevent the combustion products of the tetraethyl lead from depositing in the combustion chambers (on plugs, valves, walls, and pistons). These compounds (ethylene dibromide and ethylene dichloride, for example) change the tetraethyl lead into compounds which vaporize and exit with the exhaust gases.

⊘ 12-11 Octane Ratings Antiknock values of gasoline, or *octane ratings*, are measured in different ways (⊘ 12-8). *Research octane* is a measure of the antiknock properties under relatively mild operating conditions. A second measure is *engine octane* (also called *motor octane*), which is made under more severe operating conditions. The third octane rating now in widespread use is the *Environmental Protection Agency (EPA) octane rating*. This rating is found on the white circular stickers posted on gasoline pumps at service stations. The EPA octane rating is actually the average of the research- and motor-octane ratings. Generally, the EPA octane ratings posted on gasoline pumps are 87 for unleaded regular gasoline, 90 for leaded regular gasoline, and 95 for leaded premiums.

⊘ 12-12 Tetraethyl Lead Tetraethyl lead raises the octane number of gasoline and thus reduces knocking tendencies. It also improves valve and valve-seat life. (The lead coats the valve and valve seat and thus provides lubrication.) However, when added to gasoline tetraethyl lead has a bad effect on the engine catalysts used in catalytic converters. These converters are connected into the exhaust system to convert certain pollutants in the exhaust gases into harmless compounds (see Chap. 19). Lead from the gasoline deposits on the catalysts and stops them from doing their job, which is why the gasoline produced for late-model cars using calalytic converters has little or no lead in it. Reducing the amount of tetraethyl lead in gasoline has required a reduction in the compression ratios of automotive engines in recent years.

However, the removal of tetraethyl lead from gasoline can have a potentially harmful effect on valves and valve seats. Without the lubrication provided by the lead, the valve and valve seat can wear more rapidly. To prevent this, modern automotive-engine valves have special coatings, or facings, on their seating faces. Also, many engines use valve-seat inserts or have valve seats that are induction-hardened (⊘ 9-7).

⊘ 12-13 Mechanical Factors Affecting Knocking The shape of the combustion chamber has a great effect on engine knock. The cylinder head, intake and exhaust valves, and spark plug form the top of the combustion chamber. The pistonhead and top compression ring form the bottom of the combustion chamber (Fig. 12-4). Combustion chambers have two general shapes: wedge and hemispheric (Fig. 12-5). The shape determines turbulence, squish, and quench, and these three factors affect knock.

1. *TURBULENCE* When you stir coffee, you swirl it, or impart turbulence to it, so that the cream and sugar mix with the coffee. In the same way, imparting turbulence to the air-fuel mixture entering the combustion chamber ensures more even mixing, which makes combustion more even. Turbulence also reduces the time required for the flame front to sweep through the compressed mixture.

2. *SQUISH* In some combustion chambers, the piston squishes, or squeezes, a part of the air-fuel mixture at the end of the compression stroke. Figure 12-5 (left) shows the squish area in a combustion chamber. As the piston nears TDC, the mixture is squished, or pushed, out of the area. As it squirts out, it promotes turbulence and further mixing of the air-fuel mixture.

3. *QUENCH* It was mentioned previously that knocking results when the end-gas temperature goes too high. The end gas explodes before the flame front reaches it. However, if some heat is taken from the end gas, then its temperature will not reach the detonation point. In the cylinder shown to the left in Fig. 12-5, the squish area is also a quench area. The cylinder head is close to the piston, and these metallic surfaces are cooler than the end gas. This arrangement causes heat to be removed from the end gas. Thus, the tendency for detonation to occur is quenched. However, this causes a problem with exhaust emissions (Chap. 19).

4. *THE HEMISPHERIC COMBUSTION CHAMBER* In the hemispheric combustion chamber, the spark plug can be located near the center of the dome (Fig. 6-19). Then, when combustion starts, the flame front has a relatively short distance to travel. There are no distant pockets of end gas to detonate. The chamber has no squish or quench areas. However, there is relatively little turbulence.

5. *THE WEDGE COMBUSTION CHAMBER* In the wedge combustion chamber, the spark plug is located to one side. The flame front must travel a greater distance to reach the end of the wedge (Fig. 12-5, left). The end of the wedge has a squish and quench area which cools the end gas to prevent detonation, imparting turbulence to the mixture at the same time.

6. *CONTAMINANTS* The shape of the combustion chamber also affects the amount of contaminants that appear in the exhaust gases. The cooler metal surfaces of the cylinder head and piston top slow combustion. Therefore, the layers of air-fuel mixture next to these metal surfaces do not burn completely. The wedge combustion chamber has a larger surface area. It thus produces a greater percentage of contaminants (mostly unburned HC) than the hemispheric combustion chamber does. (See Chap. 19 for additional information.)

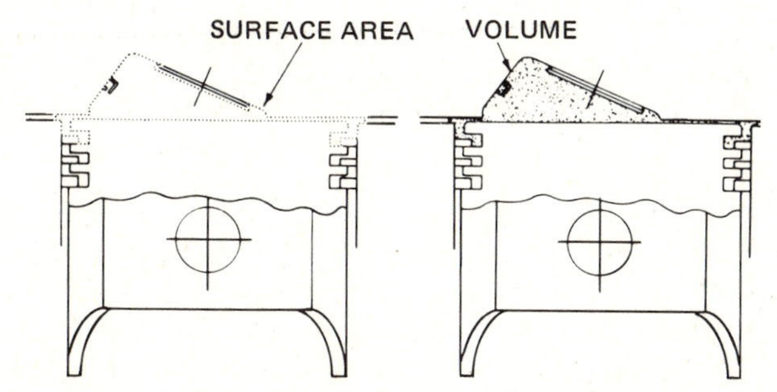

Fig. 12-4. Combustion chamber. The surface area is shown by the dotted line.

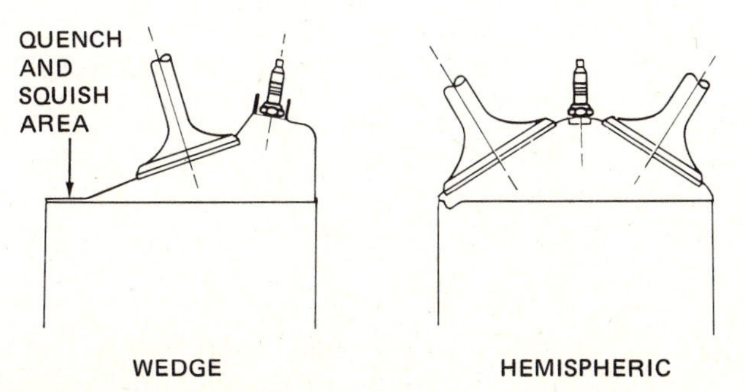

Fig. 12-5. Wedge and hemispheric combustion chambers. *(General Motors Corporation)*

⊘ 12-14 Other Factors Affecting Knocking Many operating conditions in an engine affect knocking. For example, high humidity (or damp air) as well as high altitudes (or lower-density air) reduce the tendency of the engine to knock. High air temperatures increase the tendency of the engine to knock. Engine deposits (carbon in the combustion chamber), advancing the spark, and reducing the amount of fuel in the air-fuel mixture also increase the tendency of the engine to knock.

All these factors show the need for good maintenance of the modern high-compression engine. Buildup of scale in the cooling system reduces cooling efficiency. Clogged fuel lines or nozzles in the carburetor leans out the mixture. These conditions can also increase knocking.

⊘ 12-15 Other Gasoline Additives Antiknock compounds and their related lead-compound-vaporizing substances are added to gasoline to raise its octane rating (⊘ 12-10). Other additives are also used. Major types include:

1. Oxidation inhibitors to help prevent the formation of gum while the gasoline is in storage
2. Metal deactivators to protect the gasoline from the harmful effects of certain metals which can be picked up in the refining process or in the vehicle fuel system
3. Antirust agents to protect the vehicle fuel system
4. Anti-icers to combat carburetor icing and fuel-line freeze
5. Detergents to keep the carburetor clean
6. Phosphorus compounds to combat surface ignition and spark-plug fouling
7. Dye for identification

In addition, the refining process is very carefully controlled to keep sulfur compounds and gum-forming substances to a minimum. Sulfur compounds, in excess, form sulfur acids which could seriously damage metal parts and bearings. These compounds can also contribute to air pollution as they burn with the fuel. Gum-forming substances form deposits in carburetor circuits and intake manifolds and on valves, pistons, and rings. Proper refining techniques minimize the amount of these harmful substances in gasoline.

⊘ 12-16 Chemistry of Combustion We have already noted (in ⊘ 12-1) that gasoline, in burning, forms water (H_2O) and carbon dioxide (CO_2). This occurs when enough oxygen is present to combine with all the hydrogen and carbon atoms. However, in the gasoline engine, sufficient amounts of oxygen are not usually present. As a result, the carbon does not attain complete combustion. Some atoms of carbon unite with only one atom of oxygen (instead of two). This combination produces carbon monoxide (CO). Carbon monoxide is a dangerously poisonous gas. It has no color, is tasteless, and has practically no odor. But 15 parts of carbon monoxide in 10,000 parts of air makes the air dangerous to breathe. Larger amounts may cause quick paralysis and death. An engine should never be operated in a closed space, such as a garage, without some means of exhausting the gas into the outside air. Remember this fact:

Enough carbon monoxide is produced in 3 minutes by an automobile engine running in a closed one-car garage to cause paralysis and death. Never operate an engine with the garage doors closed!

When there is incomplete combustion, the exhaust gases contain unburned hydrocarbons. These contribute to smog, a health hazard in populous places. Automotive companies are working on exhaust-system devices that will convert these compounds into harmless gases. Also, cars are equipped with closed-crankcase ventilating systems to prevent the escape of blow-by gases from the engine. Of course, with perfect combustion, there would be no problem. See Chaps. 18 and 19 for details of automotive emission controls.

⊘ 12-17 Diesel-Engine Fuels Diesel engines use fuel oil. The fuel oil is sprayed into the compressed air in the combustion chamber at the end of the compression stroke. Heat of compression ignites the fuel oil, and the combustion stroke follows (see ⊘ 6-15). Diesel oil is light, with a low viscosity and high cetane number (see ⊘ 12-19).

⊘ 12-18 Diesel-Fuel Viscosity Viscosity refers to the tendency of a liquid to resist flowing. Water has a low viscosity; it flows easily. A light oil has a higher viscosity than water, but it still flows easily. Light oil, too, has a low viscosity. Heavy oil has a high viscosity; it flows slowly. The fuel oil used in diesel engines must have a relatively low viscosity so that it flows easily through the fuel-pumping system. But it must have sufficient viscosity to lubricate the moving parts in the pumping system. However, if the viscosity is too high, the fuel will not spray, or atomize, easily and thus will not burn well.

⊘ 12-19 Cetane Number of Diesel Fuel The cetane number of diesel fuel refers to the ease with which the fuel ignites. With a high cetane number, the fuel ignites with relative ease (or at a relatively low temperature). The lower the cetane number, the higher the temperature needed to ignite the fuel. Also, the lower the cetane number, the more likely the fuel is to knock. The fuel being sprayed into the cylinder will not ignite quickly so that it tends to accumulate. Then, when ignition does take place, there will be a combustion knock as the fuel present suddenly burns. On the other hand, if the cetane number is high enough, the fuel will ignite and begin to burn as soon as the injection spray starts. There will be an even combustion-pressure rise and therefore no knock.

⊘ 12-20 Liquefied Petroleum Gas (LPG) Liquefied petroleum gas (LPG) requires a special fuel system (⊘ 13-23). There are actually two types of LPG that have been used for automotive-engine fuel: propane and butane. Of these, propane is the most widely used. Sometimes, small amounts of butane are added. Propane boils at −44 degrees Fahrenheit [−42.2°C] (at atmospheric pressure). Thus, it can be used in any climate where temperatures below this are not reached. Butane cannot be used in temperatures below 32 degrees Fahrenheit [0°C] since it is liquid below that temperature. If it remains liquid, it will not vaporize in the fuel system and thus will never reach the engine.

CHAPTER 12 CHECKUP

NOTE: Since the following is a chapter review test, you should review the chapter before taking the test.

The practicing automotive mechanic should know what gasoline is all about and how it burns in the engine. This is what we have been discussing in this chapter. To find out how well you remember and understand what you have been studying, take the test that follows.

Completing the Sentences The sentences that follow are incomplete. After each sentence there are several words or phrases, only one of which will correctly complete the sentence. Write each sentence in your notebook, selecting the proper word or phrase to complete it correctly.

1. Gasoline is made up mostly of: (*a*) hydrogen and carbon, (*b*) hydrogen and oxygen, (*c*) hydrocarbon and carbon monoxide.
2. Hydrocarbon (HC) is another name that can be used for: (*a*) water, (*b*) gasoline, (*c*) air.
3. If the gasoline in an engine burned completely, all that would remain is: (*a*) water and carbon monoxide, (*b*) carbon monoxide and hydrocarbon, (*c*) water and carbon dioxide.
4. The ease with which a gasoline vaporizes is called its: (*a*) volatility, (*b*) oxidation, (*c*) octane rating.
5. If the gasoline is too volative, it will vaporize in the fuel pump causing: (*a*) oxidation, (*b*) vapor lock, (*c*) engine runaway.
6. When the last part of the air-fuel mixture in the combustion chamber explodes before being ignited by the flame traveling from the spark plug, this condition is called: (*a*) detonation, (*b*) preignition, (*c*) vaporization.
7. A gasoline that knocks easily is called a: (*a*) high-octane gasoline, (*b*) low-octane gasoline, (*c*) blended gasoline.
8. Two ways to increase the octane rating of gasoline are: (*a*) changing the refining process and adding tetraethyl lead, (*b*) take more water out of the gasoline and add less lubricating oil, (*c*) neither (*a*) nor (*b*).
9. A high compression ratio can cause a problem because it increases: (*a*) the temperature of the air-fuel mixture, (*b*) the strain on the cylinder-head bolts, (*c*) crankshaft bearing wear.
10. The compression ratio must be kept low enough to make sure that the air-fuel mixture will not ignite from the heat of compression before the proper time, or: (*a*) a low-octane gasoline must be used, (*b*) a high-octane gasoline must be used, (*c*) gasoline cannot be used as the fuel.

Definitions and Review Questions In the following, you are asked to write the definitions of certain terms and phrases and answer questions about what you have been studying. Write in your notebook. The act of writing the information will help you remember it. It also puts good information into your notebook. Turn back into the chapter if you are not sure about an answer.

1. With perfect combustion, what two compounds would be formed when gasoline burns?
2. Name three pollutants being emitted from automobiles.
3. What is volatility and why is it important in gasoline?
4. What does the term "antiknock value" mean?
5. What is heat of compression?
6. Explain the cause of knocking produced by heat of compression.
7. What effect does increasing the compression ratio have on knocking? Why?
8. Explain one method of measuring the antiknock value of a gasoline.
9. What does octane number mean?
10. What is the difference between detonation and preignition?
11. What effect does lead have on valves and valve seats?
12. Why has lead been removed from gasoline?
13. What is quench?
14. What is squish?
15. What are the two basic combustion-chamber shapes?
16. Name six gasoline additives.
17. Why is CO dangerous?
18. Can you tell by the odor whether or not CO is present in a garage?
19. What does cetane number mean?
20. What does LPG mean?

SUGGESTIONS FOR FURTHER STUDY

Find out how gasoline is made. Go to your local library and look in one of the encyclopedias for information on oil and gasoline. Make notes on how engineers prospect for oil in the earth. Find out how the crude oil is carried to refineries and what happens to the oil when it gets there. Make notes on the important points of what you read, and file these notes in your notebook.

chapter 13

AUTOMOTIVE FUEL SYSTEMS

This chapter describes automotive fuel systems, including all fuel-system components except carburetors. (Carburetors are discussed in Chap. 14.) We describe fuel tanks, filters, gauges, pumps, and vapor-return lines. We also cover emission controls in the fuel system such as vapor-recovery systems, exhaust-gas recirculation, and positive crankcase ventilation.

⊘ 13-1 Purpose of the Fuel System The fuel system supplies a combustible mixture of air and fuel to the engine. The fuel system must vary the proportions of air and fuel for different operating conditions. When the engine is cold, for example, the air-fuel mixture must be rich (have a high proportion of fuel). This is because the fuel does not vaporize readily at low temperatures. Therefore, extra fuel must be added to the air-fuel mixture so there is enough vaporized fuel to form a combustible mixture.

⊘ 13-2 Fuel-System Components The fuel system consists of the fuel tank, fuel filter, fuel gauges, fuel pump, carburetor, intake manifold, and fuel lines. The fuel lines are tubes connecting the tank, pump, and carburetor (Figs. 13-1 and 13-2). Some gasoline engines use a fuel-injection system. In this system, a fuel-injection pump replaces the carburetor. Details of these components are discussed later.

⊘ 13-3 Fuel Tank The fuel tank (Figs. 13-3 and 13-4) is normally located at the rear of the vehicle. It is usually made of sheet metal and is attached to the frame. The filler opening of the tank is closed by a cap. The tank end of the fuel line is attached at or near the bottom of the tank. In some tanks there is a filtering element at the fuel-line connection. The tank also contains the sending unit of the fuel gauge. In older cars, the tank may also have a vent pipe to allow air to escape when the tank is being filled (Fig. 13-2).

Vaporized gasoline escaping from the fuel tank through the vent pipe or filler cap contributes to the formation of smog. Thus, cars manufactured since 1970 are equipped with a vehicle-vapor-recovery

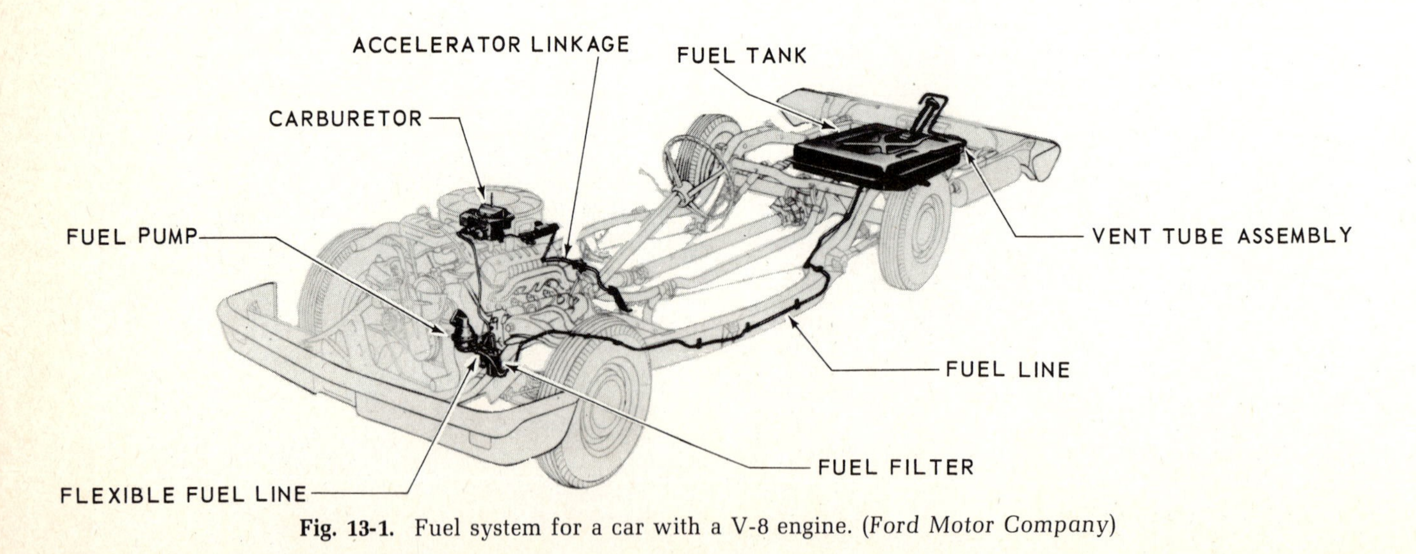

Fig. 13-1. Fuel system for a car with a V-8 engine. (*Ford Motor Company*)

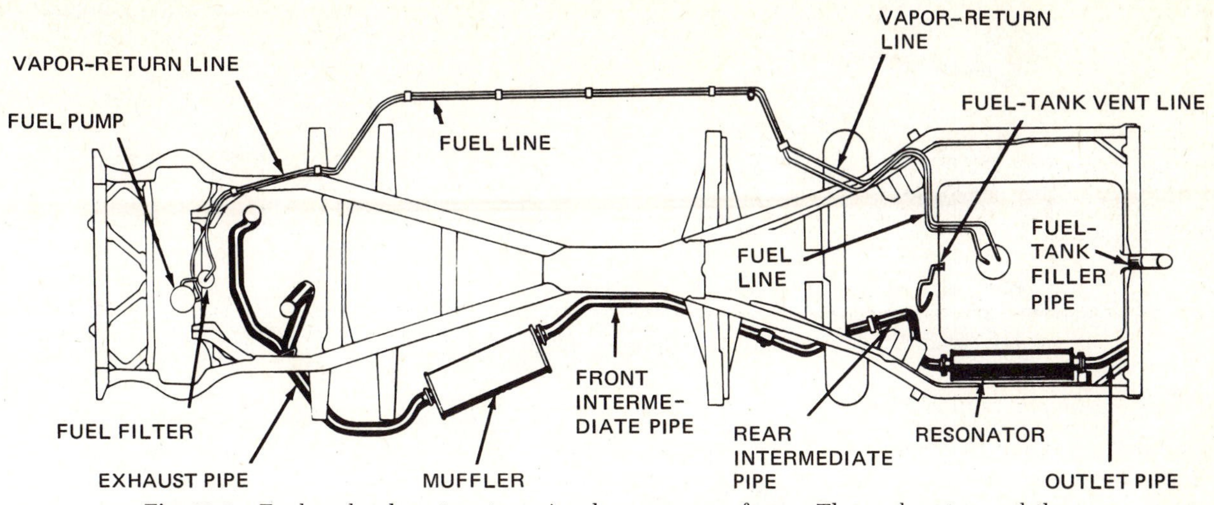

Fig. 13-2. Fuel and exhaust systems in place on a car frame. The carburetor and the engine are not shown. (*Cadillac Motor Car Division of General Motors Corporation*)

(VVR) system. In this system, the fuel-tank vent pipe is connected to a carbon canister, which retains the vapor and prevents its escape into the air (⊘ 13-8).

⊘ 13-4 Fuel Filters and Screens Fuel systems have filters and screens to prevent dirt in the fuel from entering the fuel pump or carburetor. Dirt could prevent normal operation of these units and cause poor engine performance. One type of filter is incorporated in the fuel pump (Fig. 13-9). Another type is a separate unit connected into the fuel line between the tank and fuel pump. Still another type is located between the fuel pump and carburetor (Fig. 13-1). This type may be in or on the carburetor itself. Figure 13-5 shows the type of filter that is outside the carburetor but mounted on it. The screw threads enter a tapped hole in the carburetor, and the fuel line fits on the opposite end of the filter. Figure 13-6 shows the type of filter that is installed in the carburetor itself. This filter has an element made of pleated paper.

⊘ 13-5 Fuel Gauges There are two types of fuel gauges: balancing coil and thermostatic. Each has a tank unit and a instrument-panel, or dash, unit.

1. *BALANCING-COIL GAUGE* The tank unit in this fuel gauge contains a sliding contact (Fig. 13-7). The contact slides back and forth on a resistance as the float moves up and down in the fuel tank. This action changes the amount of electric resistance the tank unit offers. Thus, as the tank empties, the float drops and the sliding contact moves to reduce the resistance. The instrument-panel unit contains two coils, as shown in Fig. 13-7. When the ignition switch is turned on, current from the battery flows through the two coils. This produces a magnetic pattern that acts on the armature to which the pointer is attached. When the resistance of the tank unit is high (tank filled and float up), the current through the E (empty) coil also flows through the F

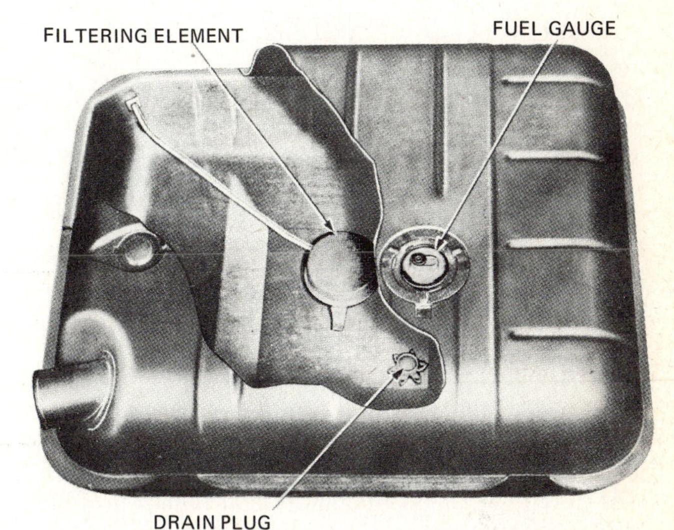

Fig. 13-3. Fuel tank partly cut away to show the filtering element and drain plug. (*Chrysler Corporation*)

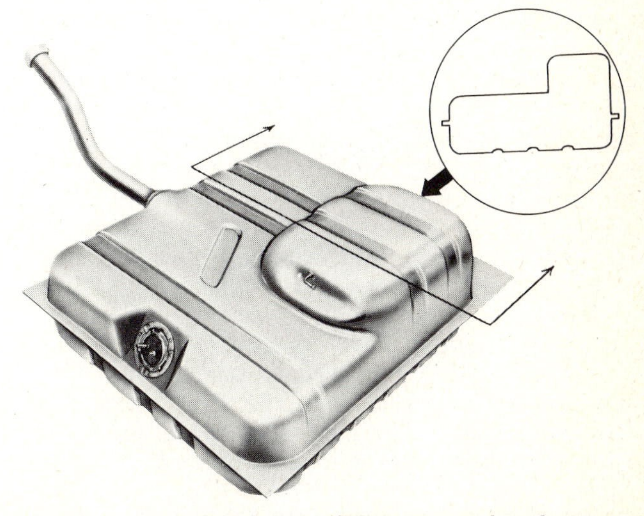

Fig. 13-4. Domed fuel tank of the type used with vapor-recovery systems. (*Chrysler Corporation*)

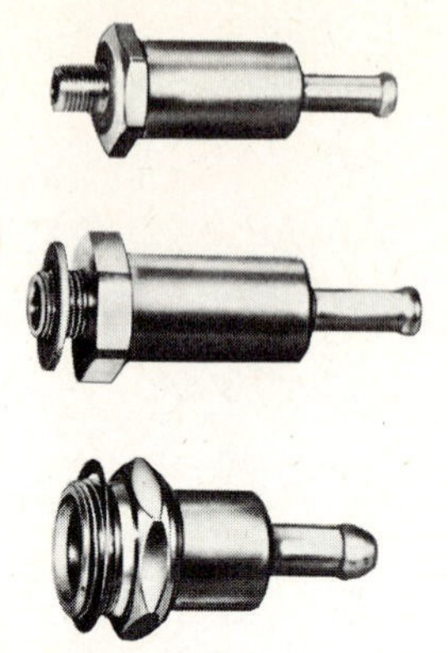

Fig. 13-5. In-line fuel filters. (*Ford Motor Company*)

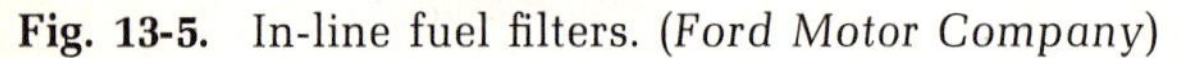

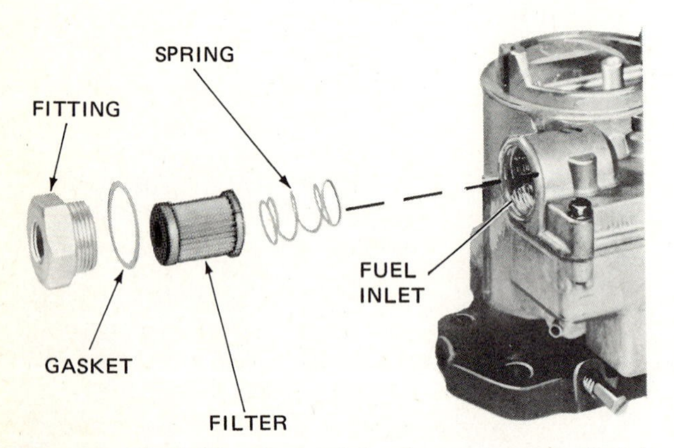

Fig. 13-6. Fuel filter located in the carburetor. (*Buick Motor Division of General Motors Corporation*)

(full) coil. Thus the armature is pulled to the right so that the pointer swings to the F (full) side of the dial. But when the tank begins to empty, the resistance of the tank unit drops. Thus, more of the current flowing through the empty coil passes through the tank unit. Since less current is flowing through the full coil, its magnetic field is weaker. As a result, the empty coil pulls the armature toward it and the pointer swings around toward the E (empty) side of the dial.

2. *THERMOSTATIC GAUGE* Figure 13-8 shows the wiring circuit of a thermostatic fuel gauge. It has a fuel-tank unit much like the balancing-coil system. That is, it has a float and sliding contact that moves on a resistor. Current flows from the battery through a heater wire in the fuel-gauge instrument-panel unit and through the resistance in the tank unit. When the tank is almost empty, most of the resistance is in the circuit. Very little current can flow. When the tank is filled, the float moves up and the sliding contact cuts most of the resistance out of the circuit. Now more current flows. As current flows through the heater coil in the instrument-panel unit, the current heats the thermostat. The thermostat blade bends because of the heat. This moves the gauge needle to the right toward the F (full) mark on the dial.

Note that the system in Fig. 13-8 has an instrument voltage regulator. This device is thermostatic;

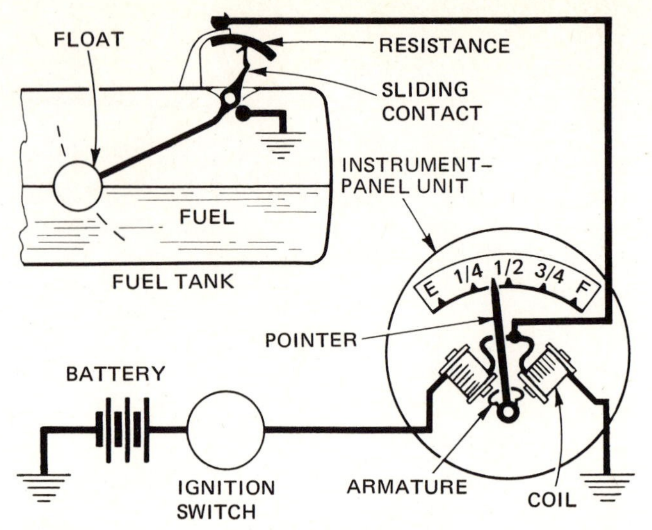

Fig. 13-7. Schematic wiring diagram for a balancing-coil fuel-gauge indicating system.

its purpose is to keep the voltage to the fuel-gauge system low.

3. *LOW-FUEL-LEVEL INDICATOR* The fuel-gauge system shown in Fig. 13-8 also has a low-fuel-level indicator. This indicator includes a thermistor assembly in the fuel tank, warning light, and warning relay. A thermistor is a special kind of resistance that loses resistance as it gets hot. As long as there are more than a few gallons of fuel in the tank, the thermistor is submerged and is cool. However, when the fuel level is low, the thermistor is exposed to air and gets hotter. Its resistance decreases and more current flows. The increased current flow that the lowered resistance allows is sufficient to operate the warning relay. It connects the warning light to the battery, and a light comes on to warn the driver that the fuel supply is getting low.

⊘ 13-6 Mechanical Fuel Pumps The fuel system uses a fuel pump to deliver fuel from the tank to the carburetor. There are two types of fuel pumps: electric and mechanical. Electric fuel pumps are discussed in ⊘ 13-9. Mechanical fuel pumps are operated by an eccentric (off-center section) on the engine camshaft, as explained in the following paragraphs (see Fig. 9-4). In in-line engines, the mechanical fuel pump is mounted on the side of the cylinder block. In some V-8 engines, the pump is mounted between the two cylinder banks. In most modern V-8 engines the fuel pump is mounted on the side of the cylinder block at the front of the engine.

The mechanical fuel pump has a rocker arm, with an end that rests on the camshaft eccentric. Many V-8 engines also use a pushrod extending from the eccentric to the rocker arm.

As the camshaft rotates, the eccentric causes the rocker arm to rock back and forth. The inner end of the rocker arm is linked to a flexible diaphragm. The diaphragm is clamped between the upper and lower pump housings (Fig. 13-9). There is a spring under the diaphragm that maintains tension on it.

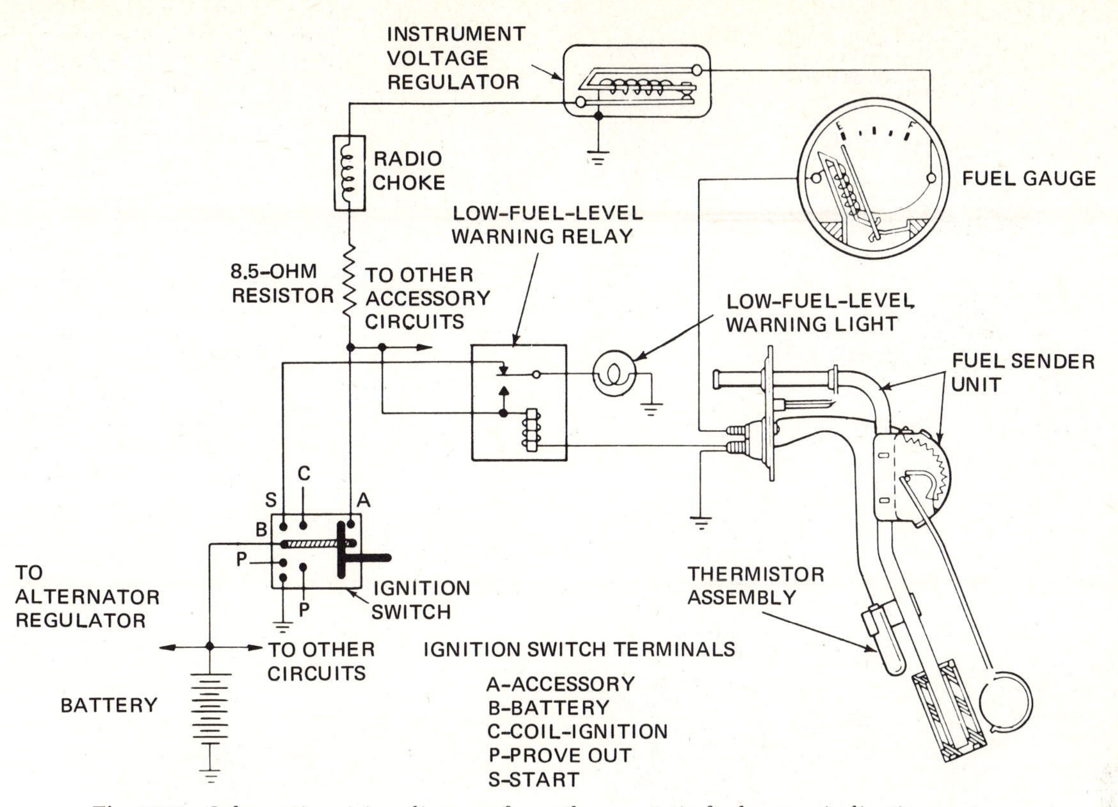

Fig. 13-8. Schematic wiring diagram for a thermostatic fuel-gauge indicating system. This system uses a variable-resistance tank unit and a thermostatic instrument-panel (dash) unit. (*Ford Motor Company*)

As the rocker arm rocks, it pulls the diaphragm down and then releases it. The spring then forces the diaphragm up. Thus, the diaphragm moves up and down as the rocker arm rocks.

This diaphragm movement produces partial vacuum and pressure in the space above the diaphragm. When the diaphragm moves down, a partial vacuum is produced. Then, atmospheric pressure, acting on the fuel in the tank, forces fuel through the fuel line and into the pump. The inlet valve in the pump opens to admit fuel, as shown by the arrows in Fig. 13-9. Note that the fuel first passes through a filter bowl and screen.

When the diaphragm is released by the return movement of the rocker arm, the spring forces the diaphragm upward. This produces pressure in the space above the diaphragm. The pressure closes the inlet valve and opens the outlet valve. Now fuel is forced from the fuel pump through the fuel line into the carburetor.

The actions in the fuel pump as the eccentric rotates are shown in Figs. 13-10 and 13-11. The fuel from the fuel pump enters the carburetor past a needle valve in the float bowl. If the float bowl is full, the needle valve closes so that no fuel can enter. When this happens, the fuel pump cannot deliver fuel to the carburetor. In this case, the rocker arm continues to rock. However, the diaphragm remains at or near its lower limit of travel. The spring cannot force the diaphragm upward so long as the carburetor float bowl will not accept fuel. However, as

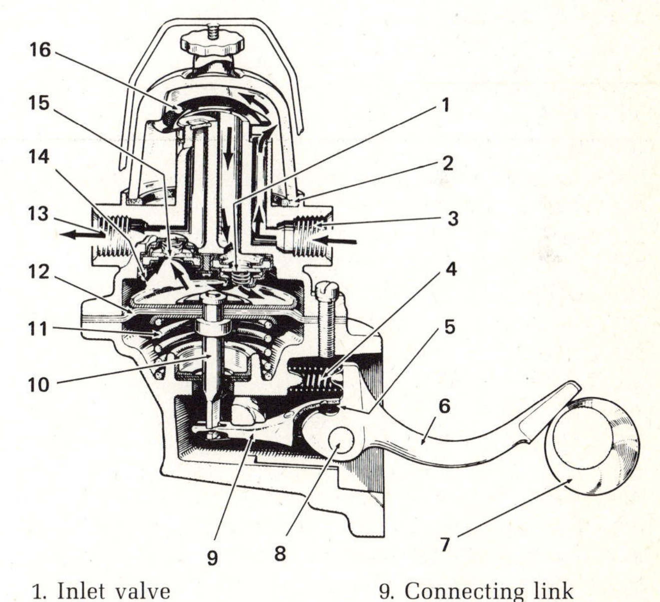

1. Inlet valve
2. Joint under cover bowl
3. Fuel inlet
4. Operating-arm return spring
5. Abutment on operating arm
6. Rocker arm
7. Eccentric on camshaft
8. Pivot
9. Connecting link
10. Pull rod
11. Diaphragm-return spring
12. Diaphragm
13. Fuel outlet
14. Pump chamber
15. Outlet valve
16. Gauze filter

Fig. 13-9. Sectional view of a fuel pump. (*Hillman Motor Car Company, Limited*)

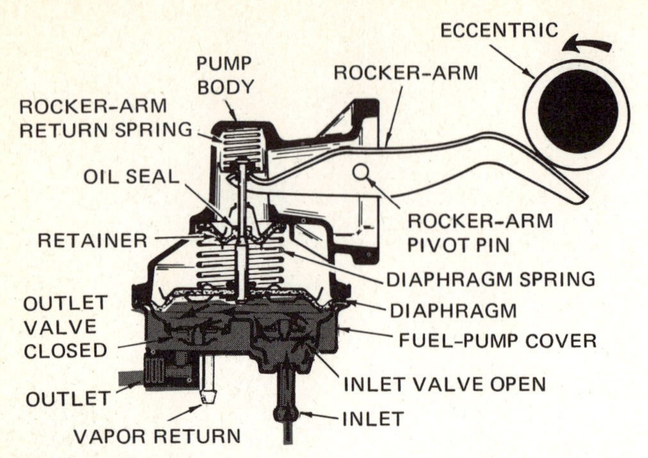

Fig. 13-10. When the eccentric rotates so as to push the rocker arm down, the arm pulls the diaphragm up. The inlet valve opens to admit fuel into the space under the diaphragm.

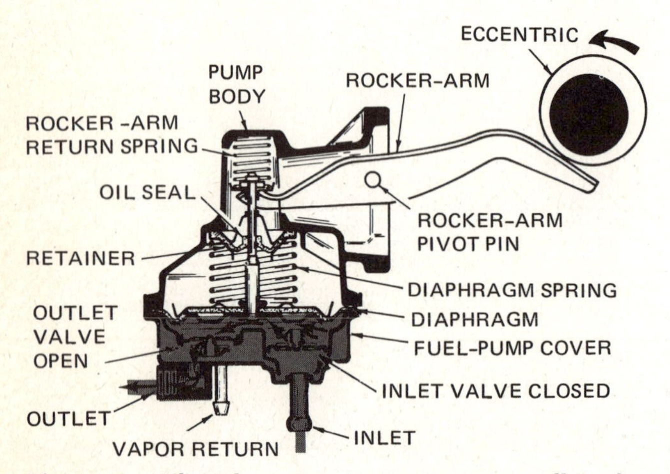

Fig. 13-11. When the eccentric rotates so as to allow the rocker arm to move up under it, the diaphragm is released so it can move down, producing pressure under it. This pressure closes the inlet valve and opens the outlet valve so that fuel flows to the carburetor.

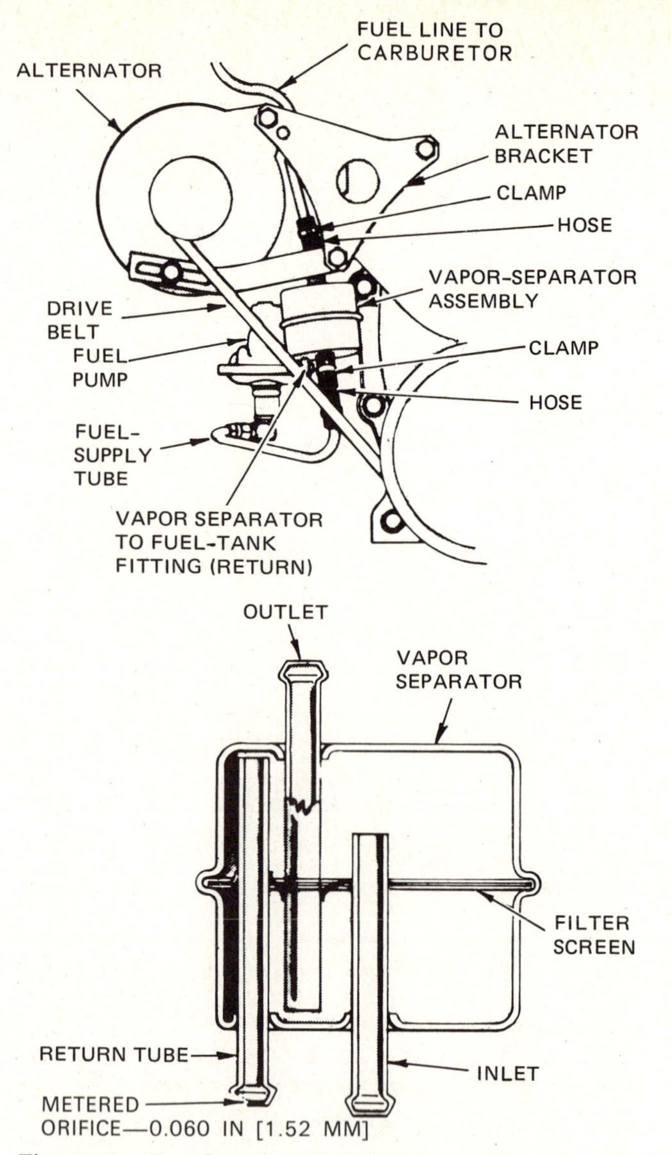

Fig. 13-12. *Top,* location of the vapor separator, in line between the fuel pump and the carburetor, on a V-8 engine. *Bottom,* enlarged sectional view of the fuel-vapor separator. (*Chrysler Corporation*)

the carburetor uses up fuel, the needle valve opens to admit fuel into the float bowl. Now the diaphragm can move up (on the rocker-arm return stroke) to force fuel into the carburetor float bowl.

⊘ 13-7 Vapor-Return Line The fuel system shown in Fig. 13-2 has a vapor-return line running from the fuel pump to the fuel tank. This line is installed on many cars having air conditioning. With air conditioning, under-the-hood temperatures are likely to be higher. The air-conditioning condenser delivers more heat under the hood. Also, during idle, the engine cooling system is not very efficient, which allows under-the-hood temperatures to increase. The higher under-the-hood temperatures tend to cause vapor to form in the fuel pump.

To understand why vapor can form in the fuel pump, note first that the pump alternately produces vacuum and pressure. During the vacuum phase, the boiling, or vaporizing, temperature of the fuel decreases. The lower the pressure, the lower the temperature at which liquid vaporizes. For example, at sea level (atmospheric pressure 14.7 psi [1.05 kg/cm^2] water boils at 212 degrees Fahrenheit [100°C] (see ⊘ 4-20). But at 16,000 feet [4.877 m] above sea level, where the pressure is around 7 psi [0.5 kg/cm^2], water boils at 185 degrees Fahrenheit [85°C].

The combination of increased under-the-hood temperature and partial vacuum in the fuel pump can cause fuel to vaporize. Vaporization produces vapor lock, a condition that prevents normal delivery of fuel to the carburetor. Thus, the engine stalls.

The vapor-return line is connected to a special outlet in the fuel pump. The line allows the vapor in the pump to return to the fuel tank. The vapor-return line also permits excess fuel being pumped

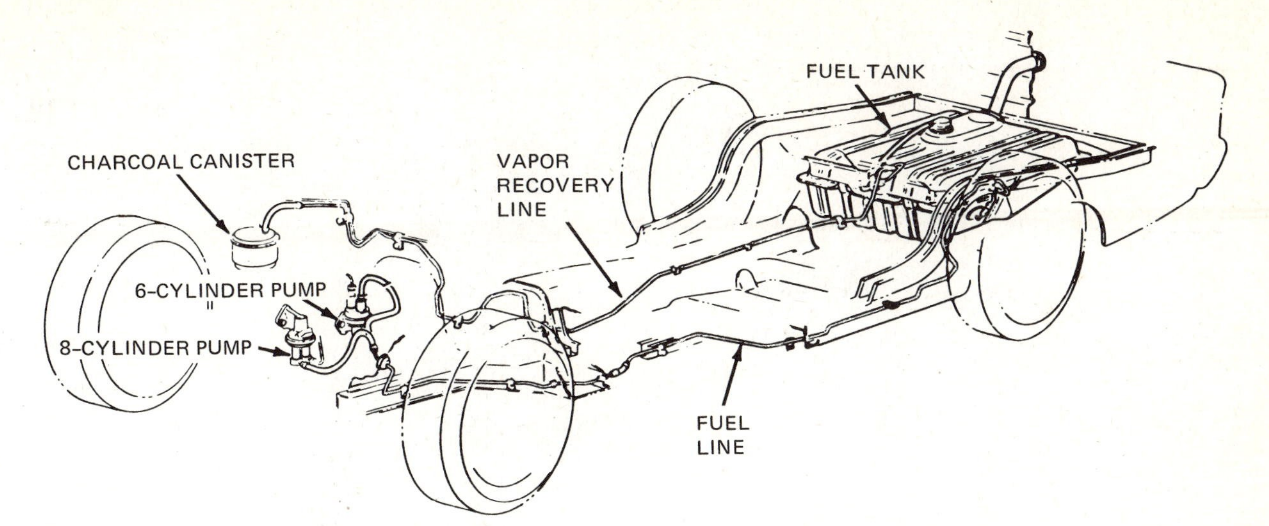

Fig. 13-13. Fuel-vapor-recovery system. (*Ford Motor Company*)

by the fuel pump to return to the fuel tank. This excess fuel, in constant circulation, helps keep the fuel pump cool. It thus prevents vapor from forming.

Some cars have a vapor separator connected between the fuel pump and the carburetor (Fig. 13-12). The separator consists of a sealed can, a filter screen, an inlet and outlet fitting, and a metered orifice, or outlet, for the vapor-return line to the fuel tank. Any fuel vapor that the fuel pump produces enters the vapor separator (as bubbles) along with fuel. These bubbles of vapor rise to the top of the vapor separator. The vapor then is forced, by fuel-pump pressure, to pass through the fuel-return line and back to the fuel tank. In the tank, the vapor condenses back into liquid fuel.

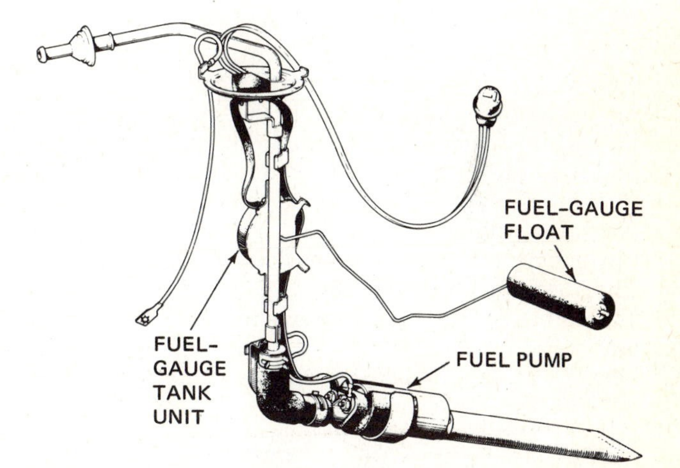

Fig. 13-14. Location of the electric fuel pump in the fuel tank. It is mounted on the same support as the fuel-gauge tank unit. (*Buick Motor Division of General Motors Corporation*)

⊘ 13-8 Fuel-Vapor-Recovery System Gasoline vapor can escape from uncontrolled fuel tanks and carburetors. This happens when a car is sitting idle and the engine is not running. At other times, when the engine is running, gasoline is being pumped from the fuel tank and carburetor so that gasoline vapor does not escape. Significant quantities of gasoline vapor (HC) can escape from parked cars without vapor controls and can add to atmospheric pollution. To prevent this loss of gasoline vapor, modern automobiles are equipped with vapor-recovery systems.

Figure 13-13 shows one fuel-vapor-recovery system. The fuel tank is sealed. Any gasoline vapor that attempts to escape has to flow through the vapor-recovery line to the charcoal canister. Gasoline vapor from the carburetor-float bowl also flows to the charcoal canister. There, the charcoal particles pick up, or *adsorb,* the gasoline vapor and hold it. Then, when the engine is started and running, air flows through the charcoal canister on the way to the carburetor. This air picks up the gasoline vapor trapped in the canister and carries it to the carburetor where it mixes with the air-fuel mixture and enters the engine. The vapor is thus burned instead of being allowed to enter the atmosphere as HC. Chapters 18 and 19, which discuss automotive emission controls, describe vapor-recovery systems in detail.

⊘ 13-9 Electric Fuel Pumps Electric fuel pumps have certain advantages over mechanical fuel pumps. Electric fuel pumps deliver fuel to the carburetor as soon as the ignition is turned on. They can deliver more fuel than the engine will require even under maximum operating conditions. Thus, the engine will never be fuel-starved. They are often used in many high-performance and heavy-duty applications.

There are various types of electric fuel pumps. One of the latest types is mounted in the fuel tank (Fig. 13-14). This type contains an impeller driven by an electric motor (Fig. 13-15). The impeller pushes fuel through the fuel line into the carburetor. Other types of electric fuel pumps are mounted in the

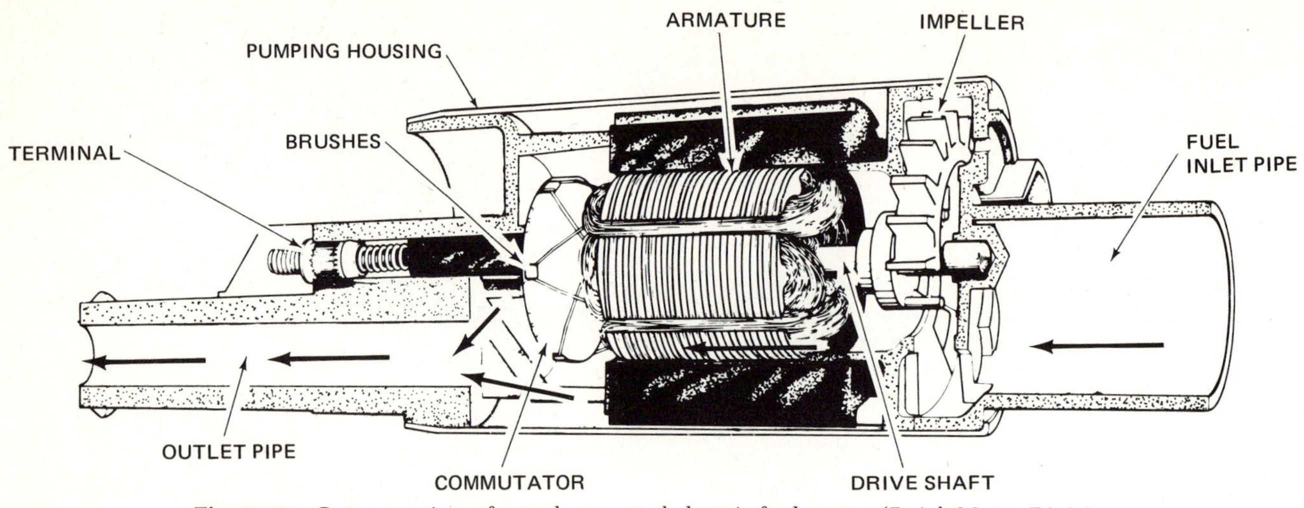

Fig. 13-15. Cutaway view of a tank-mounted electric fuel pump. (*Buick Motor Division of General Motors Corporation*)

engine compartment. One design (Fig. 13-16) contains a flexible metal bellows that is operated by an electromagnet. When the electromagnet is connected to the battery (by turning on the ignition switch), it pulls the armature down and thereby extends the bellows. This produces a vacuum in the bellows. Fuel from the fuel tank enters the bellows through the inlet valve. Then, as the armature reaches its lower limit of travel, it opens a set of contact points. This disconnects the electromagnet from the battery. The return spring therefore pushes the armature up and collapses the bellows. This forces fuel from the bellows through the outlet valve and into the carburetor. As the armature reaches the upper limit of its travel, it closes the contacts. The electromagnet is again energized and pulls the armature down. These actions are repeated as long as the ignition is on.

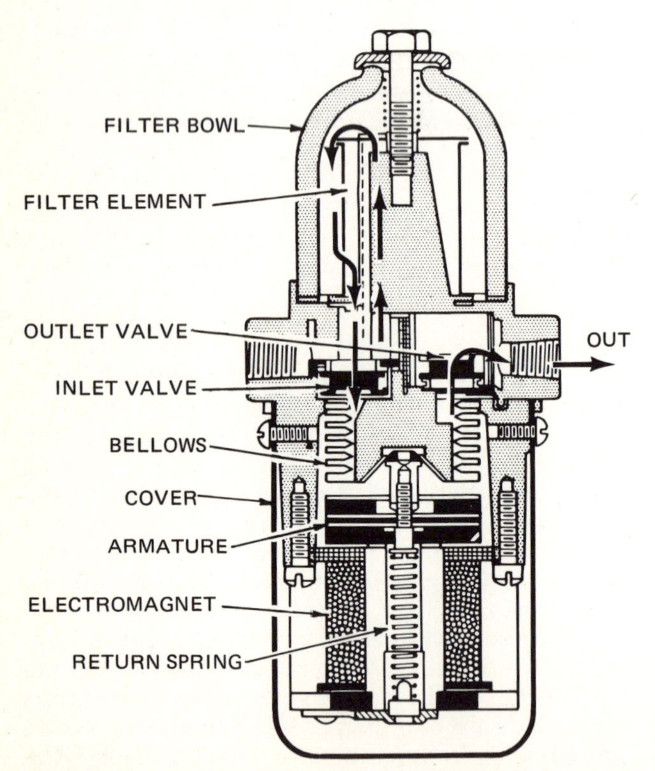

Fig. 13-16. Sectional view of an electric fuel pump.

Check Your Progress

Progress Quiz 13-1 Once again you have the chance to check up on yourself to see how well you are remembering and understanding what you are studying. If any of the following questions stump you, turn back into the chapter and reread the pages that will give you the answer.

Completing the Sentences The sentences that follow are incomplete. After each sentence there are several words or phrases, only one of which will correctly complete the sentence. Write each sentence in your notebook, selecting the proper word or phrase to complete it correctly.

1. The fuel system is responsible for getting the gasoline from: (*a*) crude oil to the engine, (*b*) the fuel tank to the engine, (*c*) the carburetor to the combustion chamber
2. Cars manufactured since 1970 have a vehicle-vapor-recovery (VVR) system to: (*a*) prevent the escape of gasoline vapors from the fuel tank, (*b*) prevent odor in the exhaust gas, (*c*) prevent vapor lock.
3. The device in the fuel system that draws gasoline from the fuel tank and delivers it to the carburetor is the: (*a*) fuel-vapor-recovery system, (*b*) fuel pump, (*c*) fuel filter.
4. The fuel pump is operated by: (*a*) an eccentric on the camshaft, (*b*) a crankpin on the crankshaft, (*c*) a rocker arm on the camshaft.
5. To prevent vapor lock, many cars are equipped with: (*a*) an antivapor lock device, (*b*) a vapor-return line, (*c*) a combination fuel pump.

6. The two types of fuel gauge are the: (*a*) thermostatic and manual, (*b*) balancing coil and electric, (*c*) thermostatic and balancing coil.
7. An electric fuel pump may be installed inside the: (*a*) engine, (*b*) fuel tank, (*c*) carburetor.
8. The thermistor is used in the: (*a*) fuel pump, (*b*) low-fuel-level indicator, (*c*) carburetor.

⊘ 13-10 Air Cleaners As already noted, the fuel system mixes air and fuel to produce a combustible mixture. A great deal of air passes through the carburetor and engine—as much as 100,000 cubic feet [2,831.7 m^3] of air every 1,000 miles [1,609.3 km]. This is a great volume of air, and it probably contains a lot of floating dust and grit. The grit and dust could cause serious engine damage if they entered the engine. Therefore an air cleaner is mounted on the air horn, or air entrance, of the carburetor to keep out the dirt (Fig. 13-17).

⊘ 13-11 Purpose of Air Cleaners All air entering the engine through the carburetor must first pass through the air cleaner. The upper part of the air cleaner contains a ring of filter material (fine-mesh metal threads or ribbons, special paper, cellulose fiber, or polyurethane). This material provides a fine maze that traps most of the dust particles in the air. Some air cleaners have an oil bath, which is a reservoir of oil that the incoming air must flow past. The moving air picks up particles of oil and carries them up into the filter. There the oil washes any dust back down into the oil reservoir. The oiliness of the filter material also improves the filtering action.

The air cleaner also muffles the noise of the intake of air through the carburetor, manifold, and valve ports. This noise would be quite noticeable if it were not for the air cleaner. In addition, the air cleaner acts as a flame-arrester in case the engine backfires through the carburetor. Backfiring may occur if the air-fuel mixture is ignited in the cylinder before the intake valve closes. When this happens, there is a momentary flashback through the carburetor. The air cleaner prevents this flame from erupting from the carburetor and possibly igniting gasoline fumes outside the engine.

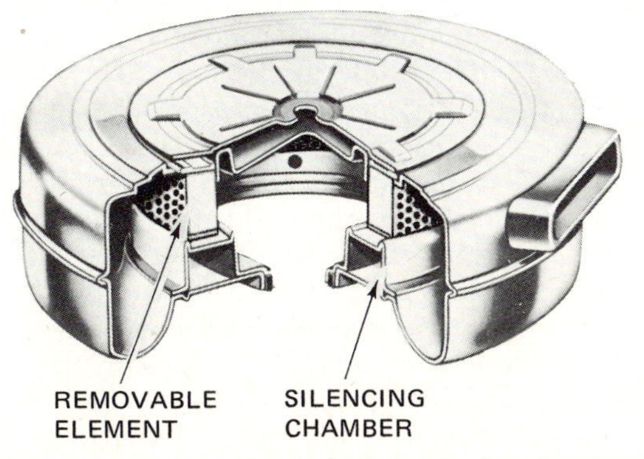

Fig. 13-17. Typical air cleaner, partly cut away to show the filter element. (*Ford Motor Company*)

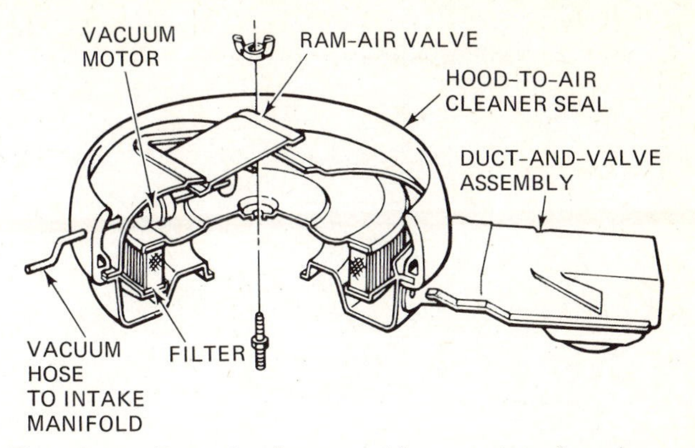

Fig. 13-18. Ram-air cleaner with ram-air valve shown open. (*Ford Motor Company*)

Some late-model cars have a ram-air air cleaner (Fig. 13-18). It allows additional air to be forced into the air cleaner during open throttle or heavy load operation. Under these conditions, a vacuum motor, connected to the intake manifold by a vacuum hose, operates to open a ram-air valve, as shown in Fig. 13-18. This valve is in line with the air scoop on the engine hood, and when it opens, extra air from the air scoop is forced into the carburetor. This improves engine performance under these conditions. At other times, the manifold vacuum is great enough to hold the ram-air valve closed, and air enters the filter through the snorkel tube or duct-and-valve assembly (see ⊘ 13-12) in the normal manner.

⊘ 13-12 Thermostatically Controlled Air Cleaners The thermostatically controlled air cleaner is part of a controlled combustion system used on late-model cars. It is one component of the emission-control equipment (details are discussed in Chaps. 18 and 19). In order to reduce the amount of HC in engine emissions, carburetors are adjusted to give leaner air-fuel mixtures at idle and part-throttle. That is, the amount of gasoline entering into the air-fuel mixture is reduced. These leaner mixtures are designed to ensure a more complete burning of the gasoline. This means that less HC is exhausted through the tail pipe.

However, these leaner air-fuel mixtures can reduce engine performance when the engine is cold. To correct for this condition, a thermostatically controlled air cleaner is used. This is also called the *heated-air system* (HAS) by General Motors. The HAS sends heated air into the carburetor during cold weather when the engine is cold (Fig. 13-19). This improves engine performance after a cold start and during warm-up. Thus, leaner carburetor calibrations can be used to reduce HC in the exhaust without affecting cold-engine performance.

One air cleaner of this type is shown in Fig. 13-20. It contains a temperature-sensing spring which reacts to the temperature of the air entering the carburetor through the air cleaner. This spring controls an air-bleed valve (see Fig. 13-21). When the air entering the air cleaner is cold, the sensing spring

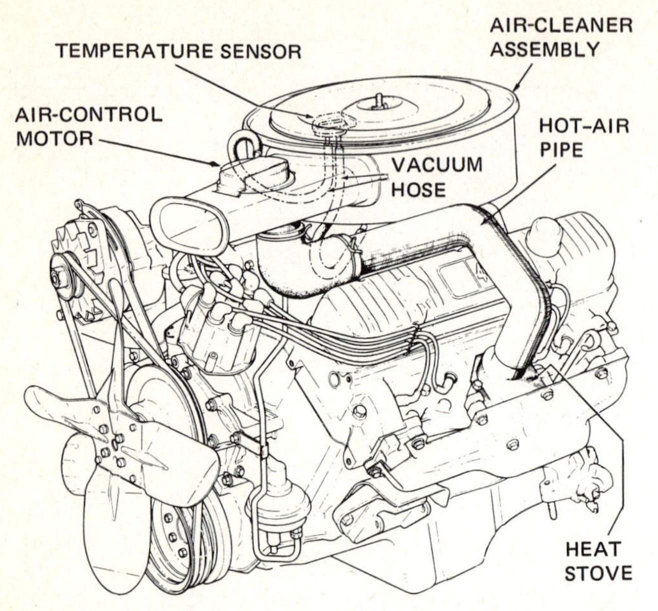

Fig. 13-19. Heated air system (HAS) installed on a V-8 engine. (*Buick Motor Division of General Motors Corporation*)

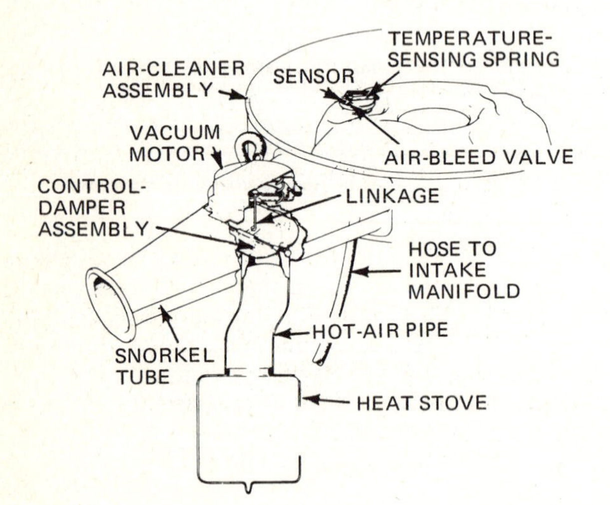

Fig. 13-20. Air cleaner with thermostatic control. (*Chevrolet Motor Division of General Motors Corporation*)

holds the air-bleed valve closed. Now, intake-manifold vacuum is applied to the vacuum chamber. The diaphragm is pushed upward by atmospheric pressure, and the diaphragm spring is compressed in this position. Linkage from the diaphragm raises the control-damper assembly. Thus, the snorkel tube is blocked off. All air now has to enter from the hot-air pipe (view B in Fig. 13-21). This pipe is connected to the heat stove on the exhaust manifold. Therefore, as soon as the engine starts and the exhaust manifold begins to warm up, hot air is delivered to the carburetor and engine. This improves engine performance after a cold start and during warm-up.

As the engine begins to warm up, the under-the-hood temperature increases. If the under-the-hood temperature goes above 128°F [53.3°C] (in the application shown), the conditions are as shown in view C in Fig. 13-21. That is, the temperature-sensing spring has bent enough to open the air-bleed valve. This action reduces the vacuum above the diaphragm so that the diaphragm spring pushes the control damper all the way down. Now all air entering the carburetor comes from under the hood; none comes from the hot air pipe.

If the temperature under the hood stabilizes between 85 and 128°F [29.4 and 53.3°C], conditions will be as shown in view D in Fig. 13-21. That is, the temperature-sensing spring will hold the air-bleed valve partly open. Some vacuum will therefore get to the vacuum chamber above the diaphragm. This vacuum will hold the control damper partly open, as shown. In this position, some air enters from under the hood and some comes up through the hot-air pipe from the heat stove around the exhaust manifold.

A similar thermostatically controlled air cleaner is shown in Figs. 13-22 and 13-23. This design, however, has a thermostatic bulb that acts directly on the valve plate. When the engine is cold, the thermostatic bulb positions the valve plate as shown in Fig. 13-23. Now, all air entering the air cleaner must come from the hot-air duct which is connected to a shroud around the exhaust manifold. As the engine warms up, the hotter air from the shroud causes the thermostatic bulb to move the valve plate. Thus, some air begins to enter from the engine compartment. With further increases of temperature, the valve plate moves farther so that more engine-compartment air enters. When the engine compartment gets hot, most or all of the air entering the air cleaner comes from the engine compartment.

The thermostatically controlled air cleaner shown in Figs. 13-22 and 13-23 includes a vacuum-override motor. The motor operates on intake-manifold vacuum. During cold-engine acceleration, when additional air is needed, the motor overrides the thermostatic control. This action opens the system to both engine-compartment and heated air so that adequate air is delivered to the carburetor through the air cleaner.

⊘ 13-13 Exhaust-Gas Recirculation The higher the combustion temperature, the more nitrogen oxides (NO_x) are formed. Nitrogen oxides contribute to smog, and so changes have been made in engines and fuel systems of late-model cars to reduce the formation of NO_x. One method uses an exhaust-gas recirculation (EGR) system. This system sends part of the exhaust gas back through the engine. The exhaust gas reduces combustion temperatures and thus the amount of NO_x coming out the tail pipe. Figure 13-24 shows one system for sending some of the exhaust gas back through the engine. The system shown has an exhaust-gas-recirculation valve that is actuated by engine vacuum. During part-throttle operation, the vacuum causes the valve to raise its diaphragm and open the port. This allows some of the exhaust gas to enter the intake manifold.

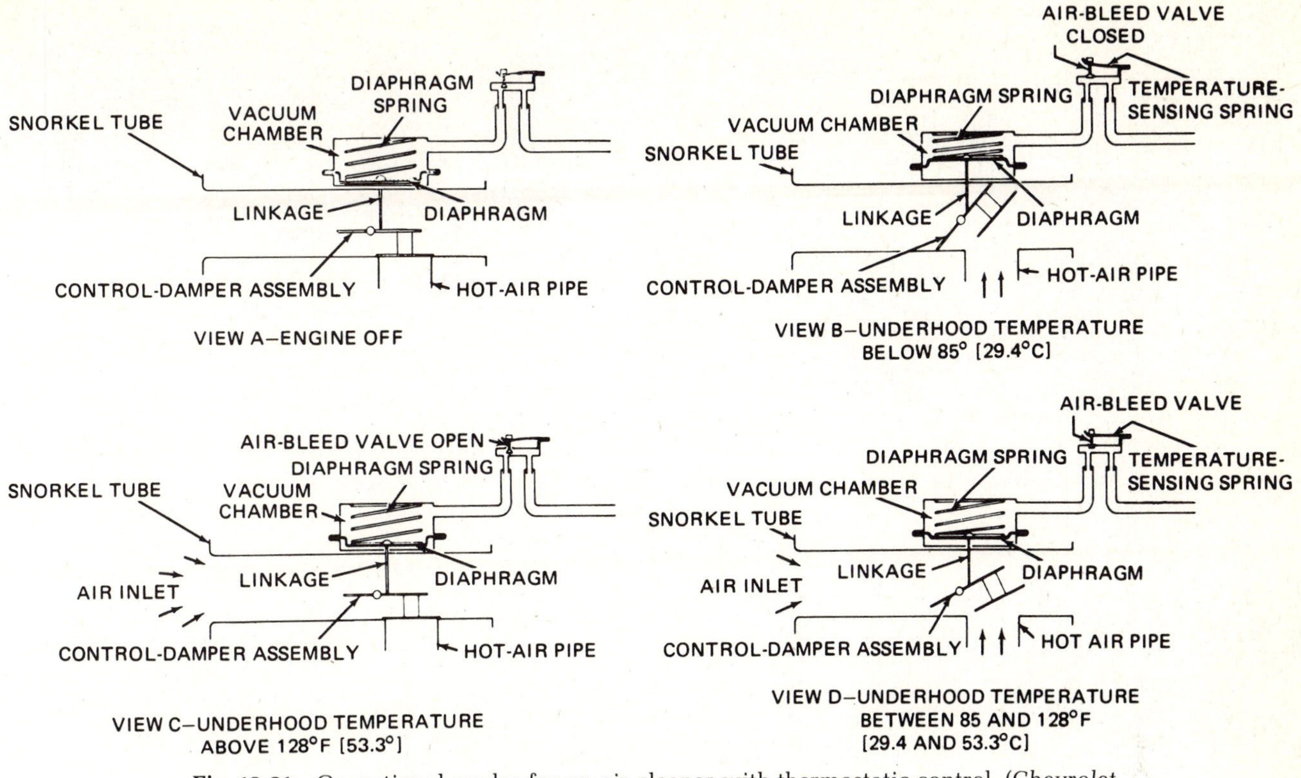

Fig. 13-21. Operational modes for an air cleaner with thermostatic control. (*Chevrolet Motor Division of General Motors Corporation*)

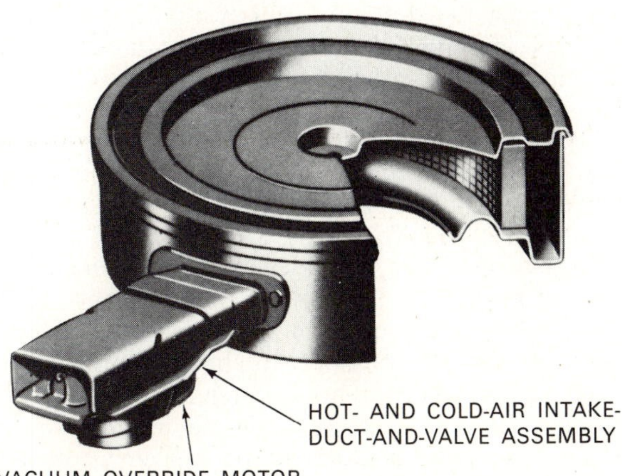

Fig. 13-22. Air cleaner with thermostatic control. (*Ford Motor Company*)

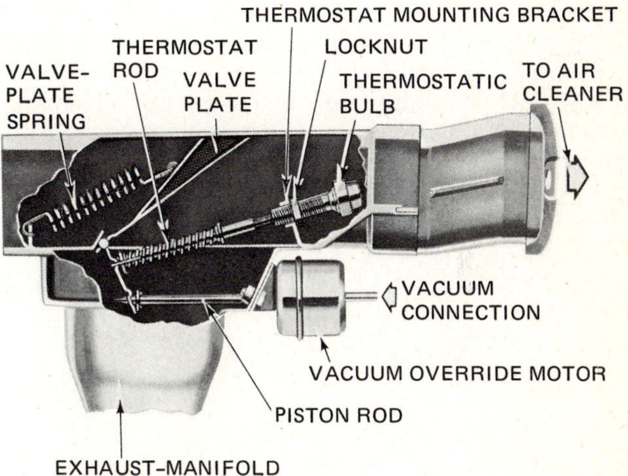

Fig. 13-23. Cutaway view of hot- and cold-air intake ducts and valve assembly for the air cleaner shown in Fig. 13-22. (*Ford Motor Company*)

Another system uses high valve overlap. That is, the camshaft cams allow the exhaust and intake valves to stay open longer at the same time. This allows more of the exhaust gases to stay behind in the cylinders and mix with the incoming air-fuel mixture. Chapters 18 and 19 describe these systems in detail.

⊘ 13-14 Crankcase Ventilation Crankcases must be ventilated. Some blow-by does get past the piston rings and enters the crankcase. In addition, water and liquid fuel appear in the crankcase during cold-engine operation. These must be cleared from the crankcase before they cause trouble. In older engines, the crankcase is ventilated by an opening at the front of the engine and a vent tube at the back. The forward motion of the car plus rotation of the crankshaft cause air to flow through and remove the blow-by, water, and fuel. However, all engines produced today have closed-crankcase ventilating systems. In this system, air circulates through the

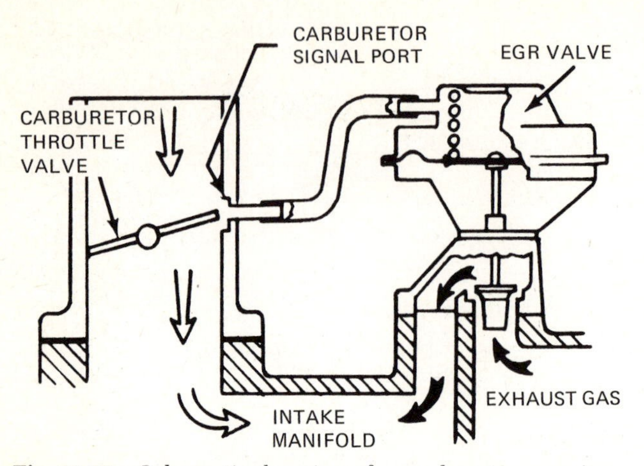

Fig. 13-24. Schematic drawing of an exhaust-gas recirculating system. (*Chevrolet Motor Division of General Motors Corporation*)

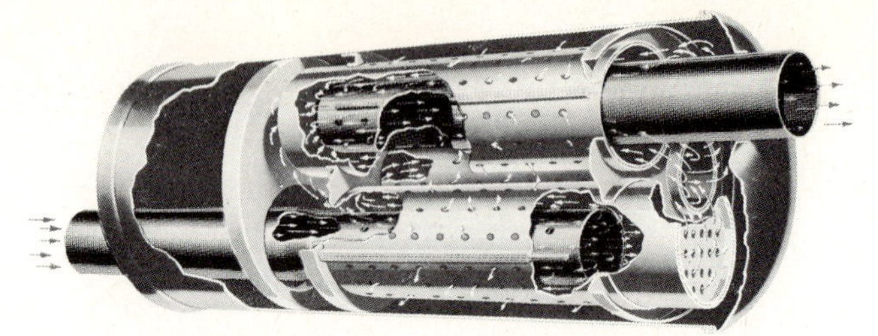

Fig. 13-25. Exhaust muffler in cutaway view. The arrows show the path of exhaust-gas flow through the muffler. (*Chevrolet Motor Division of General Motors Corporation*)

crankcase but then enters the carburetor and intake manifold. Any blow-by, water vapor, or fuel is therefore sent back through the engine instead of being emitted into the air. Closed-crankcase ventilating systems are discussed more fully in ⊘ 17-17.

⊘ 13-15 Exhaust System The exhaust system includes the exhaust manifold, muffler, exhaust pipe, and tail pipe (Fig. 13-2). Some V-8 engines have a crossover pipe to connect their two exhaust manifolds. Other V-8 engines use two separate exhaust systems (dual system), one for each cylinder bank (Fig. 7-22). This improves the "breathing" ability of the engine, allowing it to exhaust more freely and thus increase power output to some extent.

Exhaust manifolds have heat-control valves. These valves close when the engine is cold, thus directing heat to the intake manifold. The.heat helps vaporize the incoming gasoline and thus improves cold-engine operation. Manifold heat-control valves are covered in detail in ⊘ 14-25.

Some engines have exhaust manifolds equipped with air-injection systems. The system includes an air pump and a series of injection tubes in the exhaust manifold. In operation, the air pump sends a flow of air into the exhaust manifold opposite the exhaust valves. This extra air helps to burn any HC or CO still left in the exhaust gases. The air-injection system is covered in detail in ⊘ 19-16.

⊘ 13-16 Mufflers and Exhaust Pipes The muffler (Fig. 13-25) contains a series of holes, passages, and resonance chambers to absorb and damp out the high-pressure surges introduced into the exhaust system as the exhaust valves open. This quiets the exhaust. Some new exhaust systems do not use a muffler, however. Instead, the exhaust pipe has a series of scientifically shaped restrictions that damp out the exhaust noises without unduly restricting the flow of exhaust gases.

To further reduce exhaust noises, some exhaust pipes are made from laminated pipe. This type of pipe is made by covering the exhaust pipe with another layer of slightly larger pipe. Laminated pipe has very good sound-deadening properties.

⊘ 13-17 Catalytic Converters Some late-model cars have *catalytic converters* in the exhaust system. These are mufflerlike containers filled with *catalysts.* The catalysts convert the pollutants coming from the engine into harmless gases. Catalytic converters are described in detail in ⊘ 19-17.

⊘ 13-18 Diesel-Engine Fuel System In the diesel engine, air alone is compressed. Then, at the end of the compression stroke, the fuel system injects fuel oil. A typical diesel-engine fuel system is shown in Fig. 13-26. The system includes a fuel tank, filters, fuel lines, a fuel-injection-pump assembly, and fuel-injection nozzles at each cylinder. The fuel-injection-pump assembly includes a fuel-supply pump, the hydraulic head assembly, a timing-advance mechanism, a governor and fuel-control unit, and an excess-fuel device for starting.

In operation, the fuel system brings fuel from the fuel tank to the hydraulic head assembly. The hydraulic head assembly includes a pump which sends fuel, at high pressure, to the fuel nozzles in the proper firing order. This is much like the ignition distributor in the gasoline engine which sends sparks to the cylinders in the proper firing order. The diesel engine, however, fires from the heat of compression, as explained previously.

The fuel-injection-pump assembly includes a timing-advance mechanism. This mechanism is a form of governor that works against a splined sleeve to push the pump driveshaft ahead as speed increases. This sends the fuel to the cylinders earlier at high speed so that it has enough time to ignite and burn.

The fuel-injection-pump assembly also includes a fuel-control unit which varies the amount of fuel injected according to throttle position. When the throttle is pushed down, calling for more engine power, the fuel-control unit raises a metering sleeve. This action allows the pump plunger to send more fuel to the fuel nozzles.

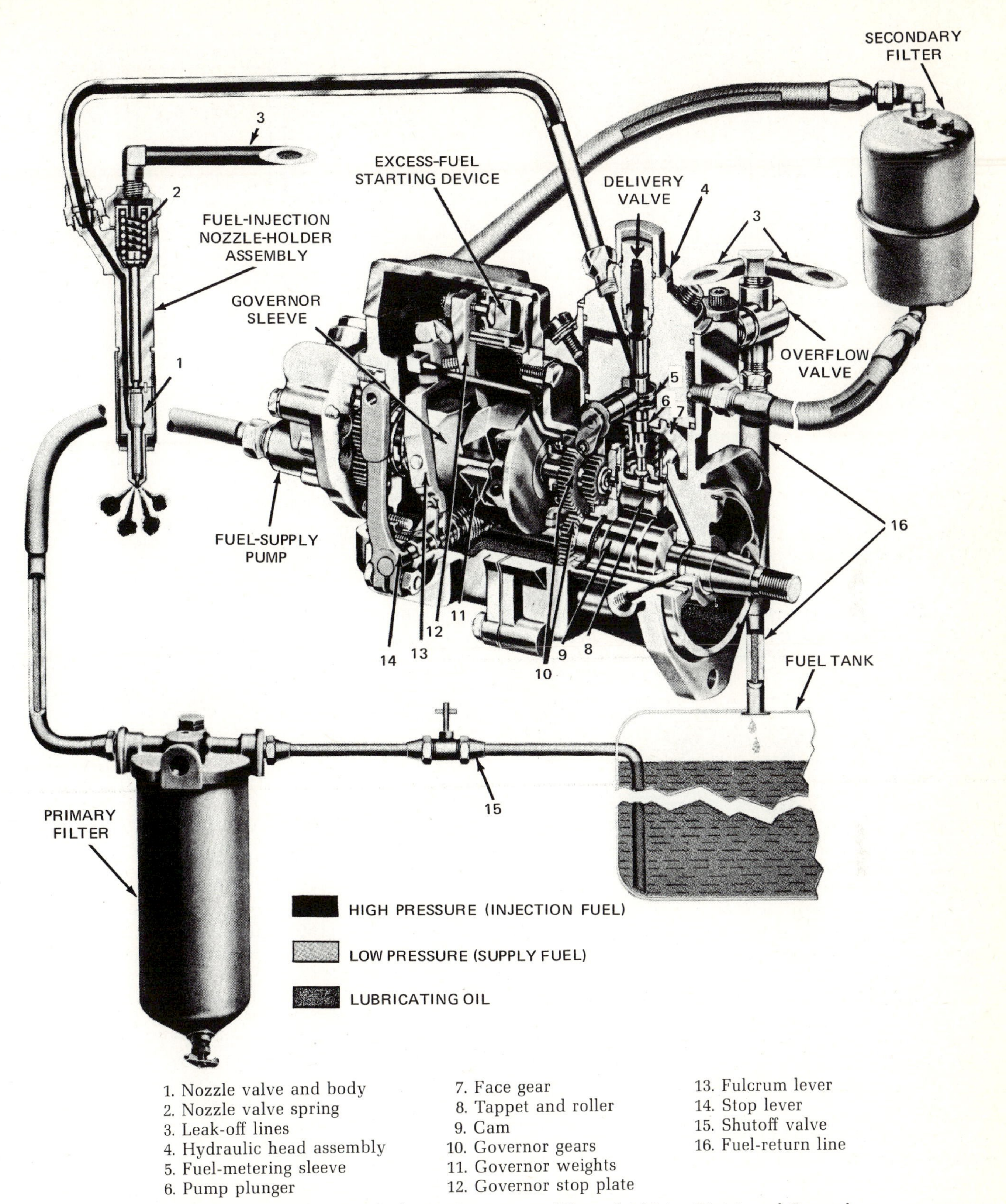

1. Nozzle valve and body
2. Nozzle valve spring
3. Leak-off lines
4. Hydraulic head assembly
5. Fuel-metering sleeve
6. Pump plunger
7. Face gear
8. Tappet and roller
9. Cam
10. Governor gears
11. Governor weights
12. Governor stop plate
13. Fulcrum lever
14. Stop lever
15. Shutoff valve
16. Fuel-return line

Fig. 13-26. Diesel-engine fuel-injection system. (*Chevrolet Motor Division of General Motors Corporation*)

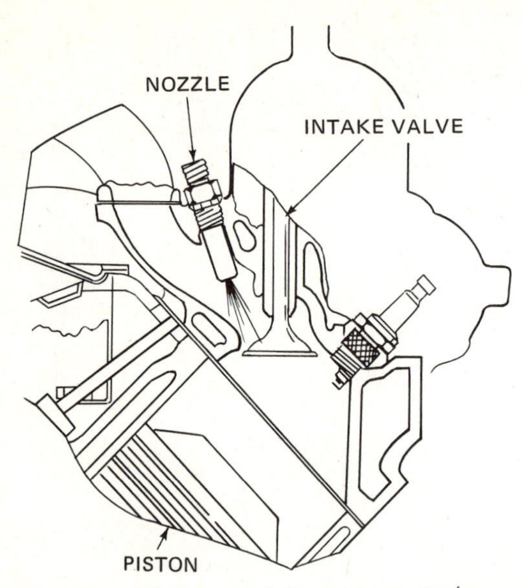

Fig. 13-27. Simplified view of the method of injecting fuel into the intake manifold just back of the intake valve.

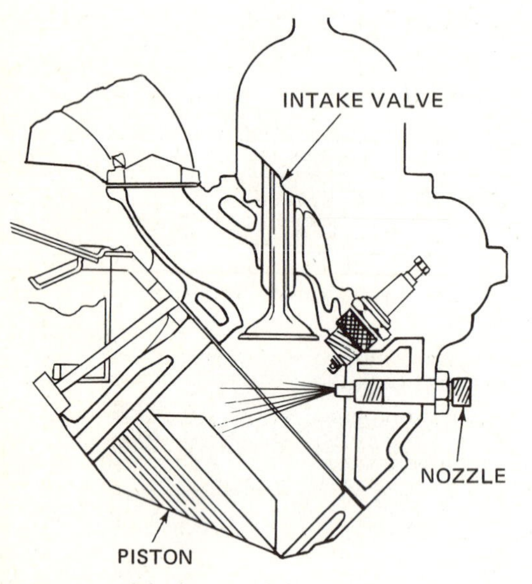

Fig. 13-28. Simplified view of the method of injecting fuel directly into the combustion chamber of the engine.

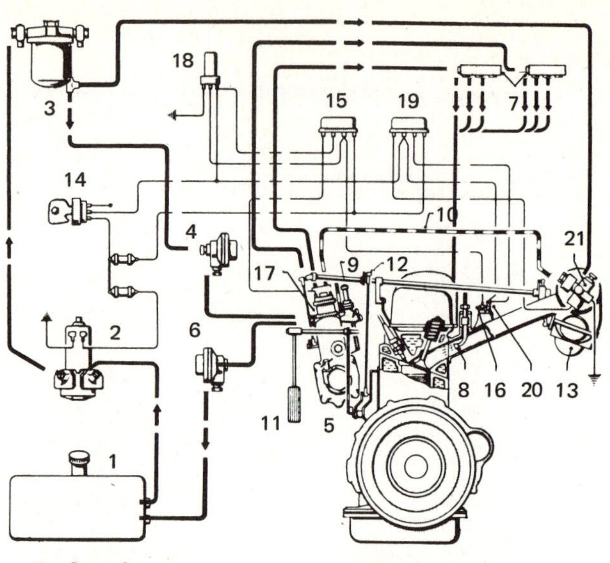

1. Fuel tank
2. Fuel-feed pump
3. Fuel filter
4. Damper container (inlet)
5. Injection pump
6. Damper container (outlet)
7. Fuel-metering units
8. Injection valves
9. Cooling-water thermostat
10. Additional air duct
11. Accelerator
12. Control linkage
13. Throttle connector
14. Ignition-starter switch
15. Relay
16. Thermo switch in cooling-water circuit
17. Magnetic switch for mixture control
18. Time switch
19. Relay
20. Thermo time switch in cooling-water circuit
21. Electromagnetic starter valve with atomizing jet

Fig. 13-29. Schematic layout of a fuel-injection system for a six-cylinder engine. (*Mercedes-Benz, Daimler-Benz Aktiengesellshaft*)

⊘ 13-19 Fuel Injection for Gasoline Engines The carburetor fuel system is the most commonly used system for automotive (gasoline) engines. In this system, the fuel is mixed with air entering the carburetor. In the fuel-injection system, air alone enters the intake manifold. Fuel nozzles are located in the intake manifold, and at the proper time they inject fuel into the air. Figure 13-27 shows, in a simplified view, how the system works. Note that the nozzle is located just back of the intake valve. This is different from the diesel-engine fuel-injection system. There, the fuel is injected directly into the engine cylinder near TDC on the compression stroke (Fig. 13-28).

The gasoline-engine fuel-injection system must supply varying amounts of fuel. The amount required changes with engine speed, throttle opening, cold starts, warm-up, and full-power running. The fuel is supplied through a pump system, which includes a metering unit that controls the amount of fuel injected. Figure 13-29 illustrates one fuel-injection system for a six-cylinder engine. The injection pump (5 in Fig. 13-29) supplies the fuel metering units (7) with fuel at high pressure. The fuel is injected through the injection valves (8) at the proper time and in the proper amount.

⊘ 13-20 Electronic Fuel-Injection System A fuel-injection system that is controlled by electronic means is illustrated in Figs. 13-30 to 13-32. This system was designed for the Volkswagen flat-four, air-cooled engine, but the basic ideas apply to all electronically controlled fuel-injection systems. The fuel is injected into the intake manifolds behind the intake valves. The injection is timed to coincide with the valve opening by triggering contacts in the ignition distributor. The amount of fuel injected is con-

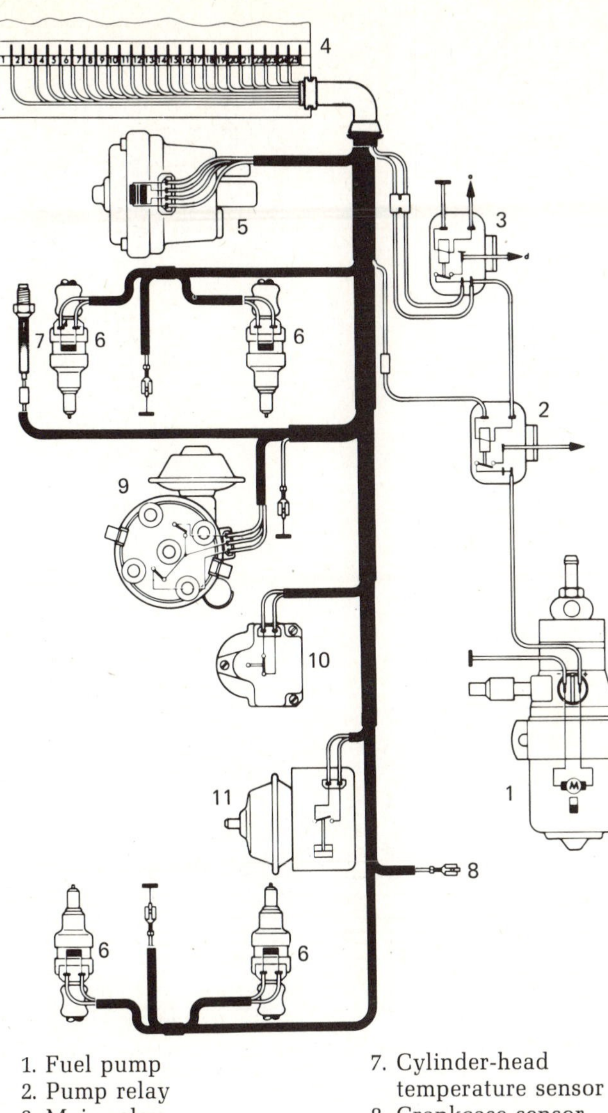

1. Fuel pump
2. Pump relay
3. Main relay
4. Control unit
5. Intake-manifold pressure sensor
6. Injector
7. Cylinder-head temperature sensor
8. Crankcase sensor
9. Ignition distributor
10. Throttle switch
11. Full-load pressure switch

Fig. 13-30. Schematic layout of the control system for the Volkswagen electronic fuel-injection system. The electronic control unit (4) receives signals from various sensors and integrates them to determine the amount of fuel to be injected. (*Volkswagen*)

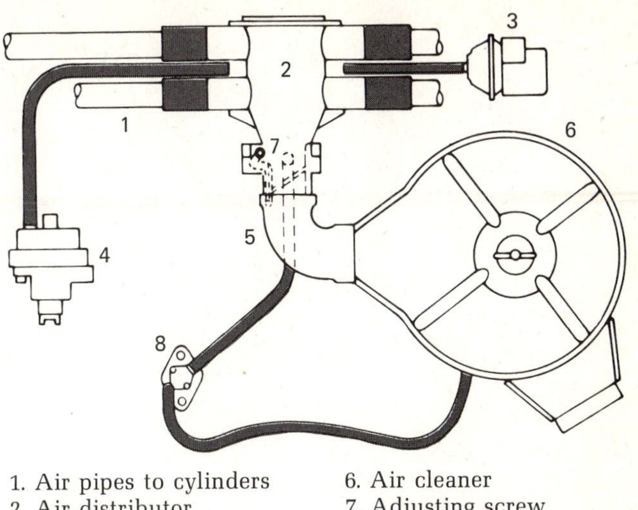

1. Air pipes to cylinders
2. Air distributor
3. Pressure switch
4. Pressure sensor
5. Idling circuit
6. Air cleaner
7. Adjusting screw
8. Auxiliary air regulator (rotary valve)

Fig. 13-31. Air-supply control for the Volkswagen electronic fuel-injection system. (*Volkswagen*)

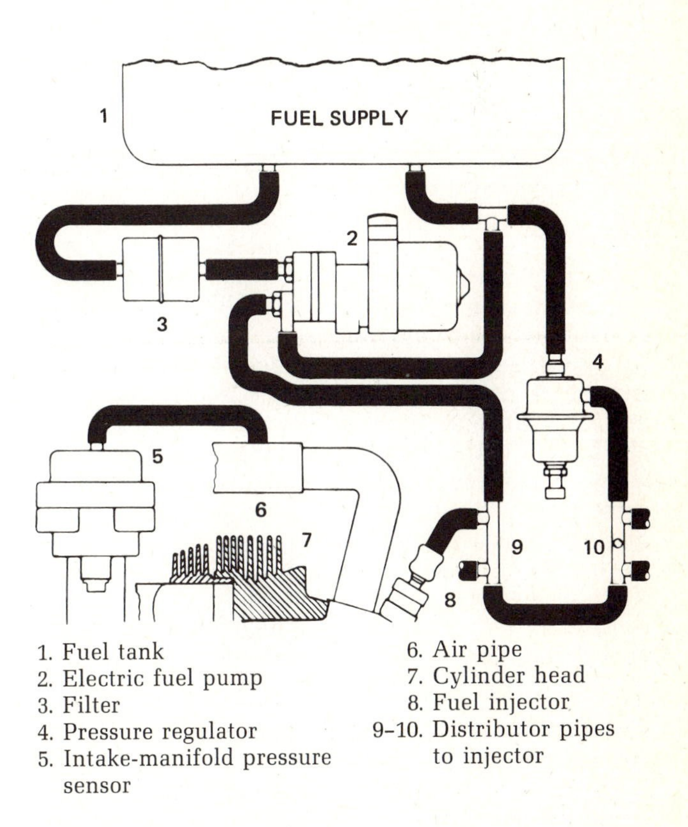

1. Fuel tank
2. Electric fuel pump
3. Filter
4. Pressure regulator
5. Intake-manifold pressure sensor
6. Air pipe
7. Cylinder head
8. Fuel injector
9–10. Distributor pipes to injector

Figure 13-32. Fuel-supply system for the Volkswagen electronic fuel-injection system. (*Volkswagen*)

trolled by the length of time the fuel injectors are open. The length of time the fuel injectors are open is, in turn, determined by sensors that send electrical signals to the transisterized control unit (Fig. 13-30). Figure 13-31 shows the air-supply system and its controls. Figure 13-32 shows the fuel-supply system.

⊘ 13-21 Superchargers and Turbochargers The word "supercharger" tells you what this device is—a mechanism for supplying the engine with a "super" charge of air-fuel mixture. The idea is simple. A centrifugal pump somewhat similar to the engine water pump is located between the carburetor and the engine cylinders (Fig. 13-33). It is driven at high speed, and it compresses the air-fuel mixture from the carburetor and delivers it to the cylinders. A greater amount of air-fuel mixture therefore enters

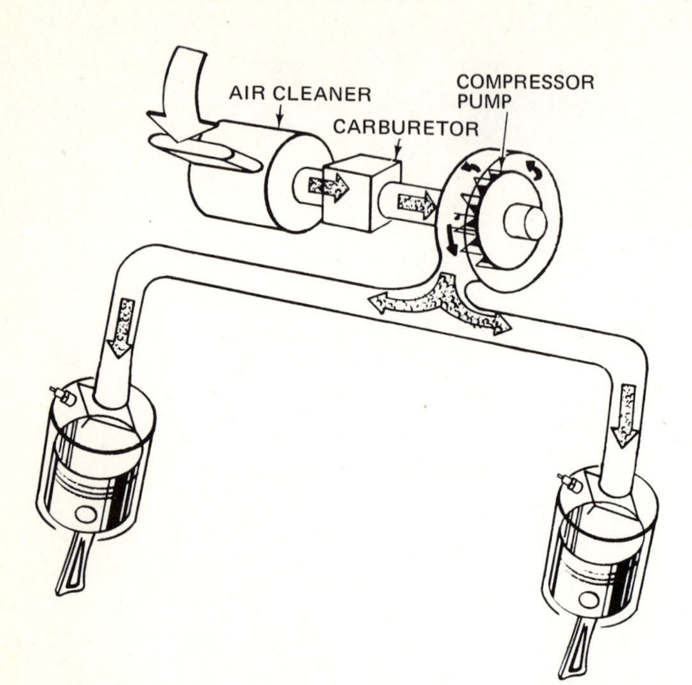

Fig. 13-33. Schematic diagram of the principle of the supercharger.

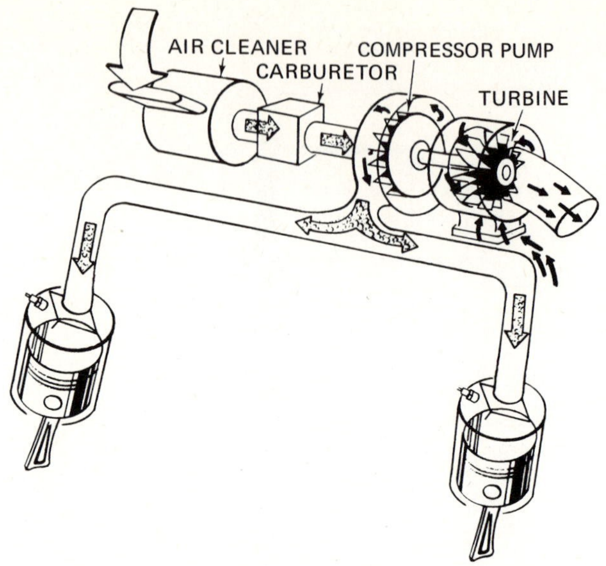

Fig. 13-34. Schematic diagram of the principle of the turbocharger.

the cylinders, and the power strokes are stronger. Thus, the engine can deliver more horsepower. The supercharger can increase the pressure on the air-fuel mixture as much as 16 psi [1.125 kg/cm^2] above normal air pressure. An increase in engine horsepower of up to 50 percent is possible.

The problem with the early supercharger was in the drive arrangement. It was driven by a belt, gears, or a chain and sprockets. These mechanical drives were put to great stress and sometimes caused trouble. Today, the supercharger is driven by the exhaust gas and carries the name "turbosupercharger," or simply "turbocharger." Figure 13-34 shows the arrangement. The exhaust gas, still under some pressure as it leaves the engine cylinders, is directed into a turbine. The turbine wheel is spun by the exhaust gas. The turbine wheel is on the same shaft as the compressor-pump rotor. So the pump rotor is driven to produce the high pressure on the air-fuel mixture entering the cylinders.

⊘ 13-22 Stratified-Charge Engine

The stratified-charge engine has a means of concentrating the rich mixture in the center of the compressed air-fuel mixture (Fig. 13-35). During combustion, the burning rich mixture in the center spreads outward and moves into areas where the mixture is lean and harder to ignite. With stratified charging, a much leaner air-fuel mixture, on the average, can be used. The combustion takes place largely in and around the concentration of rich mixture, which means that the fuel is more completely burned. The amount of pollutants, such as carbon monoxide, unburned gasoline, and nitrogen oxides, are reduced.

One way to achieve stratified charging is to give the air-fuel mixture a swirling motion as it enters the cylinder. This can be done by careful placement of the intake port.

Fig. 13-35. Principle of stratified charging.

Another method is the so-called Honda system. Here, a separate small precombustion chamber is used. This precombustion chamber has the spark plug and its own intake valve. Figure 13-36 is an outline view of the engine showing the valves, spark plug, and pistons for one cylinder. Figure 13-37 shows how the arrangement works. In operation, the carburetor delivers a very lean air-fuel mixture to the main combustion chamber and a very rich mixture to the precombustion chamber. Ignition takes place in the precombustion chamber. The rich mixture, under the high pressure of combustion, streams

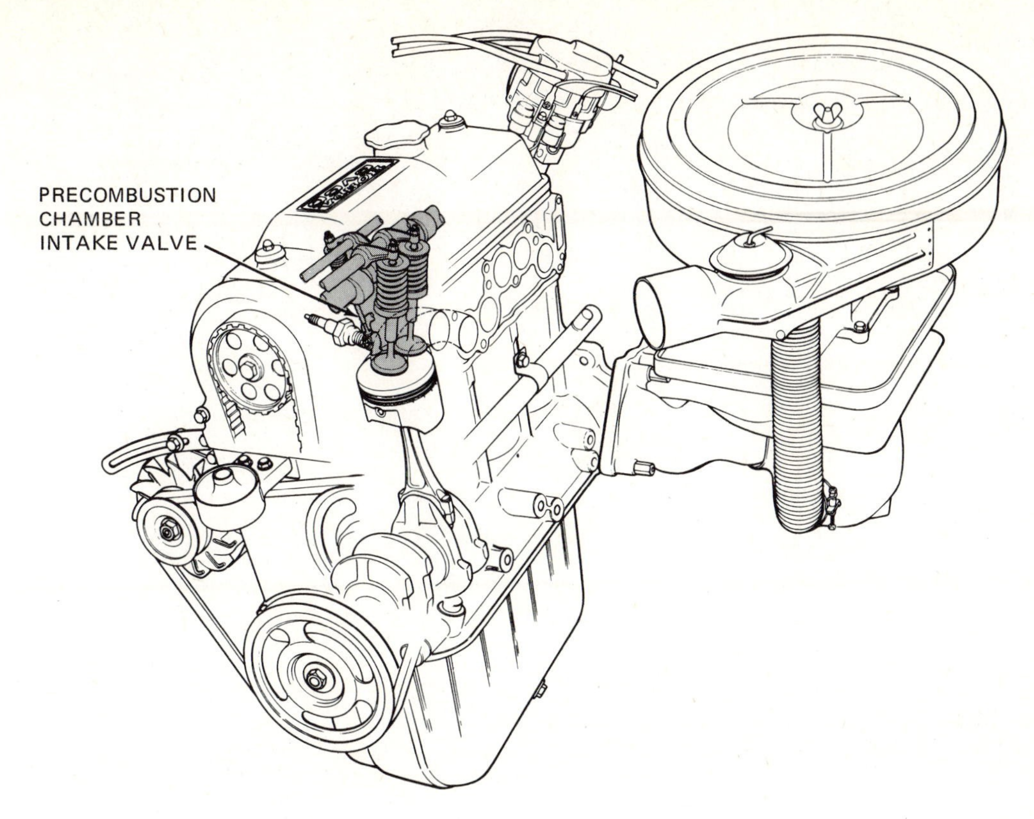

Fig. 13-36. Outline view of the Honda four-cylinder engine showing the essential working parts of one cylinder. (*American Honda Motor Company, Inc.*)

out into the main combustion chamber (3 and 4 in Fig. 13-37). There, it mixes with the lean mixture, and combustion continues. This ensures good burning of the fuel so that polluting gases—carbon monoxide, unburned gasoline, and nitrogen oxides—are kept to a low level. Figure 13-37 shows the sequence of actions.

⊘ 13-23 LPG Fuel System Liquefied petroleum gas (LPG) is a fuel that is liquid only under pressure (see Chap. 12). When the pressure is reduced, the fuel vaporizes. Thus, the LPG fuel system must have a pressure-tight fuel tank in which to store the fuel at high pressures. A typical LPG fuel system is shown in Fig. 13-38. Pressure in the tank forces fuel through the filter, high-pressure regulator, and vaporizer. The high-pressure regulator reduces the pressure so that the fuel starts to turn to vapor. This vaporizing process is completed in the vaporizer. The vaporizer has an inner tank surrounded by a water jacket through which cooling-system coolant passes. The coolant adds heat to the fuel so that it is well vaporized. The fuel then passes through the low-pressure regulator, where the pressure is further reduced to slightly below atmospheric. This prevents the fuel from flowing into the carburetor when the engine is off. The fuel then enters the carburetor. (Fuel will flow into the carburetor only when

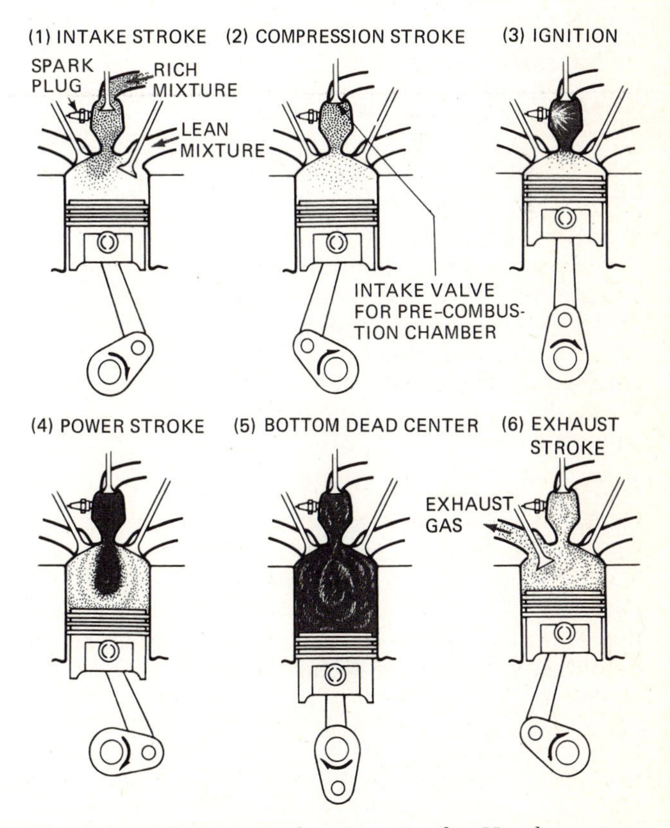

Fig. 13-37. Sequence of actions in the Honda system. (*American Honda Motor Company, Inc.*)

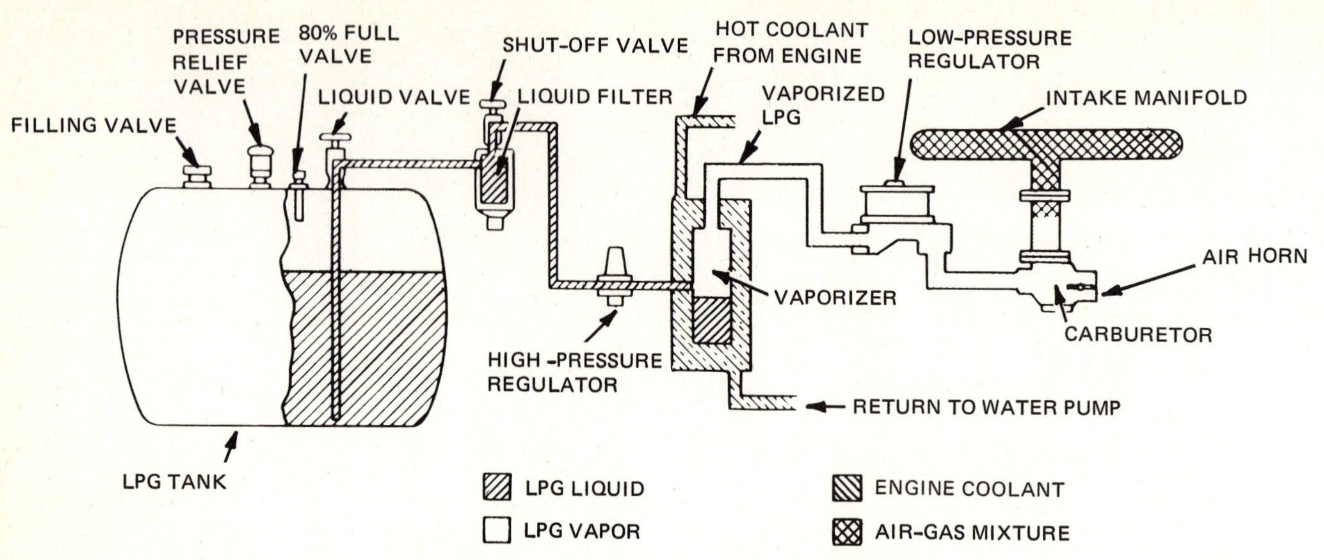

Fig. 13-38. LPG fuel system shown schematically.

the engine is running and there is a vacuum in the carburetor venturi, or air horn.) The carburetor is a mixing valve; it mixes the vaporized fuel and air as required by the engine.

LPG fuel systems are used on some cars, trucks, and buses. They are also installed in forklift and platform trucks of the type used inside factory buildings. This system is well suited for such applications because the fuel burns clean. The exhaust gases contain very little contaminants.

CHAPTER 13 CHECKUP

NOTE: Since this is a chapter review test, you should review the chapter before taking the test.

You are making good progress in your study of automotive fuels and fuel systems (except for carburetors, which are covered in Chap. 14). Now check up on yourself by taking the test that follows. If any of the questions are hard to answer, turn back into the chapter and reread the pages that will give you the information you need.

Completing the Sentences The sentences that follow are incomplete. After each sentence there are several words or phrases, only one of which will correctly complete the sentence. Write each sentence in your notebook, selecting the proper word or phrase to complete it correctly.

1. The fuel system consists of the fuel tank, fuel lines, fuel filter, fuel pump: (*a*) carburetor, and electric pump, (*b*) intake manifold, and carburetor, (*c*) carburetor, and combustion chamber.
2. The low-fuel-level indicator is attached to the: (*a*) tank unit of the fuel gauge, (*b*) instrument-panel unit of the fuel gauge, (*c*) fuel tank.
3. In the fuel-vapor-recovery system, fuel vapor escaping from the fuel tank is: (*a*) vented to the intake manifold, (*b*) adsorbed in the charcoal canister, (*c*) vented to the carburetor float bowl.
4. The filtering element in an air cleaner is made of: (*a*) fiber, special paper, or polyurethane; (*b*) steel wool or wood chips; (*c*) neither (*a*) nor (*b*).
5. The air cleaner sits on top of the: (*a*) fuel tank, (*b*) carburetor, (*c*) fuel pump.
6. When the engine is cold, all air entering the thermostatically controlled air cleaner comes from the: (*a*) intake manifold, (*b*) exhaust manifold, (*c*) heat stove on the exhaust manifold.
7. The purpose of the EGR system is to: (*a*) reduce engine temperatures, (*b*) reduce NO_x in the exhaust gas, (*c*) improve combustion of HC.
8. The purpose of the crankcase ventilator is to: (*a*) remove exhaust gases from the crankcase, (*b*) remove blow-by from the crankcase, (*c*) improve engine cooling.
9. The purpose of the catalytic converters in the exhaust system is to: (*a*) convert pollutants into harmless gases, (*b*) reduce engine temperatures, (*c*) improve engine efficiency.
10. In the fuel injection system for gasoline engines, the fuel injector: (*a*) injects fuel into the cylinders, (*b*) replaces the carburetor, (*c*) supplies an unvarying amount of gasoline.

Definitions and Explanations In the following, you are asked to write definitions, lists, and explanations relating to the fuel system. Write in your notebook.

1. Name four of the components in the fuel system.
2. What are the two types of fuel gauges?
3. Explain how a fuel pump works.
4. What is the purpose of the vapor-return line?

5. What is the purpose of the fuel-vapor-recovery system?
6. Describe the operation of two types of electric fuel pumps.
7. What is the purpose of the air cleaner?
8. What is the purpose of the thermostatically controlled air cleaner? How does it work?
9. What is the purpose of the EGR system? How does it work?
10. Why does high valve overlap reduce NO_x?
11. What is the purpose of crankcase ventilation?
12. Describe an exhaust system.
13. What is the purpose of the catalytic converters in the exhaust system?
14. What is the purpose of the muffler?
15. Explain how a diesel-engine fuel system works.
16. How does the supercharger work?
17. How is the turbocharger driven?
18. What is meant by stratified charge?
19. Explain how the Honda system works.
20. Describe how the LPG system works.

SUGGESTIONS FOR FURTHER STUDY

Keep your eyes open for articles in the automotive magazines about stratified-charge engines, superchargers, and fuel-injection systems. Some engineers tell us that these are coming things for automotive engines of the future. In fact, some of these ideas are already being used in present-day automobiles. To keep up with new developments in automotive engineering, you should first have a good understanding of the concepts behind them. That way, you will be up to date and able to understand and handle any new developments.

chapter 14

AUTOMOTIVE CARBURETORS

In previous chapters, we discussed automotive fuels and fuel systems. In this chapter, we look at automotive carburetors and find out about the various carburetor systems and how they work.

⊘ 14-1 Carburetion Carburetion is the process of mixing gasoline fuel with air so that a combustible mixture is obtained. The carburetor performs this job, supplying a combustible mixture of varying degrees of richness to suit different engine-operating conditions. The mixture must be rich (have a higher percentage of fuel) for starting, acceleration, and high-speed operation. A leaner mixture is desirable at intermediate speeds with a warm engine. The carburetor has several systems through which air-fuel mixture flows during different operating conditions. These systems produce the variations in the richness of the air-fuel mixture required for different operating conditions.

⊘ 14-2 Vaporization When a liquid changes to a vapor (undergoes a change of state), it is said to *evaporate*, or *vaporize*. Water placed in an open pan will evaporate: It changes from a liquid to a vapor. Clothes hung on a line dry: The water in the clothes turns to vapor. When the clothes are well spread out, however, they dry more rapidly than when they are bunched together. This illustrates an important fact about evaporation: The greater the surface area exposed, the more rapidly evaporation takes place. A pint of water in a tall glass takes quite a while to evaporate. But a pint of water in a shallow pan evaporates much more quickly (Fig. 14-1).

Fig. 14-1. Water evaporates from the shallow pan faster than from the glass. The greater the area exposed to air, the faster the evaporation.

⊘ 14-3 Automization In order to produce very quick vaporization of liquid gasoline, it is sprayed into the air passing through the carburetor. Spraying the liquid gasoline turns it into many fine droplets. This effect is called *atomization* (the liquid is broken up into small droplets but not actually into atoms, as the name implies). Each droplet is exposed to air on all sides so that it vaporizes very quickly. Thus, during normal running of the engine, the gasoline sprayed into the air passing through the carburetor turns to vapor, or vaporizes, almost instantly.

⊘ 14-4 Carburetor Fundamentals A simple carburetor can be made from a round cylinder with a constricted section, a fuel nozzle, and a round disk, or valve (Fig. 14-2). The round cylinder is called the *air horn*, the constricted section the *venturi*, and the valve the *throttle valve*. The throttle valve can be tilted more or less to open or close the air horn (Fig. 14-3). In the horizontal position, it shuts off, or *throttles*, the air flow through the air horn. When the throttle is turned away from this position, air can flow through the air horn.

⊘ 14-5 Venturi Effect As air flows through the constriction, or venturi, a partial vacuum is produced in the venturi. This vacuum causes the fuel nozzle to deliver a fine spray of gasoline into the passing air stream. The *venturi effect* (of producing a vacuum) can be illustrated with the setup shown in Fig. 14-4. Here, three dishes of mercury (a very heavy metallic liquid) are connected by tubes to an air horn with a venturi. The greater the vacuum, the higher the mercury is pushed up in the tube by atmospheric pressure. Note that the greatest vacuum is right in the venturi. Also, it should be remembered that the faster the air flows through, the greater the vacuum.

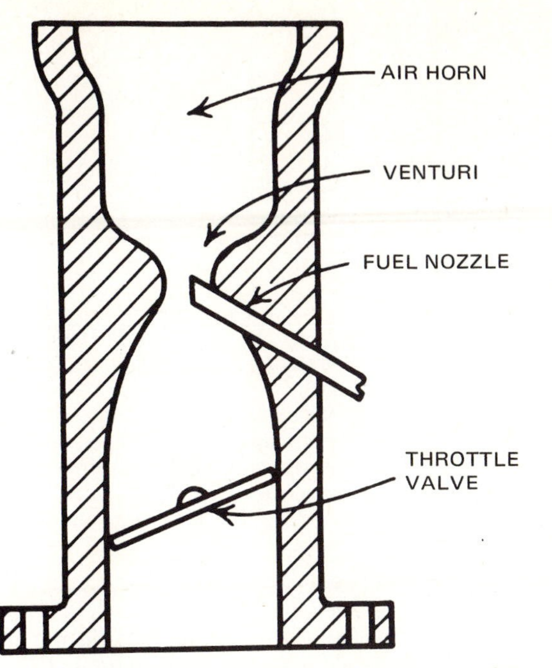

Fig. 14-2. Simple carburetor consisting of an air horn, a fuel nozzle, and a throttle valve.

Why is there a partial vacuum in the venturi? A simple explanation might be as follows: The air is made up of countless molecules. As air moves into the top of the air horn, all the air molecules move at the same speed. But if all are to get through the venturi, they must speed up and move through faster. For instance, let us see what happens to two molecules, one behind the other. As the first molecule enters the venturi, it speeds up, tending to leave the second molecule behind. The second molecule also speeds up as it enters the venturi. But the first molecule has, in effect, a head start. Thus, the two molecules are farther apart in the venturi than they were before they entered it. Now imagine a great

Fig. 14-3. Throttle valve in the air horn of a carburetor. When the throttle is closed, as shown, little air can pass through. But when the throttle is opened, as shown dashed, there is little throttling effect.

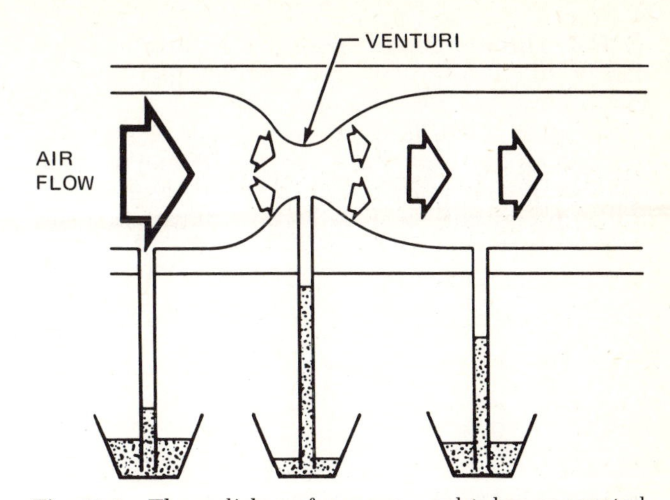

Fig. 14-4. Three dishes of mercury and tubes connected to an air horn. They show differences in vacuum by the distance the mercury rises in the tubes. The venturi has the highest vacuum.

number of molecules going through the same action. As they pass through the venturi, they are farther apart than before they entered. This is just another way of saying that a partial vacuum exists in the venturi. A partial vacuum is a thinning out of the air, where the distance between air molecules is greater than normal.

⊘ 14-6 Fuel-Nozzle Action The partial vacuum occurs just where the end of the fuel nozzle is located in the venturi. The other end of the fuel nozzle is located in a fuel reservoir, the *float bowl,* as shown in Fig. 14-5. Atmospheric pressure pushes on the fuel through a vent in the float-bowl cover. With the partial vacuum at the upper end of the nozzle, fuel is pushed up through the nozzle and enters the passing air stream. The fuel enters as a fine spray, which quickly turns to vapor as the droplets of fuel evaporate. The more air that flows through the air horn, the greater the vacuum in the venturi. The greater the vacuum in the venturi, the more fuel is delivered through the nozzle.

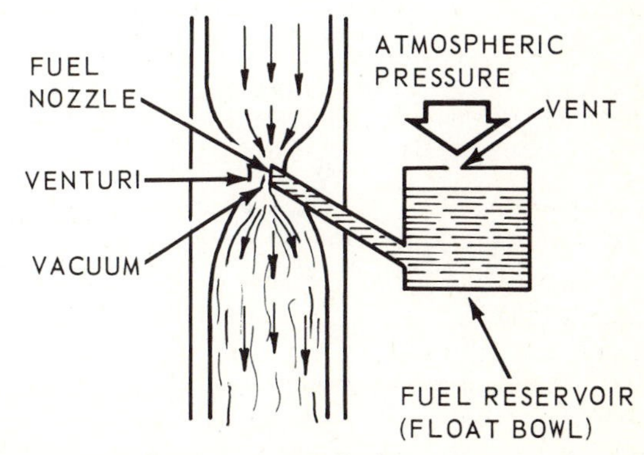

Fig. 14-5. Venturi, or constriction, causes a vacuum to develop in the air stream just below the constriction. Then atmospheric pressure pushes fuel up and out the fuel nozzle.

⊘ **14-7 Throttle-Valve Action** The throttle valve can be tilted in the air horn to allow more or less air to flow through (Fig. 14-3). When it is tilted to allow more air to flow through, larger amounts of air-fuel mixture are delivered to the engine. The engine produces more power and tends to run faster. But if the throttle valve is tilted to allow less air to flow through, then only small amounts of air-fuel mixture are delivered. The engine produces less power and tends to slow down. The throttle valve is linked with an accelerator pedal in the driver's compartment. This arrangement permits the driver to position the throttle valve to suit operating requirements (Fig. 14-6).

⊘ **14-8 Air-Fuel-Ratio Requirements** As already noted (⊘ 14-1), the fuel system must vary the richness of the air-fuel mixture, or the *air-fuel ratio,* to suit different engine-operating conditions. The mixture must be rich (have a high percentage of fuel) for starting, acceleration, and high-speed operation. It must be leaner (have a lower percentage of fuel) for part-throttle, medium-speed operation. Figure 14-7 is a graph showing typical air-fuel ratios as related to various engine speeds. Air-fuel ratios and engine speeds at which they are obtained vary with different cars. In the example shown, a rich mixture of about 9:1 (9 pounds [4.08 kg] of air for each pound [0.454 kg] of fuel) is supplied for initial starting. Then, during idle, the mixture leans out to about 12:1. At medium speeds, the mixture further leans out to about 15:1. But at higher speeds, with a wide-open throttle, the mixture is enriched to about 13:1. Opening the throttle at any speed for acceleration causes a momentary enrichment of the mixture. (This results from operation of special systems in the carburetor, which we shall study later.) Two examples are shown in Fig. 14-7 at about 20 miles per hour [32.18 km/h] and at about 30 miles per hour [48.27 km/h].

You might think that the engine itself demands varying air-fuel ratios for the different operating

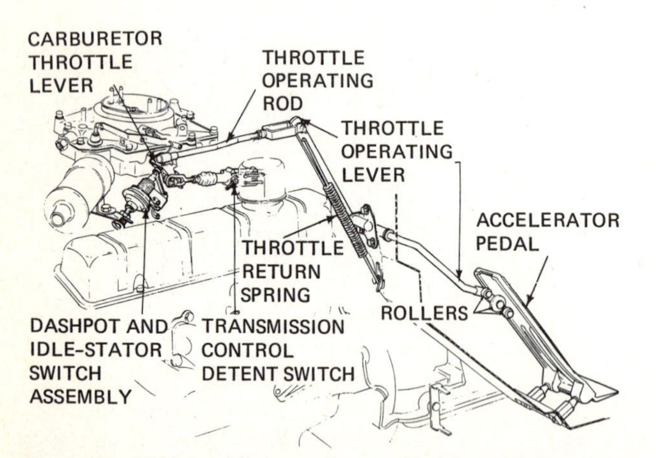

Fig. 14-6. Linkage between the accelerator pedal and the carburetor throttle lever. (*Buick Motor Division of General Motors Corporation*)

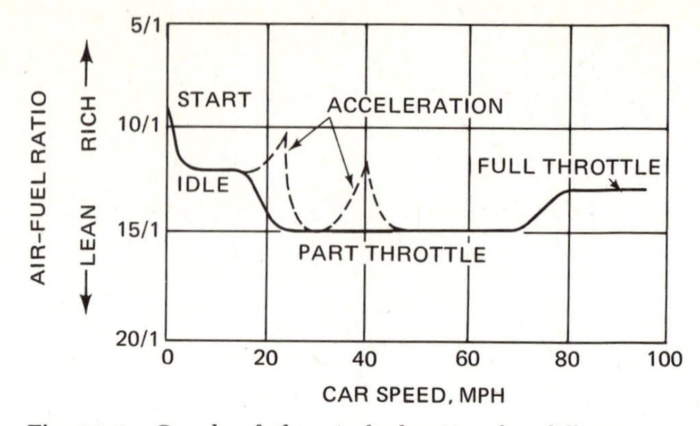

Fig. 14-7. Graph of the air-fuel ratios for different car speeds. The graph is typical. Car speeds at which the various ratios are obtained may vary with different cars. Also, there may be some variation in the ratios.

conditions. This is not quite true. For example, the mixture must be very rich for starting because fuel vaporizes very poorly under this condition. The engine and carburetor are cold, the air speed is low, and much of the fuel does not vaporize. Thus, an extra amount of fuel must be delivered by the carburetor so that enough fuel will vaporize for starting. Likewise, sudden opening of the throttle for acceleration allows a sudden inrush of air. Extra fuel must enter at the same time (that is, the mixture must be enriched). This is because only part of the fuel vaporizes and mixes with the ingoing air to provide the proper air-fuel ratio.

The following articles describe the various systems in carburetors that supply the various air-fuel mixtures required for different operating conditions.

⊘ **14-9 Carburetor Systems** The *systems* (or *circuits,* as they are sometimes called) in the carburetor are:

1. Float system
2. Idle system
3. Main metering system
4. Power system
5. Accelerator-pump system
6. Choke system

These systems are discussed in detail in the following articles.

⊘ **14-10 Float System** The *float system* includes the float bowl and a float-and-needle-valve arrangement. The float and needle valve maintain a constant level of fuel in the float bowl. If the fuel level is too high, then too much fuel will be fed from the fuel nozzle. If it is too low, too little fuel will be fed. In either event, poor engine performance will result.

Figure 14-8 is a simplified drawing of the float system. If fuel enters the float bowl faster than it is withdrawn, the fuel level rises. This causes the float to move up and push the needle valve into the valve seat. This, in turn, shuts off the fuel inlet so

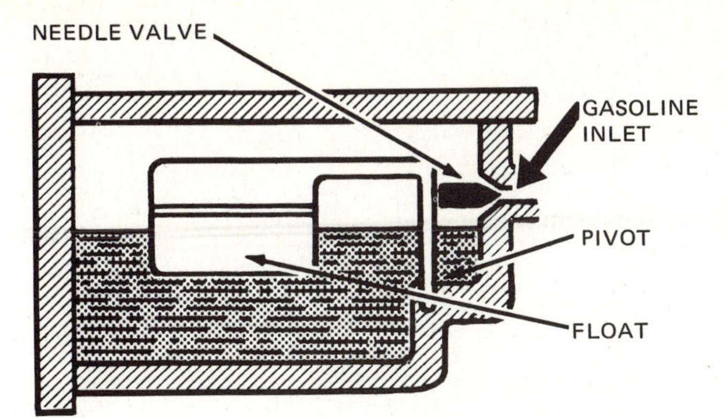

Fig. 14-8. Simplified drawing of a carburetor float system.

that no fuel can enter. Then, if the fuel level drops, the float moves down and releases the needle so that the fuel inlet is opened. Now fuel can enter. In actual operation, the fuel is kept at an almost constant level. The float tends to hold the needle valve partly closed so that the incoming fuel just balances the fuel being withdrawn.

Figure 14-9 shows an actual carburetor with a dual float assembly partly cut away so that the two floats can be seen. The carburetor has a float bowl that partly surrounds the carburetor air horn. The two floats are attached by a U-shaped lever and operate a single needle valve. Some carburetors have an auxiliary fuel valve and inlet, as shown in Fig. 14-10. During heavy-load or high-speed operation, fuel may be withdrawn from the float bowl faster than it can enter through the main fuel inlet. If this happens, the fuel level drops. The end of the float lever presses against the auxiliary valve, pushing it upward. This opens the auxiliary fuel inlet so additional fuel can enter.

A number of years ago, some four-barrel carburetors had two sets of floats (see Fig. 14-11). The four-barrel carburetor is, in effect, 2 two-barrel carburetors. As will be explained later, the primary barrels supply the engine during most operating conditions. But the secondary barrels come into the operation during acceleration and high speeds, for improved performance. The idea of using two sets of floats was to provide, in effect, separate float systems for each pair of barrels. However the two-float system requires a large float bowl and has other disadvantages not found in the single or dual floats. Therefore more recent four-barrel carburetors have a single centrally located float which responds more accurately to fuel needs. Also, the float bowl is smaller, so that there is less of a problem with fuel evaporation and the possibility of atmospheric pollution from escaping HC.

⊘ 14-11 Float-Bowl Vents The float bowl of a carburetor is vented into the carburetor air horn at a point above the choke valve (see Fig. 14-9, upper left; Fig. 14-12, upper right). The purpose of the vent is to equalize the effects of a clogged air cleaner. For example, suppose the air cleaner has become clogged with dirt. The passage of air through it is then restricted. As a result, a partial vacuum develops in the carburetor air horn. Therefore, a somewhat greater vacuum is applied to the fuel nozzle (since this vacuum is added to the venturi vacuum). However, the partial vacuum resulting from the clogged air cleaner is also applied to the float bowl (through the vent). Therefore, the only driving force that pushes fuel from the fuel nozzle is the air pressure in the air cleaner. The air pressure there is less than atmospheric pressure. Thus, the vent makes up for the effect of a clogged air cleaner. If the float bowl

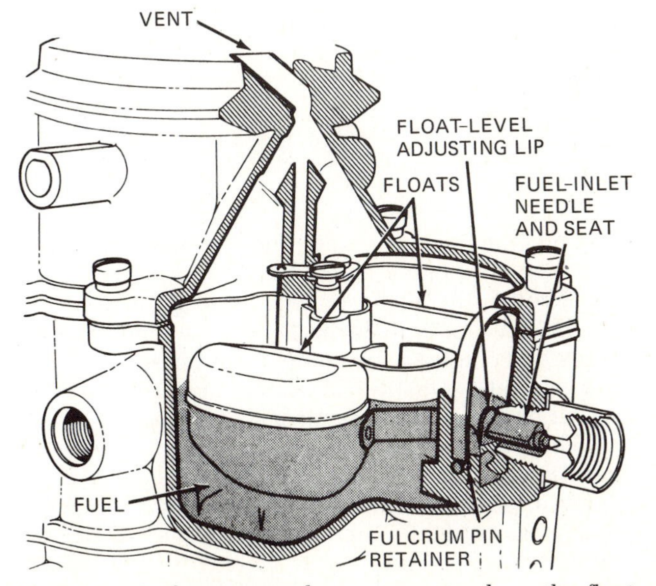

Fig. 14-9. Carburetor partly cut away to show the float system. (*Chrysler Corporation*)

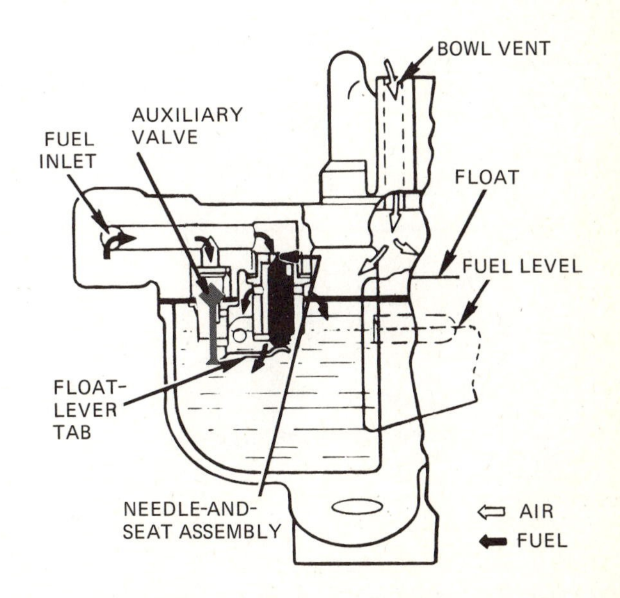

Fig. 14-10. Float system using an auxiliary fuel valve and inlet. (*American Motors Corporation*)

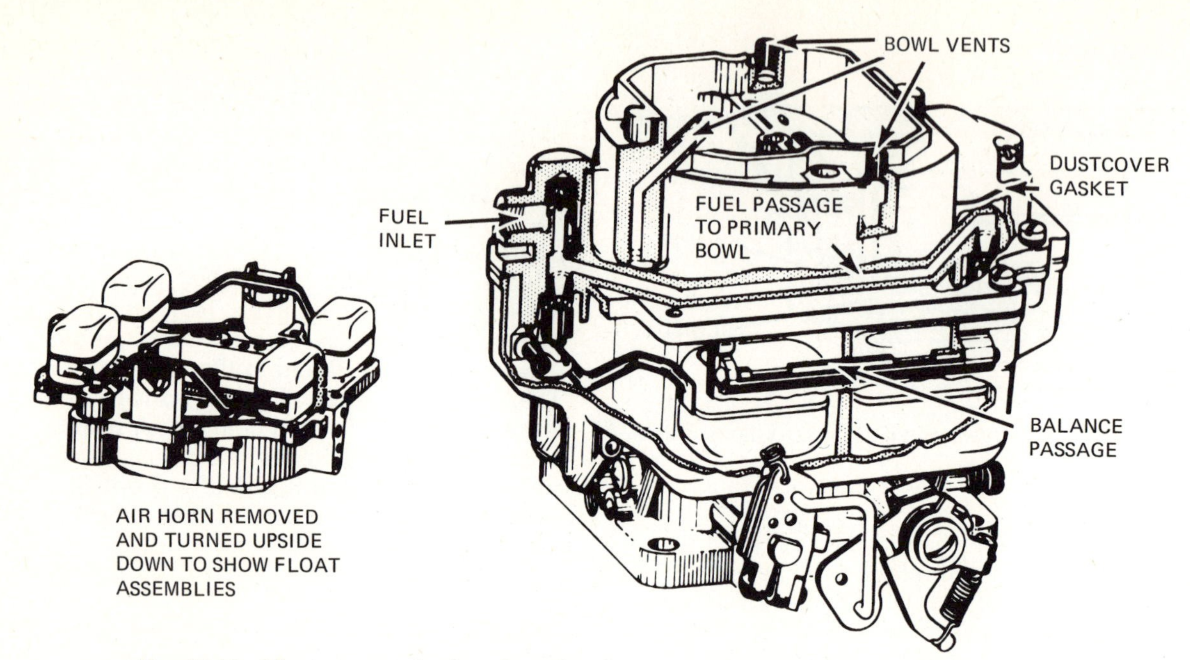

Fig. 14-11. Float system of a four-barrel carburetor using two sets of floats. (*Chevrolet Motor Division of General Motors Corporation*)

were vented to the atmosphere, then atmospheric pressure would be the driving force. This would produce a greater fuel flow from the fuel nozzle and a mixture that would be too rich.

The float bowl has another vent (Fig. 14-12, upper right). This vent is connected by a tube to the charcoal canister which is part of the fuel-vapor-recovery system (⊘ 13-8). In the carburetor in Fig. 14-12, the float bowl has a pressure-relief valve. The valve opens when vapor pressure increases within the float bowl, allowing the fuel vapor to flow to the charcoal canister. In the carburetor in Fig. 14-9, the vent to the carbon canister has a valve operated by the accelerator-pump lever. The valve opens when the engine is idling or when it has been turned off.

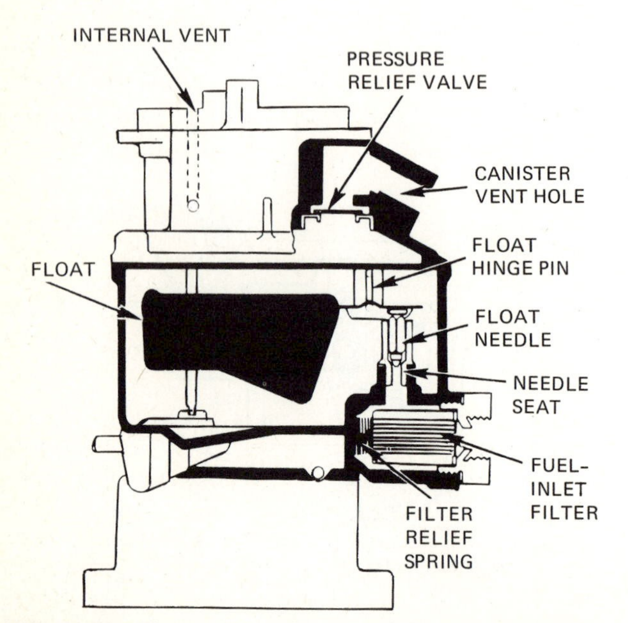

Fig. 14-12. Float system showing the two vents, one internal and the other to the charcoal canister. (*Chevrolet Motor Division of General Motors Corporation*)

⊘ 14-12 Hot-Idle Compensator Valve The internal vent could be a problem during idling or low-speed operation, especially in hot weather. Gasoline vapor from the float bowl can pass through the internal vent in sufficient amounts to upset the air-fuel ratio. In that event, gasoline vapor is added to the normal air-fuel mixture, and the mixture becomes too rich. To prevent this situation from occurring, some carburetors have a hot-idle compensator valve, as shown in Fig. 14-13 (left). This valve is operated by a thermostatic blade. When temperatures reach a preset value, the blade bends enough to open the valve port. Now additional air can flow through the auxiliary air passage, bypassing the idle system. This additional air leans out the air-fuel mixture enough to make up for the added gasoline vapor coming from the float bowl.

⊘ 14-13 Idle System When the throttle is closed or only slightly opened, only a small amount of air can pass through the air horn. The air speed is low, and very little vacuum is developed in the venturi. This means that the fuel nozzle does not feed fuel. Thus, the carburetor must have another system to supply fuel when the throttle is closed or slightly opened.

This system, called the *idle system,* is shown in operation in Fig. 14-14. It includes passages through which air and fuel can flow. The air passage is called the *air bleed.* With the throttle closed as shown, there is a high vacuum below the throttle valve from the intake manifold. Atmospheric pressure pushes air and fuel through the passages, as shown. The air

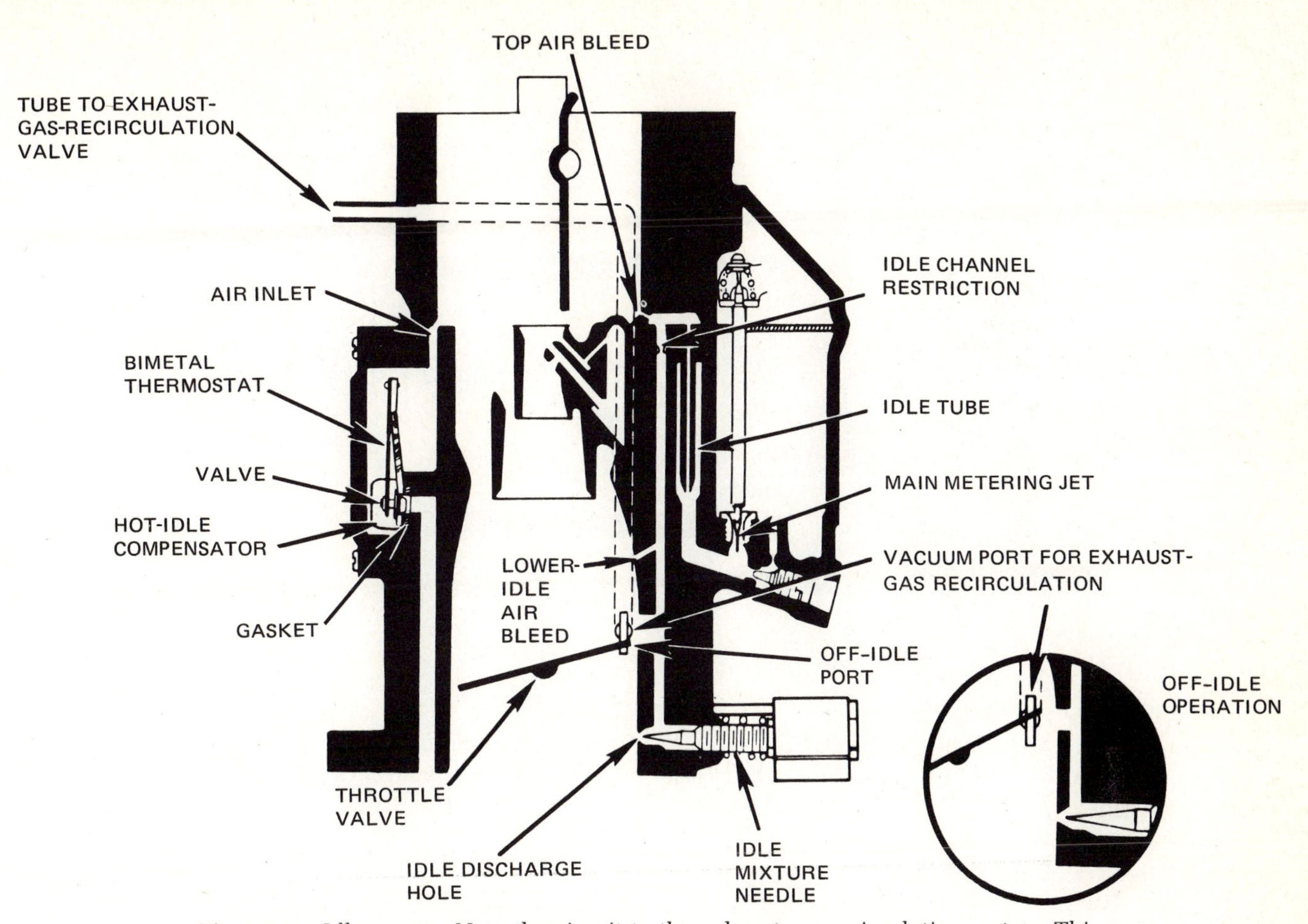

Fig. 14-13. Idle system. Note the circuit to the exhaust-gas recirculation system. This circuit allows some exhaust gas to feed into the air-fuel mixture when the throttle valve opens past the vacuum port. (*Chevrolet Motor Division of General Motors Corporation*)

and fuel mix and flow past the tapered point of the idle air-fuel-mixture adjustment screw. The air-fuel mixture has a high proportion of fuel (is very rich). It leans out somewhat as it mixes with the small amount of air that gets past the closed throttle valve. But the final mixture is still rich enough (see Fig. 14-7) for good idling. The air-fuel richness can be adjusted by turning the idle air-fuel-mixture adjustment screw in or out. This permits less or more air-fuel mixture to flow past the screw. Figure 14-15 shows a cutaway view of a carburetor with the idle system in operation.

CAUTION: In late-model cars, the idle air-fuel-mixture adjustment screw is fixed or has a locking cap on it. The reason for this is that it is illegal to adjust the idle mixture beyond the specific limits set according to federal standards.

⊘ 14-14 Low-Speed Operation When the throttle is opened slightly, as shown in Fig. 14-16, the edge of the throttle valve moves past the low-speed port in the side of the air horn. This port is a vertical slot or a series of small holes, one above the other. Additional fuel is thus fed into the intake manifold through the low-speed port. This fuel mixes with the additional air moving past the slightly opened throttle valve. Thus, a sufficient mixture richness for part-throttle low-speed operation is provided.

Some air bleeds around the throttle plate through the low-speed port when the edge of the throttle is only partway past this port. This air improves the atomization of the fuel coming from the low-speed port.

⊘ 14-15 Other Idle Systems There are many varieties of idle systems in addition to that shown in Figs. 14-14 to 14-16. In two-barrel carburetors, each barrel has its own system. In most four-barrel carburetors, only the primary barrels have idle systems (see ⊘ 14-30).

⊘ 14-16 Main Metering System Suppose the throttle valve is opened enough so that its edge moves well past the low-speed port. Now there is little difference in vacuum between the upper and lower parts of the air horn. Thus, only a small amount of air-fuel mixture is discharged from the low-speed port. But, under this condition, enough air moves through the air horn to produce a vacuum in the venturi. As a result, the fuel nozzle centered in the

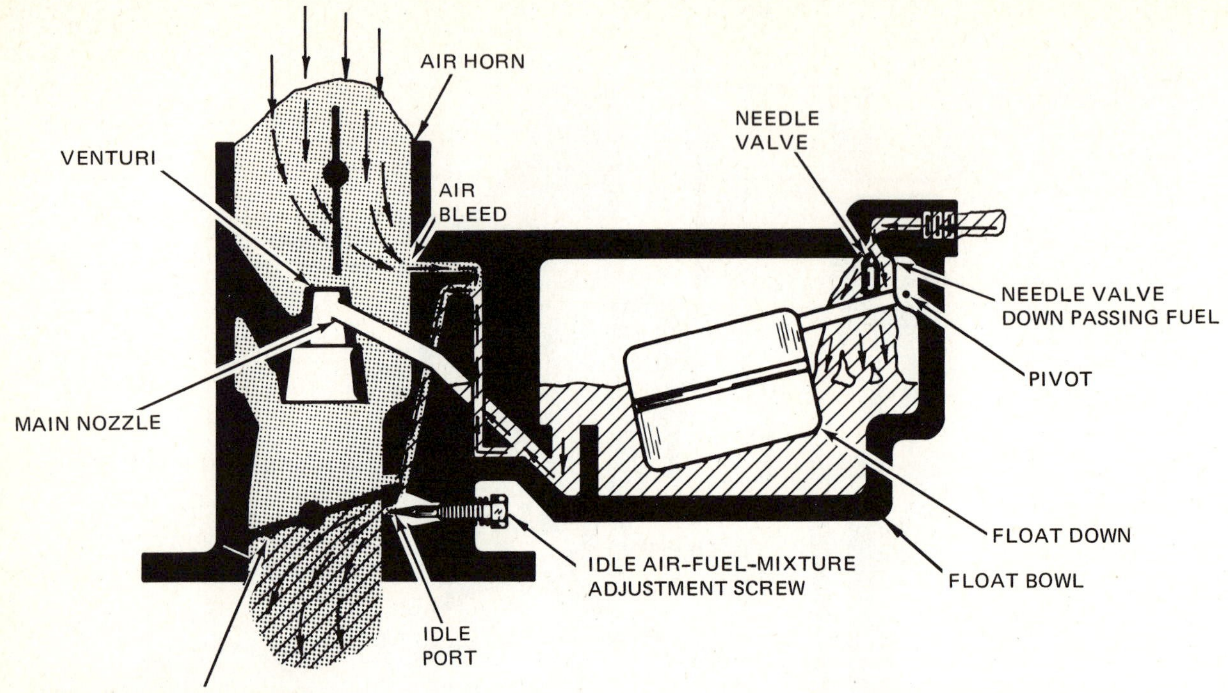

Fig. 14-14. Idle system in a carburetor. The throttle valve is closed so that only a small amount of air can get past it. All fuel is being fed past the idle adjustment screw. Arrows show the flow of air and fuel.

venturi, called the *main nozzle,* or *high-speed nozzle,* begins to discharge fuel (as explained in ⊘ 14-6). The main nozzle supplies fuel during operation with the throttle partly to fully opened. Figure 14-17 shows this action. The system from the float bowl to the main, or high-speed, nozzle is called the *main metering system.*

The wider the throttle is opened and the faster the air flows through the air horn, the greater the vacuum in the venturi. This means that additional fuel will be discharged from the main nozzle (because of the greater vacuum). As a result, a nearly constant air-fuel ratio is maintained by the main metering system from part- to wide-open throttle.

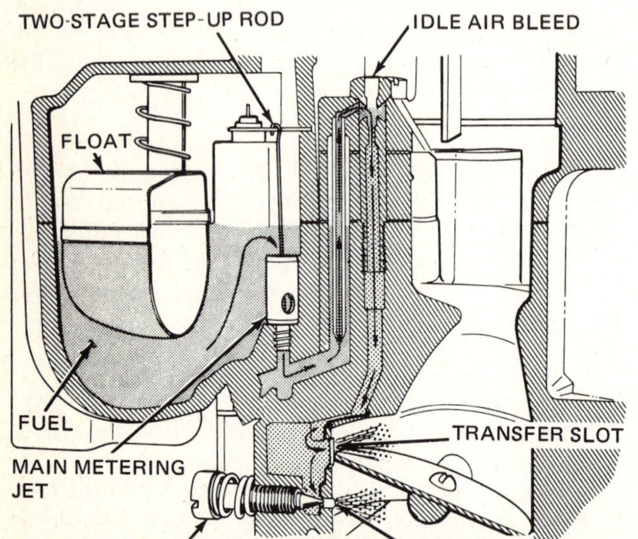

Fig. 14-15. Idle system in a carburetor. (The carburetor has been cut away to show the internal arrangement.) (*Chrysler Corporation*)

⊘ 14-17 Power System For high-speed, full-power, wide-open-throttle operation, the air-fuel mixture must be enriched (see Fig. 14-7). A power system is incorporated in the carburetor to provide this enriched mixture. This power system may be operated mechanically, by intake-manifold vacuum, or both mechanically and by vacuum.

⊘ 14-18 Mechanically Operated Power System The mechanically operated power system includes a metering-rod jet (a carefully calibrated orifice, or opening) and metering rod with two or more steps of different diameters (Fig. 14-18). The metering rod is attached to the throttle linkage (Fig. 14-19). When the throttle is opened, the metering rod is lifted. But when the throttle is partly closed, the larger diameter of the metering rod is in the metering-rod jet. This partly restricts fuel flow to the main nozzle. However, enough fuel does flow for normal part-throttle operation. When the throttle is opened wide, the metering rod is lifted enough to cause its smaller diameter, or step, to move up into the metering-rod jet. Now, the jet is less restricted, and more fuel can flow. The main nozzle is thus supplied with more fuel, and the resulting air-fuel mixture is richer.

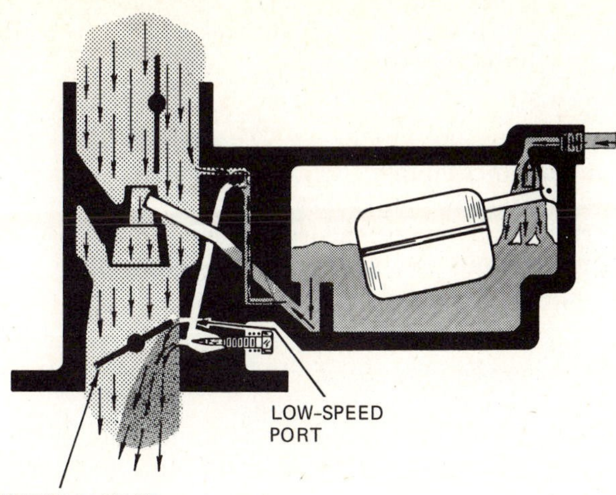

Fig. 14-16. Low-speed operation. The throttle valve is slightly open, and fuel is being fed through the low-speed port as well as through the idle port. The dark color is fuel; the light color is air.

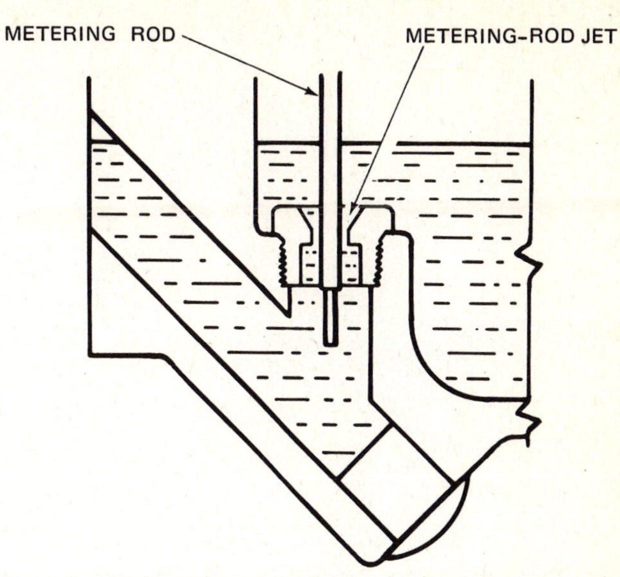

Fig. 14-18. Metering rod and metering-rod jet for better performance at full throttle.

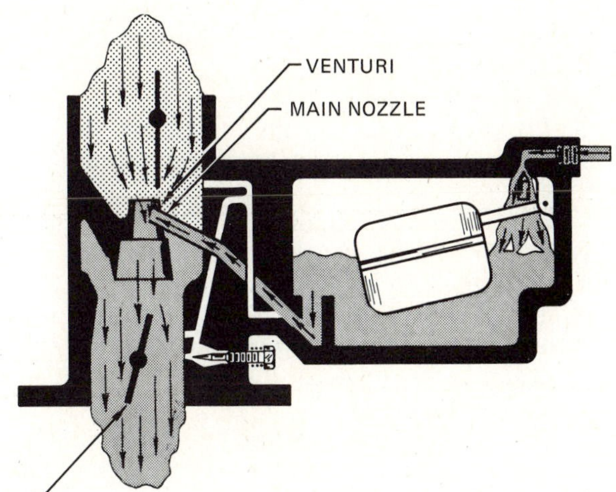

Fig. 14-17. Main metering system in a carburetor. The throttle valve is open, and fuel is being fed through the high-speed, or main, nozzle. The dark color is fuel; the light color is air.

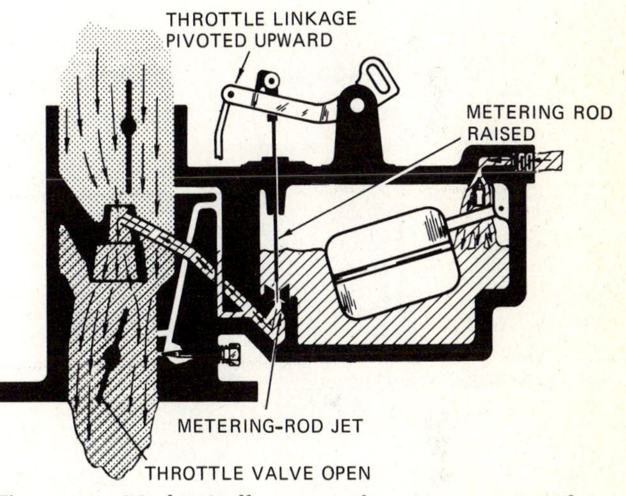

Fig. 14-19. Mechanically operated power system. When the throttle is open, as shown, the metering rod is raised so that the smaller diameter of the rod clears the jet. This allows additional fuel to flow.

⊘ 14-19 Vacuum-operated Power System The vacuum-operated power system is operated by intake-manifold vacuum. It includes a vacuum piston or diaphragm linked to a valve or metering rod similar to the one shown in Fig. 14-18. One design is shown in Fig. 14-20. During part-throttle operation, the piston is held in the lower position by intake-manifold vacuum. However, when the throttle is opened wide, the intake-manifold vacuum is reduced. This allows the spring under the vacuum piston to push the piston upward. This motion raises the metering rod so that the smaller diameter of the rod clears the metering-rod jet. Now, more fuel can flow to handle the full-power requirements of the engine.

A carburetor using a spring-loaded diaphragm to control the position of the metering rod is shown in Fig. 14-21. The action is similar to that of the carburetor using a spring-loaded piston. When the throttle is opened so that intake-manifold vacuum is reduced, the spring raises the diaphragm. This allows the metering rod to be lifted so that its smaller diameter clears the metering-rod jet. Now, more fuel can flow.

⊘ 14-20 Combination Power Systems In some carburetors, a combination full-power system is used. It is operated both mechanically and by vacuum from the intake manifold. In one such carburetor, a metering rod is linked to a vacuum diaphragm as well as to the throttle linkage (Fig. 14-21). Thus, movement of the throttle to FULL OPEN lifts the metering rod to enrich the mixture. Or loss of intake-manifold vacuum (as during a hard pull up a

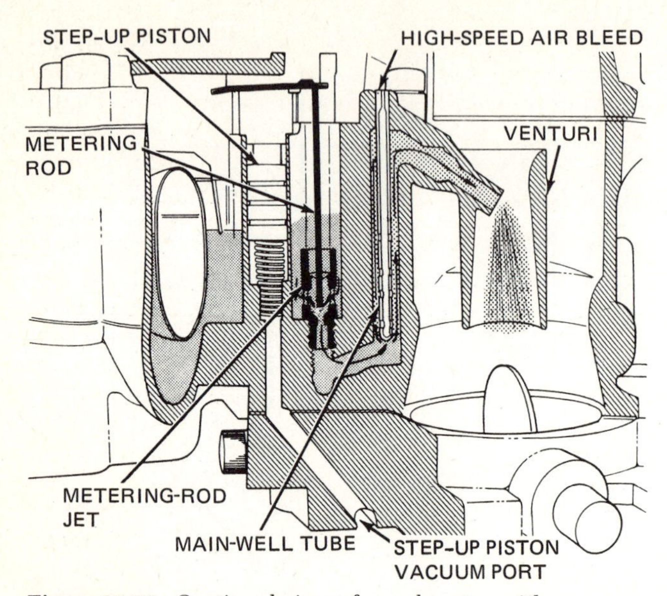

Figure 14-20. Sectional view of a carburetor with a power, or step-up, piston actuated by intake-manifold vacuum, to control the position of the metering rod. (*Chrysler Corporation*)

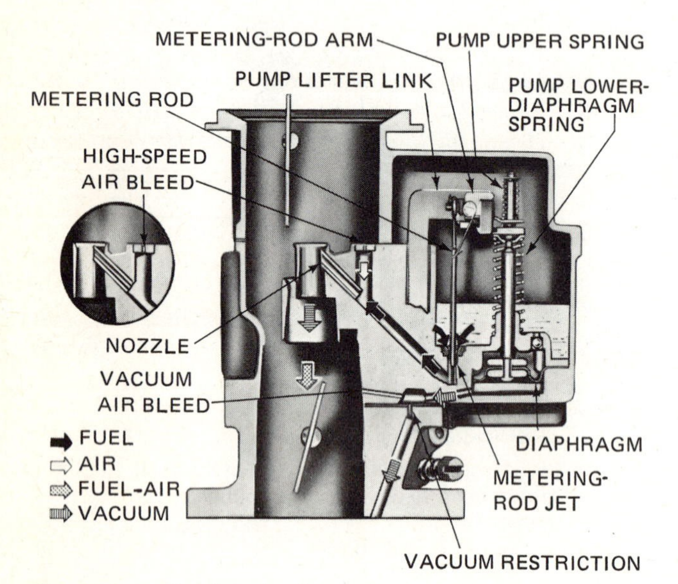

Fig. 14-21. Sectional view of a carburetor with a spring-loaded diaphragm, actuated by intake-manifold vacuum, to control the position of the metering rod. (*Ford Motor Company*)

hill or during acceleration) causes the vacuum-diaphragm spring to raise the metering rod for an enriched mixture.

Check Your Progress

Progress Quiz 14-1 Again, here is your chance to check up on yourself to find out how well you are remembering and understanding what you have been studying about the fundamentals of carburetors and carburetor systems. If any of the questions stump you, reread the past few pages to find the information you need.

Completing the Sentences The sentences that follow are incomplete. After each sentence there are several words or phrases, only one of which will correctly complete the sentence. Write each sentence in your notebook, selecting the proper word or phrase to complete it correctly.

1. The carburetor consists basically of an air horn, a fuel nozzle, and: (*a*) a choke valve, (*b*) a throttle valve, (*c*) an air cleaner.
2. The job of the carburetor is to: (*a*) mix gasoline with air, (*b*) separate gasoline from air, (*c*) increase volumetric efficiency.
3. The throttle valve: (*a*) when closed, allows little or no air to flow through the air horn, (*b*) when open, allows air to flow freely through the air horn, (*c*) both (*a*) and (*b*).
4. The reason that a richer mixture must be delivered when first starting a cold engine is that: (*a*) this allows a higher cranking speed, (*b*) only part of the gasoline will vaporize when cold, (*c*) the thick engine oil must be thinned out.
5. An air-fuel ratio of 12:1 means that the mixture has: (*a*) 12 lb of gasoline to 1 lb of air, (*b*) 12 lb of air to 1 lb of gasoline, (*c*) 1 gal of gasoline to 12 gal of air.
6. The system that maintains a constant level of gasoline in a bowl is called the: (*a*) power system, (*b*) float system, (*c*) bowl system.
7. The purpose of the float-bowl vent is to: (*a*) prevent engine overheating, (*b*) keep the level of gasoline in the bowl constant, (*c*) equalize the effects of a clogged air cleaner.
8. The purpose of the hot-idle compensator valve is to: (*a*) allow additional air to pass into the idle system, (*b*) supply additional air to prevent a rich mixture when the engine is idling hot, (*c*) increase idle speed when the engine is cold.
9. When the engine is hot and running at 600 rpm, the gasoline is supplied by the: (*a*) idle system, (*b*) low-speed system, (*c*) choke system.
10. The operating mechanism of the vacuum-operated power system includes either a: (*a*) vacuum piston or pump, (*b*) diaphragm or pump, (*c*) vacuum piston or diaphragm.

⊘ 14-21 Air-Fuel Ratios with Different Systems Figure 14-22 shows the air-fuel ratios with the different carburetor systems in operation. This is a typical curve only. Actual air-fuel ratios may be different for different carburetors and operating conditions. Note that the idle system supplies a very rich mixture to start with but that as engine speed increases, the mixture leans out. From about 25 to 40 miles per hour [40.24 to 64.37 km/h], the throttle is only partly opened and both the idle and the main metering system supply air-fuel mixture. Then, at about 40 miles per hour [64.37 km/h], the main metering sys-

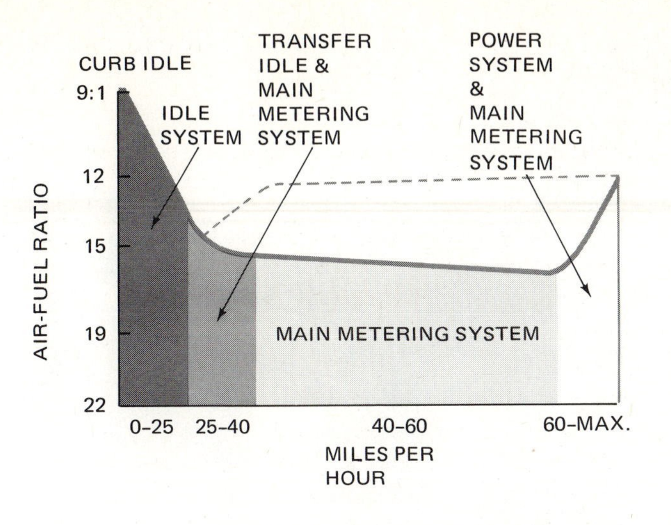

Fig. 14-22. Air-fuel ratios with different carburetor systems operating at various speeds. (*Chevrolet Motor Division of General Motors Corporation*)

tem takes over and continues by itself to about 60 miles per hour [96.56 km/h]. Note that the air-fuel ratio increases somewhat as speed increases (that is, the mixture becomes leaner). Somewhere around 60 miles per hour [96.56 km/h], the power system comes into operation (earlier, of course, if the throttle is opened wide at a lower speed). Now, the richness of the air-fuel mixture increases with higher speeds.

⊘ 14-22 Accelerator-Pump System For acceleration, the carburetor must deliver additional fuel (see ⊘ 14-8). Rapid opening of the throttle allows a sudden inrush of air. Thus, there is a sudden demand for additional fuel. Carburetors have accelerator-pump systems to provide this extra fuel. Figure 14-23 shows one type of accelerator-pump system. It includes a pump plunger which is forced downward by a pump lever that is linked to the throttle. When the throttle is opened, the pump lever pushes the pump plunger down. This forces fuel to flow through the accelerator-pump system and out the pump jet (Fig. 14-24). This fuel enters the air passing through the carburetor to supply the additional fuel needed.

However, when the throttle is opened quickly, fuel may not discharge for a long enough time to prevent engine stumble. To overcome this problem, most carburetors use a calibrated spring above the plunger cup to prolong the discharge. Figure 14-25 is a cutaway view of a carburetor using a plunger-type accelerator-pump system. In this figure note that the attachment between the pump plunger and the pump and seal is through a spring. This spring applies pressure to the pump so that the accelerator-pump system immediately begins discharging fuel through the jet. The spring maintains this pressure during the entire time that the throttle is held open until the pump plunger is all the way down, as shown in Fig. 14-24. This arrangement allows the accelerator-pump system to discharge fuel for several seconds, or until the full-power system can take over. It therefore permits smooth acceleration.

One type of accelerator-pump system uses a diaphragm instead of a plunger; it is shown in Fig. 14-26. When the throttle is opened, the lower diaphragm spring of the pump lifts the diaphragm. This forces additional fuel from the chamber above the diaphragm through the accelerator-pump system and out the pump jet.

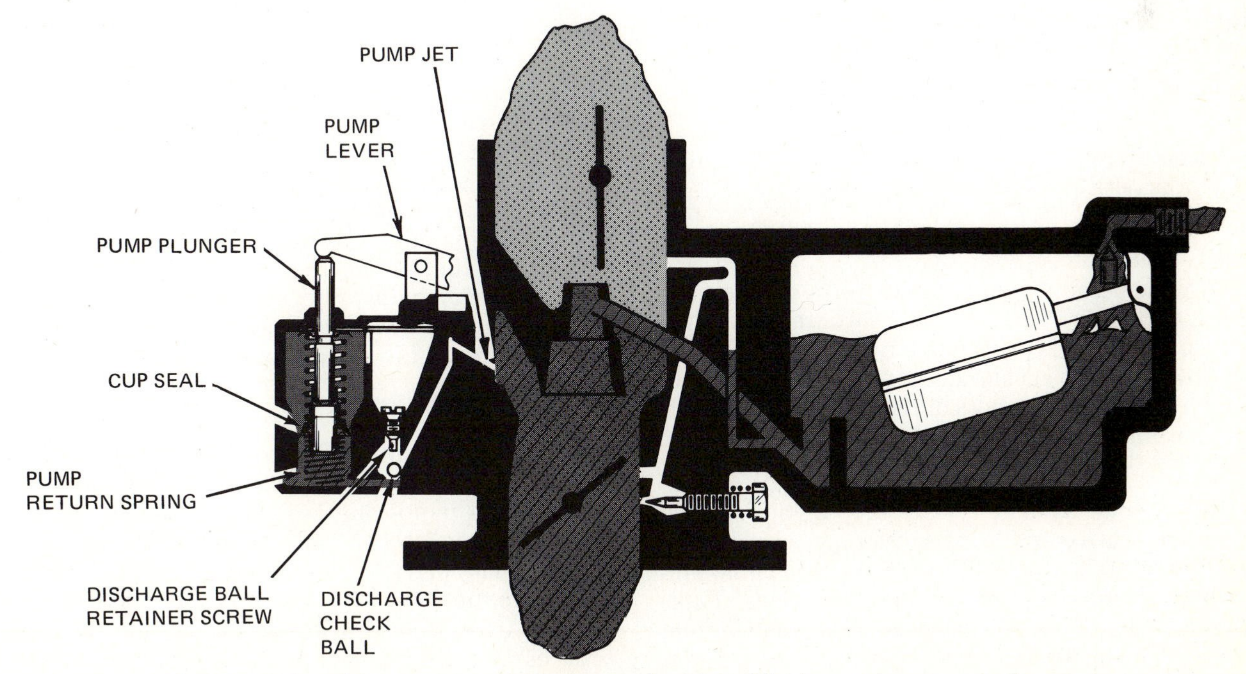

Fig. 14-23. Accelerator-pump system in a carburetor of the type using a pump plunger.

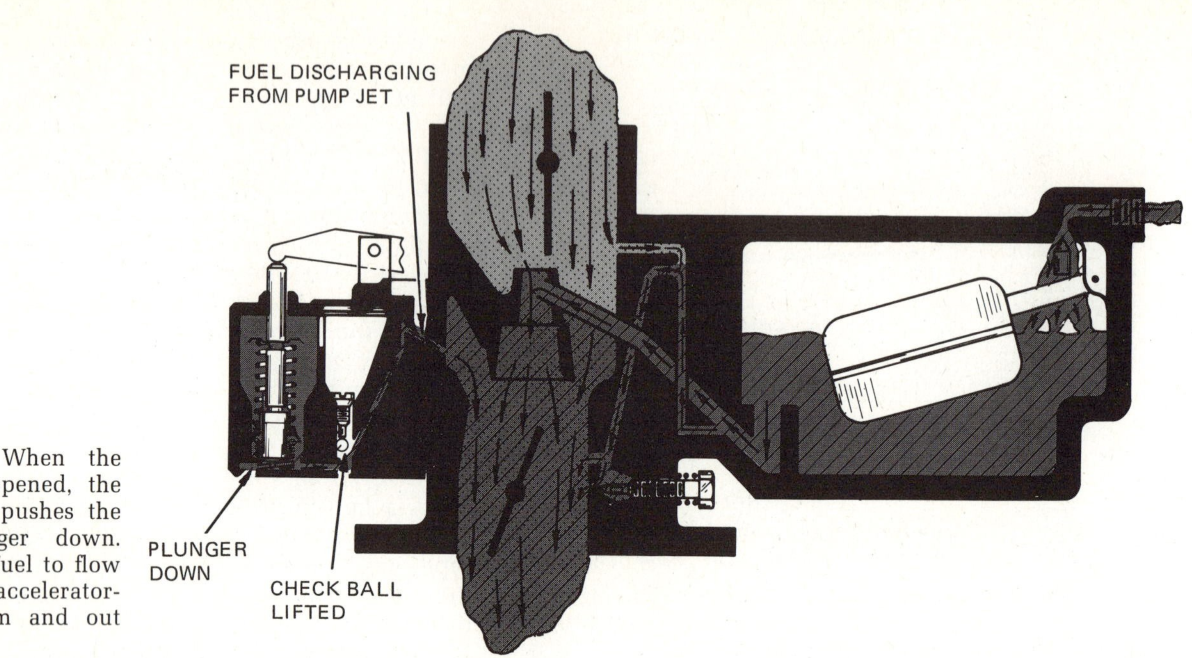

Fig. 14-24. When the throttle is opened, the pump lever pushes the pump plunger down. This forces fuel to flow through the accelerator-pump system and out the jet.

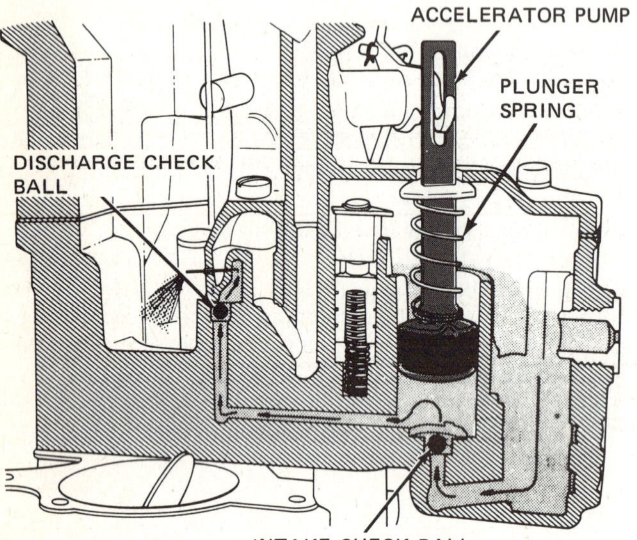

Fig. 14-25. Accelerator-pump system using a piston-type pump. (*Chrysler Corporation*)

An accelerator-pump system for a dual carburetor is shown in Fig. 14-27. This carburetor has two barrels, and there is a discharge nozzle for each. The fuel flow from the accelerator pump is split between the two barrels. Regardless of the number of barrels, most carburetors use only one accelerator pump.

⊘ 14-23 Choke System While the engine is being started, the carburetor must deliver a very rich mixture to the intake manifold. With a cold engine and carburetor, only part of the fuel vaporizes. Thus, extra fuel must be delivered so that enough fuel vaporizes to make the correct air-fuel ratio that is necessary to start the engine.

During cranking, air speed through the carburetor air horn is very low. Vacuum from the venturi action and vacuum below the throttle would be insufficient to produce adequate fuel flow for starting. Thus, to produce enough fuel flow during cranking, the carburetor has a *choke* (Fig. 14-28). The choke consists of a valve in the top of the air horn that is controlled either mechanically or by an automatic device. When the choke valve is closed, only a small amount of air can get past it (the valve "chokes off" the air flow). Then, when the engine is cranked, a fairly high vacuum develops in the air horn. This vacuum causes the main nozzle to discharge a heavy stream of fuel. The quantity of fuel produced is sufficient to produce the correct air-fuel mixture needed for starting the engine.

As soon as the engine starts, the speed increases from around 250 to 300 rpm to over 600 rpm. Now more air and a somewhat leaner air-fuel mixture are required. One method of getting more air into the engine as soon as it starts is to mount the choke valve off center on its shaft in the air horn. Then, if a spring is added to the choke linkage, the additional air the engine requires causes the valve to partly open against the spring pressure. Another arrangement includes a small spring-loaded section in the valve. This section opens to admit the additional air.

⊘ 14-24 Automatic Chokes Mechanically controlled chokes are operated by a pull rod on the instrument panel that is linked to the choke valve. When the pull rod is pulled out, the choke valve is closed. The driver must remember to push the pull rod in to the dechoked position as soon as the engine begins to warm up. If the pull rod is not pushed in, the carburetor will continue to supply a very rich air-fuel mixture to the engine. This excessive rich-

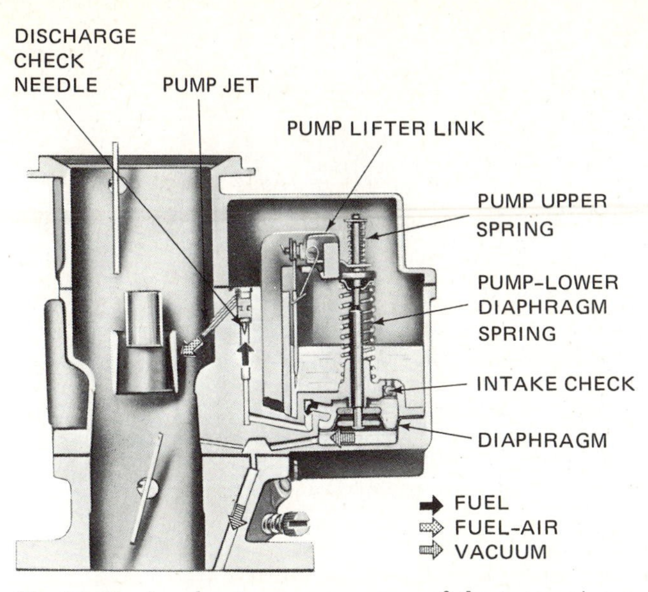

Fig. 14-26. Accelerator-pump system of the type using a spring-loaded diaphragm. Opening of the throttle allows the lower diaphragm spring to lift the diaphragm. This forces fuel through the accelerator-pump system and out through the jet. (*Ford Motor Company*)

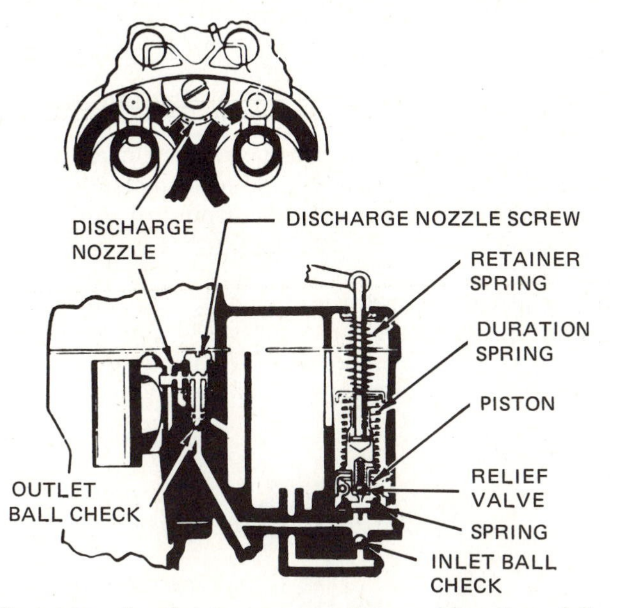

Fig. 14-27. Accelerator-pump system and location of the discharge nozzles in a dual carburetor.

ness will cause poor engine performance, high levels of exhaust emissions, fouled spark plugs, and poor fuel economy, as well as many other problems.

To prevent such troubles, most cars now have automatic chokes. Most automatic chokes operate on exhaust-manifold temperature and intake-manifold vacuum. Figure 14-29 shows a carburetor that is partly cut away so that the construction of the automatic choke can be seen. The automatic choke includes a thermostatic spring and vacuum piston, both linked to the choke valve. The thermostatic spring is made up of two different metal strips

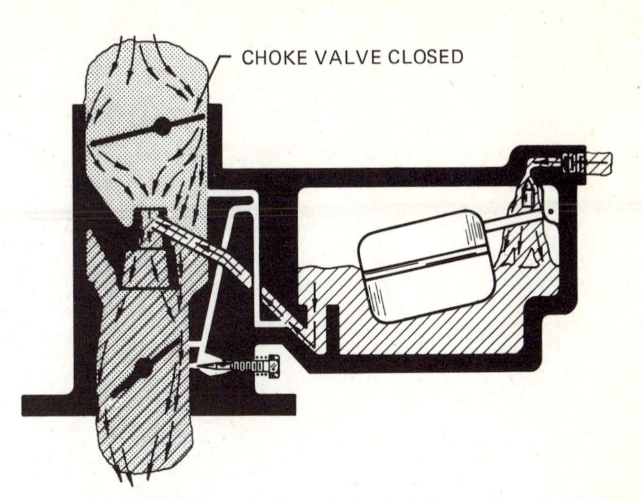

Fig. 14-28. With the choke valve closed, intake-manifold vacuum is introduced into the carburetor air horn. This causes the main nozzle to discharge fuel.

welded together and formed into a spiral. Owing to differences in the expansion rates of the two metals, the thermostatic spring winds up or unwinds with changing temperature. When the engine is cold, the spring is wound up enough to close the choke valve and spring-load it in the closed position. When the engine is cranked, a rich air-fuel mixture is delivered to the intake manifold. As the engine starts, air movement through the air horn causes the choke valve to open slightly (working against the thermostatic spring tension). In addition, the vacuum piston is pulled outward by intake-manifold vacuum. This action produces further opening of the choke valve.

The choke valve is thus positioned to let the carburetor supply the rich air-fuel mixture required for cold-engine idling operation. When the throttle is opened, the mixture must be enriched. The accelerator pump provides some extra fuel, but still more fuel is needed when the engine is cold. This additional fuel is secured by the action of the vacuum piston. When the throttle is opened, intake-manifold vacuum is lost. The vacuum piston releases and is pulled inward by the thermostatic spring tension. The choke valve therefore moves toward the closed position, causing the mixture to be enriched.

During the first few moments of operation, the choke valve is controlled by the vacuum piston. However, as the engine warms up, the thermostatic spring begins to take over. The thermostatic spring is located in a housing that is connected to the exhaust manifold through a small heat tube (not shown in Fig. 14-29). Heat passes from the exhaust manifold through this tube and enters the thermostatic spring housing. Soon, the thermostat begins to warm up. As it warms up, the spring unwinds. This causes the choke valve to move toward the opened position. When operating temperature is reached, the thermostatic spring has unwound enough to fully open the choke valve. No further choking takes place.

When the engine is stopped and cools, the thermostatic spring again winds up. This closes the

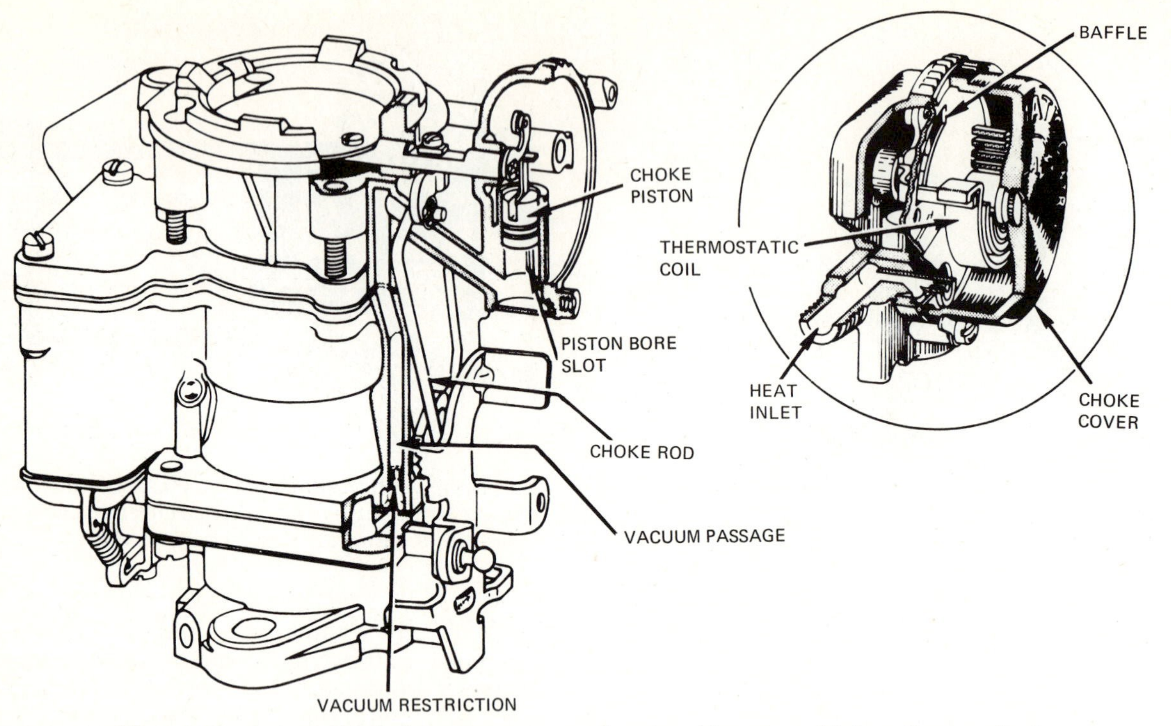

Fig. 14-29. Automatic choke system on a carburetor. (*American Motors Corporation*)

choke valve and spring-loads it in the closed position.

In many engines, the thermostat is located in a well in the exhaust manifold. There, the thermostat can react quickly to the manifold heat as the engine starts (see Figs. 14-30 and 14-34). The thermostat is connected by a link to the carburetor. Some carburetors using this arrangement have vacuum pistons (as described previously); others have vacuum diaphragms. Both work with the thermostat to control the choke-valve position during engine warm-up, as previously noted.

Some carburetors use heat from the engine coolant to operate the thermostat. In these carburetors the thermostat housing has a passage through which the coolant flows (Fig. 14-31). The action is similar to that in the automatic chokes discussed previously.

The operation of the automatic choke system which uses a vacuum diaphragm (Fig. 14-30) is quite similar to the system which uses the vacuum-piston arrangement (described previously). However, the diaphragm provides considerably more force to break loose the choke valve if it gets stuck. The linkage from the diaphragm to the choke-valve lever rides freely in a slot in the lever. During certain phases of engine warm-up, the changing vacuum causes the linkage to ride to the end of the slot in the choke lever and move the choke valve. For example, when the throttle is opened during cold-engine operation, loss of intake-manifold vacuum causes the diaphragm to move. This movement carries the choke-valve lever around so that the choke valve is moved toward the closed position. This action provides a richer air-fuel mixture for good acceleration.

Many late-model cars have electric automatic chokes. This type of choke includes an electric heating element (Fig. 14-32). The purpose of this heater is to ensure faster choke opening. This helps reduce emissions from the engine. Emissions (HC and CO) are relatively high during the early stages of engine warm-up. At low temperatures, the electric heater adds heat to the heat coming from the exhaust manifold. This action reduces the choke-opening time to

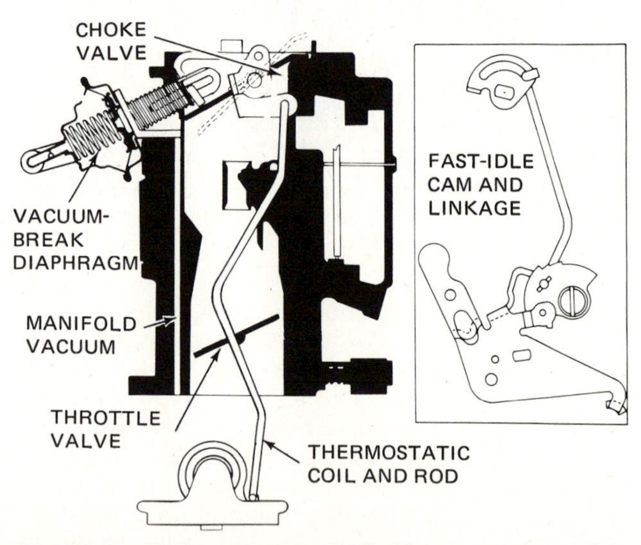

Fig. 14-30. Choke system with the thermostat located in a well in the exhaust manifold. Note the vacuum-break diaphragm. (*Chevrolet Motor Division of General Motors Corporation*)

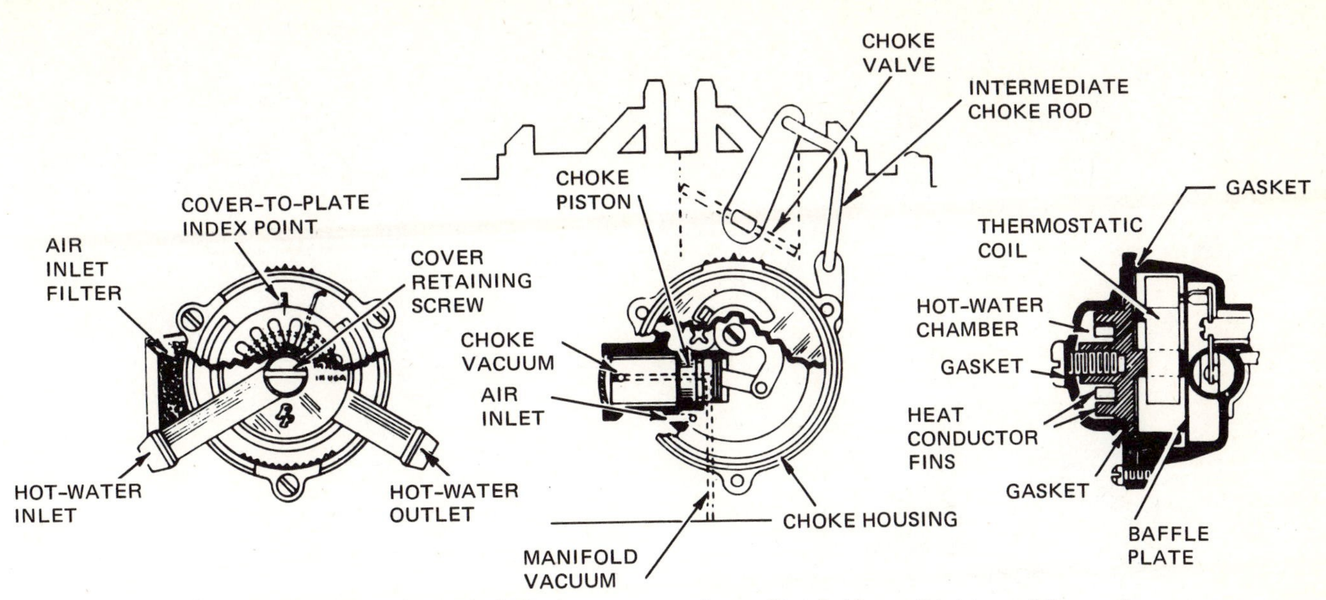

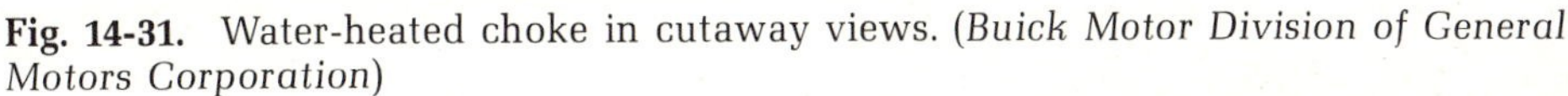

Fig. 14-31. Water-heated choke in cutaway views. (*Buick Motor Division of General Motors Corporation*)

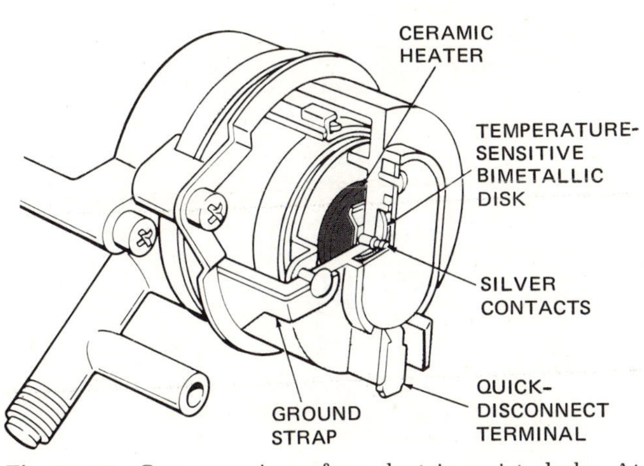

Fig. 14-32. Cutaway view of an electric-assist choke. At low temperature, the ceramic heater turns on, adding heat to the choke so that it opens more quickly. (*Ford Motor Company*)

as short as $1\frac{1}{2}$ minutes. Figure 14-33 shows the arrangement for a choke mounted in a well in the exhaust manifold. Figure 14-34 shows the choke and heating element removed from the well.

⊘ 14-25 Heat-Control Valve

During initial warm-up of the engine, just after starting, vaporization of the gasoline in the air-fuel mixture entering the engine is poor because gasoline vaporizes more slowly when it is cold (see ⊘ 12-3). To improve gasoline vaporization and therefore cold-engine operation, a device is provided to heat the intake manifold when it is cold. This device, called the *manifold heat-control valve,* is built into the exhaust and intake manifolds. Two arrangements are used, one for in-line engines and another for V-8 engines.

1. *IN-LINE ENGINES* In these engines, the exhaust manifold is located under the intake manifold. At a central point, there is an opening from the exhaust manifold into a chamber, or oven, surrounding the intake manifold (Figs. 14-35 and 14-36). A butterfly heat-control valve is placed in this opening (see Fig. 7-21). The position of this valve is controlled by a thermostat. When the engine is cold, the thermostatic spring unwinds and moves the heat-control valve to the closed position (Fig. 14-36, left). Now, when the engine is started, the hot exhaust gases pass through the opening and circulate through the oven around the intake manifold (Figs. 14-35 and 14-36). Heat from the exhaust gases quickly warms the intake manifold and helps the fuel to vaporize. Thus, cold-engine operation is improved. As the engine warms up, the thermostatic spring winds up

Fig. 14-33. Arrangement for an electric-assist choke mounted in a well in the exhaust manifold. (*Chrysler Corporation*)

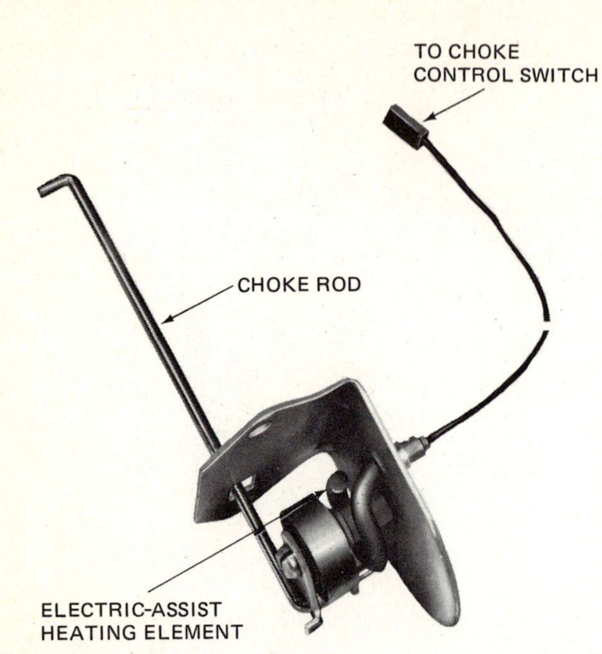

Fig. 14-34. Choke assembly removed from the well in the exhaust manifold to show the electric-assist heating element. (*Chrysler Corporation*)

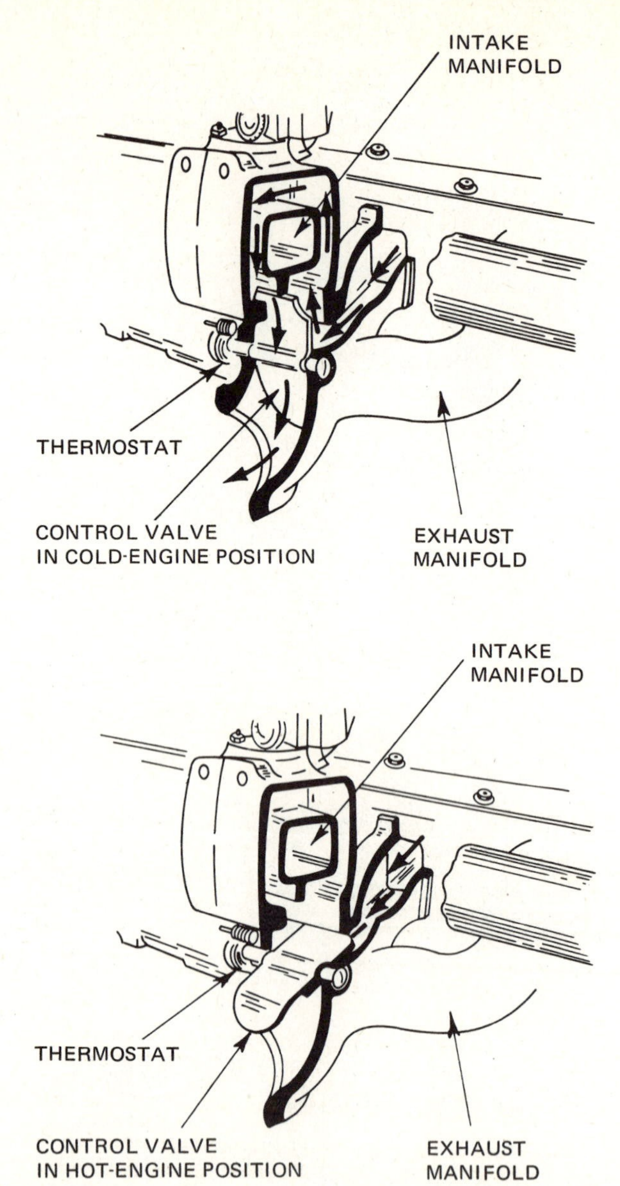

Fig. 14-35. Intake and exhaust manifolds of a six-cylinder inline engine, cut away to show the location and action of the manifold heat control. At top, the heat-control valve is in the "heat on" position. It is directing hot exhaust gases up and around the intake manifold, as shown by the arrows. At bottom, the valve is in the hot-engine position. (*Ford Motor Company*)

and moves the heat-control valve to the opened position (Fig. 14-36, right). Now, the exhaust gases no longer circulate in the oven around the intake manifold but pass directly into the exhaust pipe.

2. *V-8 ENGINES* In V-8 engines, the intake manifold is placed between the two banks of cylinders. It has a special passage (Figs. 7-24 and 14-37) through which exhaust gases can pass. One of the exhaust manifolds has a heat-control valve that is thermostatically controlled. This heat-control valve closes when the engine is cold. This causes the exhaust gases to pass from that exhaust manifold through the special passage in the intake manifold. The exhaust gases then enter the other exhaust manifold. Heat from the exhaust gases thus heats the air-fuel mixture in the intake manifold for improved cold-engine operation. As the engine warms up, the heat-control valve opens. Then, the exhaust gases from both exhaust manifolds pass directly into the exhaust pipes.

NOTE: Since the introduction of thermostatic air cleaners (⊘ 13-12), some engines no longer use a heat-control valve. To do so might add too much heat to the incoming air-fuel mixture. This would reduce the amount of air-fuel mixture entering the cylinders and thus reduce engine power.

3. *EARLY FUEL EVAPORATION (EFE) SYSTEM* Many 1975 cars came out with a vacuum-controlled manifold heat-control valve (Fig. 14-38). That is, a vacuum motor instead of a thermostat controls the position of the heat-control valve. The system is called an *early fuel evaporation* (*EFE*) *system* because of its quick action. The heat-control valve is called the EFE valve, and the vacuum to operate the vacuum motor comes from the intake manifold through a thermal vacuum switch. You can see the locations of the EFE valve and the vacuum motor on a V-8 engine in Fig. 19-21. The EFE valve does the same job as the thermostatically operated heat-control valve described in items *1* and *2* above. However, the EFE valve does the job much faster. This reduces the time that heat is going into the intake manifold and thus improves engine operation during warm-up.

When the engine is off, the heat-control valve is in the open position shown in Fig. 14-36 (for inline engines). In V-8 engines, the EFE valve is open. However, when the engine is started and intake

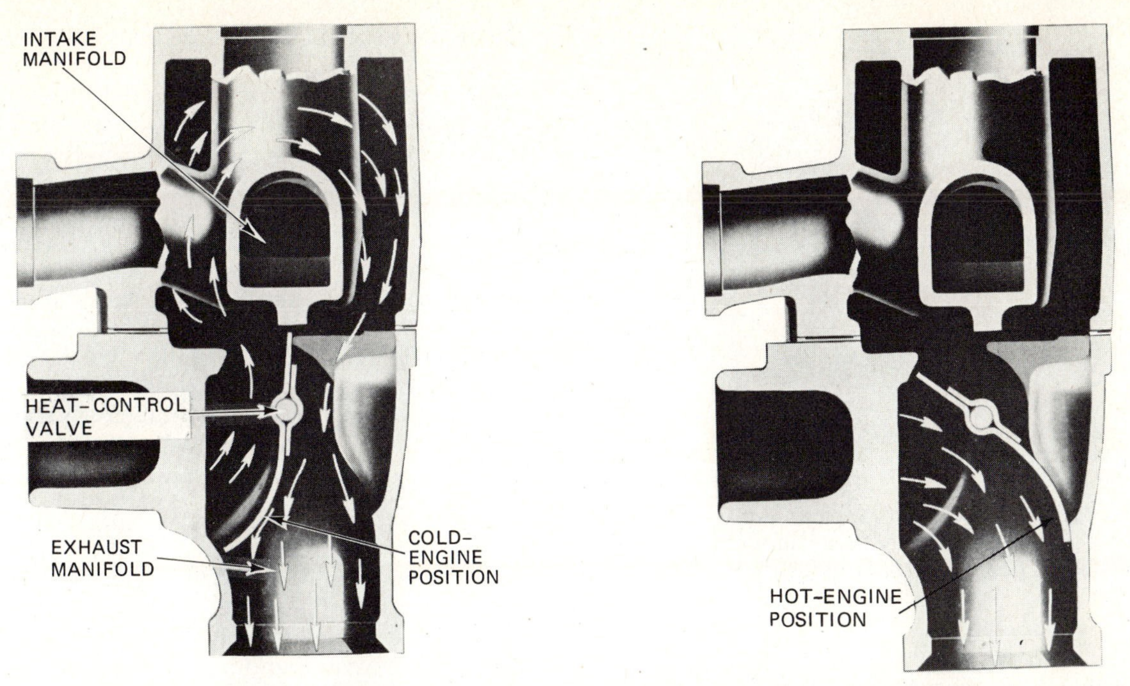

Fig. 14-36. Two extreme positions in the exhaust manifold of the manifold heat-control valve, which controls the flow of exhaust gases through the intake-manifold jacket. (*Chevrolet Motor Division of General Motors Corporation*)

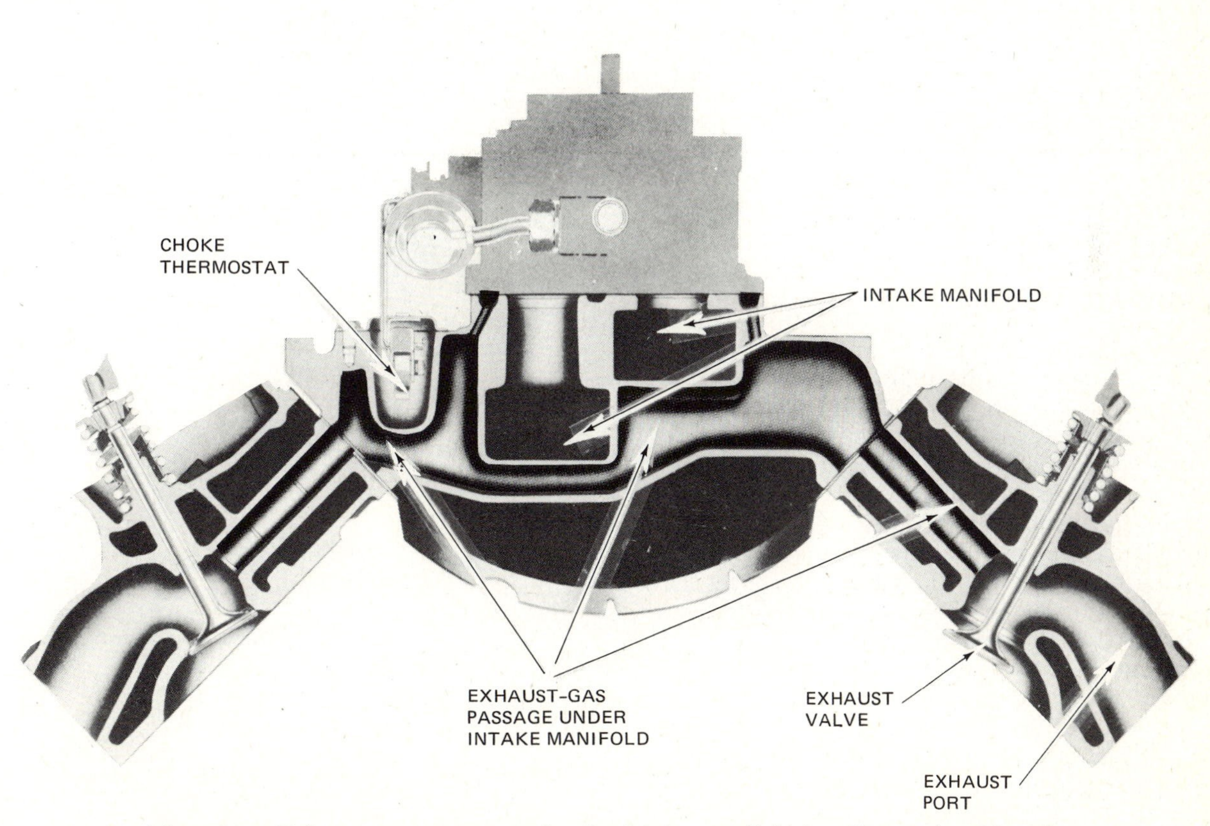

Fig. 14-37. Exhaust-gas passage under the intake manifold in a V-8 engine. Note the well in which the carburetor choke thermostat is located. (*Buick Motor Division of General Motors Corporation*)

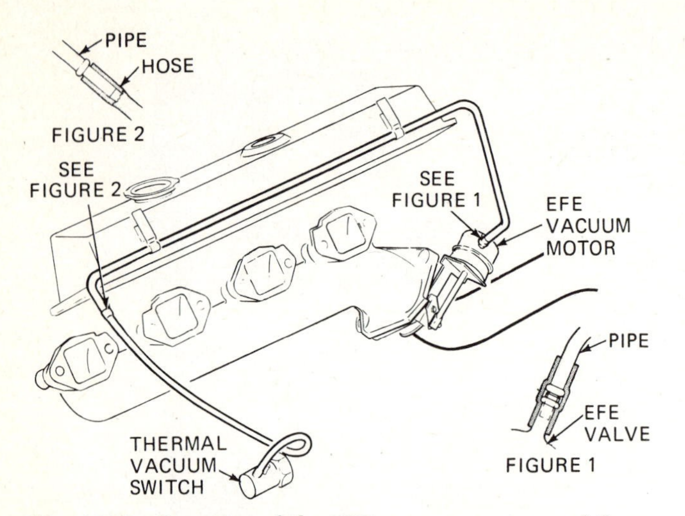

Fig. 14-38. Location of the EFE vacuum motor and thermal vacuum switch in an early fuel evaporation system. (*Cadillac Motor Car Division of General Motors Corporation*)

manifold develops, the vacuum passes through the thermal vacuum switch to the vacuum motor. The vacuum motor then operates to close the EFE valve as shown in Fig. 14-35 (for in-line engines). This sends exhaust gas up through the oven around the intake manifold. In V-8 engines, closing of the EFE valve shuts off one exhaust manifold from the exhaust pipe. In either type of engine, heat from the exhaust gases passes into the ingoing air-fuel mixture to improve fuel vaporization and cold-engine operation.

As the engine begins to warm up, the thermal vacuum switch shuts off the vacuum to the vacuum motor. Now, the motor relaxes to allow the heat-control valve (in in-line engines) to assume the hot-engine position. In V-8 engines, the EFE valve opens to permit normal movement of the exhaust gases from the exhaust manifold to the exhaust pipe.

⊘ 14-26 Anti-icing When fuel is sprayed into the air passing through the air horn, it evaporates, or turns to vapor. During evaporation, the fuel takes on heat; that is, it takes heat from the surrounding air and metal parts. This is the same effect you get when your hand gets wet and feels cold. If you blow on your hand, thus causing the water to evaporate faster, your hand will feel still colder. The faster that evaporation takes place, thus taking heat from your hand faster, the cooler your hand will feel.

Now, let us see how this principle affects the carburetor. Spraying and evaporation of the fuel "rob" the surrounding air and carburetor of heat. Under certain conditions, the surrounding metal parts are so cooled that moisture in the air will condense and actually freeze on them. If conditions are right, the ice can build up sufficiently to cause the engine to stall. This is most likely to occur during the warm-up period following the first engine start-up of the day, with high humidity and air temperatures in the range of 40 to 60 degrees Fahrenheit [4.4 to 15.6°C].

To prevent such icing, many carburetors have special anti-icing circuits. One arrangement for a V-8 engine is shown in Fig. 14-39. During the warm-up period, the manifold heat-control valve shunts hot exhaust gases from one exhaust manifold to the other (see ⊘ 14-25). Part of these hot exhaust gases circulates around the carburetor idle ports and near the throttle-valve shaft. This circulation adds enough heat to guard against ice formation. In another anti-icing arrangement, the carburetor contains water passages. A small amount of coolant from the engine cooling system passes through a special water manifold in the carburetor throttle body. This action adds enough heat to the carburetor to prevent icing.

⊘ 14-27 Fast Idle When the engine is cold, some throttle opening must be maintained so that the engine will idle faster than it does when it is warm. Otherwise, slow idle and a cold engine might cause the engine to stall. With fast idle, enough air-fuel mixture gets through and air speeds are great enough to produce adequate vaporization and a sufficiently rich air-fuel mixture. Fast idle is obtained by a fast-idle cam linked to the choke valve (Fig. 14-40). When the engine is cold, the automatic choke holds the choke valve closed. In this position, the linkage has revolved the fast-idle cam so that the adjusting screw rests on the high point of the cam. The adjusting screw prevents the throttle valve from moving to the fully closed position. The throttle valve is held partly open for fast idle. As the engine warms up, the choke valve opens. This rotates the fast-idle cam so that the high point moves from under the adjusting screw. The throttle valve closes for normal hot-engine slow idle.

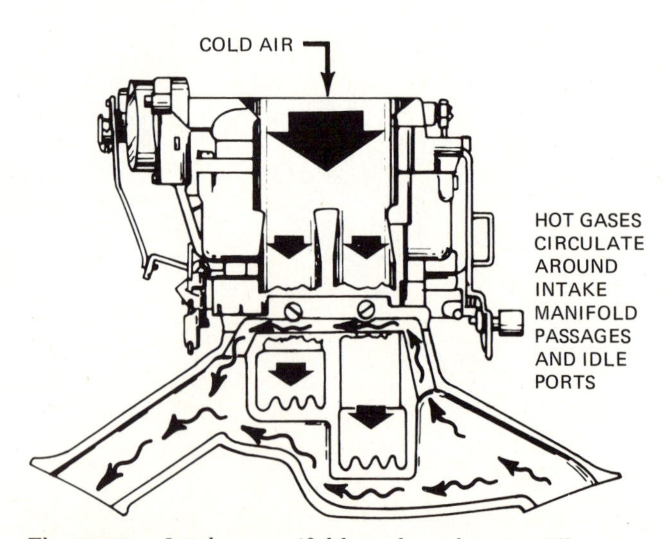

Fig. 14-39. Intake manifold and carburetor-idle-ports heating passages. Hot exhaust gases heat these areas as soon as the engine starts. (*Cadillac Motor Car Division of General Motors Corporation*)

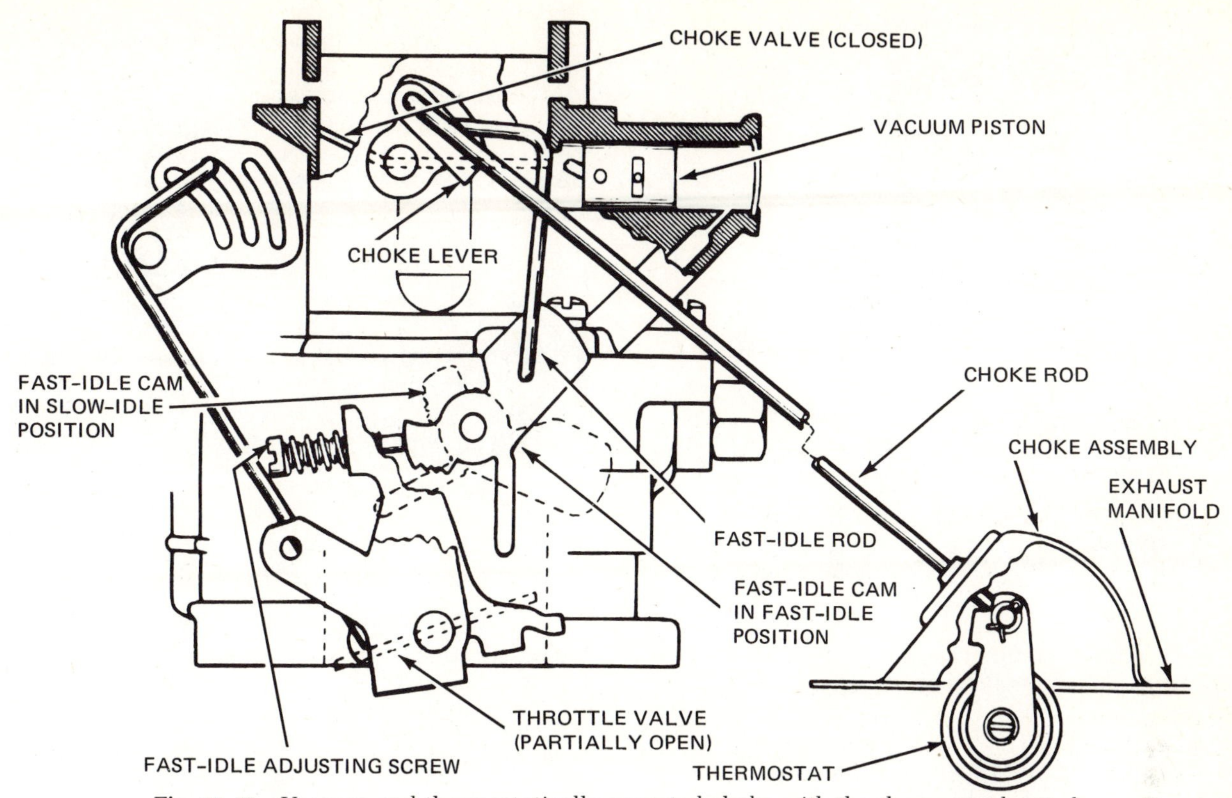

Fig. 14-40. Vacuum and thermostatically operated choke with the thermostat located in the exhaust manifold. Note two positions of the fast-idle cam. (*Chrysler Corporation*)

⊘ 14-28 Air Bleed and Antisiphon Passages In all the systems of carburetors, except the accelerator-pump system, small openings are incorporated to permit air to enter, or bleed into, the systems (Fig. 14-20). This action produces some premixing of the air and fuel so that better atomization and vaporization result. It also helps maintain a more uniform air-fuel ratio. For example, at higher speeds, a larger amount of fuel tends to be discharged from the main nozzle. But at the same time, the faster fuel movement through the high-speed system causes more air to bleed into the system. Thus, the air-bleed holes tend to equalize the air-fuel ratio.

Air-bleed passages are also sometimes called *antisiphon passages*. They act as air vents to prevent the siphoning of fuel from the float bowl at intermediate engine speeds.

If air-bleed passages become plugged, they may cause the float bowl to be emptied after the engine shuts off. When the engine is shut off, the intake manifold cools down and a slight vacuum forms as a result. With open air bleeds, air can move through the bleeds to satisfy the vacuum. But if the air bleeds are plugged, then the vacuum will cause the float bowl to empty through the idle system.

⊘ 14-29 Special Carburetor Devices Other special carburetor devices include:

1. Vacuum systems to control the ignition-distributor spark advance (see ⊘ 15-22).
2. Throttle-return checks, or dashpots (sometimes magnetically controlled) to slow throttle closing (on cars with automatic transmissions).
3. Electric kickdown switches (on some cars equipped with automatic transmissions).
4. Governors to control or limit top engine speed.
5. Idle-stop solenoids, which allow the throttle plate to close completely when the ignition is turned off. This prevents "dieseling," or continued running of the engine after the ignition is turned off (see ⊘ 22-16).

Figure 14-41 shows a throttle-return check, or dashpot, on a carburetor. The throttle-return check contains a spring-loaded diaphragm that traps air behind it when the throttle is opened. Then, when the throttle is released, the return check slows throttle movement so that the throttle closes slowly. This action guards against sudden throttle closing, which might cause the engine to stall.

The electric kickdown switch is used on some cars with automatic transmissions. It provides an electrical means of downshifting the transmission into a lower gear when the throttle is opened wide (under certain conditions).

The use of governors is largely confined to heavy-duty vehicles. A governor will prevent overspeeding and rapid engine wear. In one type, the throttle valve tends to close as rated speed is reached. In another type, a throttle plate is located between the carburetor throttle valve and the intake manifold. The throttle plate moves toward the

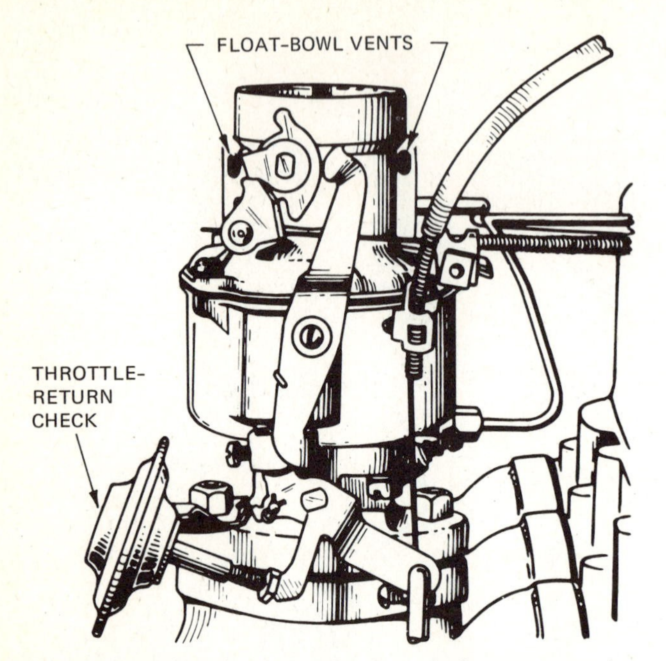

Fig. 14-41. Throttle-return check, or dashpot, on a carburetor. (*Chevrolet Motor Division of General Motors Corporation*)

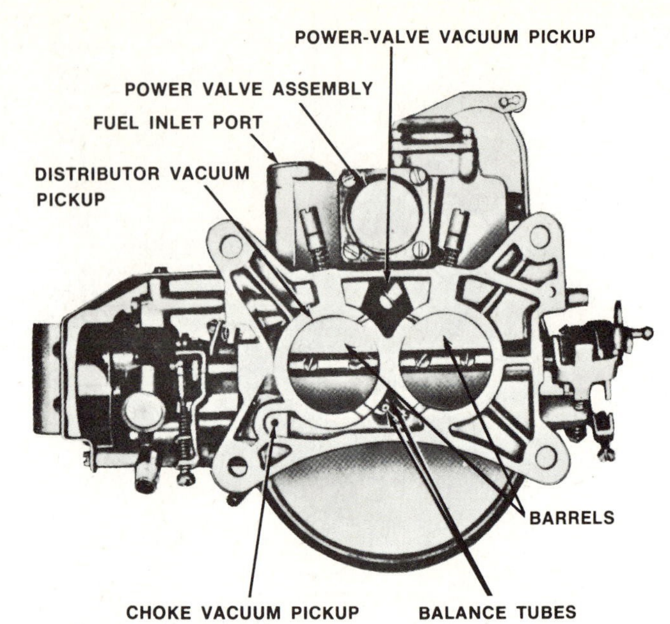

Fig. 14-42. Dual carburetor, showing the location of the two throttle valves.

closed position as rated speed is reached. This action prevents delivery of additional amounts of air-fuel mixture and any further increase in engine speed.

⊘ 14-30 Two- and Four-Barrel Carburetors Carburetors with more than a single barrel are used on many engines. Thus, many carburetors have two barrels and others have four barrels. The purpose of the additional barrels is to improve engine "breathing," particularly at high speeds. That is, the extra barrels permit more air and fuel to enter the engine. Of course, if air were the only consideration, then a single large-diameter barrel could be used. But with only a single large barrel, venturi action would be poor. Proper air-fuel ratios would be hard to achieve under varying operating conditions.

1. *TWO-BARREL CARBURETOR* The two-barrel carburetor is essentially two single-barrel carburetors in a single assembly (Fig. 14-42). Each barrel handles the air-fuel requirements of half the engine cylinders. For example, Fig. 7-24 shows the air-fuel delivery pattern (indicated by arrows) in a V-8 engine. One carburetor barrel supplies cylinders 2, 3, 5, and 8. The other barrel supplies cylinders 1, 4, 6, and 7. Each barrel has a complete set of fuel systems. The throttle valves are fastened to a single throttle shaft, so that both valves open and close together.
2. *FOUR-BARREL CARBURETOR* The four-barrel carburetor is essentially 2 two-barrel carburetors combined in a single assembly (Figs. 14-11 and 14-43). Since this carburetor has four barrels and four main nozzles, it is often called a *Quadrajet,* or *quad, carburetor.* One pair of barrels makes up the primary side, the other pair the secondary side (Fig. 14-43). Under most operating conditions, the primary side alone takes care of engine requirements. But, when the throttle is moved toward the wide-open position for acceleration or full-power operation, the secondary side comes into operation. The secondary side supplies additional amounts of air-fuel mixture. Thus, engine breathing is improved; that is, the engine receives more air-fuel mixture. Volumetric efficiency (⊘ 11-7) is higher, and the engine produces more power.

Fig. 14-43. Four-barrel, or quad, carburetor, showing the location of the four throttle valves. The small throttle valves are on the primary side. (*Carter Carburetor Division of ACF Industries, Inc.*)

There are two ways of controlling secondary-barrel action: by mechanical linkage from the primary-throttle shaft or by a vacuum device. Figure 14-44 is a sectional view of a carburetor using mechanical linkage. During part throttle, only the primary throttle valves are open. However, whenever

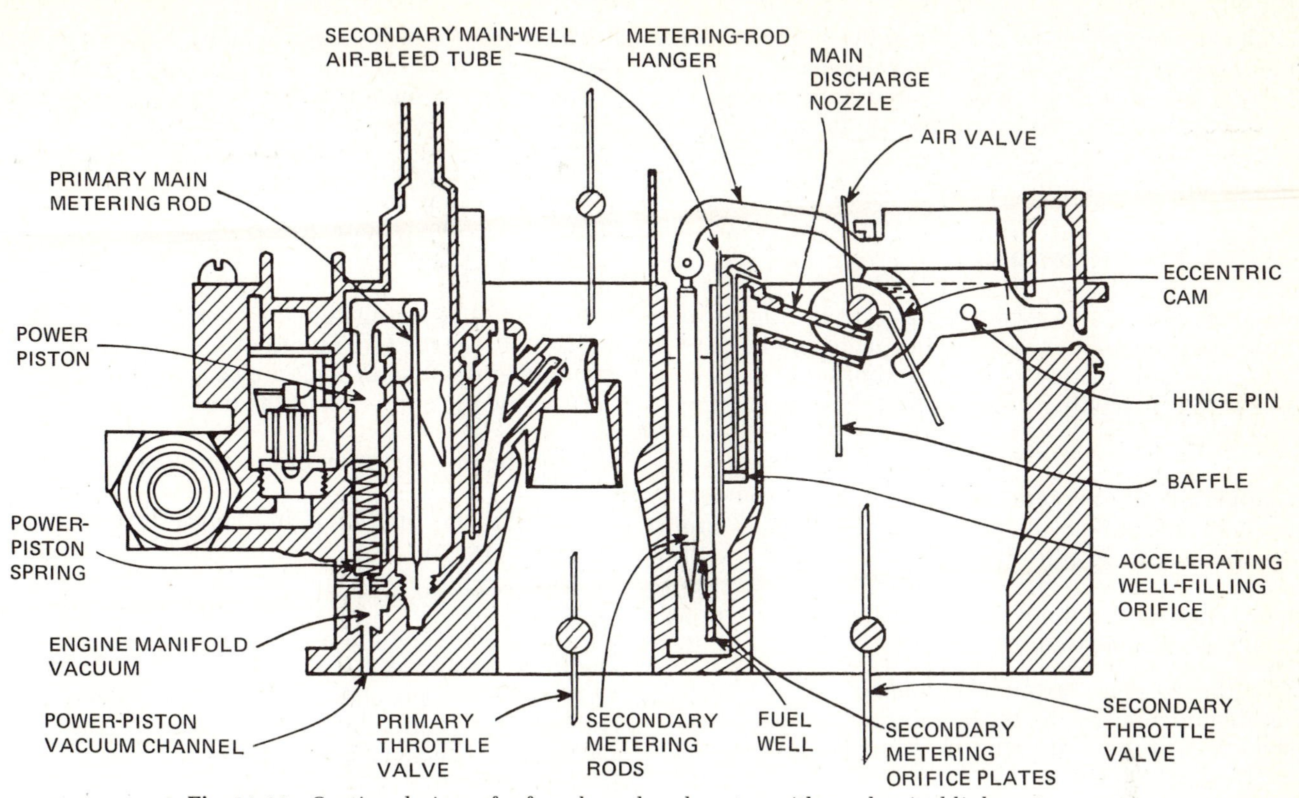

Fig. 14-44. Sectional view of a four-barrel carburetor with mechanical linkage to control the secondary throttle valves. (*General Motors Corporation*)

the throttle is opened wide for additional power, the linkage between the primary and secondary throttle valve causes the secondary throttle valves to open. The secondary barrels now supply air-fuel mixture for full-power engine operation.

In a carburetor using a vacuum device to control secondary-barrel action, there is a vacuum-operated diaphragm. The vacuum is picked up from one of the primary-barrel venturis. As air speed through the primary barrels increases, so does this vacuum. When the vacuum reaches a predetermined amount, indicating a rather high engine rpm, the vacuum actuates the diaphragm. This opens the secondary throttle valves so that the secondary barrels begin to supply air-fuel mixture.

⊘ 14-31 Multiple Carburetors To achieve still better engine breathing, some high-performance engines are equipped with more than one carburetor. The additional carburetors supply more air and fuel and thus improve high-speed, full-power engine performance. Two carburetors mounted on an engine are called *dual carburetors*. Three carburetors mounted on an engine are called *triple carburetors* or sometimes *tripower*. Figure 14-45 shows a three-carburetor installation where the three carburetors are linked together to the accelerator pedal.

For ultimate power production, each engine cylinder may be equipped with its own carburetor. Many racing and drag-strip or hot-rod cars are equipped in this manner: An eight-cylinder engine would have eight carburetors. One step beyond

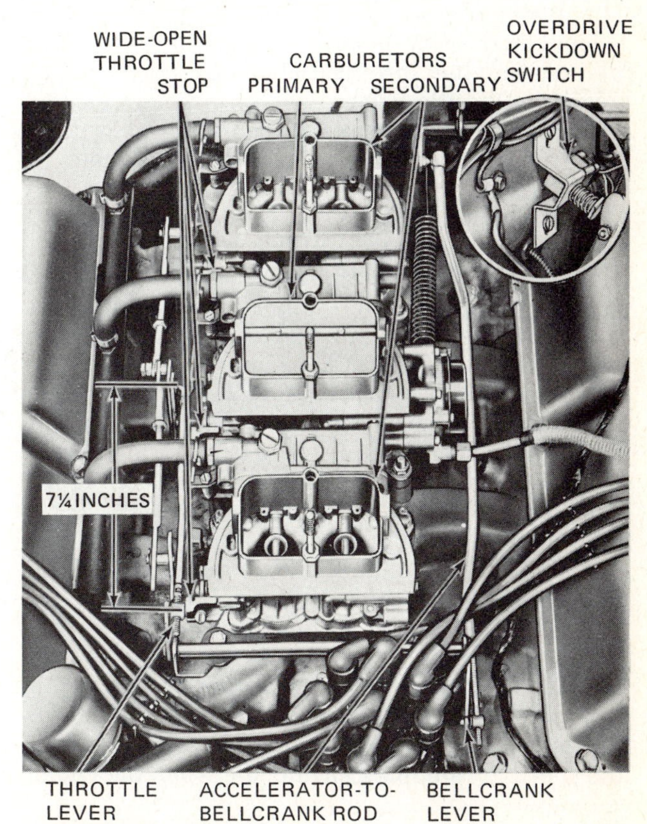

Fig. 14-45. Three-carburetor installation on a V-8 engine. (*Ford Motor Company*)

equipping each cylinder with a carburetor would be to use a fuel-injection system (see ⊘ 13-19 and 13-20). By this method fuel would feed directly to each cylinder by means of fuel nozzles in the intake manifold just opposite the intake valves. In the fuel-injection system positive hydraulic pressure forces the fuel through the nozzles. In the carburetor system, atmospheric pressure (considerably lower pressure) is the driving force.

Check Your Progress

Progress Quiz 14-2 Again, here is your chance to test your memory on how well you remember the essential facts. If any of the questions stump you, reread the preceding pages to find the information you need.

Completing the Sentences The sentences that follow are incomplete. After each sentence there are several words or phrases, only one of which will correctly complete the sentence. Write each sentence in your notebook, selecting the proper word or phrase to complete it correctly.

1. If the accelerator-pump system is not working, sudden opening of the throttle can cause: (*a*) excessive acceleration, (*b*) engine stumble, (*c*) engine overheating.
2. Most accelerator-pump systems use either a: (*a*) pump or piston, (*b*) plunger or diaphragm, (*c*) throttle valve or plunger.
3. The choke valve is: (*a*) above the throttle valve, (*b*) between the throttle valve and manifold, (*c*) below the throttle valve.
4. When the choke valve is closed, gasoline is delivered from the: (*a*) main nozzle, (*b*) idle system, (*c*) choke nozzle.
5. All automatic chokes use a: (*a*) vacuum piston, (*b*) thermostatic diaphragm, (*c*) thermostatic spring.
6. Two devices that add heat to the air-fuel mixture entering the engine are the heat-control valve and the: (*a*) choke, (*b*) air cleaner, (*c*) thermostatically controlled air-cleaner system.
7. Automatic chokes use either a: (*a*) vacuum piston or thermostatic spring, (*b*) vacuum piston or diaphragm, (*c*) thermostatic spring or electric heater.
8. The fast-idle cam is linked to the: (*a*) accelerator pedal, (*b*) throttle valve, (*c*) choke valve.
9. In the four barrel carburetor, the secondary barrels: (*a*) come into operation only during low speed, (*b*) come into operation only for acceleration or full power, (*c*) back up the primary barrels by assisting them at all times.
10. The idle-stop solenoid prevents dieseling by: (*a*) completely closing the throttle plate when the ignition is turned off, (*b*) preventing excessively high idle speed, (*c*) stopping engine idle when the engine gets hot.

CHAPTER 14 CHECKUP

NOTE: Since this is a chapter review test, you should review the chapter before taking the test.

Review Questions In this checkup, we ask you for definitions and explanations of carburetor parts and actions. Write your answers to the questions that follow in your notebook.

1. What is carburetion?
2. What is vaporization?
3. What is atomization?
4. What are the three basic parts of the simple carburetor?
5. Describe the venturi effect.
6. What drives the fuel up out of the fuel nozzle and into the air passing through the carburetor air horn?
7. Describe the operation of the carburetor float system.
8. Describe the operation of the carburetor idle system.
9. Describe the operation of the carburetor main metering system.
10. Describe the mechanically operated power system. Describe the vacuum-operated power system.
11. Why is the float bowl vented into the carburetor air horn just above the choke?
12. What is the purpose of the float-bowl vent that is connected to the charcoal canister?
13. What is the purpose of the compensator valve in the carburetor?
14. Why is it that the idle-adjustment screw is fixed or has a locking cap on it?
15. Why is an accelerator-pump system required?
16. How does the accelerator-pump system work?
17. What is the purpose of the choke? How does the automatic choke work?
18. What is the purpose of the electric choke? How does it work?
19. What is the purpose of the manifold heat-control valve?
20. What is the purpose of using two- and four-barrel carburetors?

SUGGESTIONS FOR FURTHER STUDY

Examine as many carburetors as you can. When you examine a carburetor, take a separate sheet of paper and write the facts about the carburetor. At the top of the paper write the make and model of car from which the carburetor came, together with the model and type of carburetor. Locate the venturi, main nozzle, throttle plate, and so on. Identify all the openings in the air horn. Doing all this will familiarize you with the carburetor systems.

CAUTION: Be very careful not to spill gasoline when you are examining carburetors which contain gasoline in their float bowls. Remember that gasoline is very explosive, and a spark can set off a blaze or explosion.

Study the carburetor sections in the manufacturers' shop manuals. Make sketches in your notebook of any special features you find.

chapter 15

IGNITION SYSTEMS

This chapter describes the construction and operation of automotive ignition systems and their component parts. The typical ignition system includes the battery, ignition coil, ignition distributor, ignition switch, spark plugs, and wiring (Fig. 15-1).

⊘ 15-1 Function of the Ignition System The ignition system supplies high-voltage surges of current (as high as 35,000 volts) to the spark plugs in the engine cylinders. These surges produce electric sparks at the spark-plug gaps. The sparks ignite, or set fire to, the compressed air-fuel mixture in the combustion chambers. Each spark is timed to appear at the spark-plug gap just as the piston approaches top dead center (TDC) on the compression stroke, when the engine is idling. At higher speeds or during part-throttle operation, the spark is advanced; that is, it occurs somewhat earlier in the cycle. The air-fuel mixture thus has ample time to burn and deliver its power.

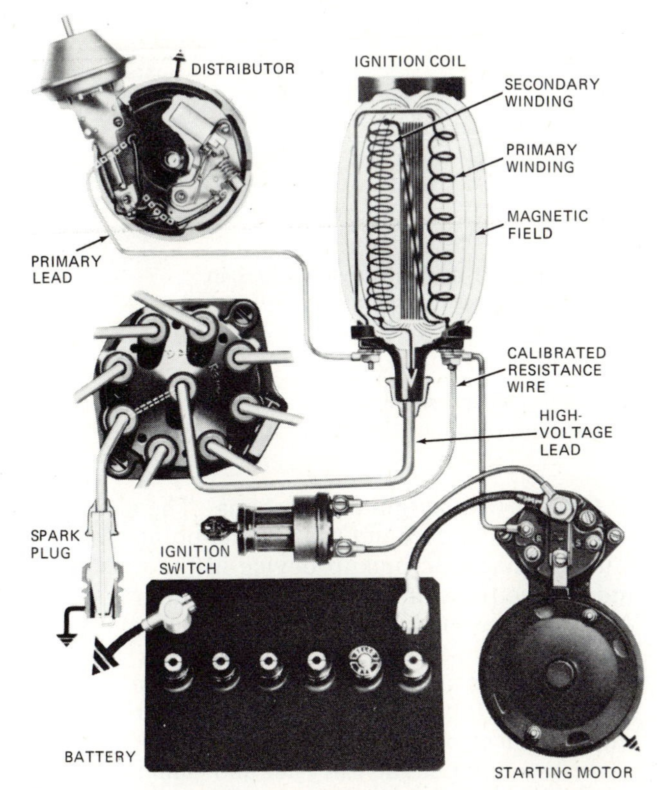

Fig. 15-1. Typical ignition system. It consists of the battery (source of power), ignition switch, ignition coil (shown schematically), distributor (shown in top view with its cap removed and placed below it), spark plugs (one shown in sectional view), and wiring. The coil is shown schematically, with magnetic lines of force indicated. *(Delco-Remy Division of General Motors Corporation)*

⊘ 15-2 Ignition Distributor The ignition distributor has two jobs. First, it closes and opens the circuit between the battery and the ignition coil. When the circuit closes, current flows in the ignition coil and builds up a magnetic field. When the circuit opens, the magnetic field in the coil collapses. The coil produces a high-voltage surge of current. (How the coil does this is explained later.) The distributor's second job is to distribute each high-voltage surge to the correct spark plug at the correct instant. It does this with the distributor rotor and cap and secondary wiring.

There are two basic types of distributors: the type using contact points to close and open the ignition-coil primary circuit and the type using a magnetic pickup device and a transistor-control unit to interrupt the current flow of the ignition-coil primary circuit. This second type of distributor is used in electronic ignition systems.

⊘ 15-3 Distributor with Contact Points This distributor (Figs. 15-2 and 15-3) consists of a housing, a driveshaft with breaker cam, an advance mechanism, a breaker plate with contact points and a condenser, a rotor, and a cap. The shaft is usually driven by the engine camshaft through spiral gears (Fig. 9-4). The shaft rotates at one-half crankshaft speed. Usually, the distributor driveshaft is coupled with a shaft that drives the oil pump.

Rotation of the shaft and breaker cam causes

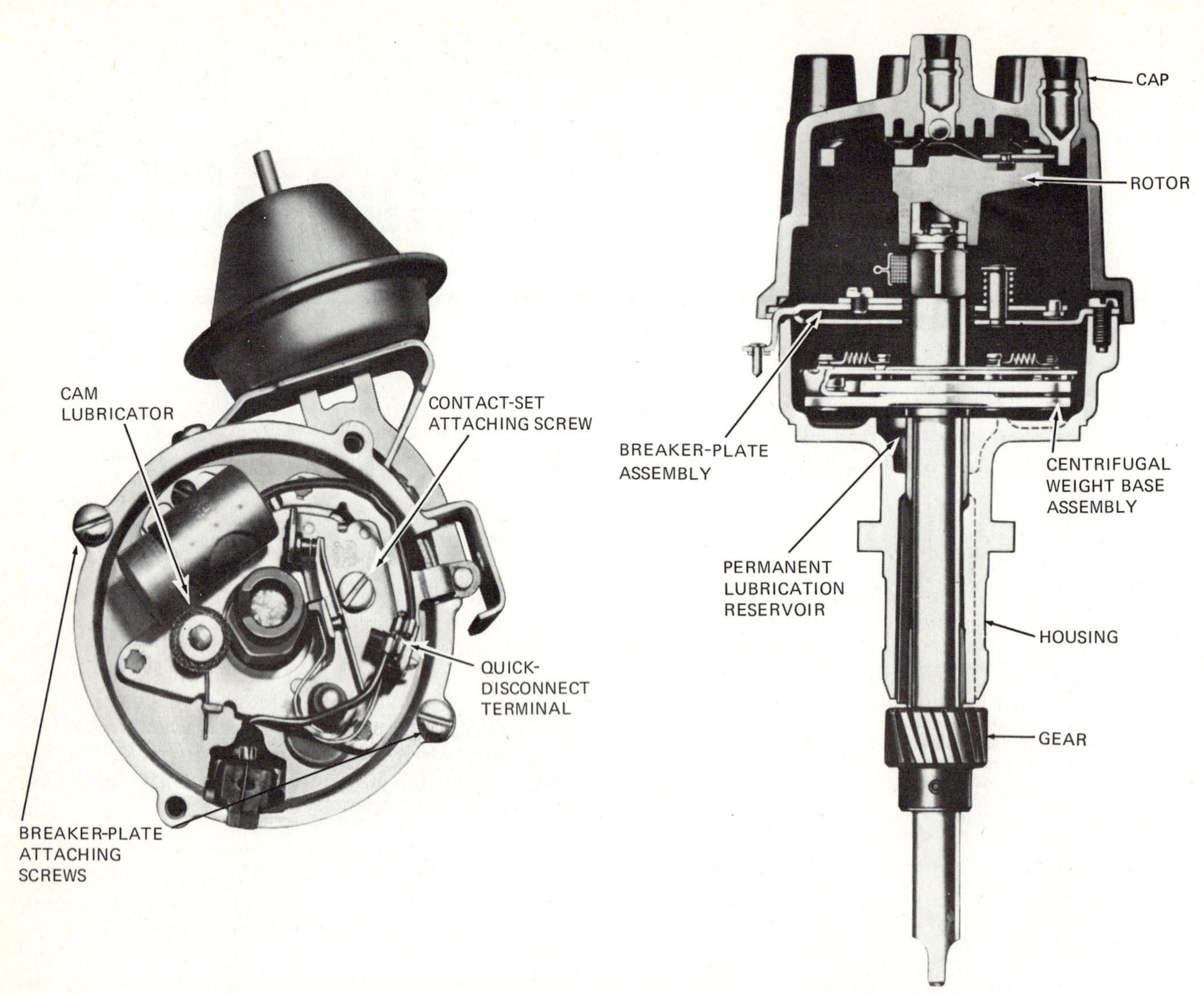

Fig. 15-2. Sectional and top views of an ignition distributor. In the top view, the cap and the rotor have been removed so that the breaker plate can be seen. (*Delco-Remy Division of General Motors Corporation*)

the distributor contact points to close and open. The breaker cam usually has the same number of lobes as there are cylinders in the engine.[1] It rotates at half crankshaft speed. The contact points close and open once for each cylinder with every breaker-cam rotation. Thus one high-voltage surge is produced by the ignition coil for each cylinder every two crankshaft revolutions. This surge ignites the air-fuel mixture compressed in each cylinder every other crankshaft revolution.

The rotor rotates with the breaker cam on which it is mounted. As it does so, a metal spring and segment or blade on the rotor connect the center terminal of the cap with each outside terminal in turn. Thus the high-voltage surges from the coil are directed first to one spark plug, then to another, and so on, according to the firing order.

[1] In some designs, the breaker cam has only one-half as many lobes as engine cylinders and there are two sets of contact points that are arranged to close and open alternately. This produces the same effect as the breaker-cam and single-contact-point-set arrangement.

⊘ 15-4 Electronic Ignition Systems The electronic ignition system does not use contact points. Instead, it uses a magnetic pickup device in the distributor and an electronic amplifying device with transistors. With the caps on, the electronic distributor looks the same as the contact-point distributor. However, when the caps are removed, the difference between the two distributors is apparent. There are various kinds of electronic ignition systems. We shall describe two of them: the Chrysler and General Motors systems.

⊘ 15-5 Chrysler Electronic Ignition System An electronic ignition system has been used in all Chrysler Corporation cars made in the United States since 1973. In this system, the distributor has a metal rotor with a series of tips on it. This rotor, called

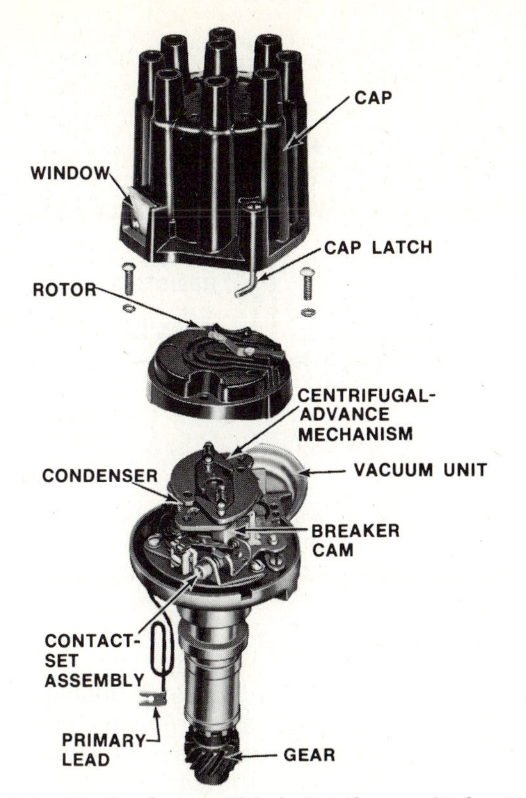

Fig. 15-3. Partly disassembled distributor. (*Delco-Remy Division of General Motors Corporation*)

the *reluctor,* is shown in Fig. 15-4. The reluctor takes the place of the breaker cam in the contact-point distributor. Notice that the reluctor in Fig. 15-4 has eight tips. This reluctor is for an eight-cylinder engine, and so there is one tip for each cylinder. Notice also that the distributor has a permanent magnet and pickup coil.

The principle of operation of Chrysler's electronic ignition system is simple. The reluctor provides a path for the magnetic lines of force from the magnet. Thus, every time a tip of the reluctor passes the pickup coil, it carries a magnetic field through the coil. This magnetic field produces a pulse of electric current in the coil. The current is very small, but it is enough to trigger the control unit into action.

The control unit uses electronic devices—diodes and transistors—to control the flow of current to the ignition coil. When the pulse of current from the pickup coil arrives at the control unit, the control unit stops the flow of current to the ignition coil. (This is the same job that the contact points do in the contact-point distributor.) When the current stops flowing in the ignition coil, the magnetic field in the coil collapses. This causes the coil to produce a high-voltage surge. This high-voltage surge is led through the distributor rotor, cap, and wiring, to the spark plug that is ready to fire.

The tip of the reluctor now rotates past the pickup coil. The pulse of current from the pickup coil ends, allowing the control unit to close the circuit from the battery to the ignition coil. Primary current flows again, and a magnetic field builds up once more in the ignition coil. Then the next tip of the reluctor passes the pickup coil, and the whole series of events is repeated.

Figure 15-5 shows the wiring circuit for the Chrysler electronic ignition system. The dual ballast is a double-resistor unit that protects the system from overload but allows a maximum current to flow during cranking. The dual ballast ensures a strong spark for good starting performance. In this system there are no contact points to adjust or wear out. Everything is automatic. The only adjustment required is the ignition timing, which we shall discuss later.

⊘ 15-6 General Motors Electronic Ignition System General Motors calls the distributor used in their electronic ignition system a *magnetic pulse distributor;* it is shown in Fig. 15-6. The General Motors distributor looks much like the Chrysler unit, and it works in about the same way. The magnetic pulse distributor has a pole piece in the form of a ring. This pole piece has a series of teeth, pointing inward. There is one tooth for each cylinder in the engine. Under the pole piece is a permanent magnet with a pickup coil. The timer core, made of iron, is placed on top of the distributor shaft. (It is placed in exactly the same way as the cam in the contact-point distributor.) The timer core also has one tooth for each cylinder in the engine.

When the engine is running, the teeth on the timer core repeatedly align with the teeth on the pole piece. They align the same number of times per timer-core rotation as there are cylinders in the engine. In an eight-cylinder engine, there would be eight teeth on the pole piece and eight teeth on the timer core. The teeth would align eight times for every revolution of the timer core. Every time the teeth align, magnetic lines of force are carried through the pickup coil. This action produces a pulse of current that flows to the ignition-pulse am-

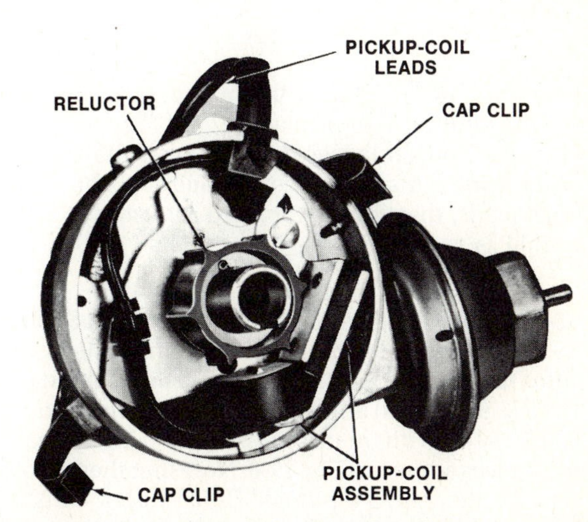

Fig. 15-4. Top view of the Chrysler electronic ignition distributor. The cap and rotor have been removed to show the reluctor and the pickup coil. (*Chrysler Corporation*)

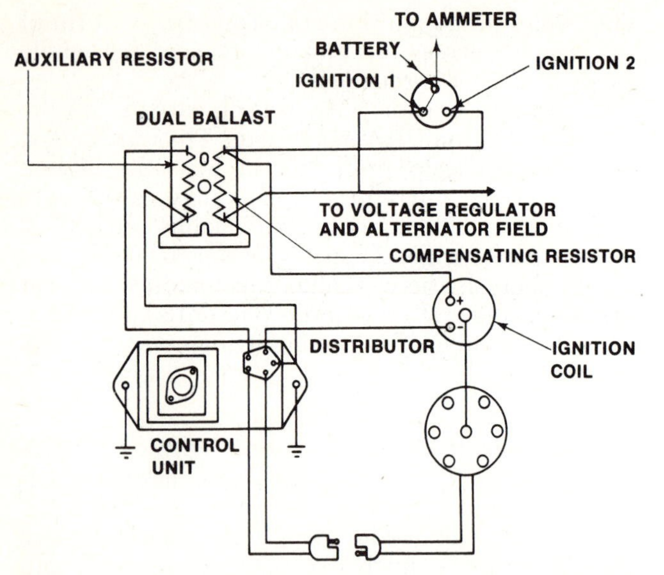

Fig. 15-5. Schematic wiring diagram for the Chrysler electronic ignition system. (*Chrysler Corporation*)

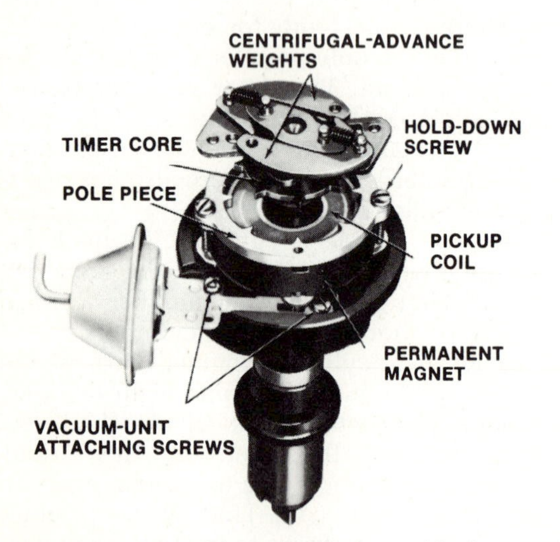

Fig. 15-6. Magnetic pickup distributor with the cap and rotor removed. (*Delco-Remy Division of General Motors Corporation*)

plifier (Fig. 15-7). There, the pulse electronically opens the ignition-coil primary circuit. The magnetic field in the coil collapses, and a high-voltage surge is produced. This surge is carried by the high-voltage leads, distributor cap, and rotor to the spark plug that is ready to fire.

⊘ 15-7 General Motors High-Energy Ignition System

In 1973 General Motors introduced an ignition distributor that has the ignition coil assembled into it (for V-6 and V-8 engines) and that produces voltages of up to 35,000 volts. With this assembly the wiring is greatly simplified, as shown in Fig. 15-8. Note that there is one lead from the battery (which Delco-Remy calls an *energizer*). This lead goes through the ignition switch to the electronic distributor. No primary resistance is used. The only other leads are the high-voltage cables

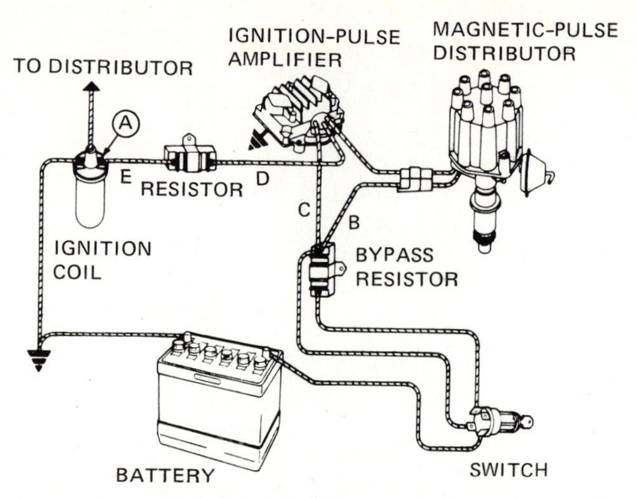

Fig. 15-7. Wiring diagram for an ignition system using a magnetic-pulse distributor and a transistorized control unit (ignition pulse amplifier) to amplify the ignition pulse. (*Delco-Remy Division of General Motors Corporation*)

going to the spark plugs. Because of the higher voltage, special silicone insulation spark-plug wires are used. These wires have a larger diameter [8 millimeters] than standard spark-plug wires. They are gray in color, have more heat resistance, and less deterioration. However, the silicone insulation is soft and must not be mishandled or allowed to rub against other parts. Special wiring-harness connectors are used to hold the spark-plug wires securely in place on top of the distributor cap.

Figure 15-9 shows the distributor assembled. It looks different from the ignition distributor in older cars, but it does the same job. All connections between the coil and distributor are inside the HEI distributor. Also, the electronic amplifier, which Delco-Remy calls the *electronic module,* is mounted inside the distributor. Figure 15-10 shows the HEI distributor with cap, rotor, and electronic module removed. The capacitor (condenser) in the distributor is for control of radio noise. It has no function in the High-Energy Ignition System. As you can see, the wiring in the High-Energy Ignition System is much simplified. The distributor uses the magnetic-pulse principle. This is explained in ⊘ 15-6.

Figures 15-11 and 15-12 show the HEI distributor in assembled and partly disassembled views. Tachometer connections to the HEI distributor are shown in Fig. 15-11.

For a short time prior to introduction of the HEI System, General Motors installed a unit-distributor system on some cars. Sort of an early version of HEI, the unit distributor and HEI distributor share many features. However, the unit distributor did not have the higher secondary voltage characteristic of the High-Energy Ignition System.

⊘ 15-8 Spark Plugs The spark plug (Fig. 15-13) includes a metal shell in which a porcelain insulator is fastened. An electrode extends through the center

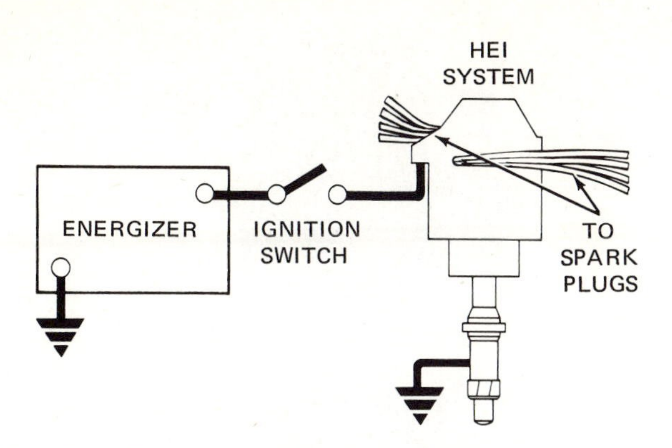

Fig. 15-8. Basic wiring diagram of the General Motors High-Energy Ignition. (*Delco-Remy Division of General Motors Corporation*)

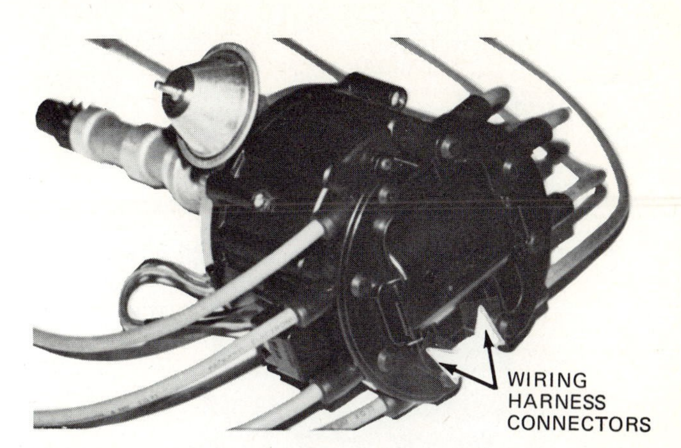

Fig. 15-9. HEI distributor, which includes the ignition coil. Wiring-harness connectors hold the silicone spark-plug wires in place. (*Delco-Remy Division of General Motors Corporation*)

of the insulator. A second electrode is attached to one side of the shell. This electrode is bent in toward the center electrode. Threads on the metal shell allow it to be screwed into a tapped hole in the cylinder head. This arrangement grounds the electrode that is attached to the shell. The two electrodes are made of special heavy wire. There is a gap of up to 0.040 inch [1.02 mm] between them.[2] The electric spark jumps this gap to ignite the air-fuel mixture in the combustion chamber. The spark jumps from the center, or insulated, electrode to the grounded, or outer, electrode. Some spark plugs have a built-in resistor (Fig. 15-13) which is part of the center electrode. This resistor reduces radio and television interference from the ignition system. It also reduces electrode erosion caused by overlong sparking.

NOTE: The High-Energy Ignition System uses spark plugs with a wider gap—as much as 0.080 inch [2.03 mm]. A standard plug cannot be used with this system because, to get the wide gap, the side electrode would have to be bent at a severe angle. Instead, special plugs made for the high-voltage system must be used.

We have been discussing the high-voltage surge from the ignition-coil secondary circuit as if it were a single powerful surge. Actually, the action is more complex. There may be a number of early surges before a full-fledged spark forms. At the end of the sparking cycle, the spark may be quenched and reformed several times. All this takes place in only a few ten-thousandths of a second. The effect is that the ignition wiring acts like a radio transmitting antenna. The surges of high voltage send out static that causes radio and television interference. However, the resistors in the spark plugs tend to concentrate the surges in each sparking cycle. They reduce the number of surges and thus the interference and its erosive effect on the electrodes.

[2] But the special spark plugs required for the high-energy ignition system shown in Figs. 15-11 and 15-12 have a gap of as much as 0.080 inch [2.032 mm].

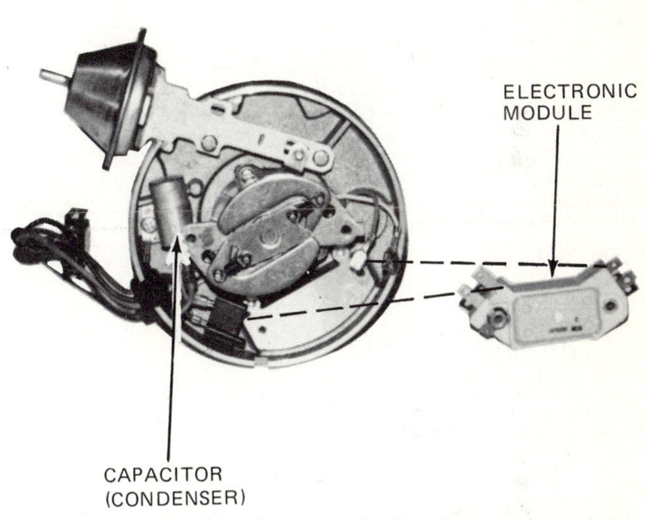

Fig. 15-10. HEI distributor with the electronic module removed. (*Delco-Remy Division of General Motors Corporation*)

⊘ **15-9 Spark-Plug Heat Range** The *heat range* of a spark plug tells how hot the spark plug becomes in operation (Fig. 15-14). The temperature that a spark plug reaches depends on how far the heat must travel. The heat path is from the center electrode to the cooler outer shell of the spark plug and then to the cylinder head. If the heat path is long, the spark plug will run hotter than if the path is short.

When a spark plug runs too cold, sooty carbon will deposit on the insulator around the center electrode. A hotter-running spark plug burns this carbon away or prevents its formation. (Carbon deposits can also be caused by too rich air-fuel mixtures or by too much oil getting into the cylinder.) When a spark plug runs too hot, the insulator may take on a white or grayish cast and appear blistered. A spark plug that runs too hot will wear more rapidly because

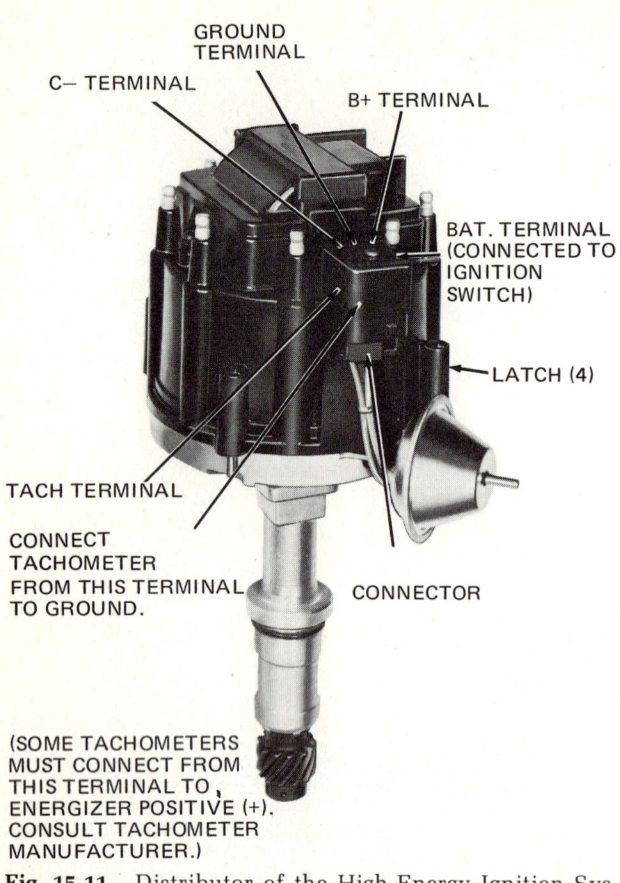

Fig. 15-11. Distributor of the High-Energy Ignition System, which includes the ignition coil. (*Delco-Remy Division of General Motors Corporation*)

the higher temperatures cause the electrodes to burn away more quickly. In addition, with a hot-running plug, there is danger of preignition (see ⊘ 12-9).

⊘ 15-10 Secondary Wiring The secondary wiring consists of the high-voltage cables connecting the distributor cap, spark plugs, and high-voltage terminal of the ignition coil. These cables carry the high-voltage surges that produce sparks at the spark-plug gaps. Therefore, the cables must be heavily insulated to contain the high voltage. The insulation must be able to withstand the effects of high temperature, oil, and high voltage.

Before 1961, the cores of these cables were made of copper or aluminum wire. However, in 1961, automotive manufacturers in the United States began to use carbon-impregnated linen cores. Carbon-impregnated linen forms a resistance path for the high-voltage surges. It produces the same effect as the resistors in the spark plugs, mentioned previously. These cables thus prevent the ignition system from interfering with radio and television.

In 1963, many automotive manufacturers began using cables with graphite-saturated fiber-glass cores. These cables work like the carbon-impregnated linen-core cables. However, it is claimed they resist breakage when pulled off spark plugs. Also, they are less apt to char from high temperatures.

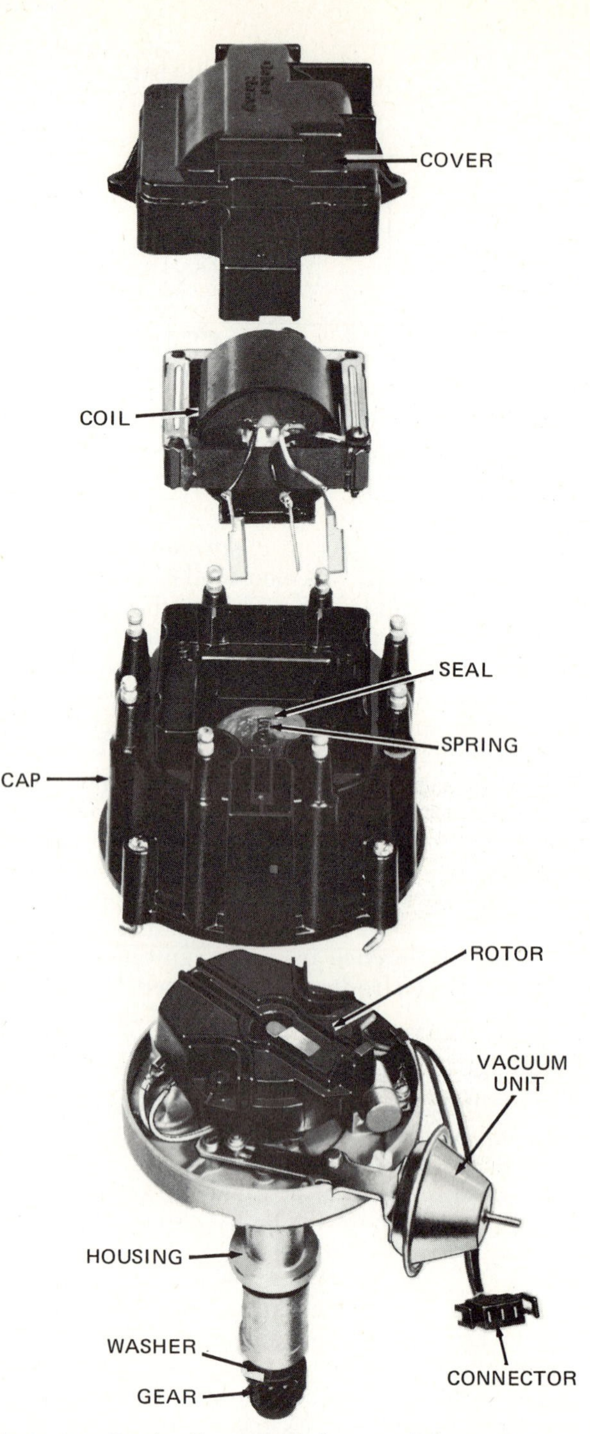

Fig. 15-12. Partly disassembled view of the High-Energy Ignition System distributor. (*Delco-Remy Division of General Motors Corporation*)

⊘ 15-11 Ignition Coil The ignition coil receives voltage from the battery. It *transforms,* or steps up, this voltage to the high voltage required to make the current jump the spark-plug gap. A high voltage is required because the air-fuel mixture between the two electrodes presents a high resistance to the passage of current. The voltage (pressure) must go very high in order to push current (electrons) from the center electrode to the outside electrode.

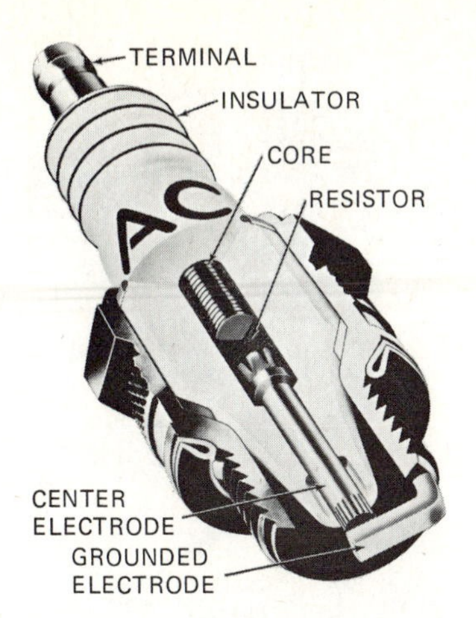

Fig. 15-13. Cutaway view of a resistor-type spark plug. (*AC Spark Plug Division of General Motors Corporation*)

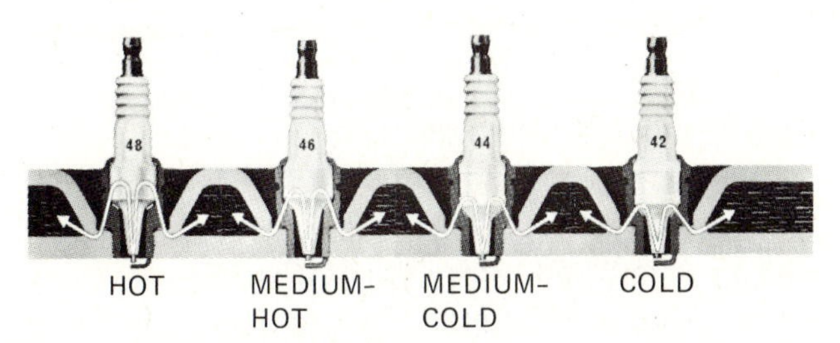

Fig. 15-14. Heat range of spark plugs. The longer the heat path (indicated by arrows), the hotter the plug runs. (*AC Spark Plug Division of General Motors Corporation*)

The ignition coil has two circuits running through it: a primary circuit and a secondary circuit (Fig. 15-1). The primary circuit is made up of a winding a few hundred turns of relatively heavy wire. The secondary circuit is made up of a winding of many thousands of turns of a fine wire. The wire of the primary circuit is wrapped, or wound, around the outside of the secondary winding, as shown in Fig. 15-15. When the distributor contact points close and current flows in the primary circuit, a magnetic field builds up. When the distributor contact points open and current stops flowing, the magnetic field collapses. The collapsing magnetic field induces high voltage in the secondary winding. This collapse produces the high-voltage surge that is conducted through the distributor rotor and cap to a spark plug.

⊘ 15-12 Primary and Secondary Circuits In order to get a clearer picture of the two interrelated circuits in the ignition system, the primary and secondary circuits, let us look at each one separately. Figure 15-16 shows the primary circuit. It consists of the battery, contact points in the distributor, primary winding in the ignition coil, ignition switch, and wiring.

Figure 15-17 is the same illustration with the secondary circuit added. The secondary circuit consists of the secondary winding in the ignition coil, distributor cap and rotor, spark plugs, and connect-

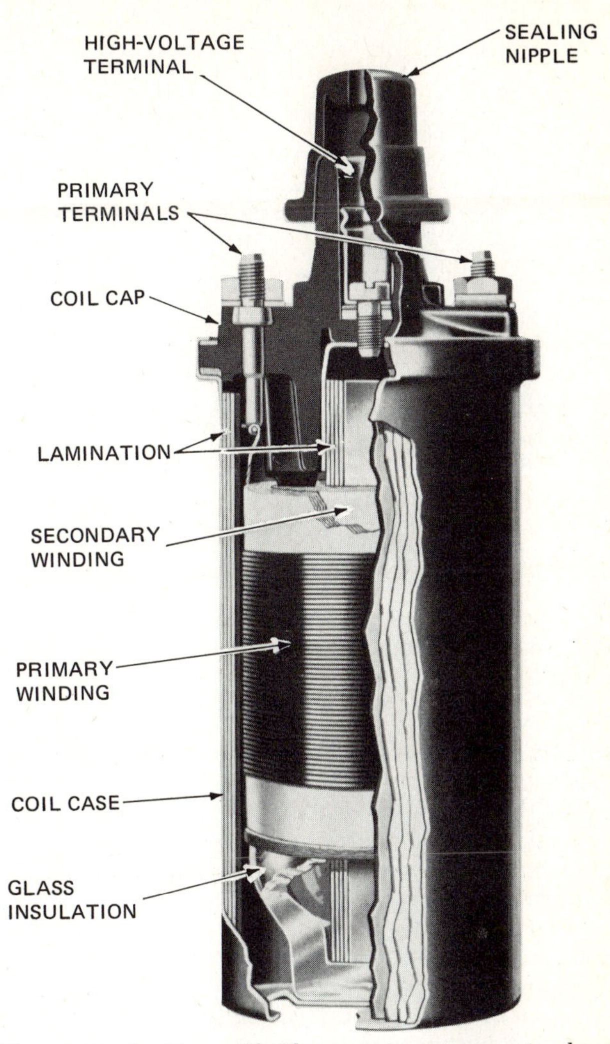

Fig. 15-15. Ignition coil. The case is cut away to show how the primary winding is wound around the outside of the secondary winding. (*Delco-Remy Division of General Motors Corporation*)

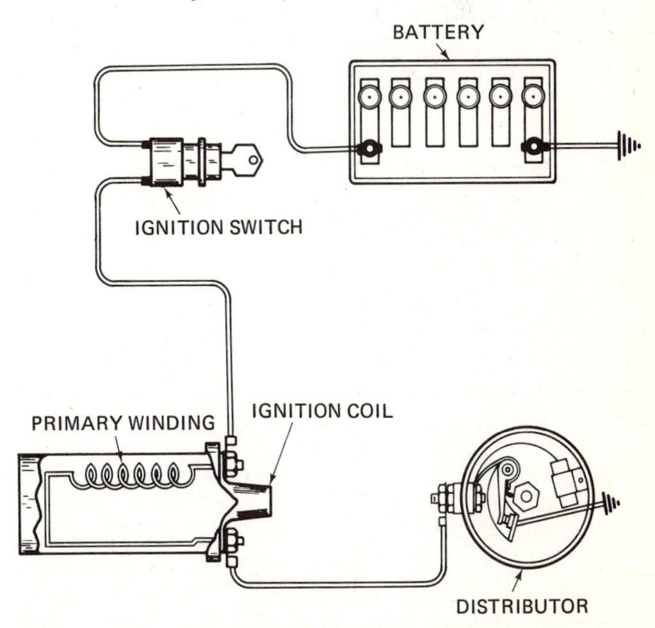

Fig. 15-16. Simplified primary circuit of the ignition system.

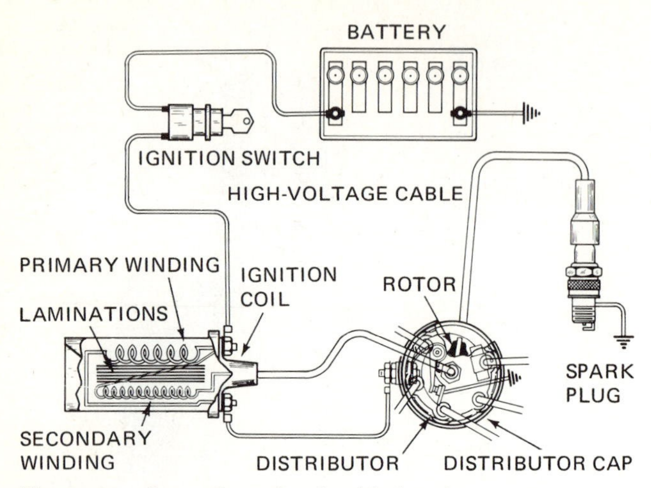

Fig. 15-17. Secondary circuit added to the primary circuit of the ignition system. Only one spark plug is shown.

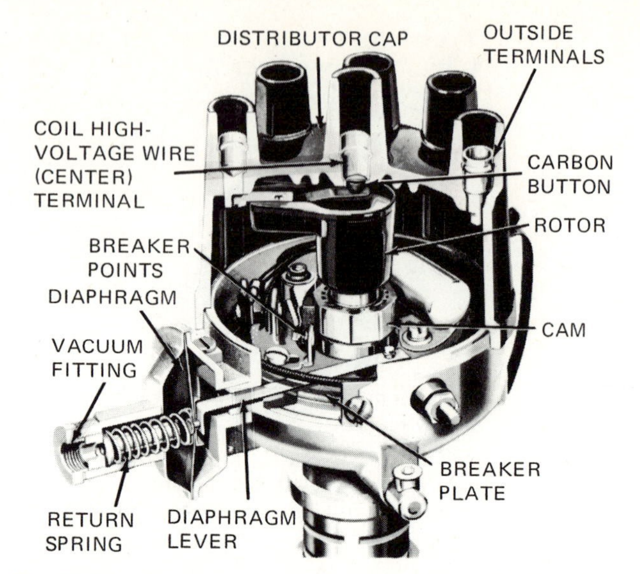

Fig. 15-19. Cutaway view of a distributor showing how the rotor is mounted on top of the cam. The construction of the vacuum-advance mechanism is also shown. (*Ford Motor Company*)

ing wires. Notice that the primary and secondary windings are connected to the same primary terminal of the coil.

⊘ 15-13 Distributor-Cap-and-Rotor Action As you can see from Figs. 15-2 and 15-3, the rotor sits on top of the cam in the distributor. Figure 15-18 shows several rotors. The purpose of the rotor is to connect the center terminal of the distributor cap to the outside terminals of the cap.

The terminals are insulated from one another and are held in place in the cap. Three terminals are shown in cutaway view in Fig. 15-19. The center terminal of the cap has a carbon button on its lower end. This button rests on one end of the rotor blade. A small spring holds the carbon button and rotor blade in continuous contact. Therefore, the rotor blade is always connected to the secondary winding of the ignition coil. Whenever the ignition-coil secondary winding produces a high-voltage surge, the rotor blade has turned so that it points to the side terminal which is connected to the spark plug that is ready to fire (Fig. 15-17). Figure 15-20 shows how the rotor and cap terminal connect the ignition-coil secondary to a spark plug.

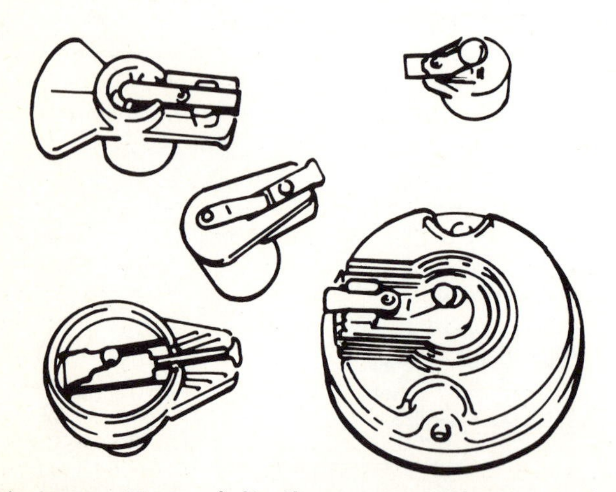

Fig. 15-18. Types of distributor rotors. The one at the lower left has a carbon resistor. The one at the lower right is attached to the advance mechanism by screws.

Here is a review of the actions: The contact points open (or the control unit in the electronic ignition system shuts off the flow of current to the ignition-coil primary winding). The magnetic field created by current flowing in the ignition-coil primary winding collapses. This collapse produces a high-voltage surge in the ignition-coil secondary winding. The high-voltage surge is led from the center terminal of the ignition coil to the center terminal of the distributor cap. From there it goes through the rotor blade and to one side terminal. This side terminal is connected to the spark plug in the cylinder in which the compression stroke is ending. The spark produced at the spark plug by the high-voltage surge ignites the compressed air-fuel mixture. The mixture burns, producing the power stroke.

Now, let us review in more detail how the magnetic field is produced, what happens when it collapses, and how the condenser enters into the action.

⊘ 15-14 Producing the Magnetic Field Current flowing through a winding causes a magnetic field. However, the magnetic field does not spring up instantly when the circuit is closed to the battery. It takes a small fraction of a second, called the *buildup time,* for the magnetic field to develop. The reason for this lies in the fact that the winding has *self-induction.* "Self-induction" expresses the effect of each turn of wire in the winding on adjacent turns.

Figure 15-21 illustrates the magnetic fields surrounding two adjacent turns of wire in the winding (seen in end view) as current flows in them. (The current is flowing away from the reader, as indicated

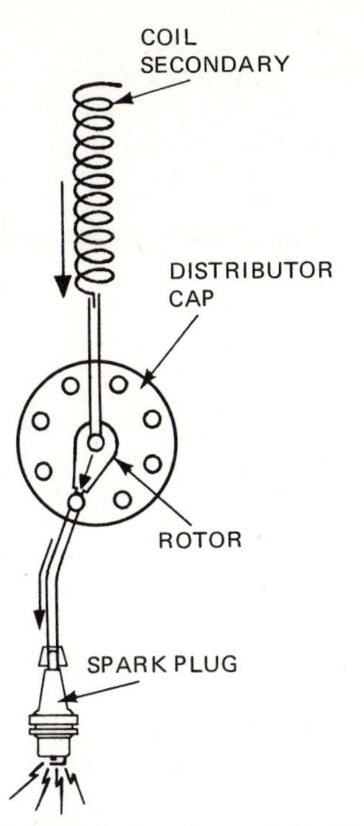

Fig. 15-20. Simplified drawing of ignition secondary showing the high-voltage surge (indicated by arrows) from coil secondary flowing through the rotor, the distributor cap, and the high-voltage lead to the spark plug.

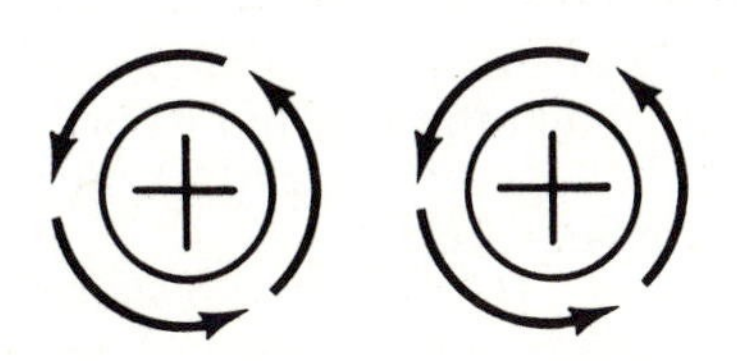

Fig. 15-21. Magnetic fields surrounding two neighboring turns of wire in a winding through which current is passing.

by the crosses.) When the current first starts to flow, the encircling magnetic fields begin to move outward from the wires. This movement is somewhat like the ripples on a pool of water moving out from where a stone has been dropped. Figure 15-22 illustrates the effect of this action on the right-hand wire: The increasing magnetic field cuts across the right-hand wire. It thus attempts to induce in that wire a flow of current in the opposite direction. (This is indicated by the dot, which means the current is flowing toward the reader.)

Actually, the current cannot flow in this direction in the right-hand wire because the battery is already forcing current through the winding and every turn of wire in the opposite direction, as shown in Fig. 15-21. But there is a tendency for a flow of current to be induced in the reverse direction in every turn of wire. The tendency is brought about by the expanding magnetic fields from adjacent turns of wire. The result is that this tendency combats any increase in current flow through the winding. This is self-induction. It takes a fraction of a

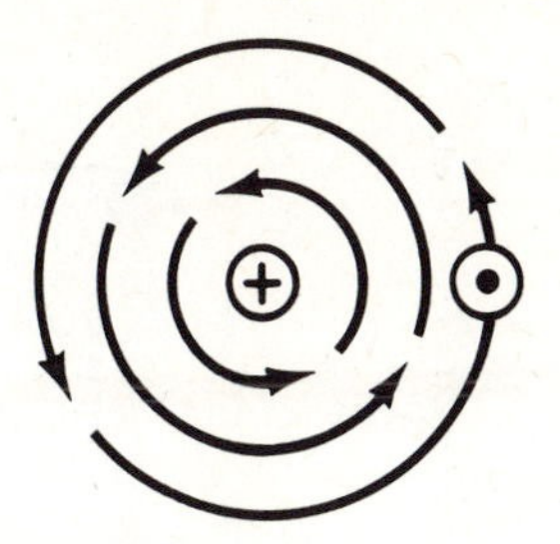

Fig. 15-22. Effect on a neighboring wire of an increasing magnetic field from one wire.

second, the buildup time, for the battery voltage to overcome this tendency and build up the magnetic field in the winding.

⊘ 15-15 Effect on Primary Winding of Collapsing the Magnetic Field When the distributor contact points open (or the control unit in the electronic ignition system is triggered), the current stops flowing. The magnetic field from the primary winding begins to collapse. This means that the magnetic field surrounding each turn of wire begins to collapse back toward the wire. Thus, instead of moving to the right, as in Fig. 15-22, the magnetic field moves to the left. This shift induces a flow of current in the right-hand wire in a direction opposite to that shown. That is, current flows in the direction in which it flowed when the winding was connected to the battery. Such action, again, is self-induction.

⊘ 15-16 Condenser Effect As the distributor contact points open (or the control unit in the electronic ignition system is triggered), the flow of current from the battery through the primary winding of the coil is stopped. Instantly, the magnetic field begins to collapse. This collapse tends to reestablish the flow of current. If it were not for the *condenser,* also called the *capacitor,* the flow of current would be reestablished. That is, a heavy electric arc would take place across the separating contact points. The points would burn, and the energy stored in the ignition coil as magnetism would be consumed by the arc. The condenser prevents this, however, by momentarily providing a place for the current to flow as the points begin to move apart.

The condenser, or capacitor, is made up of two thin metallic plates separated by an insulator. The plates are two long, narrow strips of lead or aluminum foil. They are insulated from each other by special condenser paper and wrapped on an arbor to form a winding. The winding is then installed in a container. A condenser is shown in Fig. 15-23. The two plates provide a large surface area onto which the electrons (flow of current) can move during the first instant the contact points separate. Remember, it is the massing of electrons in one place in a circuit that causes them to move and produce a flow of current. The condenser provides a large surface area. Thus, many electrons can flow onto this large

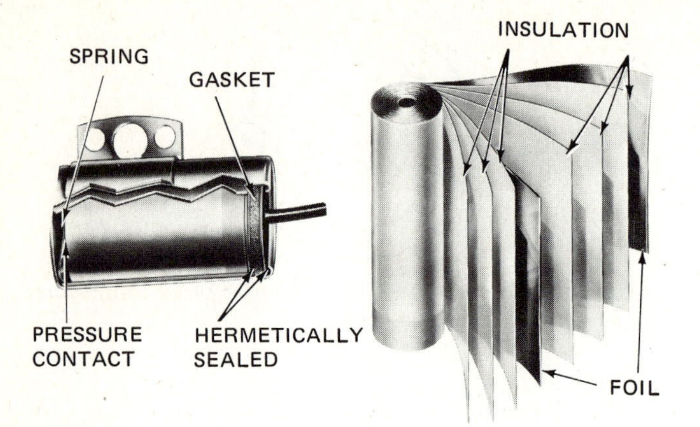

Fig. 15-23. Condenser assembled and with the winding partly unwound.

surface area without producing an excessive massing of electrons in one place.

The number of electrons the condenser can accept is limited, however, because the condenser becomes charged quickly. But, by this time, the contact points are sufficiently far apart to prevent an arc from forming between them. In effect, the condenser acts as a reservoir into which electrons flow during the first instant that the points begin to separate. By the time the reservoir is filled, the points are too far apart for the electrons to jump across them. Consequently the electrons, or current, must stop flowing in the primary circuit. It is the current flow that induces the magnetic field. Thus, the quick stoppage of the current causes the magnetic field to collapse rapidly. This rapid collapse produces a high-voltage surge in the secondary winding of the ignition coil.

⊘ 15-17 Effect on Secondary Winding of Collapsing the Magnetic Field The rapid collapse of the magnetic field causes the magnetic lines of force to move rapidly across the thousands of turns of wire in the secondary wiring. This means that each turn of wire has a voltage induced in it. All turns of wire are connected in series, so that the total voltage induced is the sum of the voltage in all the turns of wire. Thus, the secondary winding will supply a high voltage during the collapse of the magnetic field. One end of the secondary winding is connected through ground (through the cylinder head and block) to the side electrode in the spark plug. The other end of the secondary winding is connected through the cap and rotor of the distributor and through the wiring to the center electrode in the spark plug. This high voltage that is suddenly imposed on the spark plug causes electrons (current) to jump across the spark-plug gap, producing an electric spark. The timing of the spark is controlled by the spark-advance mechanisms located in the distributor (⊘ 15-20 to 15-23).

NOTE: The ignition-coil voltage output varies with different operating conditions. The ignition coil produces only enough voltage to jump the spark-plug gap. But different spark plugs in an engine will have different voltage requirements because of differences in their gaps. Also, the air-fuel-mixture richness and amount of mixture compressed with each compression stroke will vary. These variations result from changes in throttle opening and engine speed. Thus, ignition-coil voltage output must change to meet these different conditions.

⊘ 15-18 Summary of Actions in the Ignition System Let us review briefly the actions taking place in the ignition system. As the piston in one of the engine cylinders starts up on the compression stroke, one of the distributor breaker-cam lobes moves away from the contact-point breaker arm.[3] The contact points close, current flows through the primary winding of the ignition coil, and a magnetic field builds up. Then, the piston reaches the position in the cylinder at which ignition of the compressed air-fuel mixture should take place. At this instant, the next cam lobe has moved around to where it strikes against the contact-point breaker arm so that the contact points separate. The current stops flowing in the primary circuit, and the magnetic field collapses. This collapse induces high voltage in the secondary winding. In the meantime, the rotor on top of the breaker cam, has moved into position opposite the outside distributor-cap terminal connected to the cylinder spark plug. The spark plug is thus connected to the secondary winding of the ignition coil through the cap and rotor at the instant that the high voltage is induced. A spark therefore occurs at the spark-plug gap.

⊘ 15-19 Ignition-Coil Resistor Automobile engines with 12-volt systems have a resistance wire in the ignition-coil primary circuit (see Figs. 5-24 and 15-1). This wire is shorted-out by the ignition switch when it is turned to START. Now, full battery voltage is imposed on the ignition coil for good performance during cranking. After the engine is started and the ignition switch is turned to ON, the resistance is in the ignition primary circuit. The resistance wire thus protects the contact points from excessive current.

Engines equipped with the General Motors High-Energy Ignition do not use a primary resistor. On these engines, full battery voltage is always applied to the coil.

⊘ 15-20 Spark-Advance Mechanisms The spark-advance mechanisms vary the spark timing for different engine-operating conditions. There are two general types of spark-advance mechanisms: centrifugal and vacuum.

[3]This explanation is based on the actions in the contact-point-distributor ignition system. In the electronic ignition system, the results are the same except that the control unit starts and stops the flow of current in the ignition-coil primary winding (⊘ 15-4 to 15-6).

⊘ 15-21 Centrifugal Advance When the engine is idling, the spark is timed to occur just before the piston reaches top dead center (TDC) on the compression stroke. At higher speeds, it is necessary to deliver the spark to the combustion chamber somewhat earlier. Advancing the spark gives the air-fuel mixture ample time to burn and deliver its power to the piston. To provide this spark advance, a centrifugal-advance mechanism is used (Fig. 15-24). It consists of two advance weights that throw out against spring tension as engine speed increases. This movement is transmitted through an advance cam to the breaker cam (or to the timer core or reluctor of magnetic pickup distributors). (See Fig. 15-25.) This movement causes the breaker cam (or timer core) to advance, or move ahead, with respect to the distributor driveshaft. On the contact-point distributor, this advance causes the cam to open and close the contact points earlier in the compression stroke at high speeds. On the magnetic pickup distributor, the timer core is advanced so that the pickup coil advances the timing of its signals to the transistor-control unit. Since the rotor, too, is advanced, it comes into position earlier in the cycle, also.

The timing of the spark to the cylinder thus varies from no advance at low speeds to full advance at high speeds (when the weights have reached the outer limits of their travel). Maximum advance may be as much as 45 degrees of crankshaft rotation before the piston reaches TDC. Advance also varies with different makes of engines. The toggle arrangement and springs are designed to give the correct advance for maximum engine performance.

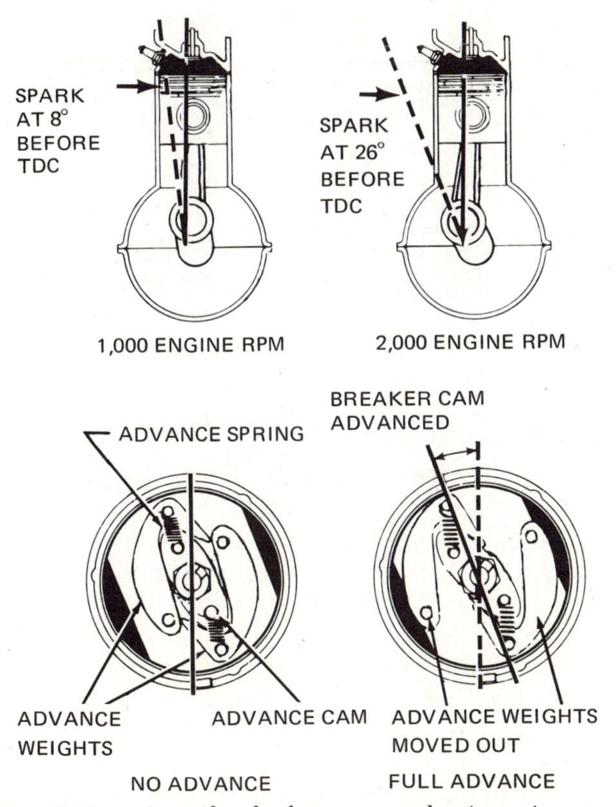

Fig. 15-24. Centrifugal-advance mechanism in no-advance and full-advance positions. In the typical example shown, the ignition is timed at 8 degrees before TDC on idle. There is no centrifugal advance at 1,000 engine rpm. There is 26 degrees total advance (18 degrees centrifugal plus 8 degrees due to original timing) at 2,000 engine rpm. (*Delco-Remy Division of General Motors Corporation*)

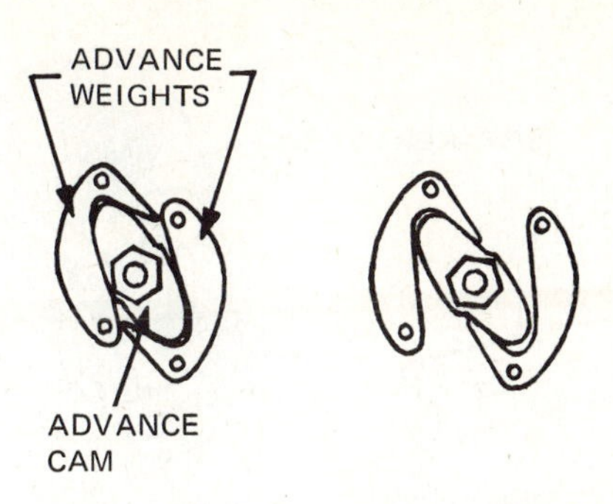

Fig. 15-25. Two extreme positions of the advance weights and advance cam.

⊘ 15-22 Vacuum Advance Under part throttle, a partial vacuum develops in the intake manifold. This means that less air and fuel will be admitted to the cylinder (volumetric efficiency is lowered). Thus, the air-fuel mixture will be less highly compressed and will burn more slowly when ignited. In order to realize full power from the air-fuel mixture, the spark should be somewhat advanced. To provide this advance of the spark, a vacuum-advance mechanism is used.

Figure 15-19 shows one type of vacuum-advance mechanism used on contact-point distributors. It contains a spring-loaded, airtight diaphragm that is connected by a linkage, or lever, to the breaker plate. The breaker plate is supported on a bearing so that it can turn with respect to the distributor housing. The breaker plate actually turns only a few degrees. The linkage to the spring-loaded diaphragm prevents any greater rotation.

The spring-loaded side of the diaphragm is connected through a vacuum line to an opening in the carburetor (Fig. 15-26). This opening is on the atmospheric side of the throttle valve when the throttle is in the idling position. There is no vacuum advance in this position.

As soon as the throttle valve is opened, however, it moves past the opening of the vacuum passage. The intake-manifold vacuum then draws air from the vacuum line and airtight chamber in the vacuum-advance mechanism. This action causes the diaphragm to move against the spring. The linkage to the breaker plate then rotates the breaker plate (or magnetic pickup assembly). This movement carries the contact points around. Thus, as it rotates, the cam closes and opens the points earlier in the cycle. The spark then appears at the spark-plug gap earlier in the compression stroke. As the throttle is

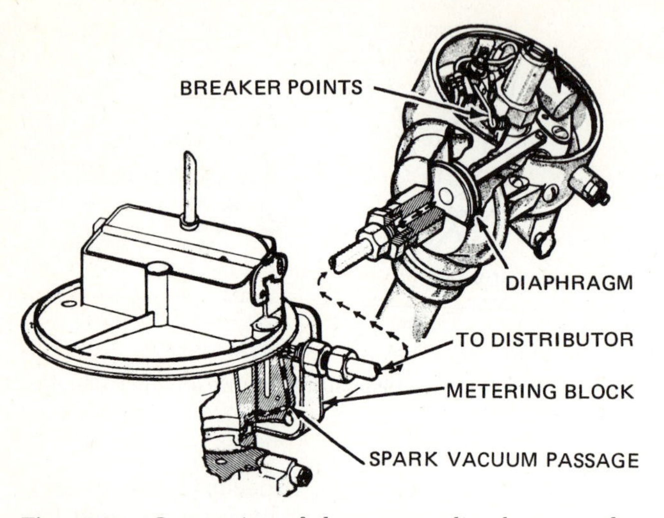

Fig. 15-26. Connection of the vacuum line between the carburetor and the vacuum-advance mechanism on the distributor. (*Ford Motor Company*)

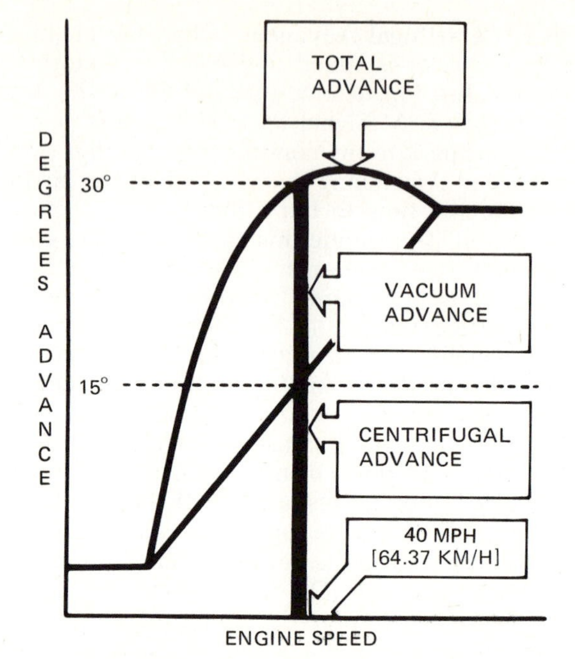

Fig. 15-27. Centrifugal- and vacuum-advance curves for one particular application.

opened wider, there is less vacuum in the intake manifold and less vacuum advance. At wide-open throttle, there is no vacuum advance at all. Under this condition, the spark advance is provided entirely by the centrifugal-advance mechanism.

On the magnetic pickup distributor, the vacuum-advance mechanism is attached to the magnetic pickup assembly (Fig. 15-6). This assembly is rotated to provide the vacuum advance.

⊘ 15-23 Combination of Centrifugal and Vacuum Advances

At any particular engine speed, there will be a centrifugal advance due to engine speed plus a possible additional spark advance due to the operation of the vacuum-advance mechanism, as shown in Fig. 15-27. At 40 mph [64.37 km/h], the centrifugal-advance mechanism provides 15 degrees spark advance on this particular application. Under part-throttle conditions, the vacuum-advance mechanism will supply up to 15 degrees of additional advance. However, if the engine is operated at wide-open throttle, no vacuum advance will be obtained. The total advance will vary somewhat between the straight line (centrifugal advance) and the curved line (centrifugal advance plus total possible vacuum advance) as the throttle is closed and opened.

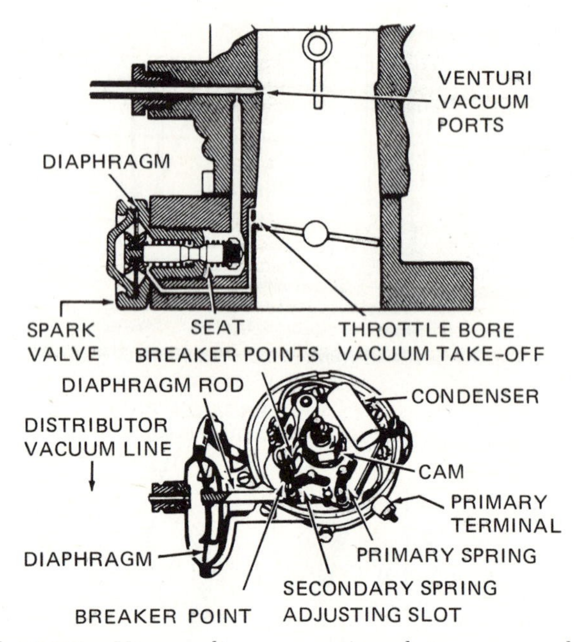

Fig. 15-28. Vacuum-line connections between a carburetor and a distributor having full vacuum control. (*Ford Motor Company*)

⊘ 15-24 Full Vacuum Control

The distributor illustrated in Fig. 15-28 does not contain a centrifugal-advance mechanism. Instead, it utilizes vacuum from the carburetor venturi and intake manifold to produce the proper advance. Full control is possible by vacuum alone because air speed through the carburetor air horn, and thus the vacuum in the venturi, is directly related to engine speed. Let us see how the system works.

In the carburetor shown in Fig. 15-28, there are two vacuum openings in the air horn. One opening is at the venturi. The other is just above the throttle when it is closed. The lower, or throttle, vacuum takeoff opening may have two ports on some models, as shown in Fig. 15-28. These two vacuum openings are connected to each other by vacuum passages and are connected to the distributor vacuum-advance mechanism by a vacuum line. Vacuum imposed on the diaphragm in the vacuum-advance mechanism causes the breaker-plate assembly to rotate. This action is very similar to that of the vacuum-advance devices discussed previously. Rotation of the breaker-plate assembly causes an advance of the spark.

As engine speed increases, the vacuum at the venturi in the carburetor increases. This is due to the increase of air speed through the venturi. The increase in venturi vacuum causes an earlier spark advance, which is related to engine speed. At the same time, under part-throttle operating conditions, there is a vacuum in the intake manifold. This vacuum condition acts at the throttle-vacuum ports in the carburetor to produce a further vacuum advance. Thus, the interrelation of the vacuum conditions at the two points in the carburetor produce, in effect, a combined speed advance (as with a centrifugal device) and vacuum advance.

⊘ 15-25 Vacuum-Advance Controls for Emission Reduction During some operating conditions, vacuum advance can increase the formation of nitrogen oxides (NO_x) during combustion. As explained in Chap. 19, nitrogen oxides form during combustion at high temperatures. Thus, part-throttle operation in lower gears can cause an increase in NO_x in the exhaust gases. To prevent this, automobiles are equipped with control systems that prevent vacuum advance under some conditions.

Car manufacturers achieve vacuum-advance control by different methods. Figure 15-29 shows one system that allows vacuum advance only in high gear with certain exceptions. The system includes a transmission switch, a solenoid vacuum switch, and a temperature-override system. For normal operation in any gear but high gear, the transmission switch is closed. This connects the solenoid vacuum switch to the battery. The solenoid pulls in its plunger, closing off the vacuum connection to the lower part of the carburetor (that is, to manifold vacuum). At the same time, the solenoid opens a connection to the upper part of the carburetor through a clean air vent. This connection releases any vacuum on the vacuum-advance mechanism at the distributor. There is no vacuum advance.

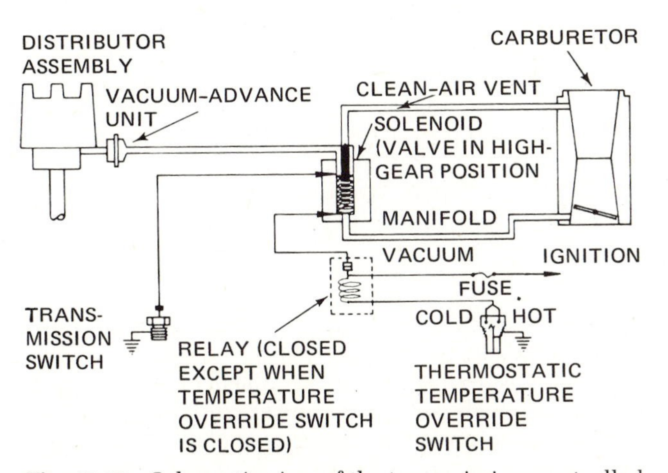

Fig. 15-29. Schematic view of the transmission-controlled spark (TCS) system. (*Chevrolet Motor Division of General Motors Corporation*)

In high gear, the transmission switch is opened, which allows the solenoid plunger to be pushed up by a spring. This action shuts off the clean air vent and opens the line from the manifold vacuum vent to the vacuum-advance unit. Now, normal vacuum advance can take place.

The temperature-override system provides full advance for better performance in all gears when the engine is cold. A thermostatic water-temperature switch is closed when the engine is cold. This connects the relay winding to the battery (through the ignition switch). The relay contact points open, thus opening the circuit to the solenoid. Now, regardless of the position of the transmission switch, the solenoid cannot operate. The vacuum-advance unit remains connected to the manifold vacuum vent in the carburetor. Thus, the system provides normal vacuum advance.

When the engine warms up, the thermostatic temperature-override switch opens. This opens the relay winding. The relay points close. The system now operates so that vacuum advance is obtained only in high gear, as explained previously.

The system also has a hot override position, which provides vacuum advance in all gear positions if the engine overheats. The vacuum advance improves engine cooling. See Chap. 19 for details of this and other vacuum-advance control systems.

⊘ 15-26 Ignition Switch In late-model cars, the ignition switch is mounted on the steering column, as shown in Fig. 15-30. This arrangement locks the steering shaft when the ignition switch is turned off and the ignition key is removed. A small gear on the end of the ignition switch rotates and releases a plunger. The plunger enters a notch in a disk on the steering shaft to lock the shaft. If a notch is not lined up with the plunger, the plunger rests on the disk. When the steering wheel, shaft, and disk are turned slightly, the plunger drops into a notch.

When the ignition key is inserted and the ignition switch is turned on, the plunger is withdrawn from the disk to unlock the steering shaft.

The ignition switch has an extra set of contacts that are used when the switch is turned past ON to START. The contacts connect the starting-motor solenoid to the battery so that the starting motor can operate. As soon as the engine is started and the switch is released, the switch returns to ON. The starting motor is then disconnected from the battery.

The alternator field circuit is connected to the battery through the ignition switch when it is turned to ON. When the ignition switch is turned to OFF, the alternator field circuit is disconnected. This action prevents the battery from running down through the field circuit.

On late-model cars, the ignition switch also operates a buzzer if the key is left in the lock when the car door on the driver's side is open. The buzzer is a reminder to the driver to remove the key from

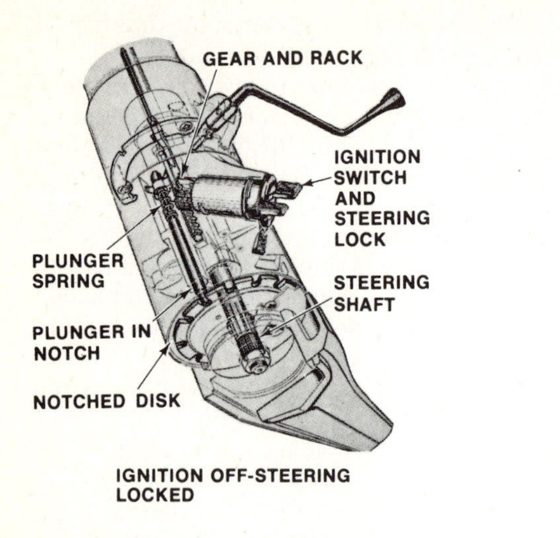

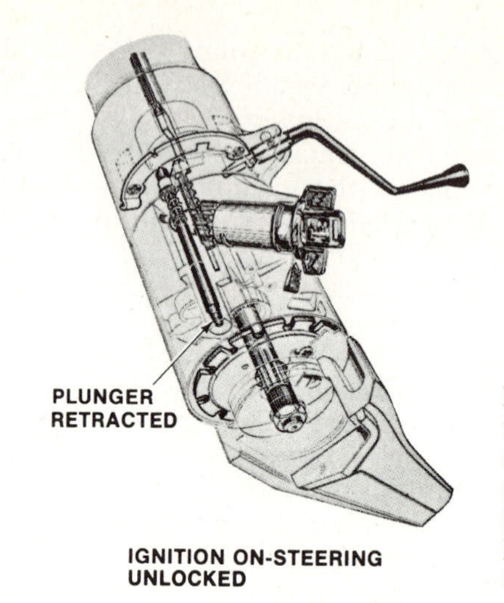

Fig. 15-30. A combination ignition switch and steering-wheel lock in phantom views showing the two positions of the lock. (*General Motors Corporation*)

the lock when leaving the car. This makes it harder for someone to steal the car.

Such accessories as the radio and car heater are also connected to the battery through the ignition switch. This arrangement prevents the driver from leaving these units running when the engine has been turned off.

CHAPTER 15 CHECKUP

NOTE: Since this is a chapter review test, you should review the chapter before taking the test.

The ignition system may seem very complex when you look at it for the first time. But if you have followed the explanations in this chapter, you know that it is really rather simple. In operation, the ignition-coil primary winding is connected to the battery momentarily (through the distributor contact points or electronic control unit), this loads the ignition coil with magnetism. Then, when the ignition coil is disconnected, it discharges this magnetism in the form of a high-voltage surge. The surge is led through the distributor cap and rotor and the wiring to a spark plug. Now, find out how well you remember the details of the ignition system by taking the test that follows. If you are not able to answer all the questions or if you are confused about any part of ignition-system action, study the chapter again. You should know the answers before you proceed to the following chapters.

Completing the Sentences The sentences that follow are incomplete. After every sentence there are several words or phrases, only one of which will correctly complete the sentence. Write each sentence in your notebook, selecting the proper word or phrase to complete it correctly.

1. The ignition coil has: (*a*) one winding, (*b*) two windings, (*c*) three windings, (*d*) four windings.
2. The primary winding of the ignition coil is connected to the battery through the: (*a*) spark-plug wiring, (*b*) distributor cap and rotor, (*c*) distributor gearing, (*d*) distributor contact points.
3. The voltage induced in a winding by self-induction is called: (*a*) reverse voltage, (*b*) reduced voltage, (*c*) countervoltage, (*d*) maximum voltage.
4. The spark-producing high voltage is produced in the secondary winding during: (*a*) magnetic buildup, (*b*) magnetic collapse, (*c*) the time points are closed.
5. The ignition condenser is connected: (*a*) across the points, (*b*) in series with the points, (*c*) in the secondary circuit.
6. The main function of the ignition condenser is to protect the contact points and produce: (*a*) quick magnetic collapse, (*b*) slow magnetic collapse, (*c*) high voltage on the points.
7. The rotating switch that connects the various spark plugs to the ignition-coil secondary in regular firing order is formed by the distributor: (*a*) points and condenser, (*b*) cap and rotor, (*c*) switch and contact points.
8. The device in many distributors that pushes the breaker cam ahead as engine speed increases is called the: (*a*) vacuum-advance mechanism, (*b*) centrifugal-advance mechanism, (*c*) full-advance mechanism, (*d*) vacuum-brake mechanism.
9. The device in many distributors that shifts the position of the breaker plate or magnetic pickup assembly to produce a change in the timing of the spark is actuated by: (*a*) intake-manifold vacuum, (*b*) engine speed, (*c*) centrifugal advance.
10. With wide-open throttle, vacuum advance will be: (*a*) at a maximum, (*b*) part to full, (*c*) at a minimum.

Reviewing the System In the following, you are asked to write in your notebook various explanations of the actions that take place in the ignition system, as well as the purposes and operation of various components. If you are not clear about some explanation, carefully review the pages in the chapter that will clear up the matter for you. Then write the explanation in your own words.

1. Describe the construction of an ignition coil.
2. Explain briefly what takes place in the ignition-coil primary circuit when the distributor contacts close and open.
3. What is meant by self-induction?
4. Explain how self-induction takes place between two adjacent turns of wire in a winding.
5. What happens in the secondary winding during buildup?
6. Explain the effect of self-induction in the primary winding after the contact points open.
7. What is meant by buildup in the ignition coil? About how long is the normal buildup time?
8. Describe the construction of a condenser.
9. Describe briefly the action of a condenser during the short interval after the contacts open.
10. Describe the action in the ignition-coil secondary winding during magnetic collapse.
11. Describe the action in the ignition-coil primary winding during magnetic collapse.
12. Summarize briefly the actions in the ignition system from the time when the breaker-cam lobe moves out of the way so that the contacts can close until the spark occurs at the spark-plug gap.
13. Why are spark advances needed?
14. Why is centrifugal advance desirable? Vacuum advance?
15. Describe two types of vacuum-advance mechanisms.
16. Explain the operation of the magnetic pickup ignition system.
17. Explain the operation of the control unit in the electronic ignition system.

SUGGESTIONS FOR FURTHER STUDY

Study the various shop manuals and bulletins issued by different automobile and electrical equipment manufacturers describing their ignition equipment. If you have a chance, examine this equipment in a local automotive electrical service shop or your school shop. Note especially the provisions for driving the distributor and the different types of advance mechanisms that distributors use. Be sure to write any important facts in your notebook.

chapter 16

ENGINE COOLING SYSTEMS

This chapter discusses the construction and operation of automotive-engine cooling systems. As mentioned in ⊘ 7-2 and 7-9, the cylinder block and cylinder head have water jackets through which coolant (water mixed with antifreeze) can circulate. This circulation of coolant between the water jackets and the cooling-system radiator removes heat from the engine.

⊘ 16-1 Purpose of the Cooling System The purpose of the cooling system is to keep the engine at its most efficient operating temperature at all engine speeds and all driving conditions. During combustion of the air-fuel mixture in the engine cylinders, temperatures as high as 6000 degrees Fahrenheit [3316°C] may be reached by the burning gases. Some of this heat is absorbed by the cylinder walls, cylinder heads, and pistons. These parts must be cooled so that their temperatures will not rise to an excessive point.

Cylinder-wall temperature must not increase beyond about 400 to 500 degrees Fahrenheit [204 to 260°C]. Temperatures higher than this will cause the lubricating-oil film to break down and lose its lubricating properties. But it is desirable to operate the engine at temperatures as close as possible to the limits imposed by oil properties. Removing too much heat through the cylinder walls and cylinder heads would lower engine thermal efficiency (⊘ 11-17). Cooling systems are designed to remove about 30 to 35 percent of the heat produced in the combustion chambers.

The engine is quite inefficient when cold. Therefore, the cooling system includes devices that prevent normal cooling action during engine warm-up. These devices shorten the inefficient cold-engine operating time and allow the working parts to reach operating temperatures more quickly. Then, when the engine reaches operating temperatures, the cooling system begins to function. Thus, the cooling system cools rapidly when the engine is hot but slowly or not at all when the engine is cold or warming up.

Two general types of cooling systems are used: *air cooling* and *liquid cooling*. Most automotive engines now employ liquid cooling, although engines for motorcycles, power lawn mowers, and so forth, are all air-cooled. Air-cooled engines have metal fins on the cylinders and heads to help radiate the heat from the engine. Cylinders are usually partly or completely separated to improve air circulation around them. Special shrouds and blowers are used on many air-cooled engines to improve air circulation around the cylinders and heads. Figures 6-9 and 6-15 illustrate air-cooled engines.

The liquid cooling systems usually have a water pump to maintain circulation in the system. Figure 16-1 shows the cooling system for a V-8 engine. The water pump is driven by a belt from the engine crankshaft. This pump circulates the cooling liquid, called the *coolant*, between the radiator and the engine water jackets. (The coolant is water to which antifreeze has been added.) The following articles describe the cooling-system components in detail.

⊘ 16-2 Water Jackets Just as we might put on a sweater or a jacket to keep warm on a cool day, so water jackets are placed around the engine cylinders. There is this difference: Water jackets are designed to keep the cylinders cool. The water jackets are cast into the cylinder blocks and heads.

⊘ 16-3 Water Pumps Water pumps are of the centrifugal type. They are mounted at the front end of the cylinder block between the block and the radiator (Fig. 16-1). The pump consists of a housing, with a water inlet and outlet, and an impeller (Figs. 16-2 and 16-3). The impeller is a flat plate mounted on the pump shaft with a series of flat or curved blades, or vanes. When the impeller rotates, the coolant between the blades is thrown outward by centrifugal force. The coolant is forced through the pump outlet and into the cylinder block. The pump inlet is connected by a hose to the bottom of the radiator, and coolant from the radiator is drawn into the pump to replace the coolant that is forced through the outlet.

The impeller shaft is supported on one or more

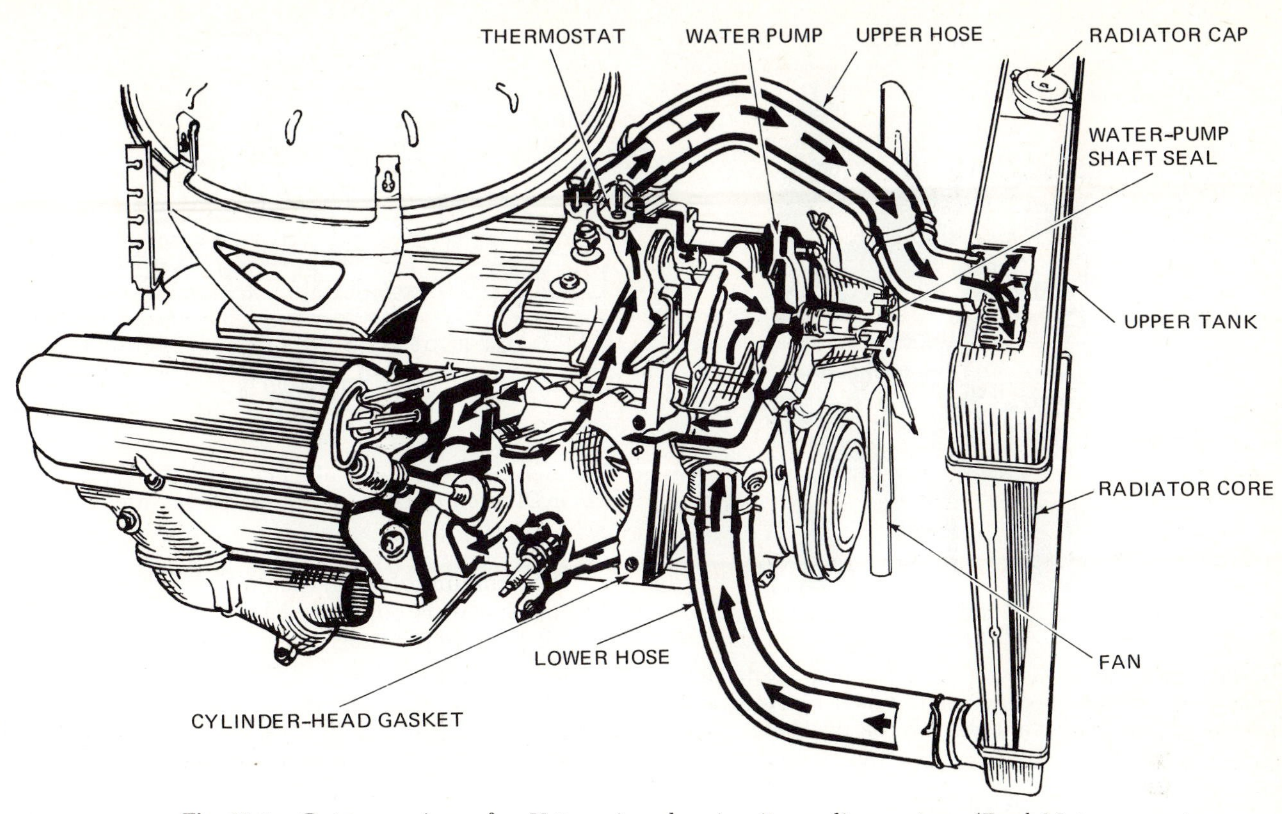

Fig. 16-1. Cutaway view of a V-8 engine showing its cooling system. (*Ford Motor Company*)

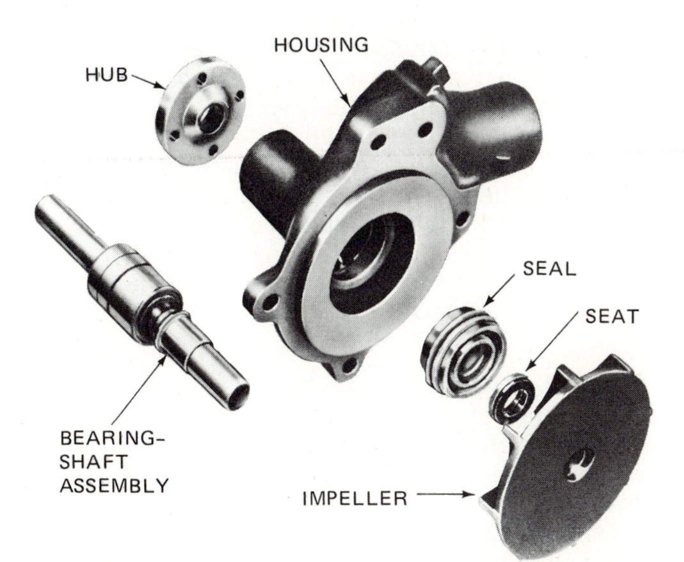

Fig. 16-2. Disassembled view of a water pump. (*Pontiac Motor Division of General Motors Corporation*)

bearings. A seal prevents coolant from leaking out around the bearing. The pump is driven by a belt to the drive pulley mounted on the front end of the engine crankshaft.

⊘ 16-4 Engine Fan The engine fan is usually mounted on the water-pump shaft. The fan is driven by the same belt that drives the pump and alternator (Fig. 16-3). The purpose of the fan is to provide a powerful draft of air through the radiator. Some fans are equipped with a fan shroud that improves their performance. The shroud increases the efficiency of the fan. It ensures that all air pulled back by the fan must first pass through the radiator.

1. FAN BELT Fan belts are of the V type. There is friction between the sides of the belt and the sides of the grooves in the pulleys. The friction causes the driving power to be transmitted through the belt from one pulley to the other. The V-type belt provides a large area of contact, so that considerable power may be transmitted. The wedging action of the fan belt as it curves into the pulley grooves aids in preventing belt slippage.

NOTE: Many engines use two V belts with double pulleys, that is, pulleys that have two belt grooves. The added belt provides the power required to drive the alternator and water pump. These belts are matched. If one belt is to be replaced, then both should be replaced at the same time. Otherwise, the new belt will take most of the driving effort and will wear rapidly.

2. VARIABLE-SPEED FAN DRIVE Many engines use a variable-speed fan drive (Figs. 16-4 and 16-5) which reduces fan speed. This fan drive conserves horsepower at high engine speeds and when cooling requirements are low. At high engine speeds, a typical fan might use up several horsepower and pro-

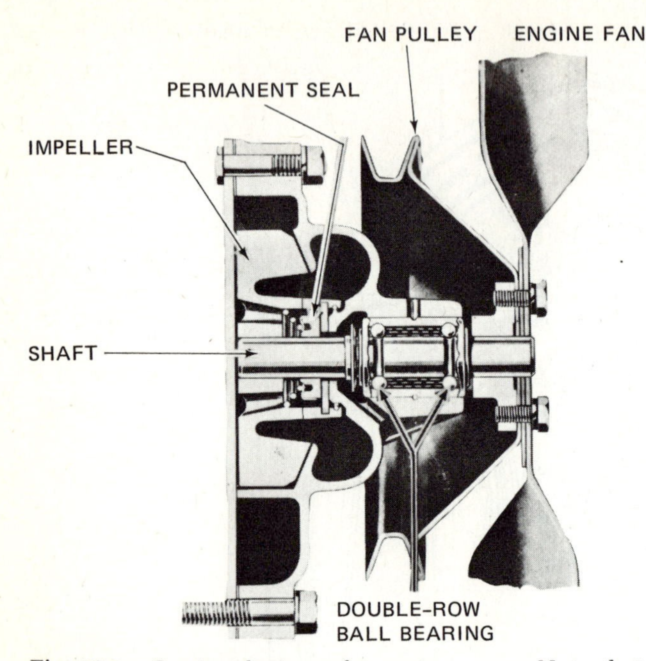

Fig. 16-3. Sectional view of a water pump. Note that the shaft is supported on double-row ball bearings, and that the fan and pulley are mounted on the shaft.

Fig. 16-4. Installation of a variable-speed fan clutch drive. The drive is positioned between the fan hub and the pulley shaft. (*Ford Motor Company*)

duce some noise. But the variable-speed fan drive contains a small fluid coupling partly filled with a special silicone oil. When engine cooling requirements are severe, as during high-temperature, high-speed operation, more oil is injected into the fluid coupling. This action causes more power to pass through the coupling. Fan speed therefore increases. When engine cooling requirements are low, as during low-temperature, intermediate-speed operation, oil is withdrawn from the fluid coupling. Less power passes through the coupling and fan speed drops off.

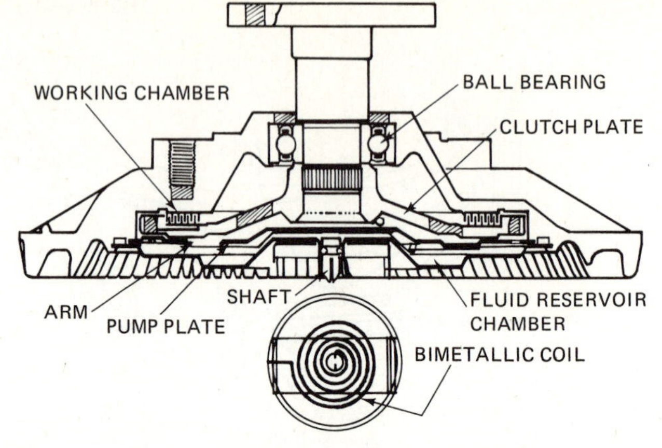

Fig. 16-5. Sectional view of a fan clutch drive. (*Chevrolet Motor Division of General Motors Corporation*)

The amount of oil in the fluid coupling, and thus the fan speed, is controlled by a thermostatic strip or coil (Fig. 16-5 shows the coil). If a strip is used, it is held at both ends by clips on the face of the fan drive. The strip bows outward with increasing under-the-hood temperatures. This motion allows a control piston to move outward. The outward-moving piston forces more oil into the fluid coupling. This action increases fan speed and engine cooling. As under-the-hood temperatures drop, the thermostatic strip straightens out, forcing the control piston in. This action causes oil to leave the fluid coupling so that fan speed drops. The bimetal thermostatic coil works in a similar manner to match fan speed with cooling requirements.

3. *FLEX FAN* Flexible blades reduce both the power required to drive the fan and noise at high speed (Fig. 16-6). With this design, the pitch of the blades decreases as fan speed increases owing to centrifugal force. The result is that each blade pushes less air. Thus, power needs and noise are lower at higher speeds.

⊘ 16-5 Radiator The radiator (Fig. 16-1) holds a large volume of coolant in close contact with a large volume of air. Thus, the radiator allows heat to transfer from the coolant to the air. The radiator core is divided into two separate and intricate compartments. Coolant passes through one, and air passes through the other. There are several types of radiator cores. Two of the more commonly used types are the tube-and-fin and ribbon-cellular cores. The tube-and-fin radiator core (Fig. 16-7) consists of a series of long tubes extending from the top to the bottom of the radiator (or from the upper to lower tank). Fins are placed around the tubes to improve heat transfer. Air passes around the outside of the tubes between the fins, absorbing heat from the coolant in passing.

The ribbon-cellular radiator core (Fig. 16-8) is made up of a large number of narrow coolant passages. The passages are formed by pairs of thin metal ribbons soldered together along their edges,

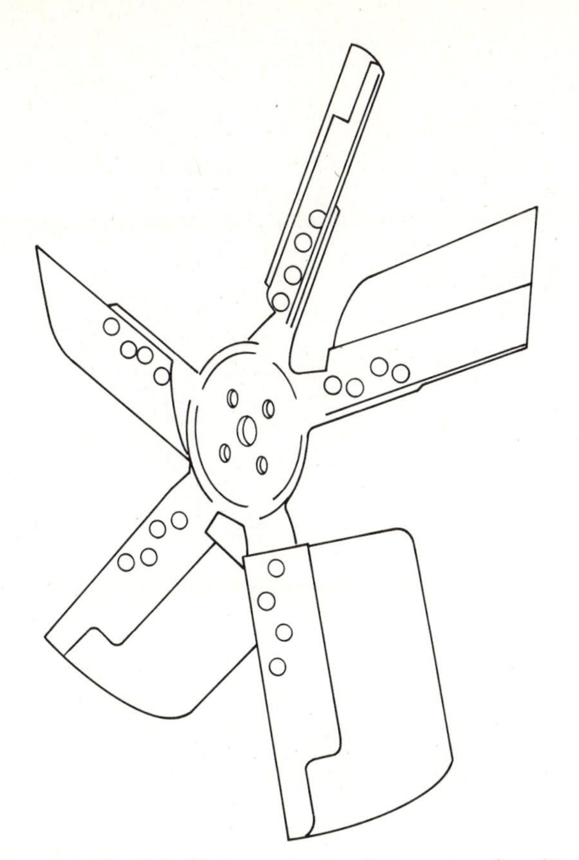

Fig. 16-6. Flexible-blade engine cooling-system fan. (*Ford Motor Company*)

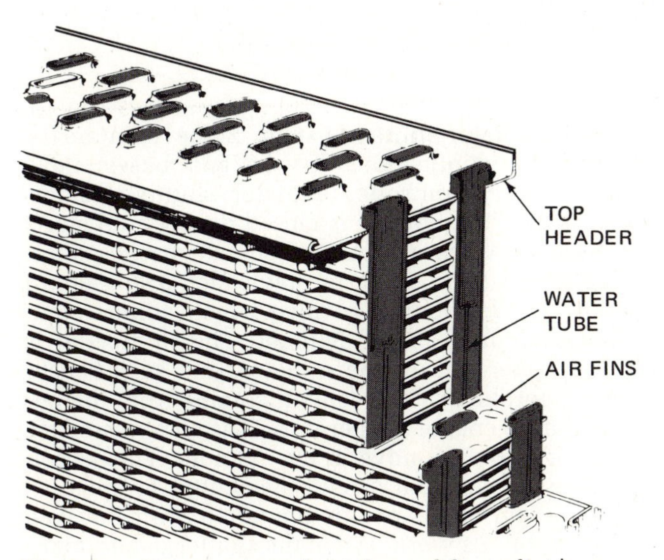

Fig. 16-7. Construction of a tube-and-fin radiator core.

running from the upper to the lower tank. The edges of the coolant passages, which are soldered together, form the front and back surfaces of the radiator core. The coolant passages are separated by air fins of metal ribbon, which provide air passages between the water passages. Air moves through these passages from front to back, taking heat from the fins. The fins, in turn, absorb heat from the water moving

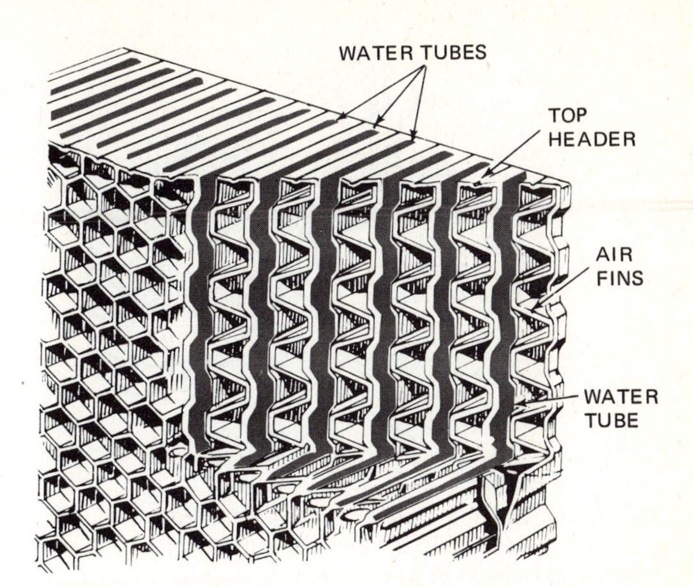

Fig. 16-8. Construction of a ribbon-cellular radiator.

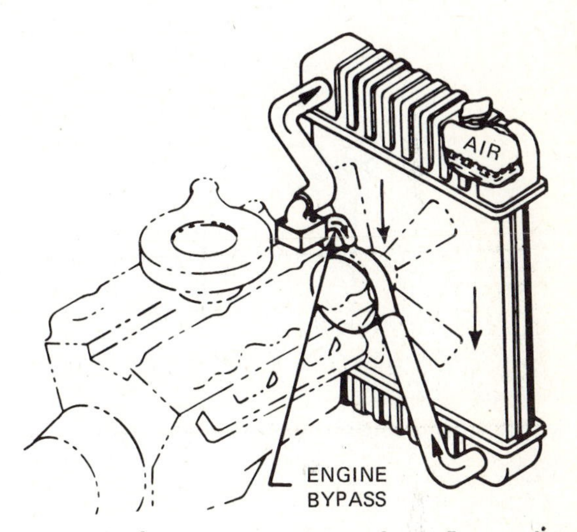

Fig. 16-9. Cooling system using a down-flow radiator. (*Harrison Radiator Division of General Motors Corporation*)

downward through the coolant passages. As a consequence, the coolant is cooled.

Radiators can be classified in another way, that is, according to the direction of coolant flow through them. In some radiators, the coolant flows from top to bottom (down-flow type). In others, the coolant flows horizontally from an input tank on one side to another tank on the other side (cross-flow type). These two types of radiators are shown in Figs. 16-9 and 16-10.

The coolant tank above or to the side of the radiator serves two purposes: It provides a reserve supply of coolant, and it provides a place where the coolant can be separated from any air that might be circulating in the system. The tank has a filler cap which can be removed for addition of coolant as necessary.

1. EXPANSION TANKS Many cooling systems have a separate expansion tank, as shown in Fig. 16-11. The mounting arrangement and connections

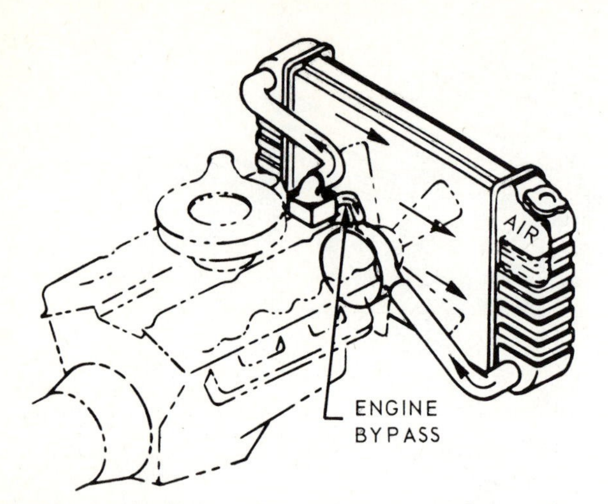

Fig. 16-10. Cooling system using a cross-flow radiator. (*Harrison Radiator Division of General Motors Corporation*)

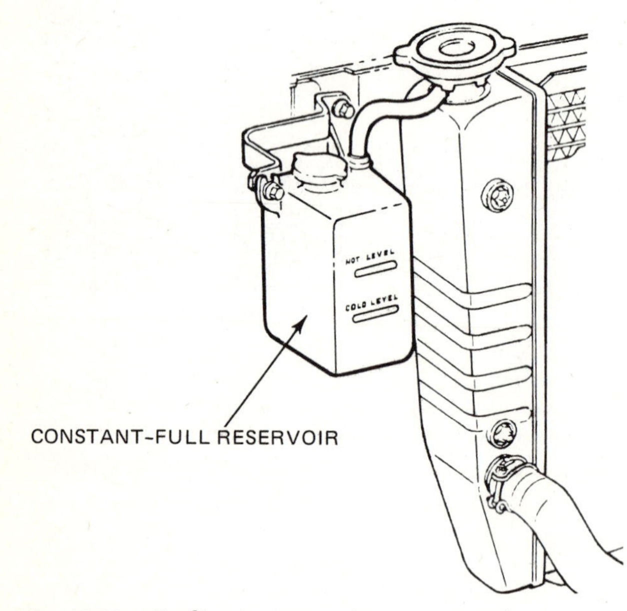

Fig. 16-11. Cooling system using an expansion, or constant-full, tank. (*Ford Motor Company*)

for the expansion tank are shown in Fig. 16-12. The expansion tank is partly filled with coolant and is connected to the radiator cap. The coolant expands in the engine as it heats up. This sends part of the coolant into the expansion tank. Then, when the engine approaches operating temperature, a valve in the radiator cap closes. The cooling system is now sealed. The pressure in the cooling system increases and thus prevents boiling. As explained in ⊘ 16-10, the increased pressure allows a higher coolant temperature and thus a more efficient cooling system.

When the engine cools off, the coolant in the cooling system contracts, producing a vacuum. Now coolant from the expansion tank flows back into the radiator. The cooling system works to keep the engine water jackets and radiator filled with coolant. This activity allows the system to operate at maximum efficiency at all times.

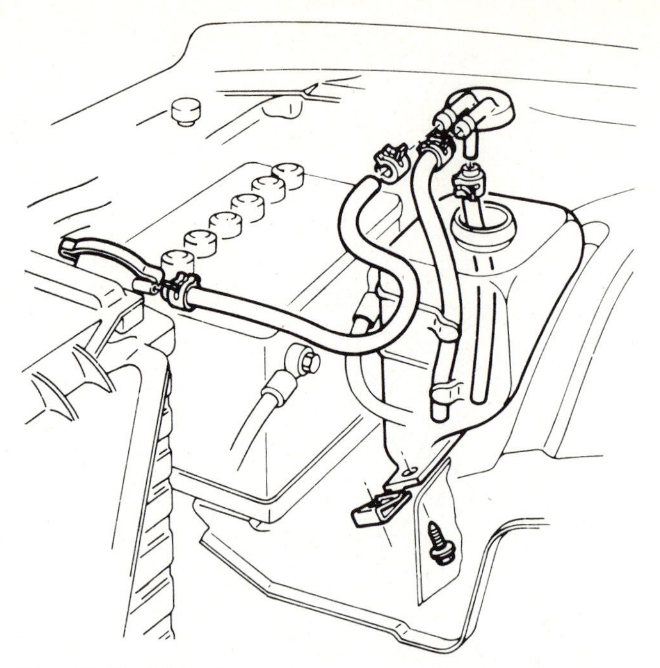

Fig. 16-12. Mounting arrangement and connections for a cooling-system expansion tank, called a *coolant-recovery system* by Chevrolet. (*Chevrolet Motor Division of General Motors Corporation*)

2. *RADIATOR GRILLS* Radiator grills, which add to the appearance of the car, place some added load on the cooling system. That is, they partially restrict the flow of cooling air through the radiator. However, the cooling system is designed to take this into account.

⊘ 16-6 Radiator Shutter System Some radiators for heavy-duty applications (trucks and buses) have automatically controlled radiator shutter systems (Fig. 16-13). The shutter is closed during cold starts and engine warm-up. As the engine reaches operating temperature, a thermostat, called a *shutterstat* in Fig. 16-13, operates a valve which admits compressed air into an air cylinder. The shutterstat is in the upper line to the radiator, and so it senses the temperature of the coolant coming from the engine, that is, engine temperature. The compressed air forces a piston to move in the cylinder. This action operates a lever that causes the shutter to open.

You can compare this operation with the way venetian blinds work. When the shutter opens, more air can flow through the radiator to increase the cooling action. During cold-weather operation, the cooling system may be working too efficiently and overcool the engine. In this case, the shutterstat, sensing the temperature of the coolant coming from the engine, causes the shutter to close. (The compressed air that operates the shutter system comes from the air-brake system, as shown in Fig. 16-13.)

⊘ 16-7 Transmission Oil Coolers Many cars with automatic transmissions are equipped with cooler

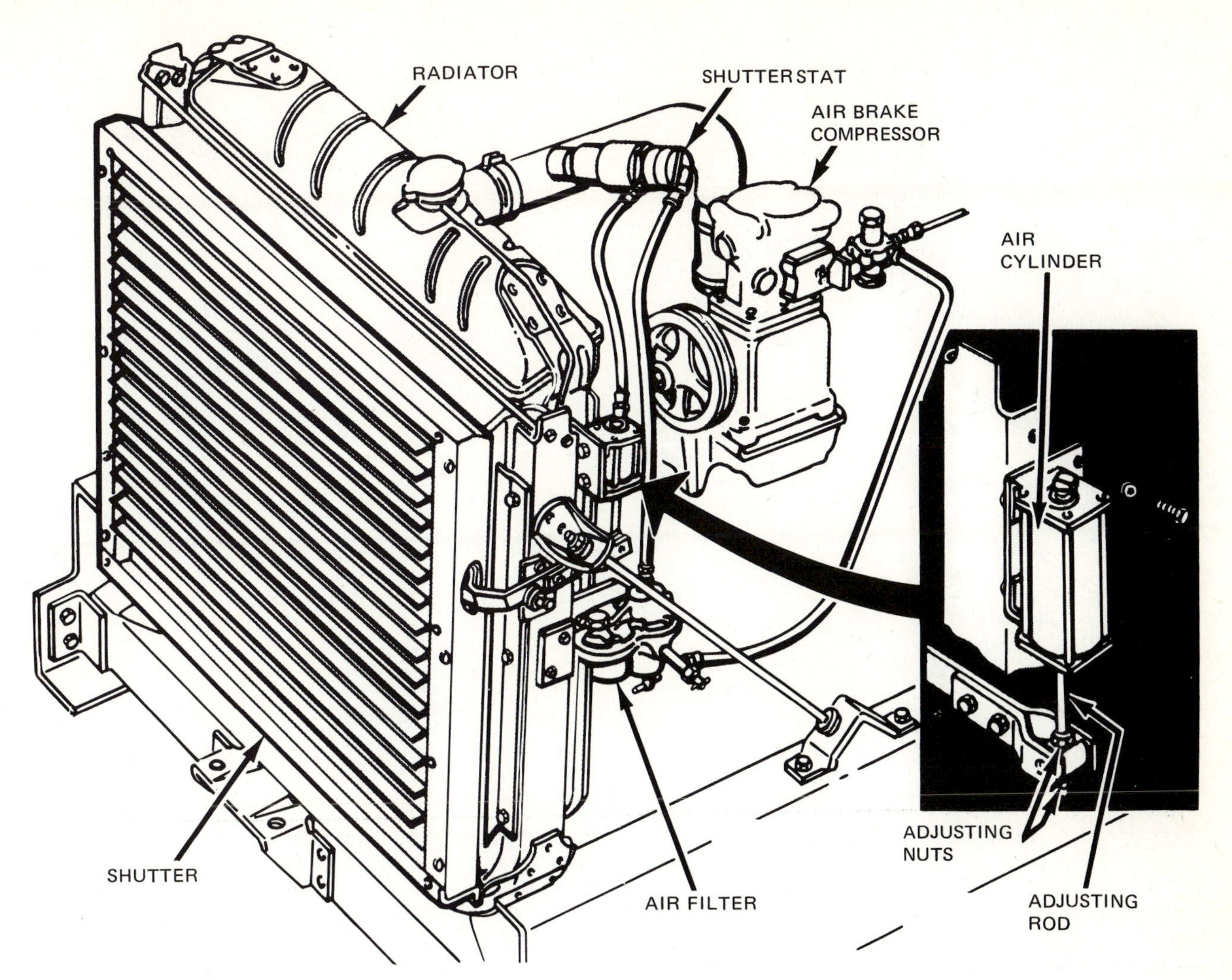

Fig. 16-13. Automatically controlled radiator-shutter system. (*Ford Motor Company*)

tubes or radiators to cool the transmission oil. The oil in automatic transmissions can get quite hot. Overheating reduces transmission performance and can damage the transmission. One oil-cooler system is shown in Fig. 16-14. The transmission is connected by two tubes to the air-cooler tube in the side or bottom tank of the radiator (Fig. 16-15). The oil-cooler tube, being immersed in the coolant in the radiator, is cooled. Thus, the transmission oil passing through it is cooled.

⊘ 16-8 Hot-Water Car Heater Many automobiles are equipped with car heaters of the hot-water type (Figs. 16-16 and 16-17). This device might be considered a secondary radiator. It transfers heat from the cooling system to the passenger compartment instead of to air passing through the main radiator. Figure 16-16 shows how hot coolant from the engine is circulated through the heater radiator. A small electric motor drives a fan that forces air through the radiator section of the heater. The air absorbs heat from the heater radiator.

A more complicated version is shown in Fig. 16-17. The principle is the same. Hot coolant from the engine circulates through a small heater radiator. An electric fan circulates air from the passenger compartment through the radiator. The system also includes a defrosting arrangement. The driver can operate controls that direct the heated air into the passenger compartment. Or, the air can be directed against the windshield to melt any frost or evaporate any mist that has formed.

In some cars, the heating system has automatic controls. These controls turn the system on when the passenger compartment is cold and turn it off when the passenger compartment is warm enough.

Some systems are fully automatic. They work with the air-conditioner system to maintain the desired temperature in summer or winter. The driver merely sets a control to the desired temperature. The system either heats or cools as required to maintain that temperature.

⊘ 16-9 Thermostat The *thermostat* is placed in the coolant passage between the cylinder head and the top of the radiator (Fig. 16-1). Its purpose is to close

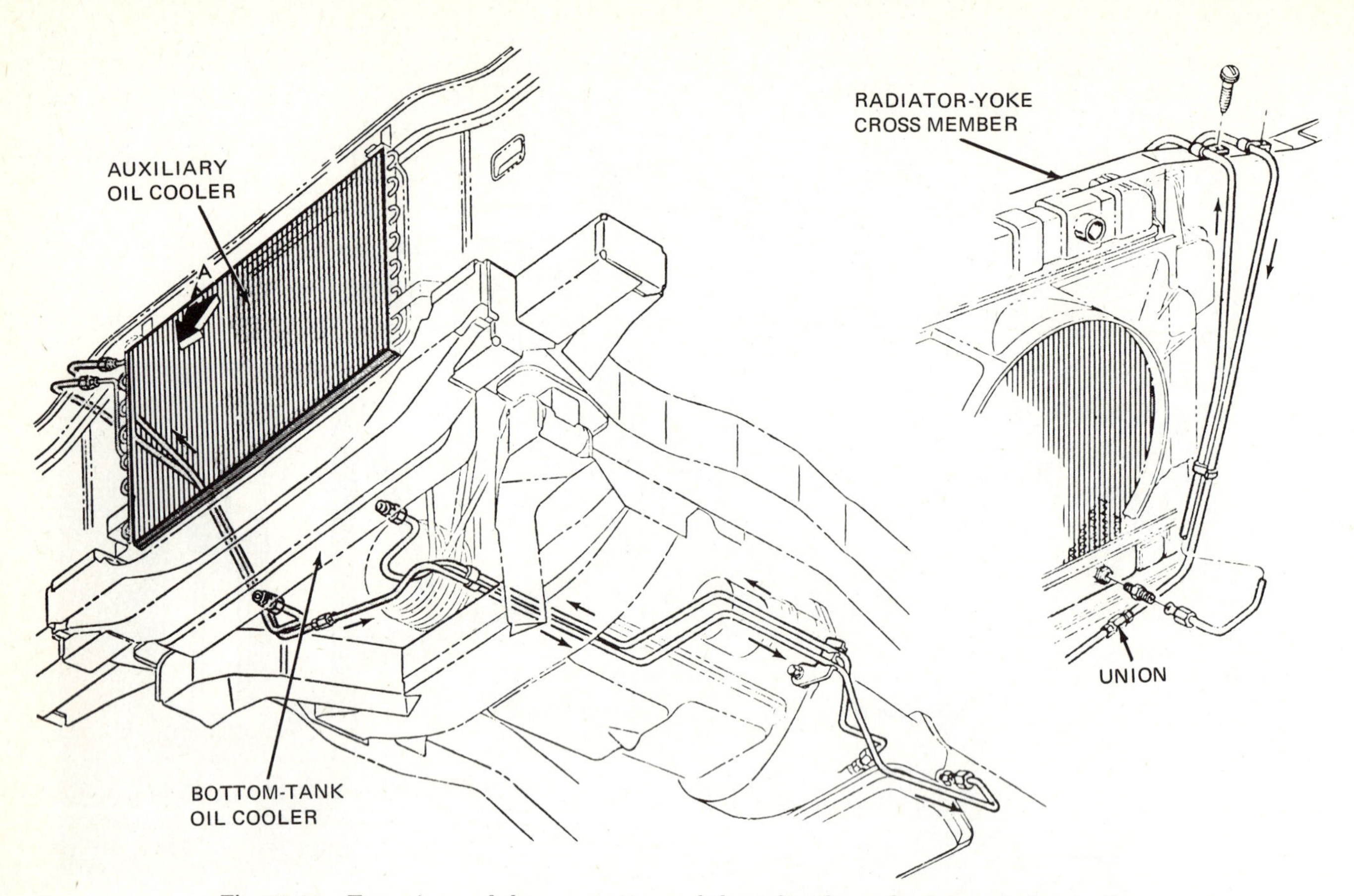

Fig. 16-14. Two views of the connections of the tubes from the transmission to the oil cooler in the bottom tank of the cooling-system radiator. (*Chrysler Corporation*)

off this passage when the engine is cold so that coolant circulation is restricted. This allows the engine to reach operating temperature more quickly. The thermostat consists of a thermostatic device and a valve (Fig. 16-18). Various thermostatic devices and valve arrangements have been used. The bellows type contains a liquid that vaporizes with increasing temperature. This vaporization increases the internal pressure and so causes the bellows to expand and raise the valve off its seat. This action permits coolant to circulate between the engine and radiator. The bimetal type uses a thermostatic coil to operate the valve. Instead of bellows or a thermostatic coil, most cooling-system thermostats are now powered with a wax pellet which expands with increasing temperature to open the valve. The thermostat shown to the right in Fig. 16-18 is of the wax-pellet type. Figure 16-19 is a sectional view of a similar thermostat.

Thermostats are designed to open at specific temperatures. For example, a thermostat designated as a 195-degree-Fahrenheit [90.6°C] unit will start to open between 192 and 197 degrees Fahrenheit [88.9 and 91.7°C] and will be fully open at 207 degrees Fahrenheit [97.2°C]. But a 180-degree-Fahrenheit [82.2°C] thermostat will operate at 15 degrees Fahrenheit [8.3°C] below these figures. Thermostats of the proper characteristics are selected to suit the operating requirements of the engine as well as the kind of antifreeze used.

With a cold engine and the thermostatic valve therefore closed, the coolant can't pass between the engine and radiator. Instead, it recirculates through the cylinder block and head. Restriction of coolant circulation in this manner prevents the removal of any appreciable amount of heat from the engine by the cooling system. Consequently the engine reaches operating temperature more rapidly. When the engine reaches operating temperature, the thermostatic valve begins to open. Then, coolant can circulate through the radiator and the cooling system operates normally, as already described.

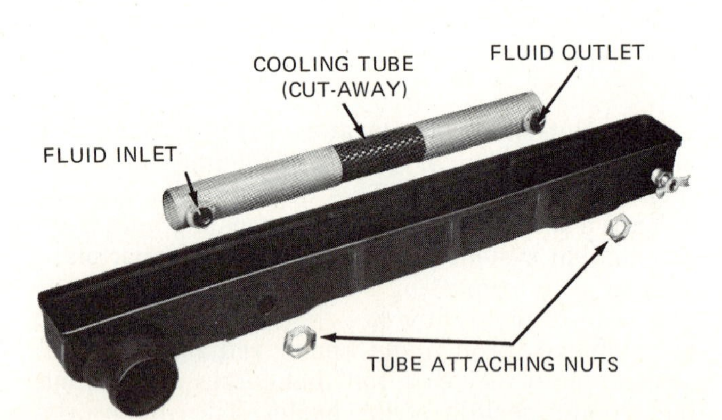

Fig. 16-15. Lower tank of the radiator and oil-cooler tube, showing how the tube fits into the lower tank. (*Chrysler Corporation*)

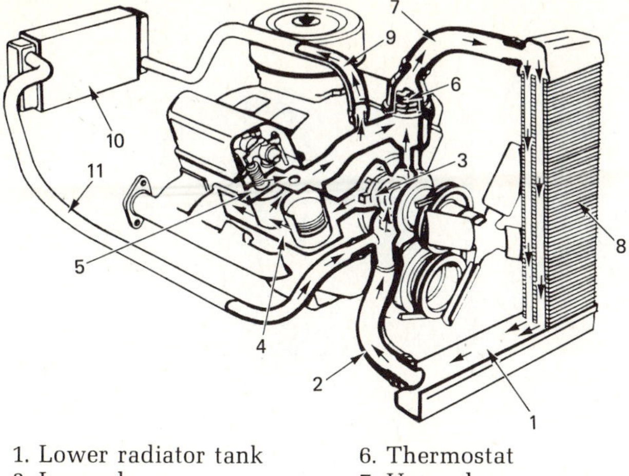

1. Lower radiator tank
2. Lower hose
3. Water pump
4. Cylinder-block water jackets
5. Cylinder-head water jacket
6. Thermostat
7. Upper hose
8. Radiator
9. Hose to heater
10. Heater
11. Return hose from heater

Fig. 16-16. Cutaway view of a V-8 engine, showing the cooling system and car hot-water heater. This is the simplest of all the car heater systems.

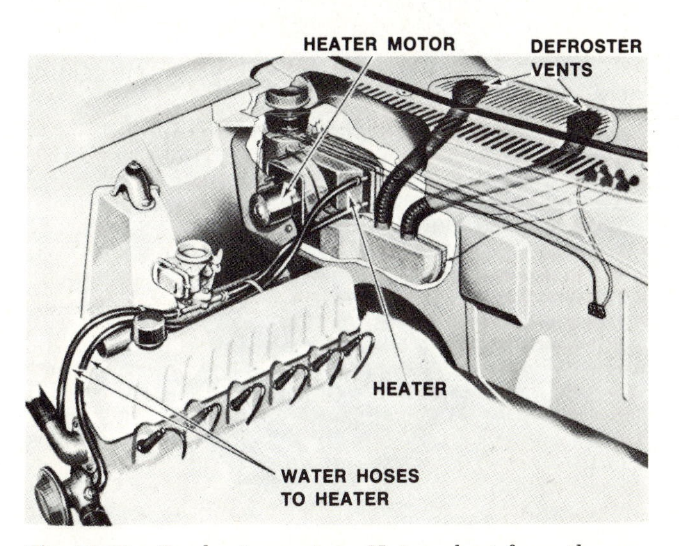

Fig. 16-17. Car heater system. Hot coolant from the engine cooling system circulates through a small radiator. The fan blows air through the radiator.

A bypass is required to permit coolant to circulate within the engine itself when the engine is cold. However, when the engine warms up, the bypass must close or become restricted. Otherwise, the coolant would continue to circulate within the engine itself and too little would go to the radiator for cooling.

One bypass system uses a small, spring-loaded valve. The valve is forced open by coolant pressure from the pump when the thermostat is closed. As the thermostat opens, the coolant pressure drops within the engine and the bypass valve closes.

Another widely used bypass system has a blocking bypass thermostat (Fig. 16-20). This thermostat operates like those already described, but it also has a secondary valve. When the primary valve is closed, the circulation to the radiator is shut off. However, the secondary valve is open, permitting coolant to circulate through the bypass. But when the primary valve opens, permitting coolant to flow to the radiator, the secondary valve closes, thus blocking off the engine bypass.

⊘ 16-10 Radiator Pressure Cap To improve cooling efficiency and prevent evaporation and surge losses, many late-model automobiles use a pressure cap on the radiator (Figs. 16-21 and 16-22). At sea level, where atmospheric pressure is about 15 psi [1.055 kg/cm^2], water boils at 212 degrees Fahrenheit [100°C]. At higher altitudes, where atmospheric pressure is less, water boils at lower temperatures. Thus, with higher pressures, the temperature required to boil water increases. Each added pound per square inch (0.0703 kg/cm^2] increases the boiling point of water about $3\frac{1}{4}$ degrees Fahrenheit [1.8°C].

The use of a pressure cap on the radiator increases the air pressure within the cooling system several pounds per square inch. Thus, the water may be circulated at higher temperatures without boiling. The water therefore enters the radiator at a higher temperature, and so the difference in temperature between the air and water is greater. Heat is then more quickly transferred from the water to the air, improving cooling efficiency. Evaporation of water is reduced by the higher pressure because the boiling point of the water is higher. The pressure cap also prevents loss of water due to surging when the car is quickly braked to a stop.

The pressure cap fits over the radiator filler tube and seals tightly around the edges. The cap contains two valves: the blowoff valve and the vacuum valve. The blowoff valve consists of a valve held against a valve seat by a calibrated spring. The spring holds the valve closed so that pressure is produced in the cooling system. If pressure rises above that for which the system is designed, the blowoff valve is raised off its seat. This action relieves the excessive pressure. Pressure caps are designed to add as much as 15 psi [1.055 kg/cm^2] in the cooling system. This capacity increases the boiling point of the water to as much as 260 degrees Fahrenheit [126.6°C].

The vacuum valve prevents the formation of a vacuum in the cooling system when the engine has been shut off and begins to cool. If a vacuum forms, atmospheric pressure from the outside causes the small vacuum valve to open, admitting air into the radiator. Without a vacuum valve, the pressure within the radiator might drop so low that atmospheric pressure would collapse it.

In cooling systems with expansion tanks (⊘ 16-5), the radiator pressure cap is more or less permanently installed. It should not be removed. If coolant must be added to the cooling system, the coolant is put into the expansion tank (see Figs. 16-11 and 16-12). In these systems, the radiator overflow tube is connected to the expansion tank.

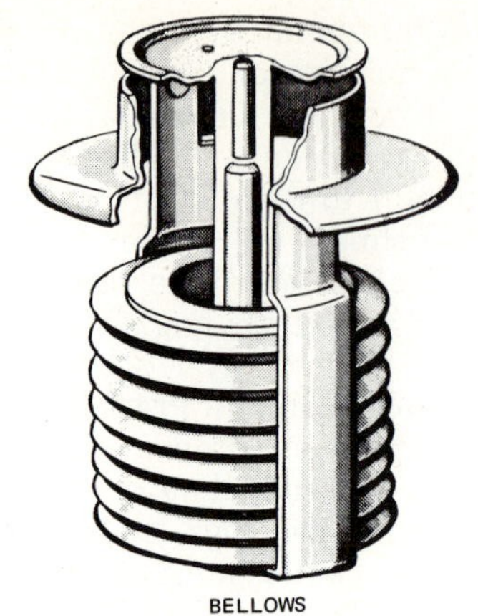

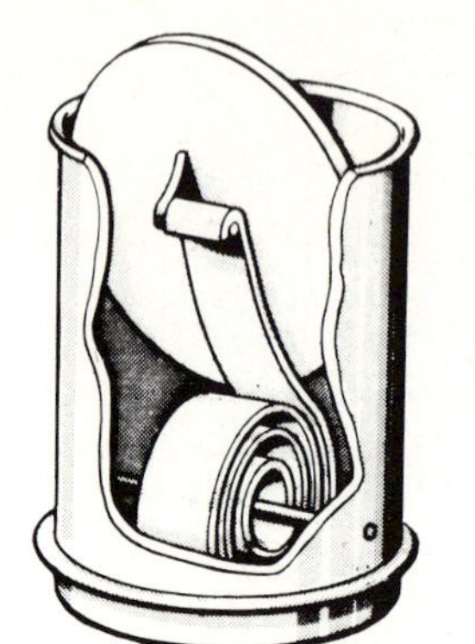

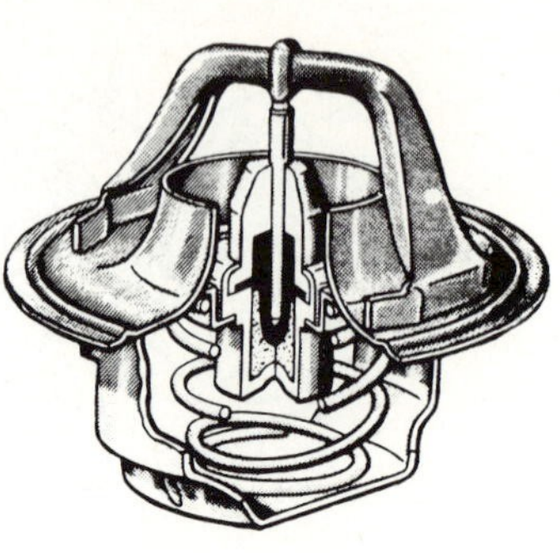

Fig. 16-18. Three types of cooling-system thermostats. (*Ford Motor Company*)

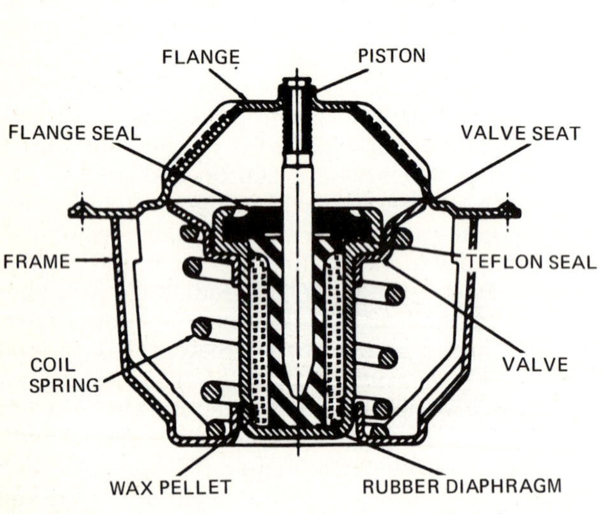

Fig. 16-19. Sectional view of a wax-pellet thermostat. (*Chevrolet Motor Division of General Motors Corporation*)

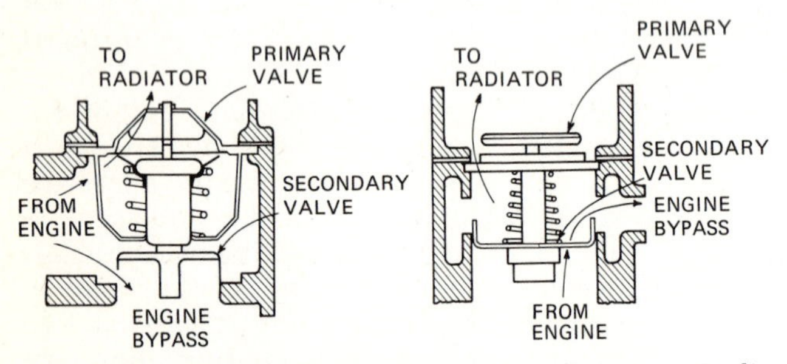

Fig. 16-20. Blocking bypass thermostats. (*Harrison Radiator Division of General Motors Corporation*)

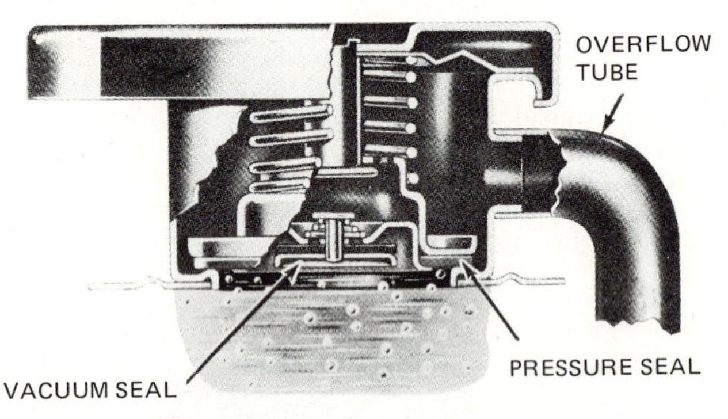

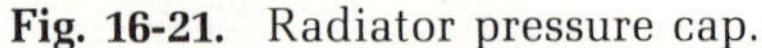

Fig. 16-21. Radiator pressure cap.

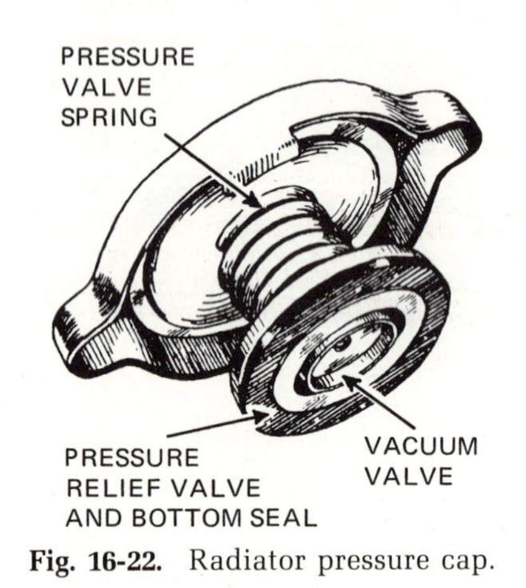

Fig. 16-22. Radiator pressure cap.

⊘ 16-11 Antifreeze Solutions Water freezes at 32 degrees Fahrenheit [0°C]. If water freezes in the engine cooling system, it stops coolant circulation. Some parts of the engine will overheat, which could seriously damage the engine. Worse, however, is the fact that water expands when it freezes. Water freezing in the cylinder block or cylinder head could expand enough to crack the block or head. Freezing water in the radiator could split the radiator seams. In either case, there is serious damage. A cracked block or head can seldom be repaired satisfactorily. A split radiator is hard to repair.

To prevent freezing of the water in the cooling system, antifreeze is added to the water to form the coolant. At one time, alcohol-base antifreeze was used, but now the most commonly used antifreeze is ethylene glycol. A mixture of half water and half ethylene glycol will not freeze above −34 degrees Fahrenheit [−36.7°C]. (This is 34 degrees Fahrenheit below zero, and it seldom gets that cold in any part of the United States.) A higher concentration of antifreeze will prevent freezing of the coolant at temperatures as low as −84 degrees Fahrenheit [−64.4°C].

Some antifreeze compounds also plug small leaks in the cooling-system radiator. These antifreeze compounds contain tiny plastic beads or inorganic fibers which circulate with the coolant. If a leak develops, the beads or fibers jam in the leak and plug it. Of course if the leak is too large, it will not be plugged. Also, these compounds cannot stop leaks in hoses, cylinder-head gaskets, or pump seals.

Corrosion protection is also built into antifreeze solutions. That is, compounds are added that fight corrosion inside engine water jackets and inside the radiator.

Antifreeze solutions also serve a purpose during hot-weather operation: They improve the hot-weather efficiency of the cooling system.

Car manufacturers recommend that the cooling system be drained, flushed out, and refilled with a fresh mixture of water and antifreeze periodically. One recommendation is that this be done every 2 years. Another is that it be done every year, preferably in the late fall just before freezing weather.

CAUTION: Cooling systems on modern cars should never be filled with water only. Indicator lights are designed not to come on to show excessive temperature until well above the boiling point of water. Thus, just plain water could boil even though the indicator light does not come on. The water-antifreeze coolant will not boil until a higher temperature is reached.

⊘ 16-12 Temperature Indicators The driver of a car should know the coolant temperature at all times. For this reason, a *temperature indicator* is installed on the instrument panel. An abnormal heat rise is a warning of abnormal conditions in the engine. The temperature indicator thus warns the driver to stop the engine before serious damage is done. Temperature indicators are of three general types: vapor-pressure, electric, and indicator-light.

1. *VAPOR-PRESSURE TEMPERATURE INDICATOR* The vapor-pressure temperature indicator consists of a metal indicator bulb and a metal tube connecting the bulb to the indicator unit. The indicator unit contains a curved, or Bourdon, tube. One end of the tube is linked to the indicator needle, and the other end is open and is connected through a tube to the indicator bulb. The indicator bulb is usually placed in the water jacket of the engine. The bulb is filled with a liquid that vaporizes at fairly low temperatures. As the engine temperature increases, the liquid in the bulb begins to vaporize. This creates pressure that is conveyed through the connecting tube to the Bourdon tube in the indicating unit. The pressure tends to straighten out the tube. The result is that the movement causes the indicating needle to move across the dial face and indicate the temperature in the water jacket.

2. *ELECTRIC TEMPERATURE INDICATOR* Electrically operated temperature indicators are of two types: the balancing-coil type and the bimetal-thermostat type. The balancing-coil oil-pressure indicator (Fig. 17-13), fuel gauge (Fig. 13-7), and temperature indicator operate similarly. The instrument-panel indicating units are practically identical. Each indicator consists of two coils and an armature to which a needle is attached (Fig. 16-23). The engine unit changes resistance with temperature. At higher temperatures it has less resistance and thus passes more current. When this happens, more current passes through the right-hand coil in the indicating unit. In this case, the armature to which the needle is attached is attracted by the increased magnetic field. The armature and the needle move around so that the needle indicates a higher temperature.

The bimetal-thermostat temperature indicator is similar to the bimetal-thermostat fuel gauge (Fig. 13-8). The instrument-panel, or dash, units are practically identical.

3. *INDICATOR LIGHTS* One indicator-light system is shown in Fig. 16-24. This system has a coolant-temperature sending unit mounted on the engine so that it is exposed to the cooling-system coolant. The sending unit is connected to two light bulbs and the battery through the ignition switch. When the ignition switch is first turned on, to start a cold engine, the sending-unit thermostatic blade is in the proper position to connect the COLD light to the battery. It comes on. The COLD light, which appears in blue on the instrument panel, remains on until the engine approaches operating temperature. As this happens, the thermostatic blade in the sending unit is bent by the increasing temperature. The blade therefore moves off the cold terminal, disconnecting the COLD light so that it turns off. If the engine overheats, the thermostat will warp further so that it moves under the hot terminal. This connects the HOT bulb to the battery so that it glows and appears in red on the instrument panel. This is a signal to the driver that the engine has overheated and should be stopped before damage results.

Some indicator-light systems do not have a COLD

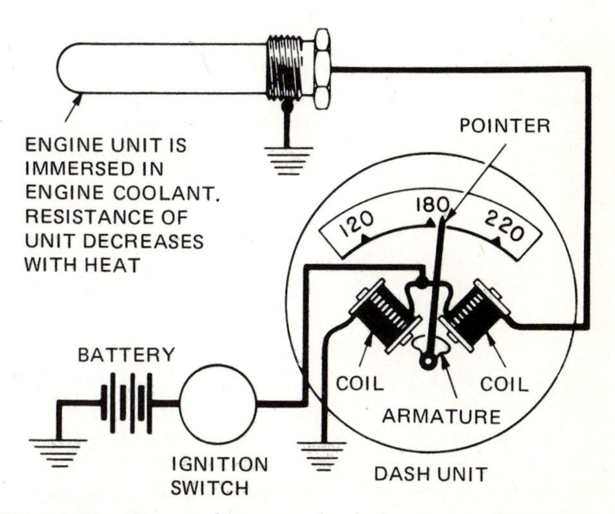

Fig. 16-23. Circuit diagram of a balancing-coil temperature-indicator system.

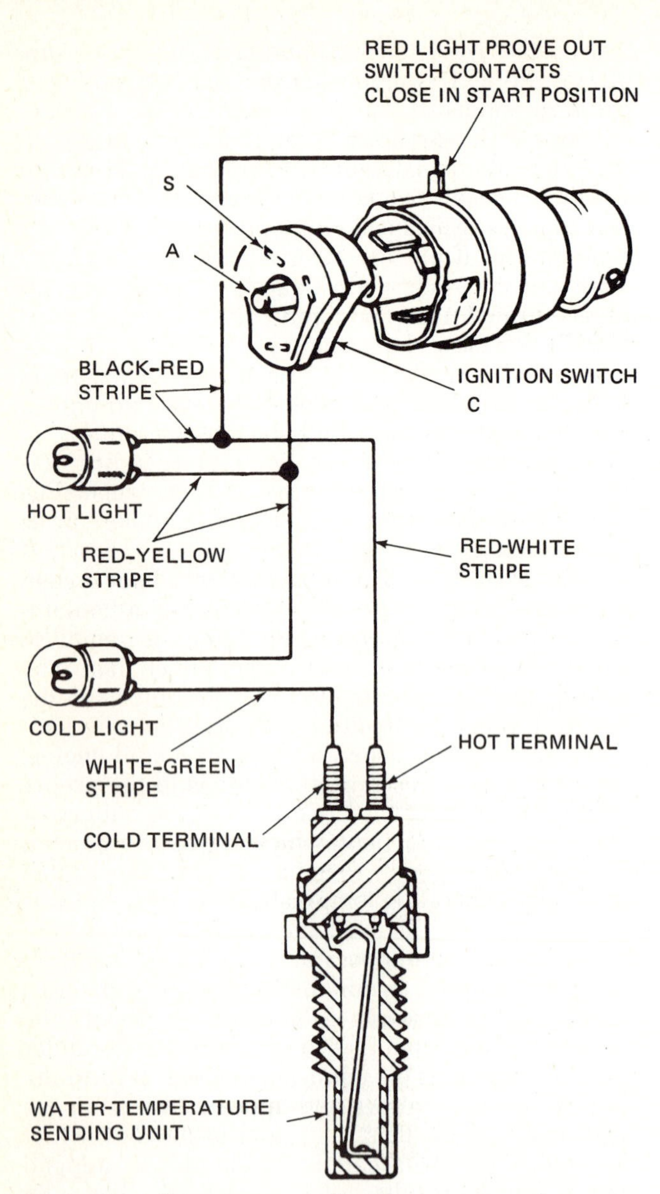

Fig. 16-24. Temperature-indicating system using COLD and HOT indicating lights. (*Ford Motor Company*)

light. Instead, the HOT light comes on during cranking and then goes off when the engine starts. The HOT light stays off unless the engine overheats. Then the light comes on to signal that the engine is getting too hot.

⊘ 16-13 Cooling-System Troubles For normal engine operation the cooling system must permit quick warm-up and then provide adequate cooling when the engine reaches operating temperature. Troubles related directly to the cooling system include *slow warm-up* and *engine overheating*. (Both of these troubles can also be caused by other conditions; for instance, the wrong ignition timing can cause the engine to overheat.) Slow warm-up could result from a thermostat's sticking open. Engine overheating could be caused by a thermostat that is stuck closed, by low water level in the system, or by accumulations of rust or scale that have clogged water passages or filled water jackets. Chapter 22 contains a trouble-diagnosis chart that relates various engine troubles to possible causes. When you study that chart, you will learn more about how cooling-system troubles and engine troubles are related.

Further information on cooling-system troubles and remedies is contained in *Automotive Fuel, Lubricating, and Cooling Systems*, another book in the McGraw-Hill Automotive Technology Series.

CHAPTER 16 CHECKUP

NOTE: Since the following is a chapter review test, you should review the chapter before taking the test.

You should have a good understanding of how the engine cooling system operates. The questions that follow will help you check yourself on how well you remember and understand what you have been reading. If the questions stump you, reread the past few pages to find the information you need.

Completing the Sentences The sentences that follow are incomplete. After each sentence there are several words or phrases, only one of which will correctly complete the sentence. Write each sentence in your notebook, selecting the proper word or phrase to complete it correctly.

1. The pump part that rotates to cause water circulation between the radiator and engine water jackets is called the: (*a*) impeller, (*b*) fan, (*c*) body, (*d*) bypass.
2. Two types of radiators discussed in the chapter are: (*a*) zigzag and fan-cooled, (*b*) horizontal and vertical, (*c*) tubular and cellular.
3. In normal operation water in the down-flow radiator circulates: (*a*) from top to bottom, (*b*) from bottom to top, (*c*) in a circular path in the radiator.
4. The part of the cooling-system thermostat that functions to open and close the valve is called the: (*a*) seater, (*b*) bellows or wax pellet, (*c*) pressure valve, (*d*) vacuum valve.
5. The device in the cooling system that raises the boiling point of the water in the system is called the: (*a*) pressure cap, (*b*) vacuum valve, (*c*) radiator, (*d*) water jackets.
6. The pressure cap contains two valves; these are the: (*a*) pressure valve and blowoff valve, (*b*) atmospheric valve and vacuum valve, (*c*) blowoff valve and vacuum valve.
7. Two types of antifreeze are: (*a*) alcohol-base and ethylene glycol, (*b*) ethylene glycol and permanent, (*c*) iso-octane and ethylene glycol.
8. Troubles of the engine related directly to the cooling system include: (*a*) hard starting and slow warm-up, (*b*) slow warm-up and overheating, (*c*) slow cranking and warm-up.
9. With reference to the direction that the water

flows through them, radiators can be classified as: (*a*) down-flow and up-flow, (*b*) right-flow and left-flow, (*c*) direct-flow and cross-flow, (*d*) down-flow and cross-flow.

10. In the blocking bypass thermostat, when the primary valve is opened, the secondary valve: (*a*) permits circulation to the radiator, (*b*) blocks off water flow to the radiator, (*c*) blocks off the engine bypass.

Review Questions When you answer a question, write it in your notebook.

1. What is the purpose of the cooling system?
2. How does the cooling system operate?
3. How does the water pump operate?
4. Describe the operation of the variable-speed engine fan.
5. What are the two general ways radiators can be classified? Describe each.
6. Describe the operation of a cooling-system thermostat.
7. Describe the operation of a cooling-system radiator pressure cap.
8. Describe a sealed cooling system.
9. Describe the indicator-light system discussed in the chapter.
10. What is the purpose of the expansion tank?
11. How does the radiator-shutter system work?
12. What is the purpose of the transmission oil cooler?

SUGGESTIONS FOR FURTHER STUDY

Automotive Fuel, Lubricating, and Cooling Systems, another book in the McGraw-Hill Automotive Technology Series, contains additional information on cooling-system operation, troubles, and servicing. You can also find out a good deal about the cooling system and its components in a friendly service shop or your school shop. Examine water-pump parts and thermostats and note the manner in which the radiator is mounted and connected with hoses to the engine and water pump. Write any interesting facts you learn in your notebook.

chapter 17

ENGINE LUBRICATING SYSTEMS

This chapter describes various types of engine lubricating systems and the operation of their component parts. In addition, the purpose and properties of lubricating oil are discussed. We have already discussed friction in the engine (⊘ 11-13), engine bearings (⊘ 7-18 to 7-27), and the action of the piston rings in controlling oil on the cylinder walls (⊘ 8-3 to 8-9). If you are not clear on any of these topics, reread the appropriate sections. They are closely related to the discussions of oil and engine lubricating systems in this chapter.

⊘ 17-1 Purpose of the Lubricating System We normally think of lubricating oil as a substance that reduces wear and friction between adjacent moving surfaces. However, the lubricating oil that circulates through the engine performs other jobs. The engine lubricating oil must:

1. Lubricate moving parts to minimize wear and power loss from friction
2. Remove heat from engine parts by acting as a cooling agent
3. Absorb shocks between bearings and other engine parts, thus reducing engine noise and extending engine life
4. Form a good seal between piston rings and cylinder walls
5. Act as a cleaning agent

1. MINIMIZING WEAR AND POWER LOSS FROM FRICTION Friction has been discussed in some detail (⊘ 4-31). Friction encountered in the engine is normally viscous friction, that is, the friction between adjacent moving layers of oil. If the lubricating system does not function properly, sufficient oil will not be supplied to moving parts. Greasy or even dry friction will result between moving surfaces. This would cause considerable power loss since much power must be used to overcome these types of friction. At most, major damage would occur to engine parts as greasy or dry friction developed. Bearings would wear with extreme rapidity. The heat resulting from greasy or dry friction would cause bearing failure, so that connecting rods and other parts would be broken. Insufficient lubrication of cylinder walls would cause rapid wear and scoring of walls, rings, and pistons. A properly operating engine lubricating system supplies all moving parts with enough oil so that only viscous friction exists in the engine.

2. REMOVING HEAT FROM ENGINE PARTS The engine oil circulates through the engine lubrication system. All bearings and moving parts are bathed in streams of oil. In addition to lubricating parts, the oil absorbs heat from engine parts and carries this heat back into the oil pan. The oil pan absorbs heat from the oil, transferring it to the surrounding air. The oil thus acts as a cooling agent.

3. ABSORBING SHOCKS BETWEEN BEARINGS AND OTHER ENGINE PARTS As the piston approaches the end of the compression stroke, the air-fuel mixture in the cylinder is ignited. Pressure in the cylinder suddenly increases many times. A load of as much as 2 tons [1,814 kg] is suddenly placed on the top of a 3-inch [76.2 mm] piston. This sudden increase in pressure causes the piston to thrust down hard through the piston-pin bearing, connecting rod, and connecting-rod bearing. There is always some space, or clearance, between bearings and journals; this space is filled with oil. When the load on the piston suddenly increases as described above, the layers of oil between bearings and journals must act as cushions. They must resist penetration, or "squeezing out." A film of oil must remain between the adjacent metal surfaces. In thus absorbing and cushioning the hammerlike effect of the sudden loads, the oil quiets the engine and reduces wear of parts.

4. FORMING A SEAL BETWEEN PISTON RINGS AND CYLINDER WALLS Piston rings must form a gastight seal with the cylinder walls. The lubricating oil that is delivered to the cylinder walls helps the piston rings to accomplish this. The oil film on the cylinder walls makes up for microscopic unevenness in the fit between the rings and walls. The film fills in any gaps through which gas might escape. The oil film also lubricates the rings, so that they move easily in the ring grooves and on the cylinder walls.

5. ACTING AS A CLEANING AGENT As it circulates the oil tends to wash off and carry away dirt, particles of carbon, and other foreign matter. The oil picks up this material and carries it back to the crankcase. There, larger particles drop to the bottom of the oil pan. Smaller particles are filtered out by the oil filter.

⊘ 17-2 Properties of Oil A satisfactory engine lubricating oil must have certain characteristics, or properties. It must have proper viscosity (body and fluidity). It must resist oxidation, carbon formation, corrosion, rust, extreme pressures, and foaming. Also, it must act as a good cleaning agent, pour at low temperatures, and have good viscosity at very high and low temperatures.

Any mineral oil, by itself, does not have all these properties. Lubricating-oil manufacturers therefore put a number of additives into the oil during the manufacturing process. An oil for severe service may have many additives, as follows:

A viscosity-index improver
Pour-point depressants
Oxidation inhibitors
Corrosion inhibitors
Rust inhibitors
Foam inhibitors
Detergent-dispersants
Extreme-pressure agents

1. VISCOSITY (BODY AND FLUIDITY) Primarily, *viscosity* is the most important property of lubricating oil. Viscosity refers to the tendency of oil to resist flowing. In a bearing and journal, layers of oil adhere to the bearing and journal surfaces. These layers must move, or slip, with respect to each other. The viscosity of the oil determines the ease with which this slipping can take place. For purposes of discussion, viscosity may be divided into two parts: body and fluidity. *Body* gives the oil resistance to oil-film puncture, or penetration, during the application of heavy loads. When the power stroke begins, for example, bearing loads sharply increase. Oil body prevents the load from squeezing out the film of oil between the journal and the bearing. This property cushions shock loads and helps maintain a good seal between piston rings and cylinder walls. The oil body maintains an adequate oil film on all bearing surfaces under load.

Fluidity has to do with the ease with which the oil flows through oil lines and spreads over bearing surfaces. In some respects, fluidity and body are opposing characteristics since the more fluid an oil is, the less body it has. The oil used in any particular engine must have sufficient body to perform as explained in the previous paragraph. But the oil must have sufficient fluidity to flow freely through all oil lines and spread effectively over all bearing surfaces. Late-model engines have more closely fitted bearings with smaller clearances. These engines require oils of greater fluidity that will flow readily into the spaces between bearings and journals.

Increasing engine temperature decreases the viscosity of oil. That is, increasing temperature causes the oil to lose body and gain fluidity. Decreasing temperature increases the viscosity of oil. The oil gains body and loses fluidity. Engine temperatures range several hundred degrees from cold-weather starting to operating temperature. Thus a lubricating oil must have adequate fluidity at low temperatures so that it will flow. At the same time it must have sufficient body for high-temperature operation.

2. VISCOSITY RATINGS Viscosity of oil is determined by a *viscosimeter*. This device determines the length of time required for a definite amount of oil to flow through an opening of a definite size. Temperature is taken into consideration during the test. (As explained previously, high temperature decreases viscosity. Low temperature increases viscosity.) In referring to viscosity, oils with the lower numbers are of lower viscosity (they are thinner). The Society of Automotive Engineers (SAE) rates oil viscosity in two different ways: for winter use and for other than winter use. Winter-grade oils are tested at 0 and 210 degrees Fahrenheit [−17.8 and 98.9°C]. There are three grades: SAE5W, SAE10W, and SAE20W, the "W" indicating winter grade. For other than winter use, the grades are SAE20, SAE30, SAE40, and SAE50 (all without the "W" suffix).

3. VISCOSITY INDEX (VI) When oil is cold, it is thicker and runs more slowly than when it is hot. It is more viscous when cold. Thus, the engine is harder to start when it is cold because the oil is more viscous. In recent years, oil chemists have developed viscosity-index (VI) improvers. These compounds tend to reduce oil viscosity when it is cold and increase oil viscosity when it is hot. Thus, the oil makes cold starting easier and yet does not thin out too much when it is hot. Oils with these characteristics are called *multiple-viscosity oils*. For example, an oil may be rated SAE10W-30. This means that the oil is the same as SAE10W when it is cold and the same as SAE30 when it is hot.

4. POUR-POINT DEPRESSANTS Pour-point depressants depress, or lower, the temperature at which oil becomes too thick to flow. Thus, this additive keeps the oil fluid at low temperatures for cold-weather starts.

5. RESISTANCE TO CARBON FORMATION Cylinder walls, pistons, and piston rings operate at temperatures of several hundred degrees. These temperatures are high enough to cause the oil to break down and form carbon. The less carbon in the engine cylinders, the better. Therefore oil chemists regulate the refining process to make sure that lubricating oil has good resistance to carbon formation.

6. OXIDATION INHIBITORS When oil is heated and then stirred up (as happens in the crankcase), the oxygen in the air tends to combine, or oxidize, with the oil. As oil oxidizes, various harmful substances can form, including some that are tarlike and

others that are like varnish. To prevent this, additives are put into the oil to prevent oxidation.
7. *CORROSION AND RUST INHIBITORS* At high temperatures, acids may form in the oil which can corrode engine parts, especially bearings. Corrosion inhibitors are added to the oil to prevent this corrosion. Also, rust inhibitors are added. These displace water from metal surfaces so that oil coats the surfaces. Rust inhibitors also neutralize combustion acids.
8. *FOAM INHIBITORS* The churning action in the engine crankcase also causes the engine oil to foam. This is like an egg beater causing an egg white to form a frothy foam. As the oil foams, it tends to overflow, or to be lost through the crankcase ventilator (⊘ 17-17). In addition, the foaming oil does not provide normal lubrication of bearings and other moving parts. Foaming oil in hydraulic valve lifters causes them to function poorly and noisily, wear rapidly, and possibly break. To prevent foaming, foam inhibitors are mixed with the oil.
9. *DETERGENT-DISPERSANTS* Despite the filters and screens at the carburetor and crankcase ventilator, dirt does get into the engine. In addition, as the engine runs, the combustion processes leave deposits of carbon on piston rings, valves, and other parts. Also, some oil oxidation may take place, resulting in still other deposits. Then, too, metal wear in the engine puts particles of metal into the oil. As a result, deposits tend to build up on and in engine parts. The deposits reduce engine performance and speed up wear of parts.

To prevent or slow down the formation of these deposits, some engine oils contain a detergent additive. The detergent acts much like ordinary soap. When you wash your hands with soap, the soap surrounds the particles of dirt on your hands. The particles become detached from your hands, and water rinses them away. In a similar manner, the detergent in the oil loosens and detaches the deposits of carbon, gum, and dirt. The oil then carries the particles away. Large particles drop to the bottom of the crankcase. Small particles remain suspended in the oil. These impurities, or contaminants, are flushed out when the oil is changed.

To prevent the detached small particles from clotting and to keep them in a finely divided state, a dispersant is added to the oil. Without the dispersant, the particles would tend to collect and form larger particles. These larger particles might then block the oil filter and reduce its effectiveness. They could also build up in oil passages and plug them, thus depriving bearings and other engine parts of oil. The dispersant prevents this. It greatly increases the amount of particles, or contaminants, the oil can carry and still function effectively.

Lubricating-oil manufacturers now place more emphasis on the dispersant qualities of an additive than on its detergent qualities. If the contaminants can be kept suspended in the oil as small particles, they will not be deposited on engine parts. There will be less need for detergent action.

10. *EXTREME-PRESSURE AGENTS* In modern automotive engines, the lubricating oil is subjected to very high pressures in the bearings and valve train. Modern valve trains have heavy valve springs and high-lift cams. This means that the valves must move farther against heavier spring loads. To prevent the oil from squeezing out, extreme-pressure agents are added to the oil. These agents react chemically with metal surfaces to form very strong, slippery films, which may be only a molecule or so thick. Thus, these additives supplement the oil by providing protection during moments of extreme pressure.

⊘ 17-3 Sludge Formation *Sludge* is a thick, creamy, black substance that often forms in the crankcase. It clogs oil screens and oil lines, preventing normal circulation of lubricating oil to engine parts. Sludge formation can result in engine failure from oil starvation.
1. *HOW SLUDGE FORMS* Water collects in the crankcase in two ways: First, water is formed as a product of combustion (⊘ 12-1). Second, the crankcase ventilating system (⊘ 17-17) carries air, with moisture in it, through the crankcase. If the engine parts are cold, the water condenses and drops into the crankcase. There, it is churned up with the lubricating oil by the crankshaft. The crankshaft acts much like a giant egg beater. It whips the oil and water into the thick, black, mayonnaiselike "goo" known as *sludge*. (The black color comes from dirt and carbon.)
2. *WHY SLUDGE FORMS* If a car is driven for long distances each time it is started, the water that collects in the crankcase while the engine is cold quickly evaporates. The crankcase ventilating system then removes the water vapor. Thus, no sludge will form. However, if the engine is operated when cold most of the time, then sludge will form. For example, the home-to-shop-to-home sort of driving, each trip being only a few miles, is sludge-forming. When a car is used for short-trip start-and-stop driving, the engine never has a chance to warm up enough to get rid of the water that collects in the crankcase. The water accumulates and forms sludge.
3. *PREVENTING SLUDGE FORMATION* To prevent sludge formation, a car must be driven long enough for the engine to heat up and get rid of the water in the crankcase. This means trips of 12 miles [19.312 km] or more in winter (but fewer miles in summer). If trips of this length are impractical, then the oil must be changed frequently. Naturally, during cold weather, it takes longer for the engine to warm up. Thus in cold weather, trips must be longer or the oil must be changed more often to prevent water accumulation and thus sludge formation.

⊘ 17-4 Service Ratings of Oil We have already mentioned that lubricating oil is rated by viscosity number (SAE10W-30, for example). Lubricating oil is also rated in another way, by its *service designa-*

tion. That is, it is rated according to the type of service for which it is best suited. There are five service ratings for gasoline-engine lubricating oils: SA, SB, SC, SD, and SE. There are four service ratings for diesel-engine lubricating oils: CA, CB, CC, and CD. The oils differ in their properties and in the additives they contain.

1. *SA OIL* SA oil is used in utility gasoline and diesel engines that are operated under mild conditions so that protection by additives is not required. This oil may have pour-point and foam depressants.
2. *SB OIL* SB oil is used in gasoline engines that are operated under such mild conditions that only minimum protection by additives is required. Oils designed for this service have been used since the 1930s. They provide only antiscuff capability and resistance to oil oxidation and bearing corrosion.
3. *SC OIL* SC oil is for service typical of gasoline engines in the 1964 to 1967 models of passenger cars and trucks. It is intended primarily for use in passenger cars. This oil provides control of high- and low-temperature engine deposits, wear, rust, and corrosion.
4. *SD OIL* SD oil is for service typical of gasoline engines in passenger cars and trucks beginning with 1968 models. This oil provides more protection from high- and low-temperature engine deposits, wear, rust, and corrosion than SC oil.
5. *SE OIL* SE oil is for service typical of gasoline engines in passenger cars and some trucks beginning with 1972 (and some 1971) models. This oil provides more protection against oil oxidation, high-temperature engine deposits, rust, and corrosion than do oils with the SC and SD ratings.

Diesel-engine oils must have different properties than oils for gasoline engines. The CA, CB, CC, and CD ratings indicate oils for increasingly severe diesel-engine operation. For example, CA oil is for light-duty service, and CD oil is for severe-duty service typical of high-speed, high-output diesel engines.

All car manufacturers recommend the use of a high-detergent engine oil such as an SE oil of the proper viscosity.

CAUTION: Do not confuse viscosity and service ratings of oil. Some people think that a high-viscosity oil is a heavy-duty oil. This is not necessarily so. Viscosity ratings refer to the thickness of the oil; thickness is not a measure of heavy-duty quality. Remember that there are two ratings: viscosity and service. Thus, an SAE10 oil can be an SC, SD, or SE oil. An oil of any other viscosity rating can have any one of the service ratings.

⊘ 17-5 Oil Changes From the day that fresh oil is put into the engine crankcase, it begins to lose its effectiveness as an engine lubricant. This gradual loss of effectiveness is largely due to the depletion, or "wearing out" of the additives. For example, the antioxidant additive becomes used up, allowing gum and varnish to form. The corrosion and rust inhibitors are gradually depleted, allowing corrosion and rust formation to begin. In addition, during engine operation, carbon tends to form in the combustion chamber. Some of this carbon gets into the oil. Also, the air that enters the engine (in the air-fuel mixture) carries a certain amount of dust. Even though the air filter is operating efficiently, it will not remove all the dust. Then, too, the engine releases fine metal particles as it wears.

All these substances tend to circulate with the oil. As the mileage piles up, the oil accumulates more and more of these contaminants. Even though the engine has an oil filter, some of these contaminants remain in the oil. Finally, after many miles of operation, the oil is so loaded with contaminants that it is not safe to use. Unless the oil is drained and clean oil is put in, engine wear will increase rapidly.

Different automotive manufacturers have different recommendations on how often engine oil should be changed. Chrysler Corporation recommends that for normal service the oil should be changed every 3 months or 4,000 miles [6.437 km], whichever comes first. For severe service (extended periods of idling and short-trip operation, dusty driving conditions, towing trailers, and so on), the oil should be changed more often—every 2 months or 2,000 miles [3.219 km].

Ford recommends that for normal service the oil should be changed every 4 months or 4,000 miles [6.437 km], whichever occurs first. For severe service, the oil should be changed every 2 months or 2,000 miles [3,219 km].

General Motors recommends that for normal service, the oil should be changed every 4 months or 6,000 miles [9,656 km], whichever comes first. For severe service, the oil should be changed every 2 months or 3,000 miles [4,828 km].

The oil filter should be changed at the first oil change on a new engine and at *every other* oil change after that. But, for severe service, when the oil is changed more frequently, the oil filter should be changed at *every* oil change.

⊘ 17-6 Oil Consumption Oil is lost from the engine in three ways: by burning in the combustion chambers, by leakage in liquid form (Fig. 17-1), and by passing out of the crankcase in the form of mist. [In the *positive crankcase ventilating* (PCV) *system* this mist will also be burned in the combustion chambers; see ⊘ 17-17.] Two main factors affect oil consumption: engine speed and wear of engine parts.

High speed produces high temperature. High temperature, in turn, lowers the viscosity of the oil. With lowered viscosity, the oil can more readily work past the piston rings into the combustion chamber, where it is burned. In addition, high speed exerts a centrifugal effect on the oil feeding through the oil lines drilled in the crankshaft to the connecting-rod journals. Thus, more oil is fed to the bearings and thrown on the cylinder walls. Also, high speed causes ring "flutter," or "shimmy." With this condi-

Fig. 17-1. Partial sectional view of an engine showing points where oil may be lost. (*Federal-Mogul Corporation*)

tion, the oil-control rings cannot function effectively. Crankcase ventilation (⊘ 17-17) causes more air to pass through the crankcase at high speed. This causes the loss of oil in the form of mist.

As engine parts wear, oil consumption increases. Worn bearings tend to throw more oil onto the cylinder walls. Tapered and worn cylinder walls prevent normal oil-control-ring action. The rings cannot change shape rapidly enough to conform with the worn cylinder walls as the rings move up and down. More oil thus gets into the combustion chamber, where it burns and fouls spark plugs, valves, rings, and pistons. Carbon formation makes the condition worse since it further reduces the effectiveness of the oil-control rings. Where the wear of cylinder-walls is not excessive, special oil-control rings can be installed. They improve the wiping action so that less oil can move past the rings. After cylinder-walls have worn beyond a certain point, however, the cylinders must be machined and new rings installed to reduce oil consumption.

Worn intake-valve guides also increase oil consumption. Oil leaks past the valve stems and is pulled into the combustion chamber along with the air-fuel mixture every time the intake valves open. Worn exhaust-valve guides can also cause high oil consumption. When the exhaust valve opens, hot exhaust gases hit the oil and it burns. Installation of new valve guides, reaming of guides and installation of valves with oversize stems, or installation of valve-stem seals will reduce oil consumption from these causes.

⊘ 17-7 Lubricating Systems for Four-Cycle Engines Two types of lubricating systems are used in four-cycle engines: the *splash* and *pressure-feed* systems. [Two-cycle engines and the Wankel engine require different kinds of lubricating systems (see ⊘ 17-8 and 17-10).]

1. *SPLASH SYSTEM* In the splash lubricating system, oil is splashed from the oil pan into the lower part of the crankcase. Usually, the connecting rod has a dipper that dips into the crankcase oil each time the piston reaches BDC. This action splashes the oil (Fig. 17-2). Some small engines also use oil slingers which are driven by the camshaft. These are gearlike parts that throw oil from the oil pan up into the moving engine parts. The splash system is used on most small four-cycle engines for power lawn mowers and similar applications.

2. *PRESSURE-FEED SYSTEM* In the pressure-feed lubricating system, many of the engine parts are lubricated by oil that is fed to them under pressure from the oil pump (Figs. 17-3 and 17-4). The oil from the pump enters an oil line (or a drilled header, channel, or gallery, as it is variously called). From the oil line, the oil flows to the main bearings and camshaft bearings. The main bearings have oil-feed holes or grooves that feed oil into drilled passages in the crankshaft. The oil flows through these passages to the connecting-rod bearings. From there, on some engines, it flows through holes drilled in the connecting rods to the piston-pin bearings.

In I-head engines, oil is fed under pressure to the valve mechanisms in the head. For example, some engines have the rocker arms mounted on hollow shafts. These shafts feed oil to the rocker arms. Some engines with independently mounted rocker arms (Fig. 9-25) have hollow mounting studs. These studs feed oil from an oil gallery in the head to the rocker-arm ball pivots. The oil spills off the rocker arms and provides lubrication for the valve stems

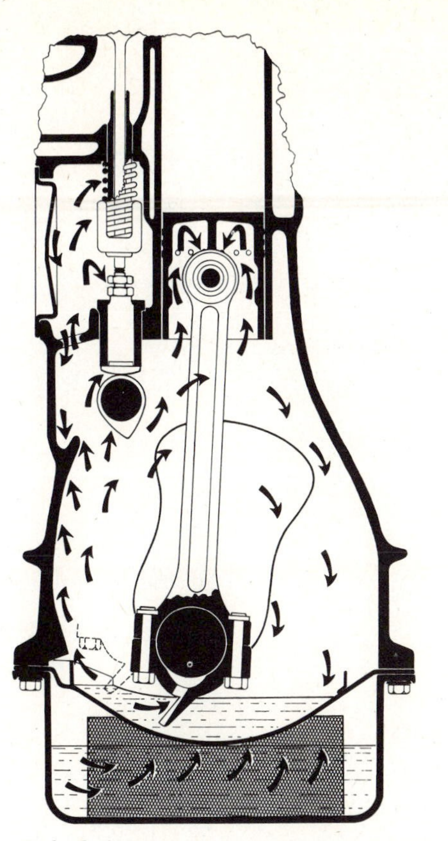

Fig. 17-2. Splash lubricating system used on an in-line engine. An oil pump maintains the proper level of oil in the tray under the connecting rods. Arrows show flow of oil splash to engine parts.

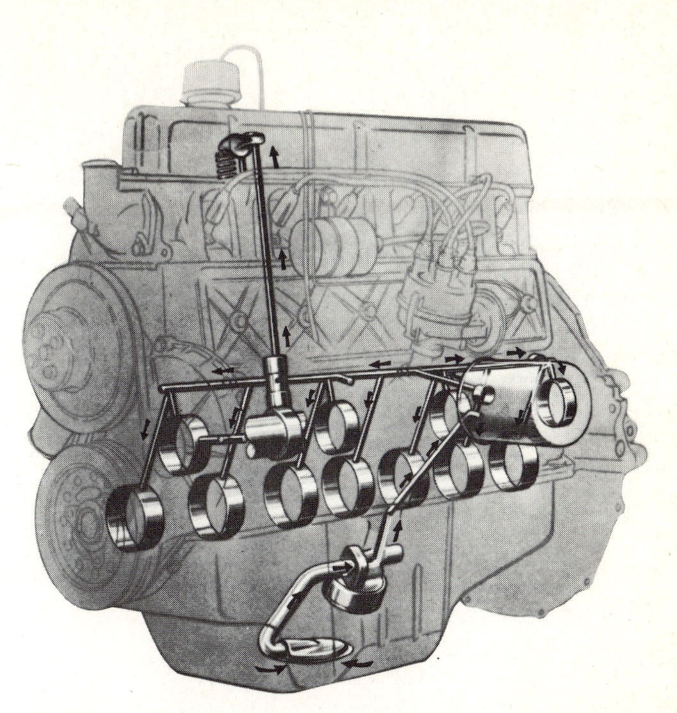

Fig. 17-3. Lubricating system for a six-cylinder, overhead-valve engine with seven main bearings. (*Ford Motor Company*)

and pushrod and valve-stem tips. Thus, all valve-mechanism parts are adequately lubricated. On other engines, the oil flows up through hollow push-rods to lubricate the valve stems and other valve-train parts.

Cylinder walls are lubricated by splashing oil thrown off from the connecting-rod bearings. Some engines have oil-spit holes or grooves in the connecting rods that index (align with) with drilled holes in the crankpin journals with each revolution. As this happens, a stream of oil is spit, or thrown, onto the cylinder walls (Fig. 17-5). On many V-8 engines, the oil-spit holes or grooves are arranged so that they deliver their jets of oil into opposing cylinders in the other cylinder bank. That is, the oil-spit holes in the connecting rods in the right-hand bank lubricate the cylinder walls in the left-hand bank, and vice versa. In many engines, the piston pins are lubricated with oil scraped off the cylinder walls by the piston rings. The pistons have grooves, holes, or slots to feed oil from the oil-control-ring groove or from oil scoops on the piston to the piston-pin bushings.

Oil passages in the block and head permit circulation of oil to bearings and moving parts (Fig. 17-6). Some engines have holes drilled in the connecting rod (as shown in Fig. 8-18) to carry oil up to the piston pin and lubricate it.

⊘ 17-8 Lubrication of Two-Cycle Engines In two-cycle engines, the air-fuel mixture passes through the crankcase on its way from the carburetor to the engine cylinders (⊘ 6-12). For this reason it is not possible to maintain a reservoir of oil in the crankcase. The oil would be picked up by the passing air-fuel mixture, carried to the engine cylinders, and burned. Therefore, to provide lubrication of two-cycle engine parts, the oil is mixed with the fuel. As the air and oil-fuel mixture enters the crankcase, the fuel, being more volatile, evaporates and passes on to the engine as an air-fuel mixture. Some of the oil is carried along with the air-fuel mixture and is burned. But enough oil is left behind to keep the moving engine parts coated with oil and thus adequately lubricated.

⊘ 17-9 Overhead-Camshaft-Engine Lubrication In the overhead-camshaft (OHC) engine, additional lubrication must be furnished to the cylinder head for the camshaft bearings. Figure 17-7 shows the lubricating system for the Opel OHC engine. Note the oil gallery running the length of the cylinder head. The oil gallery supplies oil to the camshaft bearings. The gallery also supplies oil to the hydraulic valve lifters and valve-train parts.

⊘ 17-10 Wankel-Engine Lubrication Figure 17-8 (p. 250) shows the lubricating system for the Mazda Wankel engine. The oil pump is at the bottom of the engine (lower left). The arrows show the flow of oil from the oil pump to the engine. Note that the system uses an oil cooler. The oil cooler helps remove excess heat from the engine.

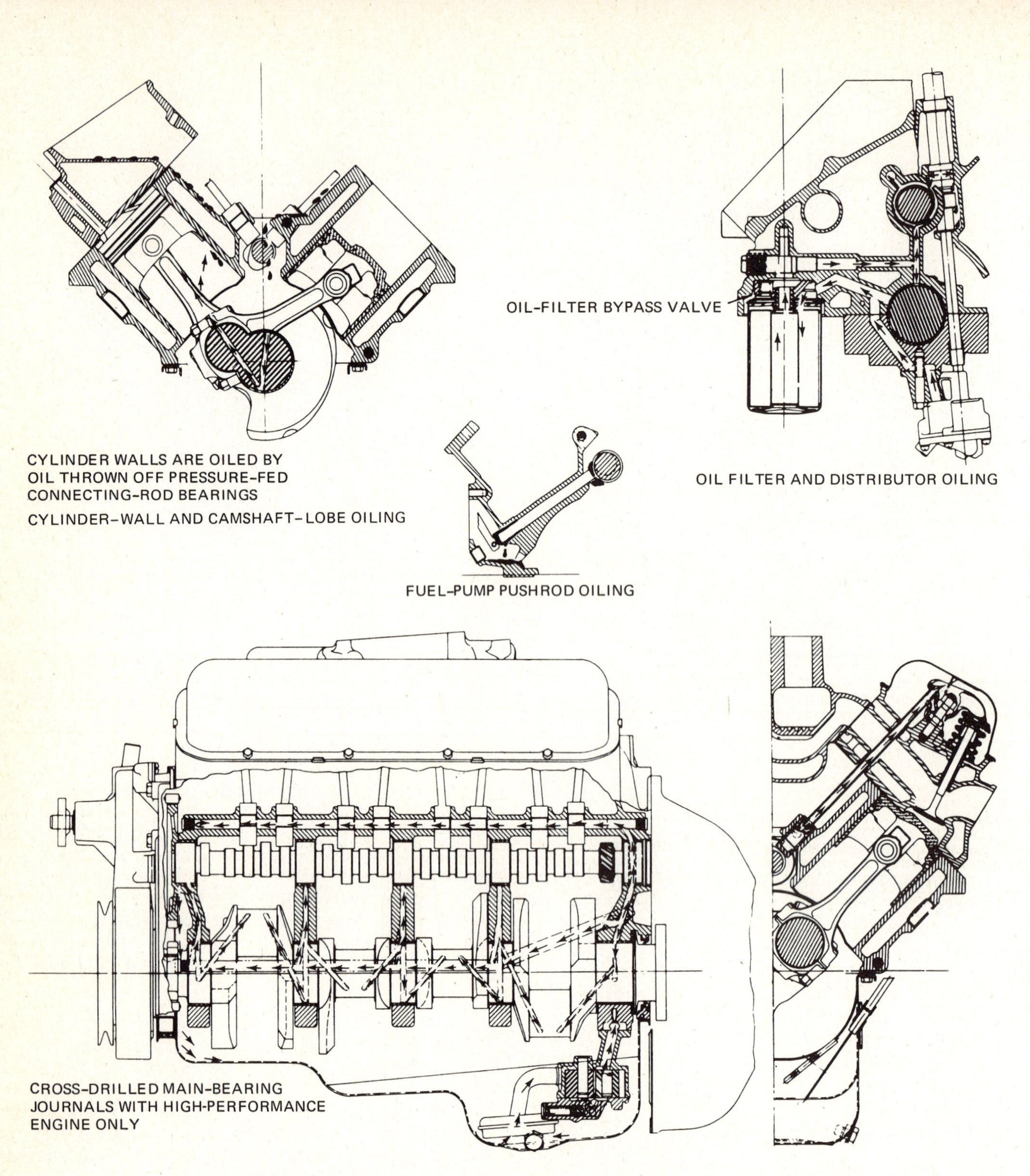

Fig. 17-4. Lubricating system of a V-8 overhead-valve engine. Arrows show the flow of oil to the moving parts in the engine. (*Chevrolet Motor Division of General Motors Corporation*)

Fig. 17-5. When a hole in the connecting rod aligns with a hole in the crankpin, oil is sprayed onto the cylinder wall, as shown. This lubricates the piston and rings. (*Ford Motor Company*)

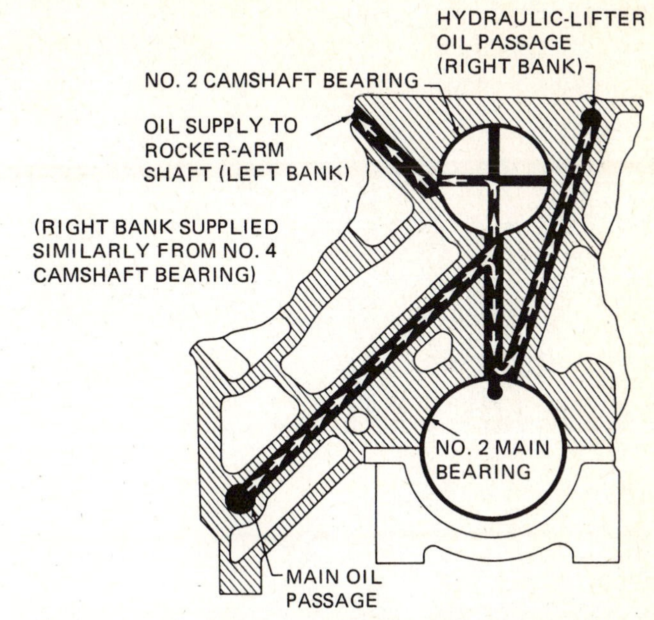

Fig. 17-6. Oil passages in the engine block carry the oil to the crankshaft main and camshaft bearings. The oil passages in the block connect with oil passages in the head so that the valve mechanisms are lubricated. (*Ford Motor Company*)

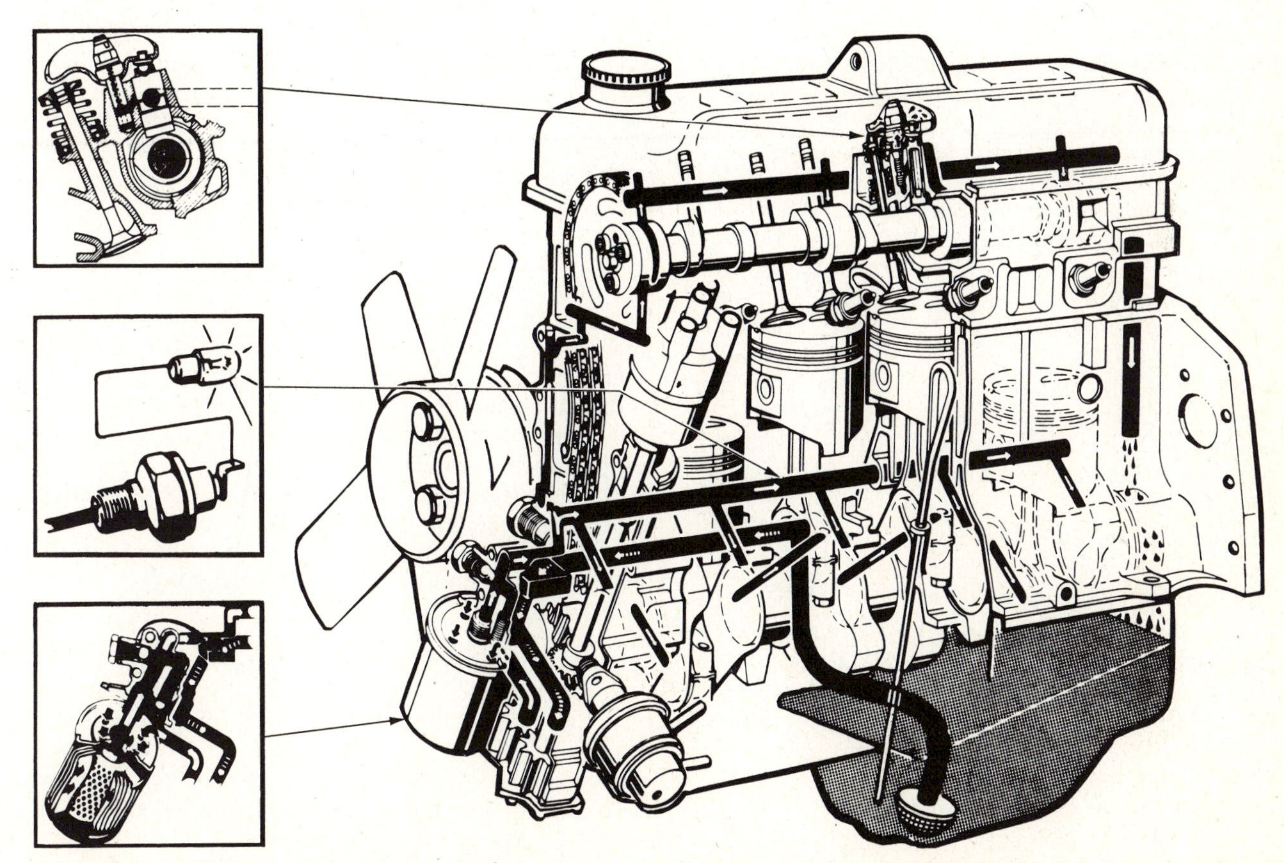

Fig. 17-7. Lubrication system for a four-cylinder, overhead-camshaft engine. Arrows show the flow of oil to the moving engine parts. (*Buick Motor Division of General Motors Corporation*)

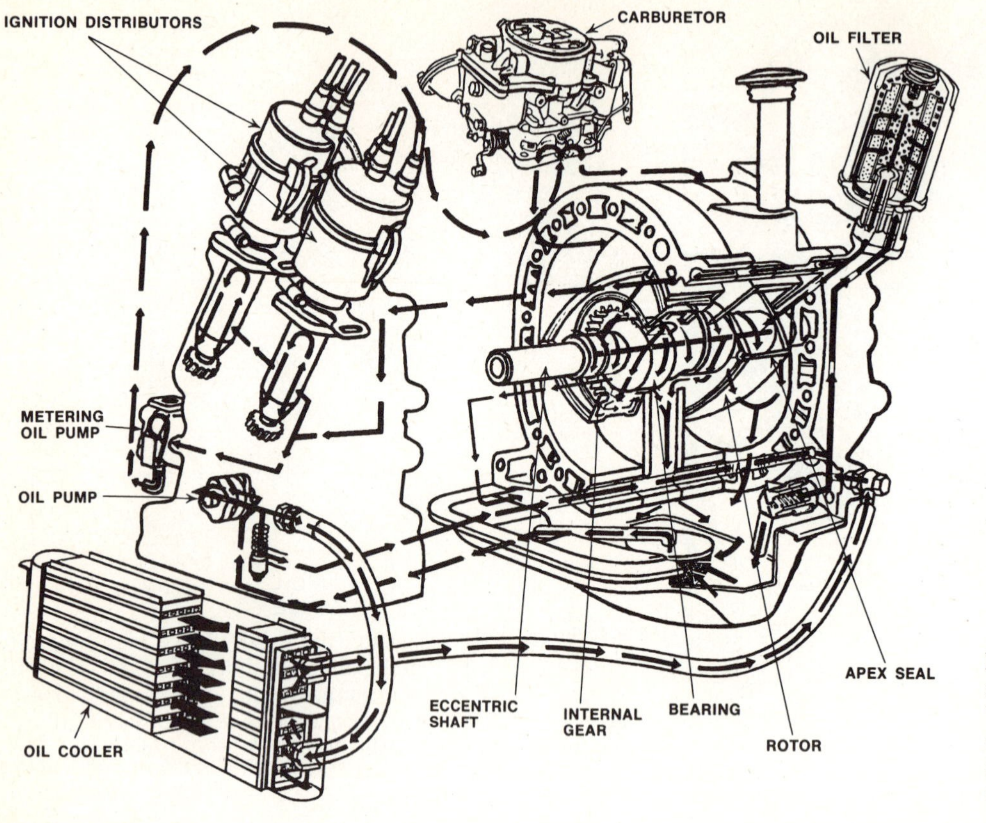

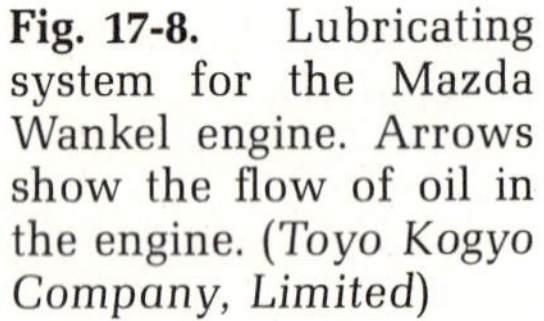
Fig. 17-8. Lubricating system for the Mazda Wankel engine. Arrows show the flow of oil in the engine. (*Toyo Kogyo Company, Limited*)

⊘ 17-11 Oil Pumps The two general types of oil pumps used in pressure-feed lubricating systems are the gear type (Fig. 17-9) and the rotor type (Fig. 17-10). The gear-type pump uses a pair of meshing gears. As the gears rotate, the spaces between the gear teeth and housing are filled with oil from the oil inlet. Then, as the teeth mesh, the oil is forced out through the oil outlet. The rotor-type pump uses an inner and outer rotor. The inner rotor is driven and causes the outer rotor to turn with it. As this happens, the spaces between the rotor lobes become filled with oil. When the lobes of the inner rotor move into the spaces of the outer rotor, oil is squeezed out through the outlet.

Oil pumps are usually driven from the engine camshaft from the same spiral gear that drives the ignition distributor (Fig. 17-11). In many engines, the oil intake for the oil pump is attached to a float. This floating intake takes only the top oil in the oil pan. (Since dirt particles sink, the top oil is cleanest.)

⊘ 17-12 Relief Valve To keep the oil pump from building up too much pressure, a relief valve is included in the lubricating system. The valve has a spring-loaded ball (Fig. 17-9) or spring-loaded plunger. When the pressure reaches the preset value, the ball or plunger moves against its spring. This action opens a port through which oil can flow back to the oil pan. Enough of the oil flows past the relief valve to prevent excessive pressure. (The oil pump can normally deliver much more oil than the engine requires. This is a safety factor that ensures delivery of enough oil under extreme operating conditions.)

⊘ 17-13 Oil Cooler Some engine lubricating systems have oil coolers. Oil coolers are used on almost all automotive air-cooled engines. One type of oil cooler consists of a small radiator mounted on the side of the engine block. Oil and water (coolant) circulates through the radiator. The coolant comes from the engine cooling system. As the coolant circulates through the oil cooler, it picks up heat. The heat is carried to the radiator, where it is passed on to the air circulating through the radiator. This process helps cool the oil and keep it at a workable temperature. Another design uses a small section of the cooling-system radiator so that an extra radiator is not required. The type used in air-cooled engines consists of a small radiator much like the radiators used in liquid cooling systems (Chap. 16).

⊘ 17-14 Oil Filters Every automotive-engine lubricating system has an oil filter. Some or all of the oil from the oil pump circulates through this filter. The oil filter is a cartridge made of material that traps particles of foreign matter. The filter thus helps keep the oil clean, preventing contaminants from entering the engine. Oil filters are of two types. Those which filter part of the oil from the oil pump are called *bypass filters*. Those which filter all the oil in circulation through the system are called *full-*

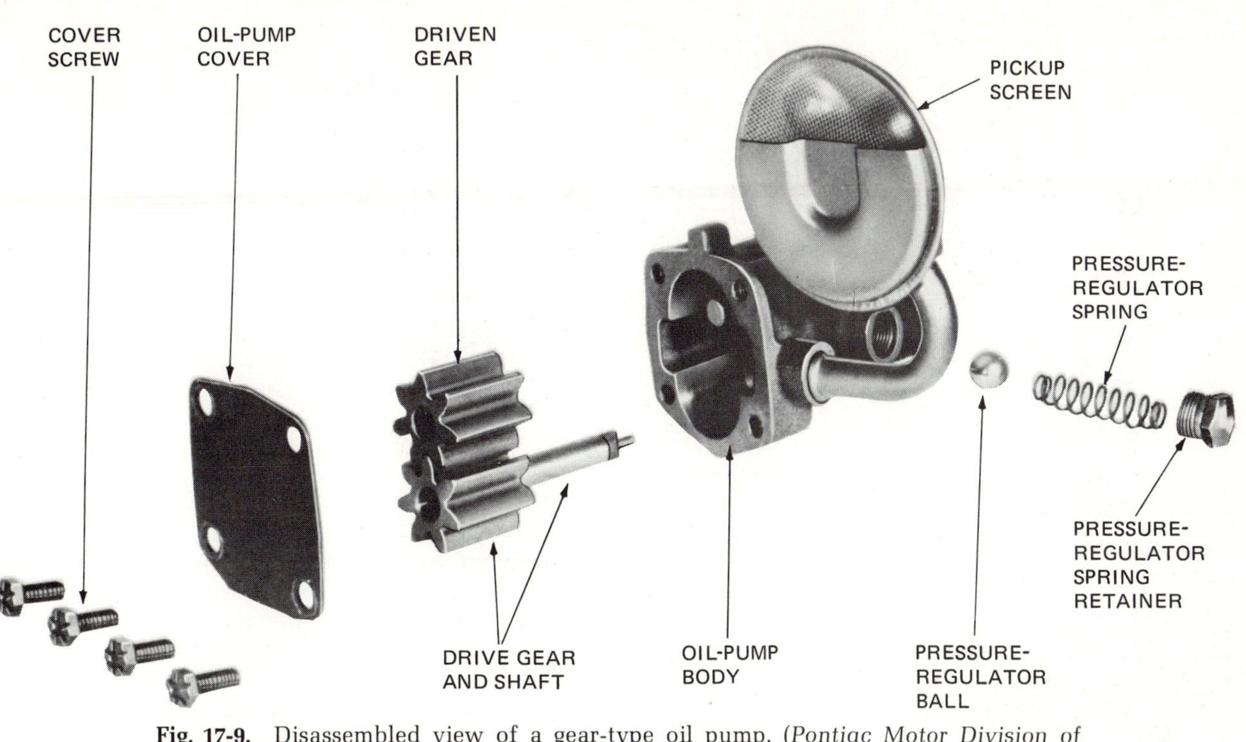

Fig. 17-9. Disassembled view of a gear-type oil pump. (*Pontiac Motor Division of General Motors Corporation*)

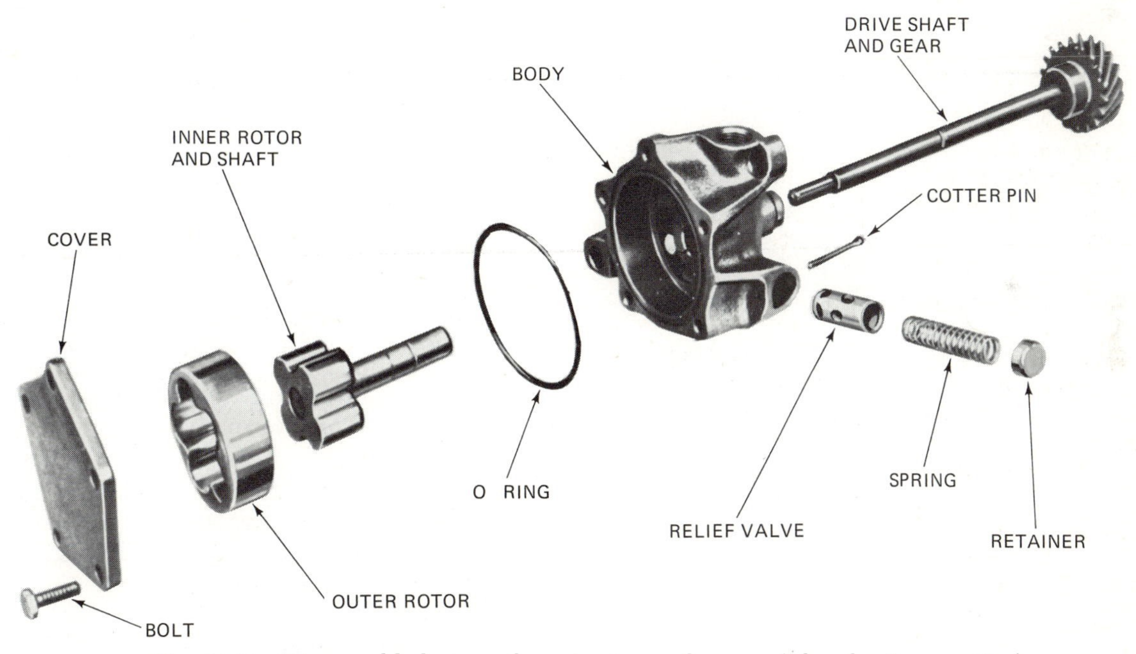

Fig. 17-10. Disassembled view of a rotor-type oil pump. (*Chrysler Corporation*)

flow filters. The full-flow filter includes a spring-loaded bypass valve. The valve protects the engine against oil starvation in case the filter becomes clogged. When this happens, the valve is opened by increased pressure from the pump trying to push oil through. With the valve opened, oil bypasses the filter. The engine is thus ensured of sufficient oil. However, the filter element should be replaced periodically so that filtering efficiency is maintained. Figure 17-12 is a cutaway view of a full-flow oil filter with a bypass valve.

⊘ 17-15 Oil-Pressure Indicators The oil-pressure indicator tells the driver what the oil pressure is in the engine. This gives warning if something occurs in the lubrication system that prevents delivery of oil to vital parts. Oil-pressure indicators are of three general types: pressure expansion, electric resistance, and indicator light. The latter two are the most commonly used.

1. *PRESSURE-EXPANSION INDICATOR* The pressure-expansion indicator uses a hollow Bourdon

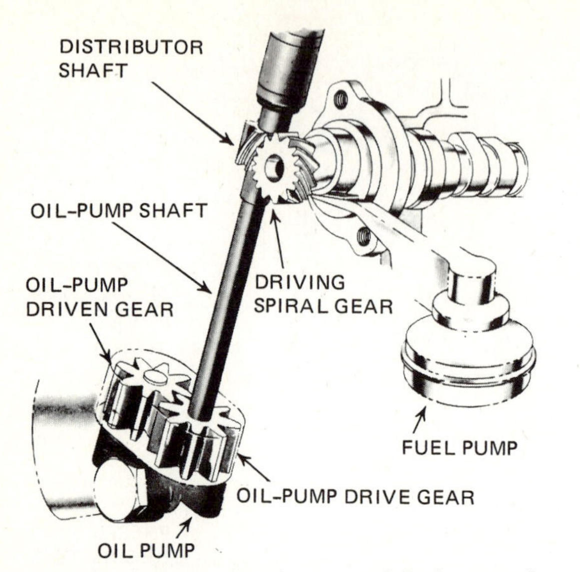

Fig. 17-11. Oil-pump, distributor, and fuel-pump drives. The oil pump is the gear type. A gear on the end of the camshaft drives the ignition distributor. An extension of the distributor shaft drives the oil pump. The fuel pump is driven by an eccentric on the camshaft. (*Buick Motor Division of General Motors Corporation*)

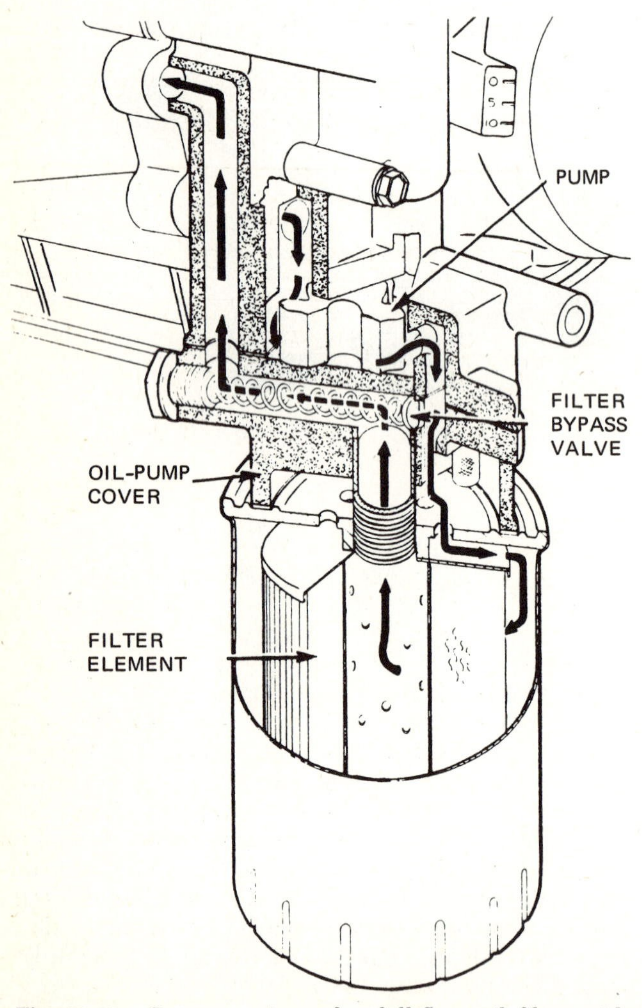

Fig. 17-12. Cutaway view of a full-flow oil filter with bypass valve. (*Buick Motor Division of General Motors Corporation*)

(curved) tube that is fastened at one end and free at the other. Oil pressure is applied to the curved tube through an oil line from the engine. The oil pressure causes the tube to straighten out somewhat as pressure increases. This movement is transmitted to a needle by linkage and gears from the end of the tube. The needle moves across the face of a dial and registers the amount of oil pressure.

2. *ELECTRIC-RESISTANCE INDICATOR* Electrically operated oil-pressure indicators are of two types: balancing coil and bimetal thermostat. The balancing-coil indicator makes use of two separate units: the engine and indicating units (Fig. 17-13). The engine unit consists of a variable resistance and movable contact. The movable contact moves from one end of the resistance to the other as oil pressure against a diaphragm varies (Fig. 17-14). As pressure increases, the diaphragm moves inward. This causes the contact to move along the resistance so that more resistance is placed in the circuit between the engine and indicating units. This reduces the amount of current that can flow in the circuit. The indicating unit consists of two coils that balance each other in a manner similar to electrically operated fuel gauges (⊘ 13-5). In fact, this type of indicator operates in the same manner as the fuel indicator. The only difference is that the fuel indicator uses a float that moves up or down as the gasoline level changes in the gasoline tank. The oil-pressure indicator uses changing oil pressure to operate a diaphragm that causes the resistance to change.

The bimetal-thermostat oil-pressure indicator is similar to the bimetal-thermostat fuel gauge (⊘ 13-5). The instrument-panel units for both are practically identical. The engine unit is similar to the engine unit of the balancing-coil oil-pressure indicator.

3. *INDICATOR LIGHT* Instead of a gauge, most vehicles have an oil-pressure indicator light. When the ignition is turned on and the oil pressure is low, the light comes on. Normally, after the engine has

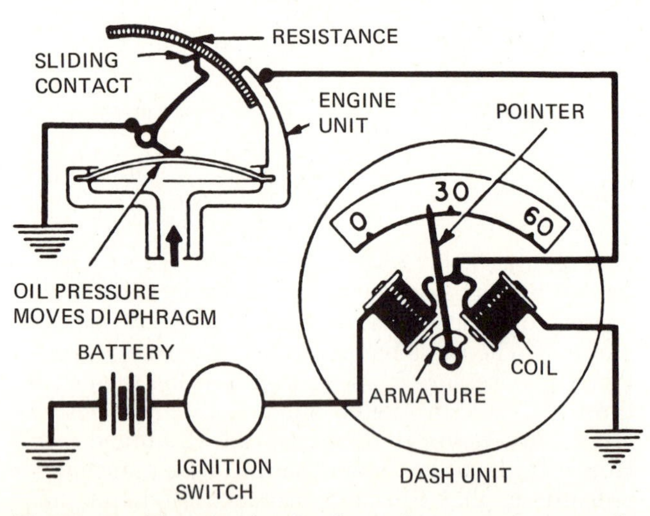

Fig. 17-13. Electric circuit of an electric-resistance oil-pressure indicator.

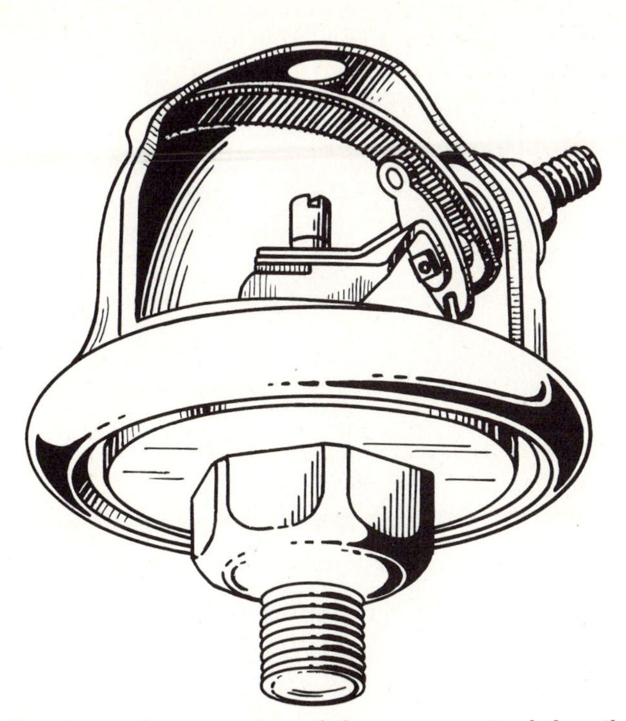

Fig. 17-14. Cutaway view of the engine unit of the oil-pressure indicator.

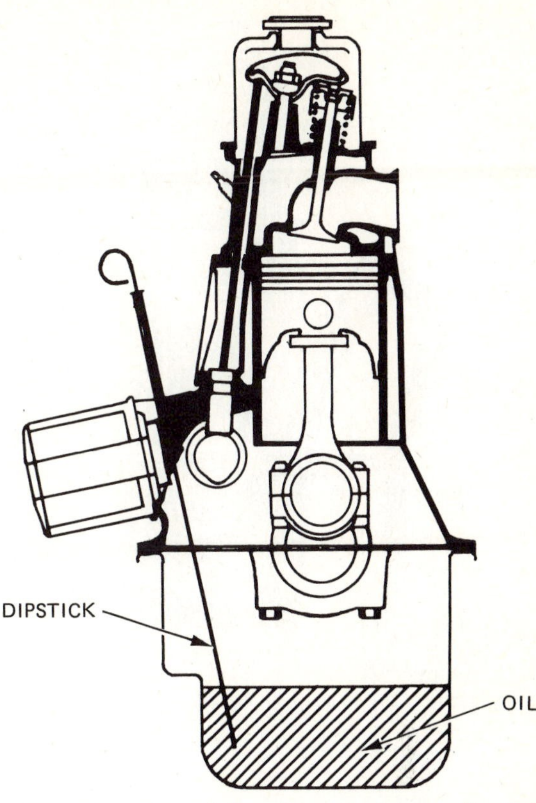

Fig. 17-15. Location of the oil-level stick, or dipstick, in the engine.

started the oil pressure has built up, the light goes off. If it does not, then the engine and lubricating system should be checked at once to find the cause of the low oil pressure. The light is connected to a pressure switch in the engine. The switch is closed except when oil pressure increases to normal values. The indicator light and pressure switch are connected in series to the battery through the ignition switch. When the ignition switch is turned on, the indicator light comes on. It stays on until the engine starts and the oil pressure builds up enough to open the pressure switch.

⊘ 17-16 Oil-Level Indicators To determine the level of the oil in the oil pan, an oil-level stick, or "dipstick," is used. The dipstick is placed so that it protrudes down into the oil (Fig. 17-15). The oil level is determined by withdrawing the dipstick and noting how high the oil rises on the dipstick. In the closed or positive crankcase ventilating system the dipstick tube is sealed at the top when the dipstick is in place. This keeps unfiltered air from entering the crankcase and crankcase gases from escaping.

⊘ 17-17 Crankcase Ventilation As mentioned in ⊘ 13-14, when the engine is running, air must circulate through the crankcase. The circulating air removes the water and liquid gasoline that appear in the crankcase when the engine is cold. Also, the air removes blow-by gases from the crankcase. Unless the water, liquid gasoline, and blow-by gases are removed from the crankcase, there will be trouble. Sludge and acids will form. Sludge can clog oil lines and starve the lubricating system, which could ruin the engine. Acids corrode metal parts, and this, too, could ruin the engine.

NOTE: The removal process requires first that the engine be heated up enough to vaporize the liquid water and gasoline. Then the circulating air can remove them, along with the blow-by gases.

In older engines, the crankcase is ventilated by an opening at the front of the engine and a vent tube at the back. The forward motion of the car and the rotation of the crankshaft move air through the crankcase, as shown in Fig. 17-16. The air passing through removes the water, fuel vapors, and blow-by. However, discharging these gases into the atmosphere produces air pollution.

To prevent atmospheric pollution, modern engines have a closed system, called a *positive crankcase ventilating* (PCV) *system*. Figure 17-17 shows a typical PCV system for a six-cylinder engine. Figure 17-18 shows the system for a V-8 engine. The principle is simple. Filtered air from the carburetor air cleaner is drawn through the crankcase. In the crankcase it picks up the water, fuel vapors, and blow-by. The air then flows back up to the intake manifold and enters the engine. There, unburned fuel is burned.

Too much air flowing through the intake manifold during idling would upset the air-fuel ratio.

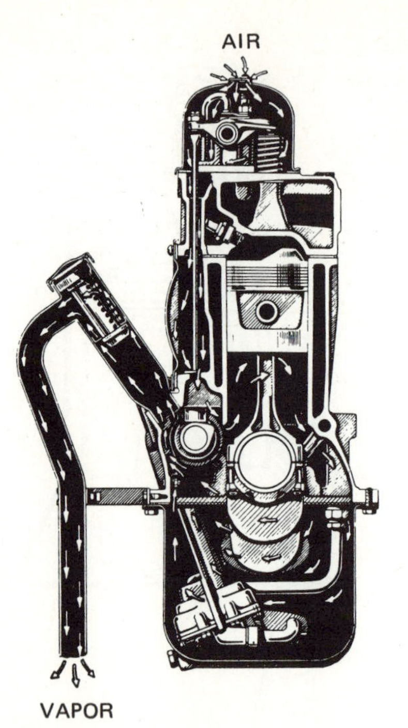

Fig. 17-16. Open-crankcase ventilating system.

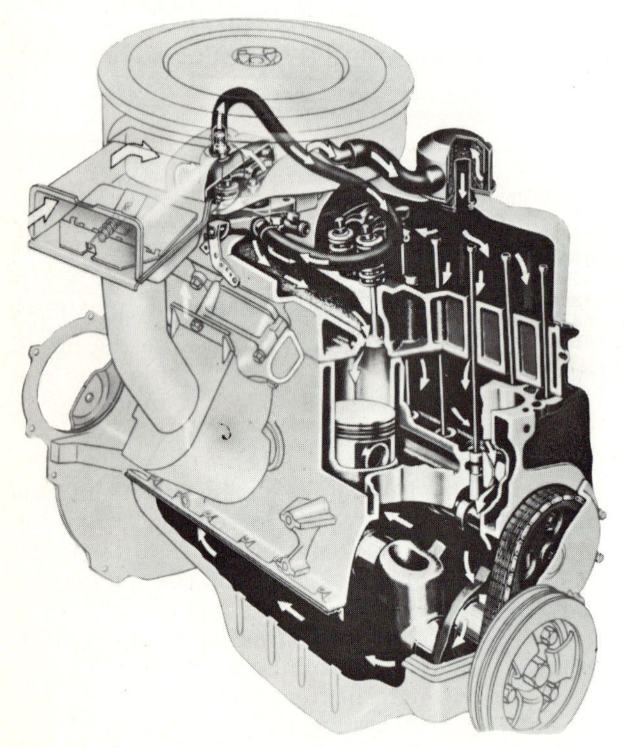

Fig. 17-17. Positive crankcase ventilating system for a six-cylinder engine. (*Ford Motor Company*)

This could cause poor idling. To prevent this, a regulator valve, called a *PCV valve,* is used. During idle, the PCV valve allows only a small amount of air to flow through. But as engine speed increases, reduced manifold vacuum allows the valve to open more. This, in turn, allows more air to flow through. Figure 17-19 shows the valve in the two positions.

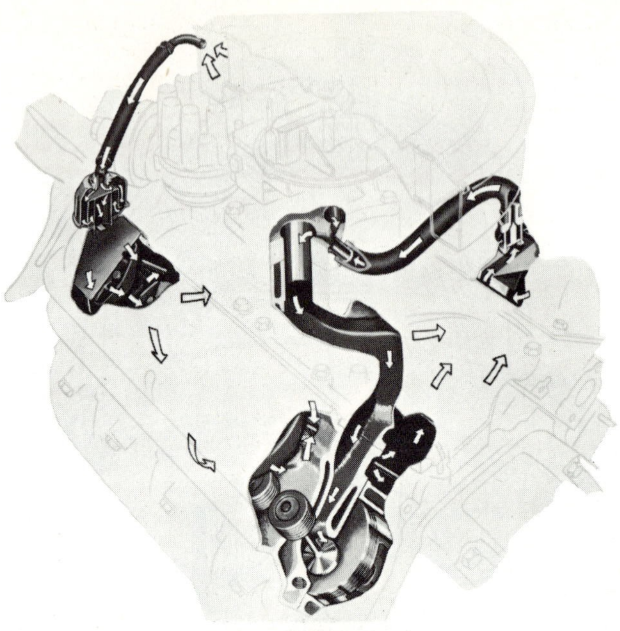

Fig. 17-18. Positive crankcase ventilating system for a V-8 engine. (*Ford Motor Company*)

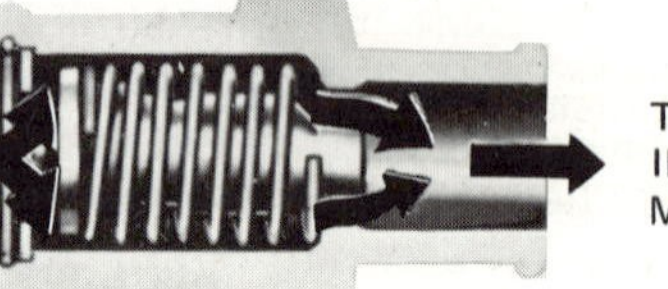

Fig. 17-19. The two extreme operating conditions of the positive-crankcase-ventilating-system regulator valve. (*Ford Motor Company*)

CHAPTER 17 CHECKUP

NOTE: Since the following is a chapter review test, you should review the chapter before taking the test.

A good understanding of the function of the lubricating system and engine oil will help you understand the various causes of engine trouble and engine wear. The questions that follow will help you check your knowledge of lubricating systems. If you have trouble with the questions, reread the chapter. It is hard to remember all the important facts the first time you read the material; most good students reread their lessons several times.

Completing the Sentences The sentences that follow are incomplete. After each sentence there are several words or phrases, only one of which will correctly complete the sentence. Write each sentence in your notebook, selecting the proper word or phrase to complete it correctly.

1. Three classes of friction are: (*a*) dry, greasy, and viscous; (*b*) dry, wet, and liquid; (*c*) dry, damp, and wet; (*d*) dry, greasy, and oily.
2. In addition to providing lubrication and acting as a cooling agent, the engine oil must: (*a*) clean, dry, and absorb shocks; (*b*) oxidize, carbonize, and burn; (*c*) absorb shocks, seal, and clean.
3. Two types of engine lubricating systems are: (*a*) pressure feed and force feed, (*b*) pressure feed and splash, (*c*) oil pump and pressure feed, (*d*) splash and nozzle.
4. The purpose of the relief valve in the pressure-feed system is to: (*a*) ensure adequate pressure, (*b*) prevent excessive pressure, (*c*) prevent insufficient lubrication, (*d*) ensure adequate oil circulation.
5. Two types of oil filters used in automotive engines are: (*a*) bypass and full flow, (*b*) open and closed, (*c*) low pressure and high pressure, (*d*) full flow and flow through.
6. The purpose of crankcase ventilation is to: (*a*) remove liquid gasoline and water, (*b*) remove vaporized water and gasoline, (*c*) cool the oil, (*d*) supply oxygen to the crankcase.
7. Viscosity can be divided into two properties: (*a*) ease of flow and fluidity, (*b*) foaming and flowing, (*c*) body and fluidity, (*d*) body and penetration.
8. The substance added to the oil which helps keep the engine clean is called a: (*a*) detergent, (*b*) soap, (*c*) grease, (*d*) thickening agent.
9. Most of the dilution of the oil in the crankcase takes place during: (*a*) high-speed operation, (*b*) long trips, (*c*) engine overheating, (*d*) engine warm-up.
10. Common causes of excessive oil consumption include: (*a*) heavy oil and tight bearings, (*b*) high speed and worn engine parts, (*c*) short trips and cold weather, (*d*) frequent oil changes and weak valve springs.

Unscrambling the Oil Jobs Following are two lists. List 1 includes the various jobs that the oil does in the engine. List 2 includes phrases that explain why or how these jobs are done (but not in the same order). To unscramble the lists take one item at a time from list 1 and write it in your notebook. Then write the item from list 2 that explains why or how the job is done. For example, the first item in list 1 is "lubricates." When you look at list 2, you will find two phrases that apply to "lubricates." One of these is "to minimize wear." So you put the two together to form "lubricates to minimize wear," which is one of the jobs the oil does. (The other phrase goes after the second "lubricates.")

LIST 1
Lubricates
Lubricates
Cools
Absorbs shocks
Seals
Cleans

LIST 2
By carrying dirt from engine parts
By carrying heat from engine parts
To minimize wear
To minimize power loss
Between bearings and other engine parts
Between piston rings and cylinder walls

SUGGESTIONS FOR FURTHER STUDY

Another book in the McGraw-Hill Automotive Technology Series, *Automotive Fuel, Lubricating, and Cooling Systems*, discusses the operation and maintenance of the lubricating system as well as the composition and action of oil in greater detail. You may be able to obtain more information on lubricating systems from service stations and oil companies. Your local service and school shops may have oil pumps and filters on hand that you can examine.

chapter 18

SMOG, AIR POLLUTION, AND THE AUTOMOBILE

In this chapter and the following one, we discuss the various special devices that are installed on engines and automobiles to reduce air pollution. These devices, called emission controls, were created because engineers realized that automobiles were causing atmospheric pollution. In addition, federal and state law-making bodies, such as the United States Congress, have passed laws that make automotive emission controls necessary.

Some emission controls are covered in other chapters. For example, the thermostatic air cleaner is discussed in ⊘ 13-12. The positive crankcase ventilating system is discussed in ⊘ 17-17. However, in this chapter and the following one, we review these devices. Thus, the complete story of emission controls will be presented in these two chapters. In this chapter we look at positive crankcase ventilating and fuel-vapor-recovery systems. In Chapter 19, we look at methods of cleaning up the exhaust gas.

⊘ 18-1 Smog The word "smog" comes from "smoke" and "fog." Smog is a sort of fog with different kinds of chemicals and other substances mixed up in it. Smog has been around for a long time. Billions of years ago, many volcanos were sending millions of tons of ashes and smoke into the air. Winds whipped up clouds of dust. Animal and vegetable matter decayed and sent polluting gases into the air.

Then people began to produce their own kind of air pollution. They discovered fire. In the Middle Ages, people in large cities such as London used soft coal to heat their homes. The smoke from these fires, combined with moisture in the air, produced dense layers of smog. The smog blanketed the cities for days, particularly in winter. The heat generated in a large city tends to produce air circulation within a domelike shape (as shown in Fig. 18-1.) This traps the smog and holds it over the city.

Because it contains chemicals and other substances, smog can be harmful, even deadly. Smog blurs vision (Figs. 18-2 and 18-3). It irritates the eyes, throat, and lungs. Smog can make people sick, and it can make sick people sicker. Smog has been linked to such human ills as eczema, asthma, emphysema, cardiovascular difficulties, and lung and stomach cancer. It also has a harmful effect on the environment. Food crops and animals suffer; paint may peel from houses. It is obvious that we must do everything possible to reduce smog.

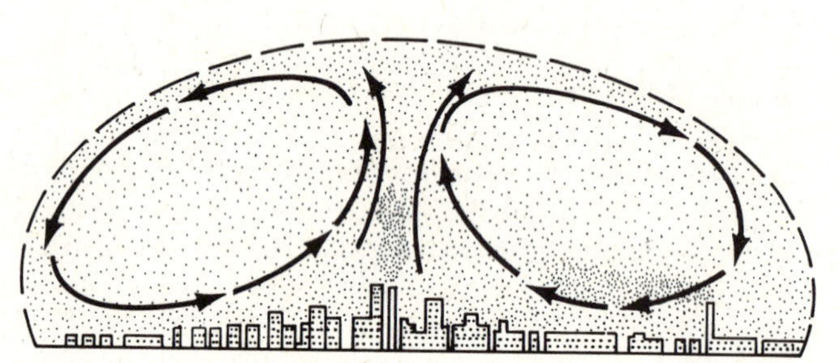

Fig. 18-1. Heat generated within a large city tends to produce a circulatory air pattern which traps smog within a dome.

⊘ 18-2 Not All Air Pollution Is Smog Smog, along with smoke, is the most visible evidence of air pollution. But some kinds of air pollution are not visible and may not become visible until they are mixed with moisture. For example, lead compounds from leaded gasoline, hydrocarbons (unburned gasoline), carbon monoxide, and other gases may pollute the air without being seen.

⊘ 18-3 Air Pollutants All air is polluted to some extent; that is, all air carries some polluting substances. Much of it is from natural causes: smoke and ash from volcanos, dust stirred up by the wind, compounds given off by growing vegetation, gases given off by decaying animal and vegetable matter, salt particles from the oceans, and so on.

Fig. 18-2. View of Los Angeles during a clear day. (*Los Angeles County Air Pollution Control District*)

Fig. 18-3. View of Los Angeles during a typical smoggy day. Note that many buildings are hidden. (*Los Angeles County Air Pollution Control District*)

People add to these pollutants by burning coal, oil, gas, gasoline, and many other substances. Altogether, it is estimated that people create 200 million tons of air pollutants every year in the United States alone. This is about 1 ton for every man, woman, and child in the country! It is these human-produced pollutants that we are concerned with in this book, especially those that come from the automobile.

Before we discuss the automobile, however, let us review what we have learned about combustion. Most fuels such as coal, oil, gasoline, and wood contain hydrogen and carbon in various chemical combinations. During combustion, oxygen unites with the hydrogen and carbon to form water (H_2O), carbon monoxide (CO), and carbon dioxide (CO_2).

In addition, many fuels contain sulfur, which burns to produce sulfur oxides (SO_x). Also, in the heat of combustion, some of the nitrogen in the air combines with oxygen to form nitrogen oxides (NO_x). Some of the fuel may not burn completely, so that smoke and ash are formed. (Smoke is simply particles of unburned fuel and soot, called *particulates,* mixed with air.)

⊘ 18-4 Los Angeles Consider Los Angeles, a large city containing about 7 million inhabitants. The city is set in a basin surrounded on three sides by mountains and on the fourth by the Pacific Ocean. When the wind blows toward the ocean, it sweeps away pollutants. But when the air is stagnant, pollutants from industry and automobiles are not blown away (Figs. 18-2 and 18-3). They just build up into a thick, smelly layer of smog. The location of Los Angeles, plus all the people and industry there, make it one of the biggest "smog centers" in the United States. And it is Los Angeles which has led in measures to reduce smog.

Los Angeles has banned unrestricted burning (for example, burning trash). Incinerators without pollution controls were outlawed. Industry was forced to change combustion processes and add controls to reduce pollutants coming from their chimneys. Laws were passed that required the addition of emission controls on automobiles. All these measures have significantly reduced atmospheric pollution in the Los Angeles area.

⊘ 18-5 Pollution from the Automobile If not controlled, the automobile can give off pollutants from four places, as shown in Fig. 18-4. Pollutants can come from the fuel tank, carburetor, crankcase, and tail pipe. Pollutants from the fuel tank and carburetor consist of gasoline vapors. Pollutants from the crankcase consist of partly burned air-fuel mixture that has blown by the piston rings. Pollutants from the tail pipe consist of partly burned gasoline, or hydrocarbons (HC), carbon monoxide (CO), nitrogen oxides (NO_x), and if there is sulfur in the gasoline, sulfur oxides (SO_x). In the pages that follow, we discuss the causes of and cures for these pollutants.

⊘ 18-6 Positive Crankcase Ventilating System We described the PCV system in ⊘ 17-17. It is illustrated in Figs. 17-17 to 17-19. Briefly, the system carries filtered air from the carburetor air filter through the crankcase. This air picks up blow-by and gasoline vapors and carries them up the intake manifold. The air then flows through the engine where the blow-by and gasoline vapors are burned.

The PCV valve is spring-loaded. At low speeds and during idle, when intake-manifold vacuum is high, the vacuum holds the valve nearly closed. In this position, the valve passes only a small amount of air. This prevents the idle-mixture ratio from

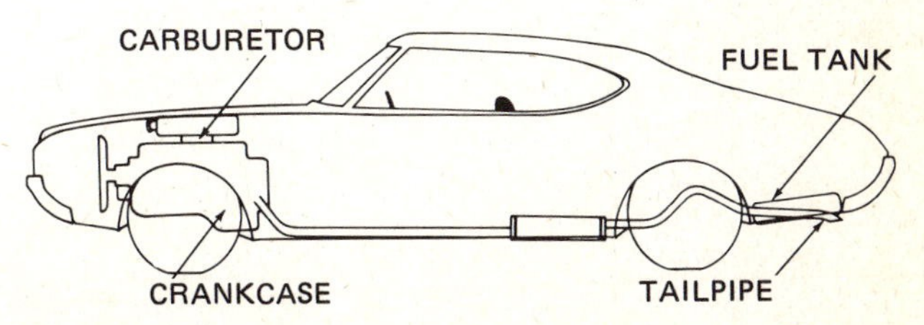

Fig. 18-4. Four possible sources of atmospheric pollution from the automobile.

being upset and producing poor idling. Then, when the throttle is opened wider and engine speed increases, the intake-manifold vacuum drops. Now, with less vacuum, the PCV valve opens, allowing more air to flow through the crankcase.

⊘ 18-7 Need for Vapor-Recovery Systems Both the fuel tank and the carburetor can lose gasoline vapor if the car does not have a vapor-recovery system. The fuel tank "breathes" as temperature changes. That is, as the tank heats up, the air inside it expands. Part of the air is forced out through the tank vent tube, or through the vent in the tank cap. This air is loaded with gasoline vapor. Then, when the tank cools, the air inside it contracts. More air enters the tank from the outside. This breathing of the tank causes a loss of gasoline. The higher the fuel-tank temperature (for instance, when the car is parked in the sun), the more gasoline vapor is lost.

The carburetor also can lose gasoline by evaporation. When the engine is running, the carburetor float bowl is full. When the engine stops, engine heat evaporates some or all of the gasoline stored in the float bowl.

If there is no vapor-recovery system, this gasoline vapor passes out into the atmosphere. However, a vapor-recovery system captures these gasoline vapors and prevents them from escaping into the air. The system thus reduces atmospheric pollution. All modern cars are equipped with vapor-recovery systems. They are called by various names: ECS (Evaporation Control System), EEC (Evaporation Emission Control), VVR (Vehicle Vapor Recovery), and VSS (Vapor Saver System). All of these systems work in the same general way.

⊘ 18-8 Vapor-Recovery Systems Figures 18-5 and 18-6 show a typical vapor-recovery system. The canister is filled with activated charcoal. Just after the engine is shut off, heat enters the carburetor. The gasoline vapor from the carburetor float bowl passes through the canister and is *adsorbed* by the charcoal. "Adsorbed" means that the gasoline vapor is trapped on the surface of the charcoal particles. (This is somewhat like the charcoal filters on cigarettes. Their purpose is to trap particles of tar and other substances to prevent their entering the mouth and lungs of the smoker). Some carburetor float bowls have a special vent (see Fig. 14-12) connected by a tube to the charcoal canister. The vent and tube carry the float-bowl vapor directly to the canister.

At the same time, vapor-laden air from the fuel tank is carried by a special emission-control pipe to the canister. As the air passes down through the canister, the gasoline vapor is adsorbed by the charcoal particles. The air exits from the bottom of the canister, leaving the HC (hydrocarbon) vapor behind. There is a filter at the bottom of the canister. It comes into action during the *purge* phase of operation. This occurs when the engine is started. Now, intake-manifold vacuum draws fresh air up through the canister. This fresh air removes, or purges, the gasoline vapor from the canister. The fresh air takes the HC through a purge line to a connection at the carburetor.

⊘ 18-9 Fuel-Return Line Notice, in Fig. 18-5, that a fuel-return line parallels the main fuel line. This fuel-return line connects the pressure side of the fuel pump to the fuel tank. Thus, any excess gasoline that is pumped by the fuel pump is returned to the fuel

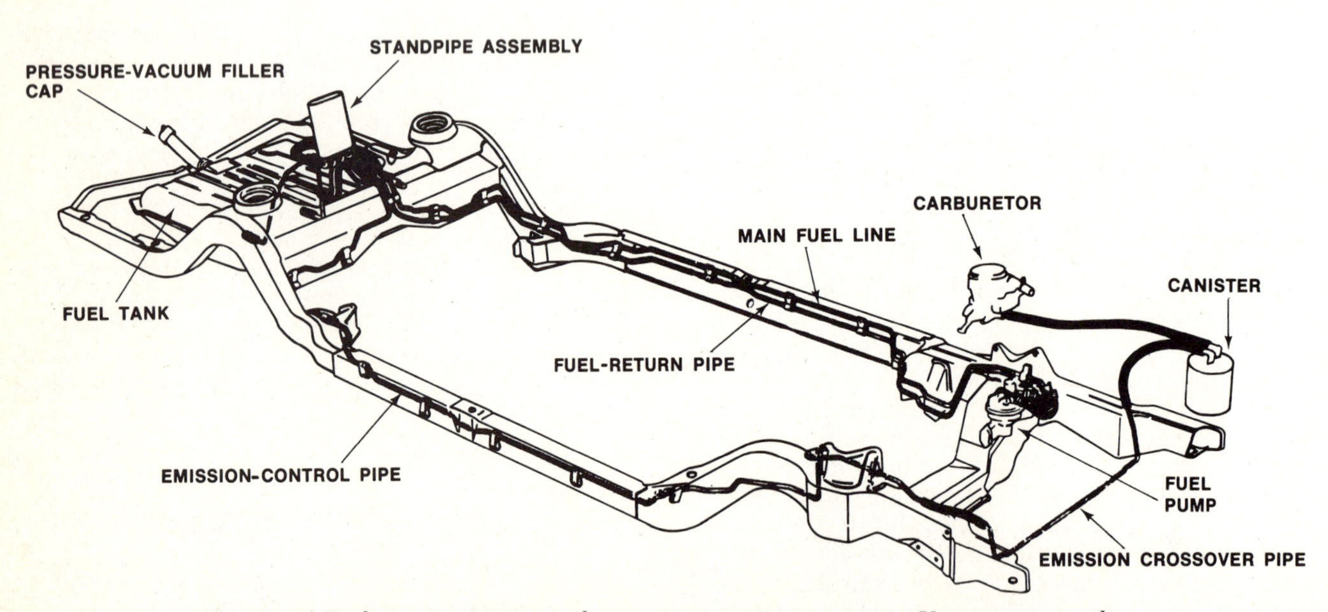

Fig. 18-5. Fuel-evaporation-control, or vapor-recovery, system. Vapor goes to the canister. The fuel-return pipe returns excess fuel (not used by the carburetor) to the fuel tank. This constant flow of excess fuel through the fuel pump helps prevent vapor lock.

tank. This action removes any vapor that might develop in the fuel pump. It also maintains a flow of fuel through the fuel pump. This flow keeps the fuel pump relatively cool and helps prevent vapor lock. There is more on fuel pumps in ⊘ 13-6.

⊘ 18-10 Charcoal Canister A charcoal canister is shown in sectional view in Fig. 18-7. An actual canister is shown in Fig. 18-8. The arrows in Fig. 18-7 show the flow of air and gasoline vapor. When the engine is turned off, vapor from the fuel tank flows into the canister. When the engine is running, intake-manifold vacuum draws air up through the canister and into the carburetor. This air cleans the gasoline vapor out of the canister.

The charcoal canister shown in Fig. 18-9 is for a six-cylinder engine. Note that it has a *restrictive valve*, also known as a *purge valve*. This valve limits the flow of gasoline vapor and air into the carburetor during idling. But it allows full air-vapor flow during part- to full-throttle operation. In the six-cylinder engine, full air-vapor flow from the canister could

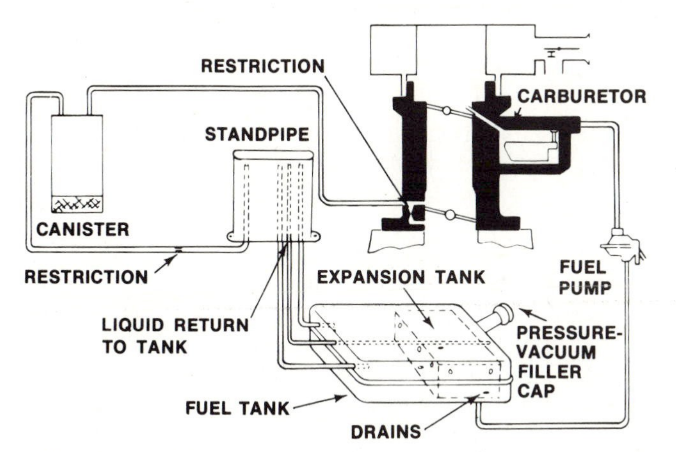

Fig. 18-6. Schematic view of a vapor-recovery system.

Fig. 18-8. Charcoal canister. Note its size.

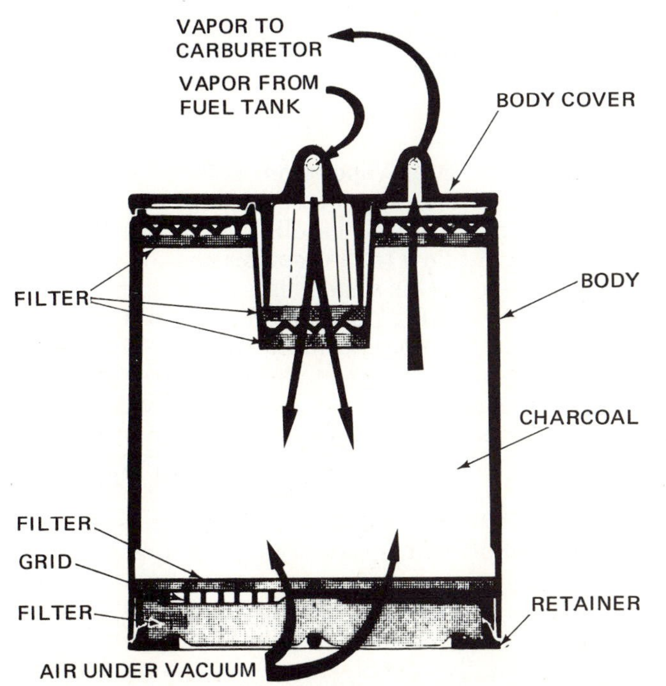

Fig. 18-7. Sectional view of a charcoal canister for a V-8 engine vapor-recovery system. (*Oldsmobile Division of General Motors Corporation*)

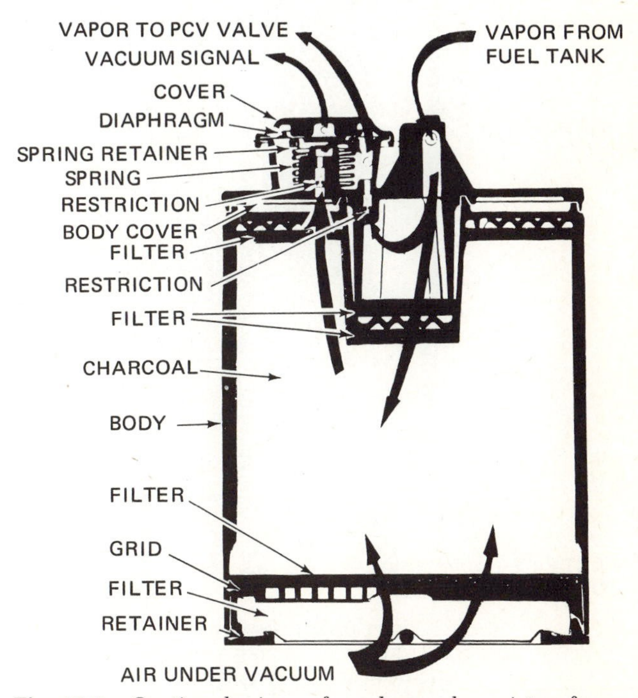

Fig. 18-9. Sectional view of a charcoal canister for a six-cylinder-engine vapor-recovery system. (*Oldsmobile Division of General Motors Corporation*)

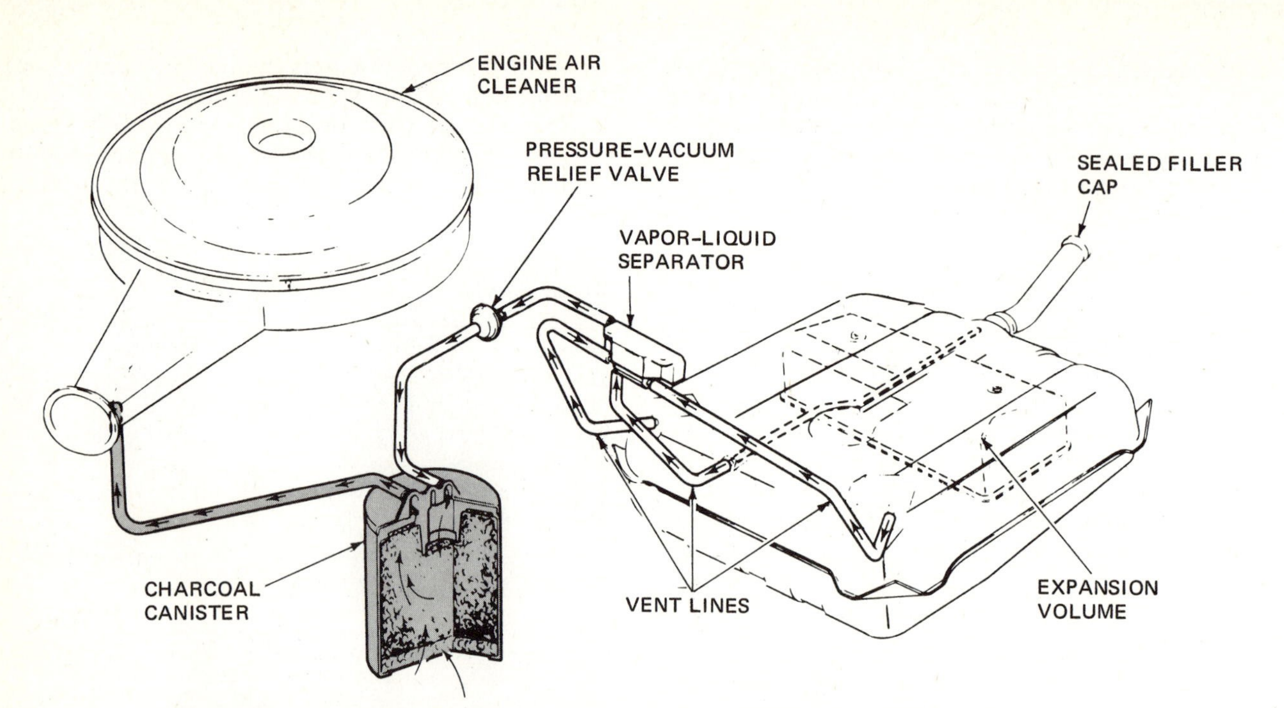

Fig. 18-10. Vapor-recovery system shown schematically. In this system, the purge line from the canister is connected to the air-cleaner snorkel. (*Buick Motor Division of General Motors Corporation*)

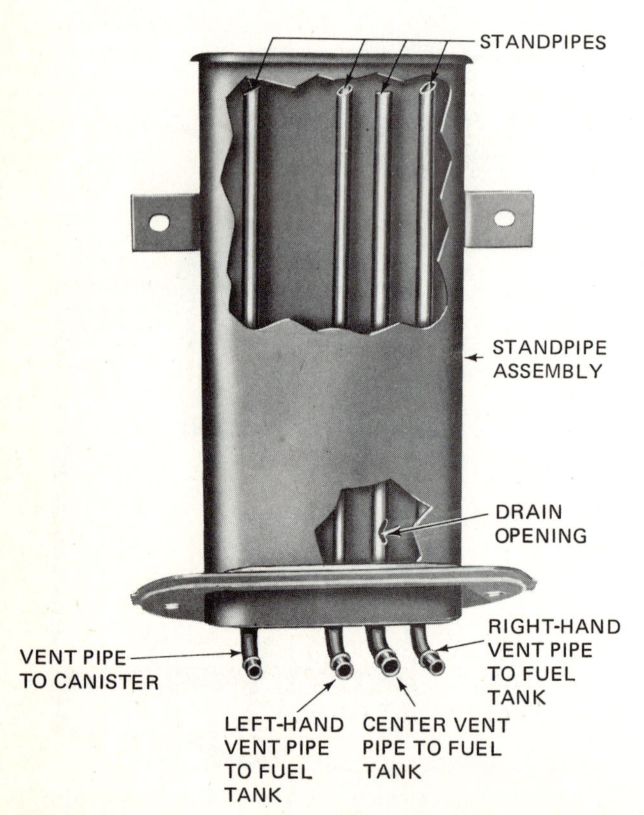

Fig. 18-11. Cutaway view of a standpipe assembly. (*Oldsmobile Division of General Motors Corporation*)

upset engine idling. At higher engine speeds, however, full air-vapor flow can be tolerated. The valve is operated by a vacuum signal from a drilled hole in the carburetor. This hole is located just below the throttle valve when the throttle valve is closed. With the throttle closed and the engine idling, a high vacuum develops below the throttle. This vacuum causes the purge valve to limit the air-vapor flow.

On some models, the purge line from the canister is connected to the air cleaner, as in Fig. 18-10. The system operates in the manner already described.

⊘ 18-11 Separating Vapor from Fuel The fuel tank must have some way of separating the gasoline vapor from the liquid gasoline. Otherwise, liquid gasoline might flow to the canister and then out into the atmosphere. One system uses a standpipe assembly, as shown in Fig. 18-5. Figure 18-11 is a cutaway view of a standpipe assembly. It contains a series of pipes with openings at the top. Three of these pipes are connected to the center and two sides of the fuel tank. These vents are shown in Fig. 18-12. In this system, at least one of the vents is always above the fuel level, regardless of the tilt of the car. Thus, air and fuel vapor can always pass up into the standpipe and through the vent line into the canister.

Figure 18-13 shows a variation of the standpipe assembly. Here, a vapor-liquid separator is positioned above the fuel tank. Four vapor-vent lines are connected from the separator to the four corners of the fuel tank. Another vapor-fuel separating arrangement is shown in Fig. 18-14. In this system, a liquid check valve is used. It is connected by tubes to the two ends of the fuel tank. The liquid check valve will pass air but not liquid gasoline.

Another type of vapor separator is shown in Fig. 18-15. This separator is mounted on top of the fuel tank. It is filled with filter material that will pass vapor but not liquid gasoline. The vapor separator is mounted at the top center of the fuel tank, thus minimizing the chances of liquid gasoline passing directly into the separator.

Some systems use a domed fuel tank (Figs. 13-4 and 18-16). The dome forms the high point of the tank, and the emission-control pipe is connected at this point.

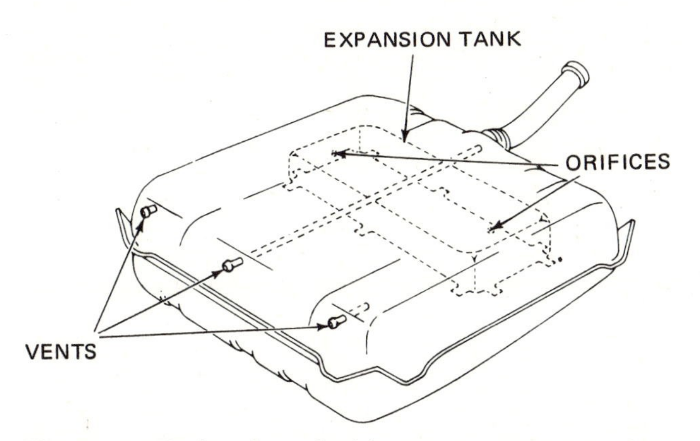

Fig. 18-12. Fuel tank used with a vapor-recovery system. The two side vents are short, but the center vent goes all the way to the rear of the tank. (*Pontiac Motor Division of General Motors Corporation*)

⊘ 18-12 Sealed Fuel Tank Older-model fuel tanks had a vent pipe or a vent in the fuel-tank cap. Modern automobiles, with vapor-recovery systems, use a sealed fuel tank with a special cap. This cap has a pressure-vacuum valve system. It is much like the pressure-vacuum cap used on the radiator of a pressurized cooling system. The cap valve opens if too much pressure develops in the tank. It also opens to admit air as fuel is withdrawn, so that a vacuum does not develop in the tank. Pressure or vacuum in the fuel tank could damage the tank.

⊘ 18-13 Carburetor Insulator Some carburetors use an insulator (Fig. 18-17) to reduce the heat flow from the engine to the carburetor. The insulator is placed between the carburetor and intake manifold and forms a heat barrier between them. This heat barrier reduces fuel evaporation from the float bowl after the engine has been turned off. Another ar-

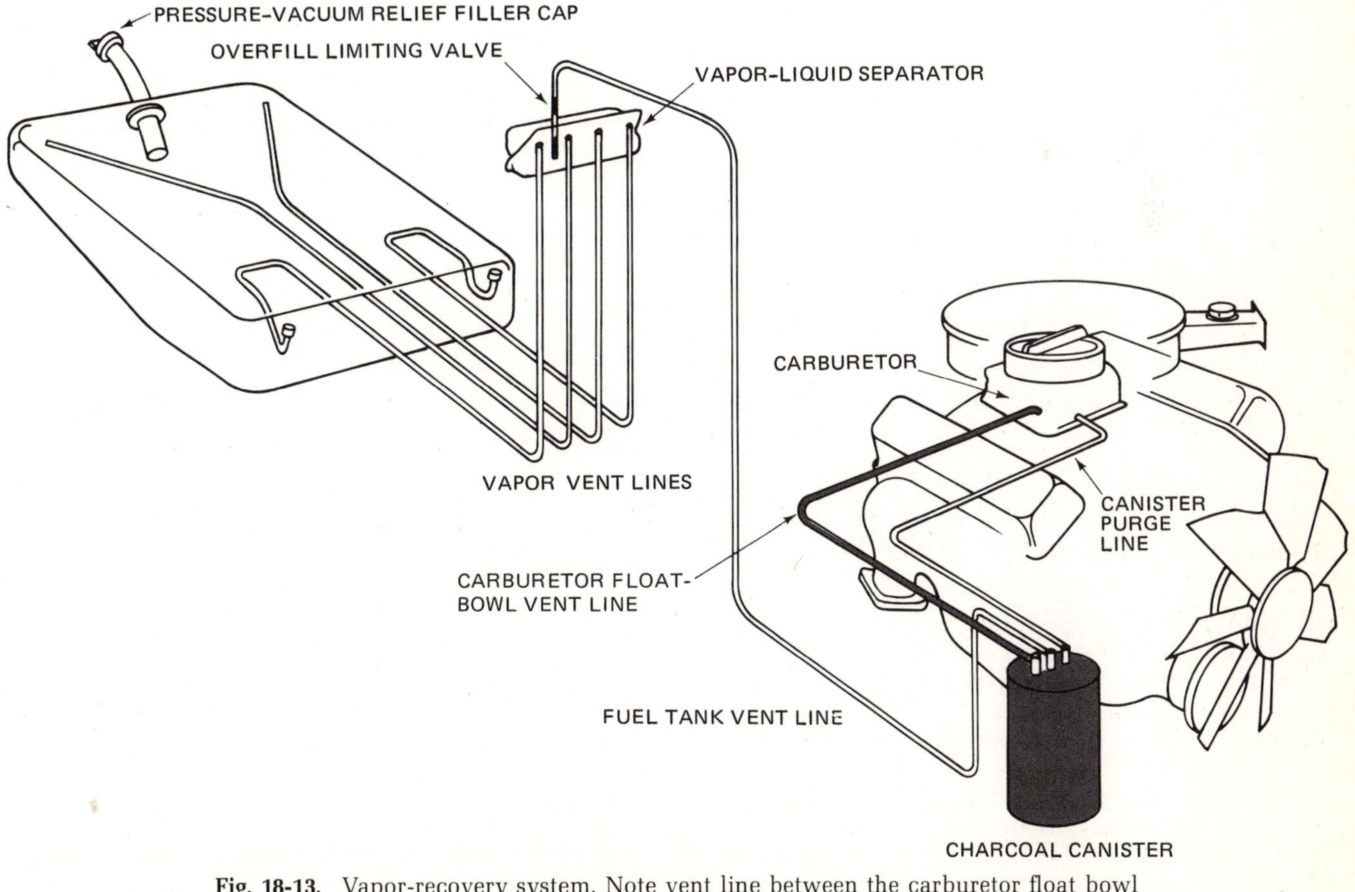

Fig. 18-13. Vapor-recovery system. Note vent line between the carburetor float bowl and the canister. (*Chrysler Corporation*)

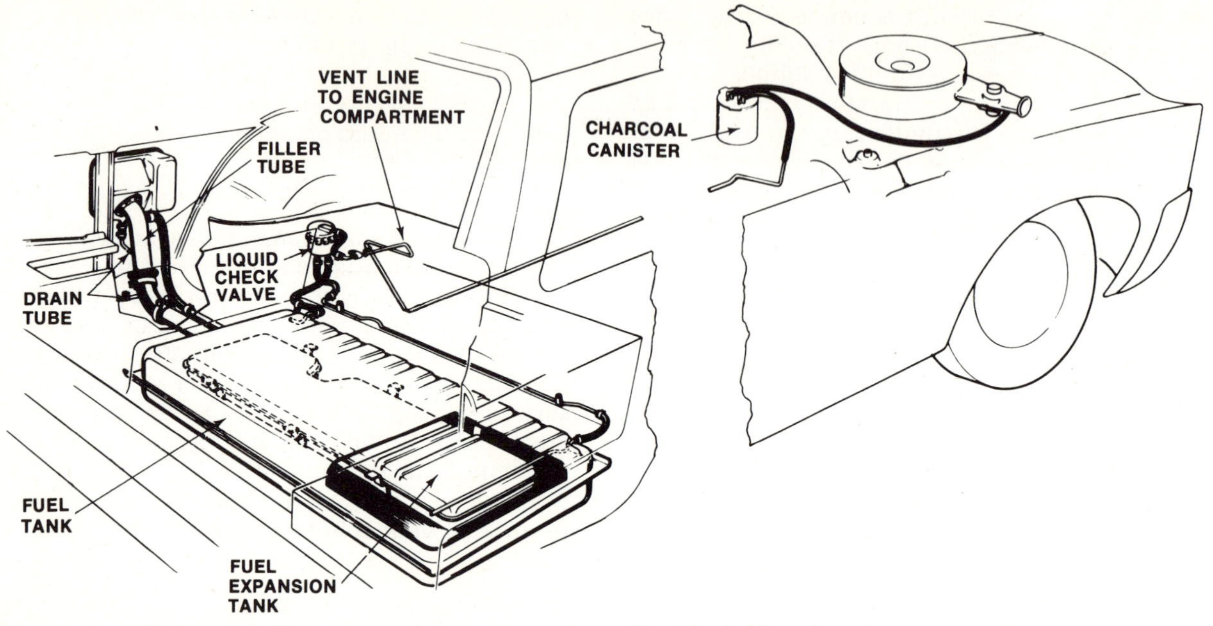

Fig. 18-14. Vapor-recovery system using a liquid check valve. (*American Motors Corporation*)

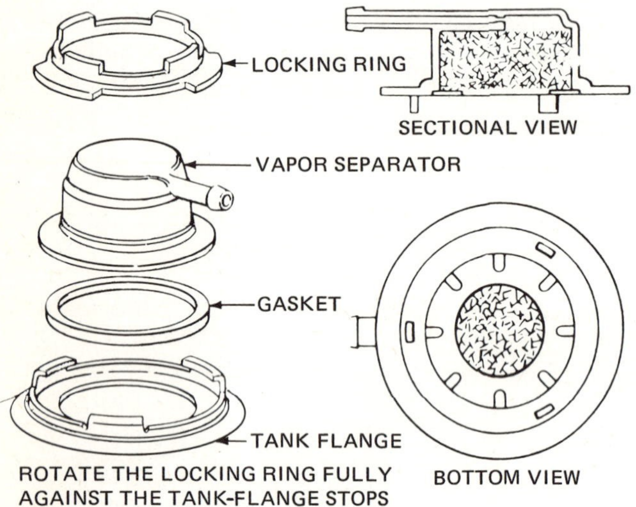

Fig. 18-15. Vapor separator using filter material. (*Ford Motor Company*)

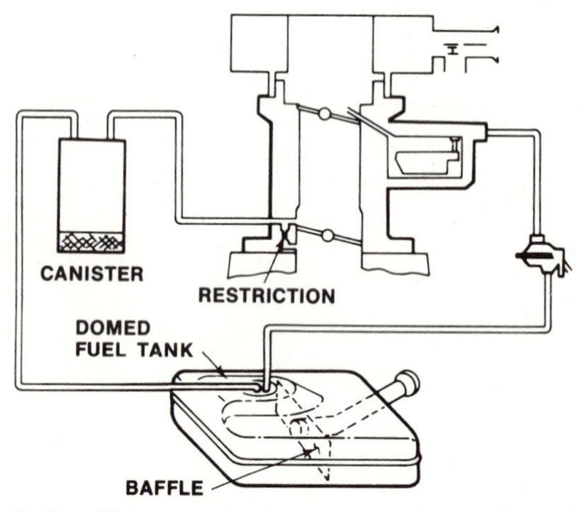

Fig. 18-16. Vapor-recovery system using a domed fuel tank. (*Pontiac Motor Division of General Motors Corporation*)

rangement uses an aluminum heat-dissipating plate which sticks out, as shown in Fig. 18-18.

⊘ 18-14 Vapor Storage in the Crankcase In some Chrysler Corporation cars, gasoline vapor from the fuel tank and carburetor is stored in the crankcase (Fig. 18-19). When the engine is turned off, gasoline vapor from the vapor separator at the fuel tank flows to the crankcase air cleaner. From there, it flows down into the crankcase. At the same time, gasoline vapor from the carburetor float bowl flows down into the crankcase. Gasoline vapor is two to four times as heavy as air. Thus it sinks to the bottom of the crankcase. Then, when the engine is started, the positive crankcase ventilating system clears the crankcase of the vapor. The vapor is carried up into the intake manifold and then into the engine, where it is burned.

⊘ 18-15 Expansion Tank You probably have noticed in Figs. 18-6, 18-10, and so on, that the fuel tanks have expansion tanks. These tanks are there in case the fuel temperature increases after the fuel tank has been filled. As fuel temperature increases, the fuel expands. The expansion tank gives the fuel a place to go.

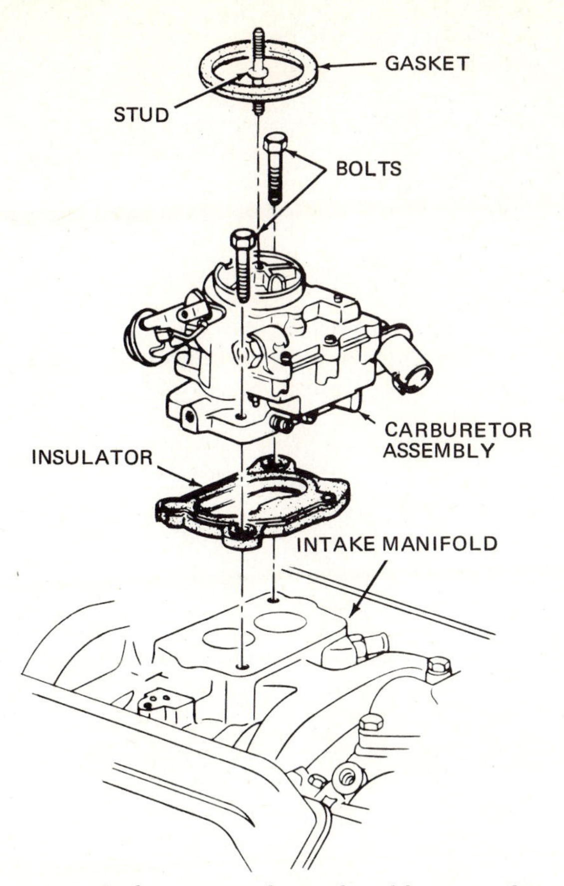

Fig. 18-17. Carburetor insulator placed between the carburetor and the intake manifold. The insulator blocks passage of heat to the carburetor. The insulator thus reduces evaporation of fuel from the float bowl. (*Chevrolet Motor Division of General Motors Corporation*)

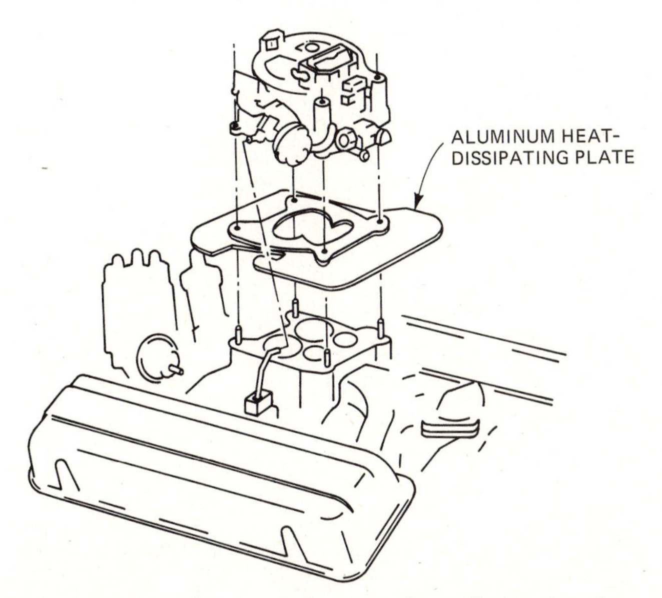

Fig. 18-18. Insulator and aluminum heat-dissipating plate between the carburetor and intake manifold. The insulator and plate reduce heat flow to the carburetor.

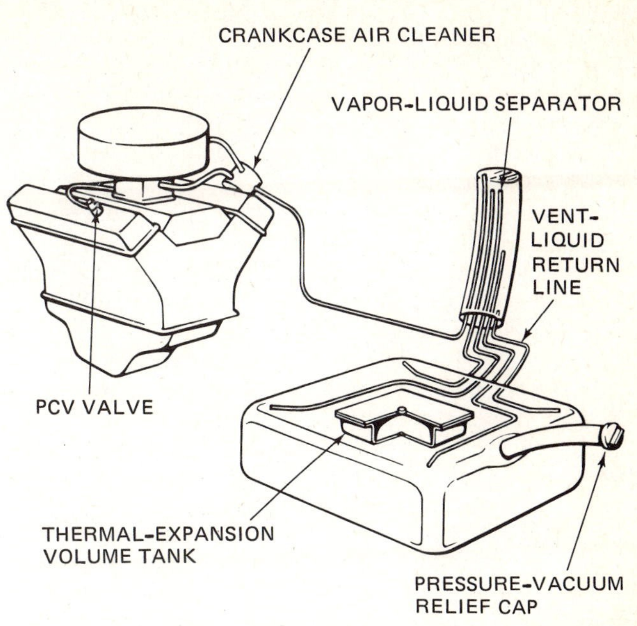

Fig. 18-19. Vapor-recovery system using the crankcase for fuel-vapor storage. (*Chrysler Corporation*)

CHAPTER 18 CHECKUP

NOTE: Since the following is a chapter review test, you should review the chapter before taking the test.

You will want to know the fundamentals of what causes air pollution and smog and how the automobile contributes to pollution. Knowing the fundamentals will make it much easier for you to understand what is being done to reduce atmospheric pollution from the automobile. The following checkup will help you review the chapter. If any of the questions are too difficult, turn back into the chapter and reread the pages that will give you the information you need.

Completing the Sentences The sentences that follow are incomplete. After each sentence there are several words or phrases, only one of which will correctly complete the sentence. Write each sentence in your notebook, selecting the proper word or phrase to complete it correctly.

1. The four places in the automobile from which pollutants can come are the fuel tank, carburetor: (*a*) engine, and manifold, (*b*) crankcase, and tail pipe, (*c*) crankcase, and manifold.
2. Three pollutants that the automobile gives off are: (*a*) HC, CO, and NO_x, (*b*) HC, CO_2, and H_2O, (*c*) H, C, and CO.

3. The system that picks up any blow-by and gasoline vapors from the crankcase and sends them through the engine is called the: (*a*) ventilator, (*b*) positive crankcase ventilating system, (*c*) antipollution system.
4. The fuel tank breathes as: (*a*) it is filled and emptied, (*b*) temperatures change, (*c*) engine speed increases.
5. "Smog" is a word made up of two separate words: (*a*) smell and fog, (*b*) smell and bog, (*c*) smoke and fog.
6. The vapor-recovery system stores gasoline vapor in: (*a*) a can, (*b*) the fuel tank, (*c*) a charcoal canister.
7. The modern automobile equipped wth a vapor-recovery system uses a: (*a*) sealed fuel gauge, (*b*) pressure-vacuum fuel-tank cap, (*c*) combination vapor and gasoline fuel pump.
8. The purpose of the expansion tank in the fuel tank is to: (*a*) allow the tank to expand to hold more fuel, (*b*) allow the gasoline to expand as its temperature increases, (*c*) prevent overfilling of the tank.

Review Questions The questions that follow give you a chance to check up on your general knowledge about pollution, smog, and the automobile. Write the answers in your notebook.

1. How did smog first start?
2. What holds smog over big cities?
3. How many tons of air pollutants are created by people in the United States each year?
4. What are the four places in the automobile from which pollutants can come?
5. Describe how the positive crankcase ventilating system works.
6. Why is the vapor-recovery system needed? How does it work?
7. What is the purpose of the fuel-return line?
8. Why is a pressure-vacuum fuel-tank cap needed?
9. What is the purpose of the expansion tank?

SUGGESTIONS FOR FURTHER STUDY

There are many articles being published in newspapers and magazines about atmospheric pollution and what should be done about it. Read and save them. Try to decide for yourself whether all the emission controls on automobiles are really necessary. Talk to other people to get their opinions. See if you can come up with some valid answers to what is really needed.

chapter 19

CLEANING UP THE EXHAUST GAS

In Chap. 18, we described the positive crankcase ventilating system and the fuel-vapor-recovery system. These systems take care of three of the four sources of automotive pollution (see Fig. 18-4). In this chapter, we consider the various steps that have been taken to clean up the fourth source of pollution—the exhaust gas from the tail pipe.

⊘ 19-1 Cleaning Up the Exhaust Gas We shall discuss three ways of cleaning up the exhaust gas:

1. Controlling the air-fuel mixture
2. Controlling the combustion process
3. Treating the exhaust gas

⊘ 19-2 Controlling the Air-Fuel Mixture Gasoline has been changed to make it burn better. One of the changes is the elimination of lead. You will recall from Chap. 12 that tetraethyl lead has been added to gasoline to control knocking and permit higher compression ratios. Lead is now being removed to permit the use of catalytic converters (which we shall discuss later in this chapter). Removing the lead has required a reduction of compression ratios. Basically, controlling the air-fuel mixture has meant (1) modifying the carburetor to deliver a leaner air-fuel mixture and (2) providing faster warm-up and quicker choke action.

⊘ 19-3 Leaner Idling Air-Fuel Mixture Modern carburetors have an idle limiter (Fig. 19-1). The idle-mixture-adjustment screw is adjusted at the factory. Then the idle-limiter cap is installed. The cap permits a small amount of adjustment of the idle-mixture-adjustment screw. But it will not permit any adjustment that goes beyond the lean-idle setting required by law. The cap can be removed, of course, if the carburetor requires a major overhaul. But it must then be reinstalled.

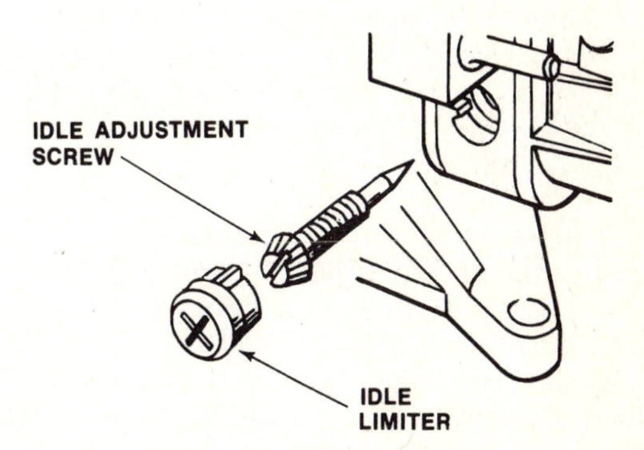

Fig. 19-1. Location of the idle limiter in one model of carburetor. (*Ford Motor Company*)

⊘ 19-4 Faster Warm-up If the air-fuel mixture coming from the carburetor is cold, only part of the fuel will vaporize. This means that an extra-rich mixture is needed. Otherwise, the engine will not get enough gasoline vapor for it to run. Of course, the situation changes as soon as the engine begins to run. Then, the hot exhaust gas circulating around the manifold-heat-control valve begins to heat the intake manifold. (See ⊘ 14-25.) However, this is too slow for the new systems. So a thermostatically controlled air cleaner is used to provide heated air quickly to the carburetor when the engine is cold. This system, called the *heated-air system* (HAS) is discussed in detail in ⊘ 13-12. Figure 19-2 shows the system installed on a V-8 engine.

As you will recall, the heated-air system includes a control-damper assembly in the snorkel of the air cleaner. Also, it has a heat stove that surrounds the exhaust manifold. When the engine is cold, the temperature sensor in the air cleaner closes the damper. In this position, all air must come from the heat stove. When the engine starts and the exhaust manifold begins to warm up, hot air is delivered to the carburetor. This improves cold and warm-up operation.

As the engine begins to warm up, the ingoing air temperature goes above 100 degrees Fahrenheit [37.8°C]. This causes the temperature sensor to admit manifold vacuum to the air-control motor. The air-control motor opens the control-damper assembly. Now air can enter from the engine compartment.

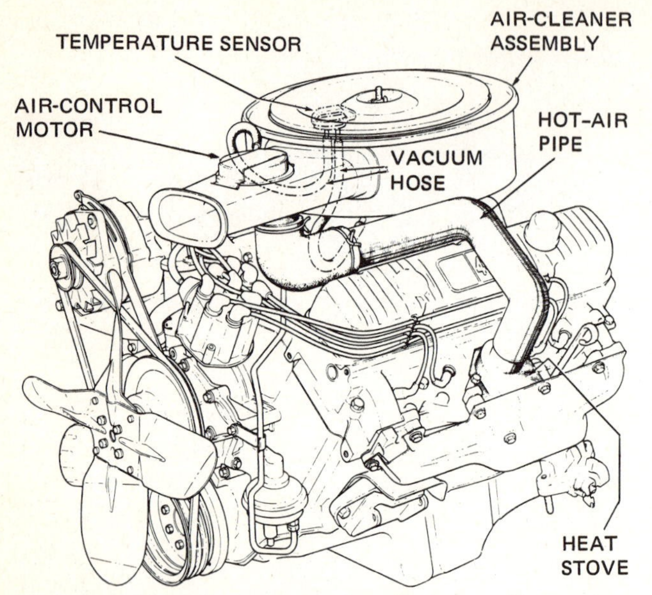

Fig. 19-2. Heated-air system installed on a V-8 engine. (*Buick Motor Division of General Motors Corporation*)

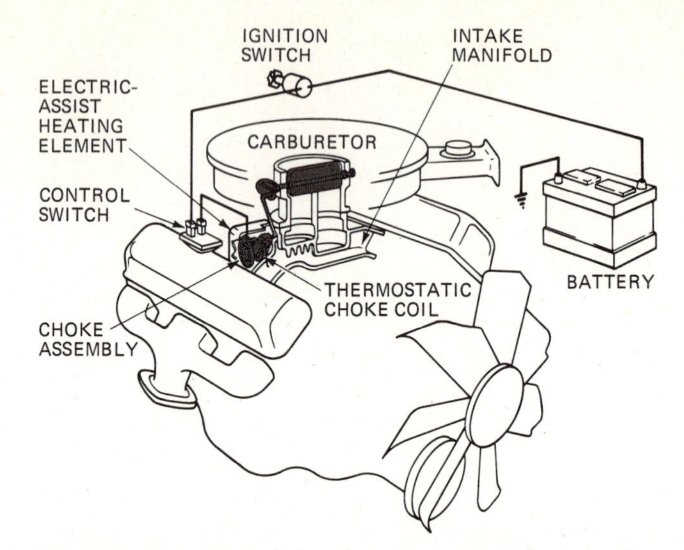

Fig. 19-3. Location of the electric-assist choke on a V-8 engine. The carburetor and choke have been cut away so that their interiors can be seen. (*Chrysler Corporation*)

This system allows the engine to start and operate satisfactorily when cold even though the idle mixture is lean.

⊘ 19-5 Faster-acting Choke In older-model cars, the automatic chokes operated only from the engine heat (⊘ 14-24). But in late-model cars, electric-assist chokes are used (Figs. 14-33, 14-34, and 19-3). The choke thermostat is subjected to heat from a heating element in the choke and heat from the exhaust manifold. This reduces the amount of time during which the engine operates in a choked condition. With the choke valve closed, the engine is fed a very rich air-fuel mixture. The exhaust gas is thus loaded with unburned HC and CO. The electric-assist choke reduces the length of time during which these pollutants are fed into the atmosphere. There is more information on the electric-assist choke in ⊘ 14-24.

⊘ 19-6 Controlling the Combustion Process The combustion process seems simple at first glance: A mixture of air and gasoline vapor is compressed in the combustion chamber. A spark ignites it. The mixture burns and produces the high pressure that pushes the piston down. However, the combustion process is actually complicated. Here are some of the factors involved:

1. The layers of air-fuel mixture next to the relatively cool cylinder head and pistonhead do not burn. The metal surfaces chill these layers below the combustion point. Thus, unburned fuel is swept out of the cylinder on the exhaust stroke. This adds polluting HC to the atmosphere. There are two methods of combating this problem. One is to use stratified charge or fuel injection. The other is to reduce the surface area surrounding the combustion chamber. We shall discuss these two methods later.

2. Increasing the combustion temperature improves combustion of the fuel. But the higher temperature produces more nitrogen oxides (NO_x), which produces another problem. More on this later.

3. Vacuum advance gives the air-fuel mixture a longer time to burn when the engine operates at part throttle. But it also gives more time for NO_x to form under certain operating conditions. Thus a means must be provided to kill the vacuum advance during these special operating conditions. We cover this in detail later.

4. Carbon buildup in the combustion chambers increases the HC in the exhaust gas. The carbon has pores that fill up during the compression and combustion strokes. Then, during the exhaust stroke, when pressure is released, the HC in the carbon pores escapes and exits with the exhaust gas.

⊘ 19-7 Reducing Combustion-Chamber Surface Area Actually, reducing combustion-chamber surface area means reducing the *S/V ratio* (Fig. 19-4). The *S/V* ratio is the ratio between the surface area and the volume of the combustion chamber. A sphere has the lowest possible *S/V* ratio. The wedge combustion chamber (Fig. 19-5) has a higher *S/V* ratio than the sphere. Thus, the hemispheric combustion chamber has a small surface area. It has less surface to chill the air-fuel mixture. Therefore, the hemispheric combustion chamber produces a lesser amount of unburned HC in the exhaust.

⊘ 19-8 Stratified Charge Stratified-charge engines are described in ⊘ 13-22. In stratified charge, the rich part of the mixture is kept away from the combustion-chamber surfaces. Thus less HC remains unburned. The Honda system (Fig. 19-6) is one special example of stratified charge. A rich mixture is delivered to the precombustion chamber and is ig-

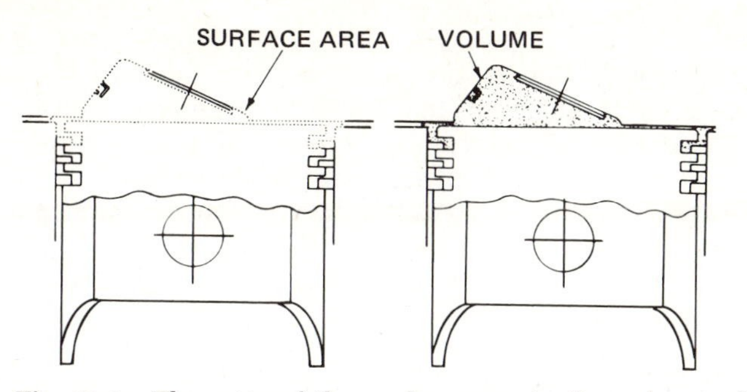

Fig. 19-4. The ratio of the surface area to the volume of the combustion chamber, or *S/V*. This ratio has an effect on the amount of unburned hydrocarbons in the exhaust gas.

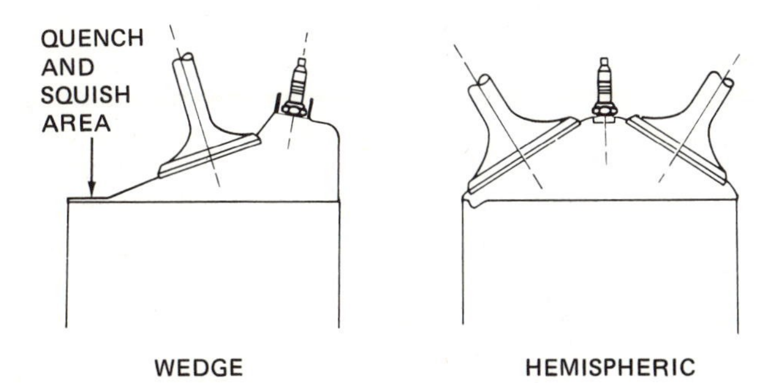

Fig. 19-5. Wedge and hemispheric combustion chambers. (*General Motors Corporation*)

nited in this chamber. It streams out into the lean mixture in the combustion chamber. Combustion takes place in the central part of the main combustion chamber.

⊘ 19-9 Fuel Injection Fuel-injection systems for gasoline engines are described in ⊘ 13-19. Fuel injection can improve combustion and reduce HC and CO in the exhaust. For one thing, the fuel-injection system more accurately meters the fuel. The system supplies the same amount of fuel to each cylinder. This is in contrast with the carburetor-type system in which some cylinders can get a richer mixture than others.

⊘ 19-10 Increasing the Combustion Temperature Increasing the combustion temperature reduces CO and unburned HC in the exhaust. But it increases the formation of NO_x. One method of reducing NO_x is to reduce the compression ratio. This reduces top combustion temperatures and thus the amount of NO_x that is formed.

Another NO_x reduction method, called the *exhaust-gas recirculation (EGR) system,* reduces NO_x during normal running of the engine. A third method reduces NO_x during acceleration in low gears and during part-throttle, high-vacuum conditions.

⊘ 19-11 Exhaust-Gas Recirculation If a small part of the exhaust gas is sent back through the engine,

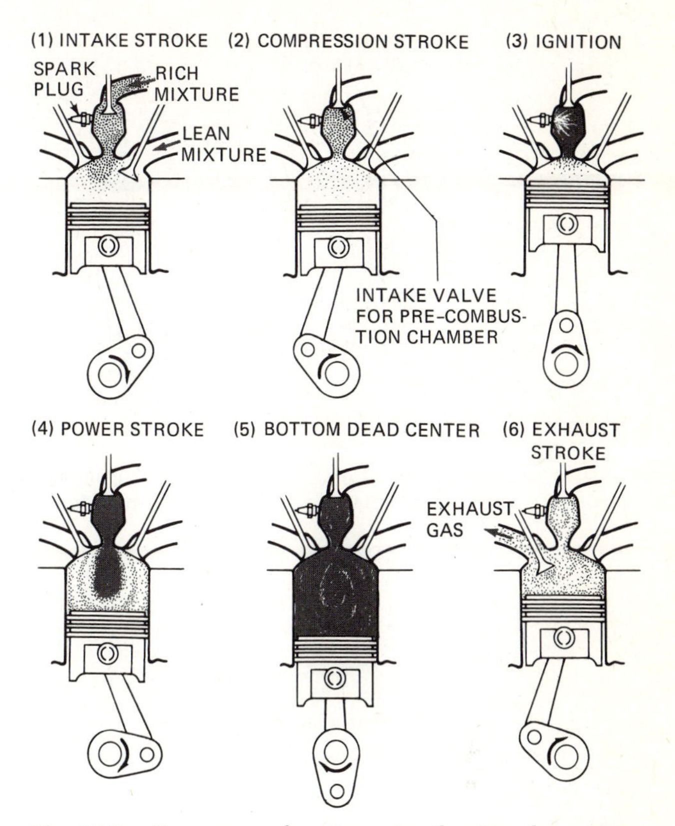

Fig. 19-6. Sequence of actions in the Honda system. (*American Honda Motor Company, Inc.*)

it reduces the combustion temperature and lowers the formation of NO_x. The amount sent through the engine should vary according to operating conditions. The simplest system is shown in Fig. 19-7. Here, a special passage connects the exhaust manifold with the intake manifold. This passage is opened or closed by a special exhaust-gas recirculation (EGR) valve. The upper part of this valve is sealed. It is connected by a vacuum line to a signal port in the carburetor, as shown. When there is no vacuum at work on the signal port, there is no vacuum in the EGR valve. The spring holds the valve closed, and no exhaust gas is recirculated. This is the situation during engine idling, when NO_x formation is near a minimum.

However, when the throttle is opened, the edge of the throttle plate passes the signal port. This allows intake-manifold vacuum to operate the EGR valve. The vacuum raises the diaphragm in the valve. This lifts the valve off the seat. Now exhaust gas can pass into the intake manifold. There, it mixes with the air-fuel mixture and enters the engine cylinders. The exhaust gas lowers the combustion temperature and thus reduces the formation of NO_x. Note that at wide-open throttle there is little vacuum in the intake manifold. Thus, the EGR valve is nearly closed. At wide-open throttle, there is less need for exhaust-gas recirculation.

Figure 19-8 shows the EGR valve in the fully opened position. On many late-model cars a thermal vacuum switch prevents exhaust gas recirculation

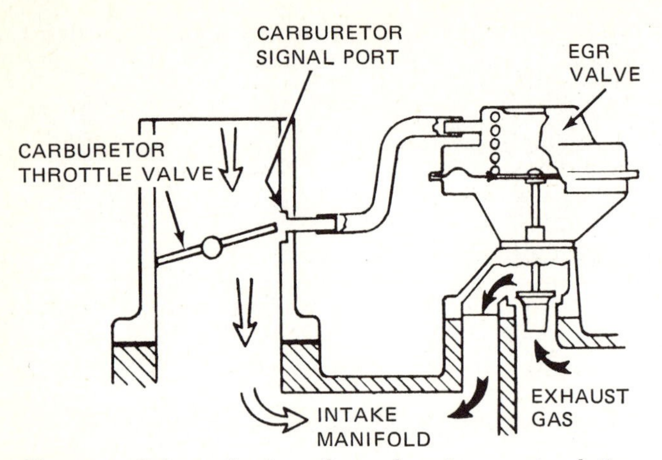

Fig. 19-7. Schematic view of an exhaust-gas-recirculation system. (*Chevrolet Motor Division of General Motors Corporation*)

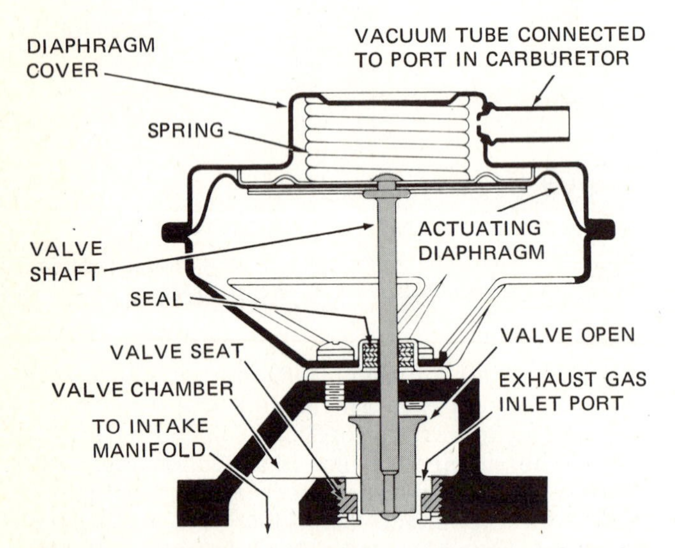

Fig. 19-8. Sectional view of the EGR valve. (*Chevrolet Motor Division of General Motors Corporation*)

until the engine temperature reaches about 100 degrees Fahrenheit [37.8°C]. The thermal switch is connected into the vacuum line between the carburetor and the EGR valve. The thermal switch is mounted in the cooling-system thermostat housing so that it senses coolant temperature. If this temperature is below 100 degrees Fahrenheit [37.8°C], the thermal switch remains closed. This prevents the vacuum from reaching the EGR valve so that exhaust gas does not recirculate. This improves cold-engine performance for the first few moments of operation. After the engine warms up to where it can tolerate exhaust-gas recirculation, the thermal valve opens. Now, vacuum can get to the EGR valve so that exhaust gas can recirculate.

There are several variations of this basic EGR system. For instance, some EGR valves have a second diaphragm. This diaphragm produces increased exhaust-gas recirculation when the engine is heavily loaded, as during hard acceleration. Also, some high-performance engines use an additional modulator system to provide additional control based on car speed. One modulator system is shown in Fig. 19-9 enclosed in dashed lines. It includes a solenoid valve that is normally open, allowing intake-manifold vacuum to pass through it. When engine temperature is high enough to open the thermal switch and the throttle is partly opened, the intake manifold can operate the EGR valve. Exhaust-gas recirculation results. However, when the engine speed reaches a certain level, the speed sensor sends a signal to the electronic amplifier. This causes the amplifier to close the solenoid valve. Now, the vacuum line is closed and exhaust-gas recirculation stops.

⊘ 19-12 Valve Overlap

One of the features of the Chrysler complete emission-control systems is additional valve overlap. The complete set of systems is shown in Fig. 19-10. Additional valve overlap does the same thing as the EGR system but in a different way. Increased valve overlap leaves more of the exhaust gases in the cylinders. That is, the intake valve opens while there is still quite a bit of exhaust gas in the cylinder. Therefore, exhaust gas mixes with the air-fuel mixture entering the cylinder. The result is that the top combustion temperatures are reduced and there is less NO_x formation. However, increased valve overlap can cause rough idling.

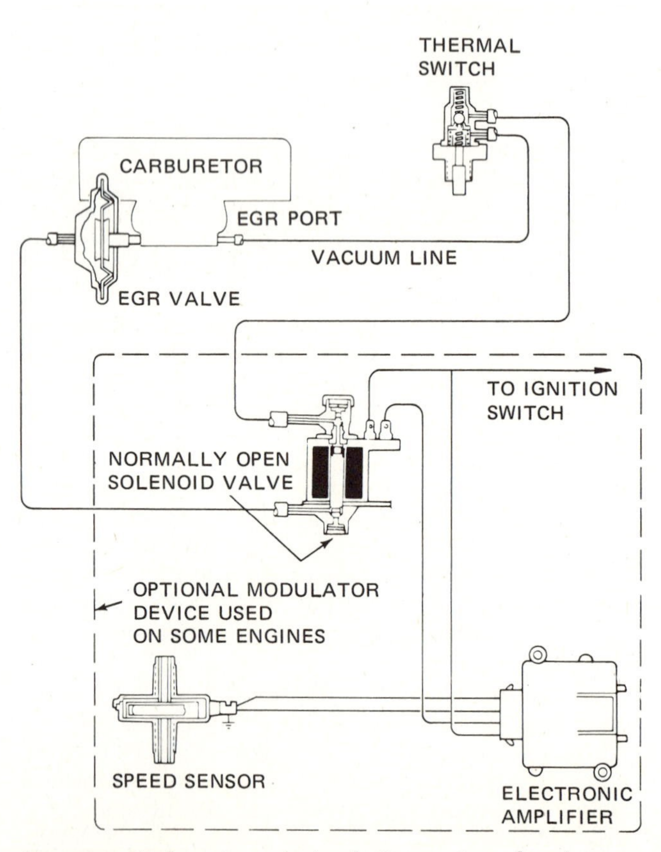

Fig. 19-9. Exhaust-gas-recirculation system showing optional modulator device for some engines. (*Ford Motor Company*)

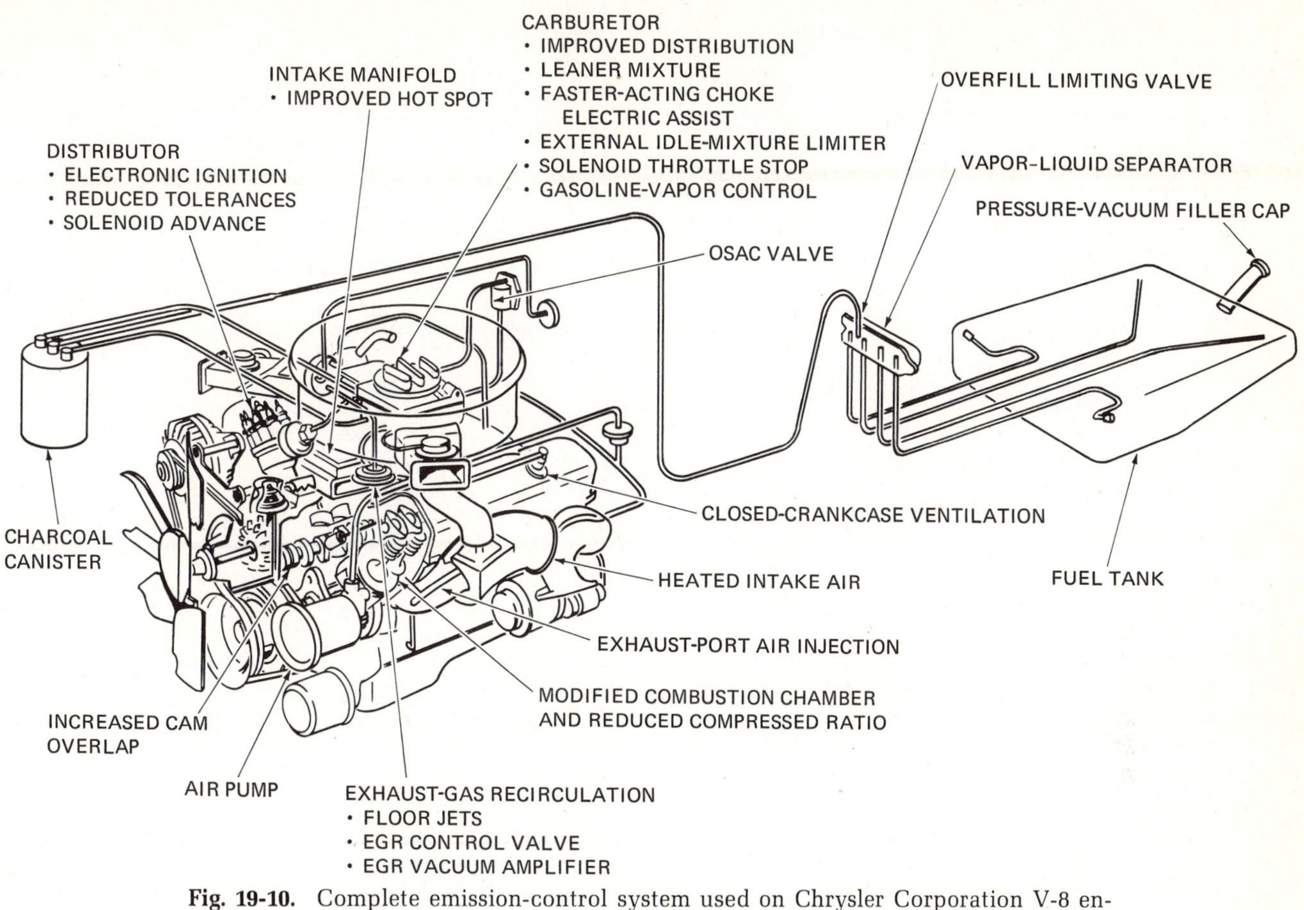

Fig. 19-10. Complete emission-control system used on Chrysler Corporation V-8 engines. The system includes a vapor-recovery system, positive crankcase ventilation, exhaust-gas recirculation (plus increased valve overlap), and other features. (*Chrysler Corporation*)

⊘ 19-13 Control of Vacuum Advance: TCS System

During part-throttle operation, the distributor vacuum advance operates. This provides more time for the leaner air-fuel mixture to burn. However, this added time also allows more NO_x to develop. Thus, a variety of controls have been used to prevent vacuum advance under certain conditions. For example, Chevrolet uses a *transmission-controlled spark* (TCS) *system* on cars with manual transmissions. The TCS system prevents vacuum advance when the car is operated in reverse, neutral, or low forward gears. Under these conditions vacuum advance could greatly increase the formation of NO_x.

Figure 19-11 shows the Chevrolet TCS system for a six-cylinder engine in a car with manual-transmission. The diagram also shows the engine temperature switch (lower left) and the idle-stop solenoid. Figure 19-12 shows the situation when first starting. Turning on the ignition switch energizes the idle-stop solenoid. The plunger extends to contact the throttle lever. This action prevents the throttle from closing completely so that idle speed stays high enough. When the engine is turned off, the idle-stop solenoid allows the throttle to close completely. This prevents "dieseling," or the engine running with the ignition off.

Now, refer to Fig. 19-12 again. Note that turning on the ignition switch completes the circuit through the vacuum-advance solenoid and the temperature-switch cold terminal. At the same time, the

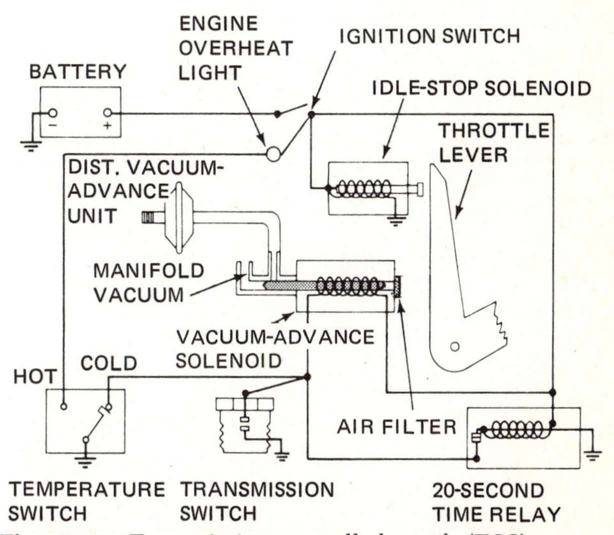

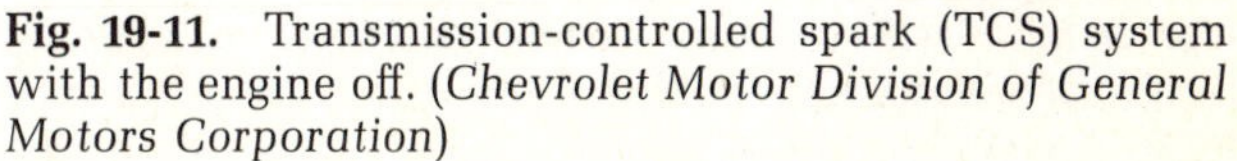

Fig. 19-11. Transmission-controlled spark (TCS) system with the engine off. (*Chevrolet Motor Division of General Motors Corporation*)

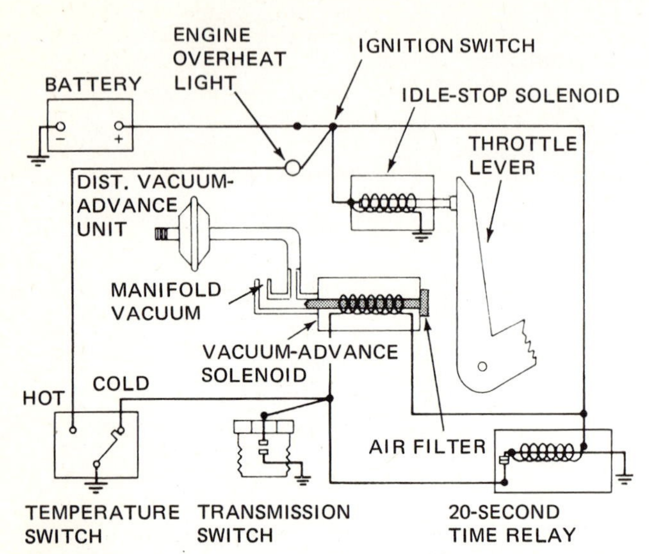

Fig. 19-12. TCS system with a cold engine running. (*Chevrolet Motor Division of General Motors Corporation*)

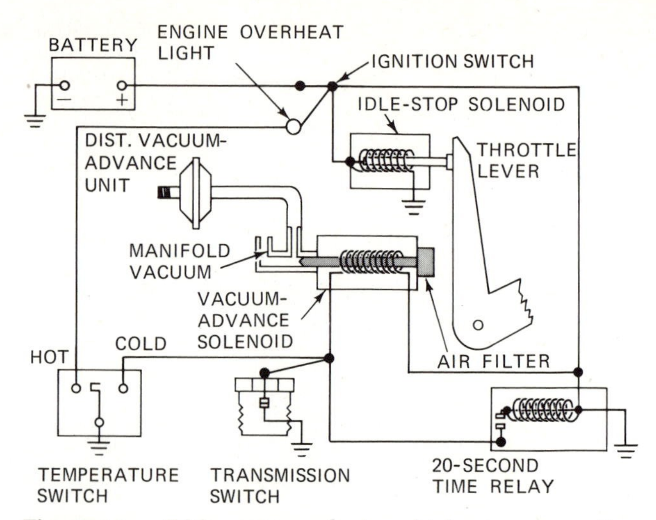

Fig. 19-14. TCS system during high-gear operation. (*Chevrolet Motor Division of General Motors Corporation*)

circuit to the 20-second time relay is completed. With either of these circuits complete, the vacuum-advance solenoid is energized. Vacuum is admitted to the distributor vacuum-advance mechanism so that vacuum advance is obtained.

Figure 19-13 shows the TCS system in low-gear operation. If the engine temperature has increased enough, the temperature-switch cold points are opened. Also, after 20 seconds, the time-relay switch points open. Thus, the circuit to the vacuum-advance solenoid is opened by either of these conditions. The solenoid plunger moves to block vacuum to the distributor vacuum advance. No vacuum advance results.

Figure 19-14 shows the TCS system in high-gear operation. The transmission switch closes its points when the transmission is shifted into high. This energizes the vacuum-advance solenoid so that vacuum is admitted to the distributor vacuum-advance mechanism. Vacuum advance can then result.

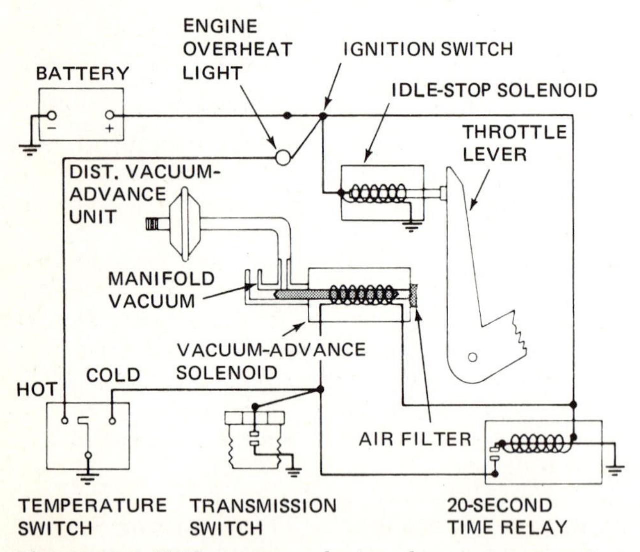

Fig. 19-13. TCS system during low-gear operation. (*Chevrolet Motor Division of General Motors Corporation*)

Some systems have a temperature-override switch. This switch causes the system to provide vacuum advance under any condition if the engine begins to overheat. This system is shown in Fig. 19-15. If the engine becomes too hot, the hot points in the temperature-override switch close. This action energizes the solenoid so that vacuum is admitted to the distributor vacuum advance. With vacuum advance, engine speed increases and improved cooling results.

⊘ 19-14 Control of Vacuum Advance: TRS System

A Ford *transmission-regulated spark* (TRS) *system* is shown in Fig. 19-16. It is for both manual and automatic transmissions. The system works in much the same way as the Chevrolet TCS system described in ⊘ 19-13. The solenoid valve is normally open, allowing vacuum advance when the transmission is in high gear. In the lower gears, the transmission switch is closed. This closes the solenoid valve. With the solenoid valve closed, vacuum is shut off from the distributor vacuum advance. Thus there is no vacuum advance.

⊘ 19-15 Other Vacuum-Advance Control Systems

There are other vacuum-advance controls that are especially tailored for the engines and vehicles in which they are used. Late-model cars produced by Chrysler Corporation use an *orifice-spark-advance control* (OSAC). This system includes a very small hole, or orifice. The orifice delays any change in the application of vacuum to the distributor by about 17 seconds when going from idle to part throttle. Therefore, there is a delay in vacuum advance until acceleration is well under way. This is a critical time during which vacuum advance could produce great amounts of NO_x.

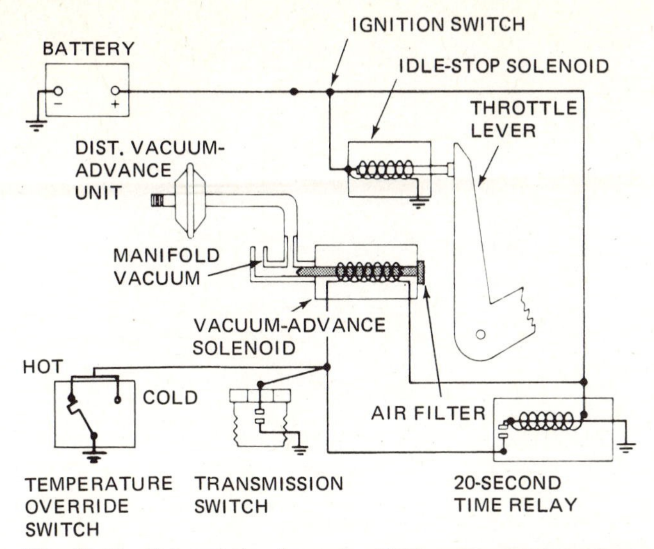

Fig. 19-15. Schematic view of a TCS system with a thermostatic temperature override switch. (*Chevrolet Motor Division of General Motors Corporation*)

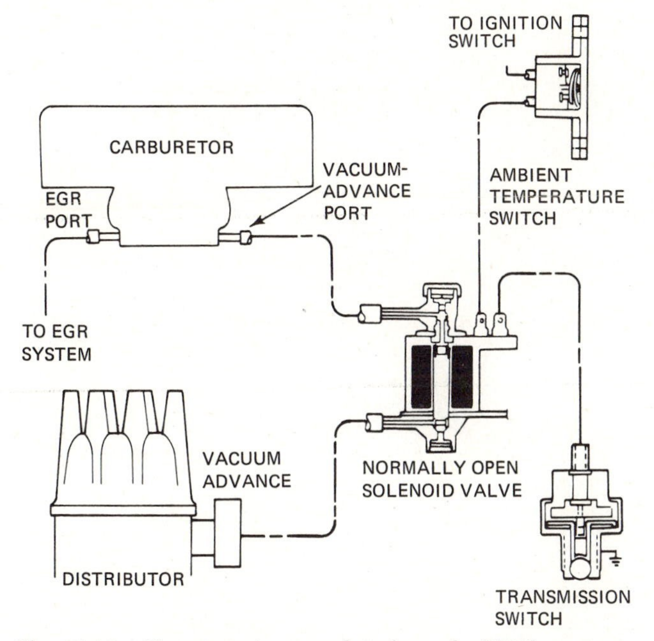

Fig. 19-16. Transmission-regulated-spark (TRS) system. (*Ford Motor Company*)

Ford has developed a similar system, called the *spark-delay-valve system* (Fig. 19-17). This system delays vacuum advance during some vehicle-acceleration conditions. The spark-delay valve is connected in series with the vacuum supply from the vacuum-advance port in the carburetor and distributor vacuum advance. During mild acceleration, the vacuum signal to the distributor can increase only gradually. This is because the spark-delay valve allows the vacuum to pass through slowly. During deceleration or heavy acceleration, the change in vacuum is great enough to open a check valve. This valve allows the vacuum to bypass the spark-delay valve. Thus vacuum advance is produced during these critical times for better engine performance. If engine tempera-

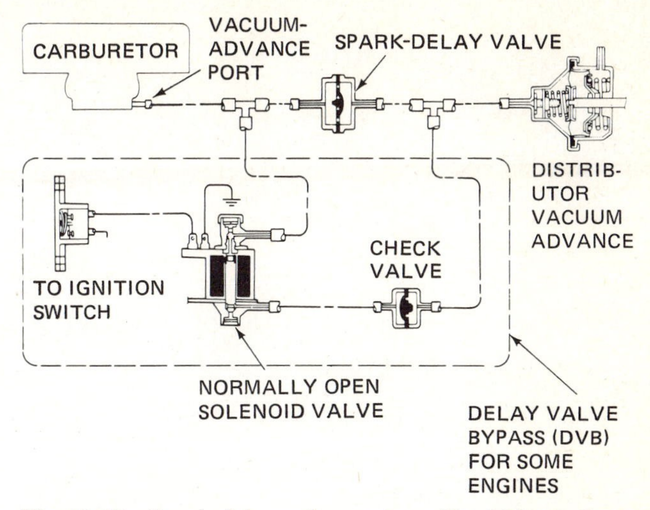

Fig. 19-17. Spark-delay-valve system. (*Ford Motor Company*)

tures are low, the temperature switch actuates the solenoid valve. The actuated valve then passes vacuum directly to the distributor vacuum advance (through the check valve). This provides vacuum advance when the engine is cold.

⊘ 19-16 Treating the Exhaust Gas by Air Injection

After the exhaust gas leaves the engine cylinders, it can be treated to reduce the HC, CO, and NO_x content. One method of treatment is to blow fresh air into the exhaust manifolds. This system is called the *air-injection system*. It provides additional oxygen to burn HC and CO coming out of the cylinders. Figure 19-18 shows the details of the system.

The air-injection pump pushes air through the air lines and air manifold into a series of air-injection tubes. These tubes are located opposite the exhaust valves in the exhaust manifold. The oxygen in the air helps burn any HC or CO in the exhaust gas in the exhaust manifold. The check valve prevents any backflow of exhaust gas to the air pump in case of backfire. The air-bypass valve operates during engine deceleration. During deceleration, intake-manifold vacuum is high. The bypass valve momentarily diverts air from the air pump to the air cleaner instead of to the exhaust manifold. This diversion tends to prevent backfiring in the exhaust system.

A variation of this system uses a special chamber called a *thermal reactor* (Fig. 19-19). In the V-8 system shown, there are two thermal reactors, one for each cylinder bank. These reactors are basically enlarged exhaust manifolds. Being larger, they hold the exhaust gas a little longer. This gives the HO and CO additional time to burn with the oxygen in the pumped-in air.

Note that the system in Fig. 19-19 includes an exhaust-gas-recirculation system, as discussed in ⊘ 19-11. Note also that the air-injection system does not treat NO_x in the exhaust gas; NO_x requires a different sort of treatment.

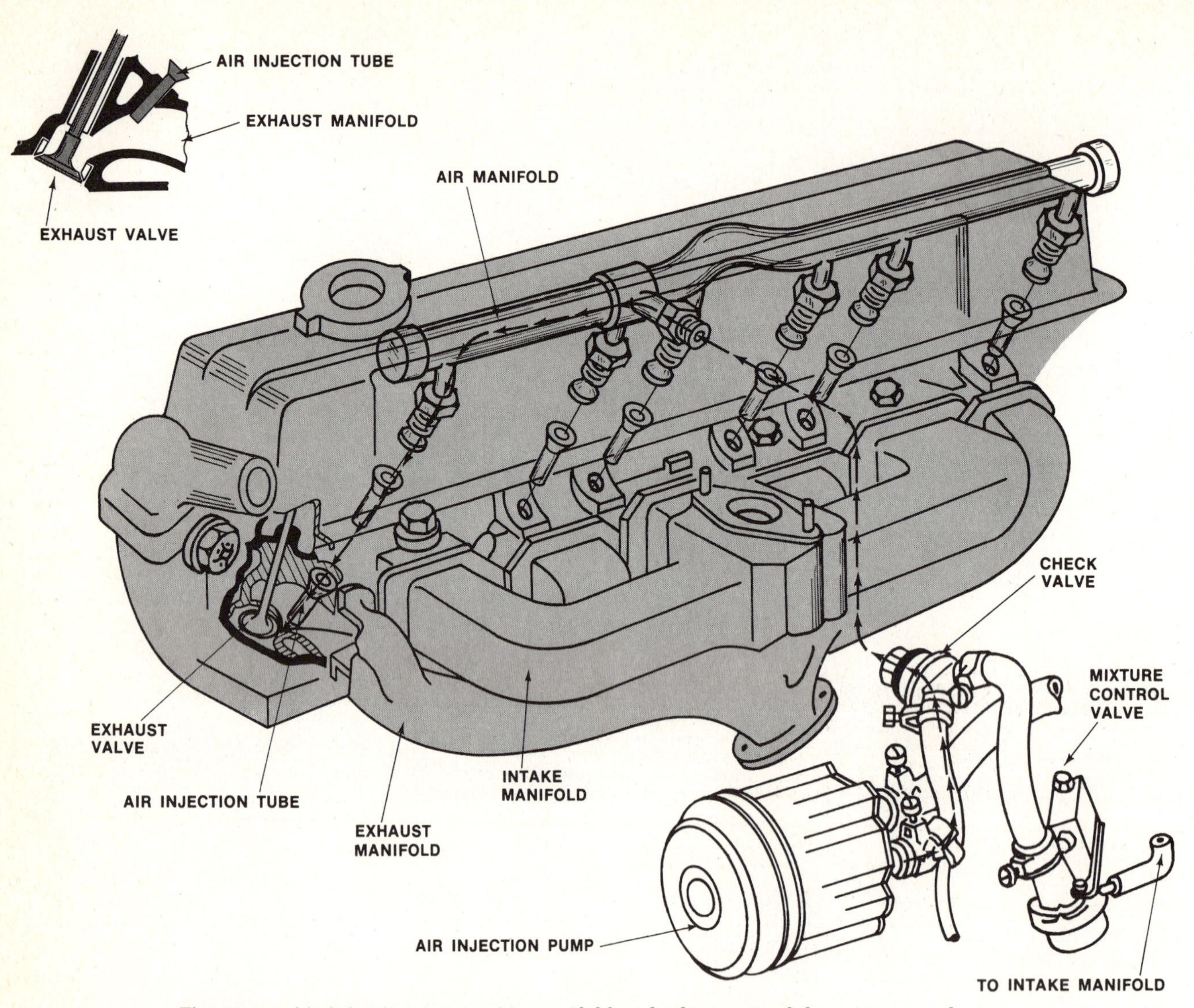

Fig. 19-18. Air-injection system. Air manifold and other parts of the system are shown detached so they can be seen better. The cylinder head has been cut away at the front to show how the air-injection tube fits into the head.

⊘ 19-17 Treating the Exhaust Gas by Catalytic Converters

Another method of treating exhaust gas is to use catalytic converters. These convert the gaseous pollutants into harmless gases. An early system using catalytic converters is shown in Fig. 19-20. Note that this is a dual-exhaust system for a V-8 engine. Each exhaust line has two catalytic converters: One handles HC and CO, the other, NO_x.

A catalyst is a material that causes a chemical change without entering into the chemical reaction. In effect, the catalyst stands by and encourages two chemicals to react. For example, in the HC/CO catalytic converter, the catalyst encourages the HC to unite with O_2 (oxygen) to produce H_2O (water) and CO_2 (carbon dioxide). In the NO_x converter, the catalyst splits the nitrogen from the oxygen. The NO_x becomes harmless O_2 and N (nitrogen).

A late-model catalytic converter engineered by General Motors is shown in Fig. 19-21. The converter is filled with pellets of metal. They are coated with a thin layer of platinum or similar catalytic metal. The pellets form a matrix through which the exhaust gas must pass. Figures 19-22 and 19-23 show one

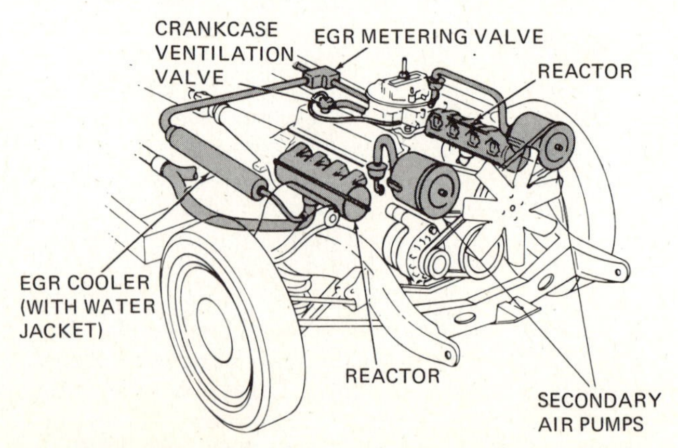

Fig. 19-19. Thermal-reactor system for exhaust-emission control.

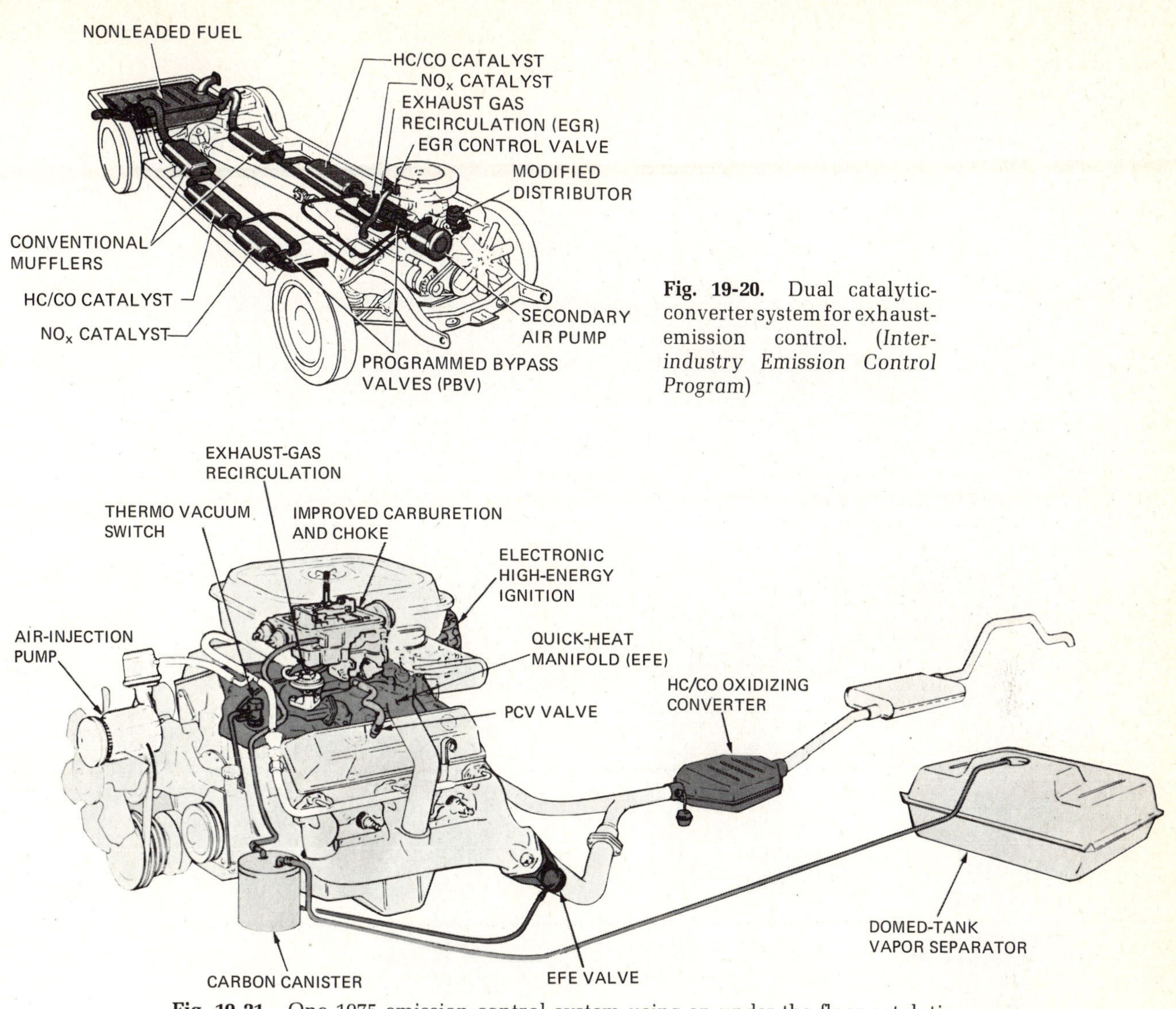

Fig. 19-20. Dual catalytic-converter system for exhaust-emission control. (*Inter-industry Emission Control Program*)

Fig. 19-21. One 1975 emission-control system using an under-the-floor catalytic converter. Note that the system uses air injection and other emission-control features described previously. (*General Motors Corporation*)

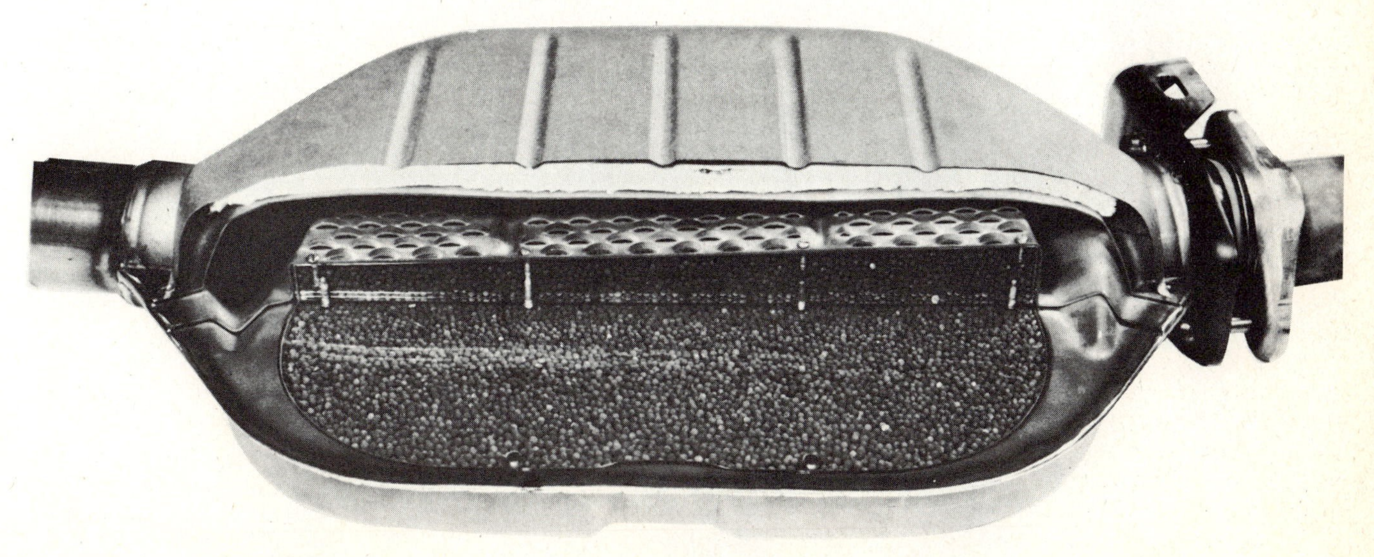

Fig. 19-22. Cutaway view of the catalytic converter shown in Fig. 19-21. (*General Motors Corporation*)

design of catalytic converter. As the exhaust gas flows through (see Fig. 19-23), the platinum or other catalyst produces the chemical reactions. Figure 19-24 shows another design.

One advantage of this system is that the pellets can be replaced when the catalyst has lost efficiency. In other words, the converter can be recharged with fresh catalyst pellets when necessary.

NOTE: The catalytic converter, in addition to converting HC and CO into harmless H_2O and CO_2, also converts the sulfur in the gasoline into sulfur oxides

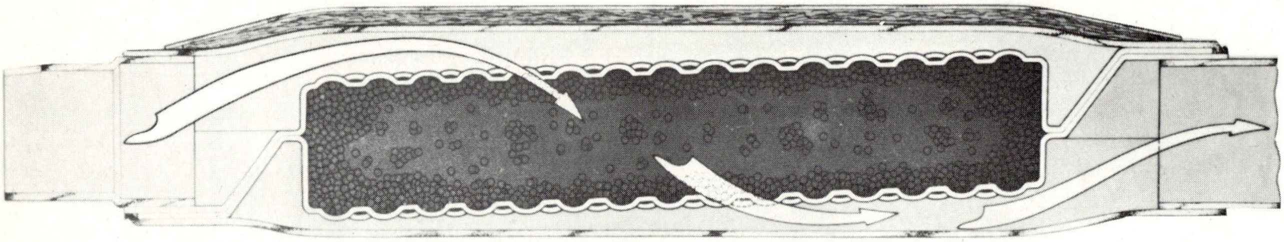

Fig. 19-23. Flow of exhaust gas through the converter (shown by arrows). (*General Motors Corporation*)

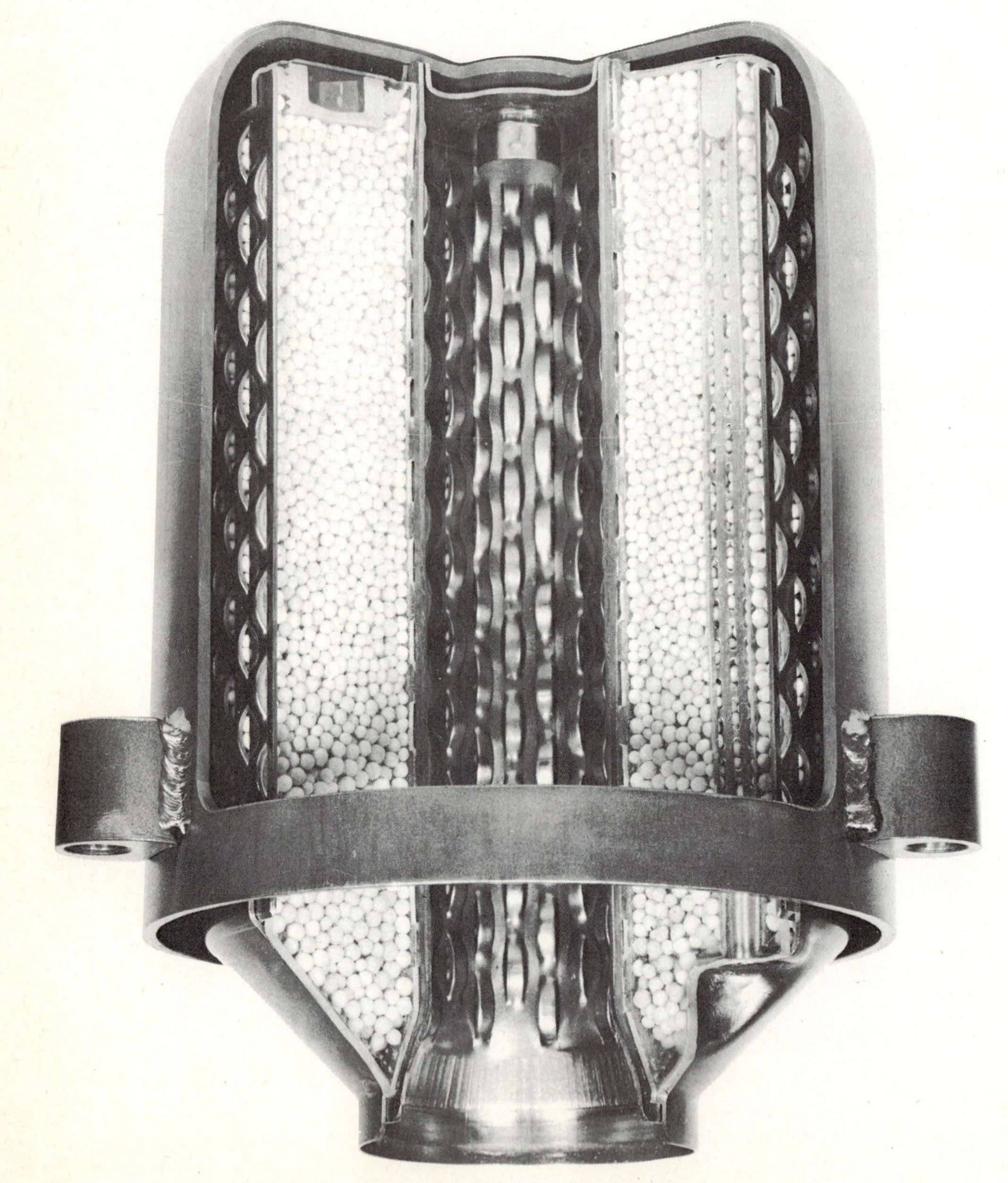

Fig. 19-24. Another design of catalytic converter cut away so that the pellets can be seen. (*General Motors Corporation*)

(SO_x). The sulfur oxides then change into sulfuric acid in the atmosphere. Opinion is divided as to how serious a threat this is to people's health. One remedy suggested is to remove the sulfur from the gasoline during the refining process. But it would take several years to rebuild the refineries so that they would remove the sulfur. And it would cost billions of dollars. An alternative would be to phase out the catalytic converter as new, low-emission engines were developed. This, also could require years and billions of dollars.

CHAPTER 19 CHECKUP

NOTE: Since the following is a chapter review test, you should review the chapter before taking the test.

The chapter you have just completed concludes our discussion of automotive emission controls. Now, take the test that follows to make sure you remember the details.

Completing the Sentences The sentences that follow are incomplete. After each sentence there are several words or phrases, only one of which will correctly complete the sentence. Write each sentence in your notebook, selecting the proper word or phrase to complete it correctly.

1. The pollutant formed by high combustion temperatures is: (*a*) HC, (*b*) CO, (*c*) NO_x.
2. Three ways of cleaning up the exhaust gas are to control the air-fuel mixture, control the combustion, and: (*a*) control the carburetor, (*b*) treat the exhaust gas, (*c*) add tetraethyl lead.
3. Two ways of treating the exhaust gas to reduce the amount of pollutants in it are by supplying additional air to the exhaust gases as they leave the cylinders and running the gases through: (*a*) an air injector, (*b*) a catalyst, (*c*) charcoal canisters.
4. One reason that the gasoline refiners are getting the lead out of gasoline is that the lead: (*a*) increases NO_x, (*b*) can ruin catalysts, (*c*) prevents normal run-on.
5. To control the air-fuel mixture and reduce atmospheric pollution, carburetors have been modified to deliver a leaner air-fuel mixture during: (*a*) idling, (*b*) acceleration, (*c*) high-speed operation.
6. Reducing the compression ratio of an engine reduces the combustion temperature, and this: (*a*) reduces the amount of NO_x formed, (*b*) increases the amount of NO_x formed, (*c*) reduces the amount of HC formed.
7. The transmission-controlled spark (TCS) system permits distributor: (*a*) centrifugal advance in high gear only, (*b*) vacuum advance in high gear only, (*c*) vacuum advance in low gears only.
8. Reducing the combustion-chamber surface area: (*a*) reduces the amount of unburned HC in the exhaust, (*b*) increases the amount of unburned HC in the exhaust, (*c*) reduces the amount of NO_x in the exhaust.
9. One method of reducing NO_x in the exhaust is to: (*a*) increase valve overlap, (*b*) reduce valve overlap, (*c*) prevent valve overlap.

Review Questions The questions that follow give you a chance to check up on your general knowledge about methods of cleaning up the exhaust gas from the automotive engine. Write the answers in your notebook. Check back into the chapter if you are not sure about your answers.

1. What are the three ways of cleaning up the exhaust gas?
2. What are the two ways of controlling the air-fuel mixture to reduce the amounts of HC and CO in the exhaust gas?
3. What is the purpose of the electric heater in the automatic choke?
4. The layer of air-fuel mixture next to the metal surfaces in the combustion chamber does not burn. Why not?
5. What effect does increasing the combustion temperature have on HC, CO, and NO_x in the exhaust?
6. What is the purpose of the EGR system? How does it work?
7. What is the purpose of increasing valve overlap?
8. What is the purpose of the TCS system? How does it work?
9. What is the purpose of the temperature-override switch in the TCS system?
10. What is the purpose of injecting air into the exhaust manifold? How does the system work?
11. What is a thermal reactor?
12. What is the purpose of the catalytic converter?

SUGGESTIONS FOR FURTHER STUDY

Continue to look for articles in the newspapers and magazines, as suggested at the end of Chap. 18. Go to your local library and look for material on automotive emissions and emission controls. Keep your eyes and ears open for any information in this field. Much is happening, and new developments are being announced periodically. File articles in your notebook and write down any important facts.

chapter 20

ENGINE-TESTING PROCEDURES AND TOOLS

The purpose of this chapter is to describe the different engine-testing procedures and the instruments used to make the tests. Later chapters in the book discuss engine troubles, or faulty conditions, as disclosed by the engine tests, and the methods for correcting these conditions.

⊘ 20-1 Engine-Testing Procedures Engine-testing procedures are of two types. One type is used when there is an obvious specific trouble that seems related to the engine. For example, if there is a miss in the engine or a complaint of excessive fuel or oil consumption, then there are definite trouble-diagnosis checks that can be made to pinpoint the cause of trouble.

The second type of engine-testing procedure is a general approach. Every engine component is tested as the procedure is carried out, and any worn condition, subnormal operation, or other defect will be detected. This general-approach procedure is often referred to as "engine tune-up." The correction of troubles found during the testing procedure "tunes up" the engine, that is, improves engine performance.

Actually, both types of engine-testing procedures have their place in the automotive business. When you encounter a specific trouble, you want to follow a specific procedure to find its cause so that you can correct it. On the other hand, it is often proper procedure to make a complete check of the engine and its components. For example, many automotive authorities recommend that the engine and its components be checked periodically (for example, every 10,000 miles [16,093 km] or at least once a year). Such an engine analysis will show up worn units, parts, or improper adjustments that soon might cause real trouble. Correction can then be made before serious trouble develops. In other words, the general-procedure diagnosis eliminates trouble before it happens. This is called *preventive maintenance*. You prevent trouble by maintaining the engine in good operating condition.

⊘ 20-2 Engine-Testing Instruments The engine-testing instruments we cover in this chapter are as follows:

1. Tachometer, which measures engine speed in revolutions per minute (rpm)
2. Cylinder compression tester, which measures the ability of the cylinders to hold compression
3. Cylinder leakage tester, which finds any places where there is compression leakage
4. Engine vacuum gauge, which measures intake-manifold vacuum
5. Exhaust-gas analyzer, which measures the amount of pollutants in the exhaust gas
6. Ignition timing light, which is used to set the ignition timing and check the spark advance
7. Oscilloscope, which shows the overall operating condition of the ignition-system circuits
8. Chassis dynamometer, which checks the engine and its operating components under actual operating conditions

There are also instruments to test the battery, starting motor, charging system, and cooling system. There are other instruments to test ignition coils, condensers, spark plugs, distributor contact-point dwell, and distributor advance mechanisms.

⊘ 20-3 Tachometer The tachometer is connected to the ignition system and operates electrically. The tachometer measures engine speed in revolutions per minute (rpm). That is, it measures the number of times the primary circuit is interrupted and translates this into engine rpm. It is a necessary instrument because the idle speed must be adjusted to a specific rpm. Also, many tests must be made at specific engine speeds. The tachometer has a selector knob that can be turned to 4, 6, or 8, according to the number of cylinders in the engine being tested. Figure 20-1 shows a tachometer connected to an engine.

Many high-performance cars have tachometers mounted on the instrument panel (Fig. 20-2). They tell the driver how fast the engine is turning. The driver can therefore keep the rpm within the range at which the engine develops maximum torque. This lets the driver get the best performance from the engine. Many of these tachometers have a red line at the top rpm on the dial. The red line marks off the danger point for engine speed.

Some tachometers are mechanical instead of electrical. They are driven off a gear on the ignition distributor shaft. They operate somewhat like the speedometer.

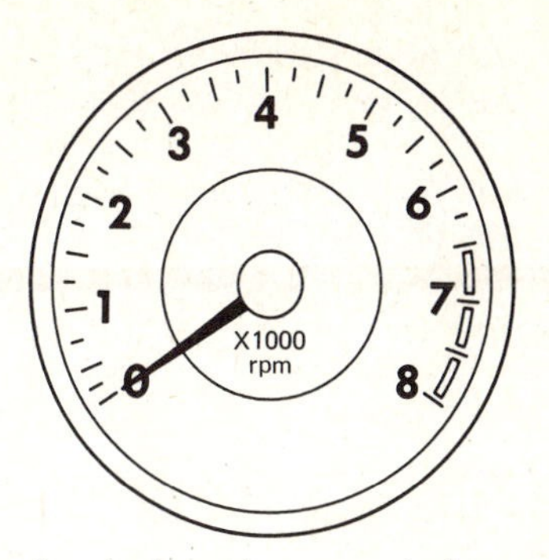

Fig. 20-2. Engine tachometer mounted on a car instrument panel.

⊘ 20-4 Cylinder Compression Tester The cylinder compression tester measures the ability of the cylinders to hold compression. Pressure, operating on a diaphragm in the tester, causes the needle on the face of the tester to move around to indicate the pressure being applied. Figure 20-3 shows a compression tester used to measure the pressure in an engine cylinder.

To use the tester, first remove all the spark plugs. A recommended way to do this is to disconnect the wires and loosen the plugs one turn. Next, reconnect the wires and start the engine. Then, run the engine for a few moments at 1,000 rpm. The combustion gases will blow out of the plug well any dirt that could fall into the cylinder when the spark plugs are removed. The gases will also blow out of the combustion chamber any loosened carbon that is caked around the exposed threaded end of the plug. This procedure prevents carbon and dirt particles from lodging under a valve and holding the valve open during the compression test.

Next, screw the compression-tester fitting into the spark-plug hole of cylinder 1, as shown in Fig. 20-3. To protect the coil from high voltage, disconnect the lead from the negative terminal of the coil. (This is the lead that goes to the distributor.) Then, hold the throttle wide open and operate the starting motor to crank the engine. The needle will move around to show the maximum compression

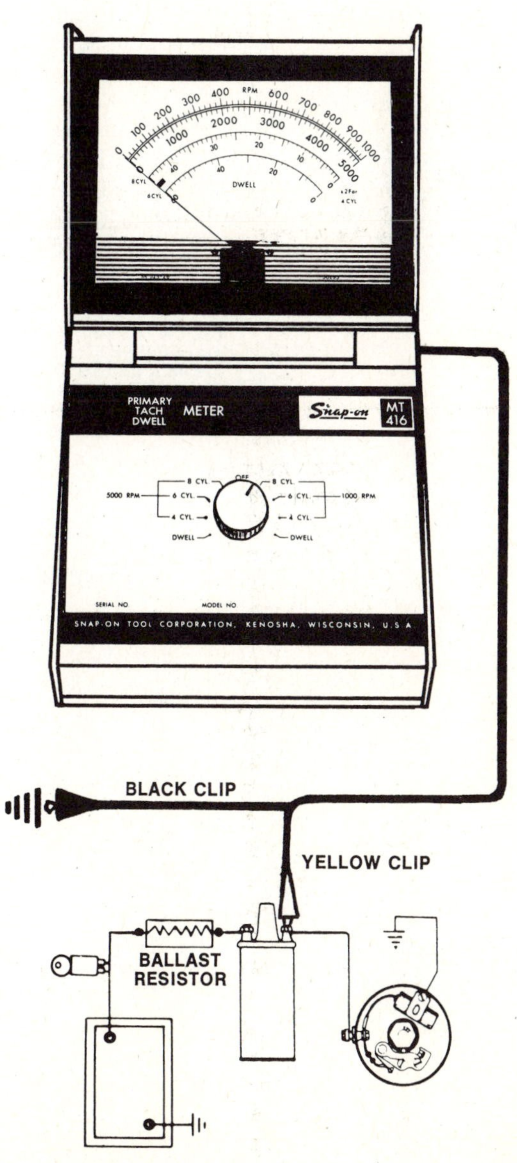

Fig. 20-1. Tachometer connected to an engine. (*Snap-on Tools Corporation*)

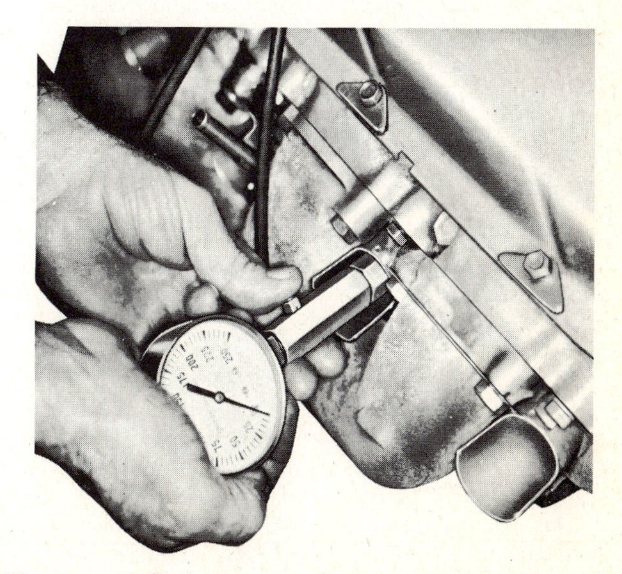

Fig. 20-3. Cylinder compression tester in use. (*Chevrolet Motor Division of General Motors Corporation*)

pressure the cylinder is developing. Write down this figure. Test the other cylinders the same way.

⊘ 20-5 Results of the Compression Test The engine manufacturer's specifications tell you what the compression pressure of the cylinders should be. If the results of the compression test show that compression is low, there is leakage past the piston rings, valves, or cylinder-head gasket. To correct the trouble, you must remove the cylinder head and inspect the engine parts.

Before you do this, however, you can make one more test to pinpoint the trouble. Squirt a small quantity of engine oil through the spark-plug hole into the cylinder. Then retest the compression. If the pressure increases to a more normal figure, the low compression is due to leakage past the piston rings. Adding the oil helps seal the rings temporarily so that they can hold the compression pressure better. The trouble in this case is caused by worn piston-rings, a worn cylinder wall, or a worn piston. The trouble could also be caused by rings that are broken or stuck in the piston-ring grooves.

If adding oil does not increase the compression pressure, the leakage is probably past the valves. This could be caused by:

1. Broken valve springs
2. Incorrect valve adjustment
3. Sticking valves
4. Worn or burned valves
5. Worn or burned valve seats
6. Worn camshaft lobes
7. Dished or worn valve lifters

It may also be that the cylinder-head gasket is "blown." This means the gasket has burned away so that compression pressure is leaking between the cylinder head and the cylinder block. Low compression between two adjacent cylinders is probably caused by the head gasket blowing between the cylinders.

Whatever the cause—rings, pistons, cylinder walls, valves, or gasket—the cylinder head must be removed so that the trouble can be fixed. (We will discuss engine service later in Chaps. 23 to 26.)

⊘ 20-6 Cylinder Leakage Tester The cylinder leakage tester does approximately the same job as the compression tester but in a different way. It applies air pressure to the cylinder with the piston at top dead center (TDC) on the compression stroke. In this position, both valves are closed. Very little air should escape from the combustion chamber. Figure 20-4 shows a cylinder leakage tester. Figure 20-5 shows the tester connected to an engine cylinder and how it pinpoints places where leakage can occur.

To use the cylinder leakage tester, first remove all plugs, as explained previously. Then remove the air cleaner, crankcase filler cap or dipstick, and radiator cap. Set the throttle wide open, and fill the radiator to the proper level. You are now ready to begin.

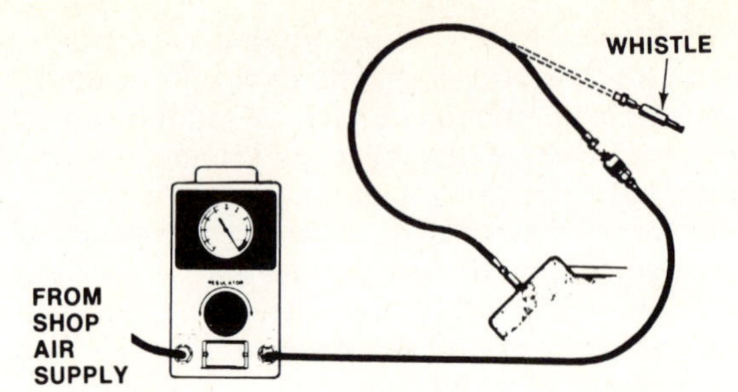

Fig. 20-4. Cylinder leakage tester. The whistle is used to locate TDC in cylinder No. 1. (*Sun Electric Corporation*)

Connect the adapter, with the whistle, to the spark-plug hole of cylinder 1. Turn the engine over until the whistle sounds. When the whistle sounds, the piston is moving up on the compression stroke. Continue to rotate the engine until the TDC timing marks on the engine align. When the marks align, the piston is at TDC. Disconnect the whistle from the adapter hose and connect the tester, as shown in Figs. 20-4 and 20-5.

Next, apply air pressure from the shop supply. Note the gauge reading, which shows the percentage of air leakage from the cylinder. Specifications vary, but a reading above 20 percent means there is excessive leakage. If the air leakage is excessive, check

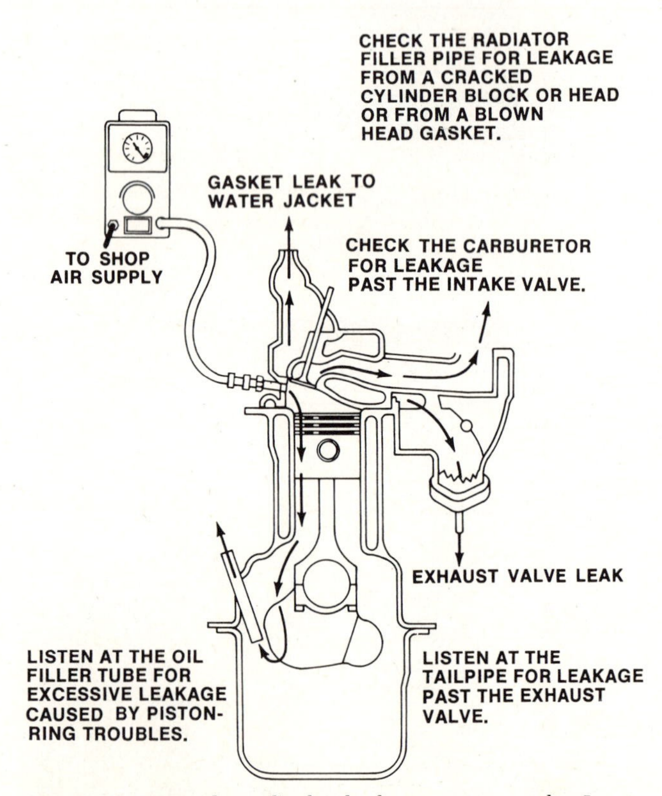

Fig. 20-5. How the cylinder leakage tester works. It applies air pressure to the cylinder through the spark-plug hole with the piston at TDC and both valves closed. Points where air is leaking can then be pinpointed, as shown. (*Sun Electric Corporation*)

further by listening at three places: the carburetor, tail pipe, and crankcase filler pipe. If the air is blowing out of an adjoining cylinder spark-plug hole, it means that the head gasket is blown between the cylinders.

Figure 20-5 shows what it means if you can hear air escaping at any of the three listening places. If air bubbles up through the radiator, then the trouble is a blown cylinder-head gasket or a cracked cylinder head. These conditions allow leakage from the cylinder to the cooling system.

Check the other cylinders in the same manner. A special adapter supplied with the tester lets you quickly find TDC on the other cylinders. When you use the tester, follow the instructions that explain how to use the adapter.

⊘ 20-7 Engine Vacuum Gauge The engine vacuum gauge is a tester for tracking down troubles in an engine that does not run as well as it should. This gauge measures intake-manifold vacuum. The intake-manifold vacuum changes with different operating conditions and with different engine defects. The way the intake-manifold vacuum varies from normal indicates what is wrong inside the engine.

Figure 20-6 shows the vacuum gauge connected to the intake manifold. With the gauge connected, start the engine. The test must be made with the engine at operating temperature. Operate the engine on idle and at other speeds as explained in the following list; the meanings of the various readings are also explained (see Fig. 20-7).

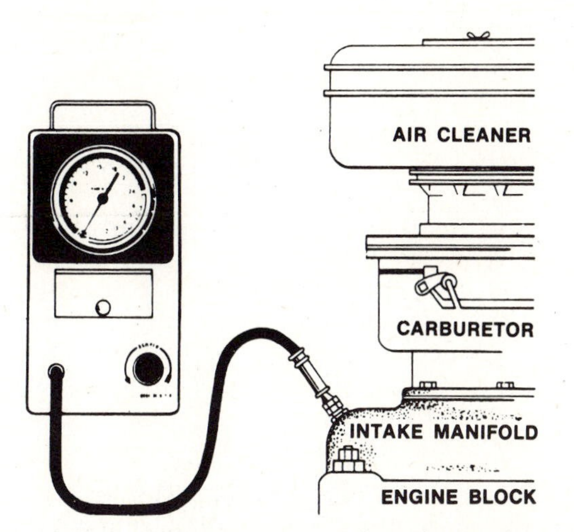

Fig. 20-6. Vacuum gauge connected to the intake manifold for a manifold vacuum test. (*Sun Electric Corporation*)

1. A steady and fairly high reading on idle indicates normal performance. Specifications vary with different engines, but a reading somewhere between 17 and 22 inches [432 and 559 mm] of mercury indicates the engine is okay. The reading will be lower at higher altitudes because of lower atmospheric pressure. For every 1,000 feet [305 m] above sea level, the reading will be reduced about 1 inch [25.4 mm] of mercury.

NOTE: "Inches or millimeters of mercury" refers to the way the vacuum gauges are set up. There is no mercury in the gauge; it is just a way of measuring vacuum.

2. A steady and low reading on idle indicates late ignition or valve timing, or possibly leakage around the pistons. Leakage around pistons—excessive blow-by—could be due to worn or stuck piston rings, worn cylinder walls, or worn pistons. Each of these conditions reduces engine power. With reduced power, the engine does not "pull" as much vacuum.
3. A very low reading on idle indicates a leaky intake manifold or carburetor gasket, or possibly leakage around the carburetor throttle shaft. Air leakage into the manifold reduces the vacuum and engine power.

NOTE: Late-model engines, with high-lift cams and more valve overlap, are likely to have a lower and more uneven intake-manifold vacuum. Also, certain automotive-engine emission controls lower the intake-manifold vacuum.

4. Back-and-forth movement of the needle that increases with engine speed indicates weak valve springs.
5. Gradual falling back of the needle toward zero with the engine idling indicates a clogged exhaust line.

TYPICAL VACUUM-GAUGE READINGS

Fig. 20-7. Vacuum-gauge readings and their meanings.

6. Regular dropping back of the needle indicates that a valve is sticking open or a plug is not firing.
7. Irregular dropping back of the needle indicates that valves are sticking only part of the time.
8. Floating motion or slow back-and-forth movement of the needle indicates that the air-fuel mixture is too rich.

A test can be made for loss of compression due to leakage around the pistons. This condition would be the result of stuck or worn piston rings, worn cylinder walls, or worn pistons. Race the engine for a moment and then quickly release the throttle. The needle should swing around to 23 to 25 inches [584 to 635 mm] as the throttle closes, indicating good compression. If the needle fails to swing around this far, there is loss of compression. Further checks should be made.

⊘ 20-8 Exhaust-Gas Analyzer At one time the major use of the exhaust-gas analyzer was to adjust the carburetor. It is still used for that purpose, but today it has the added job of checking the emission controls on the automobile. We covered emission controls in Chaps. 18 and 19, and we explained that their main purpose is to cut down on carbon monoxide (CO), hydrocarbon (HC), and nitrogen oxides (NO_x) in the exhaust gas.

Figure 20-8 shows one type of exhaust-gas analyzer. To use it, you stick a probe into the tail pipe of the car (Fig. 20-9). The probe draws out some of the exhaust gas and carries it through the analyzer. Two dials on the face of the analyzer, shown in Fig. 20-10, indicate how much HC and CO are contained in the exhaust gas. The HC meter reports in terms of parts per million. The CO meter reports in terms of a percentage. Federal and state laws set the maximum legal limits on the amount of HC and CO permitted in the exhaust gas.

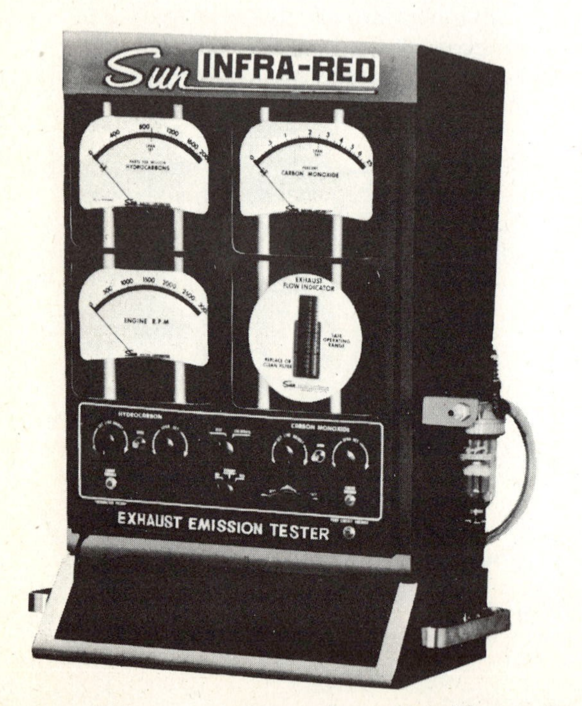

Fig. 20-8. Exhaust-emission analyzer. (*Sun Electric Corporation*)

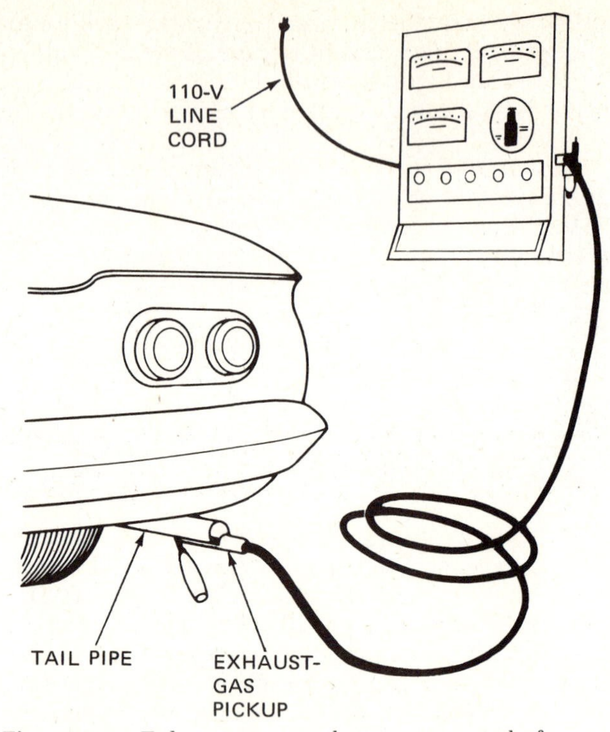

Fig. 20-9. Exhaust-gas analyzer connected for exhaust-gas test.

A different kind of tester is required for NO_x, but it works in the same general way. It draws exhaust gas from the tail pipe and runs the gas through the analyzer. The finding is reported in terms of the amount of NO_x in the exhaust gas. Generally, NO_x testers are available only in testing laboratories. They are not widely used in automotive service shops. Authorities say, however, that someday all well-equipped shops will have them.

⊘ 20-9 Ignition Timing Light The sparks must reach the spark plugs in the cylinders at exactly the right time. They must arrive a specific number of degrees before TDC (top dead center) on the compression stroke. Adjusting the distributor to make the sparks arrive at the right time is called *ignition timing*. You adjust the distributor by turning the distributor in its mounting. If you rotate the distributor in the direction opposite to normal cam rotation, you move the contact points ahead. That is, the points will close and open earlier. This action advances the sparks, and so the sparks appear at the spark plugs earlier. Rotating the distributor in the direction of normal cam rotation will retard the sparks. The sparks appear at the spark plugs later.

To time the ignition, check the markings on the crankshaft pulley with the engine running. Since the pulley turns rapidly, you cannot see the markings in normal light. But by using a special *timing light*

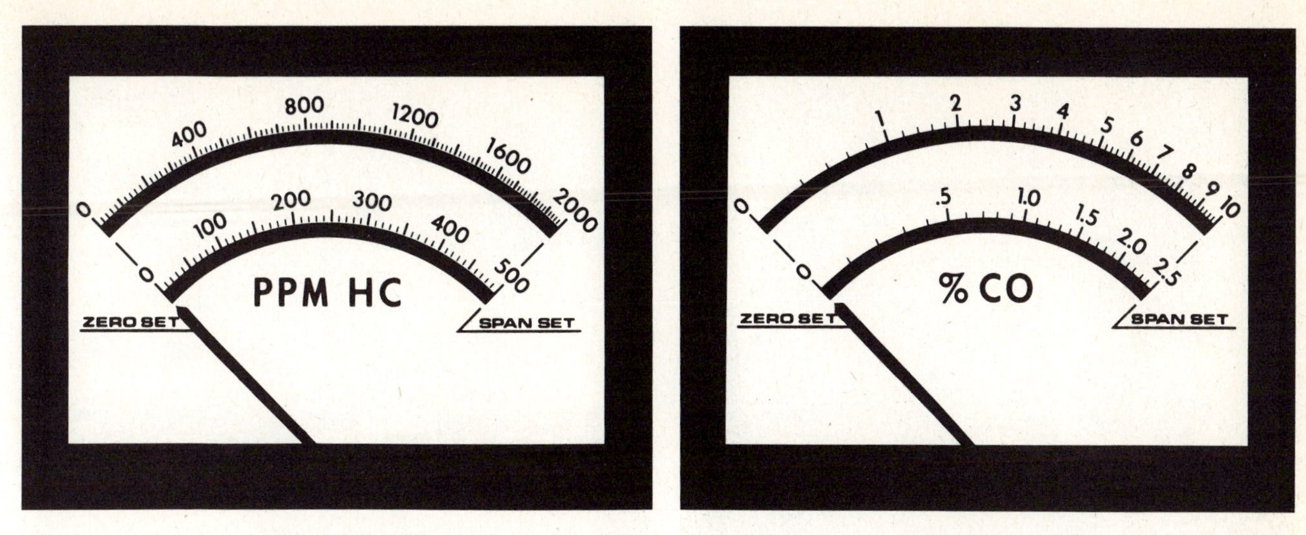

Fig. 20-10. HC and CO meter faces. (*Sun Electric Corporation*)

you can make the pulley appear to stand still. The timing light is a *stroboscopic* light. To use it, connect the timing-light lead to the No. 1 spark plug, as shown in Fig. 20-11. Every time the plug fires, the timing light gives off a flash of light (Fig. 20-12). The light lasts only a fraction of a second. The repeated flashes of light make the pulley seem to stand still.

CAUTION: When connecting a timing light, always connect the leads to the battery first. Then make the connections to the No. 1 spark plug. When disconnecting the timing light, always disconnect the timing-light lead from the No. 1 spark plug first. Then disconnect the battery leads. If you disconnect the battery leads first, you are apt to get a high-voltage shock when you touch the battery connections.

To set the ignition timing, loosen the clamp screw that holds the distributor in its mounting. Turn the distributor one way or the other. As you turn the distributor, the markings on the pulley will move ahead or back. When the timing is correct, the markings will align with a timing pointer, or timing mark, as shown in Fig. 20-13. Tighten the distributor clamp.

NOTE: There are other timing methods seldom used today. One method uses a test light connected across the points. With the engine not running but the timing marks aligned, the distributor is turned so that the points just open. This is indicated by the test light coming on. Then the distributor clamp is tightened. Another method uses a piston-position gauge. It is inserted into the spark-plug hole to determine the exact position of the piston in the No. 1 cylinder.

⊘ 20-10 Oscilloscope The oscilloscope, or "scope," is a high-speed voltmeter that uses a televisionlike picture tube to show ignition voltages. Figure 20-14 shows an electronic engine tester which includes an oscilloscope (upper left).

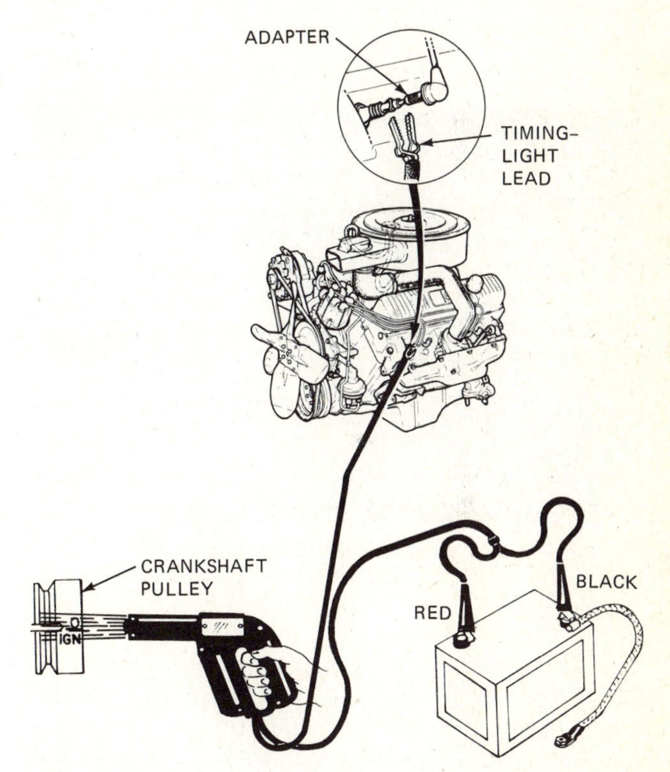

Fig. 20-11. Timing light used to check ignition timing.

The oscilloscope draws a picture of the ignition-system voltages on the face of the tube. The picture shows what is happening in the ignition system. If something is wrong, the picture will show what it is.

To understand the pictures, we shall first review the ignition system. When the distributor contact points open or the electronic control unit is triggered, the voltage in the secondary winding of the ignition coil jumps up to thousands of volts. This

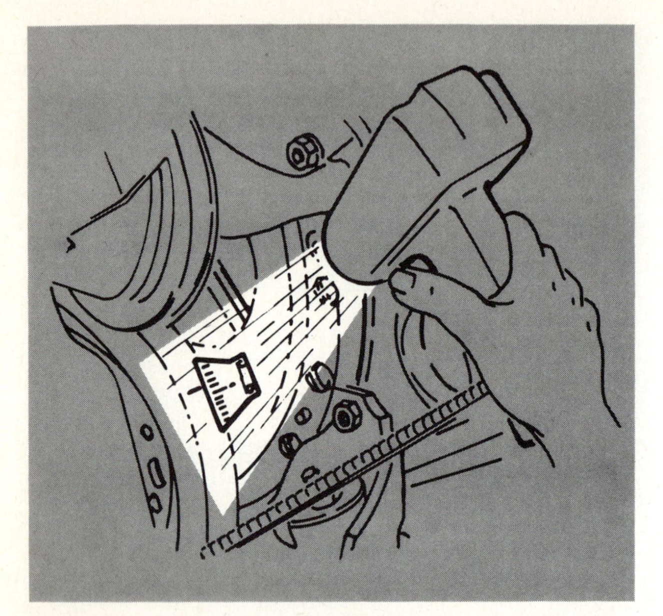

Fig. 20-12. The timing light flashes every time the No. 1 spark plug fires.

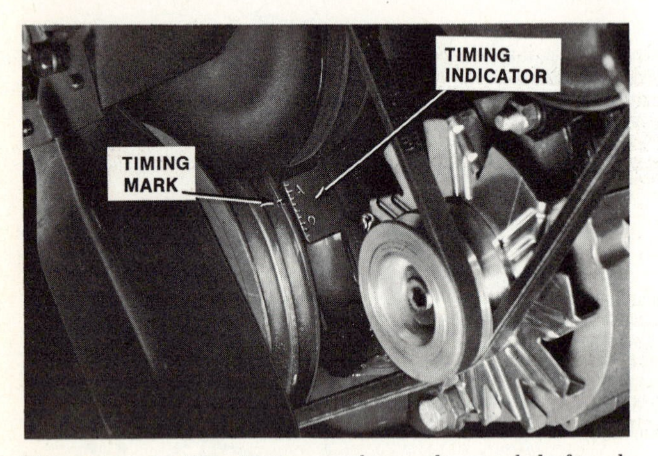

Fig. 20-13. Ignition timing marks on the crankshaft pulley.

high voltage surges to a spark plug and produces a spark. That is, the high voltage jumps the gap between the insulated and grounded electrodes of the spark plug. It takes a high voltage to start the spark. But after the spark is established, much less voltage is needed to keep the spark going. The scope can draw a picture of how and when this voltage goes up and down.

The picture is drawn on the face of the tube by a stream of electrons. This is exactly the way the picture tube in a television set works. In the scope, however, the stream of electrons draws a picture of just one thing—the ignition-system voltages feeding into the scope. Figure 20-15 shows the face of the picture tube and helps explain what we mean.

When a voltage, such as the voltage that fires a spark plug, is detected by the scope, a "spike," or vertical line, appears on the face of the tube. This is shown in Fig. 20-15. The higher the spike, the

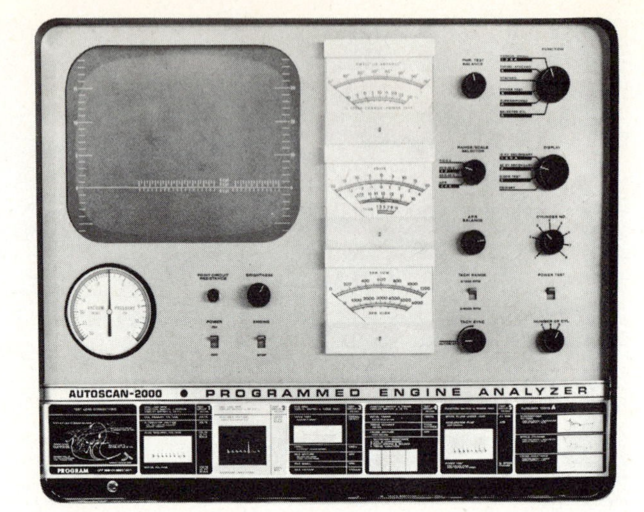

Fig. 20-14. Electronic engine tester with an oscilloscope. (*Autoscan, Inc.*)

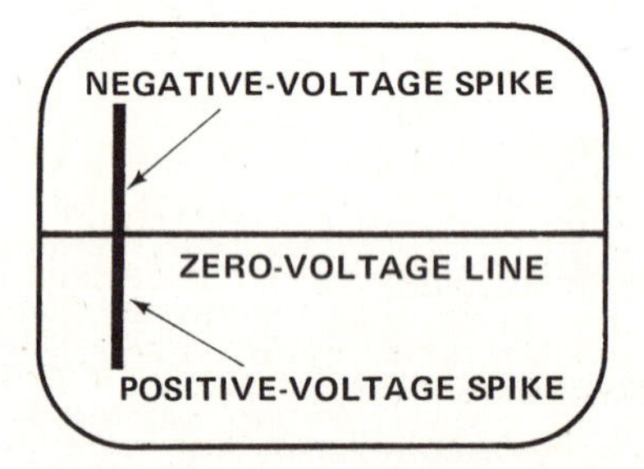

Fig. 20-15. The oscilloscope draws a horizontal zero-voltage line until a negative or positive voltage pulse enters. This causes the trace to kick up or down, as shown. The higher the voltage, the farther the trace moves up or down. These sharp up-and-down movements of the trace are called *spikes*.

higher the voltage. If the voltage spike points down, it indicates that the ignition coil or battery is connected backward.

To understand how the scope picks up the voltages and the pictures that the scope draws for us, let us first study what is called the *basic pattern* (see Fig. 20-16). The basic pattern is the picture the scope would show if it were drawing the voltage pattern for one spark plug. To start with, the contact points have opened. The high-voltage surge from the coil has arrived at the spark plug. The voltage goes up from *A* to *B*, as shown in Fig. 20-16. This line is called the *firing line*. After the spark is established, the voltage drops off considerably and holds fairly steady (from *C* to *D*). Of course, this is a very short time (measured in hundred-thousandths of a second). But the spark lasts for as long as 20 degrees of crankshaft rotation. This is long enough to ignite the compressed air-fuel mixture in the cylinder.

After most of the electromagnetic energy in the coil has been converted into electricity to make the spark, the spark across the spark-plug gap dies. However, there is still some energy left in the coil, and this produces a wavy line (from *D* to *E* in Fig.

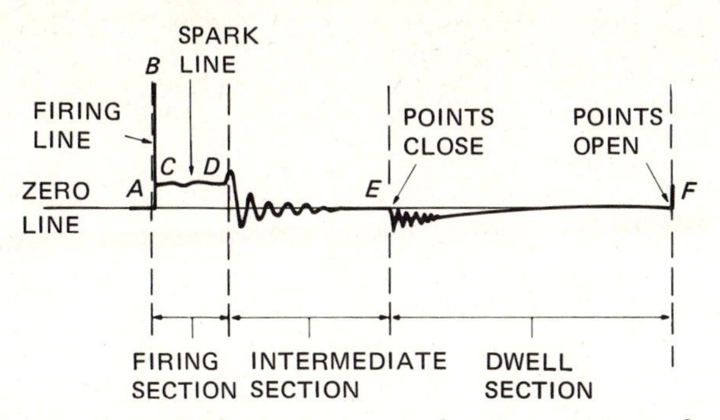

Fig. 20-16. Waveform, or trace, showing one complete spark-plug firing cycle. (*Sun Electric Corporation*)

20-16). This line is called the *coil-condenser oscillation line.* What this wavy line means is that the remaining energy is pushing electricity back and forth in the ignition-coil secondary circuit. The voltage alternates, but it is no longer strong enough to produce a spark. After a very short time, the voltage dies out. Then, at *E* (Fig. 20-16), the points close, sending current to the ignition-coil primary winding. Now, as a result of the buildup of current in the primary winding, an alternating voltage is produced in the secondary winding. This is shown by the oscillations following *E*. The section from *E* to *F* is called the *dwell section.* This is the time during which the contact points are closed. During this time, the magnetic field is building up in the ignition-coil primary winding. When the points open at *F*, we are back to *A* again. The magnetic field collapses, and the whole process is repeated.

⊘ 20-11 Oscilloscope Patterns The curves that the scope draws on the picture-tube face are called *patterns.* The patterns can be drawn in different ways. For example, the scope can be adjusted to draw a *parade pattern,* as shown in Fig. 20-17. It is called a parade pattern because the traces for the separate cylinders follow one another across the tube face like marchers in a parade. Note that the traces follow from left to right across the screen in normal firing order, with the No. 1 cylinder on the left.

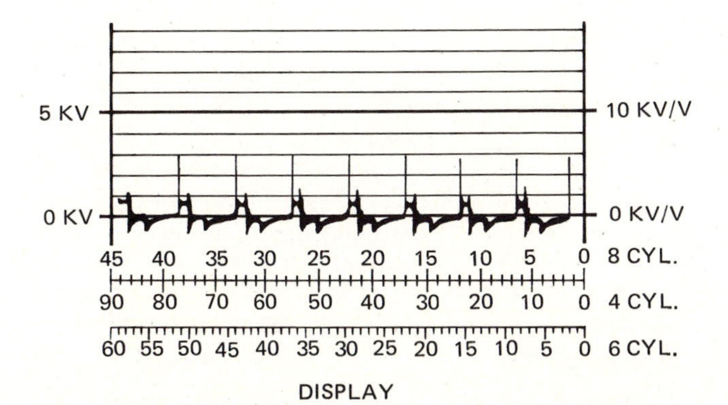

Fig. 20-17. Parade, or display, pattern of the ignition secondary voltages in an eight-cylinder engine. (*Sun Electric Corporation*)

By adjusting the scope in a different way, the traces can be stacked one above the other, as shown in Fig. 20-18. This is called a *raster pattern.* It lets you compare the traces so you can see if something is wrong in a cylinder. The raster pattern is read from the bottom up in the firing order, with the No. 1 cylinder at the bottom.

A third way to display the traces is to superimpose them, that is, place one on top of the other (Fig. 20-19). This gives a quick comparison and shows whether the voltage pattern from any one cylinder differs from the others. If everything is okay in the cylinders, only one curve appears on the tube face because all the curves fall on top of one another.

⊘ 20-12 Using the Scope There are several makes of oscilloscopes. Many are combined in consoles with other instruments for testing separate ignition components, engine rpm, intake-manifold vacuum, and so on. Figure 20-20 shows a complete console tester of this type. Figure 20-14 shows the face of a similar tester.

Scopes have pickup sensors that can be clamped onto the ignition wires, as shown in Fig. 20-21. Thus it is not necessary to disconnect and reconnect the ignition circuits. The *pattern pickup sensor* is clamped onto the wire that goes from the ignition coil to the distributor-cap center terminal. The sensor senses the high-voltage surges going to all the spark plugs. The *trigger pickup sensor* is clamped onto the wire that goes to the spark plug in No. 1 cylinder. The trigger pickup senses when the spark plug fires. This is the signal to the scope to start another round of traces.

⊘ 20-13 Reading the Patterns The patterns in Fig. 20-22 show different troubles that occur in the ignition system. The pattern that the scope draws of any cylinder's ignition-circuit voltage shows what voltages are occurring in that circuit. The way that the voltage varies from normal indicates where the electrical problem exists. For example, the scope can detect wide or narrow spark-plug gaps, open spark-plug wires, shorted coils or condensers, arcing contact points, and improper contact-point dwell. Many abnormal engine conditions change the voltage needed to fire the spark plug, and this shows up on the scope. Other abnormal engine conditions can also be identified on the scope because they change the length or the slope of the spark line. When you work in a shop that has an oscilloscope, you will be given detailed instructions on how to use it.

⊘ 20-14 Dynamometer The chassis dynamometer can test the engine-power output under various operating conditions. It can duplicate any kind of road test at any load or speed desired. The part of the dynamometer that you can see consists of two heavy rollers mounted at or slightly above floor level

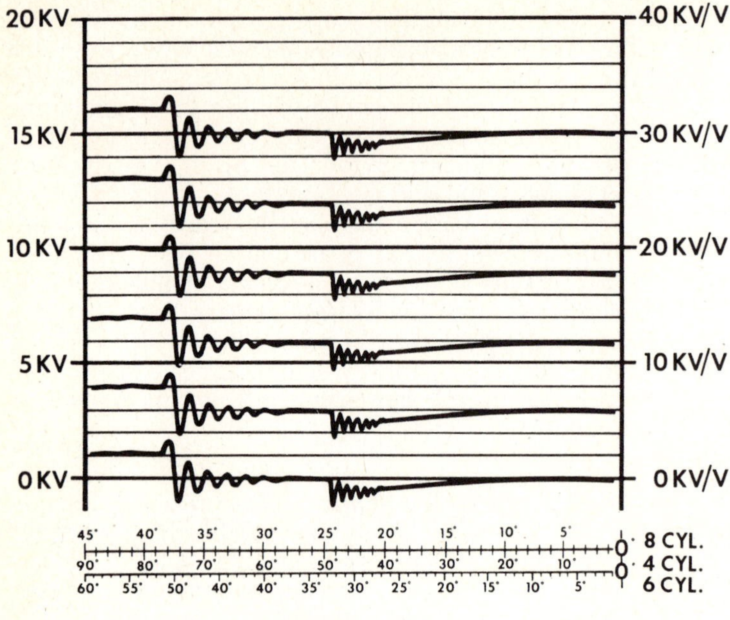

Fig. 20-18. Stacked, or raster, pattern of the ignition secondary voltages in a six-cylinder engine. (*Sun Electric Corporation*)

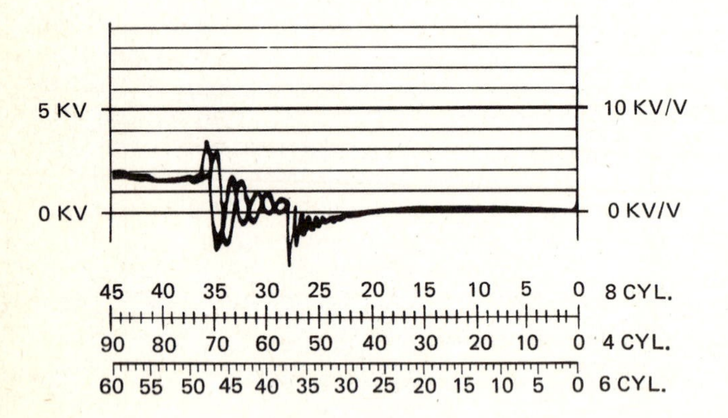

Fig. 20-19. Superimposed pattern of the ignition secondary voltages in a six-cylinder engine. (*Sun Electric Corporation*)

(Fig. 20-23). The car is driven onto these rollers, as shown in Fig. 20-24, so that the car wheels can drive the rollers. Next, the engine is started, and the transmission is put into gear. The car is then operated as though it were out on an actual road test.

Under the floor is a device that can place various loads on the rollers. This allows the technician to test the engine under various operating conditions. For example, the technician can find out how the engine would do during acceleration, cruising, idling, and deceleration. The test instruments, such as the scope, dwell-tachometer, and vacuum gauge, are hooked into the engine. These instruments then show the actual condition of the engine during various operating conditions.

The dynamometer can also be used to check the transmission and the differential. For example, the shift points and other operating conditions of an automatic transmission can be checked. Special diagnostic dynamometers are becoming more popular. These units have many instruments attached to them and have motored rollers that permit testing of wheel alignment, suspension, brakes, and steering.

⊘ 20-15 Cooling-System Testers There are three basic types of testers for the cooling system: coolant antifreeze tester, pressure tester, and belt-tension tester.

1. *COOLANT ANTIFREEZE TESTER* Antifreeze protects the engine from damage resulting from coolant freeze-up. Thus, it is important to check the coolant during cold weather to find out how much protection it has against freezing. That is, you measure the concentration of antifreeze in the coolant.

There are three types of coolant antifreeze testers: the hydrometer (Fig. 20-25), the refractometer (Fig. 20-26), and the kind with several balls in a glass tube. (The use of these testers, together with cooling-system service, is covered in *Automotive Fuel, Lubricating, and Cooling Systems,* another book in the McGraw-Hill Automotive Technology Series.)

2. *PRESSURE TESTER* The pressure tester is a small pump with a pressure gauge (Fig. 20-27). The tester is attached to the radiator-filler neck, as shown, and the pump is operated to apply pressure. If the pressure holds steady and there are no signs of leaks, the cooling system is tight.

3. *BELT-TENSION TESTER* One type of belt-tension tester is shown in Fig. 20-28. If there is not enough tension, the belt will slip. The fan and water pump will not be driven fast enough, and the engine will overheat. Also, the belt will wear out rapidly. Adjustment is made by moving the alternator out slightly. But do not overtighten. Excessive belt ten-

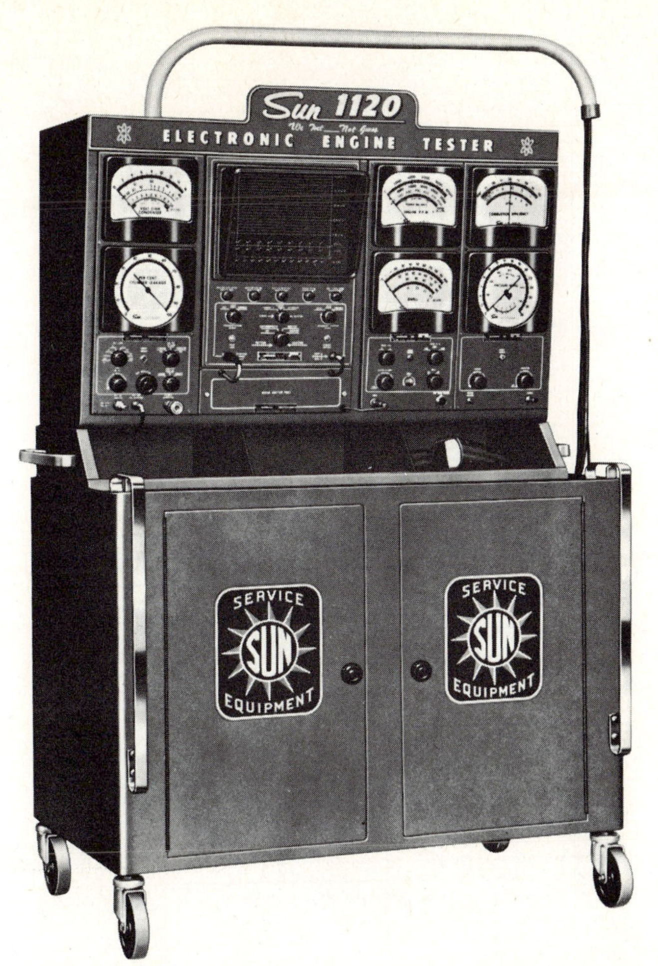

Fig. 20-20. Electronic-diagnosis engine tester. This tester includes an oscilloscope (*top center*) and other testing devices to check the condenser, distributor contact-point dwell, engine speed, and so on. (*Sun Electric Corporation*)

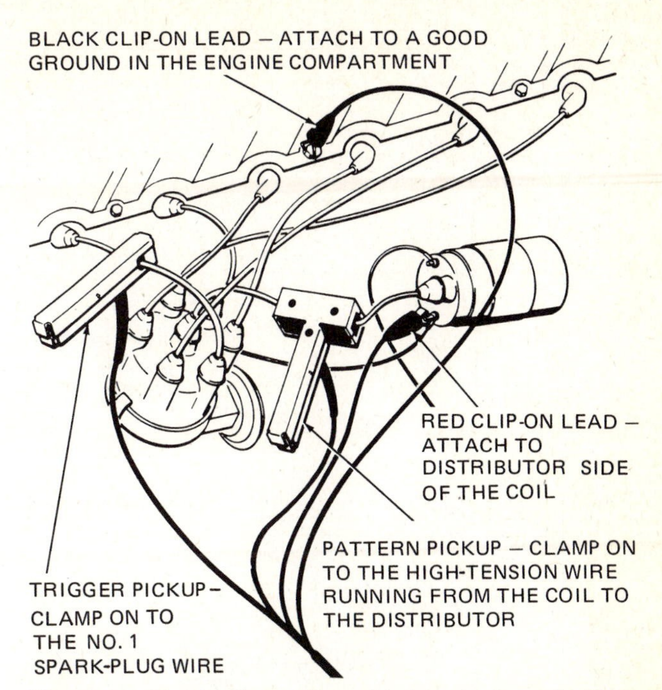

Fig. 20-21. Test leads are clipped to terminals, and pickup sensors are clamped on high-tension leads to test an ignition system. The tester is of the oscilloscope type. (*Autoscan, Inc.*)

sion can cause failure of the water-pump or alternator bearing.

⊘ 20-16 PCV-Valve Testers In Chap. 18 we mentioned that modern engines are equipped with a positive crankcase ventilating (PCV) system. This system passes air through the crankcase and then up to the intake manifold. The air passing through picks up any blow-by or hydrocarbon that has leaked down past the piston rings. The unburned and partly burned gasoline then passes through the engine again, where it has another chance to burn.

The PCV valve is designed to prevent too much air from flowing during idle. If the valve sticks open, too much air will flow into the intake manifold. The air-fuel mixture may become too lean at idle, and the engine will idle roughly and may even stall. On the other hand, if the valve sticks closed, not enough air will get through and the blow-by products can accumulate in the crankcase. This could seriously damage the engine because these products include corrosive acids and goo that could plug oil lines and cause the engine to fail from oil starvation.

The PCV valve should be checked periodically. It should be routinely replaced at stated intervals. Recommendations are:

American Motors Corporation Replace the valve every 15,000 miles [24,140 km].

Chevrolet and Ford Replace every 24 months or 24,000 miles [38,624 km] or whichever comes first.

Chrysler Corporation Check every 12 months or 12,000 miles [19,312 km], and replace every 24,000 miles [38,624 km] or 24 months.

There is a simple way to check the system and the valve. Remove the valve or valve connection with the engine running. Place your hand over the opening and feel for a slight vacuum pull against your hand. If there is no vacuum action, or if you can feel a positive pressure, then something is wrong. The PCV valve, hoses, and connections should be checked. There are special testers which can be used to check the operation of the PCV valve (Figs. 20-29 and 20-30).

Chevrolet offers a different method of checking the system. With the engine running on idle, remove the PCV valve from the rocker-cover grommet with the hose attached, and block the opening of the valve. Note the change in engine speed. A decrease of less than 50 rpm indicates a plugged ventilation valve. You should use a tachometer when making this test to get an accurate reading on the engine rpm.

Chrysler recommends this test procedure. Remove the PCV valve from the rocker-arm cover with the engine idling. The valve should hiss, and you should be able to feel a strong vacuum when you place your finger over the valve inlet. Reinstall the

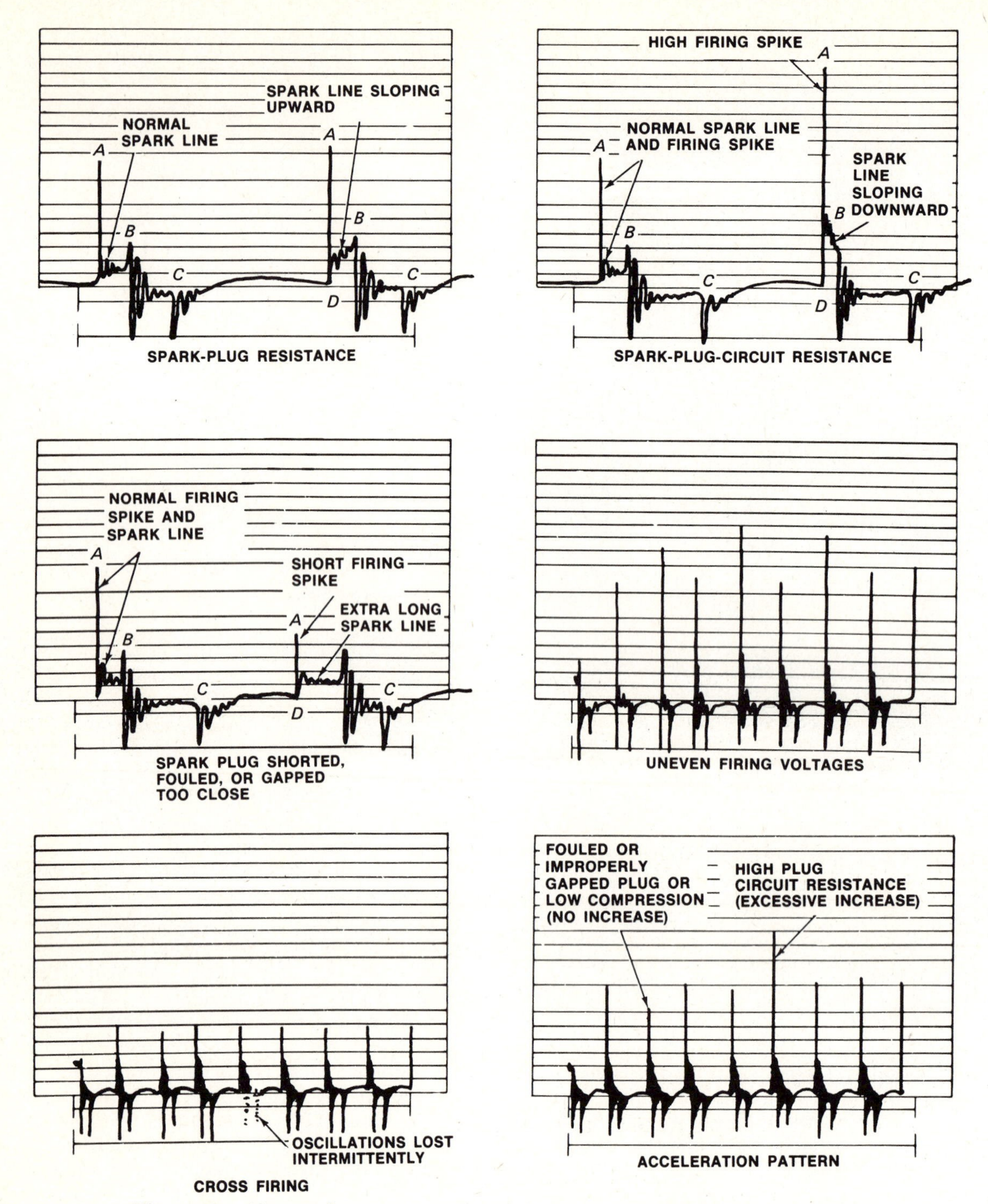

Fig. 20-22. Abnormal scope patterns and their causes. (*Ford Motor Company*)

PCV valve and remove the crankcase inlet air cleaner. Hold a piece of stiff paper over the opening of the rocker-arm cover. After a few moments, the paper should be sucked against the opening. Then stop the engine and remove the PCV valve from the rocker-arm cover and shake it. It should click, showing that the valve is free. If the system does not meet these tests, replace the PCV valve and try again. If the system still does not pass the tests, the hose may be clogged. It should be cleaned out or replaced. Or it may be necessary to remove the carburetor and clean the vacuum passage with a $\frac{1}{4}$ inch [6.35 mm] drill. Also, clean the inlet vent on the crankcase inlet air cleaner that is connected by the hose to the carburetor air cleaner.

⊘ 20-17 Fuel-System Testers The exhaust-gas analyzer, engine vacuum gauge, oscilloscope, and other instruments are used to check engine performance. An important part of the engine is the fuel system, and these test instruments also report on the fuel system. There are also fuel-pump pressure and capacity testers that check on how well the fuel pump

Fig. 20-23. Chassis dynamometer of the flush-floor type. The rollers are set at floor level. (*Sun Electric Corporation*)

Fig. 20-24. Automobile in place on a chassis dynamometer. The rear wheels drive the dynamometer rollers, which are flush with the floor. At the same time, instruments on the test panel measure car speed, engine power output, engine vacuum, and so on. (*Sun Electric Corporation*)

is doing its job. Figure 20-31 shows a fuel-pump pressure and capacity tester.

⊘ 20-18 Electrical-System Testers A variety of instruments are required to test the automotive electrical equipment. These include the distributor, coil, and condenser testers for the ignition-system components. To check the charging system, ammeters and voltmeters are required. Any of several instruments can be used to check the battery, including

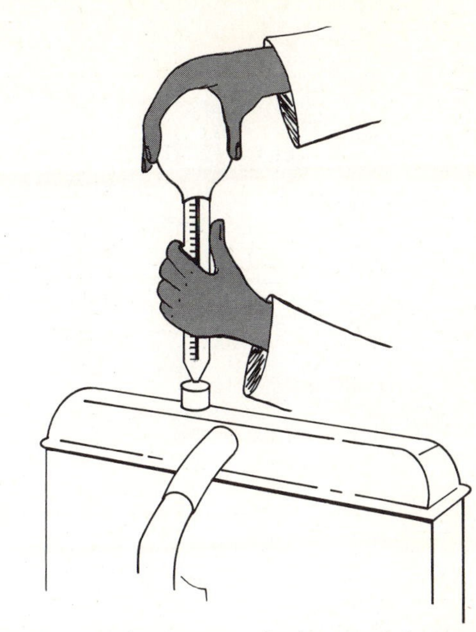

Fig. 20-25. Cooling-system hydrometer being used to check the coolant for amount of freeze-up protection. (*Ford Motor Company*)

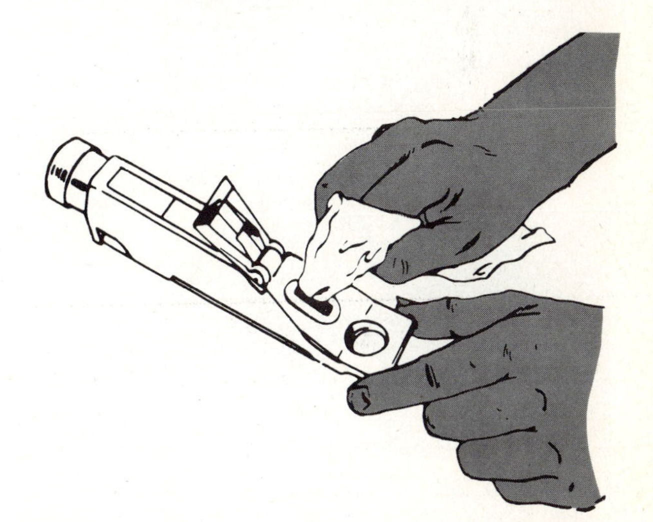

Fig. 20-26. Cleaning the refractometer-measuring window in preparation for checking coolant.

a hydrometer, voltmeter, 421 tester, and cadmium-tip tester. For details on how to use these instruments and how to service the components in the electrical system, refer to *Automotive Electrical Equipment,* another book in the McGraw-Hill Automotive Technology Series.

⊘ 20-19 Sensor Panel A recent innovation in indicating devices is the *sensor panel* introduced by Toyota in 1974 (Fig. 20-32). This panel is installed in the roof of the car above the driver (Fig. 20-33). It is connected to sensors in the light circuits, brakes, windshield washer, battery, cooling-system radiator, and engine crankcase (Fig. 20-32). The sensor panel has eleven warning lights which come on if something needs attention. If any of the four lights at the

Fig. 20-27. Pressure tester used to check the cooling system for leaks. (*Texaco Incorporated*)

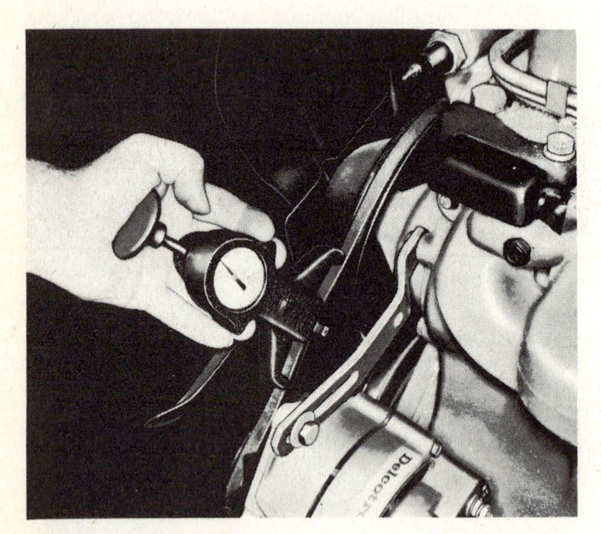

Fig. 20-28. Special tool used to check the tension of the fan belt. (*Chevrolet Motor Division of General Motors Corporation*)

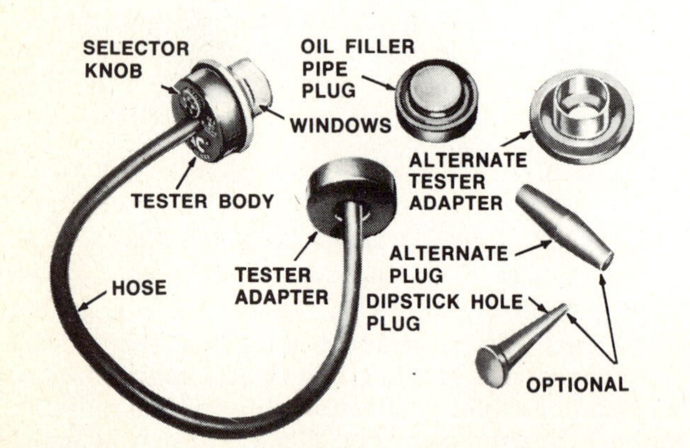

Fig. 20-29. Positive crankcase ventilating system tester. (*Ford Motor Company*)

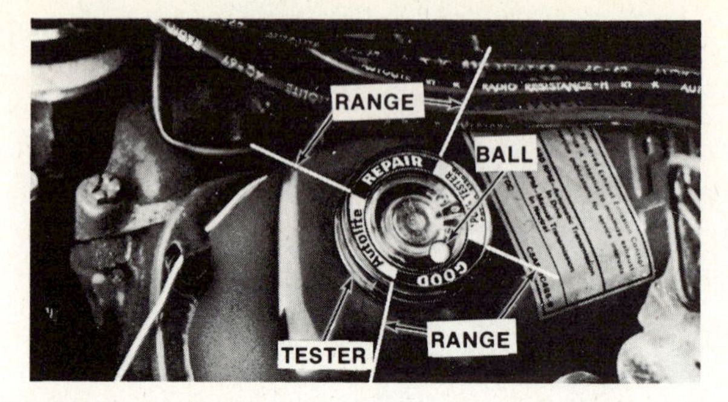

Fig. 20-30. PCV tester on a valve cover ready for a test. (*Ford Motor Company*)

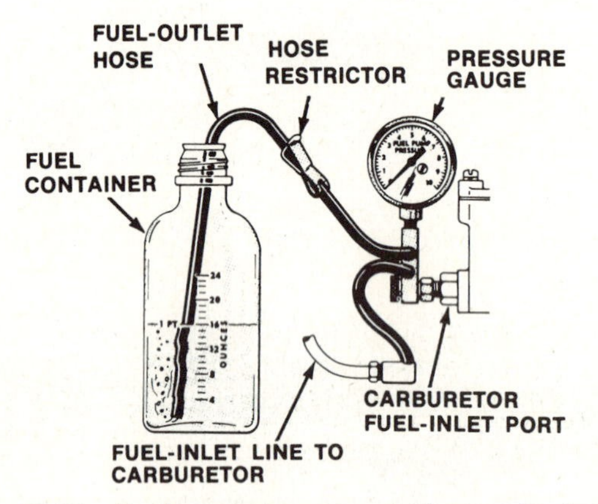

Fig. 20-31. Fuel-pump pressure and capacity tests.

top of the panel (LICENSE, BRAKE, TAIL, HEAD) comes on, it indicates trouble in that light circuit. For example, if a headlight burns out, the HEAD light comes on to warn the driver of the trouble. The four FLUID LEVEL lights (W-WASHER, BATTERY, RADIATOR, ENGINE OIL) indicate low fluid level in any of these four areas. For example, if the engine oil drops to a low level, the ENGINE OIL light comes on. The BRAKE section of the panel warns of low brake fluid, loss of vacuum in the power-brake unit, or excessive brake-lining wear. Figure 20-34 shows how the eleven warning lights are connected to sensors in the service areas.

CHAPTER 20 CHECKUP

NOTE: Since the following is a chapter review test, you should review the chapter before taking the test.

The material presented in this chapter is designed to acquaint you with the procedures and instruments used to test engines. Now find out how well you remember this material by taking the following test. If any of the questions are too difficult, turn back into the chapter and reread the pages that will give you the information you need.

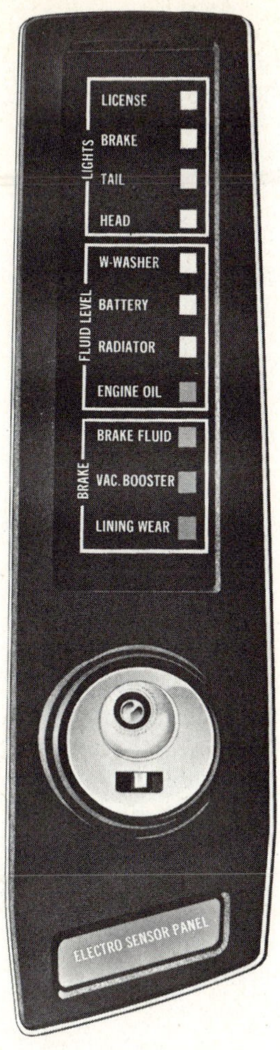

Fig. 20-32. Sensor panel, called the *Electro Sensor Panel*, or ESP, by the manufacturer. (*Toyota Motor Sales, Limited*)

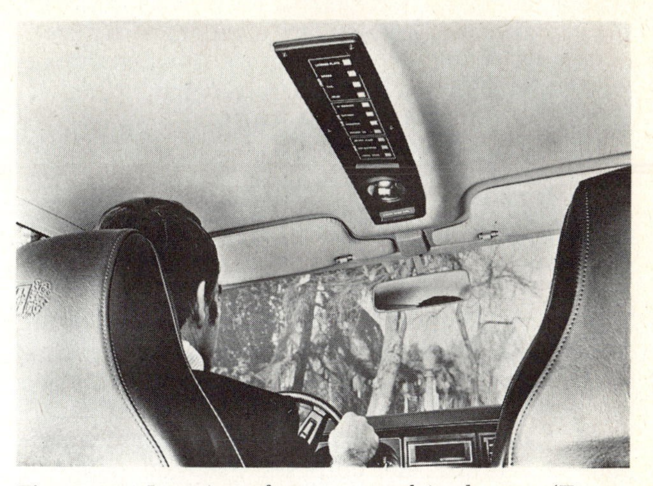

Fig. 20-33. Location of sensor panel in the car. (*Toyota Motor Sales, Limited*)

Fig. 20-34. Connections from the sensor panel to the eleven service areas. (*Toyota Motors Sales, Limited*)

Completing the Sentences The sentences that follow are incomplete. After each sentence there are several words or phrases, only one of which will correctly complete the sentence. Write each sentence in your notebook, selecting the proper word or phrase to complete it correctly.

1. Two types of engine-testing procedures are: (*a*) trouble diagnosis and fault-finding, (*b*) trouble diagnosis and tune-up, (*c*) preventive maintenance and tune-up.
2. The general procedure diagnosis which eliminates trouble before it happens is called: (*a*) preventive maintenance, (*b*) trouble diagnosis, (*c*) timing test.
3. When the tachometer is connected between the distributor primary terminal and ground, it indicates: (*a*) engine speed, (*b*) engine vacuum, (*c*) engine compression.
4. If pouring heavy oil into the cylinder increases the compression pressure, then the chances are that the loss of compression is due to leakage: (*a*) past the valves, (*b*) past the head gasket, (*c*) past the piston rings.
5. If the vacuum-gauge needle swings around to 23 to 25 in [584 to 635 mm] of mercury as the throttle is quickly closed after the engine has been raced, it indicates: (*a*) stuck valves, (*b*) low compression, (*c*) satisfactory compression, (*d*) leaky valves.
6. A steady but low vacuum reading with the engine idling indicates that the engine: (*a*) is losing power, (*b*) has a stuck valve, (*c*) exhaust line is clogged.
7. A very low vacuum reading with the engine idling indicates: (*a*) stuck valves, (*b*) air leakage into manifold, (*c*) loss of compression, (*d*) faulty piston rings.
8. A valve that sticks open or a plug that is not firing will cause the vacuum-gauge needle to: (*a*) oscillate slowly, (*b*) drop back regularly, (*c*) fall back slowly to zero, (*c*) read too high.
9. The combustion tester determines the air-fuel-mixture ratio by analyzing the: (*a*) fuel charge, (*b*) compression ratio, (*c*) compression mixture, (*d*) exhaust gas.

10. The device which can give a very close approximation of a road test in the garage is called the: (*a*) engine dynamometer, (*b*) chassis dynamometer, (*c*) tachometer, (*d*) engine tester.

Unscrambling the Test Instruments Following are two lists. The first list gives various test instruments discussed in the chapter. The second list gives the purposes of these instruments but not in the same order. To unscramble the lists, take each item in the test-instruments list in turn, and then find its purpose in the purposes list. Put the two together and write the result in your notebook. For instance, the first test instrument listed is "compression tester." When you look down the purposes list, you come to "checks cylinder compression." So you put the two together and write "compression tester checks cylinder compression."

TEST INSTRUMENTS
Compression tester
Tachometer
Vacuum gauge
Combustion tester
Timing light

PURPOSES
Analyzes exhaust gas
Checks intake-manifold vacuum
Checks ignition timing
Checks engine speed
Checks cylinder compression

SUGGESTIONS FOR FURTHER STUDY

Test-instrument manufacturers issue printed information on how to use their instruments and what the test results mean. If you can find this printed information at your local service station or garage or in your school shop, you will find it of considerable interest. You should also examine test instruments in the shop and watch carefully to note how they are used. In this connection, we give a word of caution. These instruments can be damaged by handling them carelessly or connecting them improperly. Therefore, you must know what you are doing before you attempt to use the test instruments. Study carefully the information on how to use the instruments, and make sure you know what you are doing before you attempt to use any test instrument.

chapter 21

ENGINE TUNE-UP

This chapter describes the procedure known as engine tune-up. *Tune-up includes testing the various components and accessory systems involved in engine operation. It also includes readjusting or replacing parts as required to restore engine performance. In some cases, during a tune-up, serious problems may be uncovered that will require major repair work. Chapters 23 to 26 describe the service jobs that may be performed on engines. Also, none of the various engine-testing procedures described in Chap. 20 will be repeated in this chapter. Therefore, when we refer to a specific test or repair operation, turn to the chapter in which the procedure is described.*

⊘ 21-1 What Is Tune-up? Engine tune-up means different things to different people. To some, it means a light once-over check of the engine that takes in only the obvious trouble spots. To others, it means use of the proper test instruments to do a careful, complete analysis of all engine components. In addition, it means adjusting everything to "spec" and repairing or replacing all worn parts. The latter is the proper meaning of engine tune-up, and it is the procedure outlined in this chapter.

NOTE: In this chapter, we have combined two separate programs: engine tune-up and complete car-care inspection. Engine tune-up includes checking and servicing the engine and its systems. Car-care inspection includes checking all other components on the car, such as brakes, steering, tires, and so on. Together, an engine tune-up plus a car-care inspection cover everything in and on the car that could cause trouble. Some automotive shops find that a customer who comes in for an engine tune-up can also be sold a car-care inspection. Likewise, a customer who comes in for a car-care inspection can also be sold an engine tune-up.

⊘ 21-2 Tune-up Procedure An engine tune-up follows a more or less set procedure. Many mechanics use a printed form supplied by automotive or test-equipment manufacturers (Fig. 21-1). By following the form and checking off the items listed, one by one, the mechanic is sure of not overlooking any part of the procedure. However, not all tune-up forms are the same. Different companies have different ideas about what should be done and the order in which it should be done. In addition, the tune-up procedure depends on the equipment available. If the shop has an oscilloscope or a dynamometer, it is used as part of the tune-up procedure. If these test instruments are not available, then the tune-up is performed differently.

The procedure that follows includes car-care inspection. It lists all essential checks and adjustments in what authorities believe is the most logical sequence.

⊘ 21-3 Tune-up and Car Care The tune-up procedure restores driveability, power, and performance that have been lost through wear, corrosion, and deterioration of engine parts. These changes take place gradually in many parts during normal car operation. Because of federal and state laws limiting automotive emissions, the tune-up procedure must include checks of all emission controls. Here is the procedure.

1. If the engine is cold, operate it for at least 20 minutes at 1,500 rpm or until it reaches operating temperature.
2. Connect the oscilloscope, if available, and perform an electronic diagnosis. Check for any abnormal ignition-system conditions that appear on the pattern. Make a note of any abnormality and the cylinder(s) in which it appears.
3. Remove all spark plugs. Fully open the throttle and choke valves. Disconnect the distributor lead from the coil primary terminal to prevent engine starting.
4. Check the compression of each cylinder. Record the readings. If one or more cylinders read low, squirt about a tablespoonful of engine oil through the spark-plug hole. Recheck the compression. Record the new readings.

CUSTOMER ______________ PHONE ________ DATE ________

MAKE-YEAR-MODEL ______________ MILEAGE ________

SIGNATURE ______________

PLEASE CHECK DESIRED SERVICES

Complete Infra-Red Engine Tune-Up

(CONSISTS OF ALL ITEMS LISTED BELOW)

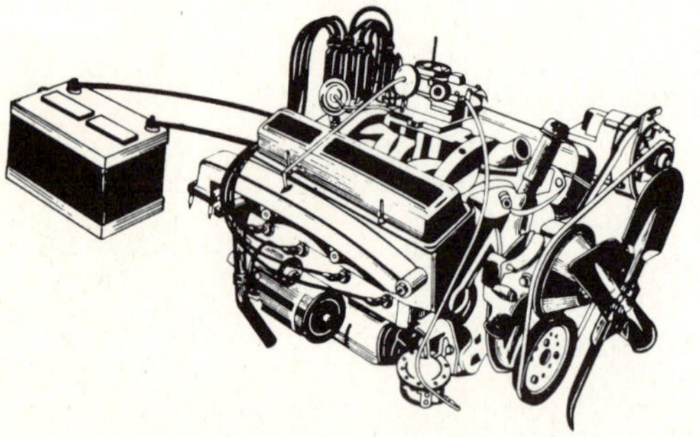

1 ☐ **USE INFRARED EXHAUST GAS ANALYSIS & VISUAL INSPECTION TO TEST & ADJUST CARBURETOR & CHOKE**
(Needed for smooth idle, economy, good performance & low exhaust emissions)

2 ☐ **TEST CYLINDER BALANCE & COMPRESSION FACTORS**
(Needed for smooth idle, economy & good performance)

3 ☐ **CLEAN OR REPLACE SPARK PLUGS AS NEEDED**

4 ☐ **REMOVE DISTRIBUTOR FROM ENGINE FOR SERVICE—TEST MECHANICAL & ELECTRICAL CONDITION—CALIBRATE ADVANCE MECHANISM**
(Has direct bearing on gasoline mileage, exhaust emissions & performance—also permits proper installation of ignition points)

5 ☐ **REPLACE IGNITION POINTS IF NEEDED & SET INITIAL TIMING**

6 ☐ **SERVICE MANIFOLD HEAT CONTROL VALVE**
(Affects warm-up & running performance)

7 ☐ **REPLACE FUEL FILTER ELEMENT IF NEEDED & TEST FUEL PUMP OPERATION**
(To guard against unpredictable road failures)

8 ☐ **SERVICE AIR POLLUTION CONTROL MECHANISM—P.C.V.**
(Affects fuel air ratio and idle performance)

9 ☐ **CLEAN, TEST & SERVICE SPARK PLUG WIRES, IGNITION COIL, DISTRIBUTOR CAP & ROTOR**
(Helps prevent starting & missing problems due to dampness and/or resistance)

10 ☐ **TEST GENERATOR OR ALTERNATOR & REGULATOR—REPLACE DRIVE BELTS IF NEEDED**
(To insure battery will be kept fully charged for starting)

11 ☐ **SERVICE & TEST BATTERY—CLEAN BATTERY TOP & CABLES—RECHARGE IF NEEDED—TEST STARTER**
(Helps prevent unpredicted starting failures)

12 ☐ **SERVICE AIR FILTER—REPLACE IF NEEDED**
(Affects general performance of engine)

13 ☐ **USE CYLINDER & CRANKCASE CHEMICAL TUNE-UP ADDITIVES**
(Helps prevent sticking valves, rings & lifters)

14 ☐ **ROAD TEST**
(For over-all automobile performance & general condition)

15 ☐ **TELEPHONE IF ADDITIONAL SERVICES ARE FOUND NECESSARY**

Sun SERVICE CONTROL SYSTEM

Fig. 21-1. Printed tune-up form. (*Sun Electric Corporation*)

NOTE: If the compression is low, indicating either bad rings or valves, tell the owner the engine is not tuneable without overhaul or repair.

5. Clean, inspect, file, gap, and test the spark plugs.[1] Discard worn or defective plugs. (Many shops install all new plugs instead of servicing the old ones.) Gap all plugs, old and new. Install the plugs.
6. Inspect and clean the battery case, terminals, cables, and hold-down brackets. Test the battery. Add water, if necessary. If severe corrosion is present, clean the battery and cables with brushes and a solution of baking soda and water.[1]
7. Test the starting voltage. If the battery is in good condition but cranking speed is low, test the starting system.[1]
8. If the battery is low or the customer complains that the battery keeps running down, check the charging system (alternator and regulator).[1] If the battery is old, it may have worn out. A new battery is required.
9. Check the drive belts and replace any that are in poor condition. If you have to replace one belt of a two-belt drive, replace both belts. Tighten the belts to the correct tension, using a tension gauge.
10. Inspect the distributor rotor, cap, and primary and high-voltage (spark-plug) wires (Fig. 21-2).
11. Clean or replace and adjust distributor contact points by setting the point gap.[1] Lubricate the distributor breaker cam if specifications call for this. On distributors with round cam lubrications, specifications call for turning the cam lubricator 180 degrees every 12,000 miles [19,312 km] and replacing the cam lubricator every 24,000 miles [38,624 km].
12. Check the distributor cap and rotor (Fig. 21-3). Check the centrifugal and vacuum advances. Set the contact dwell and then adjust ignition timing. Make sure the idle speed is not excessive because this could produce centrifugal advance during timing adjustment.
13. Use the oscilloscope to recheck the ignition system. Any abnormal conditions that appeared in step 2 should now have been eliminated.
14. Check the manifold heat-control valve. Lubricate with heat-valve lubricant. Free up or replace the valve if necessary.
15. Check the fuel-pump operation with a fuel-pump tester. Replace the fuel filter. Check the fuel-tank cap, fuel lines, and connections for leakage and damage.
16. Clean or replace the air-cleaner filter. If the engine is equipped with a thermostatically controlled air cleaner, check the operation of the control damper.
17. Check the operation of the choke and the fast-idle cam. Check the throttle valve for full opening, and the throttle linkage for free movement.
18. Inspect all engine vacuum fittings, hoses, and connections. Replace any brittle or cracked hose.
19. Clean the engine oil-filler cap if a filter-type oil-filler cap is used.
20. Check the cooling system (Fig. 21-4). Inspect all hoses and connections, the radiator, water pump, and fan clutch if used. Check the strength of the antifreeze and record the reading. Pressure-check the system and radiator cap. Squeeze the hoses to check them. Replace any defective hose (collapsed, soft, cracks, etc.).
21. Check and replace the PCV valve if necessary (see Fig. 21-5). Clean or replace the PCV filter if required. Inspect the PCV hoses and connections. Replace any cracked or brittle hose.
22. If the engine is equipped with an air pump type of exhaust-emission control, replace the pump inlet air filter if used. Inspect the system hoses and connections. Replace any brittle or cracked hose.
23. If the vehicle is equipped with a fuel-vapor-recovery system, replace the charcoal-canister filter.
24. Check the transmission-controlled vacuum-spark-advance system if the vehicle is so equipped.
25. On engines equipped with an EGR system, inspect and clean the EGR valve. Inspect and clean the EGR discharge port.
26. Tighten the intake-manifold and exhaust-manifold bolts to the proper tension in the proper sequence.
27. Adjust the engine valves if necessary.
28. Adjust the carburetor idle speed. Use an exhaust-gas analyzer and adjust the idle mixture screw. Check the amounts of CO and HC in the exhaust gas. (Many mechanics check the CO and HC both before and after the tune-up job to show how much the tune-up reduced these pollutants.)
29. Road-test the car on a dynamometer or on the

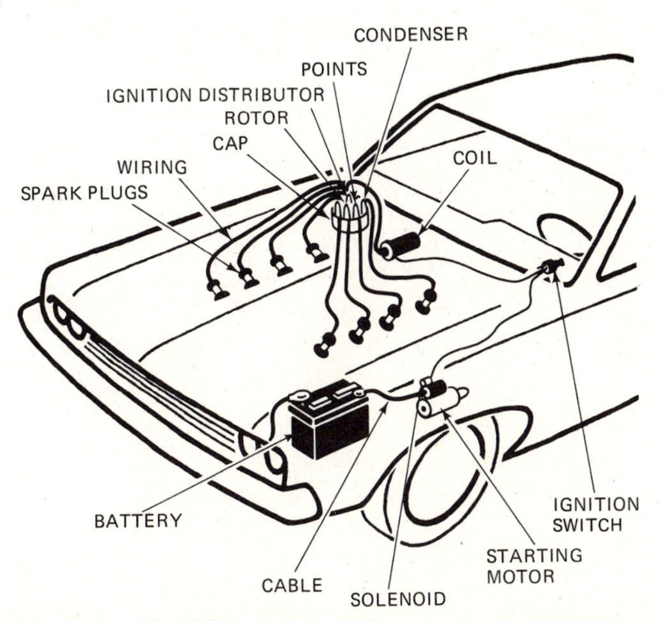

Fig. 21-2. Simplified drawing of an engine ignition system.

[1]For details of this procedure, see *Automotive Electrical Equipment,* another book in the McGraw-Hill Automotive Technology Series.

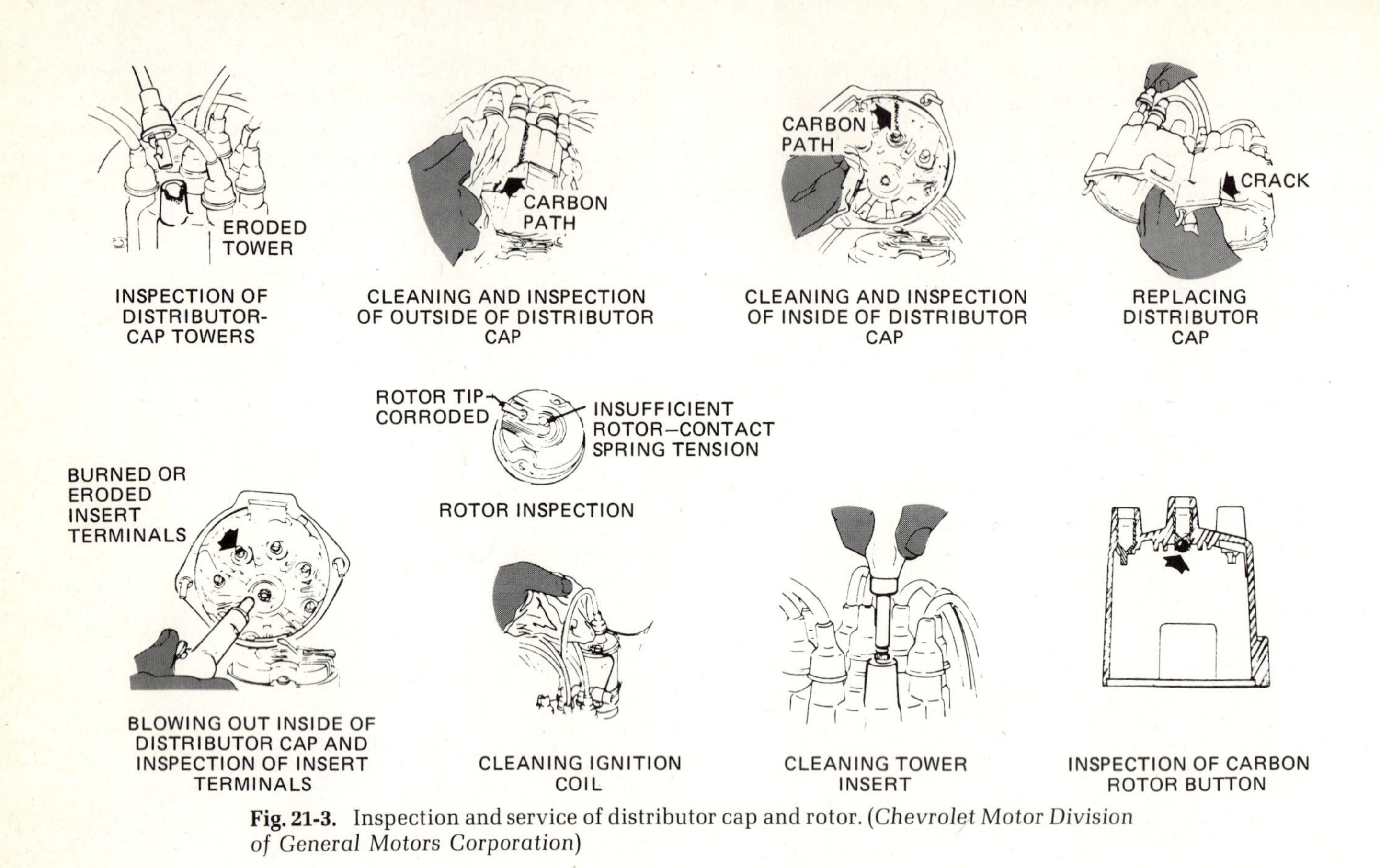

Fig. 21-3. Inspection and service of distributor cap and rotor. (*Chevrolet Motor Division of General Motors Corporation*)

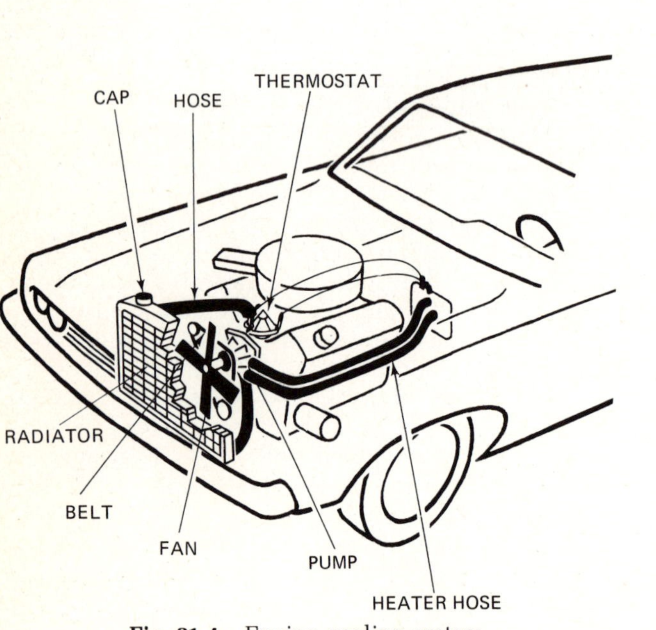

Fig. 21-4. Engine cooling system.

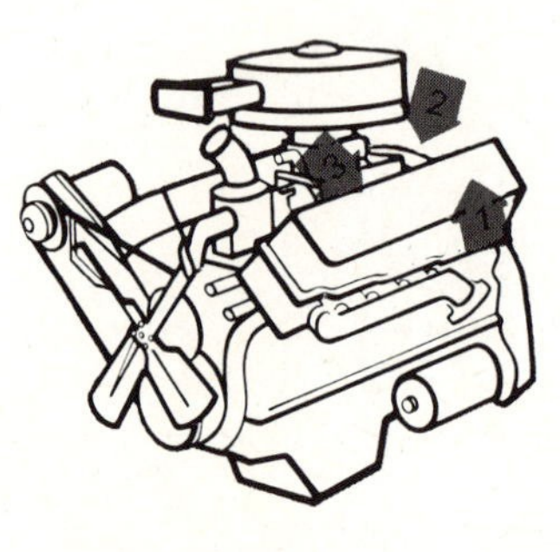
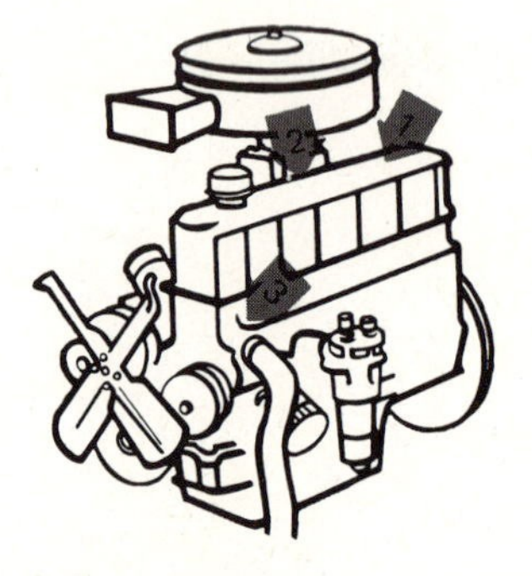

Fig. 21-5. PCV-valve locations.

road. Check for driveability, power, and idling. Any abnormal condition should be noted on the repair order before you return the car to the customer.
30. Check the doorjamb sticker to determine if an oil and oil-filter change is due. Also note the schedule for chassis lubrication. Recommend an oil change and a lube job if they are due. (See Figs. 21-6 and 21-7.) Note that car manufacturers recommend changing the oil filter every time—or every other time—the oil is changed.
31. Whenever the car is on the lift, check the exhaust system for leaks which could admit CO into the car. Also check for loose bolts, rust spots, and other under-the-car damage.

NOTE: Items *32* to *37* are not actually part of the tune-up job. They are included here so you will have the complete car-care program all in one place.

32. Check the brakes for even braking and adequate braking power.
33. Check the steering system for ease and smoothness of operation. Check for excessive play in the system. Record any abnormal conditions.
34. Check the tires for inflation and for abnormal wear. Abnormal wear can mean suspension trouble, and a front-alignment job should be recommended.
35. Check the suspension system for looseness, excessive play, and wear.
36. Check the front wheels and ball joints for excessive wear or loose bearings. Adjust the bearings if necessary.
37. Check the headlights and horns to make sure they are in good working order. Check all other lights. Replace any burned-out lights. Check headlight alignment if possible.

NOTE: As you can see, the preceding comprehensive list for tune-up and car care covers about everything on the vehicle that could cause trouble. The complete procedure therefore will uncover any problems that might affect driveability and performance. If all necessary corrective steps are taken, new-car performance will be restored to the vehicle.

Fig. 21-6. Engine oil is changed by placing the container under the oil-pan drain hole and then removing the drain plug.

⊘ 21-4 Engine Analyzers and Computer Testers Figure 21-8 shows an engine analyzer which includes an oscilloscope and other instruments for making comprehensive tests of all engine components. Once the mechanic learns how to use this equipment, a complete engine analysis can be made in a very short time.

In addition to these engine analyzers, there are still more complex testers. They run many of the tests almost automatically and then produce a printed record of the tests and their results. For example Fig. 21-9 shows a system that uses a computer and an interpreter. The various connections to be made are shown. Also shown is the remote-channel selector which can be used to select either a display device or a printer. Thus, the test results can be displayed on a screen or printed out on a form as a permanent record.

The operation of the tester shown is simple once the connections are made. You simply select the proper computer-program card (Fig. 21-10) and insert it into the computer. The punched card has the specifications for the model car under test. The tester then makes the tests listed (Fig. 21-10).

An even simpler system has been developed. In this system, the vehicle has a connector that is already attached to the various car components. The tester has a plug that is connected to all the testing instruments. Thus, all the automotive mechanic has to do is plug the tester plug into the connector on the vehicle. This completes all connections between the engine components and the test instruments. The mechanic then can run all the tests in a minimum of time.

One further refinement has been suggested—to put into the computer information on the costs of parts and repair operations. Then the computer could print out, along with the test information, the costs of fixing any troubles. That is, it would print out the costs of parts and labor. It has also been suggested that the computer could be programmed to schedule the work, depending on the availability of labor and space in the shop.

CHAPTER 21 CHECKUP

NOTE: Since the following is a chapter review test, you should review the chapter before taking the test.

Engine tune-up and car-care inspection are very important to the "health" of the automobile. Many service stations and automotive shops specialize in tune-up and car-care procedures, and so it is important to know the essentials of these procedures. Now check up on how well you remember what you have just studied by taking the following test. If any of

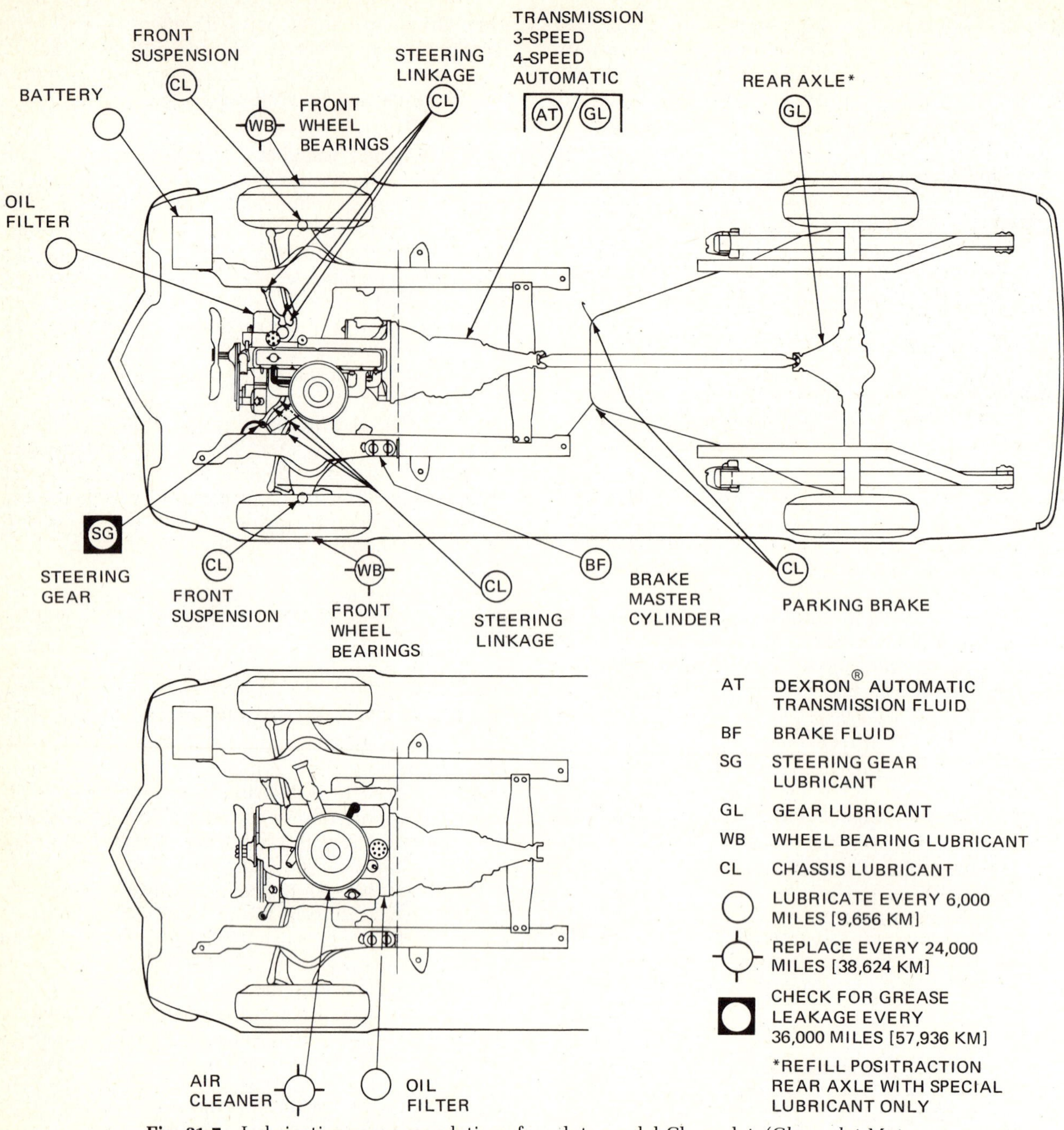

Fig. 21-7. Lubrication recommendations for a late-model Chevrolet. (*Chevrolet Motor Division of General Motors Corporation*)

the questions stump you, reread the pages in the chapter that will give you the information you need.

Completing the Sentences The sentences that follow are incomplete. After each sentence there are several words or phrases, only one of which will correctly complete the sentence. Write each sentence in your notebook, selecting the proper word or phrases to complete it correctly.

1. You determine when the engine oil was last changed by: (*a*) asking the driver, (*b*) inspecting the oil, (*c*) looking at the lube sticker.
2. A tune-up includes: (*a*) checking the air cleaner, spark plugs, and ignition system; (*b*) checking the manifold heat-control valve, PCV valve, and carburetor; (*c*) both (*a*) and (*b*).
3. If you have to replace one belt of a two-belt drive: (*a*) tighten belts based on tension of new belt, (*b*) replace both belts, (*c*) check pulleys to see what is wrong.
4. When the car is up on the lift, check the exhaust system for leaks that could cause: (*a*) carbon monoxide leaking into the car, (*b*) excess noise, (*c*) exhaust smoke.
5. A quick way to check the radiator hose is to: (*a*) squeeze the hose, (*b*) look under the car for coolant, (*c*) remove and inspect the hose.

Fig. 21-8. Electronic-diagnosis engine tester, or engine analyzer. This tester includes not only an oscilloscope (*top center*), but also other devices to check the condenser, distributor contact-point dwell, engine speed, and so on. (*Sun Electric Corporation*)

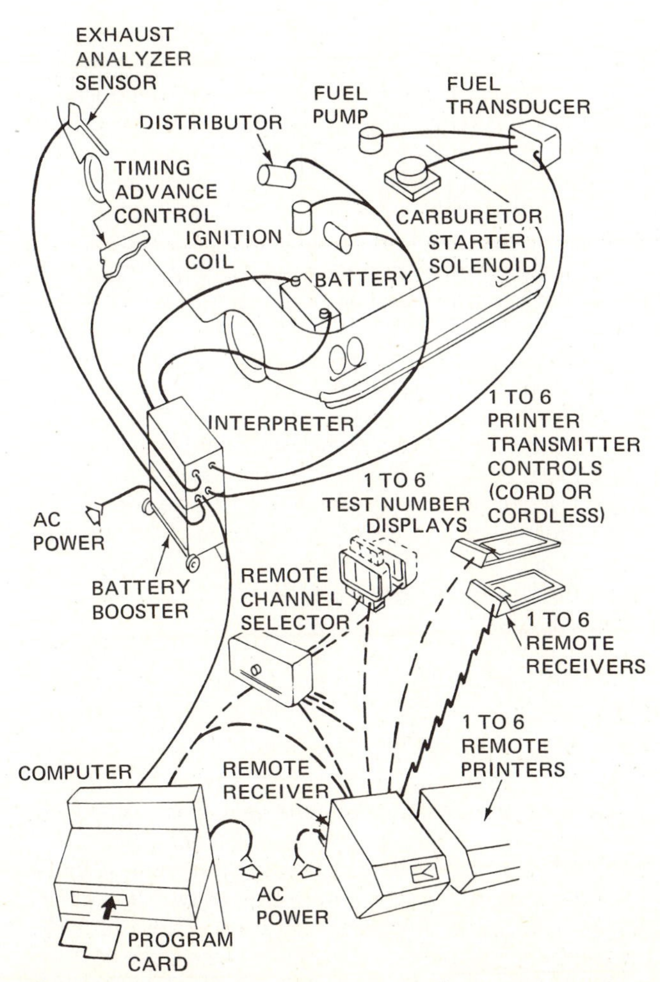

Fig. 21-9. Automotive-system test analyzer connected to make a complete analysis of the engine and accessory systems. (*Universal Testproducts, Incorporated*)

HEADER PORTION OF CARD DESIGN CHARACTERISTICS

PUNCHED CARD IMAGE

CHEVROLET 66 Air Inj Reac Biscayne-Belair Impala-Caprice 220 H P 8 Cyl
283" TRANS AUT-STD CARB 2-4 BBL

OPERATIONAL DATA

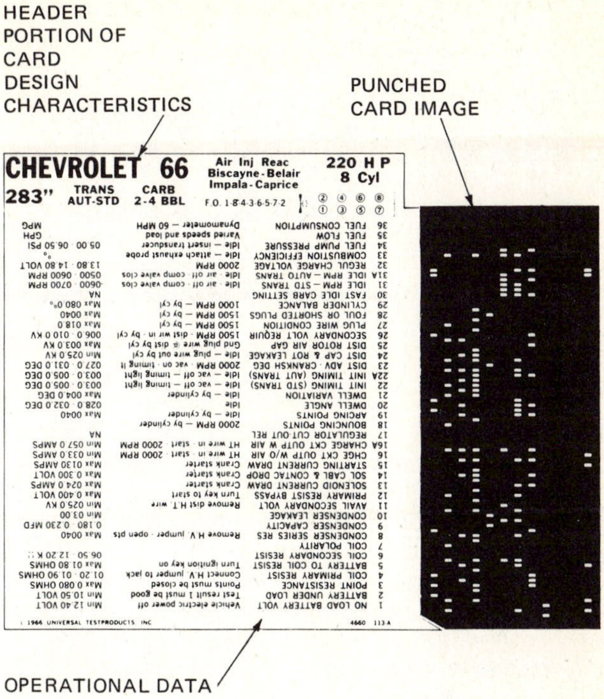

Fig. 21-10. Computer program card. (*Universal Testproducts, Incorporated*)

6. As part of the tune-up procedure, you: (*a*) clean and reinsulate the spark plugs, (*b*) clean, gap, and replace old plugs, (*c*) install all new plugs, (*d*) either (*b*) or (*c*).
7. If the vehicle is equipped with a fuel-vapor-recovery system: (*a*) replace the charcoal canister, (*b*) replace the canister filter, (*c*) replace the vapor-recovery valve.
8. To see how much the tune-up has reduced pollutants in the exhaust gases, many mechanics check for HC and CO: (*a*) after tune-up is done, (*b*) before tune-up is done, (*c*) both (*a*) and (*b*).

SUGGESTIONS FOR FURTHER STUDY

Make a collection of different tune-up printed forms supplied by automotive and test-equipment manufacturers, as shown in Fig. 21-1. Note the similarities and differences among the forms. Keep in mind that some of these forms have a dual purpose: they not only list the checks to be made, but also list them in a way that can attract potential customers. That is, they are sales sheets also. This means that the checks *may not be listed in the order in which they will actually be done.* Rather, they are listed in a way that puts the most important items first.

Also, refer to the automotive manufacturers' shop manuals and make lists of the key steps in the tune-up procedures they recommend. Then compare these to see what is common among them.

As you do all this, you are learning the basic steps in tune-up procedure. And this is important for any automotive technician to know.

chapter 22

DIAGNOSING ENGINE TROUBLES

This chapter discusses various engine troubles and relates them to possible causes and corrections; that is, it describes engine trouble-diagnosis procedures. It is not an easy chapter to study, but at the same time it is probably the most important chapter in the book. It gives you the information you need to understand the various ways in which engine trouble develops, how to determine the cause of a trouble, and how to correct it. Regardless of what you plan to do in the automotive field—whether you want to work in the service shop, plant, office, or laboratory—a knowledge of engine troubles and corrections will be of great value to you.

⊘ 22-1 How to Study This Chapter There are different ways to study this chapter. You can go through it page by page, just as you have studied the previous chapters. Perhaps a better way would be to take one complaint at a time (as listed in the Engine Trouble-Diagnosis Chart), read through the possible causes and checks or corrections, and then study the section later in the chapter that discusses the complaint. For example, you could take complaint 1, "Engine will not turn over," and after reading the causes and checks or corrections listed in the second and third columns in the chart, you would turn to ⊘ 22-4 (referred to in the first column) and study it.

Since a knowledge of trouble causes and corrections is so helpful, you will probably be referring to the Engine Trouble-Diagnosis Chart many times. One way to help yourself remember the complaints, causes, and checks or corrections is to write each complaint, with its list of causes and corrections, on a separate 3- by 5-inch card. You can also get a deck of the McGraw-Hill *Automotive Troubleshooting Cards* which list troubles and their possible causes. Carry the cards around with you. At odd moments, when you are riding a bus, eating a sandwich, or getting ready for bed, you can take out a card and read it over. Soon you will know the troubles and their causes and checks or corrections "backward and forward."

⊘ 22-2 Need for Logical Procedure After a trouble has been located in an engine, it is usually not too difficult to make the corrections necessary to eliminate the conditions causing the trouble. Chapters 23 to 26 discuss the various engine services and explain in detail the corrections to be made to eliminate different causes of engine trouble.

This chapter is devoted to trouble diagnosis, to the detective work that a mechanic is called upon to do when presented with a case of engine trouble. Careful analysis and straight thinking are often needed to find the cause of trouble. If a logical procedure is followed, the cause of the trouble can usually be spotted without delay. But haphazard guesswork wastes time and may cause you to overlook entirely the basic cause of the trouble. For example, suppose you tell a driver that the battery needs to be recharged or a new battery is necessary, blaming the run-down battery for the trouble. But you might search further to find out why the battery ran down. Perhaps it was old, or possibly the alternator or regulator was not operating properly. On the other hand, the driver might have been driving mostly at night with the lights and radio on. Also, the trouble might have been due to hard starting or frequent starts and stops which caused the driver to use the starting motor excessively, with the result that the battery ran down. In any event, unless the real cause of trouble is found and corrected, the driver will soon be in trouble again.

⊘ 22-3 Engine Trouble-Diagnosis Chart A variety of complaints bring a driver to a mechanic, but it is rare that a driver has a clear idea of what is causing the trouble. Most complaints, however, can be grouped under a few basic headings. These include: engine will not turn over; engine turns over but will not start; engine runs but misses; engine lacks power, acceleration, or high-speed performance; engine overheats; engine uses excessive oil or gasoline; or engine is noisy. The Engine Trouble-Diagnosis Chart that follows lists the various engine troubles, together with their possible causes, checks to be made, and corrections needed. (The section number given under each trouble is the section to be referred to for fuller explanations of how the troubles can be located and eliminated.) Some causes of trouble will be found to be in the engine itself; the latter part of this book is devoted to the various engine-service operations that correct these troubles. When the trouble is traced to the fuel, cooling, lubricating, or electric system, reference is made to the book in the McGraw-Hill Automotive Technology Series that explains how to correct the trouble.

NOTE: The troubles and possible causes are not listed in the chart in the order of frequency of occurrence. That is, item 1 (or item a under "Possible Cause") does not necessarily occur more frequently than item 2 (or item b).

ENGINE TROUBLE-DIAGNOSIS CHART

(See ⊘ 22-4 to 22-21 for detailed explanation of the trouble causes and corrections listed below.)

COMPLAINT	POSSIBLE CAUSE	CHECK OR CORRECTION
1. Engine will not turn over (⊘ 22-4)	a. Run-down battery	Recharge or replace battery; start engine with jumper battery and cables*
	b. Starting circuit open	Find and eliminate the open; check for dirty or loose cables*
	c. Starting-motor drive jammed	Remove starting motor and free drive*
	d. Starting motor jammed	Remove starting motor for teardown and correction*
	e. Engine jammed	Check engine to find trouble
	f. Transmission not in neutral or neutral switch out of adjustment	Check and adjust neutral switch if necessary
	g. Seat belt not fastened or interlock faulty	Check interlock*
	h. See also causes listed under item 3; driver may have run battery down trying to start	
2. Engine turns over slowly but does not start (⊘ 22-5)	a. Run-down battery	Recharge or replace battery; start engine with jumper battery and cables*
	b. Defective starting motor	Repair or replace*
	c. Bad connections in starting circuit	Check for loose or dirty cables; clean and tighten*
	d. See also causes listed under item 3; driver may have run battery down trying to start	
3. Engine turns over at normal speed but does not start (⊘ 22-6)	a. Defective ignition system	Try spark test; check timing, ignition system*
	b. Defective fuel pump or overchoking	Prime engine; check accelerator-pump discharge, fuel pump, fuel line, choke, carburetor*
	c. Air leaks in intake manifold or carburetor	Tighten mounting; replace gaskets as needed
	d. Defect in engine	Check compression or leakage (⊘ 22-7), valve action, timing
	e. Burned out coil external	Replace
	f. Plugged fuel filter	Clean or replace
	g. Plugged or collapsed exhaust system	Replace collapsed parts
4. Engine runs but misses: one cylinder (⊘ 22-8)	a. Defective spark plug	Clean or replace*
	b. Defective distributor cap or spark-plug cable	Replace*
	c. Valve stuck open	Free valve; service valve guide
	d. Broken valve spring	Replace
	e. Burned valve	Replace
	f. Bent pushrod	Replace
	g. Flat cam lobe	Replace camshaft
	h. Defective piston or rings	Replace; service cylinder wall as necessary
	i. Defective head gasket	Replace
	j. Intake-manifold leak	Replace gasket; tighten manifold bolts
5. Engine runs but misses: different cylinders (⊘ 22-8)	a. Defective distributor advance, coil, condenser	Check distributor, coil, condensor*
	b. Defective fuel system	Check fuel pump, flex line, carburetor†
	c. Cross-firing plug wires	Replace or relocate
	d. Loss of compression	Check compression or leakage (⊘ 20-4 to 20-6)
	e. Defective valve action	Check compression, leakage, vacuum (⊘ 20-4 to 20-7)

ENGINE TROUBLE-DIAGNOSIS CHART *(Continued)*

COMPLAINT	POSSIBLE CAUSE	CHECK OR CORRECTION
	f. Worn pistons and rings	Check compression, leakage, vacuum (⊘ 20-4 to 20-7)
	g. Overheated engine	Check cooling system†
	h. Manifold heat-control valve stuck	Free valve
	i. Restricted exhaust	Check exhaust, tail pipe, muffler; eliminate restriction
6. Engine lacks power, acceleration, or high-speed performance: hot or cold (⊘ 22-10)	a. Defective ignition	Check timing, distributor, wiring, condenser, coil, and plugs*
	b. Defective fuel system; secondary throttle valves not opening	Check carburetor, choke, filter, air cleaner, and fuel pump†
	c. Throttle valve not opening fully	Adjust linkage†
	d. Restricted exhaust	Check tail pipe and muffler; eliminate restriction
	e. Loss of compression	Check compression or leakage (⊘ 20-4 to 20-6)
	f. Excessive carbon in engine	Clean
	g. Defective valve action	Check with compression, leakage, vacuum testers (⊘ 20-4 to 20-7)
	h. Excessive rolling resistance from low tires, dragging brakes, wheel misalignment, etc.	Correct the defect causing rolling resistance
	i. Heavy oil	Use correct oil
	j. Wrong or bad fuel	Use good fuel of correct octane
	k. Transmission not downshifting or defective torque converter	Check transmission
7. Engine lacks power, acceleration, or high-speed performance: hot only (⊘ 22-10)	a. Engine overheats	Check cooling system†
	b. Choke stuck partly open	Repair or replace†
	c. Sticking manifold heat-control valve	Free valve
	d. Vapor lock	Use different fuel or shield fuel line
8. Engine lacks power, acceleration, or high-speed performance: cold only (⊘ 22-10)	a. Automatic choke stuck open	Repair or replace†
	b. Manifold heat-control valve stuck open	Free valve
	c. Cooling-system thermostat stuck open	Repair or replace†
	d. Engine valves stuck open	Free valves; service valve stems and guides as needed
9. Engine overheats (⊘ 22-12)	a. Lack of coolant	Add coolant; look for leak
	b. Ignition timing late	Adjust timing*
	c. Loose or broken fan belt	Tighten or replace†
	d. Thermostat stuck closed	Replace†
	e. Clogged water jackets	Clean†
	f. Defective radiator hose	Replace†
	g. Defective water pump	Repair or replace†
	h. Insufficient oil	Add oil
	i. High-altitude, hot-climate operation	Drive more slowly; keep radiator filled
	j. Defective fan clutch	Replace†
	k. Valve timing late; slack timing chain has allowed chain to jump a tooth	Retime, adjust or replace
	l. No vacuum advance in any gear	TCS system or distributor defective*†
10. Rough idle (⊘ 22-13)	a. Incorrect carburetor idle adjustment	Readjust idle mixture and speed†
	b. PCV valve stuck open	Replace
	c. See also other causes listed under items 6 to 8	
11. Engine stalls cold or as it warms up (⊘ 22-14)	a. Choke valve stuck closed or will not close	Open choke valve; free or repair automatic choke†
	b. Fuel not getting to or through carburetor	Check fuel pump, lines, filter, float, and idle systems†
	c. Manifold heat-control valve stuck	Free valve
	d. Throttle solenoid improperly set	Adjust
	e. Idling speed set too low	Increase idling speed to specified value†
	f. PCV valve stuck open	Replace
	g. Damper in thermostatic air cleaner stuck closed	Free; repair or replace control motor

ENGINE TROUBLE-DIAGNOSIS CHART *(Continued)*

COMPLAINT	POSSIBLE CAUSE	CHECK OR CORRECTION
12. Engine stalls after idling or slow-speed driving (⊘ 22-14)	a. Defective fuel pump	Repair or replace fuel pump†
	b. Overheating	See item 9
	c. High carburetor float level	Adjust†
	d. Incorrect idling adjustment	Adjust†
	e. Malfunctioning PCV valve	Replace
	f. Throttle solenoid improperly set	Adjust
13. Engine stalls after high-speed driving (⊘ 22-15)	a. Vapor lock	Use different fuel or shield fuel line†
	b. Carburetor venting or idle-compensator valve defective	Check and repair†
	c. Engine overheats	See item 9
	d. PCV valve stuck open	Replace
	e. Improperly set throttle solenoid	Adjust
14. Engine backfires (⊘ 22-15)	a. Ignition timing off	Adjust timing*
	b. Spark plugs of wrong heat range	Install correct plugs*
	c. Excessively rich or lean mixture	Repair or readjust fuel pump or carburetor†
	d. Engine overheats	See item 9
	e. Carbon in engine	Clean
	f. Valves hot or stuck	Adjust, free, clean; replace if bad
	g. Cracked distributor cap	Replace
	h. Inoperative antibackfire valve	Replace
	i. Cross-firing plug wires	Replace
15. Engine run-on, or dieseling (⊘ 22-16)	a. Incorrect idle-stop or solenoid adjustment	Adjust; fix solenoid
	b. Engine overheats	See item 9
	c. Hotspots in cylinders	Check plugs, pistons, cylinders for carbon; check valves for defects and faulty seating
	d. Timing advanced	Adjust†
16. Too much HC and CO in exhaust gas (⊘ 22-17)	a. Ignition miss	Check plugs, wiring, cap, coil, etc.*
	b. Incorrect ignition timing	Time ignition*
	c. Carburetor troubles	Check choke, float level, idle-mixture-adjustment screw, etc.,† as listed in item 20
	d. Faulty air injection	Check pump, hoses, manifold
	e. Defective TCS system	Check system
	f. Defective catalytic converters	Replace converters or catalyst
17. Smoky exhaust		
1. Blue smoke	Excessive oil consumption	See item 18 and ⊘ 22-18
2. Black smoke	Excessively rich mixture	See item 20 and ⊘ 22-20
3. White smoke	Steam in exhaust	Replace gasket; tighten cylinder-head bolts to eliminate coolant leakage into combustion chambers
18. Excessive oil consumption (⊘ 22-18)	a. External leaks	Correct seals; replace gaskets
	b. Burning oil in combustion chamber	Check valve-stem clearance, piston rings, cylinder walls, rod bearings, vacuum-pump diphragm
	c. High-speed driving	Drive more slowly
19. Low oil pressure (⊘ 22-19)	a. Worn engine bearings	Replace
	b. Engine overheating	See item 9
	c. Oil dilution or foaming	Replace oil
	d. Lubricating-system defects	Check oil lines, oil pump, relief valve†
20. Excessive fuel consumption (⊘ 22-20)	a. Jackrabbit starts	Drive more reasonably
	b. High-speed driving	Drive more slowly
	c. Short-run operation	Drive longer distances
	d. Excessive fuel-pump pressure or pump leakage	Reduce pressure; repair pump†
	e. Choke partly closed after warm-up	Open; repair or replace automatic choke†
	f. Clogged air cleaner	Clean†
	g. High carburetor float level	Adjust†
	h. Stuck or dirty float needle level	Free and clean†
	i. Worn carburetor jets	Replace†

ENGINE TROUBLE-DIAGNOSIS CHART (*Continued*)

COMPLAINT	POSSIBLE CAUSE	CHECK OR CORRECTION
	j. Stuck metering rod or full-power piston	Free†
	k. Idle too rich or too fast	Adjust†
	l. Stuck accelerator-pump check valve	Free†
	m. Carburetor leaks	Replace gaskets; tighten screws; etc.†
	n. Cylinder not firing	Check coil, condenser, timing, plugs, contact points, wiring*
	o. Automatic transmission slipping or not upshifting	Check transmission
	p. Loss of engine compression (worn engine)	Check compression or leakage (⊘ 22-4 to 22-6)
	q. Defective valve action (worn camshaft, chain slack, or jumped tooth)	Check with compression, leakage, or vacuum tester (⊘ 22-4 to 22-7)
	r. Excessive rolling resistance from low tires, dragging brakes, wheel misalignment, etc.	Correct the defects causing the rolling resistance
	s. Clutch slippage	Adjust or repair
21. Engine noises (⊘ 22-21)		
1. Regular clicking	Valve and tappet	Readjust valve clearance or replace noisy hydraulic lifters
2. Ping or knock on load or acceleration	Detonation due to low-octane fuel, carbon, advanced ignition timing, or causes listed under item 14	Use higher-octane fuel; remove carbon; adjust ignition timing
3. Light knock or pound with engine floating	Worn connecting-rod bearings or crankpin; misaligned rod; lack of oil	Replace bearings; service crankpins; replace rod; add oil
4. Light, metallic double knock, usually most audible during idle	Worn or loose pin or lack of oil	Service pin and bushing; add oil
5. Chattering or rattling during acceleration	Worn rings, cylinder walls, low ring tension, or broken rings	Service walls; replace rings
6. Hollow, muffled bell-like sound (engine cold)	Piston slap due to worn pistons or walls, collapsed piston skirts, excessive clearance, misaligned connecting rods, or lack of oil	Replace or resize pistons; service walls; replace rods; add oil
7. Dull, heavy, metallic knock under load or acceleration, especially when cold	Regular noise: worn main bearings; irregular noise: worn end-thrust bearing knock on clutch engagement or on hard acceleration	Replace or service bearings and crankshaft
8. Miscellaneous noises (rattles, etc.)	Loosely mounted accessories: alternator, horn, oil pan, front bumper, water pump, etc.	Tighten mounting

*See *Automotive Electrical Equipment*, another book in the McGraw Hill Automotive Technology Series.
†See *Automotive Fuel, Lubricating, and Cooling Systems*, another book in the McGraw-Hill Automotive Technology Series.

⊘ 22-4 Engine Will Not Turn Over If the engine will not turn over when starting is attempted, make sure the control lever is in neutral (N) or park (P). Or, if the car has a manual transmission, make sure that the clutch pedal is depressed. Check the battery and cables. A low battery can cause the solenoid plunger to repeatedly pull in and release (making a clattering noise) but will not be able to crank the engine.

If the starter spins and the drive pinion engages, the starting-motor overrunning clutch is slipping. If the solenoid plunger pulls in (a loud click) but nothing else happens, there is trouble in the solenoid (poor contacts), starting motor, or circuit. If the solenoid plunger does not pull in, the solenoid circuit is at fault. Connect a jumper lead between the solenoid battery terminal and solenoid switch terminal. If the starting motor now operates, the solenoid is okay. The trouble is in the ignition switch, in the neutral starting switch, or in the circuit between them. If the starting motor still does not operate, remove it for service.

Another method of checking for the cause of trouble when the engine will not turn over uses the headlights or dome light. This is a preliminary, or "instant," check. It is not as accurate or comprehensive as the preceding method, but some technicians use it as a first step in the diagnosis. Here is how to do the check: Turn on the headlights or dome light and try to start the engine. The lights will (1) stay bright, (2) dim considerably, (3) dim slightly, (4) go out, or (5) not burn at all.

1. If the lights stay bright, there is an open circuit in the starting motor or starting-motor circuit. Check as outlined in ⊘ 22-5. Also, the transmission may not be in neutral or the neutral switch may need adjustment. In addition, on some late-model cars, the ignition-interlock safety belts may not be fastened properly or the system may be defective.

2. If the lights dim considerably, the battery may be run down or there may be mechanical trouble in the starting motor or engine. If the battery tests okay, remove the starting motor for further checks. Try to turn the engine flywheel in the normal direction of rotation to see if the engine is jammed.
3. If the lights dim only slightly, listen for cranking action (sound of an electric motor running). If the starting motor runs, the pinion is not engaging the flywheel (Bendix type) or the overrunning clutch is slipping. If the solenoid clicks but the starting motor does not rotate, the cause could be a low battery but it is probably trouble in the starting motor. Remove the starting motor for service.
4. If the lights go out as cranking is attempted, the chances are that there is a bad connection in the main circuit, probably at a battery terminal.
5. If the lights burn dimly or not at all when they are turned on, even before cranking is attempted, the battery is probably run down.

⊘ 22-5 Engine Turns Over Slowly But Does Not Start The causes of this condition could be a run-down battery, a defective starting motor, or mechanical trouble in the engine. Check the battery, starting motor, and circuit (see *Automotive Electrical Equipment,* another book in the McGraw-Hill Automotive Technology series). If they are normal, the trouble probably is in the engine (defective bearings, rings, etc., which could produce high friction). Remember that in cold weather, cranking speed is reduced by thickening of the engine oil and reduction of battery efficiency.

NOTE: If the battery is run down, it could be that the driver has discharged the battery in attempting to start the engine. The cause of starting failure could be as noted in the following sections.

⊘ 22-6 Engine Turns Over at Normal Cranking Speed But Does Not Start This means that the battery and starting motor are in normal condition. The cause of trouble is probably in the ignition or fuel system. The difficulty could be due to overchoking.[1] Try cranking with the throttle wide open. If the engine does not start, disconnect the lead from one spark plug (or from the center distributor-cap terminal). Hold the lead clip about $\frac{3}{16}$ inch [4.76 mm] from the engine block. Crank the engine to see if a good spark occurs. If a good spark does occur, the ignition system is probably okay (the timing could be off, however).

If the ignition system seems to be operating normally, the fuel system should be analyzed. First, prime the engine by operating the carburetor accelerator pump several times.

CAUTION: Gasoline is highly explosive. Keep back out of the way while priming the engine because the engine might backfire through the carburetor. Replace the air cleaner before cranking.

If the engine now starts and runs for a few seconds, the fuel system is probably faulty. It is not delivering fuel to the engine. First make sure there is gasoline in the fuel tank. Temporarily disconnect the fuel inlet to the carburetor. Hold a container under the fuel line to catch fuel, and crank the engine to see whether fuel is delivered. If fuel is not delivered, the fuel pump is defective or the fuel line is clogged. If fuel is delivered, the fuel filter is probably at fault, the automatic choke is not working correctly, or possibly there are air leaks into the intake manifold or carburetor.

If the fuel and ignition systems seem to be okay on the preliminary checks, check the mechanical condition of the engine with the cylinder compression and leakage testers (⊘ 22-7). Also, note that a plugged or collapsed exhaust system can build up back pressure. This could prevent normal exhaust and intake so that the engine will not start.

⊘ 22-7 Cylinder Compression and Leakage Testers These testers are used to determine whether or not the cylinder can hold compression or there is excessive leakage past the rings, valves, or head gasket. The compression tester (Fig. 20-3) has been a basic engine-testing instrument for many years. Recently, the cylinder-leakage tester has come into use. Some mechanics believe that it is more accurate in pinpointing cylinder defects.

The use of compression and leakage testers has been described in detail (compression tester in ⊘ 20-4 and 20-5 and leakage tester in ⊘ 20-6).

⊘ 22-8 Engine Runs But Misses A missing engine is a rough engine. If one or more cylinders fail to fire, the engine is thrown out of balance. The result is roughness and loss of power. It is sometimes hard to track down a miss. The miss might occur at some speeds and not others. Also, the miss may skip around. The modern method of checking out a missing engine is to use an oscilloscope and a dynamometer. (The oscilloscope is discussed in ⊘ 20-10 to 20-13. The dynamometer is discussed in ⊘ 20-14.) If these testing instruments are not available, then the test can be made as follows.

Use insulated pliers to disconnect each spark-plug cable in turn to locate the missing cylinder. (Disconnecting the cable prevents the spark from reaching the spark plug, and the spark plug will not fire.) If disconnecting the cable changes the engine speed or rhythm, then the cylinder was delivering power before you disconnected the cable. But if there is no change in engine speed or rhythm, then

[1] This analysis applies to a cold engine. Failure to start with a hot engine may be caused by a defective choke that fails to open properly as the engine warms up. This condition would cause flooding of the engine (delivery of too much gasoline). Open the throttle wide while cranking (this dechokes the engine), or open the choke valve by hand while cranking.

that cylinder was missing before you disconnected the cable.

Check a missing cylinder further by holding the spark-plug lead clip close to the engine block while the engine is running. If no spark occurs, there is probably a high-voltage leak due to a bad lead or a cracked or burned distributor cap. If a good spark occurs, install a new spark plug in the cylinder (or swap plugs between two cylinders). Then reconnect the lead, and see whether the cylinder still misses. If it does, the cause of the trouble is probably defective engine parts, such as valves or rings.

If the miss is hard to locate, perform a general tune-up (Chap. 21). A tune-up will disclose and possibly eliminate various causes of missing. These could include defects in ignition system or fuel system, loss of engine compression, sticky or damaged engine valves, overheated engine, sticky manifold heat control, and clogged exhaust.

With most oscilloscopes, you can make a power-balance test that will quickly pinpoint the missing cylinder. When the oscilloscope is connected to the running engine, you turn a knob or push a button and the cylinders are shorted out, one by one, in the firing order. The scope shows which cylinder is shorted out. If shorting out a cylinder changes the engine rpm as registered on the tachometer, you know the cylinder was delivering power. But if no change in rpm takes place, then you know that the cylinder was not delivering power.

⊘ 22-9 Engine Vacuum Gauge The vacuum gauge is an important engine tester for tracking down troubles in an engine that runs but does not perform satisfactorily. It measures intake-manifold vacuum. The intake-manifold vacuum varies with different operating conditions and with different engine defects. The manner in which the vacuum varies from normal indicates the type of engine trouble. (⊘ 20-7 explains how to use the vacuum gauge.)

⊘ 22-10 Engine Lacks Power This is a general complaint that is often difficult to analyze. The best procedure is to do a tune-up job (Chap. 21). This will disclose various engine conditions that could cause loss of power. Some idea of the cause can be gained by finding out whether the engine lacks power when both hot and cold, only when hot, or only when cold. Also, find out if the problem developed suddenly or if the power gradually fell off over a period of many months or many miles of operation. A chassis dynamometer (⊘ 20-14) or an oscilloscope (⊘ 20-10 to 20-13) can be used to help locate the cause of the trouble.

1. *ENGINE LACKS POWER WHEN BOTH HOT AND COLD* The fuel system may not be enriching the mixture as the throttle is opened. This condition could be due to a faulty accelerator pump or defective main metering or power system in the carburetor. Also, the fuel system could be supplying an excessively lean or rich mixture. This condition could be due to a defective fuel pump, clogged lines, clogged filter, worn carburetor jets or lines, air leaks at the carburetor or manifold joints, malfunctioning PCV valve, and so on. Carburetors and fuel-system action can be checked with an exhaust-gas analyzer (⊘ 20-8).

Another condition could cause lack of power when the engine is hot or cold. This condition is an improper linkage adjustment that prevents full throttle opening. Also, the ignition system may be causing trouble, owing to incorrect timing, a "weak" coil, reversed polarity, wrong spark-plug heat range, and so on. The wrong fuel or oil for the engine could reduce performance. Numerous other engine conditions could cause loss of power: deposits (carbon), lack of compression (faulty valves, rings, worn cylinder walls, worn pistons), and defective bearings. A clogged exhaust (bent or collapsed exhaust or tail pipe or clogged muffler) could create back pressure that would cause poor engine performance. Also, any excessive rolling resistance would absorb engine power and hold down engine acceleration and speed. This would include dragging brakes, underinflated tires, misaligned wheels, and excessive friction in the transmission or power train. Finally, the automatic transmission may not be downshifting, or the torque converter may be defective.

2. *ENGINE LACKS POWER ONLY WHEN HOT* The engine may be overheating (⊘ 22-12). Also, the automatic choke may not be opening normally as the engine warms up. The manifold heat-control valve may be stuck. Or there may be a vapor lock in the fuel pump or line.

3. *ENGINE LACKS POWER ONLY WHEN COLD (OR REACHES OPERATING TEMPERATURE TOO SLOWLY)* The automatic choke may be leaning out the mixture too soon (before the engine warms up). The manifold heat-control valve may not be closed (so that not enough heat reaches the intake manifold). Or the cooling system thermostat may be stuck open. In this case, water circulation goes on between the engine and radiator even with the engine cold, and so warm-up is delayed. Occasionally, engine valves may stick open when the engine is cold, but as the engine warms up, the valves become free and work normally.

⊘ 22-11 Exhaust-Gas Analyzer At one time the major use of the exhaust-gas analyzer was to adjust the carburetor. Today, its major job is to check automotive emission control. If the emission controls are not working properly, there will be excessive amounts of HC and CO in the exhaust gas. The exhaust-gas analyzer measures the amount of HC and CO in the exhaust gas coming out the tail pipe. Use of exhaust-gas analyzer is discussed in ⊘ 20-8.

⊘ 22-12 Engine Overheats Most engine overheating is caused by loss of coolant due to leaks in the cooling system. Other causes include a loose or broken fan belt, a defective water pump, clogged water jackets, a defective radiator hose, and a defective thermostat or fan clutch. Also, late ignition or valve timing, lack of engine oil, overloading the engine, or high-speed, high-altitude, or hot-climate operation can cause engine overheating. Freezing of the coolant could cause lack of coolant circulation so that local hot spots and boiling develop. Then, too, if a faulty TCS system prevents vacuum advance in any gear, or if the distributor vacuum advance is defective, overheating may result.

⊘ **22-13 Rough Idle** If the engine idles roughly but runs normally above idle, chances are that the idle speed and idle mixture are incorrectly adjusted. A rough idle could also be due to other causes, such as a loose vacuum hose (or one that is disconnected from the intake manifold) or a PCV valve that is stuck open.

⊘ **22-14 Engine Stalls** If the engine starts and then stalls, note whether the stalling takes place before or as the engine warms up, after idling or slow-speed driving, or after high-speed or full-load driving. Special note should be made of the PCV valve. If this valve becomes clogged or sticks, it will cause poor idling and stalling.

1. *ENGINE STALLS BEFORE IT WARMS UP* This condition could be due to an improperly set fast or slow idle or to improper adjustment of the idle-mixture screw in the carburetor. Also, it could be due to a low carburetor float setting or to insufficient fuel entering the carburetor. This condition could result from a defective thermostatic air cleaner, dirt or water in the fuel lines or filter, a defective fuel pump, or a plugged fuel-tank vent. Also, the carburetor could be icing. Certain ignition troubles could cause stalling after starting. But, as a rule, if ignition troubles are bad enough to cause stalling, they would also prevent starting. However, burned contact points might permit starting but fail to keep the engine going. One other condition that could cause stalling before the engine warms up is an open primary resistance wire. When the engine is cranked, this wire is bypassed. Then, when the engine starts and cranking stops, this wire is inserted into the ignition primary circuit. If this wire were open, the engine would then stall.

2. *ENGINE STALLS AS IT WARMS UP* This condition could result if the choke valve sticks closed; the mixture becomes too rich for a hot engine, and the engine stalls. If the manifold heat-control valve sticks closed, the air-fuel mixture might become overheated and too lean, causing the engine to stall. If the hot-idle speed is too low, the engine may stall as it warms up because the idling speed drops too low. Also, stalling may be caused by overheating of the engine, which could cause vapor lock. In addition, if the damper in the thermostatic air cleaner sticks closed, the air-fuel mixture can become overheated and too lean, causing the engine to stall.

3. *ENGINE STALLS AFTER IDLING OR SLOW-SPEED DRIVING* This condition could occur if the fuel pump has a cracked diaphragm, weak spring, or defective valve. The pump fails to deliver enough fuel for idling or slow-speed operation (although it could deliver enough for high-speed operation), and the engine stalls. If the carburetor float level is set too high or the idle adjustment is too rich, the engine may "load up" and stall. A lean idle adjustment may also cause stalling. The engine may overheat during sustained idling or slow-speed driving. With this condition, the air movement through the radiator may not be sufficient to keep the engine cool. Overheating, in turn, could cause vapor lock and engine stalling. (See ⊘ 22-12 for causes of overheating.)

4. *ENGINE STALLS AFTER HIGH-SPEED OR FULL-LOAD DRIVING* This condition could occur if enough heat accumulates to cause vapor lock. The remedy here would be to shield the fuel line and fuel pump or use a less volatile fuel. Failure of the venting or idle-compensator valve in the carburetor may also cause stalling after high-speed or full-load operation. Excessive overheating of the engine is also a primary cause of stalling (⊘ 22-12).

⊘ **22-15 Engine Backfires** Most backfiring is caused by a faulty antibackfire valve or air bypass valve in the air-injection system (⊘ 19-16). It could also be caused by late ignition timing or ignition cross firing (caused by the spark jumping across the distributor cap or through the cable insulation). In addition, backfiring could be due to spark plugs of the wrong heat range (which overheat and cause preignition), excessively rich mixtures (caused by fuel-pump or carburetor troubles), overheating of the engine (⊘ 22-12), carbon in the engine, hot valves, or intake valves that stick or seat poorly. Carbon in the engine, if excessive, may retain enough heat to cause the air-fuel mixture to preignite as it enters the cylinder so that backfiring occurs. Carbon also increases the compression ratio and thus the tendency for knocking and preignition. Hot plugs may cause preignition; cooler plugs should be installed. If intake valves hang open, combustion may be carried back into the carburetor. Valves that have been ground excessively so that they have sharp edges, valves that seat poorly, or valves that are carboned so that they overheat often produce backfiring.

⊘ **22-16 Engine Run-on, or Dieseling** Modern engines, with their emission controls, require a fairly high, hot idle for best operation. This makes run-on, or dieseling, possible. If there are hot spots in the combustion chambers, the engine could continue to run if the throttle is not completely closed. The hot spots take the place of the spark plugs. If the throttle is slightly open, enough air-fuel mixture could get past it to keep the engine running. Ignition in the combustion chambers is caused by the hot spots.

Modern engines have an idle-stop solenoid to close the throttle completely when the ignition switch is turned off. If an engine runs on, or diesels, check the idle-stop solenoid (if present) to make sure it is releasing when the ignition is turned off. It could require adjustment to permit the throttle to close completely. Be sure the engine idle speed is not set too high. Engine run-on could also be caused by advanced ignition timing. Correction of engine overheating is covered in ⊘ 22-12. Correction of hot spots may require spark-plug service or removing the cylinder head for cleaning plus valve service.

⊘ 22-17 Too Much HC and CO in Exhaust Gas If the exhaust-gas analyzer (⊘ 20-8) discloses that there is too much HC and CO in the exhaust, correction must be made. Some states require exhaust-gas testing of all cars during state inspection. Cars that emit too much HC and CO must be repaired before they can be licensed. This restriction is designed to get the "smoggers" off the highways. Here are the possible causes. (The corrections are obvious.)

1. Missing due to ignition problems such as faulty plugs, high-tension wiring, distributor cap, ignition coil, condenser, or contact points. (It can also be caused by excessive carbon deposits in the combustion chamber or stuck or burned valves.)
2. Incorrect ignition timing.
3. Carburetor troubles such as the choke sticking closed, worn jets, high float level, and other conditions listed in ⊘ 22-20.
4. Faulty air-injection system which does not inject enough air into the exhaust manifold to completely burn the HC and CO. (This could be caused by a faulty air pump or a leaking hose or air manifold.)
5. Defective transmission-controlled spark system which permits vacuum advance in all gear positions instead of high and reverse only.
6. Defective catalytic converters which must be replaced or serviced to restore the catalytic action.

⊘ 22-18 Excessive Oil Consumption Oil may be lost from the engine in three ways: by leakage in liquid form (Fig. 17-1); by burning in the combustion chamber; and by passing out of the crankcase through the crankcase ventilating system in the form of mist or vapor.

External leakage can often be detected by inspecting the seals around the oil pan, cylinder-head cover, and timing-gear housing or at oil-line and filter connections.

Burning of oil in the combustion chamber gives the exhaust gas a bluish tinge. Oil can enter the combustion chamber through the PCV system, through the clearance between intake-valve or exhaust-valve stems and valve guides, and past the piston rings.

If the intake-valve-stem clearance is excessive, on each intake stroke oil will be "pulled" through the clearance and into the combustion chamber. The appearance of the intake-valve stem often indicates that this is occurring. Some oil remains on the underside of the valve and stem to form carbon. Oil can also seep down past the exhaust-valve stem if the clearance is excessive. The remedy is to install valve seals or a new valve guide and possibly a new valve.

Probably the most common cause of excessive oil consumption is passage of oil into the combustion chamber between the piston rings and cylinder walls. This is often called *oil pumping*. It is due to worn, tapered, or out-of-round cylinder walls or worn or carboned rings. In addition, when engine bearings are worn, excessive oil is thrown on the cylinder walls. The rings are not able to control all of it, and too much oil works up into the combustion chamber.

High speed must also be considered if there is excessive oil consumption. High speed means high temperatures and thus thin oil. More thin oil is thrown on the cylinder walls. The piston rings, moving at high speed, cannot function so effectively. More oil works up into the combustion chamber. In addition, the churning effect of the oil in the crankcase creates more oil vapor and mist at high speed. More oil is thus lost through the crankcase ventilating system. Tests show that an engine uses several times as much oil at 60 miles per hour [96.56 km/h] as at 30 miles per hour [48.28 km/h].

There is one misleading aspect of this matter of high-speed operation and oil consumption. Take, for example, the case of a car that is driven around town in start-and-stop driving where the engine never really gets warmed up. With this condition, some oil will be used, but the remaining oil may be diluted with water and unburned gasoline (see ⊘ 17-17). Thus, even though some oil is gone, the crankcase will still measure full owing to the addition of dilutes. However, suppose the car is now taken out onto the highway and driven at high speeds. The dilution elements will boil off rapidly and the car will appear to have lost perhaps a quart [0.946 l] of oil in 100 miles [160.93 km] or less.

⊘ 22-19 Low Oil Pressure Low oil pressure is often a warning of worn oil pumps or engine bearings. The bearings can pass so much oil that the oil pump cannot maintain oil pressure. Further, the end bearings will probably be oil-starved and may fail. Other causes of low oil pressure are a weak relief-valve spring, worn oil pump, broken or cracked oil line, and clogged oil line. Oil dilution, or foaming, sludge, insufficient oil, or oil made too thin by engine overheating will cause low oil pressure.

⊘ 22-20 Excessive Fuel Consumption This condition can be caused by almost anything in the car, from the driver to underinflated tires or a defective choke. A fuel-mileage tester can be used to accurately check fuel consumption (Fig. 22-1). The compression or leakage tester and the vacuum gauge (⊘ 20-4 to 20-7) will help determine whether the trouble is in the engine, fuel system, ignition system,

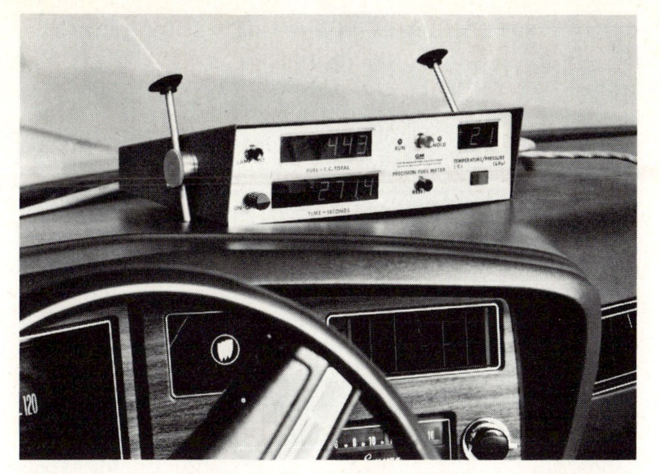

Fig. 22-1. A precision fuel-mileage indicator. (*Fluidyne Instrumentation*)

or elsewhere.[2] Also, the exhaust-gas analyzer, dynamometer, and fuel-flow meter are useful in analyzing the problem.

If the trouble seems to be in the fuel system, consider the following:

1. A driver who pumps the accelerator when idling and insists on being the first to get away when the stoplight changes will use excessive amounts of fuel.
2. Operation with the choke partly closed after warm-up will use excessive amounts of fuel.
3. Short-run operation means that the engine will operate on warm-up most of the time. This means that fuel consumption will be high.

These three conditions are due to the type of operation: Changing operating conditions is the only cure. If excessive fuel consumption is not due to any of these conditions, however, then the fuel pump should be checked for excessive pressure. High fuel-pump pressure will cause a high float-bowl level and a rich mixture. Special gauges are used to check fuel-pump pressure.

4. If excessive fuel consumption is not due to operating conditions or high fuel-pump pressure, the trouble is likely to be in the carburetor. It could be any of the following conditions:
   a. If the car is equipped with an automatic choke, the choke may not be opening rapidly enough during warm-up or may not open fully. The automatic choke should be checked by removing the air cleaner and observing choke operation during warm-up.
   b. A clogged air cleaner that does not admit sufficient air can act somewhat like a partly closed choke valve. The cleaner element should be cleaned or replaced.
   c. If the float level in the float bowl is high, it will cause flooding and delivery of excessive fuel to the carburetor air horn. The needle valve may be stuck open or may not be seating fully. The float level should be checked and adjusted.
   d. If the idle mixture is set too rich or the idle speed too high, excessive fuel consumption will result. These should be checked and adjusted as necessary.
   e. Where the accelerator-pump system has a check valve, failure of the check valve to close properly may allow fuel to feed through into the carburetor air horn. The carburetor should be disassembled for repair.
   f. The metering rod stuck in the high-speed full-throttle position or the economizer valve stuck open permits the power system to function, supplying an excessively rich mixture. The carburetor should be disassembled for repair.
   g. Worn jets permit the discharge of too much fuel. The jets should be replaced during carburetor rebuilding.
5. Faulty ignition can also cause excessive fuel consumption; the ignition system could cause engine miss and thus failure of the engine to use all the fuel. This trouble would also be associated with loss of power, acceleration, or high-speed performance (⊘ 22-10). Conditions in the ignition system that might contribute to the trouble include a "weak" coil or condenser, incorrect timing, faulty advance-mechanism action, dirty or worn plugs or contact points, and defective wiring.
6. Inferior engine action can produce excessive fuel consumption, for example, loss of engine compression from worn or stuck rings, worn or stuck valves, or a loose or burned cylinder-head gasket. Power is lost under these conditions, and more fuel must be burned to achieve the same speed. (Refer to ⊘ 20-4 to 20-6 for compression- and leakage-checking procedures.)
7. Excessive fuel consumption can also result from conditions that make it hard for the engine to move the car along the road. Such factors as low tires, dragging brakes, defective automatic transmission, and misalignment of wheels increase the rolling resistance of the car. The engine must use up more fuel to overcome this excessive rolling resistance.

[2] A rough test of mixture richness that does not require any testing instruments is to install a set of new or cleaned spark plugs of the correct heat range for the engine and operate the car for 15 to 20 minutes. Then stop the car, and remove and examine the plugs. If the spark plugs are coated with a black carbon deposit, the mixture is too rich. (See item 4, points *a* to *g*.) Black exhaust smoke is another indication of an excessively rich mixture. The mixture is too rich to burn fully, and so the exhaust gas contains "soot," or unburned fuel.

⊘ 22-21 Engine Noises Some engine noises may have little significance. Other noises may indicate serious engine trouble that requires prompt attention to prevent major damage to the engine.

A listening rod, or stethoscope, is helpful in locating the source of a noise (Fig. 22-2). The rod acts like the stethoscope that a doctor uses to listen to a patient's heartbeat or breathing. When one end is placed at the ear and the other end at some part

Fig. 22-2. Mechanic's stethoscope used to locate engine noise.

of the engine, noises from that part of the engine is carried along the rod to the ear. A long screwdriver or one of the engine stethoscopes now available can be used.

When using the listening rod to locate the source of a noise, put the engine end at various places on the engine until the noise is the loudest. You can also use a piece of garden hose (about 4 feet [1.2 m] long) to locate engine noises. Hold one end of the hose to your ear, and move the other end of the hose around the engine until the noise is loudest. By determining the approximate source of the noise, you can, say, locate a broken and noisy ring in a particular cylinder or a main-bearing knock.

CAUTION: Keep away from the moving fan belt and fan when using the listening rod.

Following are various engine noises, along with tests that may be necessary to confirm a diagnosis.

1. *VALVE AND TAPPET NOISE* This is a regular clicking sound that gets louder as engine speed increases. The cause is usually excessive valve clearance or a defective hydraulic valve lifter. A feeler gauge inserted between the valve stem and lifter or rocker arm will reduce the clearance. If the noise also is reduced, then the cause is excessive clearance; the clearance should be readjusted. If inserting the feeler gauge does not reduce noise, the noise is the result of such conditions in the valve mechanism as weak springs, worn lifter faces, lifters loose in the block, rough adjustment-screw face, and rough cams. (Possibly the noise is not coming from the valves at all; see the following conditions.)

2. *DETONATION* Spark knock, or detonation, is a pinging or chattering sound that is most noticeable during acceleration or when the car is climbing a hill. Some spark knock is normal. But when it becomes excessive, it may be due to such conditions as fuel with too low an octane rating for the engine, carbon deposits in the engine which increase the compression ratio, and advanced ignition timing. (The conditions described in ⊘ 22-15 also may cause detonation.)

3. *CONNECTING-ROD NOISES* Connecting-rod noise usually has a light knocking or pounding character. The sound is most noticeable when the engine is "floating" (not accelerating or decelerating). The sound becomes more noticeable as the accelerator is eased off with the car running at medium speed. To locate connecting-rod noise, short-out the spark plugs one at a time. The noise will be considerably reduced when the cylinder that is responsible is not delivering power. A worn bearing or crankpin, a misaligned connecting rod, inadequate oil, or excessive bearing clearances can cause connecting-rod noise.

4. *PISTON-PIN NOISE* Piston-pin noise is similar to valve and tappet noise, but it has a unique metallic double knock. It is usually most audible during idle with the spark advanced. However, on some engines, the noise becomes most audible at car speeds of around 30 miles per hour [48.28 km/h]. A check can be made by running the engine at idle with the spark advanced and then shorting-out the spark plugs. Piston-pin noise will be reduced somewhat when a plug in a noisy cylinder is shorted out. Causes of this noise are a worn or loose piston pin, a worn bushing, and lack of oil.

5. *PISTON-RING NOISE* Piston-ring noise is also somewhat similar to valve and tappet noise since it is characterized by a clicking, snapping, or rattling sound. This noise, however, is most evident on acceleration. Low ring tension, broken rings, worn rings, or worn cylinder walls produce this noise. Since the noise can sometimes be confused with other engine noises, a test can be made as follows: Remove the spark plugs and add an ounce or two of heavy engine oil to each cylinder. Crank the engine for several revolutions to work the oil down past the rings. Then replace the plugs and start the engine. If the noise has been reduced, the rings probably are at fault.

6. *PISTON SLAP* Piston slap is a muffled, hollow, bell-like sound. It is caused by the piston rocking back and forth in the cylinder. If it occurs only when the engine is cold, it is not serious. When it occurs under all operating conditions, further investigation is in order. Piston slap is caused by inadequate oil, worn cylinder walls, worn pistons, collapsed piston skirts, excessive piston clearances, or misaligned connecting rods.

7. *CRANKSHAFT KNOCK* Crankshaft knock is a heavy, dull metallic knock. It is most noticeable when the engine is under a heavy load or accelerating, particularly when cold. When the noise is regular, it probably results from worn main bearings. When the noise is irregular and sharp, it is

probably due to a worn end-thrust bearing. The latter condition, when unusually bad, will cause the noise to be produced each time the clutch is released and engaged and also when accelerating.

8. *MISCELLANEOUS NOISES* Other noises result from loosely mounted accessory parts, such as the alternator, starting motor, horn, water pump, manifolds, flywheel, crankshaft pulley, and oil pan. In addition, other automotive components, such as the clutch, transmission, and differential, may develop various noises.

CHAPTER 22 CHECKUP

NOTE: Since the following is a chapter review test, you should review the chapter before taking the test.

The chapter you have just completed is probably one of the hardest chapters in the book. At the same time, it is perhaps one of the most important chapters. For, to be an engine expert, you must know what troubles an engine might have, the causes of these troubles, and how to find the causes; that is, you need to be a good troubleshooter. The fact that you have come this far in the book shows that you have made an earnest start toward becoming an engine expert. You have done well. By the time you have finished the remaining few chapters in the book, you should have learned all the background information about engines you need. The checkup that follows will help you determine how well you have understood and remembered the information you have been studying in this chapter. If any of the questions stump you, reread the pages in the chapter that will give you the answer.

Correcting Troubles Lists In each of the following lists, you will find one item that does not belong. For example, in the list "Engine will not turn over: run-down battery, worn rings, starting circuit open, engine jammed, starting motor jammed," you would see that "worn rings" does not belong since it is the only condition that will not directly cause failure of the engine to turn over. Write each list in your notebook, but do not write the item that does not belong.

1. Engine turns over slowly but does not start: run-down battery, undersized battery cables, bad connections in the starting circuit, defective starting motor, stuck cooling-system thermostat.
2. Engine will not turn over: engine jammed, starting motor or drive jammed, starting circuit open, run-down battery, excessive carbon in engine.
3. Engine runs, but one cylinder misses: defective spark plug, stuck valve, defective fuel pump, defective rings, defective distributor cap.
4. Engine turns over at normal speed but does not start: defective ignition system, defective engine, defective fan belt, defective fuel system.
5. Engine runs, but different cylinders miss: clogged exhaust, defective rings, overheated engine, defective valve action, loss of compression, excessive vacuum in intake manifold, defective ignition.
6. Engine lacks power, acceleration, or high-speed performance when cold: stuck valves, stuck manifold heat-control valve, vapor lock, stuck automatic choke.
7. Engine lacks power, acceleration, or high-speed performance when hot or cold: defective valve action, defective ignition system, defective fuel system, loss of compression, excessive carbon in the engine, run-down battery.
8. Engine lacks power, acceleration, or high-speed performance when hot only: overheating of the engine, defective choke, vapor lock, stuck manifold heat-control valve, incorrect idle-mixture adjustment.
9. Engine overheats: ignition timing late, loose fan belt, defective water pump, high altitude, clogged water jackets, defective radiator hose, defective fuel pump, defective thermostat, lack of coolant.
10. Engine stalls as it warms up: closed choke valve, stuck manifold heat-control valve, overheated engine, idling speed too low, defective head gasket.
11. Engine stalls after idling or slow-speed drive: defective fuel pump, overheating, high float level, overcharged battery.
12. Engine stalls after high-speed drive: carburetor antipercolator defective, vapor lock, run-down battery.
13. Engine backfires: spark plugs of wrong heat range, overheated engine, hot valves, carbon in the engine, vapor lock, rich or lean mixture, ignition timing off.
14. Excessive oil consumption, blue exhaust smoke: burning oil in combustion chamber, clogged air cleaner, worn rings, worn valve guides, worn bearings.
15. Excessive fuel consumption, black exhaust smoke: clogged air cleaner, rich idle, worn carburetor jets, loss of engine compression, run-down battery, faulty ignition, defective valve action.
16. Light knock or pound with engine floating: worn connecting-rod bearing, spark knock, worn crankpin, misaligned rod, lack of oil.
17. Dull, heavy knock under load or acceleration: worn main bearings, loose piston pin, worn crankshaft end-thrust bearings.
18. Light double knock during idle: worn piston pin, lack of oil, broken rings, loose piston pin, worn piston-pin bushing.
19. Hollow, muffled, bell-like sound with engine cold: worn piston, collapsed piston skirt, worn cylinder walls, worn piston-pin bearings, lack of oil.
20. Chattering or rattling during acceleration: worn rings, worn cylinder walls, low ring tension, broken rings, misaligned rods.

Completing the Sentences The sentences that follow are incomplete. After each sentence are several words or phrases, only one of which will correctly complete the sentence. Write each sentence in your notebook, selecting the proper word or phrase to complete it correctly.

1. An engine will not turn over if it has: (*a*) a defective ignition coil, (*b*) a run-down battery, (*c*) a defective fuel pump, (*d*) valves that hang open.
2. An engine will turn over slowly because of: (*a*) a defective water pump, (*b*) vapor lock, (*c*) an undersized battery cables, (*d*) excessive fuel-pump pressure.
3. Failure of an engine to start even though it turns over at normal cranking speed could be due to a: (*a*) run-down battery, (*b*) defective starting motor, (*c*) sticking engine valve, (*d*) defective ignition.
4. Missing in one cylinder is likely to result from: (*a*) a clogged exhaust, (*b*) an overheated engine, (*c*) vapor lock, (*d*) a defective spark plug.
5. Irregular missing in different cylinders may result from: (*a*) a defective starting motor, (*b*) a defective carburetor, (*c*) an open cranking circuit.
6. Loss of engine power as the engine warms up is most likely caused by: (*a*) vapor lock, (*b*) excessive rolling resistance, (*c*) the throttle valve not closing fully, (*d*) heavy oil.
7. An engine will lose power (hot or cold) if it has: (*a*) incorrect idle adjustment, (*b*) an automatic choke valve that is stuck open, (*c*) worn rings and cylinder walls.
8. An engine may stall as it warms up if the: (*a*) ignition timing is off, (*b*) choke valve sticks closed, (*c*) battery is run down, (*d*) throttle valve does not open fully.
9. An engine will overheat if the: (*a*) automatic choke sticks, (*b*) fan belt breaks, (*c*) fuel pump is defective, (*d*) battery is run down.
10. The most probable cause of an engine stalling after a period of idling or slow-speed driving is: (*a*) loss of compression, (*b*) a defective fuel pump, (*c*) sticking engine valves.
11. Stalling of an engine after a period of high-speed driving is likely to be caused by (*a*) vapor lock, (*b*) incorrect ignition timing, (*c*) worn carburetor jets.
12. Engine backfiring may result from: (*a*) spark plugs of wrong heat range, (*b*) vapor lock, (*c*) a run-down battery, (*d*) worn piston rings.
13. A smoky blue exhaust may be due to: (*a*) an excessively rich mixture, (*b*) burning of oil in combustion chamber, (*c*) a stuck choke valve, (*d*) incorrect valve adjustment.
14. A smoky black exhaust may be due to: (*a*) worn piston rings, (*b*) worn carburetor jets, (*c*) spark plugs of wrong heat range.
15. A light knock or pound with engine floating can result from worn: (*a*) main bearings, (*b*) connecting-rod bearing, (*c*) rings.
16. A light double knock during idle can result from: (*a*) piston slap, (*b*) spark knock, (*c*) incorrect ignition timing, (*d*) loose or worn piston pin.
17. A rattling or chattering sound during acceleration may be due to: (*a*) worn main bearings, (*b*) loose valve and tappet adjustment, (*c*) worn or broken rings.
18. A hollow, muffled bell-like sound, with the engine cold, is likely to be due to: (*a*) worn or collapsed pistons, (*b*) worn main bearings, (*c*) loose oil pan, (*d*) sticking engine valves.
19. A dull, heavy knock under load or acceleration is likely to be due to worn: (*a*) rings, (*b*) main bearings, (*c*) piston pins, (*d*) pistons.
20. Loss of engine compression can result from: (*a*) worn rings or cylinder walls, (*b*) loose valve and tappet adjustment, (*c*) a defective fan belt.

Troubleshooting Engine Complaints As an engine expert you will come up against complaints of loss of power, high fuel consumption, knocking, and so on, and you must know what to do to find the causes of these troubles. The following questions are stumpers that you might actually encounter in an automotive shop. In your notebook, write the procedures you would follow to find the causes of various engine troubles. Do not copy; but write the procedures in your own words. This will help you remember the procedures. If you are not quite sure of a procedure, turn back to the pages in the chapter that will give you the information.

1. You are called out to check a car in which the engine will not turn over when the starting-motor switch is closed. You turn on the headlights and try to start the car. What are the five things that might happen to the headlights? List the probable troubles for each.
2. What are the possible causes of trouble if an engine turns over slowly but will not start, and how would you locate the actual cause?
3. A car that will not start is pulled into your shop and, on testing it, you find that the engine turns over at normal speed but will not start. What are the ignition and fuel-system checks to be made?
4. An engine misses. What check can you make to locate the missing cylinder?
5. You find that one particular cylinder is missing. What further checks can you make on this cylinder, and what are the possible causes of trouble?
6. What are the possible causes of trouble if the engine miss is irregular and cannot be traced down to any one cylinder?
7. List the possible causes of trouble and how to locate them if an engine loses power as it warms up.
8. If an engine lacks power hot and cold, what are the possible causes of the trouble and how would you diagnose the trouble?
9. If an engine lacks power only when cold but seems to run normally when hot, what could the trouble be and what would you do to make sure?
10. What would you look for if an engine overheated?
11. What are three basic conditions under which an engine will stall, and what are the causes of each condition? How can you tell which is the trouble?
12. List the possible causes of engine backfiring and describe how to locate the actual cause.

13. In what three ways is engine oil lost?
14. List the causes of excessive oil consumption due to burning of oil in the combustion chamber.
15. What are some causes of excessive fuel consumption resulting from troubles in the fuel system?
16. Describe various types of car operation that will increase fuel consumption.
17. What are causes of excessive fuel consumption due to high rolling resistance?
18. List various engine noises and explain their causes.
19. List various causes of loss of compression.
20. Explain how to use the engine vacuum gauge and make a list of the various vacuum-gauge readings, along with their causes.

SUGGESTIONS FOR FURTHER STUDY

Careful observation of trouble-diagnosis procedures in the automotive shop and examination of engine components after an engine is torn down will be of great value to you in linking causes and effects of engine troubles. For instance, if you can examine the pistons, rings, and cylinder walls of an engine which has lost compression and is using too much oil, you will see why the engine has lost compression and why it has begun to use too much oil.

It will be a great help to you in the automotive shop to know the trouble-diagnosis procedures outlined in this chapter. Thus you will want to study these procedures carefully and refer to the Engine Trouble-Diagnosis Chart over and over again. At the beginning of this chapter, we suggested that you help yourself remember the trouble-diagnosis procedures by writing them down on 3- by 5-inch cards and carrying these cards around with you. Also, you may obtain a deck of the McGraw-Hill *Automotive Troubleshooting Cards,* which list troubles and their possible causes. Whenever you get a chance, as, for instance, when you are listening to music on the radio, eating lunch, or getting ready for bed, you can take out one of these cards and read it. Soon you will know the procedures thoroughly.

Be sure to discuss with expert automotive mechanics and your teacher the various methods of locating engine troubles. Ask them about their experiences in locating troubles: how often they find that loss of compression is due to worn rings, whether they find much valve-guide wear, and so on.

chapter 23

VALVE AND VALVE-MECHANISM SERVICE

This chapter is the first of four that discuss engine-servicing jobs; it covers valves, valve seats, valve guides, valve lifters, cylinder heads, and camshafts. Note that ⊘ 23-1 and 23-2, on tools and cleanliness, apply to all servicing jobs and not merely to the valve- and valve-mechanism-servicing jobs covered in this chapter. Before you start this chapter, you may wish to review Chap. 9, which discusses various aspects of valve construction and operation.

⊘ 23-1 Engine Service Usually the correction to be made for any engine trouble is obvious once the cause of trouble has been determined. Chapters 20 and 22 explain how to test and troubleshoot an engine to find causes of various engine troubles. This and following chapters describe in detail how to service valves, camshafts, crankshafts, pistons, bearings, cylinder walls, cylinder heads, and other engine components.

Special tools are required to perform many of the engine service jobs. Such special tools are described where the service jobs are covered. In addition, numerous common hand tools are needed for engine service work.

The procedures discussed in the four engine-service chapters that follow are aimed at correcting specific engine troubles. There is another method of engine service called *engine rebuilding*. Some companies specialize in this work. They set up special engine disassembly and rebuilding lines. They bring in old, worn engines, disassemble them completely, and repair or replace all worn parts. They completely rebuild the engine, using only the old parts that are still in good condition. A good rebuilder can turn out rebuilt engines that many people consider to be almost "as good as new."

Another method of servicing engines, where there are several defects in the cylinder block, is to purchase from the manufacturer a "short block." This includes the cylinder block with all related parts, such as pistons, piston pins, rings, connecting rods and bearings, crankshaft, and main bearings. (See Fig. 23-1.)

Many engine-service jobs can be done with the engine in the vehicle. For example, valves can be adjusted and piston rings can be replaced without removing the engine from the car. For other jobs, the engine must be removed. Removing and replacing engines and replacing engine mounts are described in ⊘ 25-1 and 25-2.

In the discussions of the various engine service jobs, you will notice that the time required to do each job is usually given. These time estimates have been taken from the manufacturer's flat-rate shop manuals and are included to give you some idea of the size of the job.

NOTE: Servicing of water pumps, oil pumps, and ignition distributors is covered in other books in the McGraw-Hill Automotive Technology Series: Water-pump and oil-pump service is described in *Automotive Fuel, Lubricating, and Cooling Systems*, and ignition-distributor service is described in *Automotive Electrical Equipment*.

⊘ 23-2 Cleanliness The major enemy of good engine-service work is dirt. A trace of dirt or abrasive in a bearing, on cylinder walls, or in other working parts in the engine can ruin an otherwise good service job. When you consider that engine parts are precision-machined to tolerances of less than one-thousandth of an inch, you can understand that pieces of dirt or abrasive only a thousandth of an inch [0.0254 mm] in diameter can cause real damage. (Such fine pieces of dirt or abrasive are so small that normally you cannot see or feel them.)

Such dirt or abrasive can cause rapid wear and quick failure of engine parts. For instance, if a precision-insert main bearing is installed with dirt under it, the bearing shell will not fit snugly into the cylinder block or cap counterbore. The bearing will distort, and high spots may develop. This condition would probably cause quick bearing failure, possibly in less than 1,000 miles [1,609.3 km]. Similarly, if abrasive is left on cylinder walls after a honing job, pistons, rings, and other engine parts

Fig. 23-1. Short block. (*Ford Motor Company*)

may wear rapidly and fail in a few thousand miles. You must, therefore, be very careful to remove all dirt and abrasive produced by any service job. Keep all engine parts clean, and be doubly sure that they are clean as they go back into the engine.

Before any major engine-service job, the block should be cleaned to remove dirt and grease so that they will not get into the engine when it is opened. If steam cleaning or similar process is used, electrical parts should be covered or removed so that moisture or cleaner will not get into them.

While servicing an engine, be careful to keep all parts clean. When reassembling an engine or installing engine parts, remember that dirty parts are almost sure to cause trouble in the engine. Parts should be cleaned as explained in the sections dealing with servicing the different engine components. Several cleaning methods and materials are in use. Some methods make use of hot water mixed with solvent in which the parts are soaked. Other methods include steam jets and vapor degreasers.

CAUTION: As soon as a part is cleaned and dried, a light coating of oil should be applied to bright-finished surfaces so that rust will not form. Be sure that parts are thoroughly clean and thoroughly dry. If you use an air hose to dry parts, you should wear goggles. Remember that the airstream drives dirt particles at high velocity and that these particles could do serious damage to your eyes. Be careful where you point the air hose. Don't point it at your fellow workers. Remember that they have eyes, too!

⊘ 23-3 Valve Troubles The engine valves must open and close with definite timing in relation to the piston positions (⊘ 9-22). They must seat tightly against the valve seats and open and close promptly without lagging. The clearance between the valve stems and valve guides must be correct. Failure of the valves to meet any of these requirements means valve and engine trouble.

For example, let us see what excessive clearance between the valve stem and the valve guide might lead to. On every intake stroke, this excessive clearance will permit oil to be pulled past the intake-valve stem and opened valve into the combustion chamber, where it will burn (Fig. 23-2). The carbon from the burning oil builds up in the combustion chamber. There, it can cause excessive compression, preignition, clogged piston rings, and fouled spark plugs. The oil will tend to deposit on the valve and carbonize, thereby restricting valve action.

Furthermore, excessive clearance may allow the valve to cock and not seat properly, and it will also probably reduce valve-stem cooling. Either of these conditions will cause the valve to overheat and burn (this applies particularly to the exhaust valve). Naturally, burned valves will leak from poor seating and cause major engine power losses. And once a valve starts to burn, it burns away rapidly, so that complete valve and engine failure soon follow. As you can see, wear of an apparently minor item like the valve guide can lead to many engine troubles.

We could trace the results of other faulty valve conditions in a like manner. What we want to emphasize is that valves must work and be serviced properly if engine troubles are to be avoided.

⊘ 23-4 The Valve Trouble-Diagnosis Chart The Valve Trouble-Diagnosis Chart (p. 314) lists various valve troubles, their possible causes, and checks or corrections. Following sections supply fuller explanations of these troubles. The latter part of the chapter discusses valve service. When a possible cause or correction involves the fuel, cooling, or lubricating system, reference is made to *Automotive Fuel, Lubricating, and Cooling Systems* (another book in the McGraw-Hill Automotive Technology Series).

NOTE: The complaints and possible causes are not listed in the chart in the order of frequency of occurrence. That is, item 1 (or item a under "Possible Cause") does not necessarily occur more frequently than item 2 (or item b).

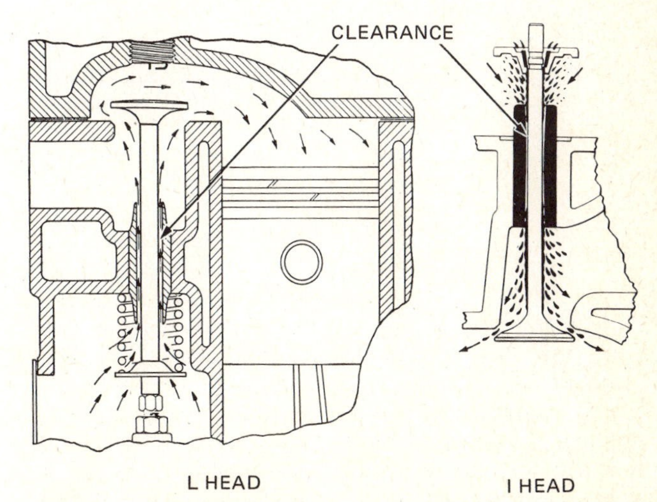

Fig. 23-2. Oil will be pulled through any excessive clearance between a worn intake-valve guide and the valve stem on the intake stroke, as shown by arrows. (*Federal-Mogul Corporation*)

VALVE TROUBLE-DIAGNOSIS CHART

(See ⊘ 23-5 to 23-10 for detailed explanations of trouble causes and corrections listed below.)

COMPLAINT	POSSIBLE CAUSE	CHECK OR CORRECTION
1. Valve sticking (⊘ 23-5)	a. Deposits on valve stem	See item 6
	b. Worn valve guide	Replace
	c. Warped valve stem	Replace valve
	d. Insufficient oil	Service lubricating system; add oil
	e. Cold-engine operation	Valves free up as engine warms up
	f. Overheating valves	See item 2
2. Valve burning (⊘ 23-6)	a. Valve sticking	See item 1
	b. Distorted valve seat	Check cooling system; tighten cylinder-head bolts
	c. Valve-tappet clearance too small	Readjust
	d. Spring cocked or weak	Replace
	e. Overheated engine	Check cooling system (see also ⊘ 22-12)*
	f. Lean air-fuel mixture	Service fuel system*
	g. Preignition	Clean carbon from engine; use cooler plugs
	h. Detonation	Adjust ignition timing (⊘ 20-9); use higher-octane fuel
	i. Valve-seat leakage	Use an interference angle
	j. Overloaded engine	Reduce load or try heavy-duty valves
	k. Valve-stem stretching from strong spring or overheated engine (see ⊘ 22-12 for causes of overheating)	Use weaker spring; eliminate overheating
3. Valve breakage (⊘ 23-7)	a. Valve overheating	See item 2
	b. Detonation	Adjust ignition timing (⊘ 20-9); use higher-octane fuel; clean carbon from engine
	c. Excessive tappet clearance	Readjust
	d. Seat eccentric to stem	Service
	e. Cocked spring or retainer	Service
	f. Scratches on stem from improper cleaning	Avoid scratching stem when cleaning valves
4. Valve-face wear (⊘ 23-8)	a. Excessive tappet clearance	Readjust
	b. Dirt on face	Check air cleaner
	c. See also causes listed under item 2	
5. Valve-seat recession (⊘ 23-9)	a. Valve face cuts valve seat away	Use coated valves and valve-seat inserts
6. Valve deposits (⊘ 23-10)	a. Gum in fuel (intake)	Use proper fuel
	b. Carbon from rich mixture (intake)	Service fuel system
	c. Worn valve guides	Replace
	d. Carbon from poor combustion (exhaust)	Service fuel, ignition system, or engine as necessary
	e. Dirty or wrong oil	Service lubricating system; replace oil

* See *Automotive Fuel, Lubricating, and Cooling Systems*, another book in the McGraw-Hill Automotive Technology Series.

⊘ 23-5 Valve Sticking Valves will stick from gum or carbon deposits on the valve stem (see Fig. 23-3 and ⊘ 23-10). Worn valve guides, which pass excessive amounts of oil, speed up the formation of deposits since the oil carbonizes on the hot valve stem. If the valve stem warps, it will stick in the valve guide. Warpage could result from overheating (⊘ 23-6), an eccentric seat (which throws pressure on one side of the valve face), or a cocked spring or retainer (which puts bending pressure on the stem). Of course, insufficient oil would also cause valve sticking. Also, sometimes valves will stick when the engine is cold but work themselves free and function normally as the engine warms up.

NOTE: When valves and piston rings have become so clogged with deposits that they no longer operate properly, it is usually necessary to overhaul the engine. However, some authorities suggest the use of special compounds in the oil and fuel which help in freeing valves and rings. One of these compounds comes in a pressurized can and is sprayed into the running engine through the carburetor (air cleaner off). When parts are not too badly worn and the major trouble seems to be from deposits, use of these compounds often postpones engine overhaul, at least for a while.

CAUTION: Before using a chemical additive in an engine equipped with a catalytic converter, make sure that the additive is harmless to the catalyst. To find this information, carefully read the label on the can and the section in the car manufacturer's service manual on the catalytic converter.

⊘ 23-6 Valve Overheating and Burning Valve overheating and burning is usually an exhaust-valve problem (Figs. 23-4 to 23-6). Any condition that causes the valve to stick so that it does not close tightly will cause valve burning. Not only does the poor seat prevent normal valve cooling through the valve seat, but it also allows hot gases to blow by, further heating the valve. The valve is cooled through both the valve seat and the valve guide (⊘ 9-5), and so poor seating or a worn guide can cause overheating and burning. Also, if the water jackets or distributing tubes in the cooling system are clogged, local hot spots may develop around valve seats. These hot spots may cause seat distortion, which then prevents normal seating and thus permits blow-by and valve burning. Valve-seat distortion can also result from improper tightening of the cylinder-head bolt. Other conditions that prevent normal seating include a weak or cocked valve spring and insufficient valve-tappet clearance. If the tappet clearance is too closely adjusted, the valve may be held open.

On engines equipped with valve rotators, check the valve-stem tip of valves that have burned to see whether or not the rotators are working. Figure 23-7 shows tip patterns indicating proper and improper rotator action. If the tip shows no rotation or a partial rotation pattern, the rotator should be replaced.

A lean air-fuel mixture may cause exhaust-valve burning since some combustion may still be going on (the lean mixture burns slowly) when the valve opens. If a lean air-fuel mixture is the cause, the fuel system should be serviced.

Preignition and detonation, both of which produce excessively high combustion pressures and temperatures, have an adverse effect on valves as well as on other engine parts. They can be eliminated by cleaning out carbon, retiming the ignition (⊘ 20-9), or using higher-octane fuel.

In some persistent cases of valve-seat leakage (especially where deposits on the valve seat and face prevent adequate sealing), the use of a so-called interference angle has proved helpful. The valve is faced at an angle $\frac{1}{2}$ to 2 degrees flatter than the valve-seat angle (Fig. 23-8). This gives greater pressure at the lower edge of the valve seat, which tends to cut through any deposits that have formed and thereby establish a good seal.

Figure 23-8 illustrates one manufacturer's recommendation for the valve and seat angles to get interference. Note that this manufacturer does not recommend interference on stellite-faced exhaust valves and induction-hardened exhaust-valve seats. These surfaces are so hard that no appreciable improvement in seating would be obtained by in-

Fig. 23-3. Gummed intake valve. Note the deposits under the valve head. (*Clayton Manufacturing Company*)

Fig. 23-4. Valve burning due to seat distortion. The valve fails to seat in one area, and hot exhaust-gas leakage in this area burns the valve. (*TRW Inc.*)

Fig. 23-5. Valve burning due to failure of the valve to seat fully. Note that the valve is uniformly burned all the way around its face. (*TRW Inc.*)

Fig. 23-6. Valve burning due to guttering. This is caused by accumulations of deposits on the valve face and seat. Parts of the deposits break off to form a path through which exhaust gas can pass when the valve is closed. This soon burns channels in the valve face, as shown. (*TRW Inc.*)

terference. (See "Note" at the end of ⊘ 9-8 for the interference angles recommended by various manufacturers.)

Any condition that causes the engine to labor hard or overheat will also overheat the valves. The causes of engine overheating are discussed in ⊘ 22-12. If the engine must be operated under heavy load and this causes valve trouble, heavy-duty valves should be installed.

In some cases valve stems have been found to stretch because of a combination of heavy springs and overheating. Lighter springs should be used and the overheating eliminated (⊘ 22-12).

⊘ 23-7 Valve Breakage Any condition that causes the valve to overheat (⊘ 23-6) or to be subjected to heavy pounding, as from excessive detonation or tappet clearance, may cause valves to break. Excessive tappet clearance permits heavy impact seating. If the seat is eccentric to the stem or if the valve spring or retainer is cocked, then the valve will be subjected to side movement or pressure every time it seats. Ultimately, this may cause it to fatigue and break (Fig. 23-9). If the stem has been scratched during cleaning, the scratch may serve as a starting point for a crack and a break in the stem.

⊘ 23-8 Valve-Face Wear In addition to the conditions discussed in ⊘ 23-6, excessive tappet clearance or dirt on the valve face or seat can cause valve-face wear. Excessive tappet clearance causes heavy impact seating that is wearing on the valve and may cause valve breakage (⊘ 23-7). Dirt may cause valve-face wear if the engine operates in dusty conditions or if the carburetor air cleaner is not functioning properly. The dirt enters the engine with the air-fuel mixture, and some of it deposits on the valve seat. The dirt will also cause bearing, cylinder-wall, and piston and ring wear.

⊘ 23-9 Valve-Seat Recession Valve-seat recession is caused by the wearing away of the valve seat. This is produced by the action of the valve face which, under certain conditions, can cut the seating surface of the valve seat. Valve-seat recession has become more of a problem in recent years because lead had been removed from gasoline. Lead additives in the gasoline form a lubricant between the valve face and the valve seat. This lubricant prevents iron particles that flake off the valve seat from sticking on the valve face. However, without this lead coating on the valve face and seat, particles of iron flake off the valve seat and tend to stick on the valve face. Gradually these particles embed in the valve face and build up into tiny bumps, or warts. This turns the valve face into a cutting surface. The valve seat is gradually cut away. The resulting valve-seat recession causes lash loss, or a decrease in valve-tappet clearance. That is, as the valve gradually cuts into the valve seat, clearance in the valve train is reduced. In an engine using mechanical valve lifters, the result can be a complete loss of clearance so that the valve can no longer close completely. The result will be valve and valve-seat burning.

To prevent valve-seat recession in engines run on lead-free gasoline, the valves are given a very thin coating of aluminum (Fig. 23-10) or other metal. This coating, less than 0.002 inch [0.051 mm] in thickness, prevents the iron particles from adhering to the valve face and thus prevents valve-seat recession.

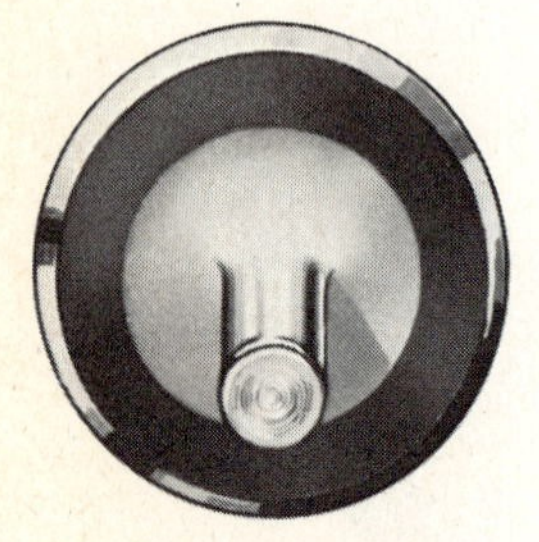

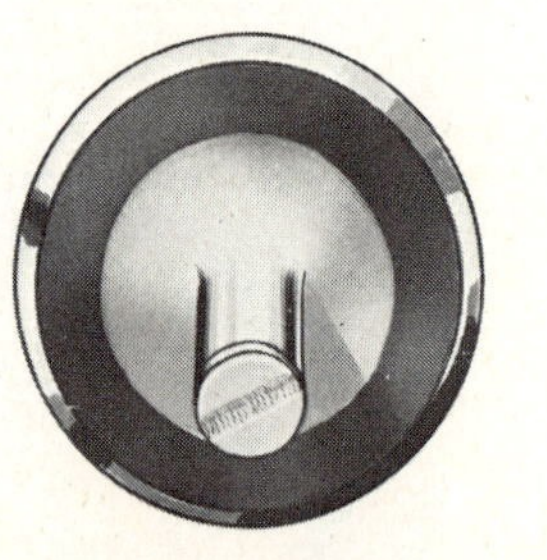

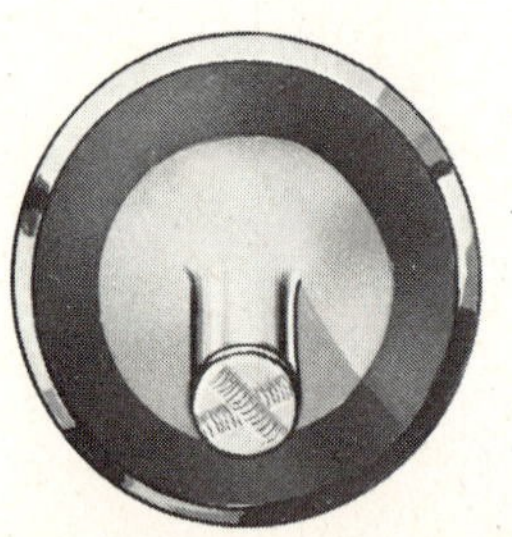

Fig. 23-7. Valve-stem-wear patterns on valves with valve rotators. (*Oldsmobile Division of General Motors Corporation*)

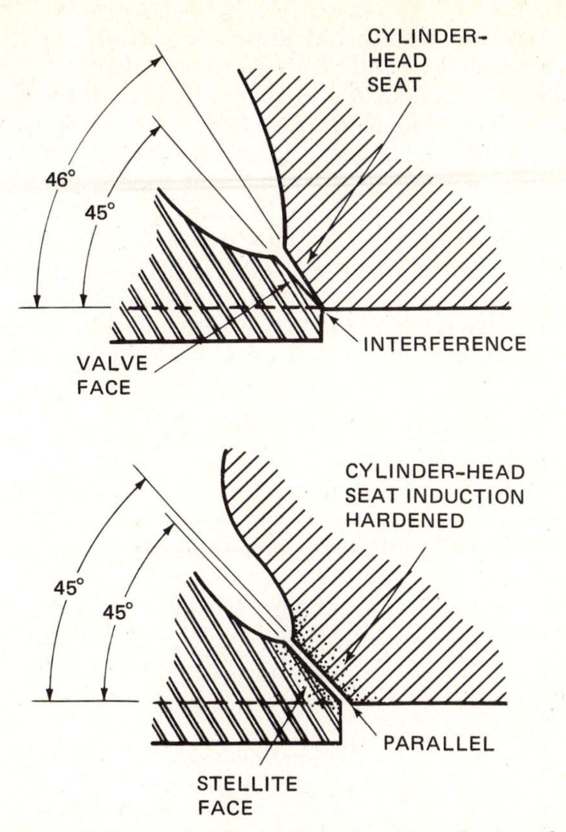

Fig. 23-8. Valve and valve-seat angles. *Top,* the interference angle recommended for many intake and exhaust valves and seats. *Bottom,* the parallel faces recommended for stellite-faced exhaust valves and induction-hardened exhaust-valve seats. (*Chevrolet Motor Division of General Motors Corporation*)

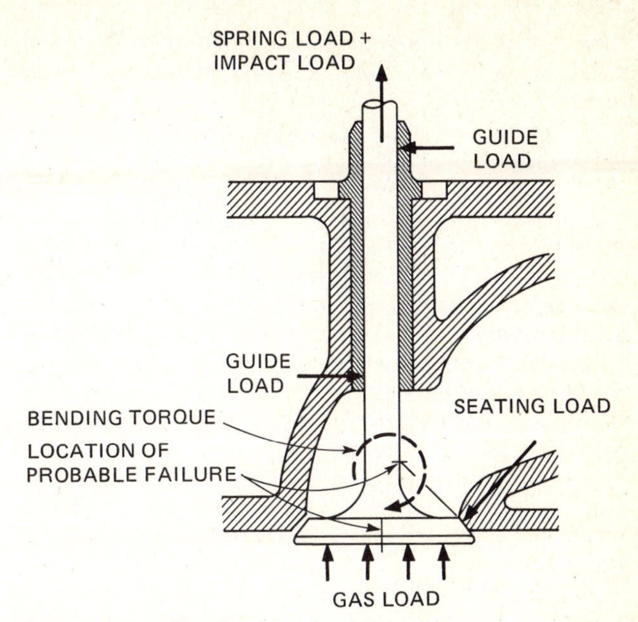

Fig. 23-9. Bending load on the valve stem can cause stem breakage. (*TRW Inc.*)

Note that the coating gives the valve face a dull, almost rough, appearance. The natural tendency for the automotive mechanic would be to reface the valve. However, *this must not be done.* Coated valves should not be refaced or lapped. Doing so would remove the coating and deny the valve seat the protection produced by the coating. The result could be very short valve and valve seat life. (Valve servicing is discussed in ⊘ 23-28.)

⊘ 23-10 Valve Deposits If the fuel has excessive amounts of gum in it, some of this gum may deposit on the intake valve as the air-fuel mixture passes the valve on the way to the engine cylinder. Carbon deposits may form from an excessively rich mixture or from oil passing a worn valve guide (intake valve). Improper combustion, due to a rich mixture, a defective ignition system, loss of compression in the engine, a cold engine, and so on, will result in carbon deposits on the exhaust valves. Dirty or improper oil will cause deposits to form on the valves.

⊘ 23-11 Valve Service We have already described methods of checking engine and valve actions with compression and leakage testers (⊘ 20-4 to 20-6) and have listed and discussed various valve troubles and their corrections. The remainder of this chapter describes valve-servicing procedures in detail.

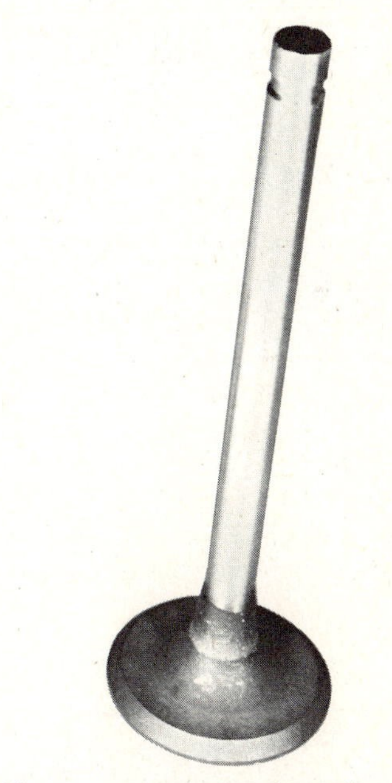

Fig. 23-10. Aluminum-coated valve. The entire head and underhood area are coated. (*TRW Inc.*)

In servicing valves, a number of components must be considered. These include valves, valve seats and guides, valve springs and retainers, rocker-arm mechanisms (in overhead-valve engines), valve tappets, camshaft, camshaft drive, and bearings.

The service jobs on valves include adjusting valve-tappet clearances (also called *adjusting valve lash*), grinding valves and valve seats, installing new seat inserts (on engines so equipped), cleaning or replacing valve guides, removing and checking the camshaft, servicing camshaft bearings, and timing the valves. A complete valve-service job, including grinding valves and seats, checking springs, cleaning guides, and tuning the engine, requires about 5 hours for a six-cylinder engine. Replacing seat inserts requires an additional ½ hour each. Replacing the camshaft requires about 8 hours; approximately 4 hours more are required to replace camshaft bearings. These operations will now be considered.

⊘ 23-12 Adjusting Valve-Lifter Clearance The procedure for checking and adjusting valve-tappet, or valve-lifter, clearance (or adjusting valve lash) varies with the type and model of engines. Some engines with hydraulic valve lifters normally require no clearance adjustment. Others require checking and adjustment whenever valve-service work is performed. The following procedures are typical.

⊘ 23-13 L-Head Engine with Mechanical Valve Lifters Some specifications call for checking the clearance with the engine cold. According to other specifications, the engine should be warmed up and idling. Remove the valve-cover plates. Use a feeler gauge to check the clearance between the valve stem and the adjusting screw in the valve lifter (Fig. 23-11). A two-step "go no-go" feeler gauge of the specified thicknesses can be used. Adjustment is correct when the "go" step fits the clearance but the "no-go" step does not.

If the clearance is not correct, the adjusting screw must be turned in or out as necessary to correct it. Some tappet-adjusting screws are self-locking; others have a locking nut. On the latter type, the locking nut must be loosened. This requires two wrenches—one to hold the screw, the other to turn the nut. On both types of tappet-adjusting screws, one wrench must be used to hold the valve lifter while a second wrench is used to turn the adjusting screw. Adjustment is correct when the feeler gauge can be removed between the screw and valve stem with some drag when the valve is closed. When a locking nut is used, it should be tightened after the adjustment is made and the clearance should be checked again. After the adjustment is completed, replace the cover plates, using new gaskets.

⊘ 23-14 I-Head Engine with Mechanical Valve Lifters Most specifications call for making the check with the engine hot and not running. First, remove the valve cover. Measure the clearance between the valve stem and rocker arm, as shown in Fig. 23-12. The clearance is measured with the valve lifter on the base circle of the cam. Turn the crankshaft by bumping the engine with the starting motor until the base circle of the cam is under the valve lifter.

NOTE: Refer to the engine manufacturer's shop manual for details of the procedure by which one crankshaft position will allow you to check half or a third of the valves. When you use this procedure, you need position the crankshaft and camshaft only two or three times to adjust all valves.

There are two kinds of rocker arms: One is shaft-mounted, and the other is ball-stud-mounted. The shaft-mounted type (Fig. 23-12) usually has an adjustment screw. This screw is normally self-locking and does not require a locking nut. Use a box-wrench to turn the adjustment screw and adjust the clearance to specifications. Do not use an open-end wrench. This could damage the screw head.

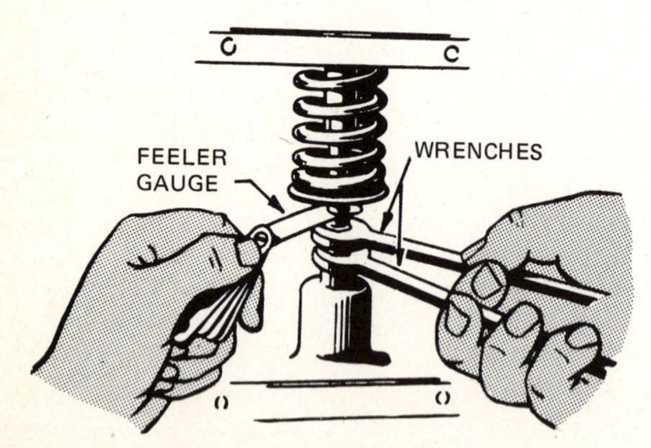

Fig. 23-11. Adjusting valve-tappet clearance on an L-head engine.

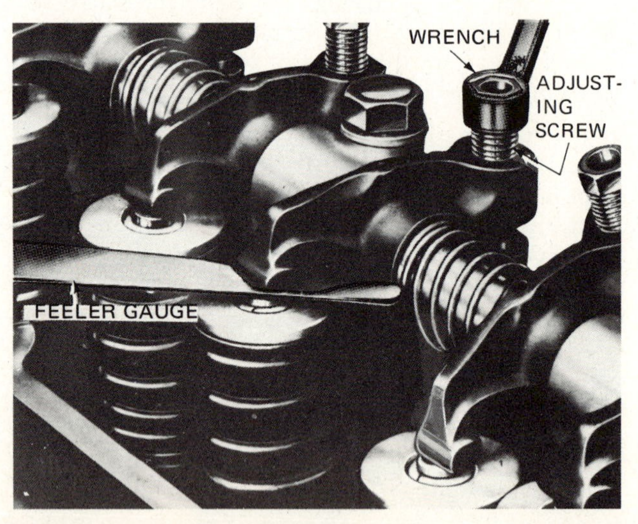

Fig. 23-12. Adjusting valve-tappet clearance on an I-head engine. (*Ford Motor Company*)

On the ball-stud-mounted rocker arms (Fig. 23-13), turn the self-locking rocker-arm-stud nut to make the adjustment. Turning the nut down reduces clearance.

⊘ 23-15 Free-Type Valve Rotators Free-type valve rotators are checked in the same way as the mechanical-lifter type. The clearance is checked between the tip cup on the valve stem and the adjusting screw in the valve lifter.

⊘ 23-16 I-Head Engines with Hydraulic Valve Lifters On some engines with hydraulic valve lifters, no adjustment is provided in the valve train. In normal service, no adjustment is necessary. The hydraulic valve lifter takes care of any small changes in the valve-train length. However, adjustment may be needed if valves and valve seats are ground. Unusual and severe wear of the pushrod ends, rocker arm, or valve stem may also require adjustment. Then some correction may be required to reestablish the correct valve-train length. Typical checking and correcting procedures follow.

⊘ 23-17 Ford Engines with Hydraulic Valve Lifters Ford engines use two types of rocker arms: shaft-mounted (Fig. 23-14) and ball-stud-mounted (Fig. 23-15). On both types the clearance in the valve train is checked with the valve lifter bled down so that the valve-lifter plunger is bottomed. First, the crankshaft must be turned so that the lifter is on the base circle or low part of the cam (rather than on the lobe). This is done by setting the piston in the No. 1 cylinder at top dead center (TDC) at the end of the compression stroke. Then check both valves in the No. 1 cylinder. The crankshaft can then be rotated as necessary to put other lifters on the base circles of their cams so that they can be checked.

To make the check, a special tool is used to apply slow pressure on the rocker arm (Figs. 23-14 and 23-15). The slow pressure gradually forces oil out of the valve lifter so that the plunger bottoms. Then, the clearance gauge is used to check the clearance between the valve stem and rocker arm. If the clearance is too small, install a shorter pushrod. If the clearance is excessive, install a longer pushrod. The clearance might become too small if valves and seats have been ground. The clearance might become excessive as a result of wear of the valve-train parts. This includes wear of the pushrod ends, valve stem, and rocker arm.

⊘ 23-18 Plymouth Engines with Hydraulic Valve Lifters The procedure for setting Plymouth valves is typical of the engines manufactured by Chrysler Corporation. It is necessary only when valves and valve seats have been ground. When this happens, the increased height of the valve stem above the cylinder head should be checked (see Fig. 23-16). With the valve seated, place the special gauge tool over the valve stem. If the height is excessive, the end of the valve stem must be ground off to reduce the height to within limits. The hydraulic valve-lifter plunger will then work near its center position rather than near the bottom, as it would with an excessively high valve stem.

Fig. 23-13. Adjusting valve-tappet clearance on an engine with rocker arms independently mounted on ball studs. Backing the stud nut out increases clearance. (*Chevrolet Motor Division of General Motors Corporation*)

⊘ 23-19 Chevrolet Engines with Hydraulic Valve Lifters The procedure for the Chevrolet engines is

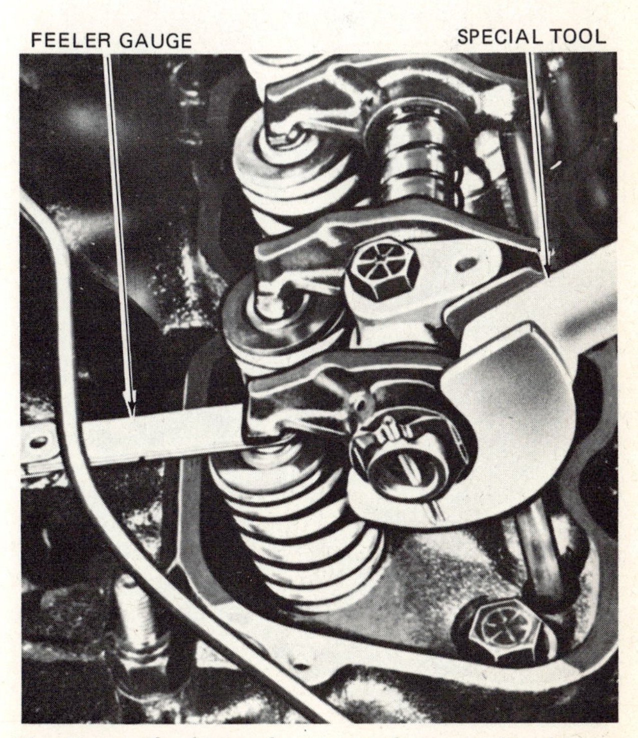

Fig. 23-14. Checking valve-tappet clearance. (*Ford Motor Company*)

Fig. 23-15. Checking valve-train clearance on a ball stud mounted on a rocker arm. The hydraulic valve lifter has been bled down with pressure from a special tool. (*Ford Motor Company*)

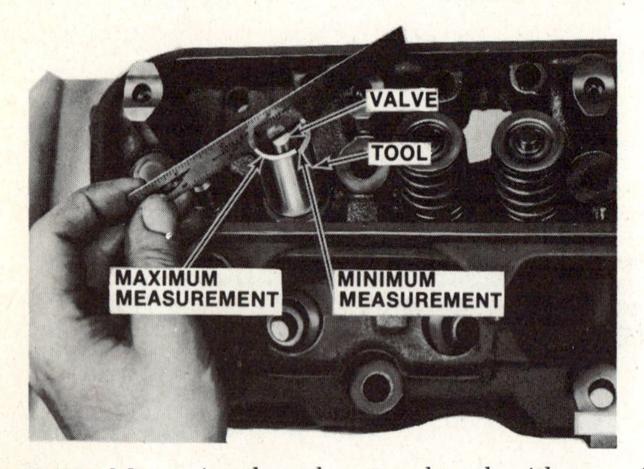

Fig. 23-16. Measuring the valve-stem length with a special tool after the valve is installed in the head. (*Chrysler Corporation*)

typical of General Motors engines using the ball-pivot type of rocker arm (Fig. 23-17). With the valve lifter on the base circle of the cam, back off the adjustment nut until the pushrod is loose. Then slowly turn the adjustment nut down. At the same time, rotate the pushrod with your fingers until the pushrod is tight. That is, until you cannot easily rotate the pushrod. Then turn the adjustment nut down one additional full turn. This places the plunger of the valve lifter in its center position.

⊘ 23-20 Overhead-Camshaft Engines Overhead-camshaft (OHC) engines have several arrangements for carrying the cam action to the valve stems. In some engines, cam action is carried directly to the valve stem through a cap, called the *valve tappet.* This cap fits over the valve stem and spring. In other engines, the cam action is carried through a rocker arm. We shall look at both arrangements. (Checks and adjustments are made with the engine cold.)

1. *CHEVROLET VEGA* The Chevrolet Vega engine is shown in Figs. 9-28 and 9-29. Figure 23-18 is a side-sectional view of the engine. Figure 23-19 shows the valve-train parts. Adjustment of valve clearance is made by turning the adjustment screw located in the valve tappet. The adjustment screw has a flat on one side. Therefore, adjustment must be made by turning the screw *full turns only.*

Turn the camshaft so the valve tappet is on the base circle. Then measure the clearance between the cam and the valve tappet with a feeler gauge (Fig. 23-20). Use the special tool, as shown, to turn the adjustment screw. The screw must be turned complete revolutions so that the flat on the screw ends up directly above the valve stem. Turning the screw in, or clockwise, decreases clearance. Each full turn of the screw changes the clearance 0.003 inch [0.076 mm].

2. *CHEVROLET LUV ENGINE* The Chevrolet LUV engine has rocker arms which are held in place by springs, as shown in Fig. 23-21. The rocker arms have dome-shaped ends which fit over ball studs in the cylinder head. The valve ends of the rocker arms fit into a depression in the valve-spring retainer and rest on the valve stems. To check the valve clearance, measure the clearance between the cam surface of the rocker arm and the base circle of the cam (Fig. 23-22). Use a flat feeler gauge. Adjust with a Phillips-head screwdriver inserted through the hole in the rocker arm dome, as shown.

3. *FORD 2,000-CC FOUR-CYLINDER ENGINE* This Ford engine has rocker arms which float between

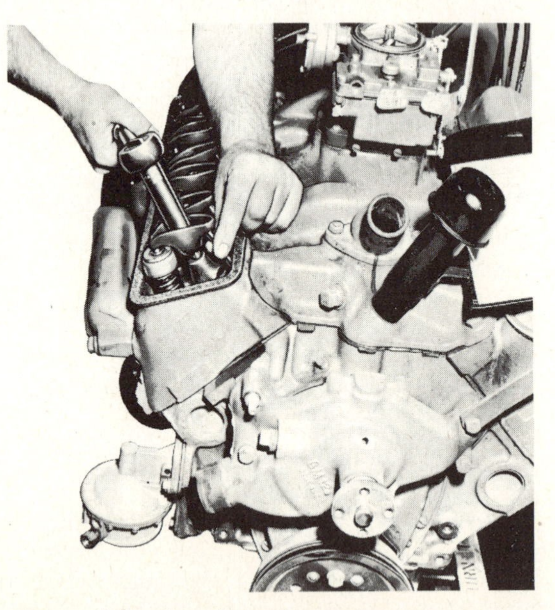

Fig. 23-17. Adjusting the valve rocker-arm stud nut to properly position the plunger of the hydraulic valve lifter. (*Chevrolet Motor Division of General Motors Corporation*)

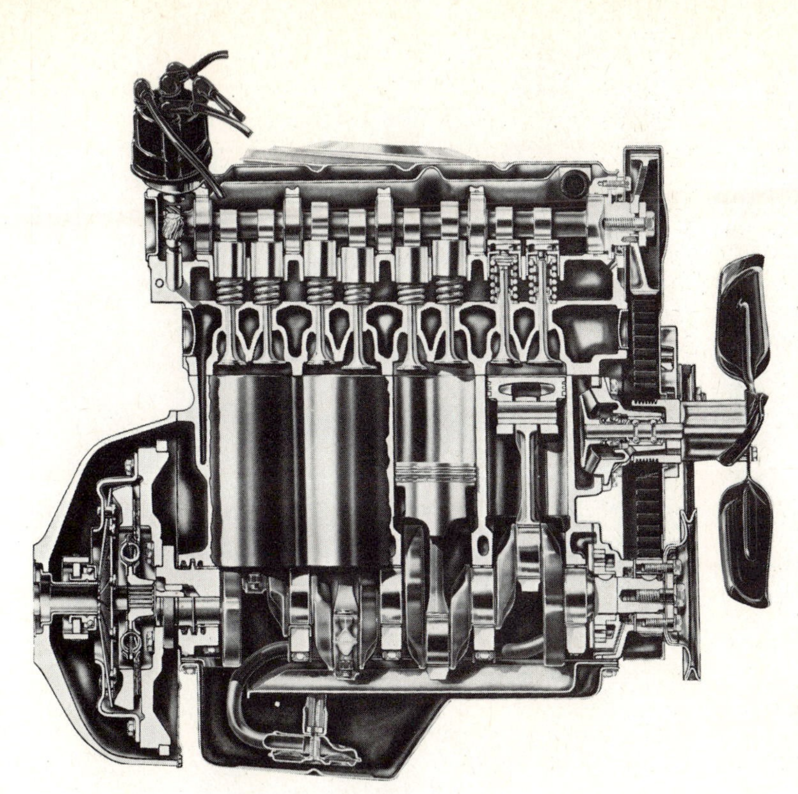

Fig. 23-18. Side sectional view showing the four-cylinder, overhead-camshaft Vega engine. (*Chevrolet Motor Division of General Motors Corporation*)

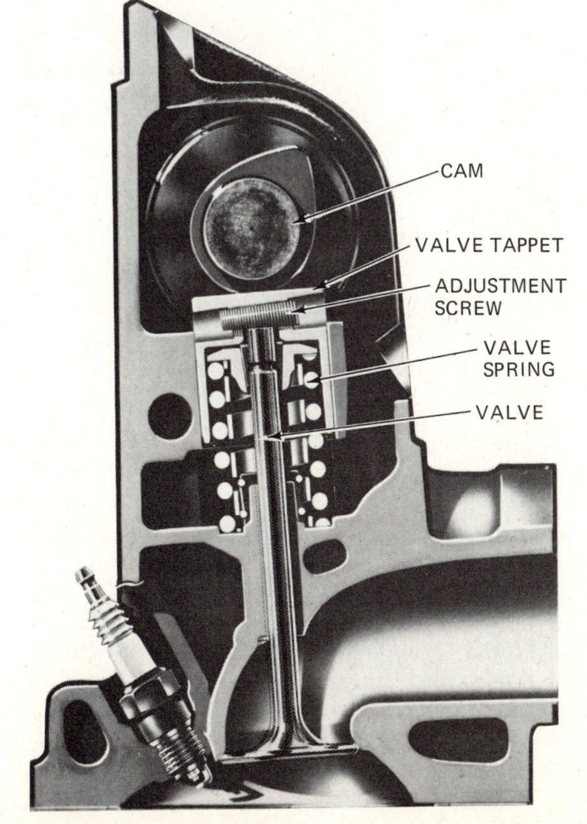

Fig. 23-19. Valve train for the Vega engine. (*Chevrolet Motor Division of General Motors Corporation*)

a stationary stud on one side and the valve stem on the other. The center of the rocker arm rests on the cam. (This is similar to the arrangement in the Chevrolet LUV engine.) Figure 23-23 shows the use of a feeler gauge to check the clearance between the base circle of the cam and the rocker arm. Adjustment is made by loosening the locknut. Use a 15-mm (0.59 inch) open-end wrench to turn the adjustment screw in or out. Turning the screw in increases the clearance. Tighten the locknut securely after the adjustment and recheck the clearance.

4. *PONTIAC OHC ENGINE* Figure 23-24 shows the valve-train arrangement in this engine. The rocker arm floats between the end of the valve stem and an automatic valve-lash adjuster. The automatic valve-lash adjuster is a special type of hydraulic valve lifter. It automatically takes up any clearance between the cam and the rocker arm. Normally, no valve adjustment is necessary on this engine.

⊘ 23-21 The Complete Valve Job A complete valve job requires the following steps. The details of valve and valve-seat service are described in ⊘ 23-22 to 23-32.

1. Drain the cooling system and disconnect the upper radiator hose from engine.
2. Remove the air cleaner and disconnect the throttle linkage, fuel line, and air and vacuum hoses from the carburetor.
3. Remove or set aside the necessary lines and hoses to get at the cylinder head.
4. Disconnect the spark-plug wires and temperature-sending unit.

Fig. 23-20. Adjusting valve-tappet clearance on the Vega engine. (*Chevrolet Motor Division of General Motors Corporation*)

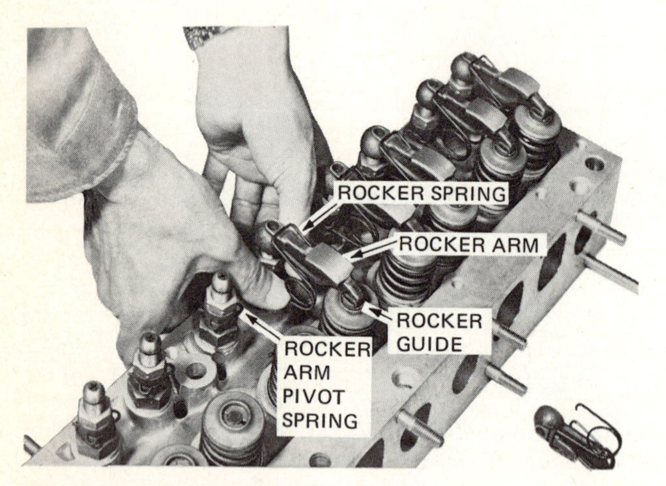

Fig. 23-21. Rocker arm and spring arrangement in the LUV engine. (*Chevrolet Motor Division of General Motors Corporation*)

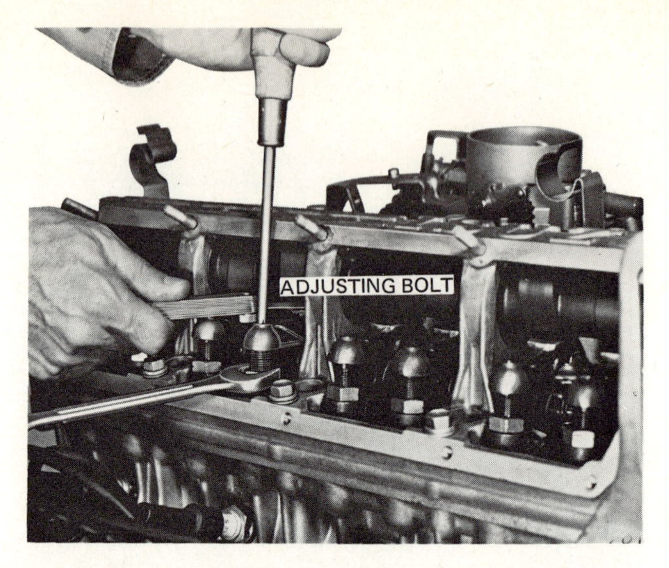

Fig. 23-22. Adjusting valve-tappet clearance on the LUV engine. (*Chevrolet Motor Division of General Motors Corporation*)

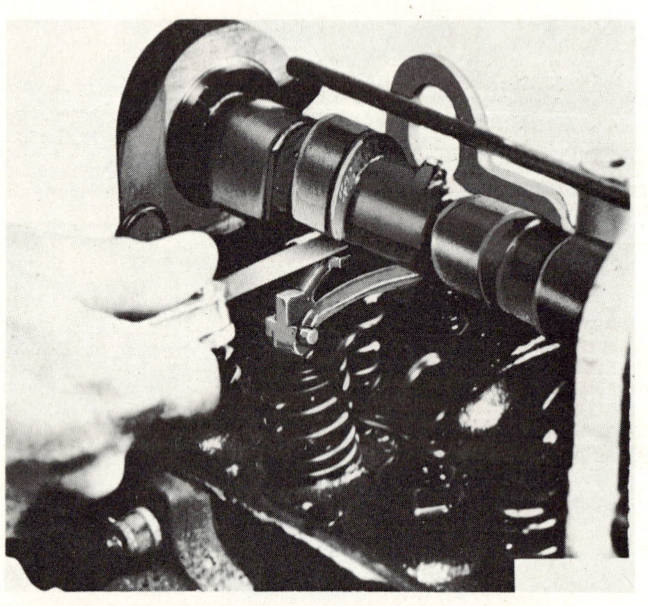

Fig. 23-23. Checking valve-tappet clearance on the Ford 2,000-cc engine. (*Ford Motor Company*)

5. Remove the crankcase ventilating system and, on air-injection systems, disconnect the air hose at the check valve. Then remove the air-supply-tube assembly.
6. On V-8 engines, remove the carburetor and intake manifold. (On many in-line engines, it is not necessary to remove the manifolds.)
7. Remove the rocker-arm cover or covers.
8. On engines with ball-stud-supported rocker arms, remove the rocker arms and pushrods at this time. If they are left on, the nuts should be loosened so that the rocker arms can be moved aside and the pushrods removed. Pushrods should be placed in a rack in proper order. Then, they can be put back into the same spot from which they were removed.
9. On engines with rocker arms supported on shafts, remove the shaft assembly or assemblies (⊘ 23-24) and then remove the pushrods in order.
10. Remove the head bolts. Take the head off the engine.
11. Remove the valves and springs from the head (keeping them in proper order so that they can be put back into their proper positions).

CAUTION: If a valve-stem end has mushroomed, the valve cannot be pulled out by hand. The mushroom must be removed, as explained in ⊘ 23-26. Otherwise, if you try to pull the valve, you can damage or break the valve guide.

12. Check valves and valve seats. Clean the valve heads and stems on a wire wheel (Fig. 23-44). Grind

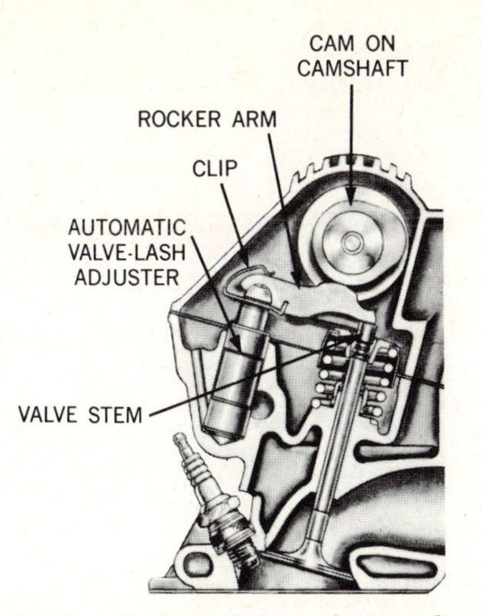

Fig. 23-24. Sectional view of the valve mechanism in an overhead-camshaft engine. (*Pontiac Motor Division of General Motors Corporation*)

the valve seats and reface valves as necessary. Check valve seating. Reface and chamfer valve-stem ends if necessary.

CAUTION: If you are installing new valves of the coated type, do not reface them. Refacing or lapping coated valves removes the protective coating and greatly shortens valve and seat life.

13. Check rocker arms for wear. Service or replace them as necessary.
14. Check valve guides for wear. Clean, replace, or knurl and ream for same-size valve stem if necessary. Or ream for a larger-diameter valve stem.
15. Replace valves and springs in head.
16. Install head, pushrods, rocker arms, rocker-arm cover, and other parts removed during head removal.
17. Check and adjust valve-stem clearance as necessary.

Check Your Progress

Progress Quiz 23-1 Here is your chance to check up on the progress you have made since starting Chap. 23. The questions that follow will help you review the material you have just covered and fix the essential points firmly in your mind. If any of the questions stump you, reread the pages in the chapter that will give you the answer.

Correcting Troubles Lists In each of the following lists, you will find one item that does not belong. For example, in the list "Excessive oil consumption: worn valve guides, worn piston rings, worn bearings, worn fuel pump," you would see that "worn fuel pump" does not belong since it is the only condition that is not directly related to excessive oil consumption. Any of the other conditions could cause excessive oil consumption. Write each list in your notebook, but do not write the item that does not belong.

1. Valve sticking: deposits on valve stem, worn valve guide, warped valve stem, insufficient oil, excessive tappet clearance, overheated valves.
2. Valve burning: insufficient tappet clearance, cocked spring, distorted seat, lean mixture, preignition, detonation, seat leakage, excessive tappet clearance, stretched valve stem, overloaded engine.
3. Valve breakage: overheated valve, detonation, excessive tappet clearance, idle too high, seat eccentric to stem, cocked spring, scratches on stem.
4. Valve-face wear: excessive tappet clearance, dirt on face, interference angle, conditions causing valve burning.
5. Valve deposits: gum in fuel, rich mixture, poor combustion, heavy valve spring, worn valve guides, dirty oil.

Completing the Sentences The sentences that follow are incomplete. After each sentence there are several words or phrases, only one of which will correctly complete the sentence. Write each sentence in your notebook, selecting the proper word or phrase to complete it correctly.

1. When the valve is faced at an angle flatter than the seat angle, the difference is called the: (*a*) interference angle, (*b*) seat angle, (*c*) flat angle, (*d*) face angle.
2. On L-head engines valve-tappet clearance is measured between the adjustment screw in the valve lifter and the: (*a*) rocker arm, (*b*) valve stem, (*c*) valve retainer.
3. On overhead-valve engines valve-tappet clearance is measured between the valve stem and the: (*a*) rocker arm, (*b*) valve retainer, (*c*) adjustment screw in lifter, (*d*) adjustment screw.
4. To adjust clearance in the L-head engine, an adjustment screw is turned in the: (*a*) rocker arm, (*b*) pushrod, (*c*) valve lifter, (*d*) valve stem.
5. To adjust clearance in the overhead-valve engine, an adjustment screw is turned in the: (*a*) rocker arm, (*b*) pushrod, (*c*) valve lifter, (*d*) valve stem.
6. In the Vega overhead-camshaft engine, valve clearance is adjusted by: (*a*) grinding the valve tip, (*b*) adjusting the screw in the rocker arm, (*c*) adjusting the screw in the valve tappet.
7. On Ford engines with hydraulic valve lifters, insufficient clearance after bleed-down requires correction by: (*a*) turning the adjustment screw, (*b*) installing shorter pushrods, (*c*) grinding the valves.
8. On Chevrolet engines with hydraulic valve lifters, clearance adjustment is made by backing off the adjustment nut until the pushrod is loose, tightening it until looseness is gone, and then: (*a*) backing off the nut one turn, (*b*) installing longer pushrods, (*c*) turning the nut down one turn.
9. The type of engine that requires no valve-tappet-clearance adjustment uses: (*a*) free valves, (*b*) hydraulic valve lifters, (*c*) F-head valves, (*d*) I-head valves.

10. If the clearance between the tip cup and the valve stem on the free valve is insufficient, correction can be made by grinding the: (*a*) tip cup, (*b*) valve stem, (*c*) valve retainer, (*d*) valve lock.

⊘ 23-22 Removing, Cleaning, and Replacing Cylinder Heads On some engines, the manifolds must be removed before the cylinder heads can be taken off (⊘ 23-35). On other engines, the manifolds may be left in place.

1. REMOVING THE CYLINDER HEAD Follow the general instructions in ⊘ 23-21 for removing the cylinder head. Slightly loosen all cylinder-head bolts first to ease the tension on the head. Then remove the bolts. If the head sticks, carefully pry it loose. Do not pry hard. Do not insert the pry bar too far between the head and block. This could mar the mating surfaces and lead to leaks. Lift the head off and place it in a headholding fixture (Fig. 23-25).

CAUTION: Never remove a cylinder head from a hot engine. Wait until the engine cools. If the head is removed hot, it can be distorted so that it cannot be used again.

2. CLEANING THE CYLINDER HEAD After the valves and other parts are removed (as explained later), clean and inspect the cylinder head. Clean carbon from the combustion chambers and valve ports. Use a wire brush driven by a drill motor (Fig. 23-26). An air-powered motor is best. Keep the wire brush away from valve seats because it could scratch the seating surfaces. Scratched seats can cause poor valve seating and serious engine trouble. Blow out all dust with an air hose.

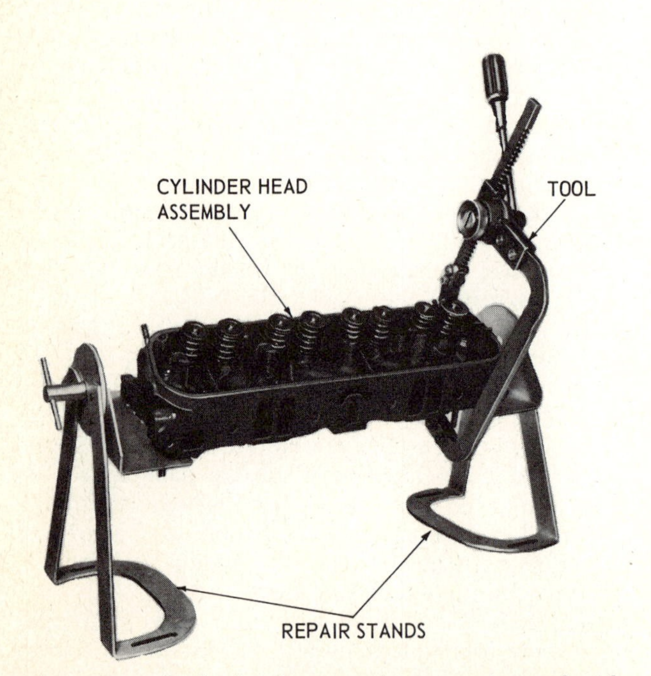

Fig. 23-25. Cylinder head mounted in a repair stand with a spring-compressor tool in place. (*Chrysler Corporation*)

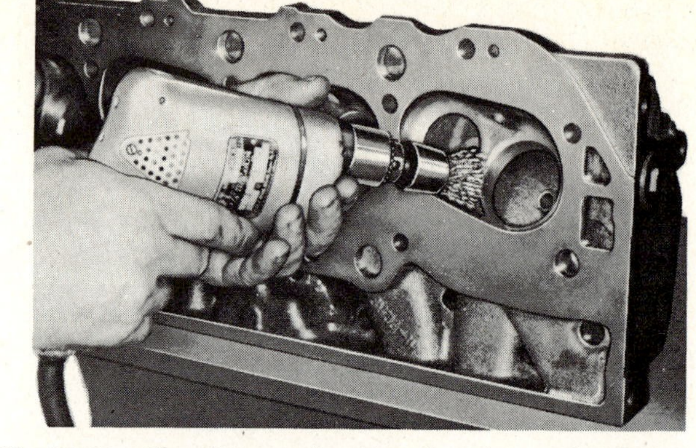

Fig. 23-26. Cleaning combustion chamber and valve ports with a wire brush. (*Chevrolet Motor Division of General Motors Corporation*)

CAUTION: Always use goggles when using a wire brush, compressed air, or similar equipment. They will protect your eyes from flying particles.

Clean gasket surfaces with a flat scraper. Be very careful not to scratch the gasket surface. All traces of old gasket material and sealer should be removed.

Remove dirt and grease from the cylinder head. Then clean the water jackets and passages by soaking the head in a boil tank. Flush the water jackets as recommended by the manufacturer of the cleaning agent.

3. INSPECTING THE CYLINDER HEAD As you remove the cylinder head, examine the gasket and mating surface for traces of leakage or cracks. A blown gasket or coolant leakage could result from a warped head or improper gasket installation. In the head, cracks usually occur between valve seats. If they are not too bad, they can often be repaired by *cold welding*. First, drill a small hole in the crack and thread it. Then screw in a threaded rod and cut it off. Next, drill a second hole overlapping the first one and thread it. Screw in the threaded rod and cut it off. Repeat until the crack is completely treated. Make sure to get to both ends of the crack to relieve the stress that caused the crack. Sometimes it is desirable to install a seat insert when the crack runs into the valve seat. This is done by making an undercut in the head and pressing in the insert (⊘ 23-31).

Clean and inspect valve guides. Note the condition of valve seats and ball studs (on heads using them). We cover servicing of valve guides, ball studs, and seats later.

Check the cylinder head for cracks, warpage, and rough gasket surfaces. If cracks are suspected, have the head checked with Magna-Flux equipment.

Warpage is detected by laying a straightedge against the gasket surface of the head (Fig. 23-27). Check crossways and longways. One specification calls for not more than 0.005 inch [0.137 mm] maximum out of straight. More than this requires either

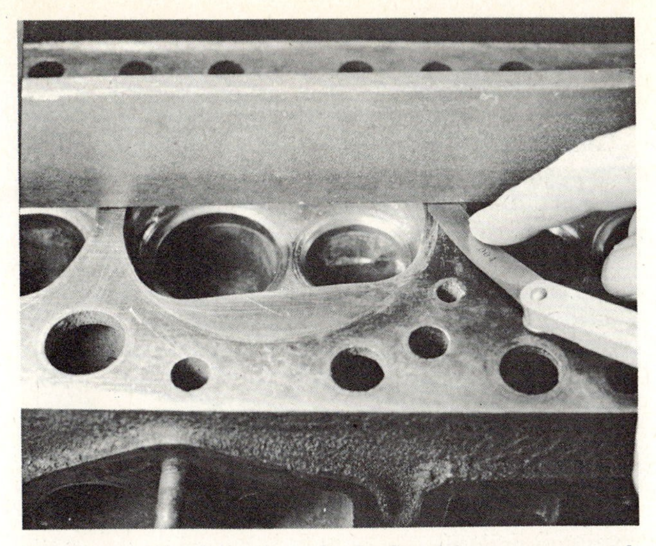

Fig. 23-27. Checking the cylinder head for warpage with a straightedge and feeler gauge. (*TRW Inc.*)

a new head or machining the head so that the gasket surface is back to straight again.

Check the gasket surface of the head for nicks or rough spots. These can be removed with a fine-cut mill file (Fig. 23-28).

NOTE: If one head from a V-8 engine requires machining to remove gasket-surface roughness, then the other head should be machined a like amount. Otherwise, uniform compression and manifold alignment will be lost. Also, remember that removing metal from the gasket surface lowers the head with respect to the intake manifold (Fig. 23-29). Therefore, a compensating amount may have to be machined from the manifold to restore alignment.

4. REPLACING THE CYLINDER HEAD Reassemble the cylinder head (as explained later) so

Fig. 23-28. Removing nicks on cleaned gasket surface with a fine-cut file. (*TRW Inc.*)

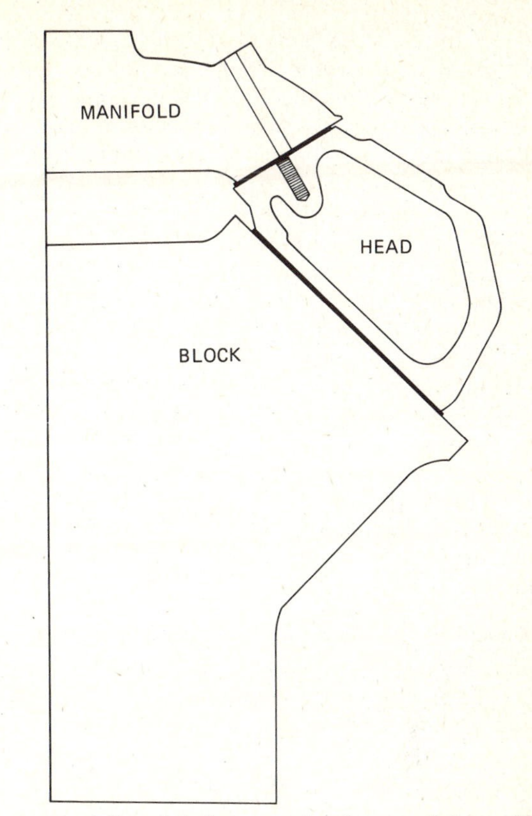

Fig. 23-29. Mismatch between intake manifold and head after machining of head-gasket surface. (*TRW Inc.*)

that valve springs, rocker arms, and other parts are in place. Then install the head, as follows. Always use a new gasket.

Before installing the head, check the cylinder block. Make sure the gasket surface is flat and in good condition. Make sure all traces of gasket material are removed from the block. Bolt holes in the block (where present) should be cleaned out. Cylinder-block studs (where present), should be in good condition. Cylinder-head bolts should be cleaned with a wire brush or wheel (Fig. 23-30).

Fig. 23-30. Cleaning cylinder-head bolts with a wire wheel. (*Dana Corporation*)

Cylinder-block studs (where present) can be cleaned up with a thread chaser if the threads are damaged.

Use care when handling the gasket. If it is of the lacquered type, do not chip the lacquer. If the block has studs, put the gasket into place, right side up. Some are marked "TOP" so that you know which side goes up. Also, some are marked "FRONT" so that you know which end goes to the front. Use gasket cement only if specified by the manufacturer. For example, one manufacturer says to use cement on steel gaskets but not on composition steel-asbestos gaskets.

If the block does not have studs, use two pilot pins set into two bolt holes to ensure gasket alignment. Then lower the head into position. Substitute bolts for pilot pins (if used). Run on the nuts or bolts finger tight.

CAUTION: Make sure that all bolt holes in the block have been cleaned. If they are not clean and the bolts bottom on the foreign material, the head will not be tight.

Use a torque wrench to tighten the nuts or bolts. They must be tightened in the proper sequence and to the proper tension. If they are not, head or block distortion, gasket leakage, or bolt failure may occur. Refer to the sequence chart for the engine being serviced and note the torque called for. Figure 23-31 shows the sequence for tightening the cylinder-head bolts on one head of a V-8 engine. Each bolt should be tightened in two or more steps. In other words, the complete circuit should be made at least twice, with each bolt or nut being drawn down little by little. After engine assembly is completed, the engine should be run until it is warm. Then the tensions should be checked again. Also, some engines using aluminum heads must be turned off and allowed to cool. Then the tensions should be checked again.

NOTE: Some torque specifications call for clean, dry threads. Others call for lightly lubricated threads. Antiseize or sealing compounds are often used on bolts in aluminum blocks.

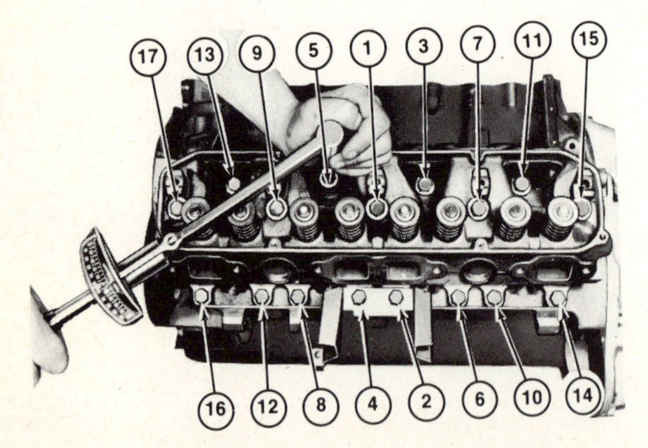

Fig. 23-31. Sequence chart for tightening cylinder-head bolts on a V-8 engine. (*Chrysler Corporation*)

CAUTION: If the rocker arms are in place, tighten the bolts slowly. This gives the hydraulic valve lifters time to bleed down to their operating length. If the bolts are tightened too rapidly, excessive pressure will be put on the lifters. They could be damaged, and the pushrods could be bent. On most engines, the head bolts cannot be tightened if the rocker arms are in place.

CAUTION: On overhead-valve cylinder heads with the rocker arms and shaft in place, make sure that the pushrods are in position. That is, make sure the lower ends of the pushrods are in the valve-lifter sockets.

⊘ 23-23 Ball-Stud Service If a ball stud is loose in the cylinder head, has damaged threads, or has begun to pull out, it should be replaced. The old stud is removed with a special puller (Fig. 23-32). The stud hole is then reamed to a larger size to take the oversize stud (Fig. 23-33). For example, Chevrolet supplies 0.003- and 0.013-inch [0.0763 and 0.330 mm] oversize studs. If the 0.003-inch oversize stud is to be installed, the special 0.003-inch oversize reamer must be used.

CAUTION: Always ream the stud hole oversize before installing an oversize stud. Otherwise, you may crack the head.

To install the new stud, use the special tool, as shown in Fig. 23-34. Drive the stud down into place.

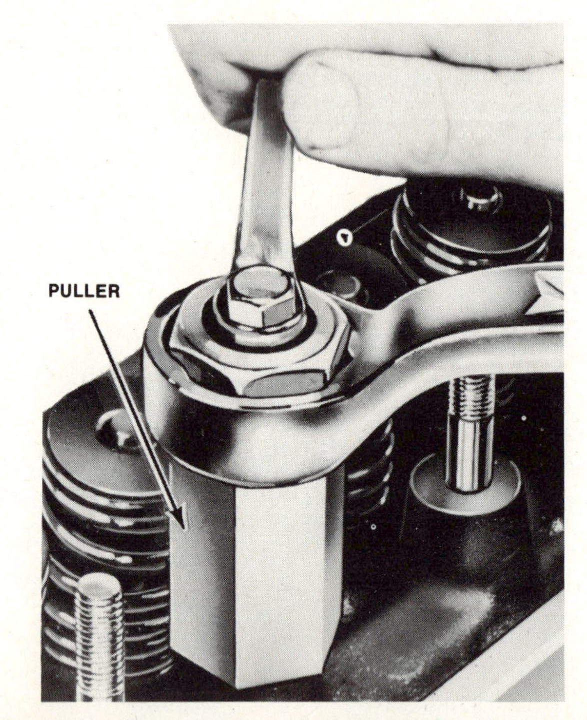

Fig. 23-32. Removing a rocker-arm stud with a special puller. The sleeve is held stationary, and the puller is turned. The screw-thread action pulls the stud out of the head. (*Ford Motor Company*)

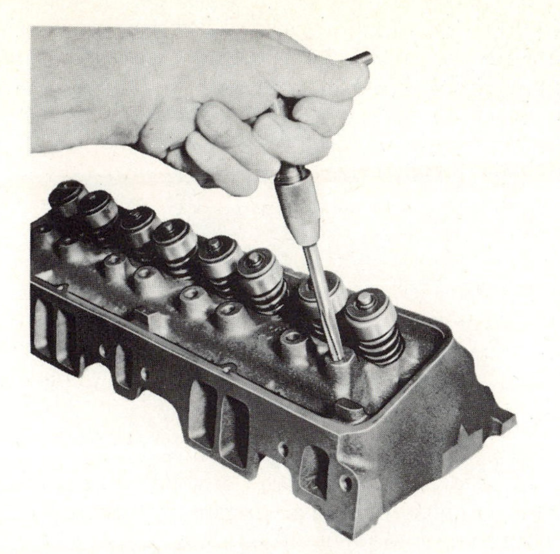

Fig. 23-33. Reaming a stud hole to take a ball stud. (*Chevrolet Motor Division of General Motors Corporation*)

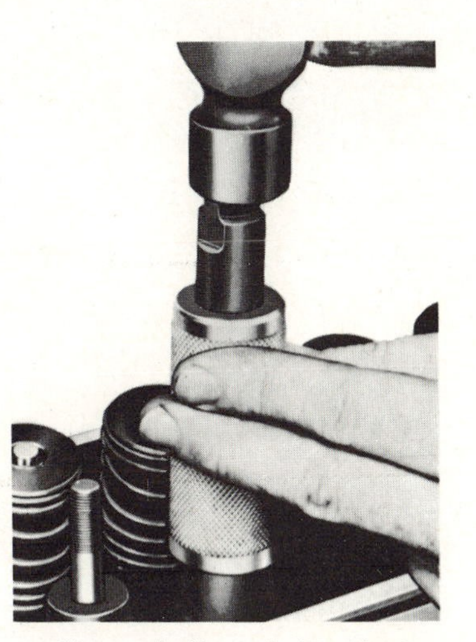

Fig. 23-34. Installing a ball stud. (*Ford Motor Company*)

When the tool is firmly driven down to the cylinder head, the stud is in the correct position.

NOTE: Some studs come threaded. On these, the stud hole must be tapped to take the stud.

⊘ 23-24 Servicing Rocker-Arm Assemblies There are two methods of attaching the rocker arms to the cylinder head. In one method, the rocker arms are mounted on a common shaft. In the other, the rocker arms are mounted individually on separate studs. The rocker arms move on ball pivots instead of a shaft.

There are several variations of the shaft-mounted arrangement. Figure 23-35 shows one of these. With this design, remove the rocker arms by first taking out the shaft-locking plug. Then slide the shaft out from the five supporting shaft struts on the head. In the design shown in Fig. 23-36, the shaft is removed with the rocker arms on it. First remove

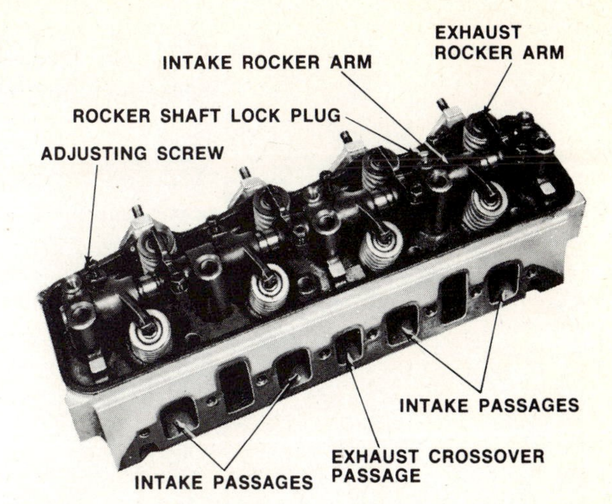

Fig. 23-35. Cylinder-head assembly. (*Chrysler Corporation*)

the five bolts that hold the five shaft support brackets and shaft on the head. Then slip the rocker arms, brackets, and spacers off the shaft. A similar design is shown in Fig. 23-37. One difference in this design is that there are four shaft supports. Also, the end rocker arms are held in place by washers and cotter pins. Several other variations can be found on different engines.

On engines with independently mounted ball-pivot rocker arms (Fig. 23-17), remove the rocker arms by taking off the adjusting nuts. The rocker-arm studs in the head can be replaced if they have become loose or if the stud threads are damaged (⊘ 23-23).

After rocker arms are removed, they should be inspected for wear or damage. Rocker arms with bushings can be rebushed if the old bushings are worn. On some rocker arms, the valve ends, if worn, can be ground down on the valve-refacing machine (Fig. 23-38). Excessively worn rocker arms should be discarded.

When reinstalling rocker arms and shafts on the cylinder head, make sure that the oilholes (in shaft so equipped) are on the underside. Otherwise, they will not feed oil to the rocker arms. Be sure that all springs and rocker arms are in their original positions when the shafts are reattached to the head.

⊘ 23-25 Servicing Pushrods Pushrods should be inspected for wear at the ends. Roll the rods on a flat surface to check for straightness. Replace defective rods. On certain engines, rods have one tip hardened and are marked with strips of color. The pushrod should be installed so that the hardened end is toward the rocker arm. Always make sure that the lower end of the pushrod is seated in the valve-lifter socket.

Special short-length pushrods are available. These may be used in engines after valves and valve seats have been reground. Regrinding may result in excessive lengthening of the valve train since the

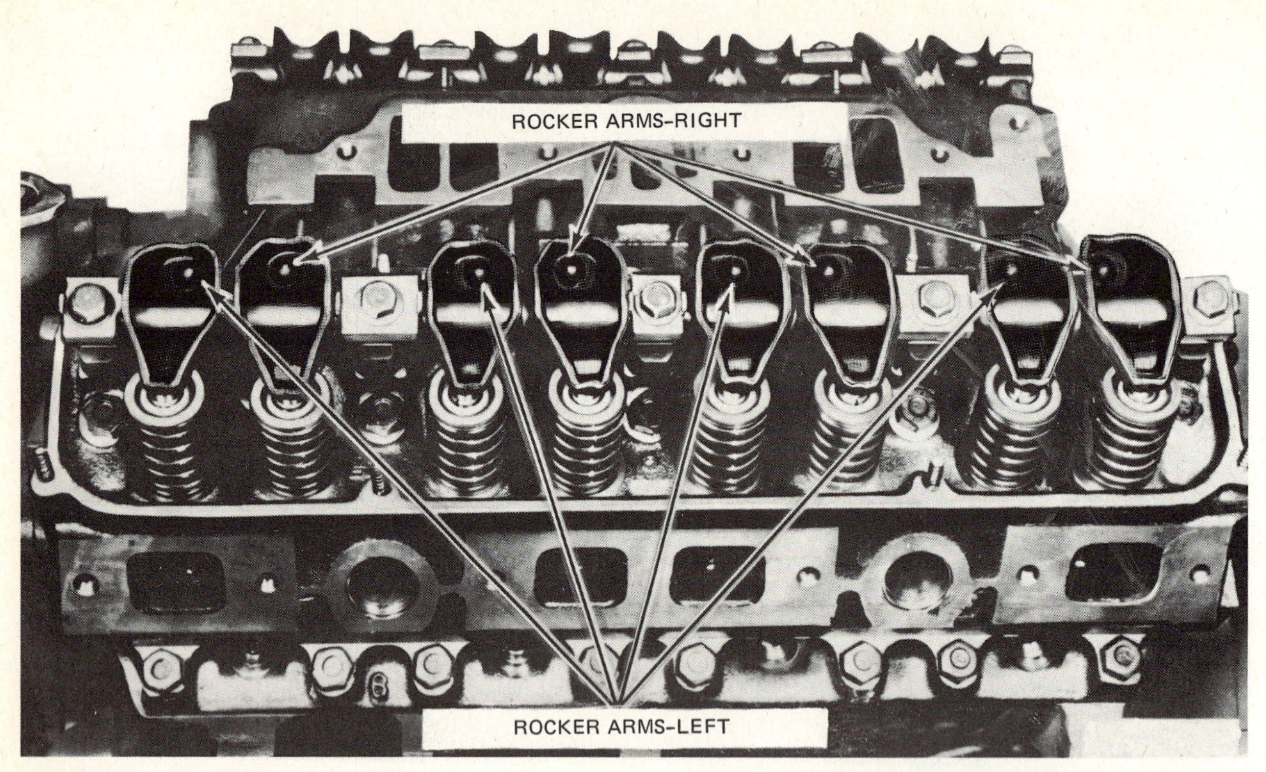

Fig. 23-36. Cylinder-head assemblies installed on the block of a V-8 engine. (*Chrysler Corporation*)

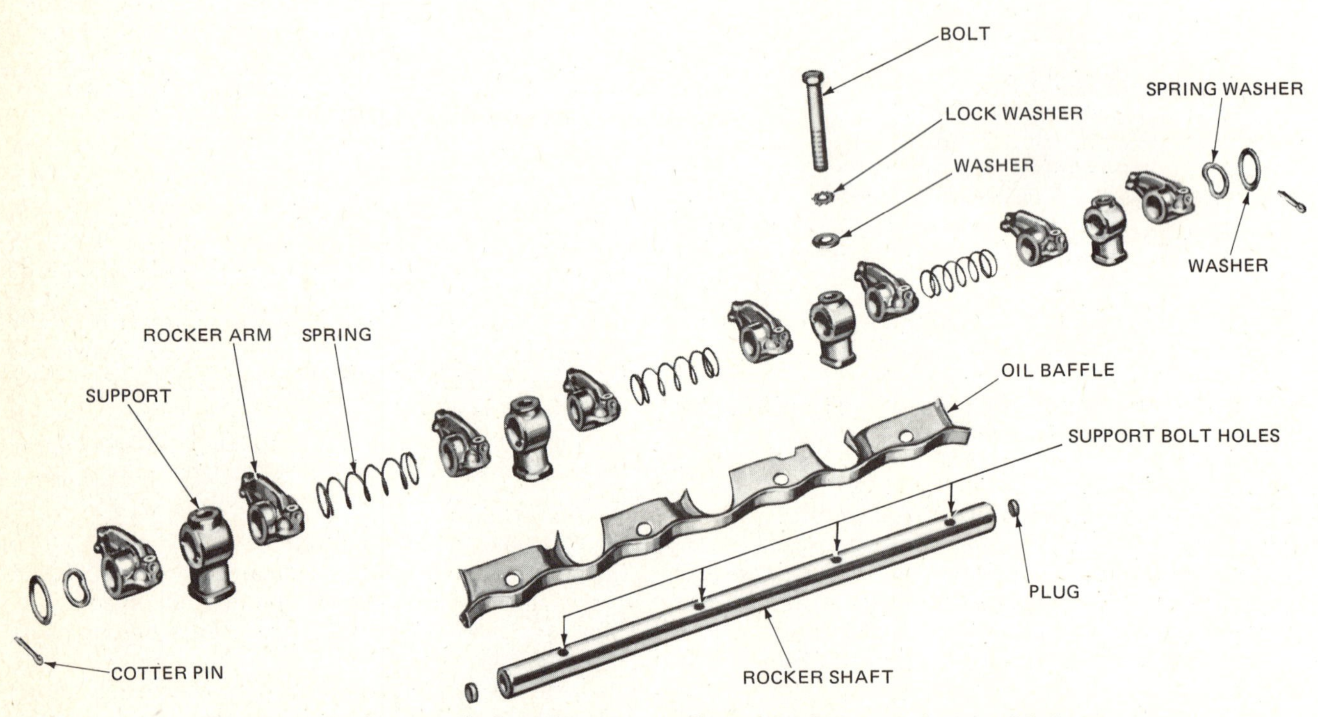

Fig. 23-37. Rocker-arm shaft, rocker arms, and associated parts in disassembled view. (*Ford Motor Company*)

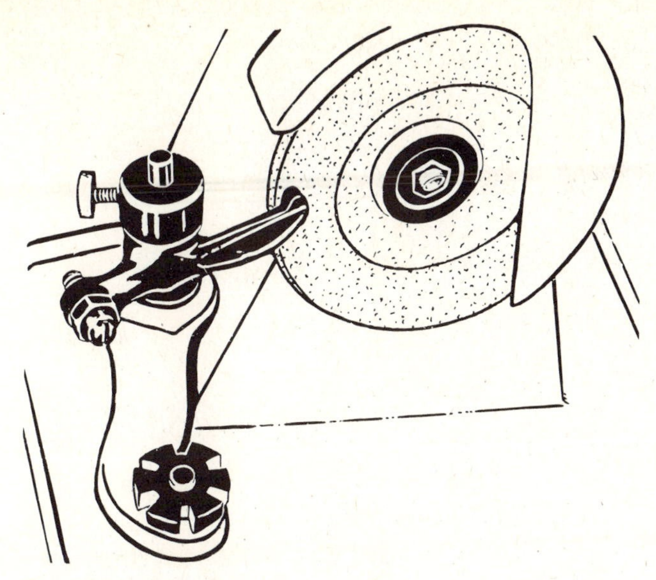

Fig. 23-38. Grinding the valve-end tip of a rocker arm. (*Snap-on Tools Corporation*)

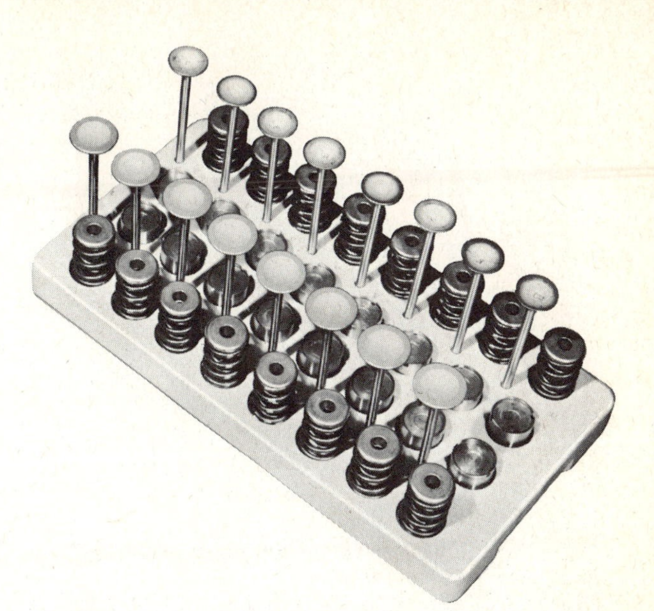

Fig. 23-39. Valve rack for holding valves and associated parts. (*Chevrolet Motor Division of General Motors Corporation*)

valve stem rises higher out of the cylinder head. Therefore, the plunger may almost bottom in the hydraulic valve lifter. Using a shorter pushrod corrects this condition (see ⊘ 23-17). Instead, some engine manufacturers recommend grinding off the end of the valve stem. This procedure brings the valve-train length back to normal after a valve and valve-seat grinding job (⊘ 23-18).

⊘ 23-26 Removing Valves After the cylinder head is removed and interfering rocker-arm mechanisms are removed from it, the valves should be taken out. Valves and valve parts must not be interchanged. Each valve, with its own spring, retainer, and lock, should be reassembled in the same valve port from which it was removed. For this reason, a special valve rack is recommended (Fig. 23-39). Likewise, each rocker arm and pushrod should be replaced in its original position.

CAUTION: Before removing a valve from the cylinder head, examine the valve stem. Look for burrs at the retainer-lock grooves and for mushrooming on the end. Burrs and mushrooming must be removed with a file or small grinding stone in a drill motor. If they are not removed, the valve guide could be badly damaged or broken when the valve is forced out. This would mean extra work in cleaning up the valve guide.

1. *L-HEAD ENGINES* For the benefit of those relatively few who may work on a flat-head engine, here is how to remove the valves. First remove any manifolds that may be in the way. Then use a valve-spring compressor to compress the valve spring. This releases the retainer lock so that the lock and the retainer can come off. Do not drop the lock down into the crankcase. It could jam in moving parts and ruin the engine. Then remove the spring and take out the valve.

2. *OVERHEAD-VALVE ENGINES* After everything else is removed from the head, the valves can be removed. This requires a spring compressor. One type is shown in Fig. 23-25. A similar type is shown in Fig. 23-40. Figure 23-41 shows how a spring compressor is used. As the handle is pressed, the spring is compressed. This allows removal of the retainer lock. Then the spring can be released so that the retainer and spring can be removed. Valve-stem

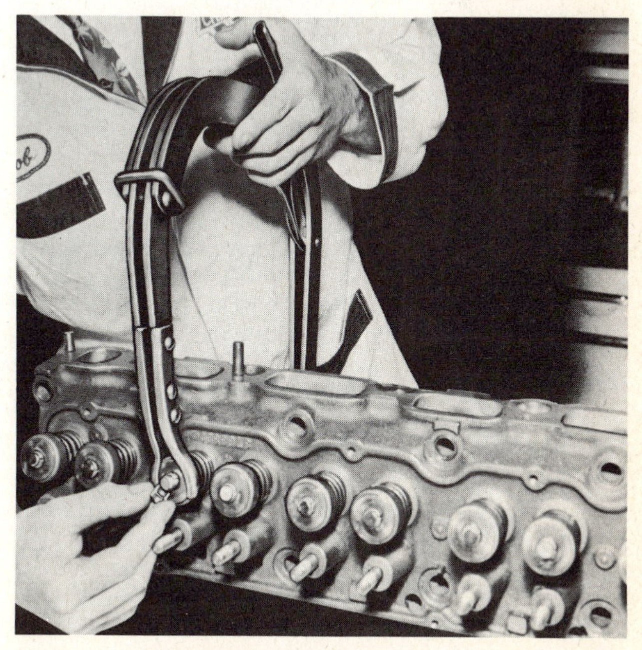

Fig. 23-40. Using a valve-spring compressor on the valve assemblies in a cylinder head. (*Chevrolet Motor Division of General Motors Corporation*)

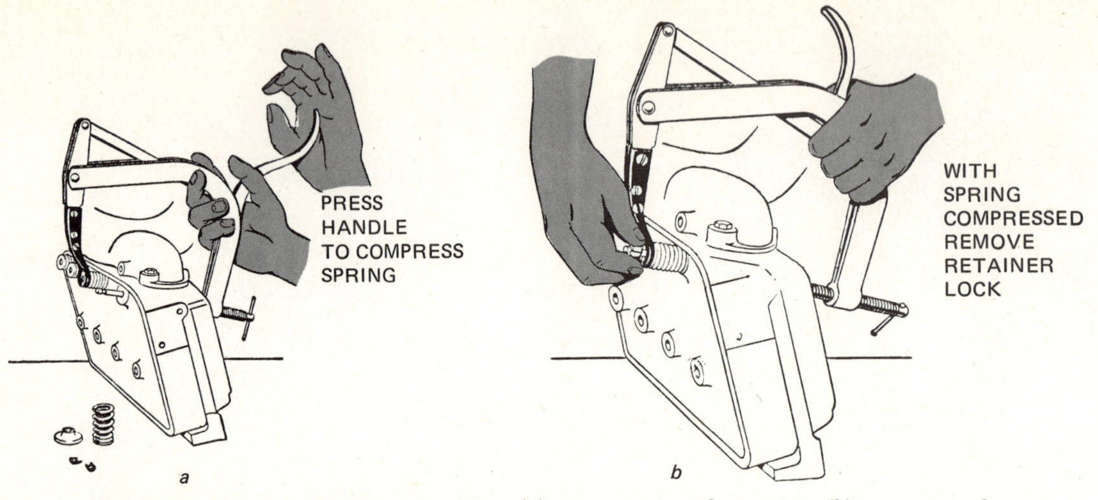

Fig. 23-41. Using a spring compressor: (*a*) compressing the spring; (*b*) removing the retainer lock.

seals or shields (Figs. 9-37 to 9-39) should be carefully inspected and replaced if worn or damaged. Many manufacturers recommend installation of new seals or shields whenever valves or valve springs are removed.

NOTE: Some mechanics use a 6-inch [152.4 mm] length of ¾-inch [19.05 mm] water pipe instead of a spring compressor to break loose the retainer lock. The pipe is centered on the retainer and hit on the end with a hammer. This forces the retainer down against the spring tension enough to break the retainer lock loose.

On many models, the valve springs, stem seals, or shields can be replaced without removing the cylinder head. For instance, on the engine using the ball-pivot-type rocker arm, a special spring compressor is installed in place of the rocker arm to compress the spring. To hold up the valve while the spring is being compressed, compressed air from the shop air supply is introduced into the cylinder through the spark-plug hole (Fig. 23-42). A special air-hose adaptor, which can be screwed into the spark-plug hole, is required. The air pressure will hold the valve closed while the spring is compressed. If the air pressure does not hold the valve closed, then the valve is stuck or damaged. The cylinder head should be removed for a closer look.

NOTE: The air pressure may push the piston to BDC.

On some engines with the rocker arms mounted on a shaft, it is possible to bleed down the hydraulic valve lifter. To do this, you apply pressure with a special tool (Fig. 23-14). Then you can remove the pushrod, and move the rocker arm to one side. With the rocker arm wired out of the way (Fig. 23-43), use a valve-spring compressor to compress the spring. Then you can remove the retainer, spring, and seal. Air pressure must be applied to the cylinder (as explained in the previous paragraph) to hold the valve on its seat when the spring is compressed.

⊘ 23-27 Inspecting Valves As you take the valves out of the head, inspect each one. Decide whether it can be serviced and used again. (See ⊘ 23-1 to 23-10.) If the valve looks good enough to use again, put it into its proper place in the valve rack (Fig. 23-39). If it looks too bad to be cleaned up for further service, discard it. Put a new valve into the appropriate place in the valve rack.

⊘ 23-28 Servicing Valves Once all the valves are out of the cylinder head, remove them one by one from the valve rack. Clean each of them. Clean the

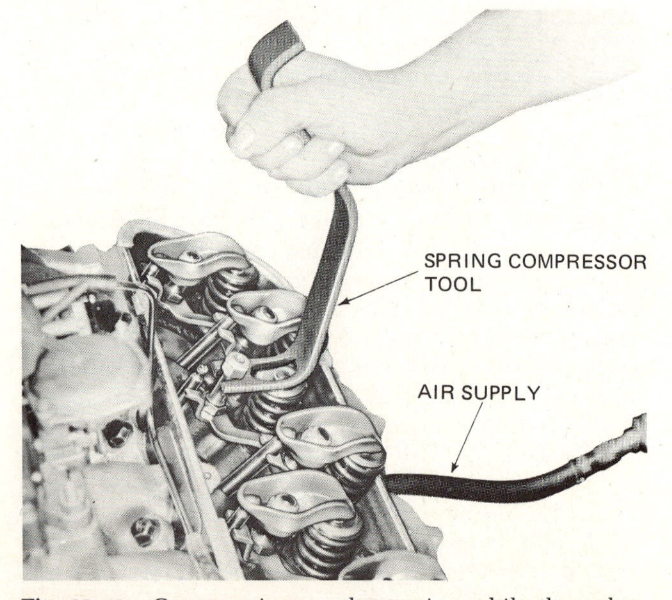

Fig. 23-42. Compressing a valve spring while the valve is held closed with air pressure. (*Chevrolet Motor Division of General Motors Corporation*)

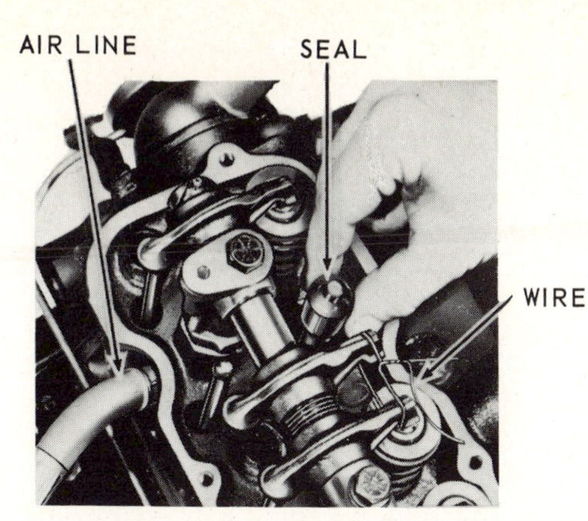

Fig. 23-43. Removal of a valve-stem seal after the rocker arm has been wired out of the way. (*Ford Motor Company*)

carbon off the valves with a wire wheel, as shown in Fig. 23-44. (Wear goggles to protect your eyes from flying particles of metal and dirt!) Polish the stems, if necessary, with a fine grade of emery cloth. Do not take off more than the dirty coating on the surface. Do not take metal off the stems!

CAUTION: Do not scratch the valve-seating surface or valve stem with the wire brush or emery cloth.

As you clean the valves, reexamine them to make sure all are usable. Small pits or burns in the valve-seating face can be removed by grinding the valve. Larger pits or burns are hopeless; new valves will be required. Figure 23-45 shows specific parts of the valve to be examined. Some engine manufac-

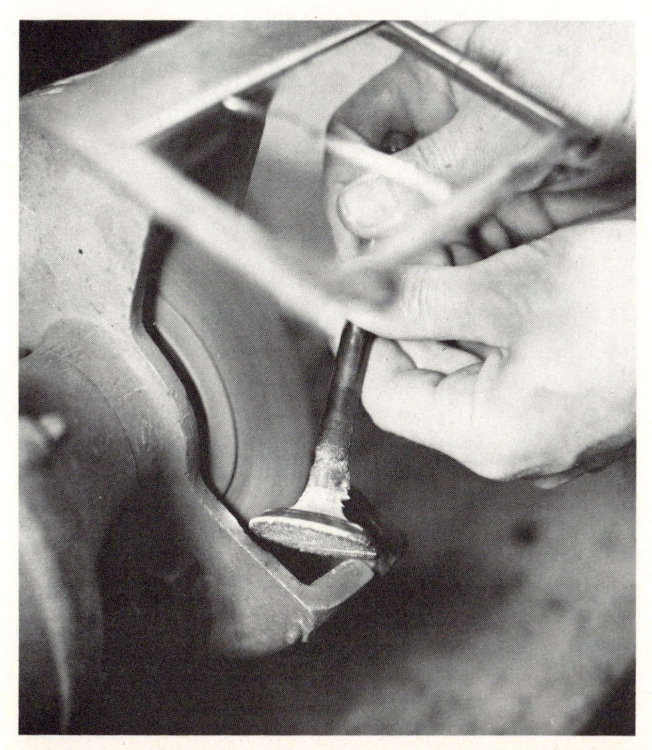

Fig. 23-44. Cleaning a valve on a wire wheel. (*TRW Inc.*)

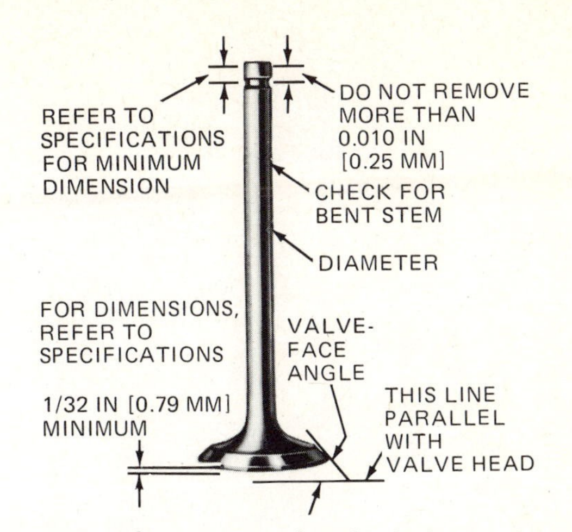

Fig. 23-45. Valve parts to be checked. On the valve shown, the stem is hardened at the end. Therefore, not more than 0.010 inch [0.25 mm] should be removed. (*Ford Motor Company*)

turers recommend the use of a *run-out gauge* (Fig. 23-46) to check for a bent valve stem. Eccentricity can also be checked in the valve grinder. If the run-out, or eccentricity, is excessive, discard the valve.

After cleaning the valves, replace them temporarily in their valve guides to check for guide wear. This procedure and valve-guide service are covered in ⊘ 23-30.

1. REFACING, OR GRINDING, VALVES If the valves are good enough to reuse, the next step is to reface, or grind, them. This process requires a valve-refacing machine (Fig. 23-47). The machine has a ginding wheel, coolant delivery system, and chuck that holds the valve for grinding. Set the chuck to grind the valve face at the specified angle. [This angle must just match the valve-seat angle or be the interference angle recommended by the engine manufacturer. (See ⊘ 9-8 and Fig. 23-8.)] Then put the valve into the chuck and tighten it. The valve should be deep in the chuck, so that not too much sticks out. Otherwise, the valve can whip during grinding. The result will be a poor grinding job.

To start the operation, align the coolant feed so

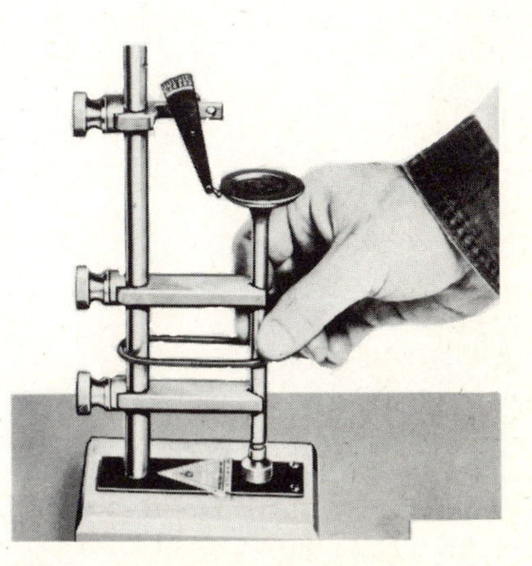

Fig. 23-46. Using a run-out gauge to check for valve-face eccentricity. (*Ford Motor Company*)

Fig. 23-47. Valve-refacing machine. (*Black and Decker Manufacturing Company*)

that it feeds coolant on the grinding wheel. Then start the machine. Move the lever to carry the valve face across the grinding wheel. The first cut should be a light one. If this cut removes metal from only one-half or one-third of the face, the valve may not be centered in the chuck. Or the valve stem may be bent, in which case the valve should be discarded. Cuts, after the first one, should remove only enough metal to true up the surface and remove pits. Do not take heavy cuts. If so much metal must be removed that the margin is lost, discard the valve. (See Fig. 23-48.) Loss of the margin causes the valve to run hot, and it will soon fail.

NOTE: If new valves are required, they will not need to be refaced. Seating should be checked, however, as already explained. *Never reface or lap coated valves!*

NOTE: Follow the operating instructions of the refacer manufacturer. In particular, dress the grinding wheel as necessary with the diamond-tipped dressing tool. As the diamond is moved across the rotating face of the grinding wheel, it cleans and aligns the grinding face.

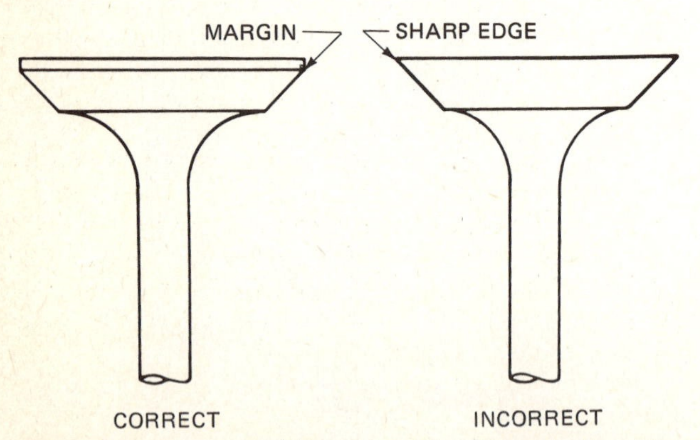

Fig. 23-48. Correct and incorrect valve-face grinding. The valve to the right, having no margin, would soon fail. (*TRW Inc.*)

2. *REFACING VALVE-STEM TIPS* If the tip of a valve stem is rough or worn unevenly, it can be ground lightly. Use the special attachment furnished with the valve-refacing machine (Fig. 23-49). The attachment allows you to swing the valve slightly and rotate it. In this way, the tip can be ground to produce a slightly crowned, or rounded, end. One recommendation is to grind off as much from the stem as you ground off the valve face. That way, you make up for the amount the valve sinks into the seat owing to face grinding.

CAUTION: The ends of some valve stems are hardened. These should have no more than a few thousandths of an inch ground off them (see Fig. 23-45). Excessive grinding exposes soft metal so that the stem wears rapidly in service.

⊘ **23-29 Replacing Valves** As the valves are refaced and cleaned, they should be returned to the valve rack (Fig. 23-39). They are now ready for installation in the cylinder head. First, however, the valve guides and valve seats must be serviced (⊘ 23-30 and 23-31). Also, the other components of the valve train—pushrods, rocker arms, and valve lifters—must be checked and serviced as necessary.

The valve-assembly sequence for an overhead-valve engine is shown in Fig. 23-50. New shields or seals should be installed if the old ones are worn or if the manufacturer recommends them. To avoid damage to some types of seals, special plastic caps can be placed over the ends of the valve stems. The seals will then slip on without being damaged by the sharp edges of the stem end or lock grooves.

NOTE: If valves and seats have been ground, the effective length of the valve spring will not be great enough. In order to restore normal spring tension, spring shims will be required.

Using a spring compressor (Figs. 23-40 and 23-41), install the springs, spring retainers, and locks. Measure the installed spring height (Fig. 23-51). Note that the end of the steel scale has been cut away (Fig. 23-52). If the spring height is excessive, a spring shim is required. The shim is installed between the spring and cylinder head.

NOTE: The procedure of installing the spring and then measuring the installed spring height is slow and can ruin stem seals if the spring has to be removed to install shims. Some mechanics do a faster job, as follows. Put the valve in the guide and install the retainer and locks. Pull the stem down until the retainer hits the cylinder head. Tap the end of the stem with a plastic hammer to seat the locks. Then

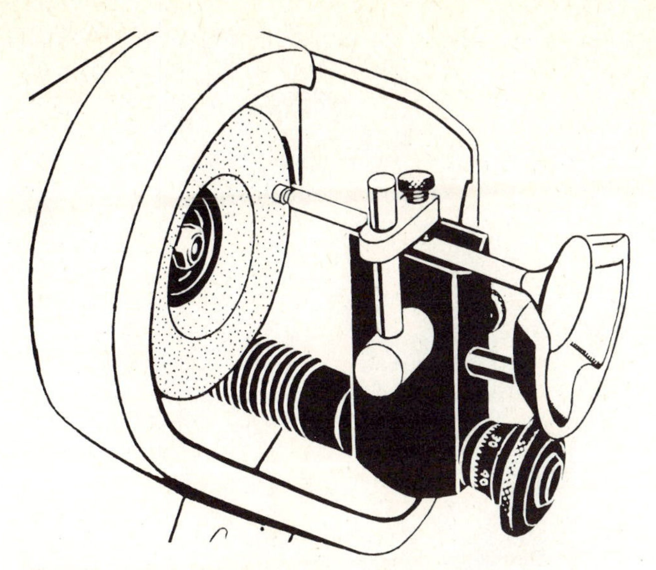

Fig. 23-49. Grinding a valve tip. (*Snap-on Tools Corporation*)

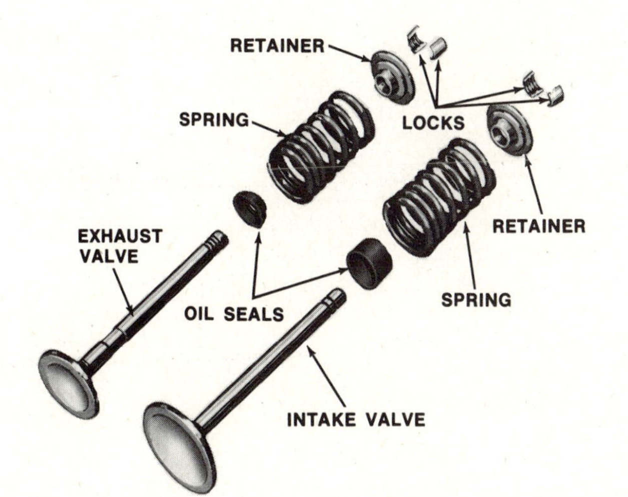

Fig. 23-50. Exhaust and intake valves and associated parts in the proper assembly relationship. (*Chrysler Corporation*)

pull the valve up to close it. Measure the distance between the head and retainer. Compare it with the spring height specified. Add shims if necessary to reduce the height to specifications.

CAUTION: Do not install a shim that would reduce the spring height below the specified minimum. This results in excessive spring pressure and rapid wear of valve-train parts.

Install the valve springs with the proper side against the cylinder head. As a rule, the close-spaced coils go next to the head (on springs with differential spacing of coils). Also, a damper spring (where used) must be placed inside the valve spring in an exact relationship with the spring coils. One typical example is that the coil end of the damper spring should

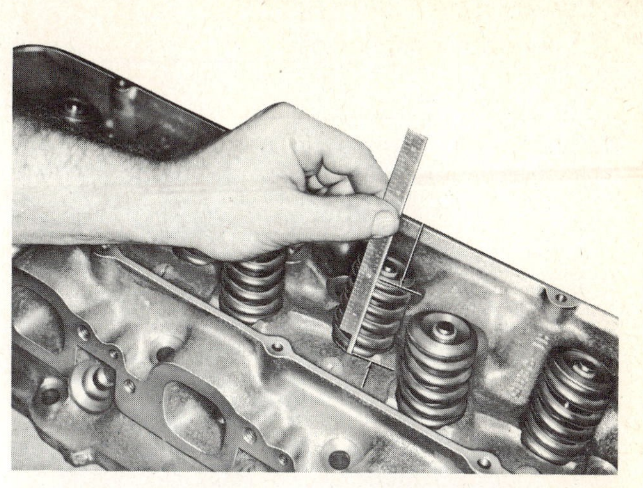

Fig. 23-51. Measuring valve-spring installed height. (*Chevrolet Motor Division of General Motors Corporation*)

be 135 degrees counterclockwise from the coil end of the valve spring.

⊘ 23-30 Servicing Valve Guides The valve guide must be clean and in good condition for normal valve seating. It must be serviced before the valve seats are ground if grinding is required. As a first step, the valve guide should be cleaned with a wire brush or adjustable-blade cleaner (Fig. 23-53). Then, it should be checked for wear. If it is worn, it requires service. The type of service depends on whether the guide is replaceable or integral. If it is replaceable, the old guide should be pressed out. Then a new guide should be installed and reamed to size. If the guide is integral, it can be serviced in either of two ways: (1) by reaming it to a larger size and installing a valve with an oversize stem and (2) by knurling and reaming it. All these services are described in the following paragraphs.

1. TESTING VALVE GUIDES FOR WEAR One method of testing valve guides uses a dial indicator (Fig. 23-54). With the valve in place, attach the dial indicator as shown. The button just touches the edge of the valve head. Use a special tool, as shown in Fig. 23-55, to hold the valve off its seat. Then rotate the valve and move it sideways to determine the amount of guide wear. On some engines, the recommendation is to check valve movement for the stem with the valve seated (Fig. 23-56).

Another method of testing the valve guide is to insert a tapered pilot into the guide until it is tight. Then, pencil-mark the pilot at the top of the guide and remove it. Measure the pilot diameter ½ inch [12.7 mm] below the pencil mark. This gives the guide diameter, which can then be compared with the valve-stem diameter.

Neither of these two methods will accurately show valve-guide eccentricity and bellmouthing, however. The valve guide may wear bellmouthed or oval-shaped because the valve tends to wobble as it opens and closes. The bellmouth wear shown in

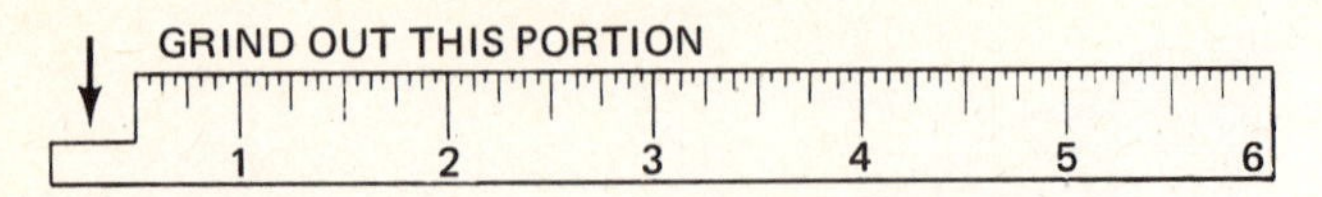

Fig. 23-52. Scale cut away to measure valve-spring installed height. (*Chevrolet Motor Division of General Motors Corporation*)

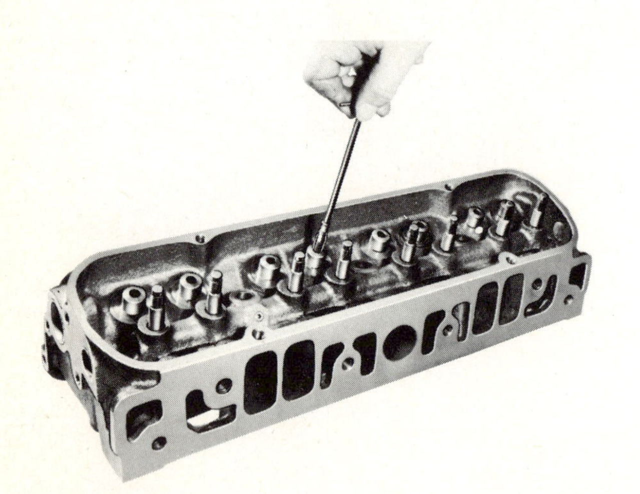

Fig. 23-53. Cleaning the valve guide with an adjustable-blade cleaner. (*Pontiac Motor Division of General Motors Corporation*)

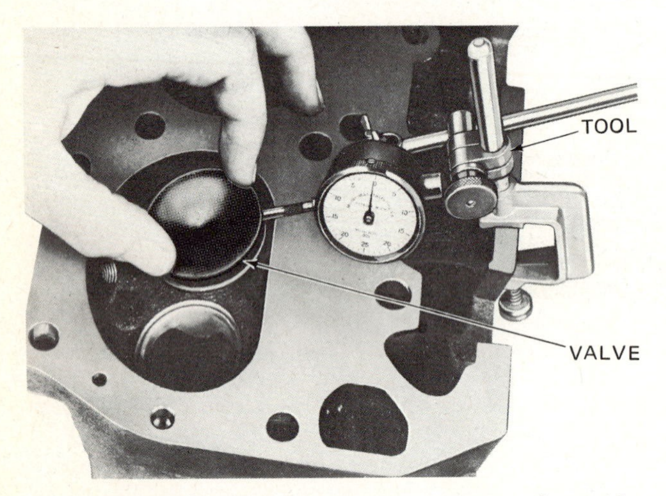

Fig. 23-54. Dial indicator set up to measure valve-guide wear. (*Chrysler Corporation*)

Fig. 23-57 is exaggerated. A small-hole gauge, shown in Fig. 23-57, will detect oval or bellmouth wear. It is used as shown: The split ball is adjusted until it is a light drag fit at the point being checked. Then, the split ball is measured with a micrometer. By checking the guide at various points, any eccentricity will be detected.

2. *REMOVING VALVE GUIDES (REPLACEABLE TYPE)* A valve-guide puller (Fig. 23-58) is handy for removing an old valve guide. As the nut is turned on the screw, the guide is pulled out. On some L-head engines, the guide can be driven down into the valve-spring compartment. On I-head engines,

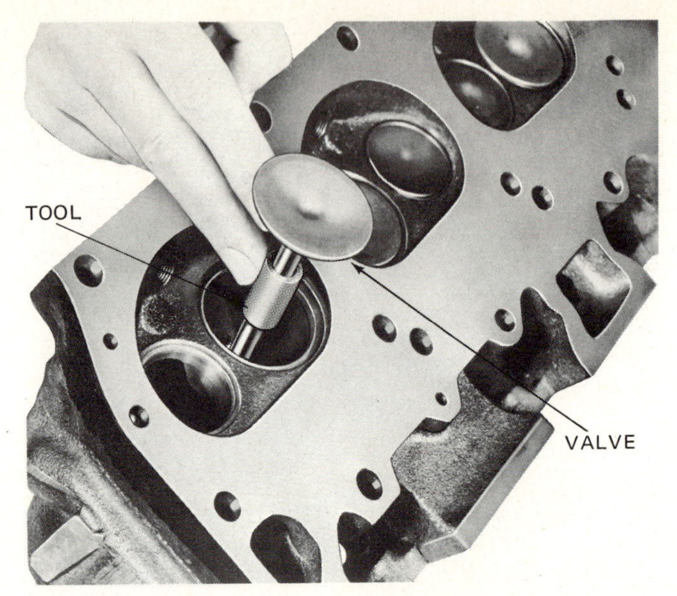

Fig. 23-55. Using a tool on a valve stem to hold the valve off its seat in the cylinder head. (*Chrysler Corporation*)

the valve guide can be pressed out of the head with an arbor press.

3. *INSTALLING VALVE GUIDES (REPLACEABLE TYPE)* New valve guides can be installed with a special driver, or replacer (Fig. 23-59). (On I-head engines, valve guides can be installed with an arbor press.) Valve guides must be installed to the proper depth in the block or head. Then they must be reamed to size. This is usually done in two steps: a rough ream and then a second, or final, finishing ream. Figure 23-60 illustrates both the depth of assembly of valve guides and the reaming dimensions on one engine.

4. *KNURLING THE VALVE GUIDE* In the knurling operation, a knurling tool is run down into the valve guide. One type of knurling tool has a small wheel set in a rod, or arbor (Fig. 23-61). It is rotated by a slow-speed electric drill as it is pushed down into the guide. This action causes the wheel to form a spiral groove in the guide (Fig. 23-62). The metal displaced from the groove is pushed inward, as shown. Then the guide is reamed. The procedure usually takes several steps: for example, first knurl, first ream, second knurl, final ream. When the job is finished, the valve guide should be the right size to provide a good fit for the valve stem. Also, the grooves that are left (Fig. 23-63) are filled with oil. This provides excellent lubrication of the valve stem.

5. *CHECKING CONCENTRICITY WITH THE VALVE SEAT* After the valve guide is serviced and checked for size, check its concentricity with the valve seat. As a rule, the seat is ground as part of the service job (⊘ 23-31).

⊘ 23-31 Valve-Seat Service For effective valve seating and sealing, the valve face must be concentric with the valve stem. Also, the valve guide must

Fig. 23-56. Valve-stem clearance in the valve guide being checked from the stem end of the valve. (*Ford Motor Company*)

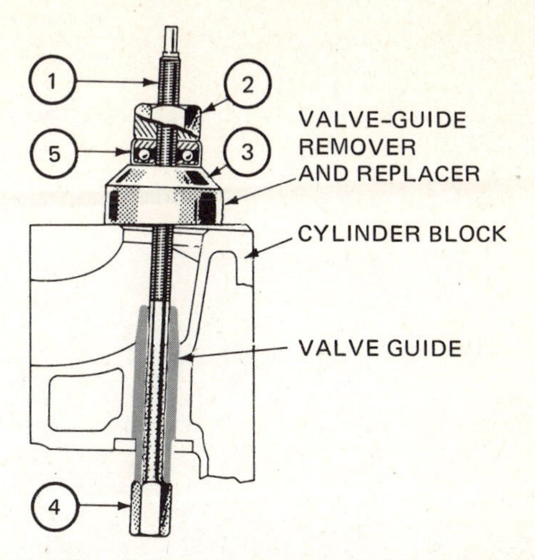

Fig. 23-58. The valve guide is removed from the cylinder block with a special puller: (1) screw, (2) nut, (3) spacer, (4) nut, (5) bearing.

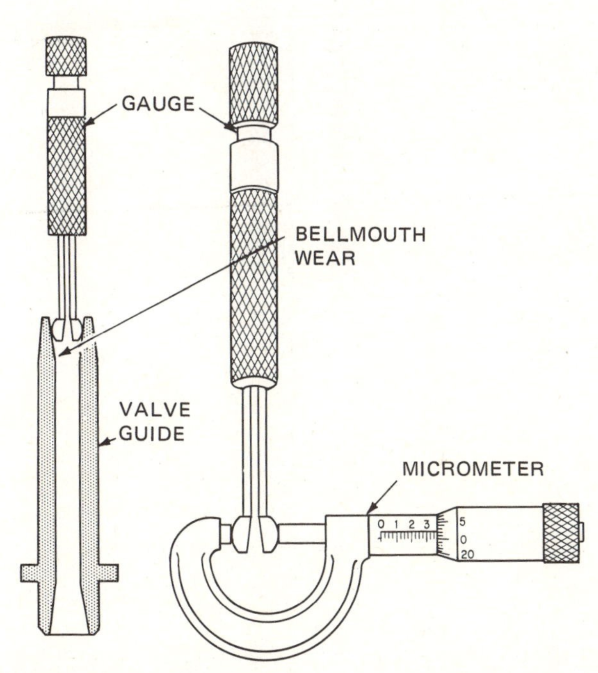

Fig. 23-57. A small-hole gauge is the most accurate device for inspecting valve-guide wear. The gauge is adjusted so that the split ball is a drag fit in the guide (*left*). Then the split ball can be measured with a micrometer (*right*).

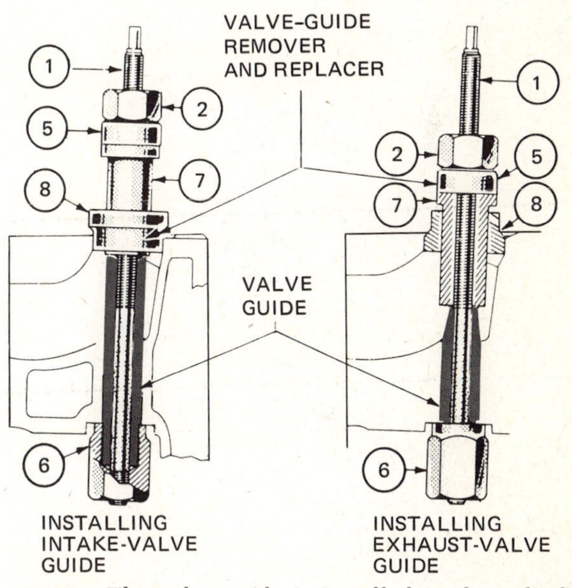

Fig. 23-59. The valve guide is installed in the cylinder block with a special replacer: (1) screw, (2) nut, (5) spacer, (6) recessed nut, (7) sleeve, (8) collar.

be concentric with the valve face. In addition, the valve-face angle must match the valve-seat angle (or have an interference angle). Thus, as a first step in valve-seat service, the valve guides must be cleaned and serviced (⊘ 23-30).

Valve seats are of two types. The *integral* type is actually the cylinder block or head. The *insert* type is a ring of special metal set into the block or head (Fig. 9-14). Replacing seat inserts and grinding seats are described below.

1. *REPLACING VALVE-SEAT INSERTS* A valve-seat insert may be badly worn. Or it may have been ground down on previous occasions so that there is insufficient metal for another grind. In either case, it must be replaced. The old seat must be removed with a special puller. If a puller is not available, the insert is punch-marked on two opposite sides and an electric drill used to drill holes almost through the insert. Then, a chisel and hammer can be used to break the insert into halves so that it can be removed. Care must be taken that the counterbore is

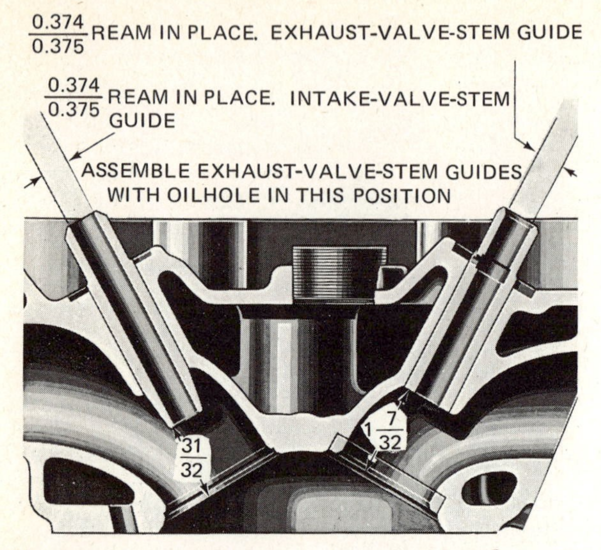

Fig. 23-60. Sectional view of a cylinder head from a V-8, overhead-valve engine. The reaming dimensions and correct locations of the intake and exhaust valve-stem guides are shown. (*Chrysler Corporation*)

Fig. 23-62. Sectional view of the valve guide showing the effect of knurling before reaming. (*United Tool Processes*)

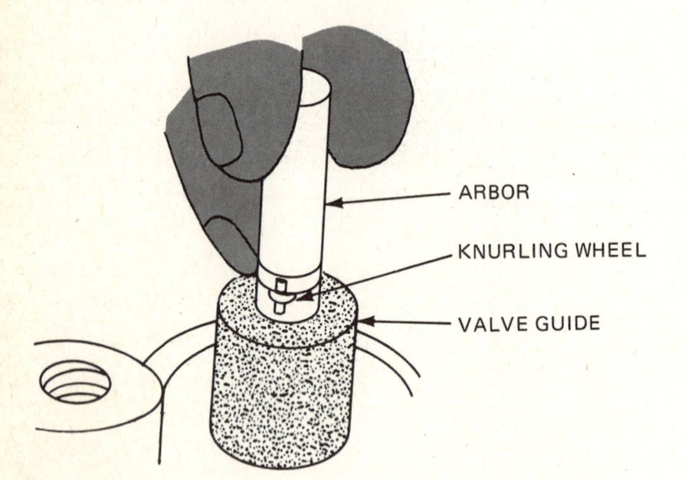

Fig. 23-61. Insertion of arbor, with knurling wheel in place, in readiness to knurl valve guide. (*United Tool Processes*)

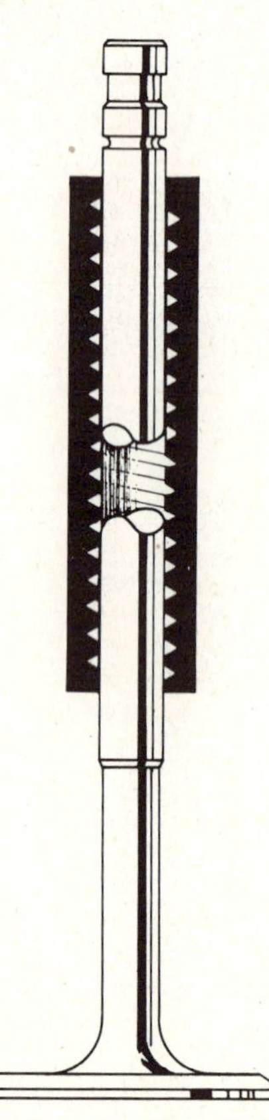

Fig. 23-63. Valve in place in a knurled and reamed valve guide. The valve stem is cut away to show the grooves in the valve guide. (*United Tool Processes*)

not damaged. If the new inserts fits too loosely, the counterbore must be rebored oversize. Then an oversize insert is installed. The new insert should be chilled in dry ice for 15 minutes to shrink it so that it can be driven into place. Then, the valve seat should be ground.

2. *GRINDING VALVE SEATS* Two types of valve-seat grinders are used: *concentric* and *eccentric* grinders. The concentric grinder rotates a grinding stone of the proper shape on the valve seat (Fig. 23-64). The stone is kept concentric with the valve seat by a pilot installed in the valve guide (Fig. 23-65). This means that the valve guide must be cleaned and serviced (⊘ 23-30) before the valve seat is ground. In the unit shown in Fig. 23-64, the stone is automatically lifted about once a revolution. This permits the stone to clear itself of grit and dust by centrifugal force. After the valve seat is ground, it may be too wide. It must be narrowed with upper and lower grinding stones to grind away the upper and lower edges of the seat. A typical valve seat is shown in Fig. 23-66. A gauge that measures valve-seat width is shown in Fig. 23-67. (A steel scale can also be used to measure valve-seat width.)

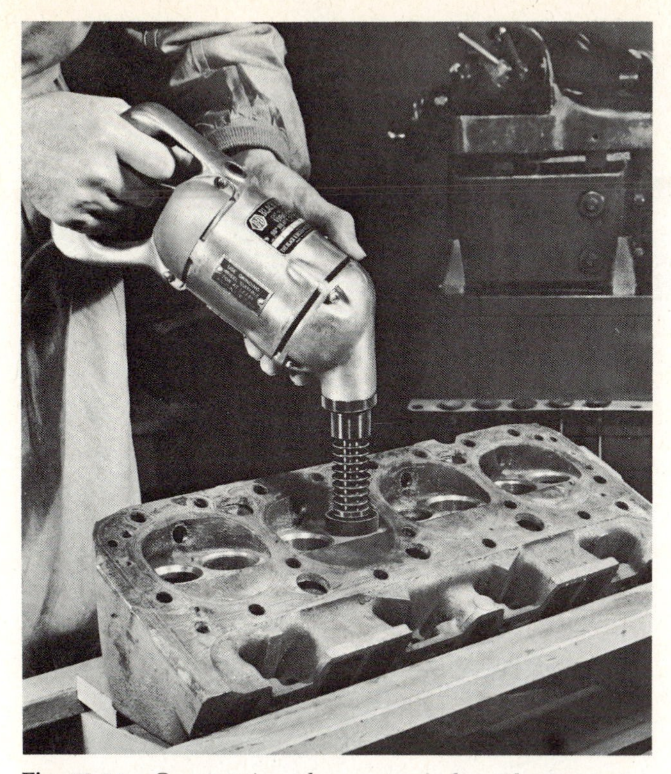

Fig. 23-64. Concentric valve-seat grinder. The stone is rotated at high speed. About once every revolution, it is automatically lifted off the valve seat so that it can throw off loosened grit and grindings. (*Black and Decker Manufacturing Company*)

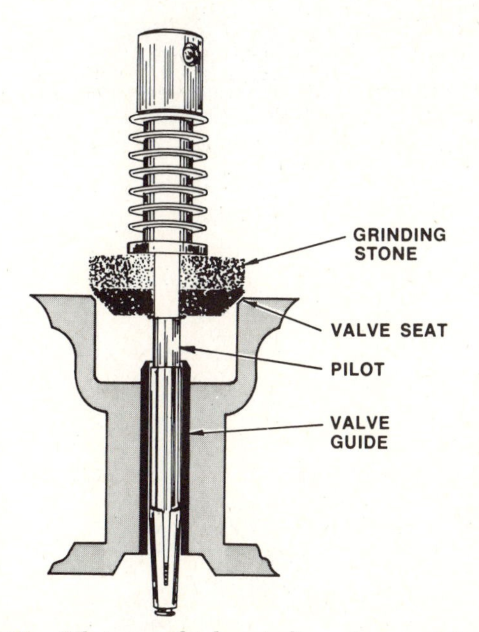

Fig. 23-65. Pilot on which grinding stone rotates. The pilot keeps the stone concentric with the valve seat. (*Black and Decker Manufacturing Company*)

In the eccentric valve-seat grinder (Fig. 23-68), the grinding stone is offset from the center of the valve seat. It makes only line contact with the valve seat. As the stone revolves, its center rotates slowly on an eccentric shaft. The slow rotation permits the line contact at which grinding is taking place to progress evenly around the entire valve seat. The eccentric valve-seat grinder also pilots the valve guide.

CAUTION: Be sure to follow instructions furnished by the grinder manufacturer. Note that the grinding stone must be dressed frequently with the diamond-tipped dressing tool.

3. *CHECKING VALVE GUIDES AND SEATS FOR CONCENTRICITY* After the valve guides are serviced and valve seats ground, the concentricity of both can be checked with a valve-seat dial gauge (Fig. 23-69). The gauge is mounted in the valve guide and is rotated so that the indicator finger sweeps around the valve seat. Any eccentricity (or run-out) of the seat is thus registered on the gauge dial.
4. *TESTING VALVE SEATING* Contact between the valve face and seat may be tested as follows: Mark lines with a soft pencil about $\frac{1}{4}$ inch [6.35 mm] apart around the entire valve face. Then put the valve into place and, with light pressure, rotate it half a turn to the left and then half a turn to the right. If this removes the pencil marks, the seating is good.

The seating can also be checked with prussian blue. Coat the valve face lightly with prussian blue. Put the valve on its seat and turn it with light pressure. If blue appears all the way around the valve seat, the valve seat and guide are concentric with each other. Now, check the concentricity of the valve face with the valve stem by removing the prussian blue from the valve and seat. Lightly coat the seat with prussian blue and then lightly rotate the valve on the seat. If blue transfers all the way around the valve face, the valve face and stem are concentric. This is a check similar to the run-out check (Fig. 23-46).

⊘ 23-32 Valve-Spring Inspection Valve springs should be checked for proper tension and squareness. A special fixture, such as shown in Fig. 23-70, checks tension. To check for squareness, stand the spring, closed-coil end down, on a flat surface. Hold a steel square next to it, as shown in Fig. 23-71. Rotate the spring slowly to see if the top coil moves away from the square more than $\frac{1}{16}$ inch [1.587 mm] (Ford). If the spring is excessively out of square, or has lost tension, discard it. (One manufacturer's recommendation is to replace all valve springs during the complete valve job. Then you are sure of good spring action.)

⊘ 23-33 Camshaft Service Camshaft removal varies somewhat from engine to engine. It is less complex in overhead-camshaft engines. In an overhead-valve engine the general procedure begins with removal of the radiator. Then take the pulley from the crankshaft and remove the gear or timing-chain cover. Detach the camshaft thrust plate (where

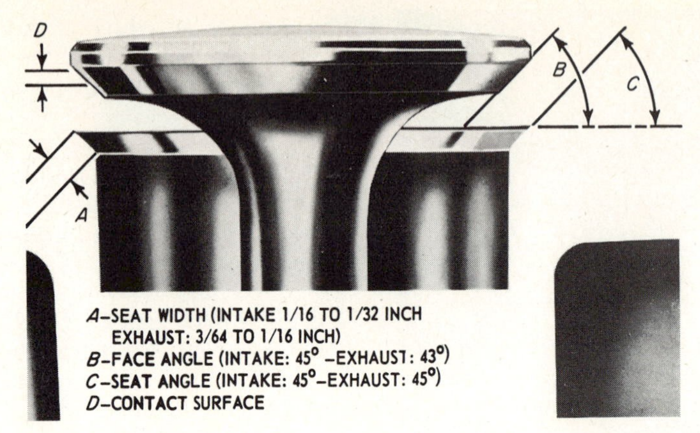

Fig. 23-66. Angles to which the valve seat and upper and lower cuts must be ground on one engine. The dimensions and angles vary with different engines. (*Chrysler Corporation*)

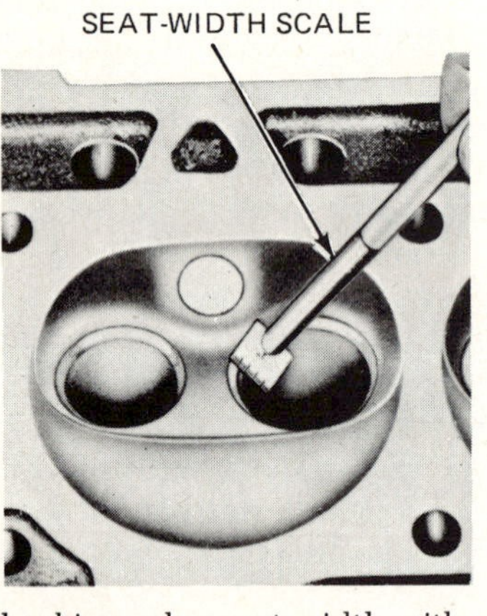

Fig. 23-67. Checking valve-seat width with a special tool. (*Ford Motor Company*)

Fig. 23-69. Valve-seat dial gauge. The unit shown is mounted in the valve guide. (*Oldsmobile Division of General Motors Corporation*)

Fig. 23-68. Eccentric valve-seat grinder installed on a cylinder block ready to grind a valve seat. The micrometer feed permits accurate feeding of the grinding wheel into the valve seat. (*Hall Manufacturing Company*)

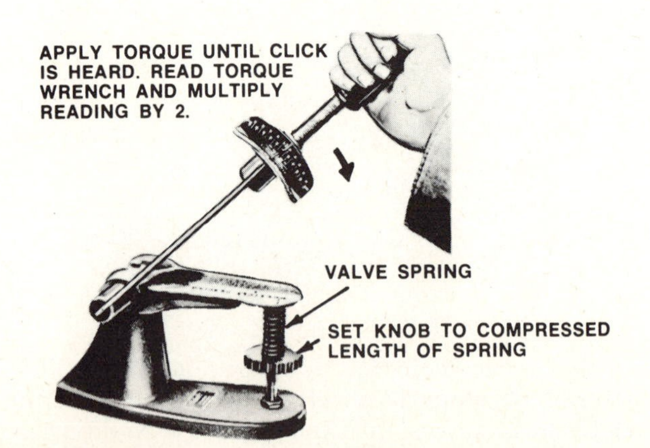

Fig. 23-70. Testing a valve spring for the proper tension in a special fixture. (*Ford Motor Company*)

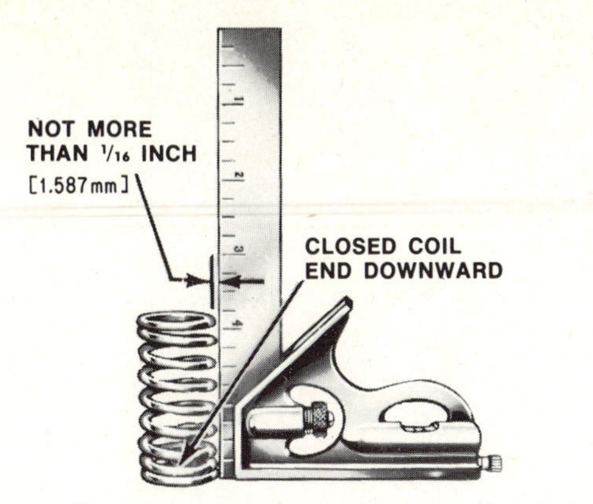

Fig. 23-71. Checking spring squareness. (*Ford Motor Company*)

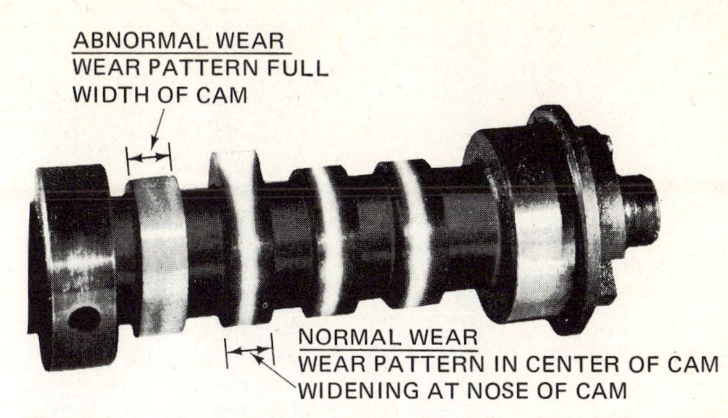

Fig. 23-73. Normal and abnormal cam wear. (*Oldsmobile Division of General Motors Corporation*)

present). Take off the camshaft sprocket and chain (where used). Remove the distributor or oil pump (whichever has the driven gear) so that the gear will not interfere with camshaft removal.

Remove the pushrods so that the valve lifters can be raised up out of the way. Now, the camshaft is free and can be pulled forward and out. Be very careful to keep the journals and cams from scratching the camshaft bearings. Support the rear of the camshaft as it is pulled out so that the bearings are not damaged.

NOTE: Supporting the camshaft is easier said than done. On most engines, it is necessary to remove the crankshaft in order to get to the camshaft and support it.

1. *CHECKING THE CAMSHAFT* Check the camshaft for alignment by rotating the camshaft in V blocks and using a dial indicator (Fig. 23-72). Journal diameters should be checked with a micrometer, and the bearings with a telescope gauge. The two dimensions can be compared to determine whether bearings are worn and require replacement. Pits or wear on the journals or cam lobes requires camshaft replacement.
2. *CHECKING FOR CAM WEAR* Figure 23-73 shows normal and abnormal cam wear. Normal cam wear is close to the center of the cam, as shown. The reason for this is that the cam is slightly tapered in most engines (Fig. 23-74). Also, the lifter foot is slightly spherical, or crowned, in shape. Therefore, when all is well, the contact pattern is as shown in

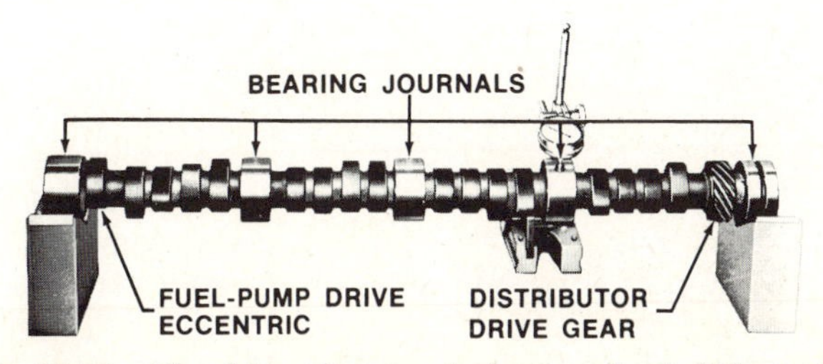

Fig. 23-72. Checking alignment of a camshaft. (*Chevrolet Motor Division of General Motors Corporation*)

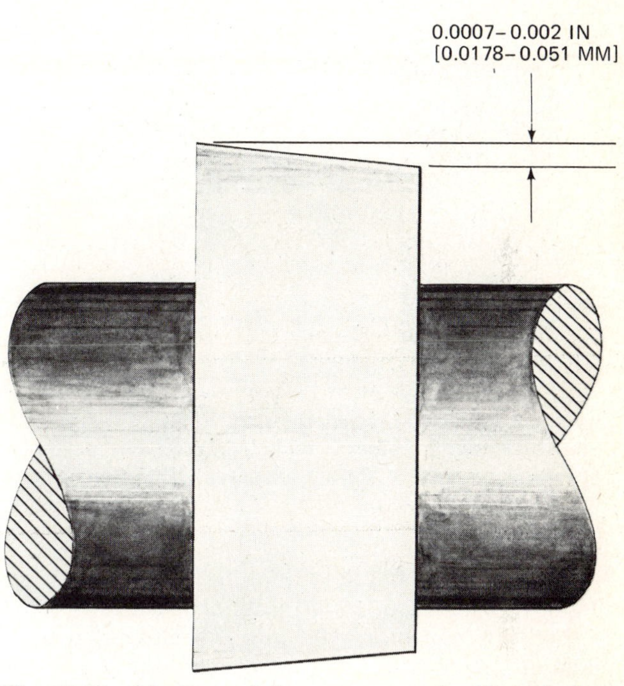

Fig. 23-74. Most cam lobes are tapered, although they appear to be flat. Taper is greatly exaggerated. (*Service Parts Division of Dana Corporation*)

Fig. 23-75. If wear shows across the full width of the cam, a new crankshaft is required. The lifter should also be checked (⊘ 23-34).

The cam-lobe lift can be checked with the camshaft in or out of the engine. Figure 23-76 shows the check with the camshaft in the engine. The setup in Fig. 23-72 can be used to measure the lobe lift with the camshaft out of the engine.
3. *REPLACING CAMSHAFT BEARINGS* A special bearing remover and replacer bar is required to replace the camshaft bearings. On some engines, the puller bar is threaded and a nut is turned to remove the bearings. On others, a hammer is used to drive against the bar and force the bearings out. Figure 23-77 shows a screw-type puller used to remove center bearings in a V-8 and an in-line engine. (When using this puller, you must work from one end of the block and then the other.) The two end

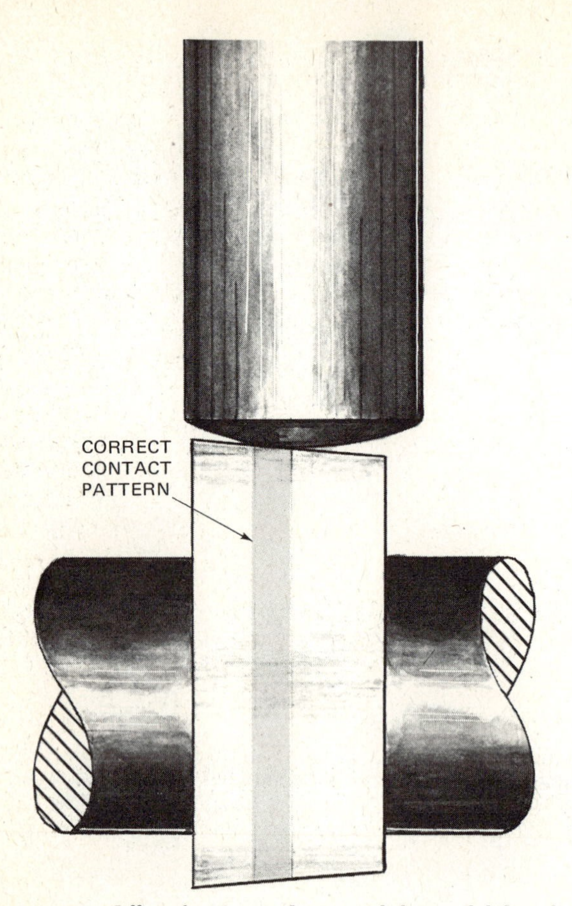

Fig. 23-75. Offset between the cam lobe and lifter face (which is crowned) gives a wide, centered, contact area. Taper and crown are shown exaggerated. (*Service Parts Division of Dana Corporation*)

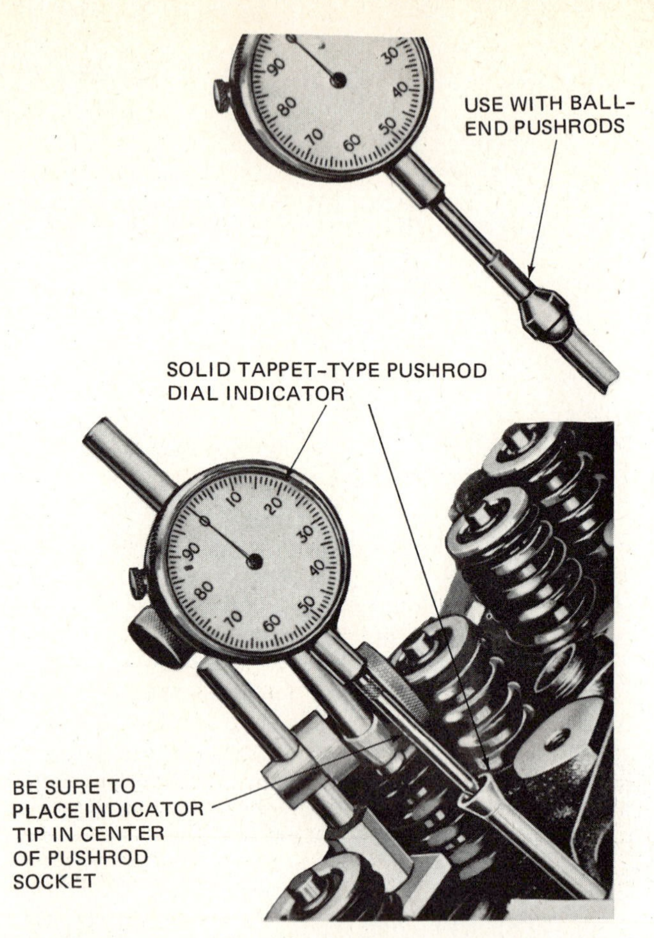

Fig. 23-76. Cam-lobe lift check on one engine. (*Ford Motor Company*)

bearings are removed with a different tool (Fig. 23-78). It is used to drive the end bearings inward so that they can be forced into the block and then taken out. In reinstalling the bearings, the end bearings must be driven in first (Fig. 23-78). These bearings serve as pilots for pulling new bearings into place with the screw-type puller.

Some engine manufacturers recommend a drive-type remover and replacer bar. The bar is put into position (Fig. 23-79), and a hammer is used to drive out the bearings, one at a time. The same tool is used to drive the new bearings into place (Fig. 23-80). Oilholes in the new bearings should align with the oilholes in the block. Also, new bearings should be staked in place if the old bearings were staked. If the new bearings are not of the precision type, they will require reaming to establish the proper fit.

4. *TIMING THE VALVES* The timing gears, or sprockets and chain, are marked for proper positions and correct valve timing (Figs. 9-5 to 9-7). To get to these markings, however, the front of the car must be partly torn down. Thus, some engines have another marking system for checking valve timing. This marking is on the flywheel or vibration damper near the ignition-timing markings. When this marking is visible or registers with a pointer, a designated valve should be just opening or should have opened a specified amount. Valve action is observed by removing the valve cover.

When the flywheel or vibration damper is not marked, piston position can be measured with a special gauge. The gauge is inserted through a special hole in the cylinder head. The relationship of the piston with the valves can thus be established.

5. *TIMING GEAR AND CHAIN* Gear run-out can be checked by mounting a dial indicator on the block. The indicating finger should rest on the side of the gear (Fig. 23-81). Run-out will then be indicated as the gear is rotated. Gear backlash is measured by inserting a narrow feeler gauge between the meshing teeth. Excessive run-out or backlash requires gear replacement. Excessive slack in the timing chain indicates a worn chain and possibly worn sprockets.

⊘ 23-34 Valve Lifters Solid and hydraulic valve lifters require different servicing procedures.

1. *SOLID VALVE LIFTERS* On some engines, the solid valve lifters are removed from the camshaft side. This procedure requires camshaft removal as a first step. In most engines, solid valve lifters are removed from the valve or pushrod side. As they are removed, valve lifters should be kept in order so that they can be restored to the bores from which

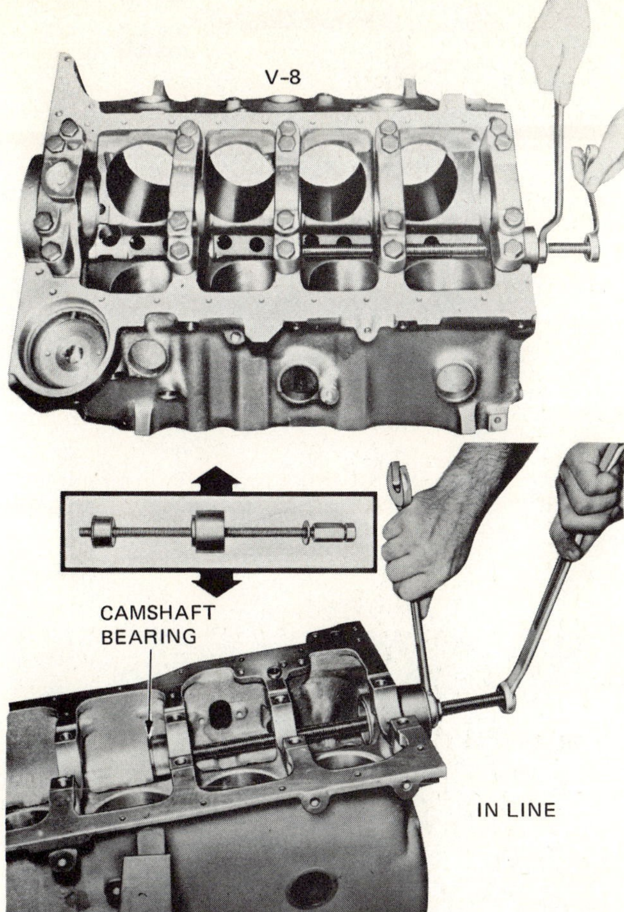

Fig. 23-77. Screw-type camshaft-bearing remover and replacer in use on a V-8 and an in-line engine. (*Chevrolet Motor Division of General Motors Corporation*)

they were removed. On many engines, oversize valve lifters may be installed if the lifter bores have become worn. Before this is done, however, the lifter bores must be reamed oversize.

NOTE: If the lifters have mushroomed on the bottom, they must be removed from the camshaft side.

2. HYDRAULIC VALVE LIFTERS On some engines, a "leak-down" test is used to determine the condition of the hydraulic valve lifters. One way to make this test is to insert a feeler gauge between the rocker arm and valve stem. Then, note the time it takes for the valve lifter to leak enough oil to allow the valve to seat. As the valve seats, the feeler gauge becomes loose. This indicates the end of the test. If the leak-down time is too short, the valve lifter is defective.

A more accurate leak-down test is made with the valve lifter out of the engine and installed in a special tester (Fig. 23-82). With this tester, the time required for a uniform pressure (from the weight on the end of the lever) to force the lifter plunger to bottom is measured. If it is too short, the lifter is defective.

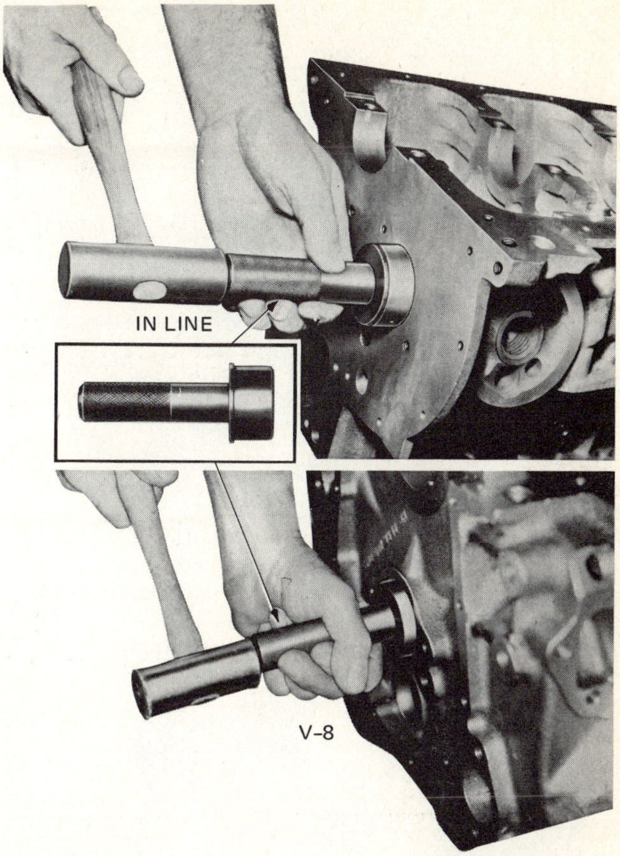

Fig. 23-78. Driver-type camshaft-bearing remover and replacer in use on the end camshaft bearings of an in-line and a V-8 engine. (*Chevrolet Motor Division of General Motors Corporation*)

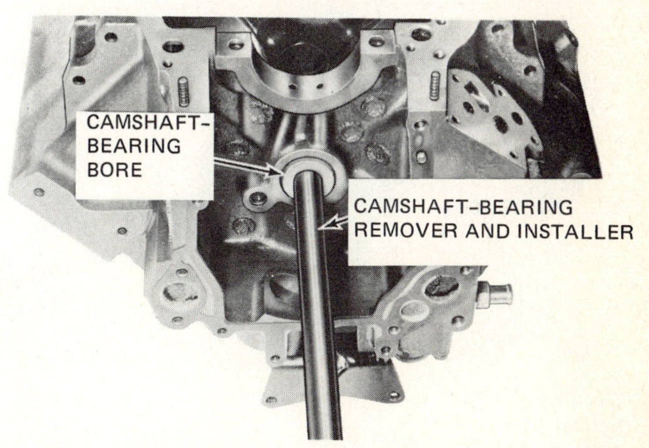

Fig. 23-79. Using a driver-type camshaft-bearing remover-and-replacer tool to remove camshaft bearings. (*Cadillac Motor Car Division of General Motors Corporation*)

To remove the hydraulic valve lifters from some engines, the pushrod cover and rocker-arm assembly must be removed first. Then the pushrods are taken out. On some engines with shaft-mounted rocker arms, a rocker arm can be moved by compressing the spring so that the pushrod can be removed. Thus, the rocker-arm assembly does not have to be taken off.

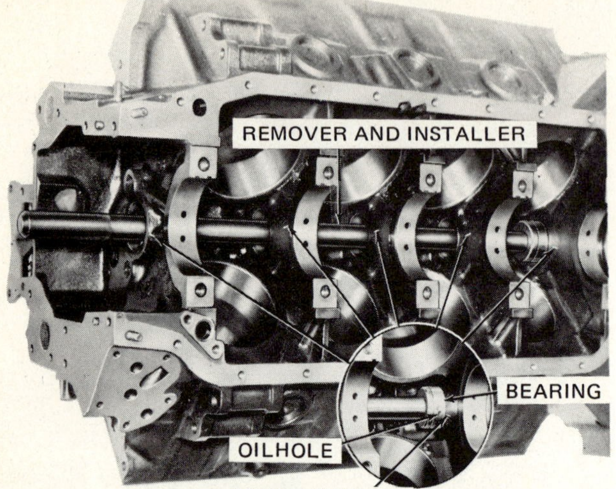

Fig. 23-80. Using a special tool to replace the camshaft bearings. Note the scribe marks on the faces of the bores to ensure the proper alignment of the bearing oilholes with the oilholes in the cylinder block. (*Cadillac Motor Car Division of General Motors Corporation*)

Fig. 23-81. Checking timing-gear run-out, or eccentricity, with a dial indicator. (*Buick Motor Division of General Motors Corporation*)

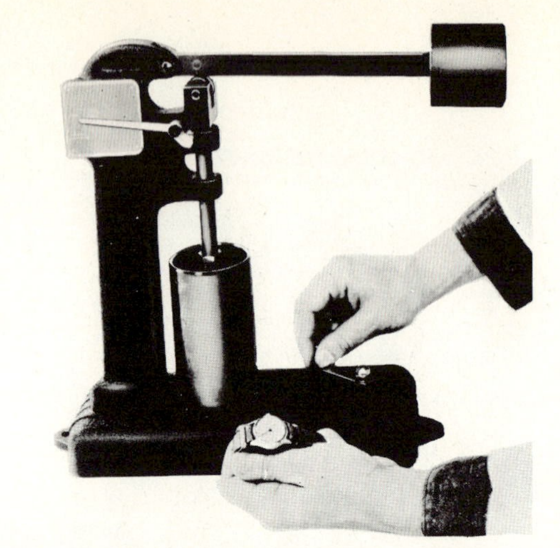

Fig. 23-82. Checking the leak-down rate of a hydraulic valve lifter removed from an engine. (*Pontiac Motor Division of General Motors Corporation*)

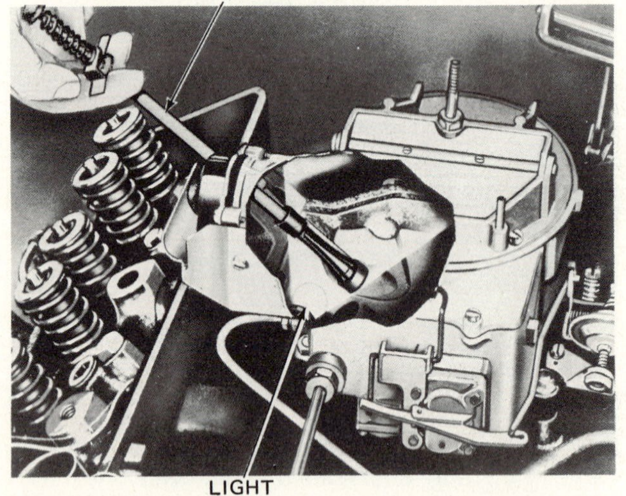

Fig. 23-83. Removing a valve lifter with a special tool. (*Ford Motor Company*)

The use of one valve-lifter-removing tool is shown in Fig. 23-83. The tool is inserted through the pushrod opening in the block and is seated firmly on the end of the valve lifter. The valve lifter is then removed through the pushrod opening.

3. *SERVICING HYDRAULIC VALVE LIFTERS* If a hydraulic valve lifter is defective, it is usually cheaper to replace it than to disassemble and service it. The labor required would probably cost more than the new lifter. However, if you prefer to spend the time, you can service the valve lifter as follows: Disassemble the lifter, and clean all parts in solvent. If any part is defective, replace the lifter. On reassembly, fill the lifter with clean, light engine oil.

Work on only one valve lifter at a time so that you do not mix lifter parts. Also, make sure each lifter goes back into the bore from which it was removed.

CAUTION: When servicing and handling hydraulic valve lifters, be extremely careful to keep everything clean. It takes only one tiny particle of dirt to cause a lifter to malfunction.

4. *CHECKING THE VALVE-LIFTER FOOT* As shown in Fig. 23-75, the foot of the valve lifter should be slightly spherical, or crowned. If it is worn or pitted, it can often be reground and reused. Figure 23-84 shows the setup for grinding: The crown is produced by rocking and rotating the lifter during finish-grind.

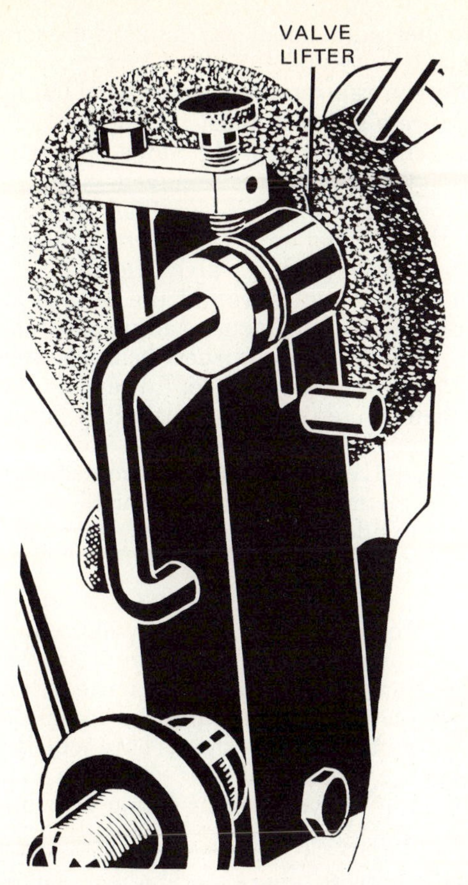

Fig. 23-84. Grinding the foot of a valve lifter. (*Snap-on Tools Corporation*)

Fig. 23-85. Sequence chart for tightening the intake-manifold attaching bolts on a V-8 engine. (*Chevrolet Motor Division of General Motors Corporation*)

⊘ 23-35 Removing and Replacing Manifolds Take the carburetor off. (Handle it with care to avoid damaging it or spilling gasoline from the float bowl.) Disconnect vacuum lines, exhaust pipes, pollution control hoses, and any other pipe or wire connected to the manifolds. Remove the nuts or bolts and take the manifolds off.

When reinstalling manifolds, be sure all old gasket material has been removed from the manifolds and cylinder head. Use new gaskets. Tighten nuts or bolts to the proper tension and in the proper sequence (Fig. 23-85).

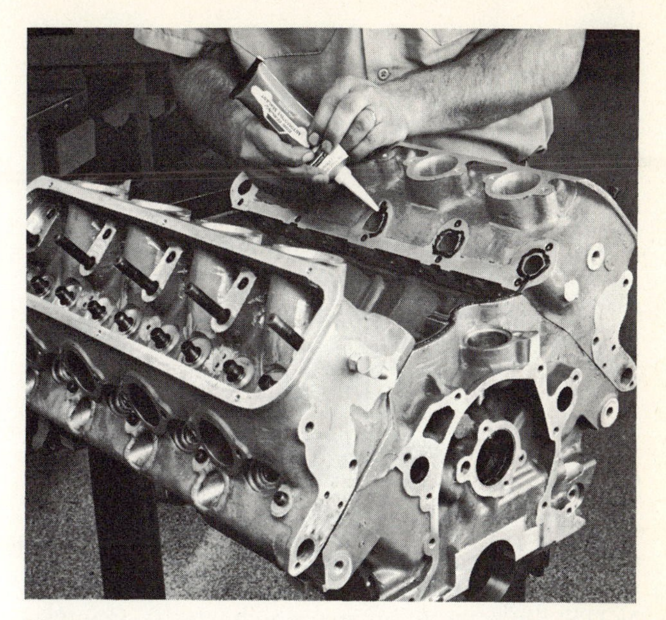

Fig. 23-86. Using plastic rubber to form a gasket. (*Dow Corning Corporation*)

NOTE: Some mechanics say they have success with a plastic gasket compound (silicone rubber) that comes in a tube. It is squeezed out of the tube onto the gasket surfaces (Fig. 23-86), and it spreads to form the gasket.

Check Your Progress

Progress Quiz 23-2 Once again you can find out how well you are understanding and remembering the material you have been studying. The following quiz, covering the second half of Chap. 23, will help you review the material and fix the important points clearly in your mind. If any of the questions stump you, turn back into the chapter and reread the pages that will give you the answer.

Completing the Sentences The sentences that follow are incomplete. After each sentence there are several words or phrases, only one of which will correctly complete the sentence. Write each sentence in your notebook, selecting the proper word or phrase to complete it correctly.

1. If the cylinder-block-bolt holes are not cleaned out, the head bolts may bottom in the bolt holes and prevent: (*a*) head-to-block sealing, (*b*) normal valve action, (*c*) adequate oil-ring action, (*d*) loss of compression.
2. The proper method of servicing valve faces is: (*a*) lapping, (*b*) grinding, (*c*) coating.
3. The spring compressor is used to install the: (*a*) valves, (*b*) spring retainers and locks, (*c*) rocker arms and pushrods.
4. If a valve is ground down so much that its outer edge is sharp (that is, its margin is lost), it will: (*a*) run too hot, (*b*) run too high, (*c*) run too cool.
5. The difference between the valve-face angle and the valve-seat angle is called the: (*a*) differential

angle, (*b*) angular difference, (*c*) interference angle.
6. The tip end of the valve stem should be: (*a*) rounded, (*b*) curved, (*c*) flat, (*d*) convex, (*e*) concave.
7. Valve seats may be serviced by either: (*a*) reaming or cutting, (*b*) cleaning or grinding, (*c*) grinding or replacing seat insert.
8. The two types of valve-seat grinding equipment are: (*a*) concentric and eccentric, (*b*) centrifugal and concentric, (*c*) valve-guide piloted and eccentric.
9. Valve guides are likely to wear: (*a*) round or oval-shaped, (*b*) out-of-round or oval-shaped, (*c*) oval-shaped and bell-mouthed.
10. On many engines, valve timing can be checked by means of markings on the: (*a*) valve guide or valve cover, (*b*) valve lifter or engine flywheel, (*c*) crankshaft and camshaft gears or sprockets.

CHAPTER 23 CHECKUP

NOTE: Since the following is a chapter review test, you should review the chapter before taking the test.

You are now well into the part of the book that is designed to give you practical guidance in actual work on automotive engines. Nearly all servicing procedures on modern engines are discussed on these pages. You should, of course, remember the essentials of these various procedures so that when you are in the shop, you will have a good idea of what to do and why you should do it. The following checkup will give you a chance to test yourself on how well you remember these procedures.

Completing the Sentences The sentences that follow are incomplete. After each sentence there are several words or phrases, only one of which will correctly complete the sentence. Write each sentence in your notebook, selecting the proper word or phrase to complete it correctly.

1. Among other things, excessive clearance between the valve stems and guides is apt to cause: (*a*) excessive compression, (*b*) spark knock, (*c*) tappet noise, (*d*) excessive oil consumption.
2. With the proper tools a complete valve-servicing job on six-cylinder engine, including grinding valves and seats, checking springs, and cleaning guides, should require about: (*a*) 2 hr, (*b*) 4 hr, (*c*) 6 hr, (*d*) 5 hr.
3. When you measure valve-tappet clearance on an I-head engine, you measure the clearance between the: (*a*) valve stem and tappet-adjustment screw, (*b*) rocker arm and pushrod, (*c*) rocker arm and valve stem, (*d*) valve stem and seat.
4. When you measure valve-tappet clearance on the Vega overhead-camshaft engine, you measure the clearance between the: (*a*) cam and valve tappet, (*b*) rocker arm and pushrod, (*c*) valve stem and tappet-adjustment screw, (*d*) valve stem and seat.
5. Engines equipped with hydraulic valve lifters normally require: (*a*) valve-tappet adjustment, (*b*) no valve-tappet adjustment, (*c*) adjustment-screw adjustment.
6. Before you can remove a valve from the cylinder head, the retainer lock must be removed after the: (*a*) retainer is removed, (*b*) valve spring is compressed, (*c*) valve guide is removed.
7. In some engines, grinding valves and seats may require the installation of: (*a*) shorter pushrods, (*b*) longer pushrods, (*c*) new valve lifters.
8. Before a valve seat is ground, the valve guide must be: (*a*) replaced, (*b*) reamed, (*c*) adjusted, (*d*) cleaned.
9. If you find that the camshaft bearings are excessively worn, you will have to: (*a*) replace them, (*b*) rebore them, (*c*) replace the camshaft, (*d*) grind camshaft journals.
10. After a valve seat has been smoothed and trued, upper and lower cutting reamers or stones must be used to: (*a*) widen valve seat, (*b*) narrow valve seat, (*c*) produce interference angle.

Service Procedures In the following, you should write the procedures asked for in your notebook. Do not copy from the book, but try to write in your own words just as you would explain it to another person. Give a step-by-step story. This will help you remember the procedures later when you go into the shop. In addition, you will be filling your notebook with valuable information.

1. Explain how to check and adjust valve-tappet clearance on an overhead-valve engine. On an overhead-camshaft engine.
2. Explain how to remove the valves from an overhead-valve engine.
3. Explain why mushroom ends on valve stems should always be removed before the valves are taken out of the cylinder head.
4. Explain how to grind valves.
5. Explain how to grind a valve seat.
6. Explain how to remove and replace valve guides in an overhead-valve engine.
7. Explain how to remove and check a camshaft.
8. Explain how to remove and replace the manifolds.
9. Explain how to remove and replace an overhead-valve cylinder head.
10. Make a list of the various valve troubles and their causes and their corrections.

SUGGESTIONS FOR FURTHER STUDY

When you are in the engine service shop, keep your eyes and ears open so that you can learn more about how various engine jobs are done. Study the operating manuals supplied by the service-equipment manufacturer so that you can learn how to operate valve-refacing machines, seat grinders, and so on. Also study the shop manuals issued by automobile manufacturers. These manuals include specific information on how to do various service jobs.

chapter 24

ENGINE SERVICE: CONNECTING RODS, PISTONS, AND RINGS

This chapter continues the discussion of engine service. It covers the servicing of connecting rods, connecting-rod bearings, pistons, and piston rings. Figure 24-1 is a disassembled view of a cylinder block for a V-8 engine. Only one rod-and-piston assembly is shown. Note that the piston pin is of the free floating type.

Connecting Rods and Rod Bearings

⊘ **24-1 Engine-Bearing Prelubricator** The engine bearing prelubricator (Fig. 24-2) provides initial lubrication to all bearings after a service job. That is, it is used to charge the reconditioned engine with oil before the engine is operated. This prevents damage to the bearings during initial startup.

The bearing prelubricator is used in many engine building and rebuilding shops. To use it, fill it with SAE20 or SAE30 oil and connect it to an air hose. Then connect the prelubricator hose to the engine lubricating system—at the oil filter, for example. Then, when air pressure is applied, oil is forced through the lubricating system.

Many automotive technicians believe that the best prelubrication is to put plenty of oil on engine parts during assembly. Additional prelubrication may be required, for example, to fill the hydraulic valve lifters and oil filter and prime the oil pump. If so, this should also be done: after the engine is assembled, add oil to the oil pan. Then use a slow-speed drill motor to turn the oil pump until the system pressurizes. Finally, install the ignition distributor.

⊘ **24-2 Using the Engine-Bearing Prelubricator to Check Bearings** The engine-bearing prelubricator can also be used to check for bearing wear before starting a service job. It is used with the oil pan off (Fig. 24-3). Worn bearings pass much more oil. This means that more oil gets on the cylinder walls and works up into the combustion chambers where it is burned. A normal bearing will leak between 20 and 150 drops of oil per minute when the prelubricator is used. If the bearing leaks more, it is worn. If it leaks less than 20 drops per minute, then either the bearing clearance is too small or the oil line to the bearing is stopped up.

NOTE: When oilholes in the crankshaft and in the bearing align, considerable oil is forced through the bearing. This will give the appearance of excessive wear. In such a case, the crankshaft should be turned a few degrees to move the oilholes out of register.

⊘ **24-3 Preparing to Remove Rods** Connecting rods and pistons are removed from the engine as assemblies. Removing, servicing, and replacing connecting rods requires about 5 to 8 hours, depending on the type of engine. About 3 additional hours are required to install new piston rings. Additional time is needed for such services as piston-pin bushing replacement. On most engines, the piston-and-rod assemblies are removed from the top of the egine. Thus the first step is to remove the cylinder head (⊘ 23-21). Cylinders should be examined for wear. If wear has taken place, there will be a ridge at the top of the cylinder. This ridge, called the *ring ridge,* marks the upper limit of piston-ring travel. If the ring ridge is not removed, the top ring could jam under it as the piston is moved upward. This could break the rings or the piston-ring-groove lands (Fig. 24-4). Thus, the ridge, if present, must be removed.

A quick way to check for a ring ridge is to see if your fingernail catches under it. If your fingernail catches on the ring ridge, so will the piston rings. A more accurate check is to use an inside micrometer. Measure the diameter on the ring ridge and then immediately below it. If the difference is more than 0.004 inch [0.102 mm], remove the ring ridge.

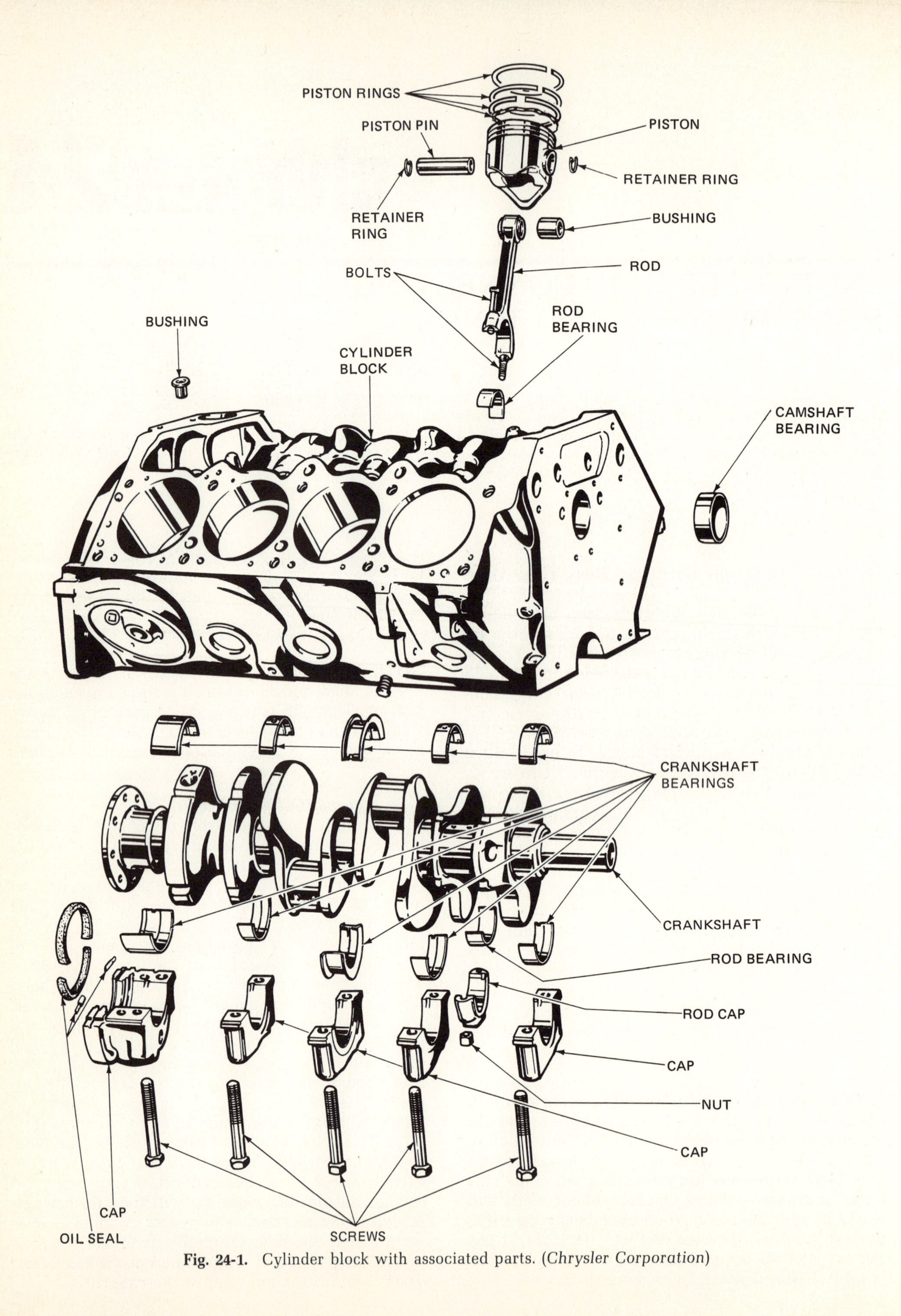

Fig. 24-1. Cylinder block with associated parts. (*Chrysler Corporation*)

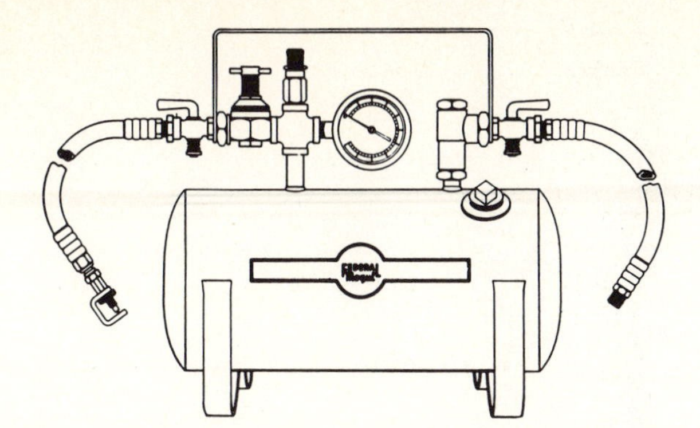

Fig. 24-2. Engine prelubricator. (*Federal-Mogul Corporation*)

Fig. 24-3. Oil-leakage test using the engine prelubricator. (*Federal-Mogul Corporation*)

⊘ 24-4 Removing the Ring Ridge To remove a ring ridge, use a special ring-ridge remover, as shown in Fig. 24-5. With the piston near BDC, stuff cloths into the cylinder and install the ridge remover.

NOTE: There are several different kinds of ridge removers. Be sure to read and follow the instructions carefully when you use a ridge remover.

Adjust the cutter blades to take off just enough metal to remove the ridge. Cover the other cylinders to keep cuttings from getting into them. Rotate the tool to cut the ridge away.

CAUTION: Turn the ridge remover by hand, not with an impact wrench! Do not remove too much metal. That is, do not undercut the top of the cylinder deeper than the material next to the ring ridge. Do not run the cutting tool above the cylinder. This would taper the edge.

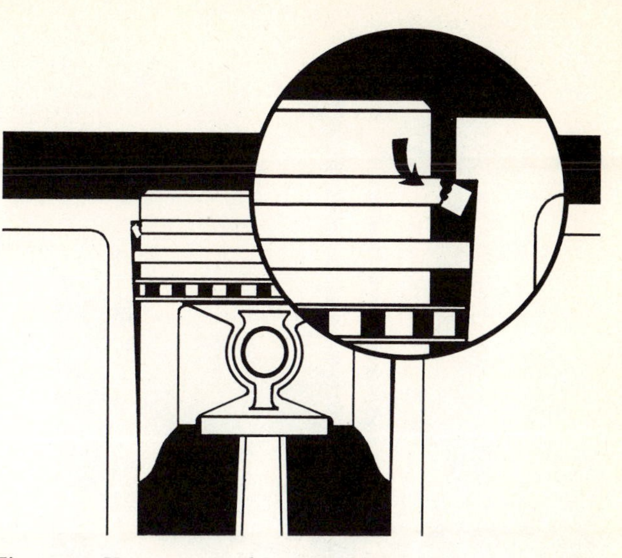

Fig. 24-4. How a ring ridge caused by cylinder wear might break the ring if the piston is withdrawn without removing the ridge. (*Sealed Power Corporation*)

Remove the tool, take the cloth out, and wipe the cylinder clean. Repeat the process for the other cylinders.

⊘ 24-5 Removing the Oil Pan The oil pan must be removed so that the connecting rods can be detached from the crankshaft. First, remove the drain plug to drain the engine oil (see Fig. 21-6). On many cars, the steering idler or other steering linkage must be removed. In such case, note how the linkage is attached. Note the number and location of shims (if used). On some cars, the oil pan is easier to remove if the engine mounting bolts are removed and the engine is raised slightly. Other parts may require removal before the oil pan can be taken off. These include exhaust pipe, oil-level tube, brake-return spring, and starting motor.

Next, the nuts or bolts holding the oil pan to the engine cylinder block can be removed. Steady the pan as the last two nuts or bolts are removed so that it does not drop. If the oil pan strikes the crankshaft so that the pan does not come free, turn the crankshaft over a few degrees. If the oil pan does not break loose when the last bolt is out, tap the sides of the oil pan with a rubber mallet. If it still doesn't come loose, carefully force the claw or flat edge of a pry bar or scraper between the edge of the pan and cylinder block (Fig. 24-6). Try to get the flat edge on the pan side of the gasket to avoid scratching the block. Tap the pry bar with a hammer to help free the oil pan.

NOTE: Many engines have metal reinforcements under the corner bolts of the oil pan. These reinforcements help seal around the rear main bearing and bottom of the timing cover. Don't let these get away from you when you remove the bolts.

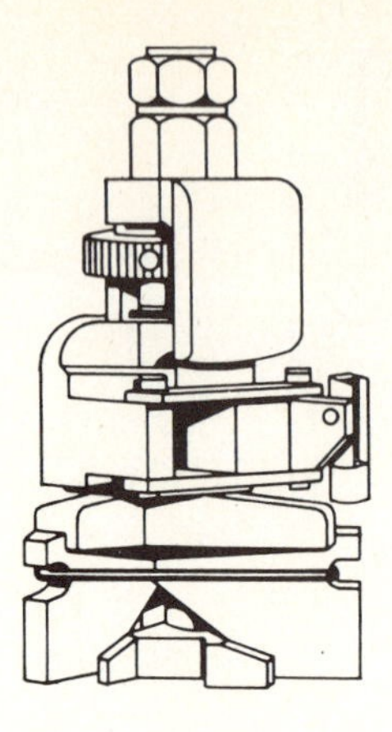

Fig. 24-5. Ridge-removal tool in place in the top of a cylinder. Cutters remove the ridge as the tool is turned in the cylinder.

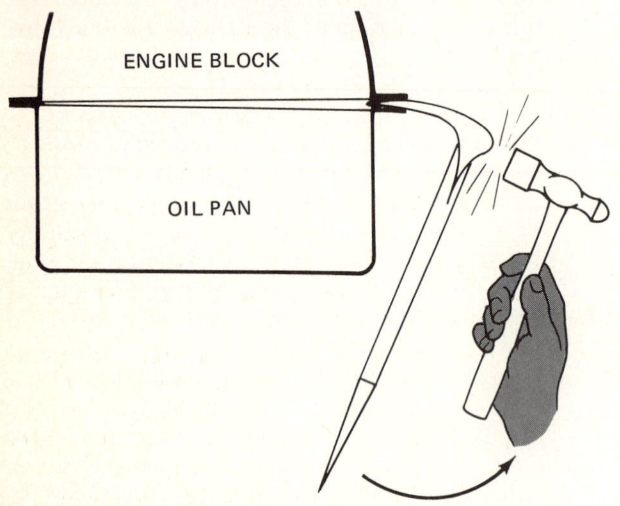

Fig. 24-6. Using a pry bar and hammer to break the oil pan loose.

Clean the oil pan, oil screen, and oil pump thoroughly before replacing the oil pan. Make sure that the gasket material is scraped off the pan and block gasket surfaces. Then check the flatness of the oil-pan gasket surfaces. Make sure that the bolt holes have not been dished in by overtightening of the bolts. The gasket surfaces can be straightened by laying the oil pan on a flat surface and tapping the gasket-surface flanges with a hammer.

Apply new gasket cement, if specified. Lay the gasket (or gaskets) in place on the oil pan. Be sure that the bolt holes in the gasket and oil pan line up. Install the oil pan and tighten the bolts or nuts to the proper tension.

NOTE: Some oil pans are installed with plastic gasket material (see Fig. 23-86).

⊘ 24-6 Removing Piston-and-Rod Assemblies After the preliminaries are out of the way, as noted in previous sections, remove the piston-and-rod assemblies, as follows.

CAUTION: Handle pistons and rods with care because they can be easily damaged. Never clamp a rod tightly in a vise. This can bend the rod and ruin it. Never clamp a piston in a vise. This can nick or break the piston. Do not allow the pistons to hit against each other or against other hard objects or bench surfaces. Distortion of the piston or nicks in the soft aluminum piston material may result from careless handling. These conditions will ruin the piston.

With the cylinder head and oil pan removed, crank the engine so that the piston of the No. 1 cylinder is near the bottom. Examine the rod and cap for identifying marks (Fig. 24-7). If none can be seen, use a small watercolor brush and a little white metal paint to mark a "1" on the rod and rod cap. Marks are needed to make sure that the parts go back into the cylinders from which they were removed. Each piston should also be numbered.

CAUTION: Do not mark the rods and rod caps with metal numbering dies or a center punch and hammer. This can distort and ruin the rods and caps.

Remove the rod nuts and caps. Slide the rod-and-piston assembly up into the cylinder away from the crankshaft. Use guide sleeves on the rod bolts, as shown in Fig. 24-8. These prevent the bolt threads from scratching the crankshaft journals. Also, the long handle permits easy removal and replacement of the piston-and-rod assembly. Short pieces of rub-

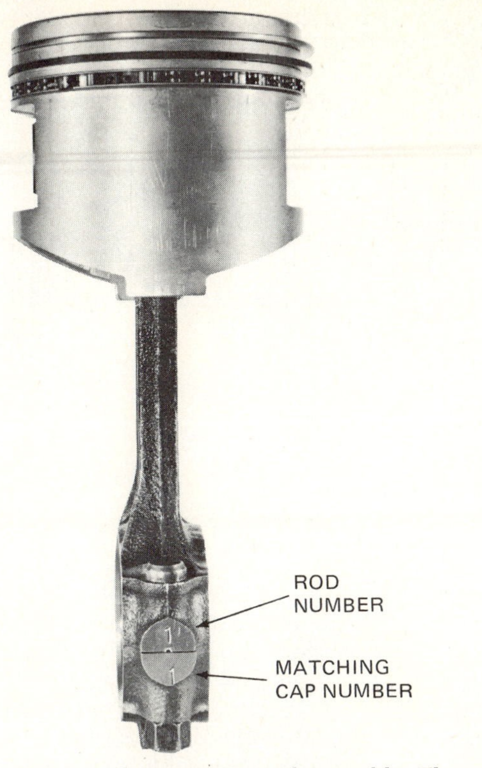

Fig. 24-7. Piston-and-connecting-rod assembly. The numbers on the rod and rod cap indicate that the assembly is to be replaced in the No. 1 cylinder. (*American Motors Corporation*)

Fig. 24-8. Using a special short guide and long guide to remove a connecting rod. (*Buick Motor Division of General Motors Corporation*)

ber hose, split and slipped over the rod bolts, will also protect the crankshaft journals.

Turn the crankshaft as you go from rod to rod so that you can reach the rod nuts. When all rods are detached and the assemblies have been moved up to the top of the cylinders, remove the assemblies. Lay the assemblies out in order on a cloth spread out on the bench or set them in a wooden piston box. Make sure all rods, rod caps, and pistons are marked with the number of the cylinder from which they were removed.

⊘ 24-7 Separating Rods and Pistons There are five basic piston-rod-bushing arrangements, as shown in Fig. 24-9. The rods and pistons are separated by removing the piston pins. If the pin is free-floating (Fig. 24-9A), remove the retainer ring and slide the pin out. If the pin is locked to the connecting rod or piston with a lock bolt (Fig. 24-9B, C, E), loosen the lock bolt and slide the pin out.

If the pin is a press fit in the connecting rod (Fig. 24-9D), it must be pressed out. This requires a special tool assembly, as shown in Fig. 24-10. The tool is put together in the arrangement shown. Then an arbor press is used to press the pin out (Fig. 24-11).

CAUTION: Be very careful to avoid nicking or scratching the pistons, piston rings, or rods. (See the Caution at the beginning of ⊘ 24-6.) Check the rods, rings, and pistons, as explained below.

⊘ 24-8 Attaching Connecting Rods and Pistons After connecting rods and pistons have been cleaned, serviced, and checked, lay them out on a clean bench in their engine order. Make sure parts match as in the original assembly. The pistons should then be attached to the rods with the piston pins, as follows. Make sure the piston is in the correct position on the rod as the two are attached. On many engines, the piston notches face to the front of the engine. Also, the rod oil hole faces toward the inside of the block.

To attach the piston and connecting rod where a lock bolt is used, simply put the pin through the piston and rod. Then tighten the lock bolt to hold the parts together. On the free-floating type (Fig. 24-9A), install the pin and retainer rings.

On the type with a press fit of the pin in the rod (Fig. 24-9D), a special tool assembly is required to press the pin in. The tool shown in Fig. 24-12 is properly arranged to install the pin. Figure 24-13 shows the tool put together and an arbor press used to install the pin. The parts are put together and the nut placed on the end of the main screw to hold everything together. Then pressure is applied in an abor press to push the pin into place. Plymouth recommends a fit test after the pin has been installed. This is done by placing the assembly in a vise, as shown in Fig. 24-14. Then a torque wrench is used to apply 15 pound-feet [2.073 kg-m] of torque to the nut on the end of the tool. If this amount of torque causes the connecting rod to move down on the piston pin, the press fit is too loose. The connecting rod must be discarded. If the rod does not move, the fit is satisfactory.

⊘ 24-9 Replacing Piston-and-Rod Assemblies After rods are reattached to their pistons and the pis-

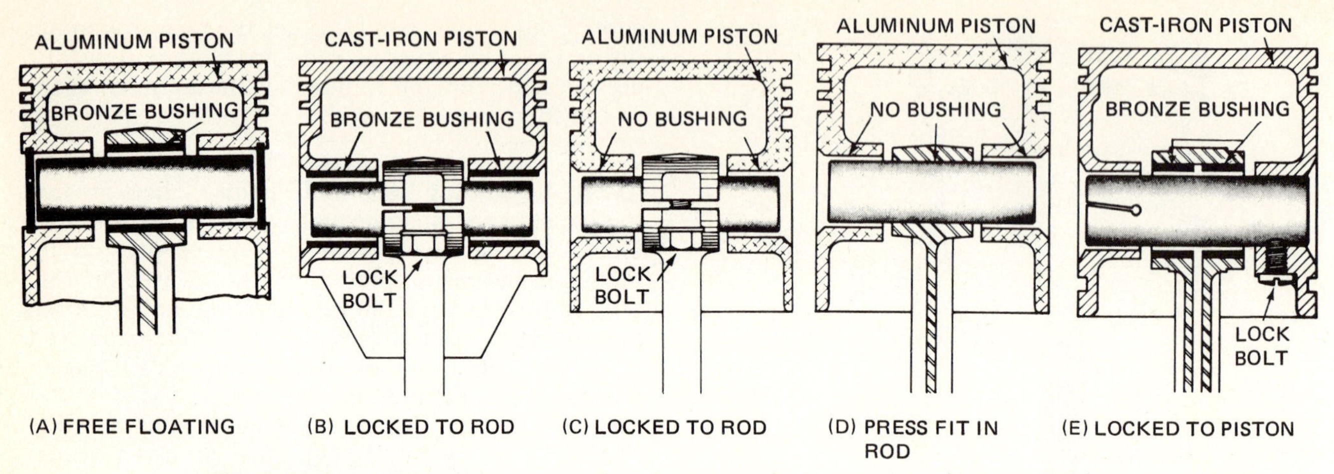

Fig. 24-9. Five piston, piston-pin, and connecting-rod arrangements. (*Sunnen Products Company*)

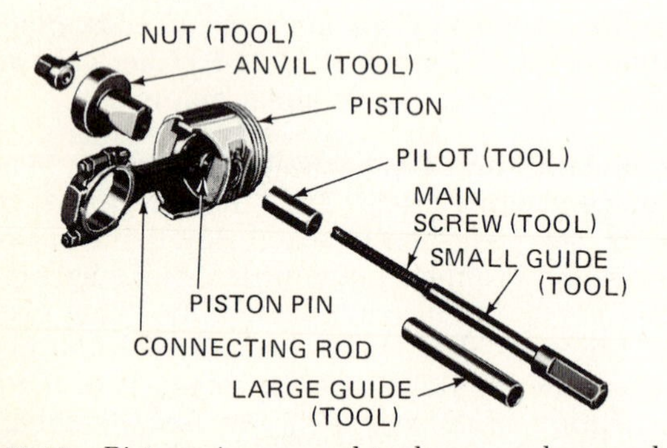

Fig. 24-10. Piston-pin-removal tool arranged properly to remove a pin. (*Chrysler Corporation*)

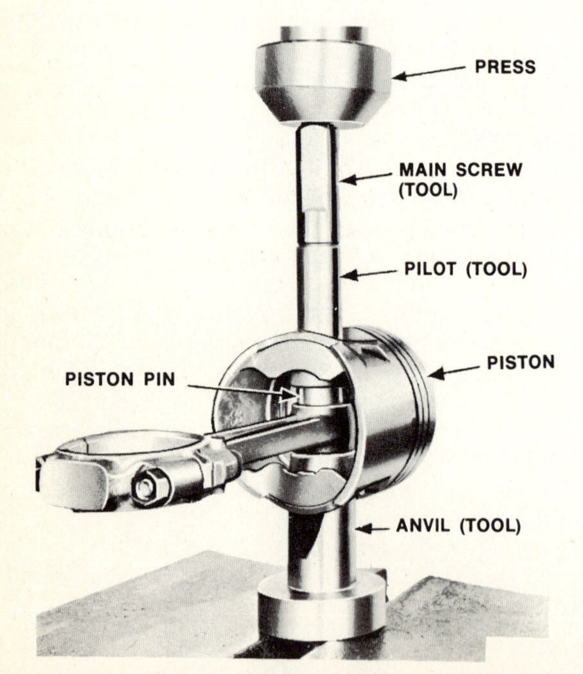

Fig. 24-11. Piston-pin-removal tool in use. (*Chrysler Corporation*)

ton rings are installed (⊘ 24-26), the assemblies go back into the engine. Rings should be positioned so that the ring gaps are uniformly spaced around the piston or as specified by the manufacturer. For example, Fig. 24-15 shows a recent Ford recommendation. Note that the gaps in the compression rings are toward the front, while the gaps in the oil ring parts are toward the back.

Dip the piston assembly above the piston pin in SAE30 oil (Fig. 24-16). Drain out excess oil. Use a piston-ring compressor or a loading sleeve to compress the rings into the piston-ring grooves (Fig. 24-17). Install guide sleeves on the rod bolts (Fig. 24-8), or cover the rod bolts with rubber hose. Then push the piston down into the cylinder. Tapping the head of the piston with the wooden handle of a hammer helps get the piston started. Make sure the assembly is installed with the piston facing in the right direction. Many pistons have a notch or other mark that should face toward the front of the engine (Figs. 24-15 and 24-17).

NOTE: ⊘ 24-28 explains in detail how to use the loading sleeve that comes with some ring sets. See also Fig. 24-55.

Attach the rod cap with the nuts turned down lightly. Then tap the cap on its crown lightly to help center it. Tighten the nuts to specifications with a torque wrench.

NOTE: Bearing clearances must be checked (⊘ 24-15).

⊘ 24-10 Checking Rod Side Clearance Make sure that the rods are centered on the crankshaft crankpins. If a rod is offset to one side, the rod-and-piston assembly probably has been put in backward. It probably has been turned 180 degrees from its correct position. Also, offset could mean a bent rod (see ⊘ 24-11). Clearance between connecting rods on V-8 engines should also be checked (Fig. 24-18). Incorrect side clearance means a bent rod.

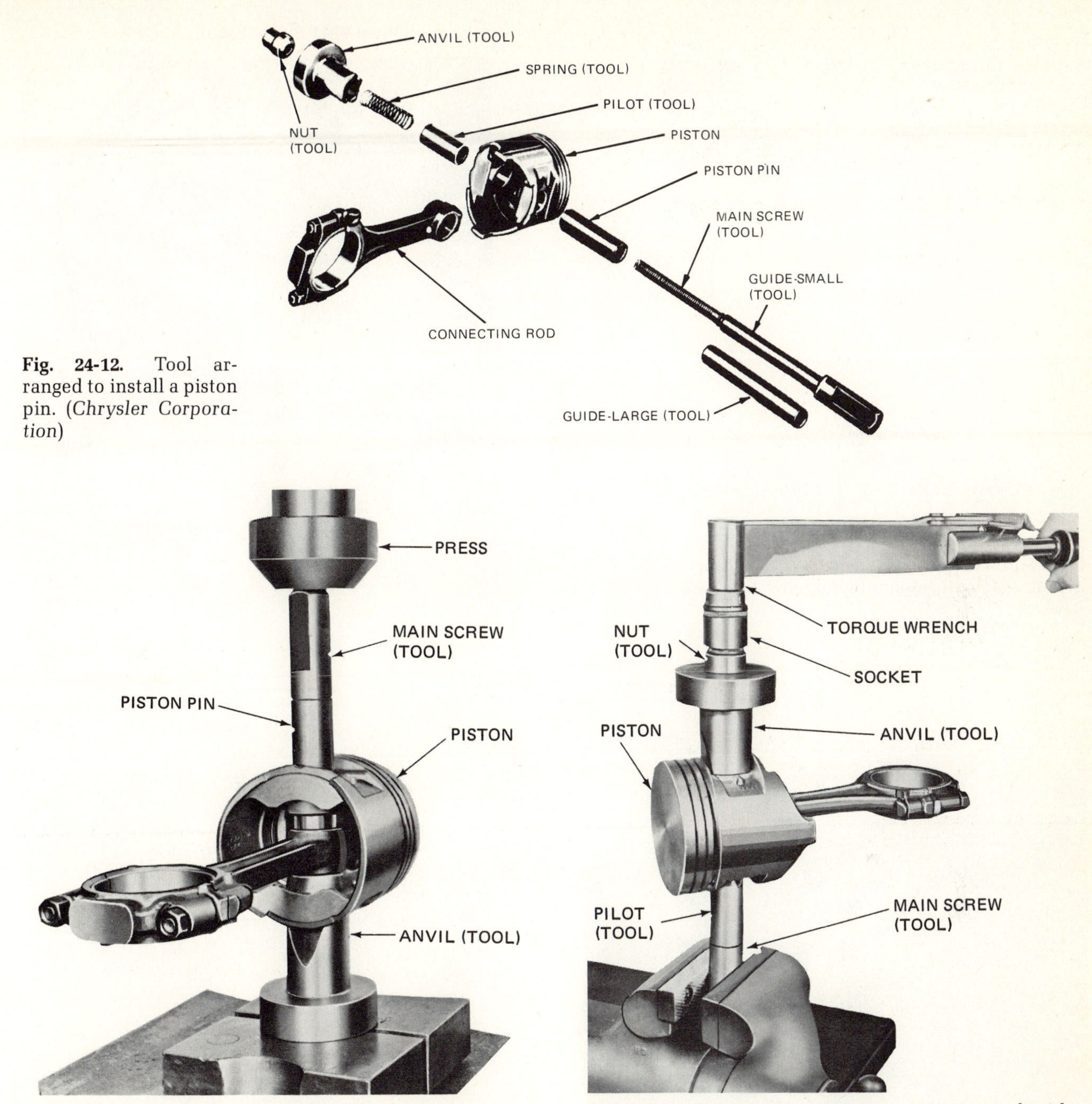

Fig. 24-12. Tool arranged to install a piston pin. (*Chrysler Corporation*)

Fig. 24-13. Installing a piston pin. (*Chrysler Corporation*)

Fig. 24-14. Testing piston fit in a connecting rod with a torque wrench. (*Chrysler Corporation*)

⊘ 24-11 Checking Connecting Rods After connecting rods are detached from the pistons, the rods and rod caps should be cleaned and inspected. Make sure to clean out the oil holes in the rods. Blow them out with compressed air.

Inspect the rod big-end bearings (see ⊘ 24-14). If the rod has a bushing in the small end, check its fit with the piston pin. If it is not correct, service is required (see ⊘ 24-12).

Check rod alignment. Figure 24-19 is an exaggerated view of the effects of a misaligned connecting rod. Heavy loading at points *A* and *B* on the bearing would cause bearing failure at these points. The heavy-pressure spots *C* and *D* on the piston cause heavy wear and possibly scoring of the piston and cylinder wall. A basic inspection check recommended by engine manufacturers is to look for uneven wear or shiny spots on the pistons. If any are found, the piston, pin, and rod should all be replaced.

A rough check for rod alignment can be made by detaching the oil pan and watching the rod while the engine is cranked. The rod should stay centered on the pin. If the rod moves back and forth on the piston pin or is not centered, the rod is out of line.

To check rod alignment out of the engine, reinstall the piston pin in the rod. Then mount the rod on the arbor of the special fixture by attaching the

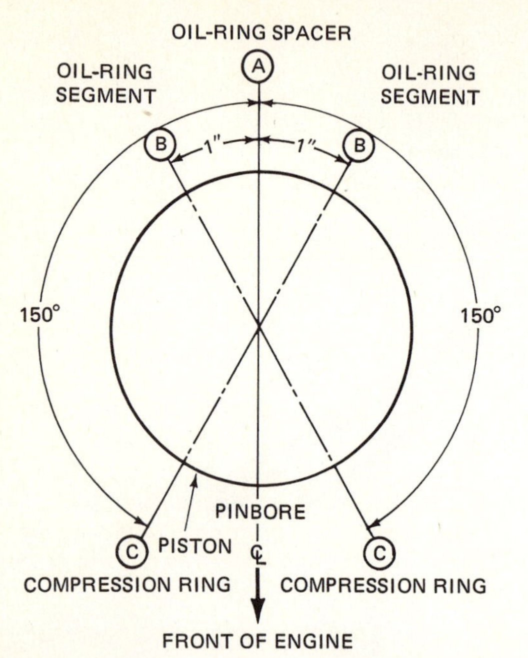

Fig. 24-15. Proper arrangement of ring gaps, as specified by one automotive manufacturer. (*Ford Motor Company*)

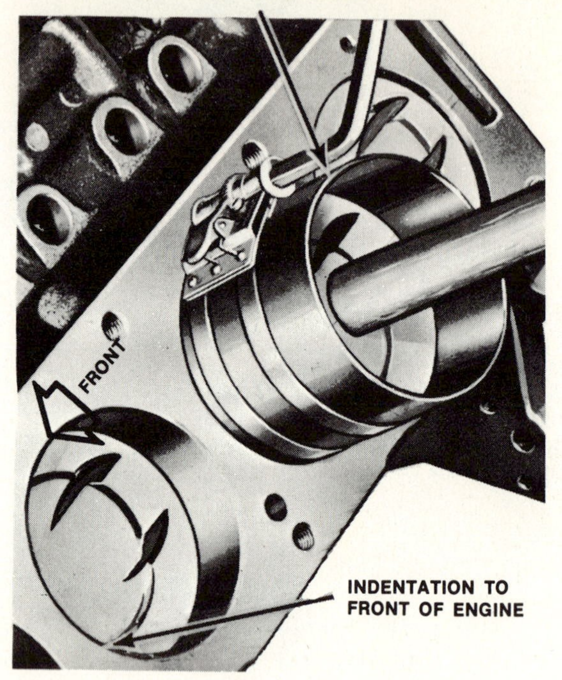

Fig. 24-16. Dipping the assembly into engine oil to cover the piston rings and pistonhead.

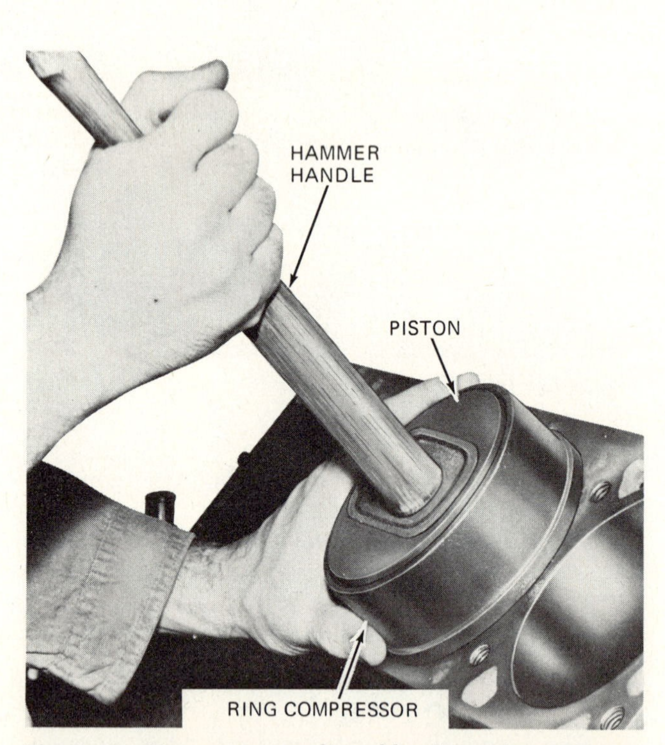

Fig. 24-17. *Top*, using an adjustable piston-ring-compressor tool to install a piston with rings; *bottom*, using a loading sleeve type of ring compressor. (*Chevrolet Motor Division and Cadillac Motor Car Division of General Motors Corporation*)

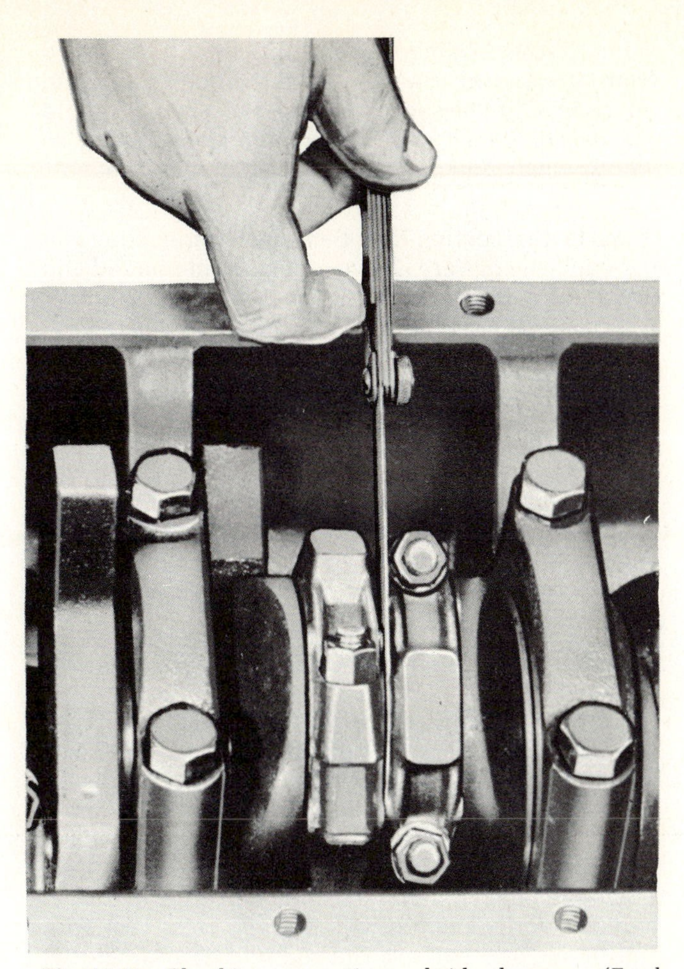

Fig. 24-18. Checking connecting-rod side clearance. (*Ford Motor Company*)

rod cap (Fig. 24-20). Put the V block over the piston pin and move it in against the faceplate. If the V block does not fit squarely against the faceplate, the rod is out of line. (This same fixture can be used to check alignment of the rod-and-piston assembly before the rings are installed.)

If the rod is out of line, check the crankpin for taper (⊘ 24-16). A tapered crankpin causes the rod to be subjected to bending stress.

Bent rods must be straightened or replaced. To straighten a rod, use a straightening bar inserted into the piston-pin hole. Bend the rod a little past straight and then back to straight again. This relieves the stress set up by the bending process.

NOTE: Experience has shown that bent connecting rods tend to take on a permanent set. Even if they are straightened, they may drift back to the bent condition. Most engine manufacturers require that bent connecting rods be replaced.

⊘ 24-12 Piston-Pin Bushings in Rods When the connecting rod has a piston-pin bushing (Fig. 24-9A and E), check the fit of the pin. If the fit is correct, the pin will not drop through the bushing of its own weight when held vertical. It will require a slight push to force it through. If the fit is too loose, the bushing should be reamed or honed for an oversize pin or replaced.

NOTE: Aluminum pistons have no bushings. They are supplied with prefitted piston pins as a matched set. If the pin is worn or is too loose a fit in the piston, a new pin-piston set is required. However, some automotive shops hone the piston-pin holes and install oversize pins, provided the piston is otherwise in good condition. This is not a recommended

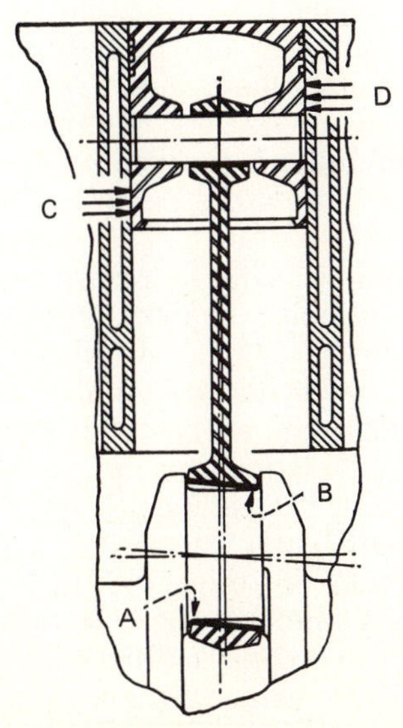

Fig. 24-19. Heavy-pressure areas caused by a bent rod. (The bent condition is exaggerated.) Areas of heavy pressure (A, B, C, and D) wear rapidly so that early failure results. (*Federal-Mogul Corporation*)

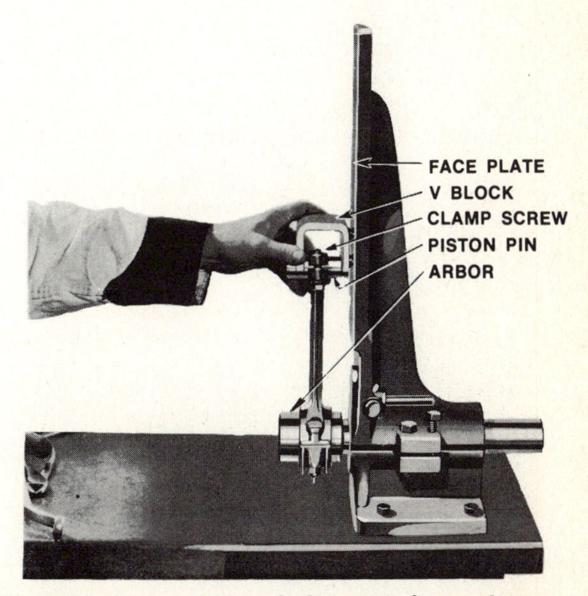

Fig. 24-20. Connecting-rod alignment fixture being used to check the alignment of a connecting rod. (*Chevrolet Motor Division of General Motors Corporation*)

practice, according to most automotive-engine manufacturers.

On some rods, the bushing cannot be replaced; if the bushing is so worn that it cannot be reamed or honed for an oversize pin, the complete rod must be replaced. On other rods, worn bushings can be replaced. The new bushings can be reamed or honed to fit the old pins (if they are in good condition) or new standard-sized pins. Pins that are worn, pitted, or otherwise defective should be discarded. To replace a bushing, press out the old bushing in an arbor press. Burrs on the edges of the bushing bore in the rod should be removed with a hand scraper or tapered burring reamer. Then, a new bushing can be pressed in with an arbor press. A tapered mandrel should then be used to expand the edges of the bushing. This procedure swages, or expands, them firmly in the rod. Make sure the oilholes in the bushing and rod align. Ream or bore the new bushing to size.

When reaming a set of bushings, proceed slowly on the first rod. Use an expansion reamer. Expand the reamer in easy stages, taking off a little metal each time. Try the pin fit after each reaming operation. This guards against overreaming. Then, after the first rod is reamed, all other rods may be rough-reamed quickly by reducing the reamer diameter about 0.0005 inch [0.0127 mm]. Then expand the reamer to take the final cut. At this stage, check the pins with a micrometer so that any slight variation in size can be taken care of. Suppose you find that one pin is slightly larger than the others. Then the bushing into which the pin will fit can be reamed slightly larger to provide a good fit. This ensures a good matching fit of pins to bushings.

To hone a set of bushings, follow the same two-step procedure. Rough-hone the bushings to within about 0.0005 inch [0.0127 mm] of the proper size. Then change hones, and finish-hone to size. Check pins with a micrometer during finish-honing so that variations in pin size can be taken care of. Figure 24-21 shows a clamp that holds the connecting rod. It also shows the method of holding the rod and clamp during honing. The bushing should be moved from one end of the stone to the other and should not be held in one spot. However, the bushing should not be moved past the end of the stone. Doing so would wear the edges of the bushing bell-shaped.

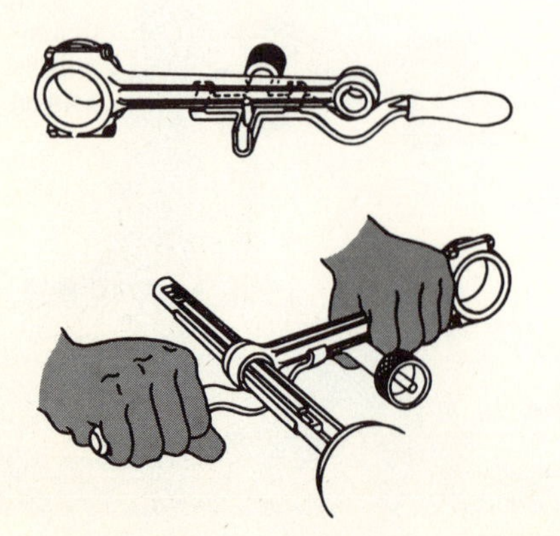

Fig. 24-21. *Top,* connecting rod in a clamp; *bottom,* method of holding the rod and clamp while honing, or grinding, the piston-pin bushing.

⊘ 24-13 Connecting-Rod Bearings Connecting-rod big-end bearings are of two types: *direct-bonded* and *precision-insert.* Some adjustment is possible on the direct-bonded type (⊘ 24-15), but if this bearing is worn, the complete rod and cap must be replaced. The precision-insert bearing is not adjustable. However, this type of bearing can be replaced without difficulty provided the rod, crankpin, and other engine components are in good condition. Whenever a rod bearing fails, an analysis should be made to determine the cause. Then the cause can be eliminated so that the failure will not be repeated quickly (⊘ 24-14).

⊘ 24-14 Analysis of Bearing Failures The following are the various types of bearing failure.

1. *BEARING FAILURE DUE TO LACK OF OIL (A IN FIG. 24-22)* When insufficient oil flows to a bearing, actual metal-to-metal contact results. The bearing overheats, and the bearing metal melts or is wiped out of the bearing shell. Welds may form between the rotating journal and bearing shell. There is a chance that the engine will "throw a rod." In other words, the rod will "freeze" to the crankpin and break, and parts of the rod will go through the engine block. Oil starvation of a bearing could result from clogged oil lines, a defective oil pump or pressure regulator, or insufficient oil in the crankcase. Also, bearings with excessive clearance may pass all the oil from the pump so that other bearings are starved and thus fail.
2. *BEARING FAILURE DUE TO FATIGUE (B IN FIG. 24-22)* Repeated application of loads on a bearing will ultimately fatigue the bearing metal. It starts to crack and flake out. Craters, or pockets, form in the bearing. As more and more of the metal is lost, the remainder is worked harder and fatigues at a faster rate. Ultimately, complete bearing failure occurs.

Fatigue failure seldom occurs under average operation conditions. However, certain special conditions will cause this type of failure. For instance, if a journal is worn out of round, the bearing will be overstressed with every crankshaft revolution. Also, if the engine is idled or operated at low speed much of the time, the center part of the upper rod-bearing half will carry most of the load and will "fatigue out." On the other hand, if the engine is operated at maximum torque with wide-open throttle (that is, if the engine is "lugged"), then most or all of the upper bearing half will fatigue out. High-speed operation tends to cause fatigue failure of the lower bearing half.

3. *BEARING SCRATCHED BY DIRT IN THE OIL (C IN FIG. 24-22)* Embeddability (⊘ 7-23) enables

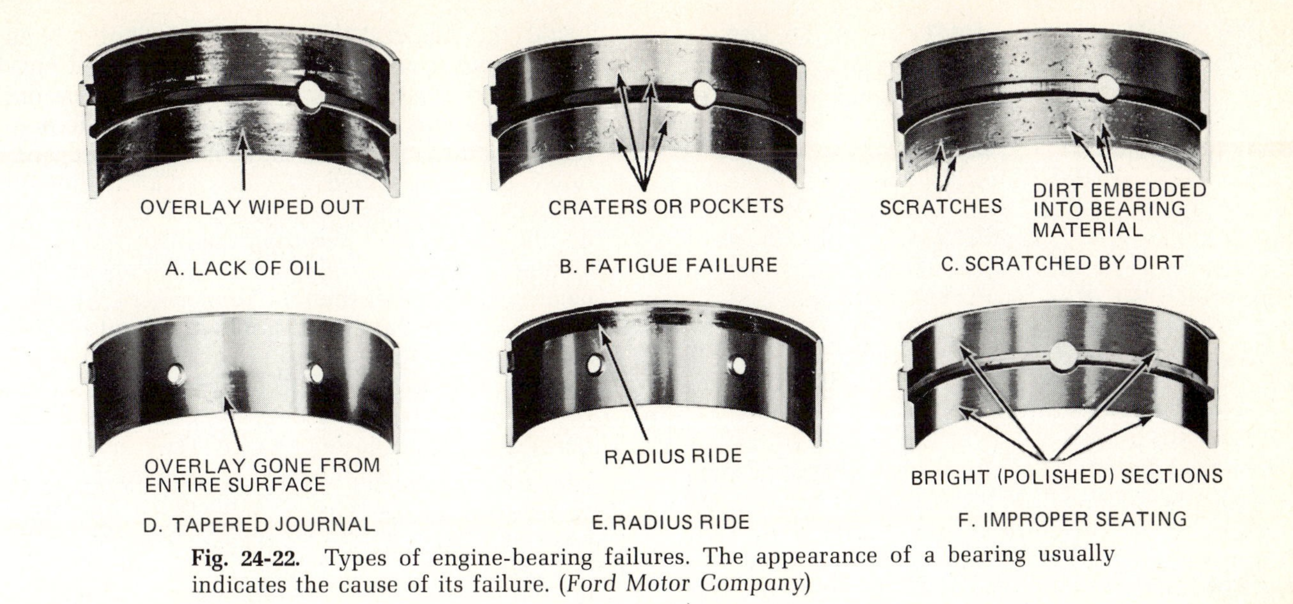

Fig. 24-22. Types of engine-bearing failures. The appearance of a bearing usually indicates the cause of its failure. (*Ford Motor Company*)

a bearing to protect itself by allowing dirt particles to embed so that they will not gouge out bearing material or scratch the rotating journal. Figure 24-23 shows, in exaggerated view, what happens when a particle embeds. The metal is pushed up around the dirt particle, reducing oil clearance in the area. Usually the metal can flow outward enough to restore adequate oil clearance. However, if the dirt particles are too large, they do not embed completely and are carried with the rotating journal, gouging out scratches in the bearing. Also, if the oil is very dirty, the bearing becomes overloaded with particles. In either case, bearing failure soon occurs.

4. BEARING FAILURE DUE TO TAPERED JOURNAL (D IN FIG. 24-22) If the journal is tapered, one side of the bearing carries most or all of the load. This side will overheat and lose its bearing metal. Do not confuse this type of failure with the failure that would result from a bent connecting rod. With a tapered journal, both bearing halves will fail on the same side. With a bent rod, failure will be on opposite sides (A and B in Fig. 24-19).

5. BEARING FAILURE FROM RADII RIDE (E IN FIG. 24-22) If the journal-to-crank-cheek radius is not cut away sufficiently, the edge of the bearing rides on this radius. This condition causes cramming of the bearing, possibly poor seating, rapid fatigue, and early failure. This trouble would be most likely to occur after a crankshaft-grind job during which the radii were not sufficiently relieved.

6. BEARING FAILURE FROM POOR SEATING IN THE BORE (F IN FIG. 24-22) Poor seating of the bearing shell in the bore causes local high spots where oil clearances are too low. Figure 24-24 shows, in exaggerated view, what happens when particles of dirt are left between the bearing shell and the counterbore. This reduces oil clearance (as at X). Also, an air space exists which prevents proper cooling of the bearing. The combination can lead to quick bearing failure.

7. BEARING FAILURE FROM RIDGING Crankpin ridging, or "camming," may cause failure of a partial-oil-groove type of replacement bearing installed without removal of the ridge. The ridge forms on the crankpin because of uneven wear between the part of the crankpin in contact with the partial oil groove and the part that runs on the solid bearing. The original bearing wears to conform to this ridge. However, when a new bearing is installed, the center zone may be overloaded at the ridge and may soon fail. A ridge so slight that it can hardly be detected (except with a carefully used micrometer) may be enough to cause this failure. Failures of this sort have been reported in engines having ridges of less than 0.001 inch [0.025 mm].

⊘ 24-15 Inspecting the Connecting-Rod-Bearing Fit
Precision-insert bearings must be inspected in one way, and the direct-bonded bearings in another.

CAUTION: Before installing new bearings, the crankpins should always be checked for taper or out-of-roundness (⊘ 24-16).

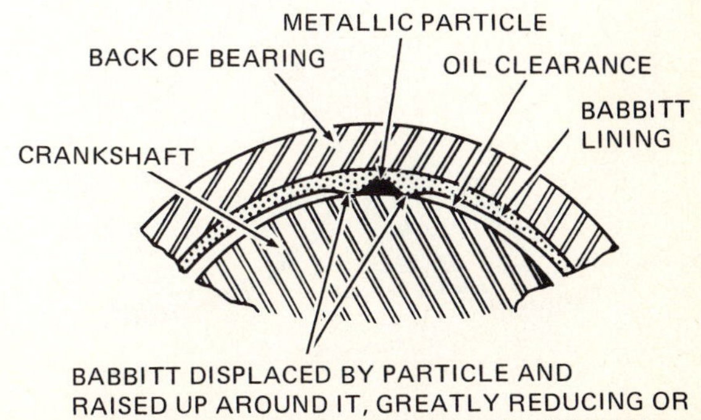

Fig. 24-23. Effect of a metallic particle embedded in bearing metal (the babbitt lining). (*Federal-Mogul Corporation*)

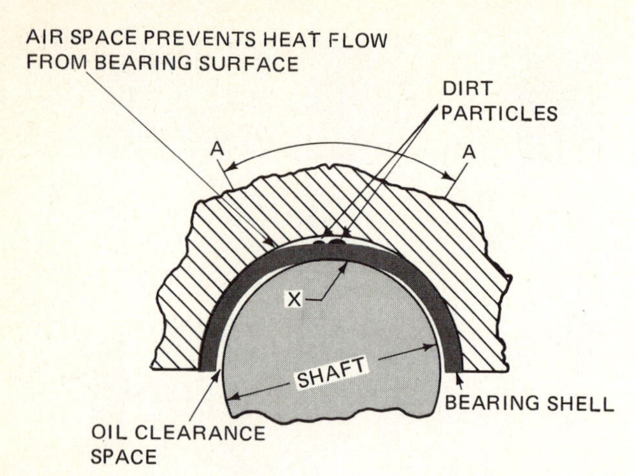

Fig. 24-24. Effect of dirt particles under the bearing shell, caused by poor installation. (*Federal-Mogul Corporation*)

1. *PRECISION-INSERT BEARINGS* The fit of these bearings can be inspected with Plastigage shim stock or with micrometer and telescope gauge.

a. *Plastigage* Plastigage is a plastic material that comes in strips and flattens when pressure is applied to it. Put a strip of the material into the bearing cap. Install the cap and tighten the rod nuts to the specified tension. Then, remove the cap and measure the amount of flattening. If the Plastigage is flattened only a little, then oil clearances are large. If it is flattened considerably, oil clearances are small. Actual clearance is measured with a special scale supplied with the Plastigage (Fig. 24-25).

The bearing cap and crankpin should be wiped clean of oil before the Plastigage is used. The crankshaft should be turned so that the crankpin is about 30 degrees back of BDC. Do not move the crankshaft while the cap nuts are tight. This would flatten the Plastigage further and throw off the clearance measurement.

b. *Shim Stock* Lubricate a strip of 0.001-inch [0.0254 mm] shim stock ½ inch [12.70 mm] wide. Lay it lengthwise in the center of the bearing cap.

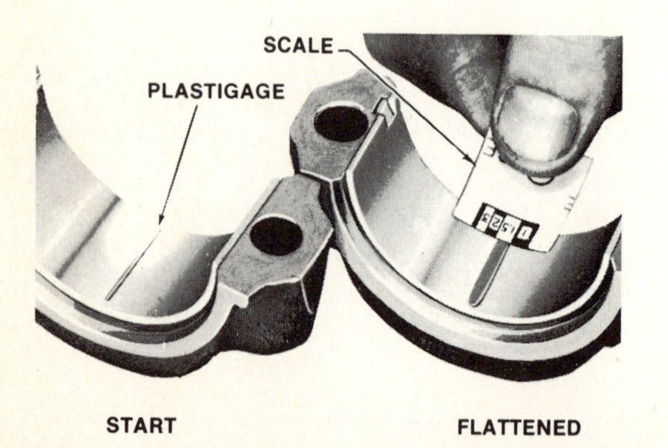

Fig. 24-25. Bearing clearance being checked with Plastigage. *Left,* Plastigage in place before tightening the cap; *right,* measuring the amount of flattening (or bearing clearance) with a scale. (*Buick Motor Division of General Motors Corporation*)

Install the cap, and tighten cap nuts lightly. Note the ease with which the rod can be moved endwise on the crankpin. If the rod moves easily, tighten the nuts a little more and recheck. Repeat the procedure until the nuts have been drawn down to the specified tightness or until the rod tightens up on the crankpin. If the rod tightens up, the clearance is less than the thickness of the shim stock. If the rod does not tighten up, the clearance is greater than the thickness of the shim stock. With the latter condition, lay an additional strip of shim stock on top of the first and repeat the checking procedure. If the rod still does not tighten up, keep adding shim stock until the actual bearing clearance is determined. Then remove the cap, take out the shim stock, replace the cap, and torque the cap nuts to specifications. Excessive clearance requires bearing replacement (⊘ 24-16).

c. *Micrometer and Telescope Gauge* Inspect the crankpin diameter with a micrometer. Inspect the bearing diameter (cap in place) with a telescope gauge and micrometer (or an inside micrometer). Compare the two diameters to determine the difference, or bearing clearance. At the same time, inspect the crankpin for taper or eccentric wear. Measure the diameter at several places along the crankpin (to check for taper). Also measure around the crankpin (to check for eccentricity, or out-of-roundness).

2. *DIRECT-BONDED CONNECTING-ROD BEARINGS* On these bearings, adjustment is made by installation or removal of shims under the cap. Shims placed between the cap and rod (at the bolt bosses) hold the cap away from the rod when the nuts are tightened. This increases the bearing clearance. Clearance can be checked with a micrometer and telescope gauge, as noted previously. It can also be checked by attempting to snap the rod back and forth on the crankpin with one hand. If the rod moves easily, take off the rod cap. Remove one shim only from each side of the cap. Replace the cap and try to move the rod. If the rod still moves easily, take off another pair of shims. Repeat this procedure until the rod will not move. Then add one shim to each side of the cap, replace and tighten the cap, and retest.

If the bearing is worn, pitted, scored, chipped, or otherwise damaged, replace the rod and cap as a unit. Rebabbitting of this type of rod should not be attempted in the field. Special equipment is required to do this job.

⊘ 24-16 Installing Precision Connecting-Rod Bearings New precision connecting-rod bearings are required if the old ones are defective (⊘ 24-14) or have worn so much that clearances are excessive. They are also required if the crankpins have worn out of round or have tapered so much that they must be reground. In this case, new undersize bearings are required. Engine rebuilders usually replace the bearings in an engine when it is torn down whether or not the old bearings are in bad condition. Their reasoning is that when the engine is torn down for rebuilding, it costs only slightly more to put in new bearings. However, if the engine had to be torn down especially for bearing installation, the cost would be high. They believe it is cheap insurance against failure to install new bearings during an engine-rebuilding job.

1. *INSPECTING CRANKPINS* Crankpins should always be inspected with a micrometer for taper and concentricity. Measurements should be taken at several places along the crankpin to check for taper. Diameter should be checked all the way around for out-of-roundness. If crankpins are out of round or tapered more than 0.0015 inch [0.037 mm], the crankshaft must be replaced or the crankpins reground (⊘ 25-14). Bearings working against taper or out-of-roundness of more than 0.0015 inch [0.037 mm] will not last long. And when bearings go, there is the chance that the engine will be severely damaged.

2. *INSTALLING NEW BEARINGS* When new bearings are to be installed, make sure your hands, the workbench, tools, and all engine parts are clean. Keep the new bearings wrapped up until you are ready to install them. Then handle them carefully. Wipe each with a fresh piece of cleaning tissue just before installing it. Be very sure that the bores in the cap and rod are clean and not excessively out of round.[1] Then put the bearing shells in place (Fig. 24-26). If they have locking tangs, make sure that the tangs enter the notches provided in the rod and cap. Note the following comments about bearing spread and crush. Check clearance after installation (see ⊘ 24-15).

CAUTION: Do not attempt to correct clearance by filing the rod cap. This destroys the original relationship between cap and rod and will lead to early bearing failure.

3. *BEARING SPREAD* Bearing shells are usually manufactured with "spread." That is, the shell diameter is somewhat greater than the diameter of the rod cap or rod bore into which the shell will fit (Fig. 24-27). When the shell is installed into the cap or rod, it snaps into place and holds its seat during later assembling operations.

4. *BEARING CRUSH* In order to make sure that the bearing shell will "snug down" into its bore in the rod cap or rod when the cap is installed, the bearings have "crush" (Fig. 24-28). They are manufactured to have some additional height over a full half. This additional height must be crushed down when the cap is installed. Crushing down the additional height forces the shells into the bores in the cap and rod.

[1] Some manufacturers recommend a check of bore symmetry with the bearing shells removed. The cap should be attached with nuts drawn up to specified tension. Then a telescope gauge and micrometer or special out-of-round gauge can be used to check the bore.

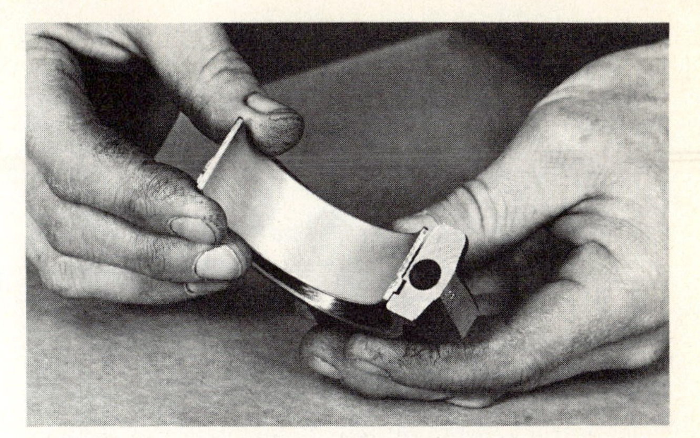

Fig. 24-26. Inserting a new bearing in the connecting-rod cap. (*Service Parts Division of Dana Corporation*)

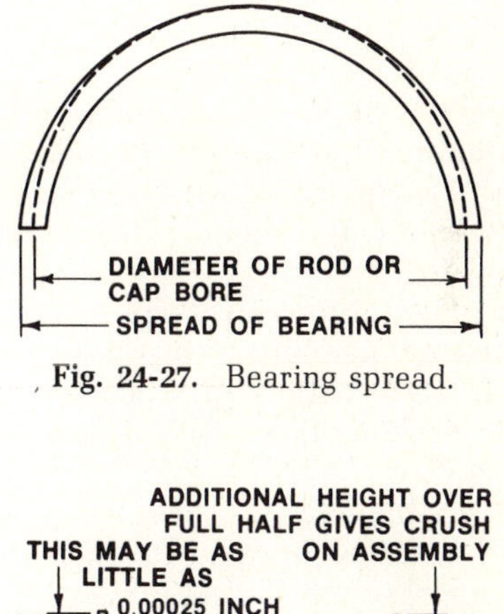

Fig. 24-27. Bearing spread.

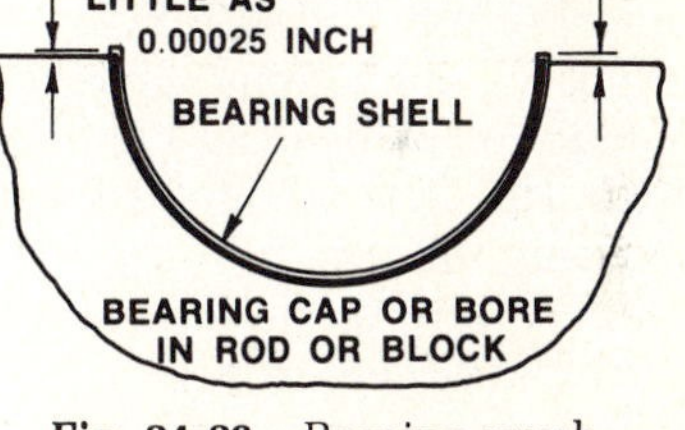

Fig. 24-28. Bearing crush.

It ensures firm seating and snug contact with the bores.

CAUTION: Never file off the edges of the bearing shells in an attempt to remove crush. When you select the proper bearings for an engine (as recommended by the engine manufacturer), you will find that they have the correct crush. Precision-insert bearings must not be tampered with in any way to make them "fit better." This usually leads only to rapid bearing failure.

Check Your Progress

Progress Quiz 24-1 You are now deep in the part of the book that deals with engine service, and these periodic progress checks are more important to you than ever. The proof of your training is your ability to handle engine service; this part of the book is designed to give you the facts you need to be successful in engine-service work. Check yourself on how well you are remembering those facts by answering the following questions.

Completing the Sentences The sentences that follow are incomplete. After each sentence there are several words or phrases, only one of which will correctly complete the sentence. Write each sentence in your notebook, selecting the proper word or phrase to complete it correctly.

1. With the bearing oil-leak detector in use, a bearing in normal condition should leak about: (*a*) 2 to 15 drops a minute, (*b*) 20 to 150 drops a second, (*c*) 20 to 150 drops a minute.
2. One of the major reasons for removing the ridge from the top of the cylinder before taking out the piston is to keep from: (*a*) damaging piston pin, (*b*) breaking piston rings, (*c*) scratching the cylinder wall, (*d*) breaking the connecting rod.
3. If piston-pin bushings have become worn so that they must be reamed or honed, then it will be satisfactory to use: (*a*) original piston pins, (*b*) new undersize piston pins, (*c*) new oversize piston pins.
4. When honing piston-pin bushings, you should first rough-hone to within: (*a*) 0.0005 in [0.0127 mm], (*b*) 0.005 in [0.127 mm], (*c*) 0.05 in [1.27 mm] of the finished size.
5. When an engine "throws a rod," it means that the rod has: (*a*) broken, (*b*) gone through the cylinder head, (*c*) gone through the oil pan.
6. The material which flattens varying amounts to indicate the amount of bearing clearance is called: (*a*) feeler stock, (*b*) shim stock, (*c*) Plastigage.
7. If both bearing halves fail on the same side, chances are the failure is due to: (*a*) a bent rod, (*b*) a tapered journal, (*c*) lack of oil.
8. Where clearance is excessive but the crankpin is not excessively tapered or out of round, a repair can be made by installing: (*a*) new undersize bearings, (*b*) new oversize bearings, (*c*) a new crankshaft.
9. The amount that the bearing-shell diameter is greater than the diameter of the bore into which it is placed is called the bearing: (*a*) crush, (*b*) spread, (*c*) diameter, (*d*) bore.
10. The additional height over a full half that the bearing shell has is called bearing: (*a*) crush, (*b*) spread, (*c*) diameter, (*d*) bore.

Unscrambling Causes of Bearing Failure Following are two lists. The first list gives various types of bearing failure. The second list gives various causes of bearing failure but not in the same order. To unscramble the lists, take each item in the bearing failures list in turn, and then find the cause in the causes list. Put the two together and write the result in your notebook. For example, the first bearing failure listed is "scratches in bearing metal." When you look down the causes list, you come to "dirt in oil," which will cause this sort of bearing trouble. So you put the two together and write "scratches in bearing metal—dirt in oil."

BEARING FAILURES
Scratches in bearing metal
Bearing metal wiped out uniformly
Bearing metal wiped out on one side
Bearing metal wiped out at center
Bright spots on bearing metal
Craters flaked out
Bearing failure at edge

CAUSES
Tapered Crankpin
Fatigue
Dirt under shell
Radius ride
Lack of oil
Dirt in oil
Crankpin ridged

Pistons and Rings

⊘ 24-17 Piston Service After the piston-and-rod assemblies are removed from the engine, the pistons and rods should be separated (⊘ 24-3 to 24-7). Then the rings can be removed from the pistons. (The rings can also be removed from the pistons before the pistons and rods are separated.) A special ring-expander tool can be used for ring removal. The tool has two small claws that catch under the ends of the ring (Fig. 24-29). When pressure is applied to the tool handles, the ring is sprung enough so that it can

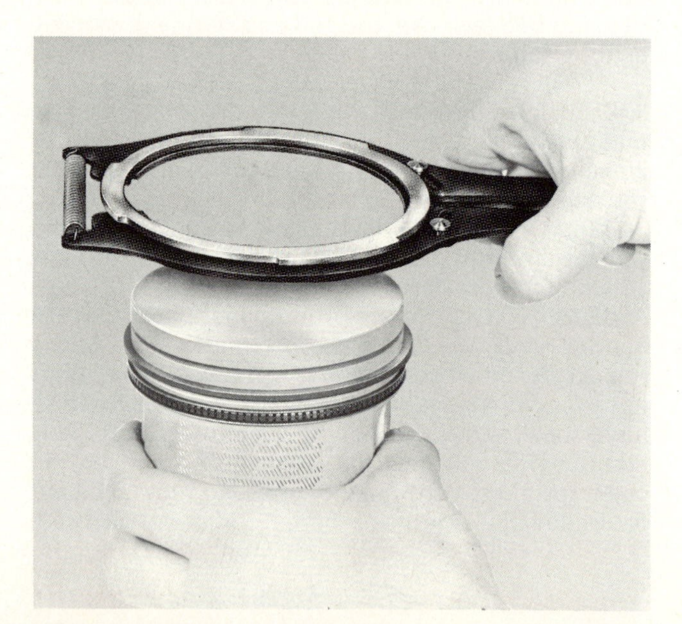

Fig. 24-29. Using a piston-ring-expander tool to remove or install a compression ring on a piston. (*Service Parts Division of Dana Corporation*)

Fig. 24-30. A groove-cleanout tool is used to clean the piston-ring grooves. (*Ford Motor Company*)

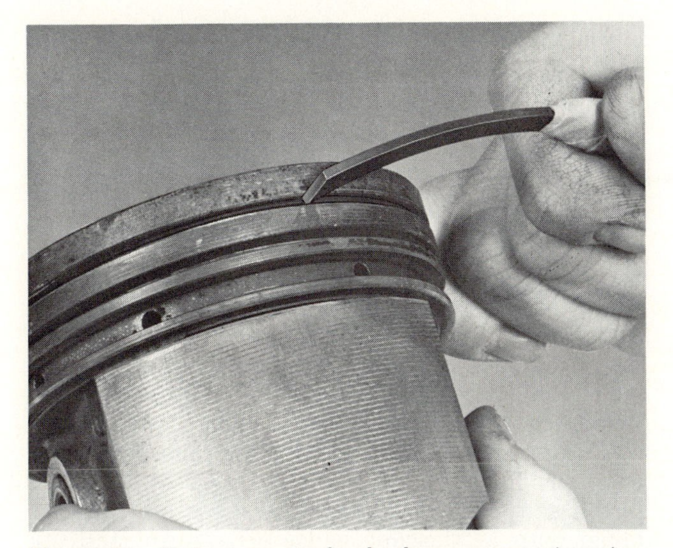

Fig. 24-31. Using a part of a broken compression ring, which has been sharpened on one end, to clean out the ring grooves. (*Service Parts Division of Dana Corporation*)

be lifted out of the ring groove and off the piston. Discard the old piston rings. As a rule, expert mechanics replace the old rings with new ones during an engine overhaul. Once the ring break-in coating and tool marks are worn off, the ring will not reseat itself if it is reinstalled.

⊘ 24-18 Piston Cleaning Remove carbon and varnish carefully from piston surfaces. Do not use a caustic cleaning solution or wire brush! These could damage the piston-skirt finish. You may decide to reinstall the pistons in the engine; therefore, you should not damage them. Use the cleaning method provided in your shop to clean the pistons. Clean out ring grooves with a clean-out tool (Fig. 24-30). You can also use the end of a broken piston ring filed to a sharp edge (Fig. 24-31). Oil-ring slots, or holes, must be clean so that oil can drain back through them. Use a drill of the proper size. Do not remove metal when cleaning the slots or holes.

⊘ 24-19 Piston Inspection Examine the pistons carefully for wear, scuffs, scored skirts, worn ring grooves, and cracks. Defects such as excessive wear, scuffs (Fig. 24-32*a*), scores, or cracks require piston replacement. Pistons that have failed due to preignition (Fig. 24-32*b*) or detonation (Fig. 24-32*c*) obviously must be replaced. Look for cracks at the ring lands, skirts, bushing bosses, and heads. Any defects require replacement of the piston, with these exceptions: Worn ring grooves can sometimes be repaired

Fig. 24-32. Typical piston failures. (*a*) Piston that has failed from scuffing. Note the scratch or scuff marks that run vertically on the piston skirt. (*b*) Piston that has failed due to preignition. Note how the ring lands, particularly the top one, appear to have been "nibbled" away. (*c*) Piston that has failed due to detonation. The excessive pressure and temperature have caused a hole in the pistonhead. (*TRW, Inc.*)

a

V-8
PISTON

B

45°

A

THE ELLIPTICAL SHAPE OF THE PISTON SKIRT SHOULD BE 0.008 TO 0.010 IN. FOR 273 IN³ AND 0.010 TO 0.013 IN. FOR 318 IN³ LESS AT DIAMETER A THAN ACROSS THE THRUST AT DIAMETER B

b

Fig. 24-33. (*a*) Using a micrometer to measure piston diameter. (*b*) Piston measurements. On these Plymouth pistons, measurements should be within specifications shown. (*Pontiac Motor Division of General Motors Corporation; Chrysler Corporation*)

by cutting the grooves larger and using ring-groove spacers (⊘ 24-20). Piston-skirt wear or collapse (reduction in skirt diameter) can sometimes be corrected by knurling the piston skirt (⊘ 24-21).

Inspect the fit of the piston pins to the pistons or piston bushings. One way to do this is to use a small-hole gauge to check the piston bearing bores and a micrometer to measure the pin diameter. On the type of piston without a bushing in which the pin oscillates (Fig. 24-9), the piston and pin are supplied in matched sets. If the fit is too loose or there are other pin or piston defects, the pin and piston are replaced as a matched set. (Chevrolet specifies a fit no looser than 0.001 inch [0.0254 mm].)

NOTE: Some engine shops hone the piston-pin holes and install oversize piston pins, provided the piston is otherwise in good condition. However, this is not generally recommended by most automotive-engine manufacturers.

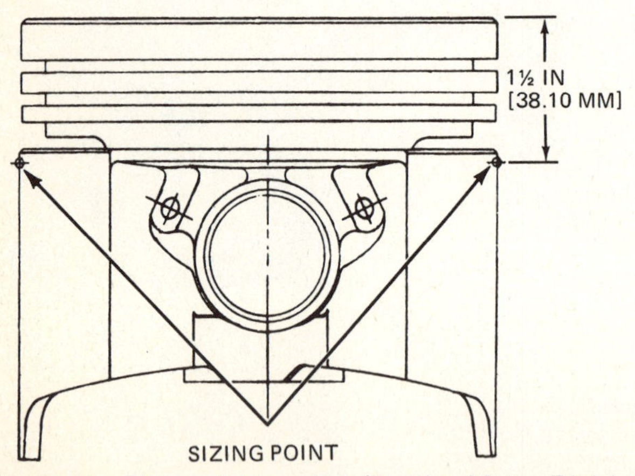

Fig. 24-34. Piston sizing point. (*Pontiac Motor Division of General Motors Corporation*)

Check the piston with a micrometer (Fig. 24-33*a*). Some manufacturers specify taking various measurements (Fig. 24-33). It is important to take the measurement at the sizing point (Fig. 24-34) because this is the point of maximum wear.

Compare the sizing-point measurement with the cylinder diameter, measured 90 degrees from the piston pin (Fig. 24-35). This measurement may be made with a telescope gauge (Fig. 24-35) and micrometer (Fig. 24-36). Measurement may also be made with a dial indicator (Fig. 24-37) and a micrometer (Fig. 24-38).

TELESCOPE GAUGE
90° FROM PISTON PIN

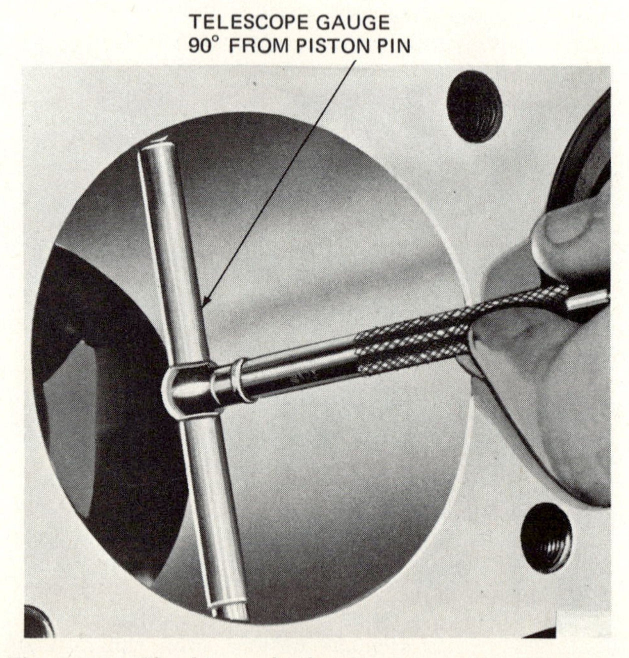

Fig. 24-35. Checking cylinder diameter with a telescope gauge. (*Buick Motor Division of General Motors Corporation*)

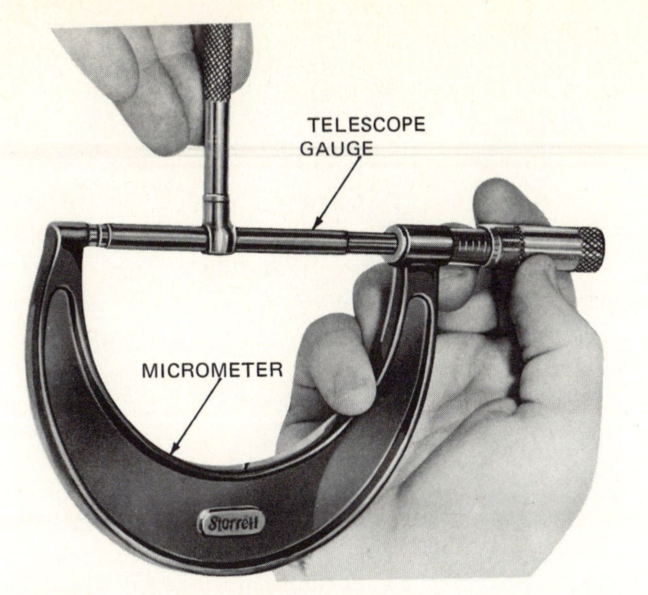

Fig. 24-36. Measuring the setting of a telescope gauge set at cylinder diameter. (*Buick Motor Division of General Motors Corporation*)

If the cylinder wall is excessively worn or tapered, it will require refinishing (⊘ 25-19 to 25-24). If the cylinder wall is refinished, then a new, oversized piston will be required.

Piston fit can also be checked another way. Some manufacturers specify using a feeler ribbon and spring scale (Fig. 24-39). If only a light pull is required to pull the ribbon out, the fit is the thickness of the ribbon. If the feeler ribbon comes out too easily, the fit is too loose.

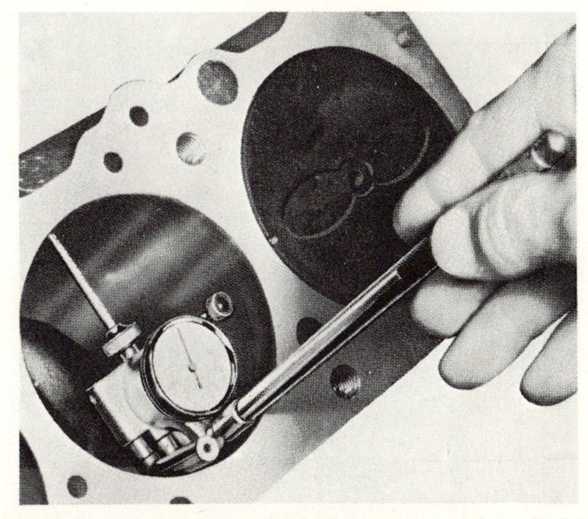

Fig. 24-37. Measuring cylinder diameter with a dial indicator. Once the reading is taken, the dial should be set at zero and the reading measured with a micrometer (see Fig. 24-38). (*Pontiac Motor Division of General Motors Corporation*)

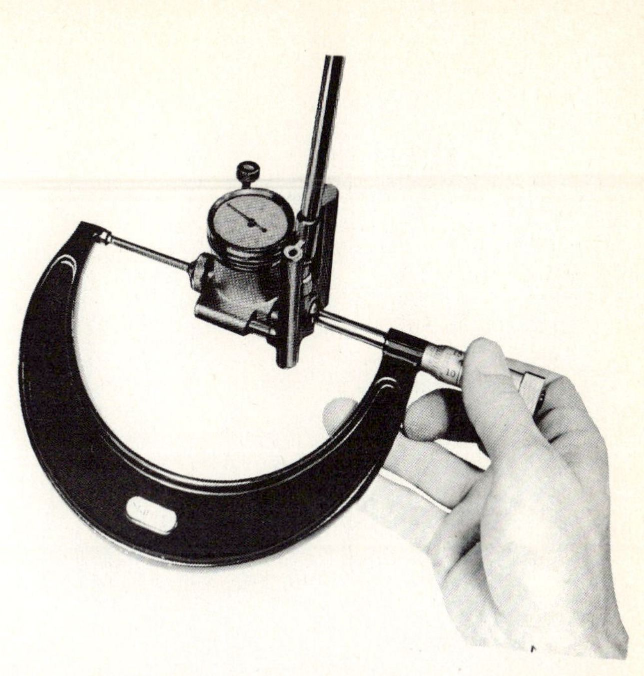

Fig. 24-38. Measuring the dial-indicator reading with a micrometer. (*Pontiac Motor Division of General Motors Corporation*)

⊘ 24-20 Ring-Groove Repair If a piston is in good condition except for excessive ring-groove wear, it can often be repaired. The top ring groove is the one that wears most because it gets the highest temperatures and pressures (Fig. 24-40). One piston-ring manufacturer states that almost all aluminum pistons checked at the time of overhaul have excessively worn top ring grooves. The ring groove may be checked with a special gauge, as shown in Fig. 24-41. If the ring groove is excessively worn (as much as 0.006 inch [0.152 mm] or more), the groove can be machined to a larger width with a special

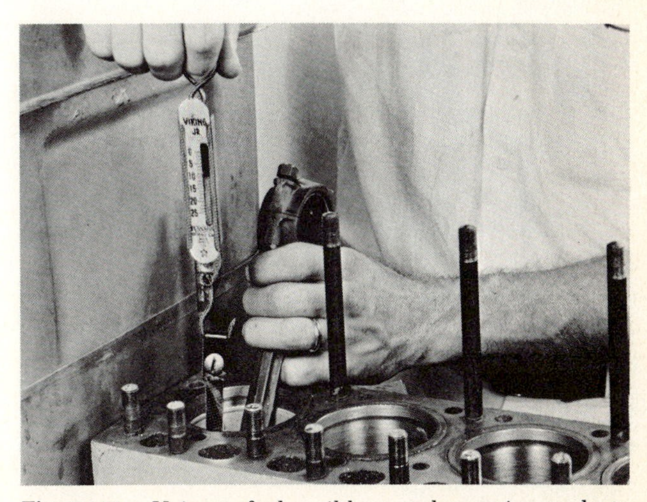

Fig. 24-39. Using a feeler ribbon and a spring scale to check piston clearance. (*Federal-Mogul Corporation*)

Fig. 24-40. Piston cut away to show wear of the ring groove. (*Federal-Mogul Corporation*)

Fig. 24-41. Using a special gauge to check the top ring groove for wear. (*Federal-Mogul Corporation*)

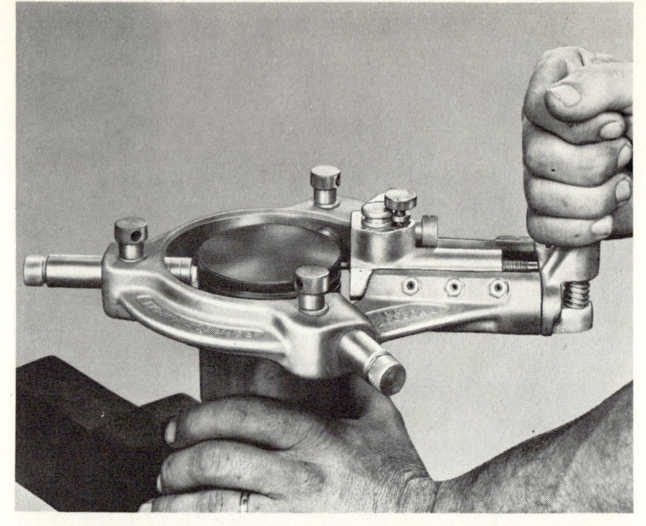

Fig. 24-42. Using a manually operated lathe, called a *Manulathe* by the manufacturer, to widen and square up the top ring groove. (*Federal-Mogul Corporation*)

hand-operated lathe (Fig. 24-42). The lathe squares up the top and bottom sides of the ring groove. Then the ring is installed with a spacer, as shown in Fig. 24-43.

⊘ 24-21 Piston Resizing Resizing of modern pistons is not recommended by automotive manufacturers. The procedure can damage the piston finish. One piston-ring manufacturer has developed a knurling procedure, called *nurlizing*. The piston skirt is run between a supporting wheel and a nurlizing tool, as shown in Fig. 24-44. The result is shown in Fig. 24-45. The procedure displaces metal and expands the diameter of the piston skirt. Also, the indentations form little pockets that can hold lubricating oil.

⊘ 24-22 New Pistons New pistons are of the finished type, ready for installation. They are available in a number of sizes. When these are used, the cylinders are finished to fit the pistons. Engine manufac-

Fig. 24-43. Top ring-groove spacer in place above the ring. (*Federal-Mogul Corporation*)

Fig. 24-44. Piston skirt cut away to show how the nurlizer operates to increase the diameter of the piston skirt. (*Federal-Mogul Corporation*)

turers supply oversize pistons of the same weight as the original pistons. Thus it is not necessary to replace all pistons when only some of the cylinders require service. There is no problem of balance if all pistons are of the same weight even if some are oversize.

NOTE: Aluminum pistons are usually supplied with piston pins already fitted. This ensures factory specifications on the pin fit to the piston.

Fig. 24-45. Nurlized piston. (*Federal-Mogul Corporation*)

CAUTION: Finished pistons have a special finish. They must not be buffed with a wire wheel or finished to a smaller size. This would remove the finish and cause rapid piston wear after installation.

⊘ 24-23 Fitting Piston Pins in Pistons On pistons with piston-pin bushings, worn bushings may be replaced. The new bushings are honed to size to fit the piston pins (Fig. 24-46). (Figure 24-9B shows the type of piston we are discussing.)

NOTE: As mentioned in ⊘ 24-12, aluminum pistons have no bushings. They are supplied with prefitted piston pins as a matched set. If a pin is worn or has too loose a fit in the piston, a new pin-piston set is required. However, as mentioned in ⊘ 24-19, some engine shops hone the piston-pin holes and install oversize piston pins, provided the piston is otherwise in good condition. But this is not recommended by most engine manufacturers.

⊘ 24-24 Rod-and-Piston Alignment After the rod and piston have been reassembled, but before the rings are installed on the piston, alignment should be checked. The alignment tool shown in Fig. 24-20 is used for this check. The V block is held against the piston. If the V block does not line up with the face plate as the piston is moved to various positions, the connecting rod is twisted.

⊘ 24-25 Piston-Ring Service If an engine is torn down for service, the old piston rings should be discarded. Rings that have been used, even for only short mileage, will not seat properly to provide sealing.

NOTE: If the engine trouble is due to rings sticking in the piston-ring grooves, a special compound can be introduced into the intake manifold and engine oil before engine disassembly. This will sometimes free the rings (⊘ 23-5).

CAUTION: Before adding any chemical to an engine equipped with a catalytic converter, be sure that the chemical will not harm the catalyst.

Proper selection of new piston rings depends on the condition of the cylinder walls and whether they are to be reconditioned. (⊘ 25-20 describes the inspection of cylinder walls for wear and taper.) If the cylinder walls are only slightly tapered or out of round (consult the manufacturer's specifications for maximum allowable deviations), then standard-type rings can be used. Where the cylinder walls have some taper but not enough to warrant the extra expense of a rebore or hone job, special "severe," or "drastic," rings should be used. These rings have greater tension and are more flexible. This enables them to expand and contract as they move up and down in the cylinder. Thus, they follow the changing

Fig. 24-46. Piston-pin bushings being honed. (*Sunnen Products Company*)

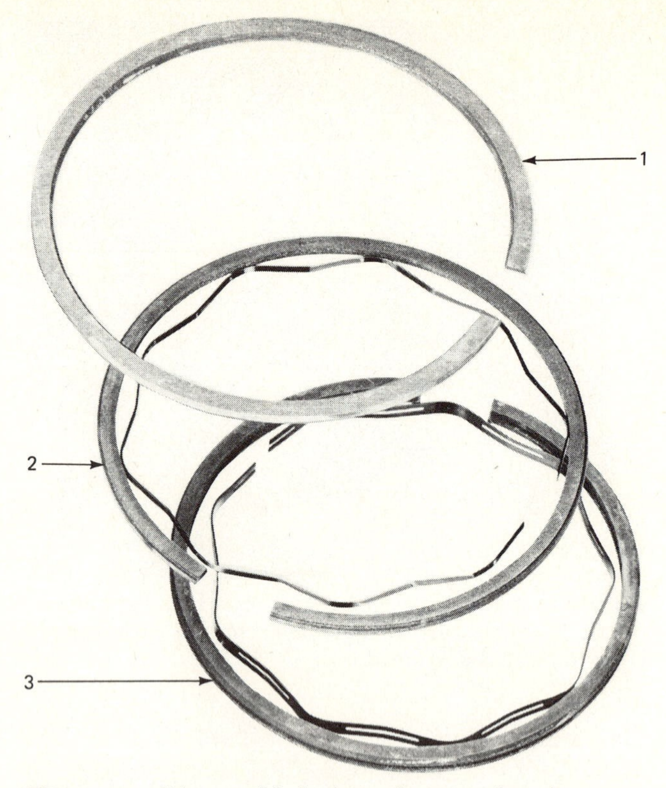

Fig. 24-47. Disassembled view of a set of replacement piston rings: (1) top compression rings; (2) second compression ring, which includes an expander spring; (3) oil-control ring, which includes an expander spring. (*Grant Piston Rings*)

contours of the cylinder wall and provide adequate sealing (preventing blow-by) and oil control. Figure 24-47 shows a set of replacement rings for tapered cylinder walls.

Automotive manufacturers generally recommend honing the cylinder walls lightly (⊘ 25-21 and 25-22) before piston-ring installation to "break the glaze." Cylinder walls take on a hard, smooth glaze after the engine has been in use for a while. In some automotive shops it is the practice to knock off this glaze by running a hone or glaze-breaker up and down the cylinder a few times before putting in new rings. However, at least one piston-ring manufacturer says that honing does not have to be done on cast-iron cylinder walls, provided the walls are not wavy or scuffed. The glaze is a good antiscuff material and will not unduly retard the wear-in of new rings. However, the walls should be reasonably concentric and in relatively good condition.

The best honing job leaves a cross-hatch pattern (Fig. 24-48) with hone marks intersecting at a 60-degree angle, as shown. This leaves the best surface for proper seating of new rings. (See ⊘ 25-21 and 25-22 on honing procedures.)

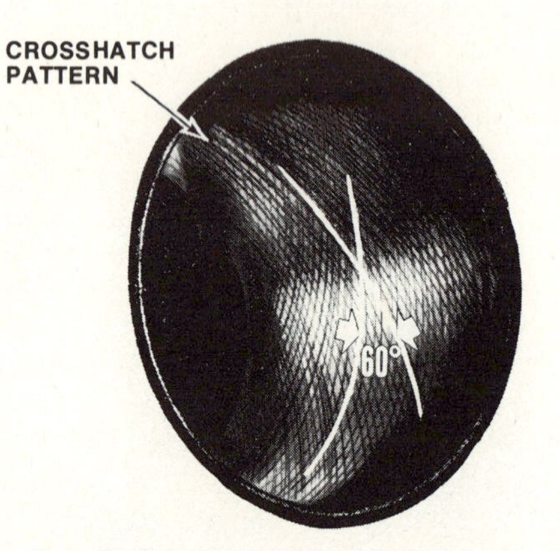

Fig. 24-48. Crosshatch pattern that remains on a cylinder wall after a good honing job. (*Chrysler Corporation*)

⊘ 24-26 Fitting Piston Rings Piston rings must be fitted to the cylinder and to the ring grooves in the piston. Rings come in packaged sets in graduated sizes to fit various sizes of cylinders. All packages have instruction sheets that describe in detail exactly how to install the rings. These instructions should be carefully followed.

CAUTION: Never throw the instructions away until you have finished the ring-installation job.

As a first step in fitting a piston ring, the ring should be pushed down into the cylinder with a piston and the ring gap measured. The ring gap is the space between the ends of the ring. It is measured with a feeler gauge (Fig. 24-49). Figure 24-49 shows the gap being measured with the ring at the

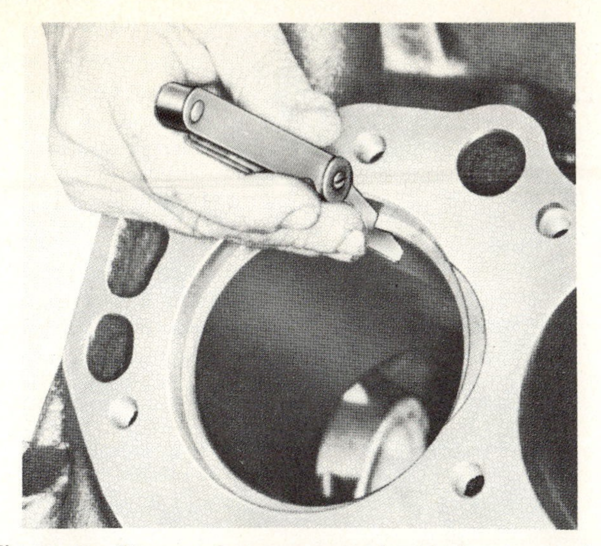

Fig. 24-49. Ring gap being measured with the ring in the cylinder. (*Chevrolet Motor Division of General Motors Corporation*)

top of the cylinder. This makes it easier for you to see what is going on. Actually, the gap is measured with the ring pushed down to the lower limit of ring travel. If the cylinder is worn, that is where the ring gap will be smallest. If the ring gap is too small, check the package that the ring came in. The ring set may be wrong for the job. (Rings come in sets in graduated sizes.) If the ring gap is wrong, then either you have the wrong rings, you have incorrectly measured the cylinder diameter, or the wrong-size rings were inserted in the package.

On older-model engines, the recommendation was to file the ends of the ring with a fine-cut file. The file was first clamped in a vise. Then the ring was worked back and forth on the file (with the ring ends on the two sides of the file). This procedure is no longer recommended. Filing the ring ends can remove some of the ring coating and cause early ring failure.

CAUTION: Remember that if the cylinder is at all tapered, the diameter at the lower limit of ring travel (in the assembled engine) will be smaller than the diameter at the top (Fig. 25-24). This means the ring must be fitted to the diameter at the lower limit of ring travel. If it is fitted to the upper part of the cylinder, as shown in Fig. 24-49, the ring gap will not be great enough at the lower limit of ring travel. As a result, the ring ends will come together. The ring will be broken, and the cylinder wall will be scuffed. Always measure the ring gap with the ring pushed down to the point of minimum diameter at the lower limit of ring travel.

If the ring gap is correct, insert the outside surface of the ring into the proper ring groove in the piston (Fig. 24-50). Then roll the ring around in the groove to make sure the ring has a free fit around the entire piston. An excessively tight fit probably

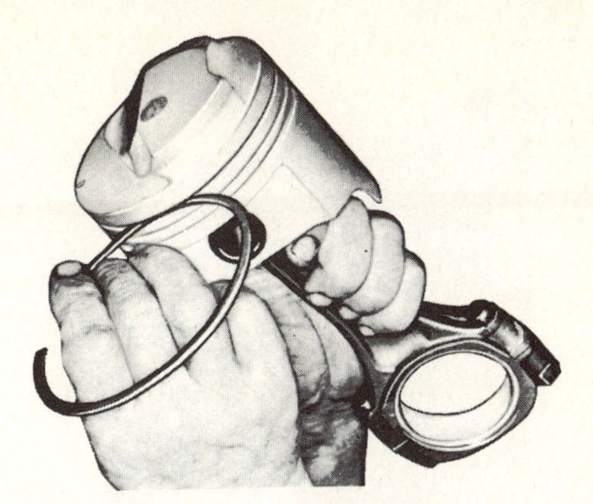

Fig. 24-50. Checking the fit of the ring in the ring groove. (*Chevrolet Motor Division of General Motors Corporation*)

means that the ring groove is dirty. Another possibility is that the ring groove has been nicked or burred with the blade of the ring tool (see Fig. 24-30). Some authorities recommend using the end of a broken ring which has been filed to a sharp edge to clean the ring grooves (Fig. 24-31). This is preferred by many technicians because the piece of ring will not cause nicks or burrs.

Install the rings in the ring grooves, using the ring tool, as shown in Fig. 24-51. Then, recheck the fit. Insert a feeler gauge of the proper size between the ring and the side of the groove (Fig. 24-52).

⊘ 24-27 Cautions on Installing Rings The three-part oil-control ring is installed one part at a time, as shown in Fig. 24-53. Various types of compression rings and their proper installation are shown in Fig. 24-54. One special caution to observe is never to spiral the compression rings into the grooves. (Spi-

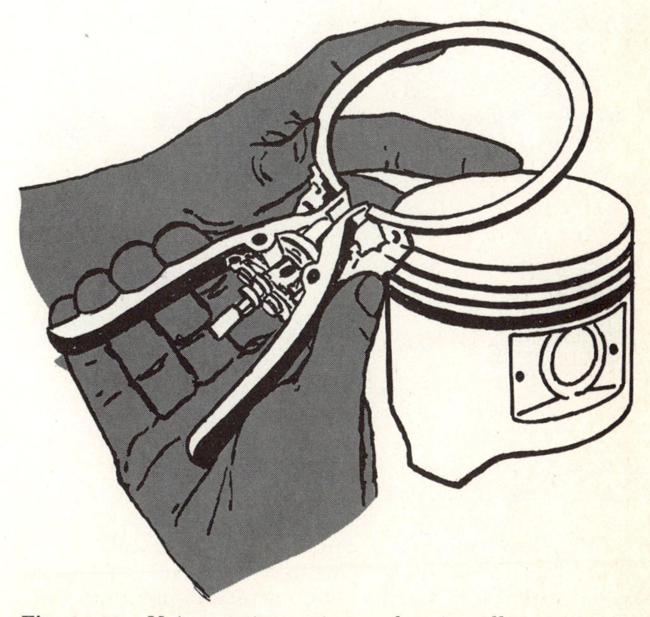

Fig. 24-51. Using a piston-ring tool to install a compression ring on a piston. (*Federal-Mogul Corporation*)

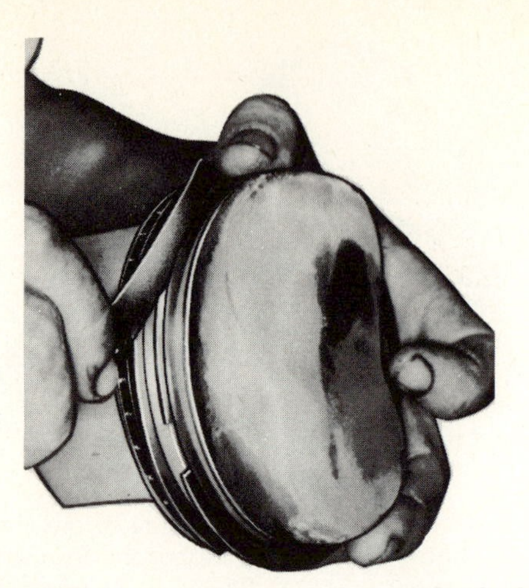

Fig. 24-52. Piston-ring clearance being checked with a feeler gauge. (*Chevrolet Motor Division of General Motors Corporation*)

raling the rails of the oil-control rings is shown in Fig. 24-53.) This could distort or break the compression ring and cause major loss of compression and blow-by. Instead, *always* use a ring expander tool, as shown in Fig. 24-51. Also, never overexpand the compression rings.

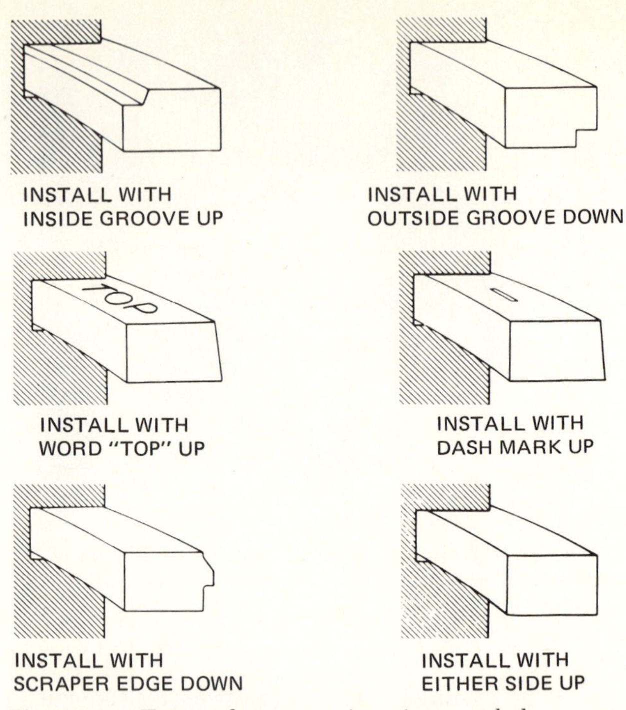

Fig. 24-54. Types of compression rings, and the proper way to install them. (*Federal-Mogul Corporation*)

⊘ 24-28 Installing the Piston-and-Rod Assembly We have already covered this procedure in ⊘ 24-9. When installing the piston-and-rod assembly, be sure to use a ring compressor. Install the assembly with the correct side facing forward (Fig. 24-17).

Some ring sets are now supplied with a special ring compressor, or loading sleeve which is used as shown in Fig. 24-55. The loading sleeve has a larger inside diameter at the top than at the bottom. Therefore, when the piston, with rings, slides down through the loading sleeve, it compresses the rings in the piston-ring grooves. Note the four steps. First, make sure the rings are free in the grooves. Lubricate the piston and rings. Insert the assembly through the large end of the loading sleeve. Next, center the rings on the piston, and slide the sleeve up as shown in step 2. Put the assembly into the cylinder bore and press downward on the piston until the loading sleeve rests on the cylinder block. (Be sure to use protective guide sleeves, as explained in ⊘ 24-9.) Then hold the loading sleeve firmly against the block. With a hammer handle, thrust downward with *one fast motion*, as shown in step 4. This pushes the piston-and-rod assembly into the cylinder bore.

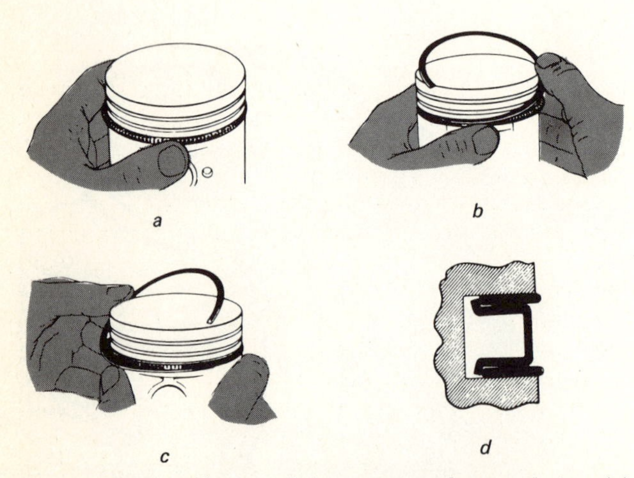

Fig. 24-53. Installation of three-piece oil-control ring. (*a*) Place expander spacer in oil-ring groove with ends of spacer above a solid part of the groove bottom. (*b*) Hold ends of spacer together, and install steel rail above the spacer. (*c*) Install other spacer on lower side of spacer. Make sure ends of spacer are not overlapping. (*d*) Sectional view of the three parts fitted into groove. (*Federal-Mogul Corporation*)

CAUTION: If the piston-and-rod assembly does not enter the bore easily—STOP! Remove the assembly and push the assembly with the rings through the loading sleeve by hand. Examine the rings for damage and if everything looks okay, repeat the procedure. Examine the ring grooves in the piston. If these are not cleaned out properly, they may prevent the rings from compressing enough to enter the cylinder.

Check Your Progress

Progress Quiz 24-2 Here is your checkup quiz on the last half of Chap. 24. Answering the following questions will tell you how well you have remembered the material you have just read. If any ques-

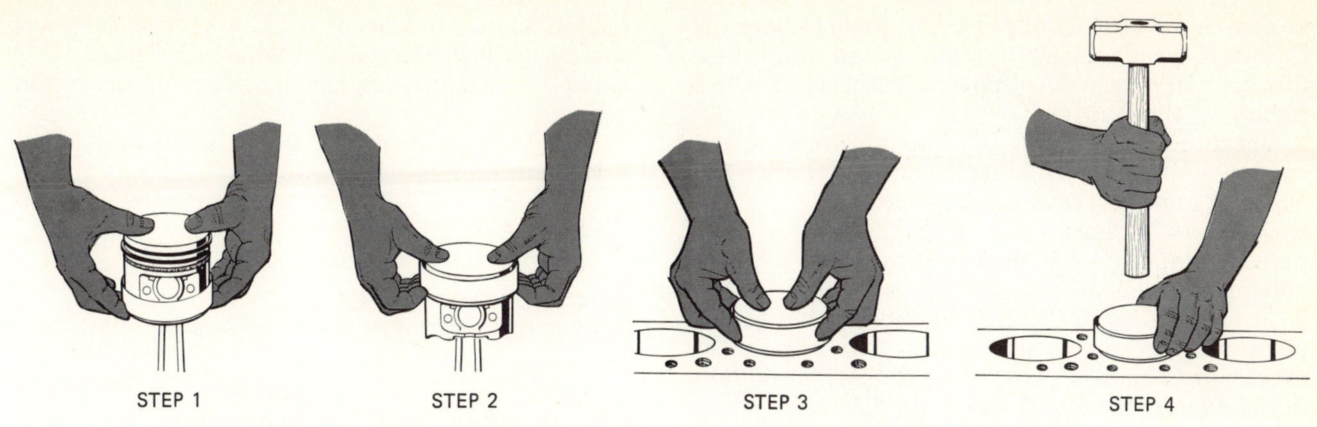

Fig. 24-55. Using the loading sleeve which comes with the new ring set to install the piston-and-ring assembly in the cylinder bore. (*Chrysler Corporation*)

tion stumps you, review the pages that will give you the answer.

Completing the Sentences The sentences that follow are incomplete. After each sentence there are several words or phrases, only one of which will correctly complete the sentence. Write each sentence in your notebook, selecting the proper word or phrase to complete it correctly.

1. Aluminum pistons are supplied with: (*a*) prefitted bushings, (*b*) prefitted piston pins, (*c*) connecting rods as an assembly.
2. Finished pistons: (*a*) should be ground to fit the cylinders, (*b*) should never be buffed or finished to a smaller size, (*c*) have special finishes that must not be removed, (*d*) both (*b*) and (*c*).
3. After the ring groove has been machined to a larger width: (*a*) a special spacer should be installed with the ring, (*b*) a reinforcing ring should be used, (*c*) a wider ring should be used.
4. Chevrolet specifies that the fit of the pin to the piston should be no looser than: (*a*) 0.10 in [2.54 mm], (*b*) 0.010 in [0.254 mm], (*c*) 0.001 in [0.025 mm].
5. When new pistons are installed: (*a*) the cylinder walls are finished to fit the pistons, (*b*) the pistons are finished to fit the cylinders, (*c*) both the pistons and cylinder walls are finished.
6. Whenever an engine is torn down for service: (*a*) the old rings should be discarded, (*b*) the old rings should be refinished for reinstallation in the engine, (*c*) neither (*a*) nor (*b*).
7. If the ring gap is too small: (*a*) file the ring ends, (*b*) the ring is the wrong size for the cylinder, (*c*) the cylinder will require refinishing.
8. The ring gap should be checked with the ring at the: (*a*) top of the cylinder, (*b*) bottom of the cylinder, (*c*) lower limit of ring travel.
9. The three-part oil-control ring should be installed: (*a*) one part at a time, (*b*) as a unit, (*c*) with the ring expander.
10. When installing a compression ring on a piston: (*a*) never use a ring expander, (*b*) always use a ring expander, (*c*) install it one part at a time.

CHAPTER 24 CHECKUP

NOTE: Since the following is a chapter review test, you should review the chapter before taking the test.

You have made excellent progress in the engine-servicing part of the book. The material you are now studying is really the heart of the subject since it deals with the various service jobs that are required to correct different engine troubles. It is important for you to remember the details of these various service jobs. The following checkup will give you the chance to test yourself and find out how well you remember the ways in which these service jobs are done.

Completing the Sentences The sentences that follow are incomplete. After each sentence there are several words or phrases, only one of which will correctly complete the sentence. Write each sentence in your notebook, selecting the proper word or phrase to complete it correctly.

1. Items that must be removed before the pistons can be taken out of the engine are: (*a*) rings, piston pin, and head; (*b*) head, oil pan, and ridge (if present); (*c*) head, block, and oil pan.
2. One group of parts removed from the engine as an assembly includes the: (*a*) piston, pin, rings, and rod; (*b*) piston, rod cap, rings, and valve; (*c*) head, rod, valves, and lifters.
3. A ring compressor should be used when reinstalling the piston-and-rod assembly in the engine to compress the piston rings so that they will: (*a*) not catch on rod, (*b*) enter cylinder, (*c*) slip into piston, (*d*) mate with piston pin.
4. When checking connecting rods, items to be considered include: (*a*) ring fit, bushings, pins, and alignment; (*b*) bushings, bearings, pins and rings; (*c*) bushings, bearings, and alignment.
5. Two methods of checking clearance of a precision-insert bearing are to use: (*a*) Plastigage and taper shim, (*b*) Plastigage and adjuster, (*c*) Plastigage and shim stock.

6. Two characteristics of precision bearings that help to install them properly are: (*a*) spread and crush, (*b*) bore and spread, (*c*) pitch and crush, (*d*) bore and pitch.
7. The direct-bonded type of connecting-rod bearing can be adjusted by: (*a*) filing the shells, (*b*) filing the caps, (*c*) removing or installing shims.
8. Severe, or drastic, rings are for use: (*a*) in tampered cylinders, (*b*) with tapered pistons, (*c*) with collapsed pistons, (*d*) in tapered cylinders.
9. In fitting rings to tapered cylinders, ring gap is measured with the ring located at the: (*a*) point of maximum diameter, (*b*) top of cylinder, (*c*) bottom of cylinder, (*d*) point of minimum diameter.
10. As a rule, before the pistons are removed, it is necessary to remove the: (*a*) crankshaft, (*b*) piston rings, (*c*) ring ridges.

Service Procedures In the following, write the procedures asked for in your notebook. Do not copy them from the book, but try to write them in your own words. Give a step-by-step account. This will help you remember the procedures when you go into the engine shop.

1. Explain how to remove a piston-and-rod assembly from an engine.
2. List various connecting-rod-bearing troubles and their causes.
3. Explain how to install and ream, or hone, a set of piston-pin bushings in the connecting rods.
4. Explain how to inspect the fit of a precision-insert bearing with Plastigage and with feeler stock.
5. List the steps in replacing a precision-insert rod bearing.
6. Explain how to inspect piston fit in a cylinder.
7. Explain how to fit piston rings.
8. List important points of selecting new rings for a job.

SUGGESTIONS FOR FURTHER STUDY

Always keep your eyes and ears open while you are in the engine shop. Notice how the different jobs are done, and pay particular attention to the methods of testing, removing, servicing, and replacing engine parts. Always examine all the defective parts you can find, and notice particularly exactly why the parts are defective. If you are not sure, ask someone why a part was discarded. Soon you will learn to spot defects quickly and easily.

Study the operating manuals of the different service equipment. For instance, the honing-equipment manufacturer supplies an operating manual that explains what the machine will do and how to operate it. You will find such manuals of great interest. Keep a notebook and jot down every important fact that you learn while you are in the shop or while you are studying the different manuals. Jotting down these facts will help you remember them. Also, it will increase the value of your notebook.

chapter 25

ENGINE SERVICE: CRANKSHAFTS AND CYLINDER BLOCKS

This chapter concludes our discussion of piston-engine service. It discusses the servicing of crankshafts, main bearings, and cylinder blocks. If the cylinder block, crankshaft, and main bearings all need service, it is often cheaper to buy a short block than to invest the time required for the services. A short block is shown in Fig. 23-1.

Removing and Replacing Engines and Mounts

⊘ **25-1 Removing an Engine** Many engine-service jobs can be performed with the engine in the vehicle. Other jobs, such as boring cylinders or main-bearing bores, require removal of the engine from the vehicle. Specific removal procedures vary, so always check the manufacturer's shop manual for the car you are servicing before starting the job. What follows is a typical procedure.

1. Drain the cooling system and remove the hood.
2. Disconnect the battery ground cable and the alternator ground cable from the cylinder block.
3. Remove the air cleaner.
4. Disconnect the radiator hoses from the head and water pump.
5. Disconnect the automatic transmission oil-cooler lines (where present) from the radiator.
6. Remove the fan shroud (if present) and the radiator.
7. Disconnect the oil-pressure and coolant-sending-unit wires.
8. Disconnect the fuel-pump hoses and plug the hoses to prevent leaks.
9. Disconnect the throttle linkage from the carburetor and the engine ground strap.
10. Such units as the alternator, air-conditioner compressor, and power-steering pump need not be removed from the engine compartment. Instead, they can be detached from the engine and moved to one side out of the way.

CAUTION: Do not disconnect the air-conditioner pressure hoses unless necessary. These hoses hold Freon-12 under pressure, and disconnecting them would allow Freon-12 to escape. This can be dangerous. In addition, it would then be necessary to purge and recharge the system with Freon-12.

11. Disconnect the heater hoses from the water pump, intake manifold, and choke housing (if present).
12. Disconnect the primary wire from the ignition coil. Remove the wiring harness from the top of the engine.
13. Raise the vehicle. Drain the engine oil.
14. Disconnect the starting motor cable from the starting motor. (The starting motor may also have to be removed.)
15. Disconnect the exhaust pipe (or pipes) from the exhaust manifold (or manifolds). Wire up the exhaust system to support it.
16. Disconnect the engine mounts from the frame brackets.
17. Disconnect the clutch or automatic transmission. Make sure the converter (automatic transmission) is secure in its housing.

NOTE: It is sometimes easier to remove the transmission with the engine because it is easier and prevents damage to the seal.

18. Lower the vehicle. Support the transmission with a floor jack.
19. Attach the engine-lift cable to the engine. Raise the engine enough to remove the mount bolts (Fig. 25-1). Make a final check to make sure all wiring, hoses, and other parts are free and clear of the engine.
20. Raise the engine enough to clear the mounts. Then alternately raise the transmission jack and engine until the engine separates from the transmission.
21. Carefully pull the engine forward from the

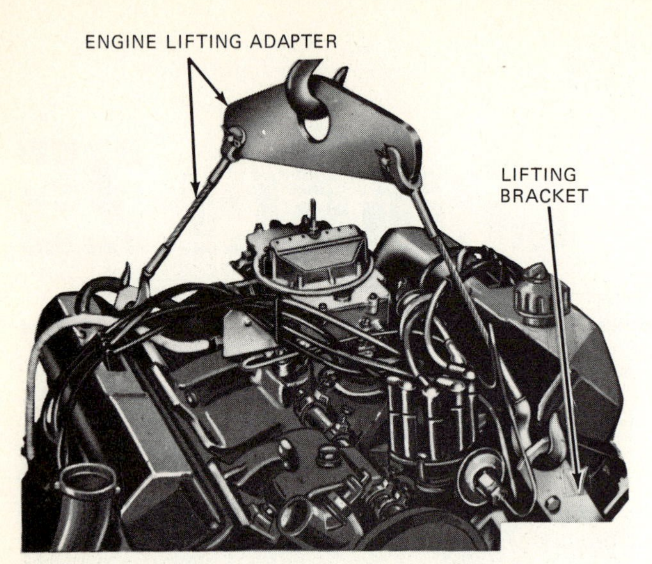

Fig. 25-1. Hoisting an engine from a car. (*Ford Motor Company*)

transmission. Then lift the engine out of the engine compartment.

NOTE: Engine installation is essentially the reverse of the preceding procedure.

⊘ 25-2 Replacing Engine Mounts Typical engine mounts are shown in Fig. 25-2. Broken or deteriorated engine mounts put extra strain on other mounts and the drive line. They should be replaced with new mounts. Always check the manufacturer's shop manual before attempting to replace an engine mount. Typical procedures follow.

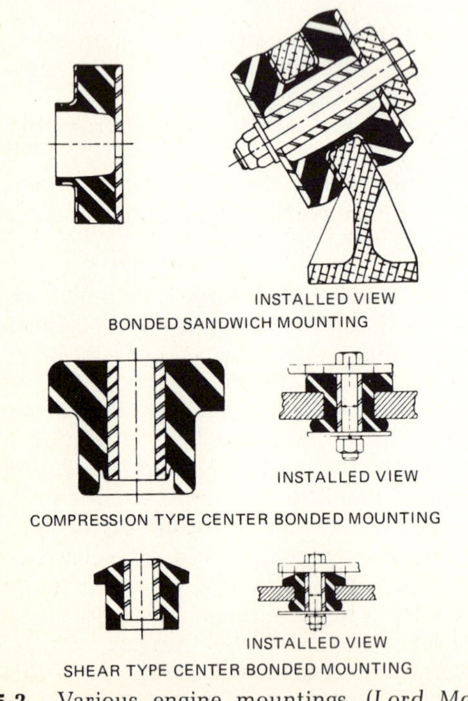

Fig. 25-2. Various engine mountings. (*Lord Manufacturing Company*)

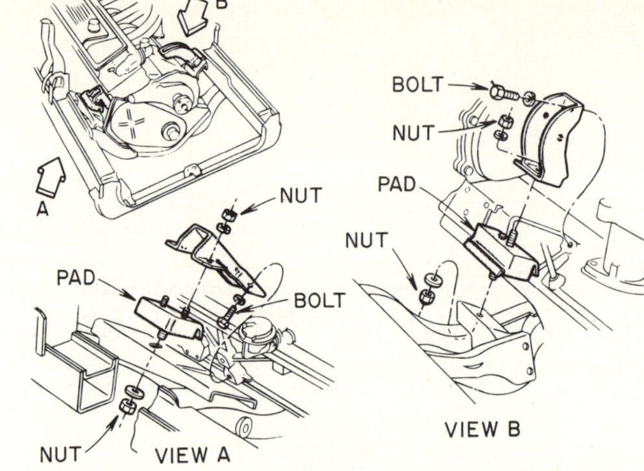

Fig. 25-3. Front engine flexible mounts used to support the engine on the frame. (*Chrysler Corporation*)

1. *REPLACING A FRONT ENGINE MOUNT* Support the engine with a jack and wood block under the engine oil pan. Raise the engine enough to take the weight off the mount. Remove the nut and through bolt. Remove the mount-and-frame-bracket assembly from the cross member, as shown in Fig. 25-3. Install the new mount on the cross member. Use new self-locking bolts and nuts. Do not tighten the bolts until you have lowered the engine. Then tighten them to the specified tension.
2. *REPLACING A REAR ENGINE MOUNT* Support the engine with a jack and wood block under the oil pan. Remove the cross-member-to-mount bolts and remove the mount locknut and bolt.
3. *RAISE THE TRANSMISSION TO RELEASE THE ENGINE WEIGHT FROM THE MOUNT* Remove tle mount-to-transmission bolts and remove the mount (Fig. 25-4). Install the new mount with bolts and nuts.

Crankshafts and Main Bearings

⊘ 25-3 Crankshaft and Bearing Service Modern automotive engines have precision-insert main bearings that can be replaced without removing the crankshaft. Many main-bearing difficulties can be taken care of by this method of bearing replacement. However, bearing replacement will not fix stopped-up oil passages, worn crankshaft journals, a damaged crankshaft, or a block in which a bearing has spun. That is, the bearing and crankshaft journal can become so hot, owing to lack of oil, that they weld momentarily. Then the bearing spins with the crankshaft and gouges the bearing bore in the cylinder block. Bearing spin will damage the cylinder block, and block replacement will be required.

If all bearings have worn fairly evenly, then probably only crankshaft-journal inspection and bearing replacement will be required. It should be noted that usually all bearings do not wear the same amount. Some bearings will wear more than others.

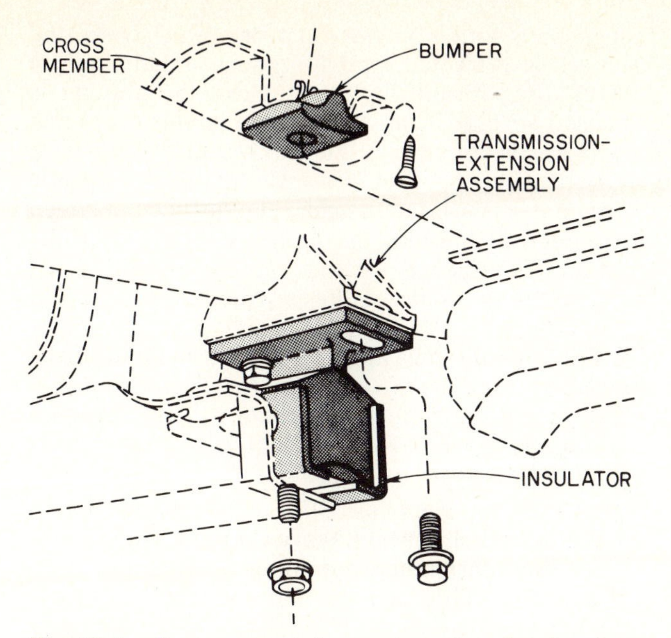

Fig. 25-4. Rear support for engine, showing insulator arrangement. (*Chrysler Corporation*)

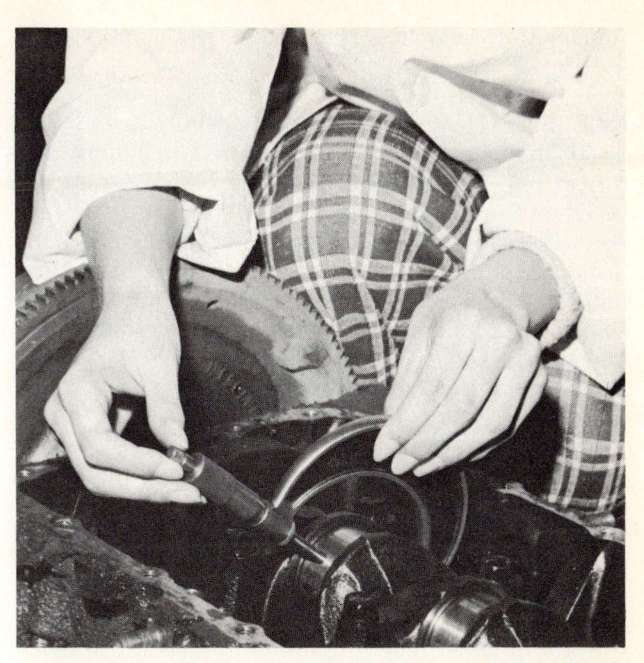

Fig. 25-5. Checking crankpin with a micrometer.

This is acceptable provided none of the bearings wear beyond manufacturer's specifications. The lower bearing half wears the most. It takes the weight of the crankshaft and combustion pressures through the rods and cranks. Uneven wear can result from normal aging of the oil pump. As the oil pump wears, oil pressure and circulation drop. The main and rod bearings farthest from the oil pump get less oil. Thus they wear the most. Also, a clogged oil passage will starve bearings, and when this happens, the bearings fail (they may also spin, as noted previously.)

If main bearings have worn very unevenly, the best service procedure is to removed the crankshaft from the engine block. Then the block and crankshaft can be checked separately for damage and clogged oil lines.

However, as we said, many main-bearing problems can be taken care of by replacing the bearings with the crankshaft still in the engine, as we shall explain later. Replacing the main bearings (or mains) without removing the crankshaft requires about 5 hours. Removing and replacing the crankshaft requires several additional hours.

⊘ 25-4 Inspecting Crankshaft Journals Both the crankpins and crankshaft main journals should be inspected whenever the bearing caps are removed. Inspecting crankpins was discussed in ⊘ 24-16. Figure 25-5 shows a crankpin being checked with a micrometer. Main journals can be checked with a special crankshaft gauge or with a special micrometer. Measurements should be taken in several places along the journal to check for taper. Also, the crankshaft should be rotated by one-quarter or one-eighth turns to check for out-of-round wear. (See ⊘ 24-14 for a discussion of what a tapered, ridged, or out-of-round journal will do to the bearing.) If journals are tapered or out of round by more than 0.003 inch [0.076 mm], they should be reground. Some authorities consider 0.0015 inch [0.037 mm] the outside tolerable limit for wear. They point out that any noticeable out-of-round or taper shortens bearing life.

To check journals, remove the oil pan (⊘ 24-5) and bearing caps (⊘ 25-5). It is not necessary to detach the connecting rods from the crankshaft. However, the spark plugs should be removed so that the crankshaft can be turned over easily.

⊘ 25-5 Removing Bearing Caps Sometimes, it is difficult to remove all bearing caps with the engine in the car. In some cars, the front cross member is so close to the engine that the engine must be lifted to get enough clearance to remove the cap. Also, the rear main-bearing cap may be hard to remove because of interference with other parts.

If you are planning to merely check journals and bearings, then remove bearing caps one at a time to make the checks. If the crankshaft is to come out of the engine, the connecting rods should be detached and all main bearing caps removed.

Caps should be marked so that they can be replaced on the same journals from which they were removed. (See ⊘ 24-6 on marking rod caps.) To remove a cap, remove the nuts or bolts. Bend back the lock-washer tangs (if used). Disconnect oil lines where necessary. (Use new lock washers on reassembly.)

If a cap sticks, work it loose carefully to avoid nicking or cracking it. In some engines, a screwdriver or pry bar can be used to work the cap loose. Sometimes tapping the cap lightly on one side and then the other with a brass or plastic hammer will loosen it.

CAUTION: Heavy hammering or prying can nick or crack the cap, bend the dowel pins, or damage the dowel holes. In such a case, the bearing may not fit when the cap is replaced, and early bearing failure will occur. Also, remember that the bearing caps are made of cast iron and so are brittle. A hard blow can crack or break a cap. A damaged cap will have to be discarded, and a new cap used. A cracked cap can break if it is reinstalled on the engine. Such a break usually means a ruined engine.

When a bearing cap is damaged or lost, a new cap is required. Some new caps are supplied with a shim pack so that the cap can be shimmed into alignment. However, it is difficult to get a good fit this way. So it may become necessary to take the engine out of the car, disassemble it, and install the caps, without bearings, back in the block. Then the block will have to be line-bored to reestablish bearing-bore alignment. (This procedure is discussed in ⊘ 25-18.)

⊘ 25-6 Measuring Main Journals with Crankshaft Gauge The special crankshaft gauge shown in Fig. 25-6 is used as shown in Fig. 25-7. The journal and gauge pads and plunger must be clean. Then, the plunger is retracted, and the gauge is held tightly against the journal (Fig. 25-7). Next, the plunger is released so that it moves out into contact with the journal. The plunger is then locked in this position by tightening the thumbscrew. Finally, an outside micrometer is used to measure the distance between D, the end of the plunger, and C, the button on the bottom of the gauge. This measurement, multiplied by 2, is the diameter of the journal.[1]

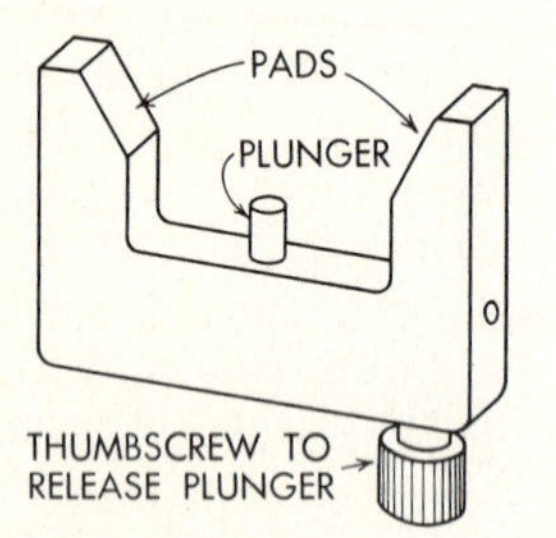

Fig. 25-6. Special gauge for checking journal diameter. (*Federal-Mogul Corporation*)

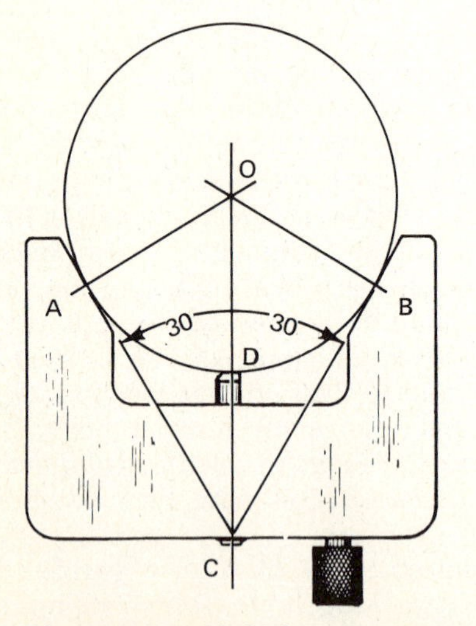

Fig. 25-7. Checking crankshaft main journal with a crankshaft gauge. (*Federal-Mogul Corporation*)

NOTE: Take the measurements from one end of the journal to the other. Rotate the crankshaft by one-eighth turns to repeat the checks. This repeated measurement will detect journal taper and out-of-round. Write down the readings as you take them.

⊘ 25-7 Measuring Main Journals with Micrometer To use the micrometer, the upper bearing half must be removed. This is done with a special roll-out tool, as explained in ⊘ 25-10. Then the micrometer can be used as shown in Fig. 25-8. Take measurements from one end of the journal to the other. Rotate the crankshaft by one-eighth turns to repeat the check. This procedure will detect journal taper and out-of-round.

⊘ 25-8 Inspecting Main Bearings Main, or crankshaft, bearings, should be replaced if they are worn, burned, scored, pitted, rough, flaked, cracked, or otherwise damaged. (See ⊘ 24-14 on bearing failures.) It is important to inspect the crankshaft journals (⊘ 25-4 to 25-7) before installing new bearings. If the journals are not in good condition, the new bearings may soon fail. Also, bearings may have worn unevenly; that is, some bearings may be worn considerably more than others. If so, or if a bearing is damaged, the possibility of a bent crankshaft or clogged oil passages should be considered. This means the crankshaft should come out so it and the oil galleries in the block can be inspected. The following sections describe checking bearing fit, replacing bearings, and servicing crankshafts.

NOTE: If one main bearing requires replacement, then all main bearings should be replaced even though the others appear to be in good condition. If only one main bearing is replaced, crankshaft alignment might be lost. This would overload some bearings, causing them to fail rapidly.

[1] If you are interested in the geometry of the gauge, note the following, and refer to Fig. 25-7.

$2AO = OC$ [In a 30° right triangle, the hypotenuse (OC) is twice the side opposite the 30° angle]

$AO = OD$

$2OD = OC = OD + DC$

$OD = DC$

Since DC is therefore equal to the radius of the journal, $2DC$ equals the diameter of the journal.

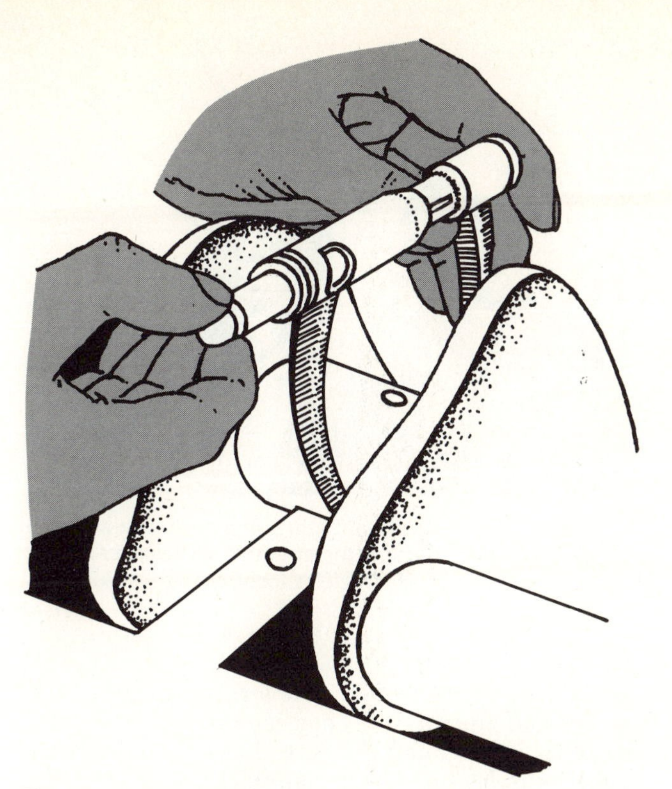

Fig. 25-8. Crankshaft-journal diameter being measured with a special micrometer.

⊘ 25-9 Inspecting Main-Bearing Fit Bearing fit (or oil clearance) should always be inspected after new bearings are installed. The fit should also be inspected whenever the condition of the bearings is being determined. (Crankshaft-journal condition should also be inspected at the same time.)

1. *PRECISION-INSERT MAIN BEARINGS* Bearing clearance can be checked with shim stock or Plastigage.

a. *With Shim Stock* Put a piece of shim stock of the right size and thickness in the bearing cap after the cap has been removed (Fig. 25-9). Coat the shim stock lightly with oil. Replace the cap. Tighten the cap nuts or bolts to the specified tension. Note the ease with which the crankshaft can be turned.

CAUTION: Do not attempt to rotate the crankshaft, as this could damage the bearing. Instead, see whether it will turn about 1 inch [25.4 mm] in one direction or the other.

If the crankshaft is locked or drags noticeably, the bearing clearance is less than the thickness of the shim stock. If it does not drag, an additional thickness of shim stock should be placed on top of the first. The ease of crankshaft movement should again be checked. Clearance normally should be about 0.002 inch [0.05 mm] (see the engine manufacturer's specifications for exact clearance). After clearance is determined, remove the cap, take out the shim stock, replace the cap, and torque the cap nuts according to specifications.

b. *With Plastigage* Wipe the journal and bearing clean of oil. Put a strip of Plastigage lengthwise in the center of the journal (Fig. 25-10). Replace and tighten the cap. Then remove the cap and measure the amount the Plastigage has been flattened (Fig. 25-11). Do not turn the crankshaft with the Plastigage in place. (See ⊘ 24-15 for more detailed information on Plastigage.)

Fig. 25-9. Main-bearing clearance being checked with shim stock. (*Chrysler Corporation*)

CAUTION: The crankshaft must be supported so that its weight will not cause it to sag. A sagging crankshaft could result in an incorrect measurement. One way to support it is to position a small jack under the crankshaft. Let the jack bear against the counterweight next to the bearing being checked. Another method is to put shims in the bearing caps of the two adjacent main bearings.

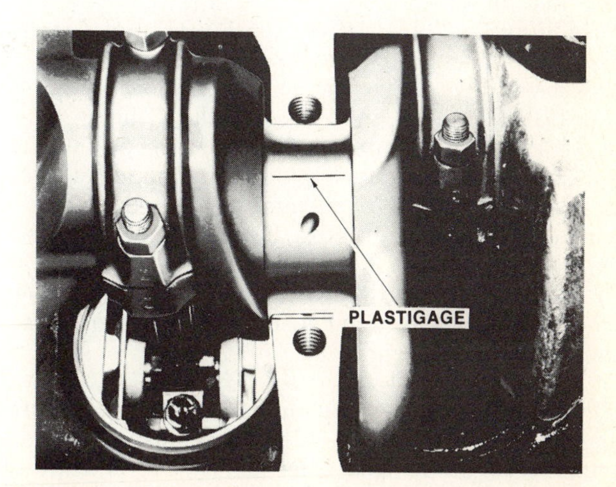

Fig. 25-10. Plastigage strip in place for bearing-clearance check. (*Chevrolet Motor Division of General Motors Corporation*)

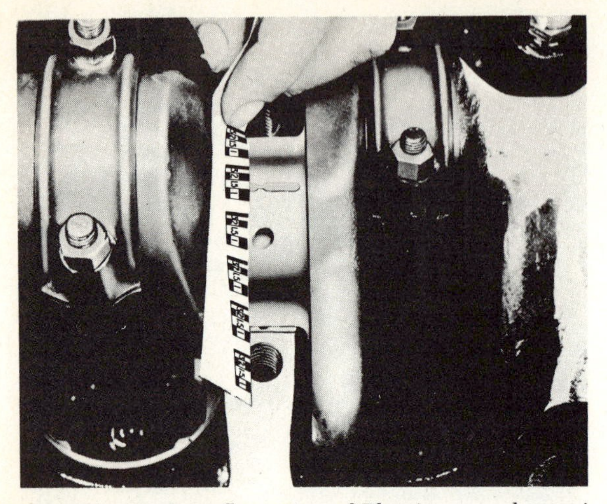

Fig. 25-11. Checking flattening of Plastigage to determine bearing clearance. (*Chevrolet Motor Division of General Motors Corporation*)

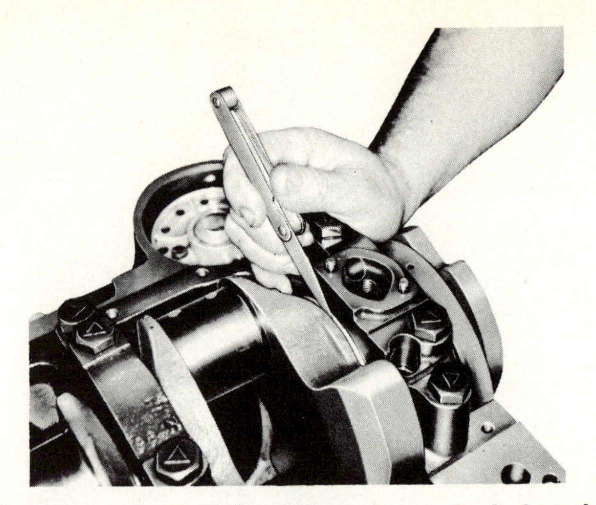

Fig. 25-12. Crankshaft end play being checked at the end-thrust bearing with a feeler gauge. (*Chevrolet Motor Division of General Motors Corporation*)

Then tighten the cap bolts. This lifts and supports the crankshaft. (Of course, if the engine is out of the car and inverted, this is not necessary.)

2. *SHIM-ADJUSTED MAIN BEARINGS* Loosen all bearing caps just enough to permit the crankshaft to turn freely. Take off the rear main-bearing cap. Remove one shim from each side of the cap. Replace and tighten the cap bolts or nuts to the specified tension. Rotate the crankshaft to see whether it now drags. If it does not drag, remove additional shims (in pairs). Check for drag after each pair is removed. When a drag is felt, replace one shim on each side of the cap. If the crankshaft now turns freely when the cap is tightened, the clearance is correct. Loosen the cap bolts or nuts. Go to the next bearing and adjust its clearance in the same way. Finally, when all bearings are adjusted, tighten all cap bolts or nuts to the proper tension. Then recheck for crankshaft drag as it is turned. If if drags, then recheck and readjust the bearings.

3. *INSPECTING CRANKSHAFT END PLAY* Crankshaft end play will become excessive if the end-thrust bearings are worn. This condition produces a noticeably sharp, irregular knock. If the wear is considerable, the knock will occur every time the clutch is released and applied. This action causes sudden endwise movements of the crankshaft. Check end play by forcing the crankshaft endwise as far as it will go. Then measure the clearance at the end-thrust bearing with a feeler gauge (Fig. 25-12). Consult the engine manufacturer's shop manual for allowable end play.

⊘ 25-10 Replacing Precision Main Bearings Before replacing bearings, crankshaft journals should be checked (⊘ 25-4). Also, after bearings are installed, bearing fit should be checked (⊘ 25-9). Precision-insert main bearings can be replaced without removing the crankshaft. However, some authorities do not recommend this. They say that you are working blind. You cannot be sure that the counterbore in the cylinder block is perfectly clean and that the shell is seating tightly. Furthermore, neither the crankshaft nor the block can be checked for alignment. As previously noted (⊘ 25-3), with uneven bearing wear the crankshaft should be removed for further inspection.

To install a precision-insert main bearing without removing the crankshaft, use a special roll-out tool as shown in Figs. 25-13 and 25-14. The tool is inserted into the oilhole in the crankshaft journal, as shown. Then the crankshaft is rotated. The tool forces the bearing shell to rotate with the crankshaft so that it is turned out of the bore. The crankshaft must be rotated in the proper direction so that the lock, or tang, in the bearing is raised up out of the notch in the cylinder block.

To install a new bearing half, coat the bearing surface with engine oil. Leave the outside of the bearing dry. Make sure that the bore, or bearing seat, in the block is clean. Do not file the edges of the shell (this would remove its crush). Use the tool to slide the bearing shell into place as shown in Fig. 25-14. Make sure that the tang on the bearing shell seats in the notch in the block. Then place a new bearing shell in the cap. Install the cap, and tighten the cap bolts or nuts to the specified tension. Tap the crown of the cap lightly with a brass hammer while tightening it. This helps to align the bearings properly. After all bearings are in place, check bearing fit.

NOTE: If the crankshaft is removed, it is easier to install main bearings (Fig. 25-15). Also, you can wipe the bearing bores in the cylinder block and make sure they are in good condition. Then, the bearing inserts can be slid into position, as shown.

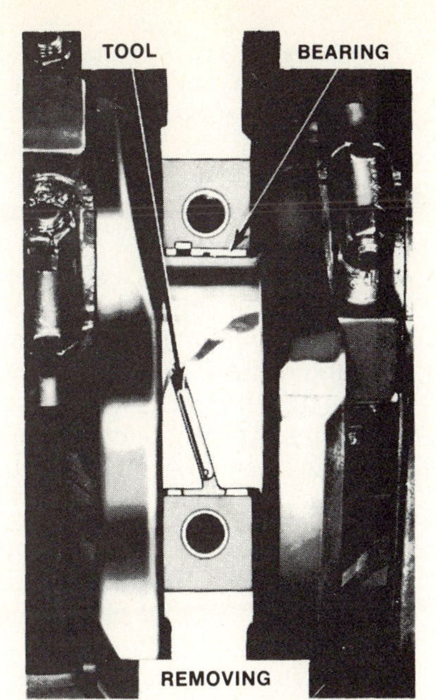

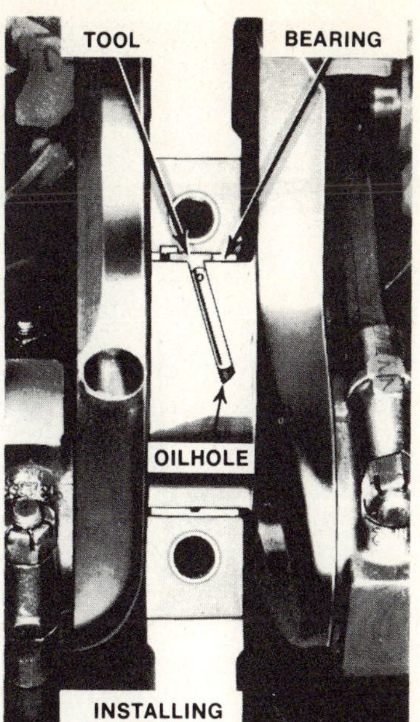

Fig. 25-13. Removing and installing an upper main bearing. The crankshaft journal is partly cut away to show the tool inserted in the oil hole in the journal. (*Chrysler Corporation*)

Fig. 25-14. Using a roll-out tool to install a new upper main bearing. Arrow shows the direction in which crankshaft is turned to roll the bearing into place. (*Federal-Mogul Corporation*)

Fig. 25-15. Installing a main bearing with the crankshaft out. (*Federal-Mogul Corporation*)

Some bearing sets have annular (or ring) grooves in only one bearing half; others have grooves in both halves. Some do not use grooves. Be sure to check the service manual for the engine you are servicing to determine what kind of bearing half goes where.

Some crankshaft journals have no oilhole. For example, the rear main journals of many in-line engines do not. To remove and replace the upper bearing half on these journals, first start the bearing half with a small pin punch and hammer. Then use a pair of pliers with taped jaws to hold the bearing half against the oil slinger. Rotate the crankshaft (Fig. 25-16). This movement will pull the old bearing out. The new bearing is put into position in the same manner. The last fraction of an inch can be pushed into place by holding only the oil slinger with the pliers while rotating the crankshaft. Or the bearing may be tapped down with a pin punch and hammer. Be careful that you do not damage the bearing.

NOTE: While removing and replacing the upper bearing shell of a rear main bearing, hold the oil seal

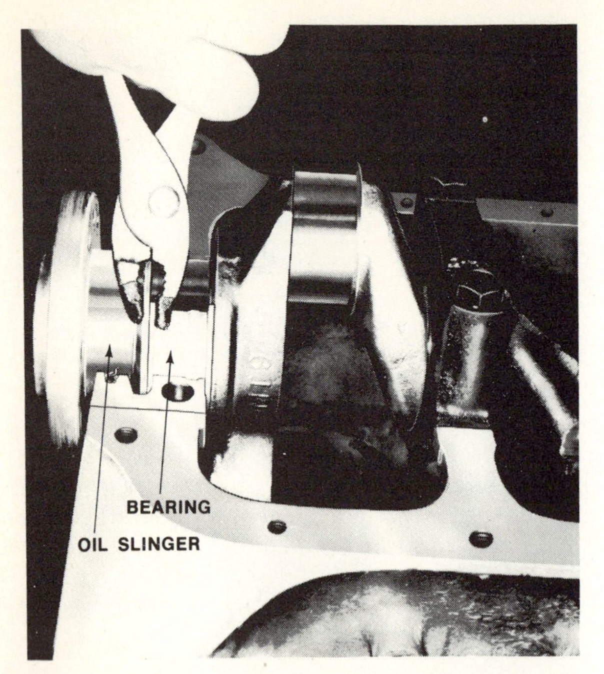

Fig. 25-16. Replacing a rear main-bearing half with pliers. On this engine, the crankshaft has no oilhole in the rear journal. (*Chevrolet Motor Division of General Motors Corporation*)

in position in the cylinder block; otherwise, it may move out of position (see ⊘ 25-11).

Bearing fit should be checked after all bearing caps have been replaced, as already explained (⊘ 25-9). If excessive clearances are found with the new bearings, the journals are worn. This means the crankshaft must be removed for service (⊘ 25-12 and 25-13). Then, undersize bearings must be installed. (Bearings are available in several undersizes.) The journals should be ground down enough to remove imperfections and then ground an additional amount to fit the next undersize bearings.

On all but a very few engines, precision-insert bearings are installed without shims. Never use shims on these bearings unless the engine manufacturer specifies them. Similarly, bearing caps must not be filed in an attempt to improve bearing fit.

⊘ 25-11 Replacing the Main-Bearing Oil Seal An oil seal is required at the rear main bearing to prevent oil leakage at that point (Fig. 25-17). When main-bearing service is being performed, or whenever leakage is noted at the rear main bearing, the oil seal must be replaced.

Replacement of the main-bearing oil seal varies with different constructions. On some engines using a split-type oil seal, the crankshaft must be removed. A special oil-seal compressor or installer is then used to insert the new seal in the cylinder-block bearing. The seal should then be trimmed flush with the block, as shown in Fig. 25-18. The oil seal in the cap can be replaced by removing the cap, installing the oil seal, and trimming it flush. On other engines it is not necessary to remove the crankshaft since removal of the flywheel will permit access to the upper oil-seal retainer. The retainer cap screws can then be removed along with the retainer for oil-seal replacement.

Some engines use a one-piece rubber-type oil seal. It can be pulled from around the crankshaft with a pair of pliers. Then a new oil seal can be worked into place. The new oil seal should be coated with oil. Then coat the crankshaft contact surfaces of the seal with a suitable grease. Install the seal, using a seal-installing tool if required.

⊘ 25-12 Removing the Crankshaft Such parts as the oil pan, timing-gear or timing-chain cover, crankshaft timing gear or sprocket, interfering oil lines, and oil pump must be removed before the crankshaft can be taken off. Also, on some engines, the flywheel must be detached from the crankshaft. With other parts off, the bearing caps are removed to release the crankshaft.

CAUTION: The crankshaft is heavy! Support it adequately as you remove the bearing caps.

NOTE: For a complete engine overhaul, the cylinder head and piston-and-rod assemblies must be removed. However, if only the crankshaft is being removed, the piston-and-rod assemblies need not be removed. Instead, they can be detached from the crankshaft and pushed up out of the way. *Be very careful not to push them up too far!* If you do, the top piston ring may move up beyond the cylinder

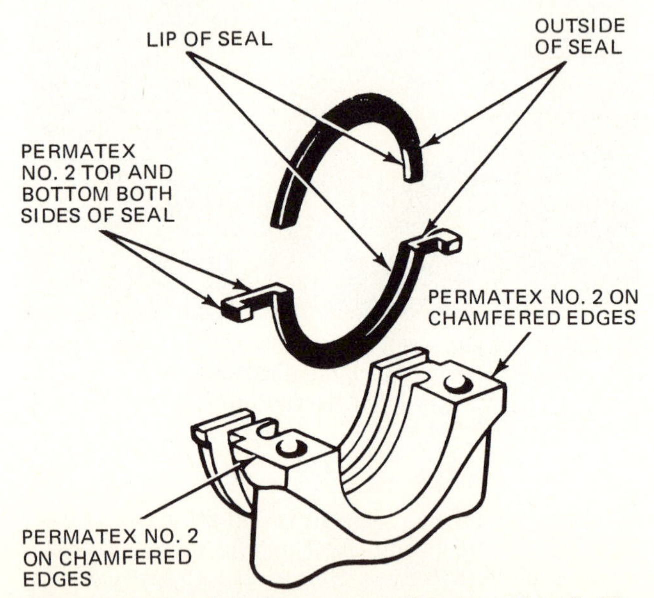

Fig. 25-17. Crankshaft rear main-bearing oil seal. The upper part fits in a groove in the block. (*American Motors Corporation*)

block. In this case, it can catch on the top of the block. You will then be unable to pull the piston-and-rod assembly back down. You will have to remove the cylinder head and use a ring compressor to get the ring back down into the block again.

⊘ 25-13 Inspecting and Servicing the Crankshaft Inspect the crankshaft for alignment and for main journal and crankpin wear. Alignment can be checked with the setup shown in Fig. 25-19. As the crankshaft is rotated in the V blocks, the dial indicator will show any misalignment. Make sure the V blocks are clean. Oil them lightly so they do not scratch the crankshaft journals rotated on them.

CAUTION: Do not leave the crankshaft supported only at the ends, as shown in Fig. 25-19. This could cause the crankshaft to sag and go out of alignment from its own weight. To prevent misalignment, set the crankshaft on end or hang it from one end (Fig. 25-20).

If the crankshaft is out of line, a new or reground crankshaft should be used. It is very difficult to straighten a bent crankshaft.

Inspection of the journals and crankpins for taper or out-of-round has already been discussed (⊘ 24-16 and 25-4 to 25-7). If journals or crankpin taper or out-of-round exceeds safe limits, or if they are rough, scratched, pitted, or otherwise damaged, they must be ground (⊘ 25-14). Then new undersize bearings must be installed. Journals and crankpins must be ground down to fit the next undersize bearings available.

Fig. 25-18. Using a special tool to install a rear main-bearing oil seal in a cylinder block. (*Ford Motor Company*)

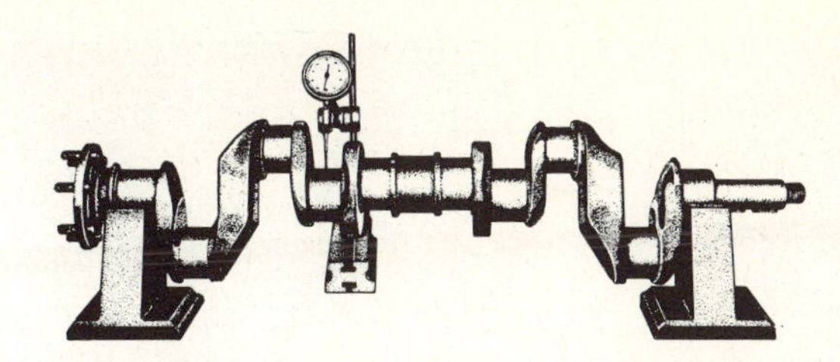

Fig. 25-19. Checking a crankshaft on V blocks with a dial indicator. This check is made to determine if the crankshaft is out of line. (*Federal-Mogul Corporation*)

NOTE: It is possible to "metalize" journals and crankpins and then regrind them to their original sizes. This is done by first rough-turning the journals and crankpins in a lathe. Then, a high-temperature flame is used to spray liquid metal onto the prepared surfaces. This metal adheres and can be ground to form a new journal surface. The procedure is useful for crankshafts from large, heavy-duty engines. A second procedure uses an electric arc to add metal to the journal. The cost of these crankshafts often makes saving the old crankshaft worthwhile.

⊘ 25-14 Finishing Main and Crank Journals A special *lathe*, or crankshaft grinder, is required to service main and crank journals. This machine is found in automotive machine shops specializing in crankshaft service and in engine-rebuilding shops (Fig. 25-21). The bearing surface must be finished to extreme smoothness. Some authorities recommend a final polishing with a long strip of oiled crocus cloth (a special cloth used to polish metal). Wrap the strip halfway around the journal and pull it back and forth, being careful to work all the way around the journal.

CAUTION: Be sure to relieve (grind back) the journal and crankpin radii (where they curve up to the

Fig. 25-20. Completed crankshafts, hanging in racks, ready for installation in engines. (*Automotive Rebuilders, Inc.*)

Fig. 25-21. Crankshaft in a crankshaft lathe. The picture was taken at high speed to show the grinding wheel grinding the journal. (*Automotive Rebuilders, Inc.*)

crank cheeks). This procedure will guard against bearing failure from radii ride (see ⊘ 24-14, item 5). Also, use care in touching up the thrust faces on the journal that takes the flanged or end-thrust bearing. Make sure that these faces are smooth and square with the crankshaft journal.

1. CLEANING THE CRANKSHAFT Any time that a crankshaft is out of the engine or after grinding journals and crankpins, the crankshaft should be thoroughly cleaned. Journals and crankpins can be cleaned with a long strip of oiled crocus cloth, as explained previously. Then the crankshaft should be washed in a suitable solvent. A valve-guide or rifle cleaning brush should be used to clean out oil passages (Fig. 25-22). Remember that any trace of abrasive left in an oil passage can work out onto a bearing surface and cause early bearing failure. Oil the bearing surfaces immediately after they have been cleaned to keep them from rusting.

2. GRINDING CRANKPINS ON ENGINE A special grinder (Fig. 25-23) can be used to grind crankpins with the crankshaft still in the engine. The grinder is attached to a crankpin, and the crankshaft is rotated by a driving device at the rear wheel (the transmission in gear). *This is not a recommended procedure.*

Check Your Progress

Progress Quiz 25-1 You are nearing the end of your studies of *Automotive Engines*. If you have remembered the essentials of what you have read in this book, you have become well acquainted with how automobile engines are constructed, how they operate, what can go wrong with them, and how troubles can be eliminated. Regardless of what your plans are for your future in the automobile business, you will find such information valuable. These progress quizzes give you a good chance to check up on your

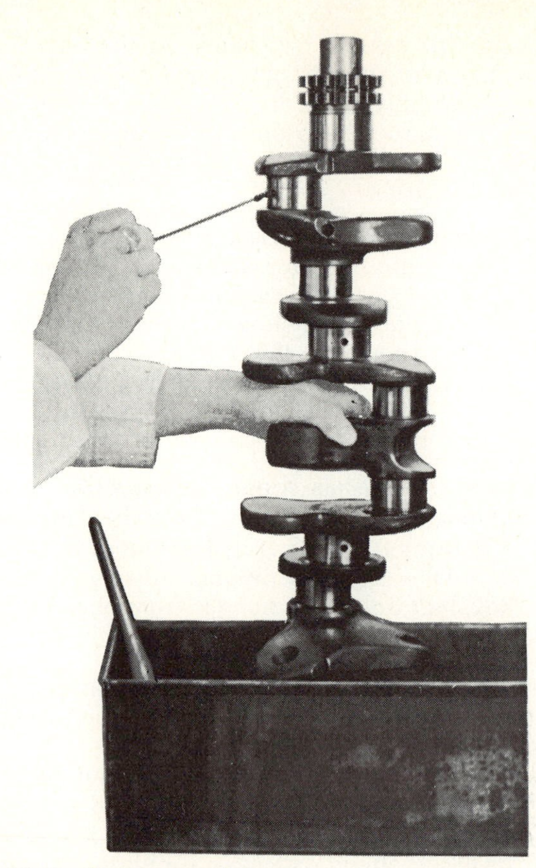

Fig. 25-22. Using a small brush to clean out oil passages in the crankshaft. (*Federal-Mogul Corporation*)

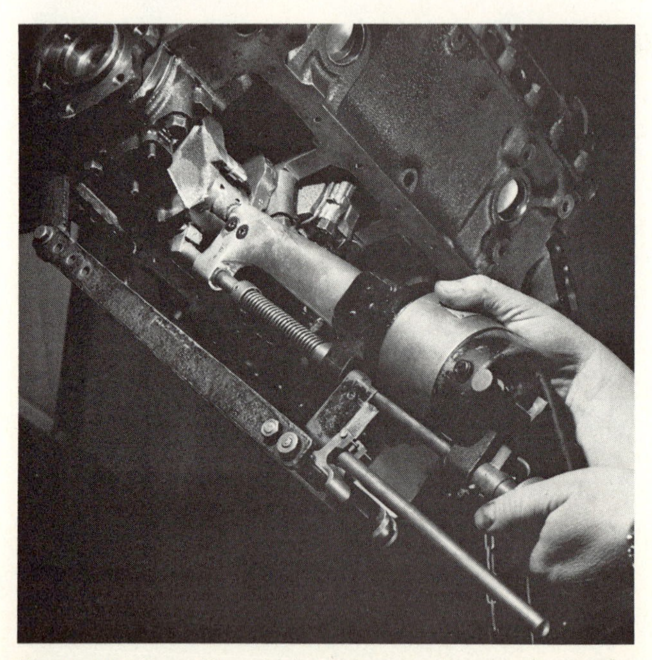

Fig. 25-23. Portable grinder being used to grind a crankpin on a crankshaft in an engine. (*Federal-Mogul Corporation*)

understanding of what you have been reading. The following quiz covers the past few pages.

Completing the Sentences The sentences that follow are incomplete. After each sentence there are several words or phrases, only one of which will correctly complete the sentence. Write each sentence in your notebook, selecting the proper word or phrase to complete it correctly.

1. Crankshaft main journals can be checked on the engine by either of two special measuring devices: (*a*) a micrometer or a dial indicator, (*b*) a crankshaft gauge or a dial indicator, (*c*) a micrometer or a crankshaft gauge.
2. Crankshaft journals should be reground if they are tapered or out of round by more than: (*a*) 0.0003 in [0.0076 mm], (*b*) 0.003 in [0.076 mm], (*c*) 0.030 in [0.76 mm], (*d*) 0.300 in [7.60 mm].
3. If one main bearing requires replacement, you should replace: (*a*) adjacent main bearings, (*b*) all main bearings, (*c*) camshaft bearings, (*d*) rod bearings.
4. If the Plastigage flattens more at one end than at the other when it is used to check main-bearing clearance, it means that the: (*a*) bearing cap is loose, (*b*) bearing shell is loose in bore, (*c*) journal is tapered, (*d*) shaft is bent.
5. Excessive end-thrust-bearing wear will cause excessive: (*a*) crankshaft end play, (*b*) crankshaft deflection, (*c*) crankshaft torsional vibration.
6. Before installing new main bearings, you should always: (*a*) check crankshaft journals, (*b*) check connecting-rod journals, (*c*) grind crankshaft journals, (*d*) replace the crankshaft.
7. After installing new main bearings, you should always: (*a*) check crankshaft journals, (*b*) check bearing fit, (*c*) check bore eccentricity, (*d*) shim-adjust bearings.
8. Main-bearing clearances can be checked with either: (*a*) shims or dial indicator, (*b*) feeler stock or micrometer, (*c*) micrometer or Plastigage, (*d*) feeler stock or Plastigage.
9. The crankshaft should not be supported by V blocks at its two end journals for any length of time since this could cause the: (*a*) journals to scratch, (*b*) journals to rust, (*c*) crankshaft to sag.
10. If all bearings have worn evenly, all that is normally required is to: (*a*) grind journals and replace bearings, (*b*) replace the bearing caps, (*c*) check the journals and install new bearings.

Cylinder Blocks

⊘ 25-15 Cylinder Wear The piston-and-ring movement, the high temperatures and pressures of combustion, the washing action of gasoline entering the cylinder—all these tend to cause cylinder-wall wear. At the start of the power stroke, pressures are the greatest. The compression rings are forced with the greatest pressure against the cylinder wall. Also, at the same time, the temperatures are highest. The oil film is therefore least effective in protecting the cylinder walls. Thus, the most wear will take place at the top of the cylinder. As the piston moves down on the power stroke, the combustion pressure and temperature decrease. Thus, less wear takes place. The cylinder thus wears irregularly, as shown in Fig. 25-24. This is *taper wear,* which leaves a large ridge at the top and a smaller ridge at the bottom. These ridges mark the limits of ring travel.

In addition to taper wear, the cylinder tends to wear oval-shaped because the piston tends to push sideways against the cylinder wall as it moves up and down in the cylinder. These side thrusts of the piston are due to the tilting of the connecting rod. For example, on the power stroke, in the piston position shown in Fig. 25-25, the total push on the piston is as shown at *A*. Most of this push is carried downward through the connecting rod, as at *B*. But a small component of the total push *C* thrusts the piston sideways against the cylinder wall. Although this sideward force does tend to cause wear, it is actually not a major factor in cylinder wear because it is, after all, relatively small.

It is easy to be fooled by "oval wear" because cylinders have different shapes at different temperatures. When you remove the cylinder head of a cold engine and carefully measure cylinders for eccentricity, you are quite likely to find that they are indeed oval. What happens, however, when you put the head back on and run the engine? First, in replacing the head and drawing the bolts or nuts down right you are introducing certain stresses into the block. These stresses alter the shape of the cylinder. Then, when the engine is started and the block

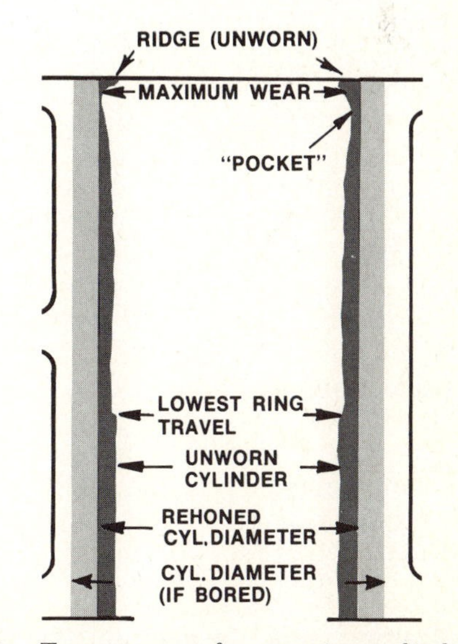

Fig. 25-24. Taper wear of an engine cylinder (shown exaggerated). Maximum wear is at the top, just under the ring ridge. Honing the cylinder usually requires removal of less material than boring, as indicated. Material to be removed by honing is shown solid. Material to be removed by boring is shown both solid and shaded. (*Sunnen Products Company*)

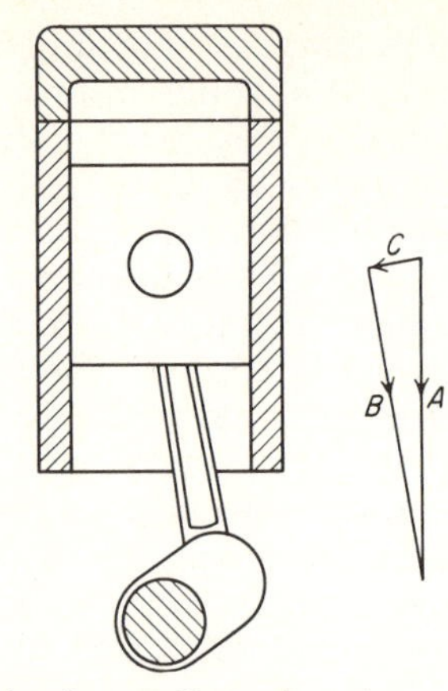

Fig. 25-25. Side thrust *C* as the piston and connecting rod reach the position shown during the power stroke. *A* represents the component of that pressure that is transmitted through the connecting rod.

warms up, the expansion of the metal alters the cylinder shape still further. In many cases these changes in cylinder shape reduce the out-of-roundness; that is, replacing the head and warming the block reduce the eccentricity of the cylinders. Here is how this condition comes about. When the block is first manufactured, the bores are all machined to be perfectly round when cold. But installing the head and warming the new block (by running the engine) cause the cylinders to be distorted out of round. However, as the engine operates, the piston-ring action tends to wear the cylinders round. In other words, the cylinders wear round when hot. After that, when the block cools and the head is removed, the stresses are changed so that the cylinders distort out of round. But replacing the head and warming the block reduce this distortion; the cylinders become more nearly round.

Another type of cylinder wear results from the washing action of entering gasoline. This wear is most likely to occur on the cylinder wall opposite the intake valve. At times the air-fuel mixture is not perfectly blended, and small droplets of gasoline, still unvaporized, enter the cylinder. These droplets of gasoline strike the cylinder wall and tend to wash away the protective film of lubricating oil. With reduced protection, some additional wear takes place. A closed choke, which excessively enriches the air-fuel mixture, hastens this sort of wear because quantities of unvaporized gasoline are likely to enter the cylinder as long as the closed choke causes the carburetor to supply a rich mixture.

⊘ 25-16 Cleaning and Inspecting the Cylinder Block As a first step, make a visual inspection of the block. Major damage, resulting from a main bearing spinning or a broken connecting rod going through the block means the block must be discarded. If the engine has overheated and there are cracks in the cylinder walls, discard the block.

Inspect the cylinder bores for cracks, grooves, scratches, or discoloration. Inspect for cracks across the top of the block between cylinders, between bolt holes, on the outside of the block, and in the main-bearing-bore webs. (See ⊘ 25-26 on block repair.) If everything looks okay, measure the cylinder bores with a micrometer. If the bores are not too badly worn and they can be honed or rebored within specified limits, then clean the block. In other words, if you have decided to reuse the block, you should clean and service it.

To clean the block, first remove the cam bearings and expansion (freeze) plugs (see ⊘ 25-27). Then use the cleaning method that is available in the shop. One method uses a jet of steam from a steam cleaner directed by a nozzle onto the block to wash away dirt, sludge, and oil. Another method is to boil the block in a hot solution of caustic soda or similar chemical. A good solvent can also be brushed on to clean the block. Be sure that all old gasket material is removed from machined surfaces. All pipe plugs that close off oil passages should be removed so the passages can be blown out with compressed air. Long rods of the proper diameter, or rifle (valve-guide cleaning) brushes (Fig. 25-22) can be pushed through the oil passages. This will clear out sludge that does not easily blow out. Remember that clogged oil passages prevent normal bearing lubrication so that the bearings wear rapidly and fail. Make sure the passages are clean.

Threaded holes in the block should be blown out with compressed air. If the threads are not in good condition, use a tap or thread chaser of the correct size to clean them up. Then blow out the holes. Dirty or battered threads may give false torque readings that will prevent normal tightening of bolts on reassembly. This could cause engine failure from loose engine parts. Dirt in the bottom of a bolt hole may prevent normal tightening of the bolt. In addition, it could cause the block to break

Fig. 25-26. Checking the cylinder block for warpage with a straight edge.

Fig. 25-27. Checking for main-bearing-bore alignment in the block with bearing caps in place. The alignment-checking bar should turn with all caps tightened to specified tension. (*Federal-Mogul Corporation*)

because of the pressure applied at the bottom of the hole as the bolt is tightened. Repeairing damaged threads with thread inserts is described in ⊘ 25-28.

If there is any question of cracks, the block should be Magna-Fluxed. In this procedure, a strong magnetic field is put on the block. Powdered iron is applied to the block. If there is a crack, the powdered iron is attracted to it and the crack is shown plainly.

Machined gasket surfaces should be inspected for burrs, nicks, and scratches. Minor damage can be removed with a fine-cut file. Check the head end of the block for warpage by laying a long straightedge against the sealing surfaces (Fig. 25-26). If there is any clearance between the straightedge and the cylinder block, measure it with a feeler gauge, as shown. The surface of the block should be flat and true within the manufacturer's specifications. A typical specification is that the surface should not be more than 0.003 inch [0.08 mm] out of true for each 6 inches [152.40 mm] of surface. If the surface is out of true more than this, it should be refaced in a surface grinder. This is a job for the engine machine shop. No more than the minimum amount of metal should be removed as required to true up the surface.

Inspect the main-bearing bores for alignment and out-of-roundness (⊘ 25-17). Bearing shells must be removed, but caps must be in place for this check.

If expansion plugs have not been removed (that is, if the block has not been cleaned), they should be checked for leaks. Any plugs that look as though they have leaked should be replaced (⊘ 25-27).

⊘ 25-17 Inspecting Bearing Bores Very uneven bearing wear with some bearings worn much more than others may means out-of-round bearing bores or a warped block. Inadequate oiling could also be a factor, possibly resulting from a worn oil pump or clogged oil passages (⊘ 25-3).

To check for out-of-round bores, the crankshaft and main bearings must be removed. The bearing bores must be clean and the caps installed with the bolts torqued to specifications. Then an inside micrometer can be used to check for out-of-round. If the bores are out of round or out of true, they must be line-bored to restore roundness (⊘ 25-18).

To check bores for alignment, a special alignment bar is installed in place of the crankshaft with the bearings removed (Fig. 25-27). The bar is 0.001 inch [0.03 mm] smaller than the diameter of the bores. When the cap bolts are torqued to specifications, the bar should turn with the help of an extension handle on the bar. If it does not turn, the block is out of line and the bores must be machined (⊘ 25-18).

NOTE: If a bearing cap is damaged or lost, a new cap is required. It is sometimes hard to get good cap alignment when a new cap is installed. To secure good alignment, the bearing bores may require line boring (⊘ 25-18).

⊘ 25-18 Line-boring the Bearing Bores As a first step, all bearing caps should have about 0.015 inch [0.38 mm] removed from their *parting faces*. The parting face is the surface that meets the cylinder block when the cap is installed. One method is to gang the caps together in a special fixture. Then use a surface grinder, as shown in Fig. 25-28, to grind all the caps at once.

Next the caps are installed and the cap bolts torqued to specifications. Then, the block is clamped in the boring machine (Fig. 25-29). Bores should be measured to determine how much metal has to be

Fig. 25-28. Removing metal from the parting faces of the bearing caps. (*Federal-Mogul Corporation*)

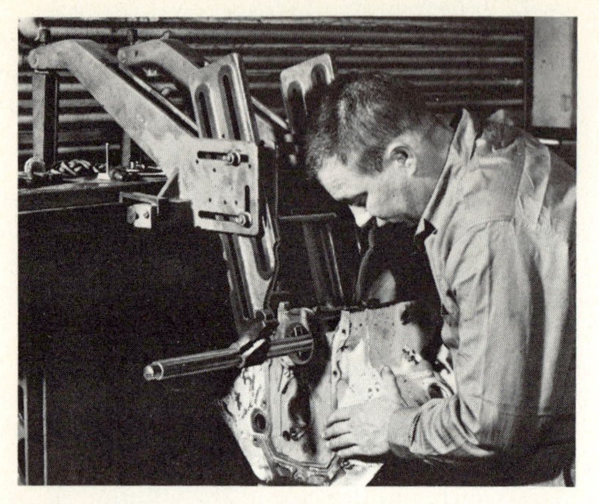

Fig. 25-29. Mounting the cylinder block in a boring machine. (*Federal-Mogul Corporation*)

removed. Enough metal should be removed to allow the bores to take standard bearings. Then the boring tool is adjusted to cut the bores to their original diameters (Fig. 25-30). The adjustment must allow most of the material to be removed from the caps. If any amount of material is removed from the bore half in the block, the distance between the crankshaft and camshaft centerlines will be shortened. This could cause timing-gear meshing problems.

After the bore job is finished, remove the caps. Scrape off any burrs from the edges of the caps and block bores. Then clean the block and cap, making sure you remove all traces and metal particles. The block and oil passage *must be clean!*

⊘ 25-19 Cylinder Service Up to certain limits, cylinders may wear tapered or out of round and not require refinishing. As noted in ⊘ 24-25, special drastic replacement rings will control compression

Fig. 25-30. Boring tool ready to cut the bearing bores to their original diameter. (*Federal-Mogul Corporation*)

and oil in cylinders with some taper and out-of round wear. But when wear goes byond a certain point, even the severest rings cannot hold compression and control oil. Loss of compression, high oil consumption, poor performance, and heavy carbon accumulations in the cylinders will result. Then, the only way to get the engine back into good operating condition is too refinish the cylinders. New pistons (or resized pistons) and new rings must be installed at the same time.

Refinishing cylinders requires 12 to 20 hours, according to the type of engine. This includes fitting and installing new pistons, rings, piston pins, and connecting rods. When new bearings are fitted, about 10 additional hours are required. Grinding valves would require several more hours. These various times are mentioned because an engine that requires cylinder refinishing is usually in need of a general overhaul. These other services will also be required.

⊘ 25-20 Inspecting Cylinder Walls Wipe walls and examine them for scores and spotty wear (which shows up as dark, unpolished spots). Hold a light at the opposite end of the cylinder so that you can see the walls better. Scores or spots mean the walls must be refinished. Even drastic rings cannot give satisfactory performance on such walls.

Next, measure the cylinders for wear, taper, and out-of-roundness (see Fig. 25-31). This can be done with an inside micrometer, a telescope gauge and an outside micrometer, or a dial indicator (Fig. 24-37). The dial indicator is moved up and down in the cylinder and rotated at various positions to detect wear. To measure taper, start with the dial set to zero and with the dial indicator at the top of ring travel. Then push the dial indicator to the bottom of the cylinder. The movement of the indicator needle will tell you how much the cylinder has worn tapered. To check for out-of-round wear, rotate the dial indicator at various positions in the cylinder.

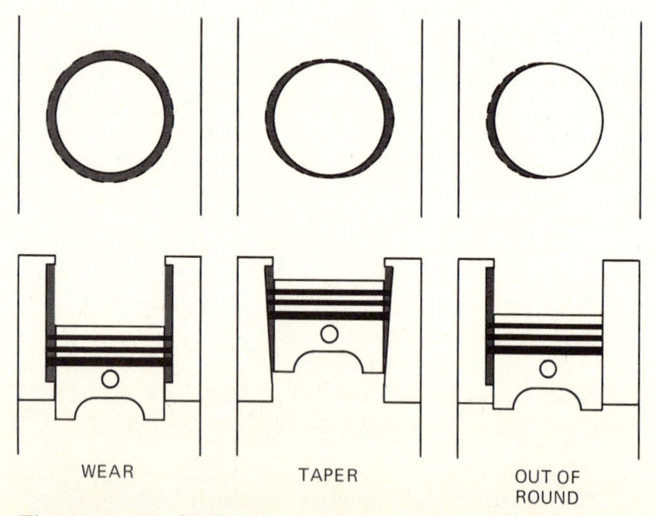

Fig. 25-31. Cylinder wear, taper, and out-of-roundness.

Note the amount of needle movement. Write down the wear measurements for each cylinder. Compare them with the allowable maximum specified by the engine manufacturer to decide what has to be done.

A quick way to measure cylinder taper is to push a compression ring down to the lower limit of ring travel. Measure the ring gap with a feeler gauge (Fig. 25-32). Then pull the ring up to the upper limit of ring travel. Remeasure the ring gap. The top gap minus the bottom gap divided by three gives you the approximate cylinder taper.

⊘ 25-21 Refinishing Cylinders As a first step, the block should be cleaned (⊘ 25-16). A decision must be made on whether the cylinders are to be honed or bored. This decision depends on the amount of cylinder wear. Figure 25-24 illustrates the amount of metal removed by the hone and by boring. The hone (Fig. 25-33) uses a set of abrasive stones which are turned in the cylinder. The boring machine uses a revolving cutting tool. Where cylinder wear is not too great, only honing is necessary. But if wear has gone so far that a considerable amount of material must be removed, then honing will not do the job and the cylinder must be rebored.

NOTE: If cylinders are within specifications and do not require honing or boring, a glaze breaker should be run through the cylinders. Three or four strokes of the glaze breaker will "break the glaze" on the cylinder wall. This practice produces a slightly roughened surface which will let the new rings wear in properly.

CAUTION: If the crankshaft has not been removed, the main bearings and crankshaft must be protected from grit and cuttings. Stuff clean rags down in the cylinders to catch the cuttings.

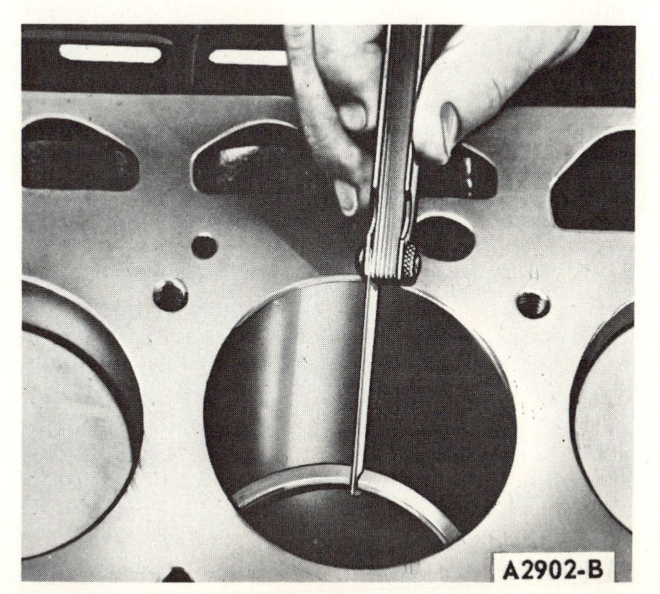

Fig. 25-32. Checking the ring gap of a ring pushed down into the cylinder to the lower limit of ring travel. (*Ford Motor Company*)

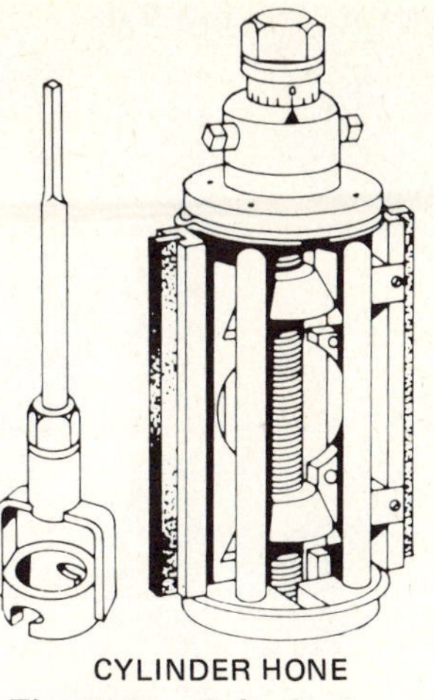

Fig. 25-33. Cylinder hone.

⊘ 25-22 Honing Cylinders If wear, taper, or out-of-round are not too great, only honing will be required. Figure 25-34 shows a cylinder hone in place ready for the honing operation. Figure 25-35 shows a honing operation in progress in an engine rebuilding shop. Note that the cylinder is flooded with honing lubricant, a cutting oil, during the honing operation.

NOTE: Dry honing is also done. With dry honing, a vacuum device removes the dust produced by the operation.

If a considerable amount of material must be removed, start with coarse stones. Sufficient material must be left, however, so that all rough hone marks can be removed with fine, or finishing, stones. (See ⊘ 24-25 and Fig. 24-48 for information about how the cylinder wall should look after final honing.) The proper honing pattern is important in obtaining good wearing-in of new piston rings. This pattern is obtained by moving the hone up and down as it rotates.

Final honed size should be just right to take a standard oversize piston and rings. In other words, the cylinder should be finished so that the piston and rings will fit it. Here is the way it is done. First, the piston is miked. Then the cylinder is honed to take the piston plus the proper clearance. Usually, the clearance is about 0.001 inch [0.03 mm]. That is, the cylinder is finished 0.001 inch [0.03 mm] larger than the piston diameter.

During the later stages of honing, the hone should be withdrawn periodically and the cylinder measured. This guards against overhoning.

Cylinders must be cleaned after honing (see ⊘ 25-24).

NOTE: Only those cylinders requiring service need be honed. Oversize pistons can be installed in some cylinders of an engine, and standard pistons in other

Fig. 25-34. Cylinder hone in place in a cylinder. In operation, the hone revolves in the cylinder. The abrasive stones in the hone remove material from the cylinder wall.

Fig. 25-35. Close-up view of a honing machine in action. This is a heavy-duty machine used by companies specializing in engine rebuilding. (*Automotive Rebuilders, Inc.*)

cylinders. Manufacturers supply oversize pistons that weigh the same as the standard pistons. Thus, there is no problem of balance when pistons of different sizes are used in an engine. Actually, if you have to service only one or two cylinders, you can leave the pistons in the other cylinders. That way, you would not have to install new rings on all pistons. And you wouldn't have to break the glaze on the cylinders not requiring honing. However, it is best to have all cylinder bores the same size.

⊘ 25-23 Boring Cylinders If cylinder wear is too great to be cleaned up by honing, the cylinder must be bored. The size to which it is rebored depends on the amount of material that must be removed from the cylinder wall. It also depends on the size of oversize pistons available. For instance, Chevrolet supplies pistons 0.010, 0.020, and 0.030 inch [0.25, 0.51, and 0.76 mm] oversize.

CAUTION: Modern thin-wall engines are limited as to the amount of material that can be taken from the cylinder walls. If you take more than the maximum specified, the cylinder wall will be so thin that it will not hold up in service.

NOTE: First, refinish the cylinder that has the most wear. If you cannot clean up that cylinder when refinishing it to take the maximum-sized piston available, discard the cylinder block. Or, if the block is otherwise in good condition, you could consider installing cylinder sleeves (⊘ 25-25).

Figure 25-36 shows one type of boring bar. Figure 25-37 shows the centering fingers of the cutting head extended so that the head will be centrally located in the cylinder. The centering should be done with the cutting head at the bottom of the cylinder. Then, when the centering fingers are extended, the cutting head will be centered in the original, unworn, center part of the cylinder. Figure 25-38 shows the cutting tool extended and cutting as the head revolves. This is one of several types of boring bars.

Several cautions must be observed in the use of the boring bar. First, the top of the cylinder block must be smooth and free of any nicks or burrs. Nicks or burrs upset the alignment of the bar. They cause the cylinder to be bored at an angle to the crankshaft. Nicks and burrs should be removed with a fine-cut file. All main bearing caps should be in place. The cap bolts or nuts should be tightened to the specified tension. If this is not done, the bearing bores may become distorted from the cylinder-boring operation.

The boring operation should remove just enough material to clean up the cylinder wall so that all irregularities are removed. One recommended procedure is to bore the cylinder to the same diameter as the oversize piston to be installed. The final honing will remove enough additional material to

Fig. 25-36. Cylinder-boring machine. The cutting tool is carried in a rotating bar that feeds down into the cylinder as it rotates. This causes the rotating tool to remove material from the cylinder wall. (*Rottler Boring Bar Company*)

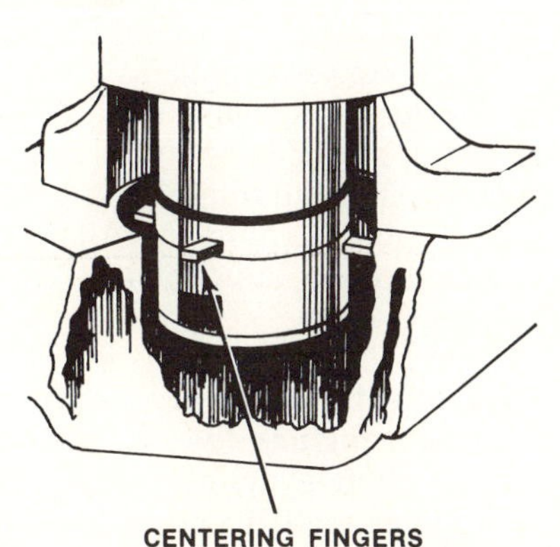

Fig. 25-37. Centering fingers extended to center the cutting head of the boring bar in the cylinder.

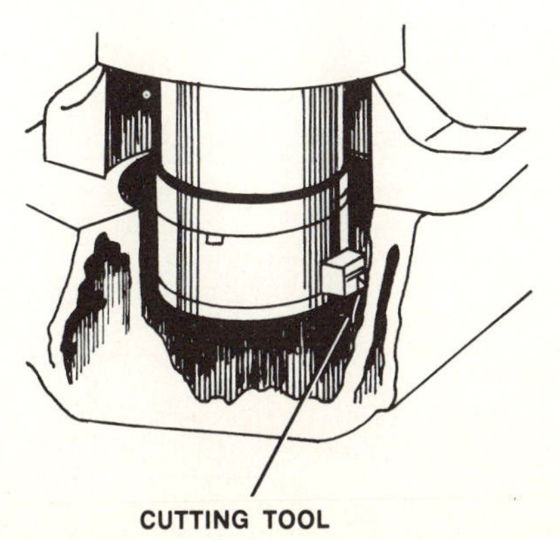

Fig. 25-38. Revolving head of a cylinder-boring machine, showing the cutting tool.

give the recommended clearance. This is usually 0.001 to 0.002 inch [0.03 to 0.05 mm] depending on the engine model.

NOTE: Only those cylinders requiring it need to be rebored and final honed. (See the note at the end of ⊘ 25-22 about standard and oversize pistons of the same weight.) However, it is recommended practice to rebore all cylinders in an engine to the same oversize.

The last step in the boring operation is to use a hone or glaze breaker to roughen the cylinder walls. This roughs up the cylinder walls enough to provide a surface suitable for ring wear-in. The boring bar can leave a very smooth surface, particularly if a light final cut has been taken. This surface can be too smooth to properly seat the rings. So it should be roughed up with a glaze breaker or hone.

⊘ 25-24 Cleaning Cylinders Cylinders must be cleaned thoroughly after the honing or boring operation. Even slight traces of grit or dust on the cylinder walls may cause rapid ring and wall wear and early engine failure. As a first step, some engine manufacturers recommend wiping down the cylinder walls with very fine crocus cloth. This loosens embedded grit and also knocks off "fuzz" left by the honing stones or cutting tool. Then use a stiff brush and hot soapy water to wash down the walls. It is absolutely essential to clean the walls of all abrasive material. If not removed, such material causes rapid wear of pistons, rings, and bearings.

After washing down the walls, swab them down several times with a cloth dampened with light engine oil. Wipe off the oil each time with a clean, dry cloth. At the end of the cleaning job, the cleaning cloth should come away from the walls showing no trace of dirt.

Clean out all coolant and oil passages in the block, as well as stud and bolt holes, after the walls are cleaned (see ⊘ 25-16).

NOTE: Gasoline and kerosene will not remove all the grit from cylinder walls. Their use to clean grit or dust off cylinder walls is not recommended.

⊘ 25-25 Replacing Cylinder Sleeves There are two types of cylinder sleeves: wet and dry. The wet sleeve is sealed to the block at the top and bottom. It is in direct contact with the coolant. The dry sleeve is pressed into the cylinder. This type is in contact with the cylinder wall from top to bottom.

Cracked blocks, scored cylinders, cylinders worn so badly that they must be rebored to an excessively large oversize—all these can often be repaired by the installation of cylinder sleeves (Fig. 25-39). As a first step, the cylinders are bored oversize to take the sleeves. Then the sleeves are pressed into place.

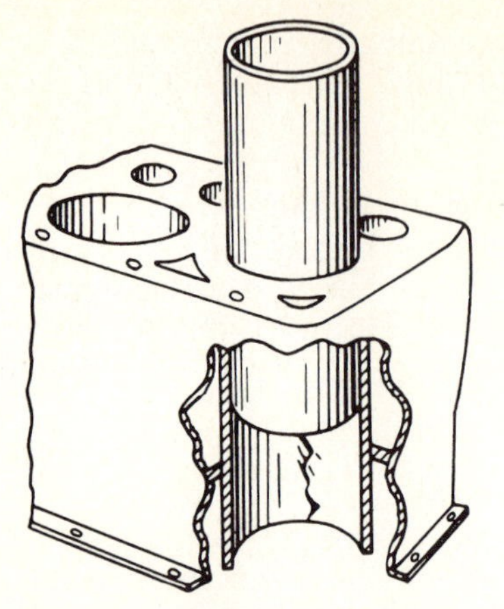

Fig. 25-39. Cracked blocks and badly scored or worn cylinder bores can sometimes be repaired by installing cylinder sleeves. (*Sealed Power Corporation*)

Figure 25-40 shows an operator using a pneumatic hammer to drive sleeves into place. The cylinder block has been prepared, and the sleeves have been positioned on the cylinder block. The pneumatic hammer uses compressed air, which operates the driving head. It hammers the sleeves down into place in the cylinder block.

The sleeves are then finished to the proper size to take a standard piston and set of rings.

⊘ 25-26 Repairing Cylinder-Block Cracks or Porosity Sometimes a cylinder block is in good condition except for some cracks or sand holes (left in the block during casting). It may then be worthwhile to repair the block. Areas which are not subjected to temperatures of more than 500 degrees Fahrenheit [260°C] or pressure (from coolant, oil, or cylinders) can be repaired with a metallic plastic or epoxy.

Fig. 25-40. Installing sleeves in a previously prepared V-8 engine block. (*Automotive Rebuilders, Inc.*)

Permissible repair areas for one manufacturer's engines are shown in Fig. 25-41.

The epoxy repair is started by cleaning the porous or cracked area down to bright metal with a grinder. Chamfer or undercut the crack or holes. If a hole is larger than $\frac{1}{4}$ inch [6.35 mm], drill, tap, and plug it. Smaller holes or cracks can be repaired with epoxy. Mix the two ingredients according to the directions on the package. Apply the mixture with a putty knife or similar tool. Fill cracks and holes and smooth the surface as much as possible. Allow the mixture to cure according to directions on the package. Then sand or grind the surface smooth and paint it.

A crack in a critical area, as for instance between bolt holes or cylinders on the gasket surface, may often require cold welding to repair it. First, drill a small hole in the crack and thread it. Screw in a threaded rod, and cut it off. Then drill a second hole overlapping the first, and thread it. Screw in a threaded rod, and cut it off. Repeat until the whole crack is treated. Make sure to get to both ends of the crack to relieve the stress that caused the crack. Then resurface the block in a surface grinder. [This is the same procedure used to repair cracks in cylinder heads (⊘ 23-21).]

⊘ 25-27 Expansion-Core Plugs You may have to remove an expansion plug from the block (because of coolant leakage, for example). To do this, put the pointed end of a pry bar against the center of the plug. Tap the end of the bar with a hammer until the point goes through the plug. Then press the pry bar to one side to pop the plug out. Another method is to drill a small hole in the center of the plug and then pry the plug out.

CAUTION: Do not drive the pry bar or drill past the plug. On some engines the plug is only about $\frac{3}{8}$ inch [9.52 mm] from a cylinder wall. You could damage the cylinder wall if you drove the pry bar or drill too far. Do not drive the plug into the water jacket. You would have trouble getting it out, and it could block coolant circulation.

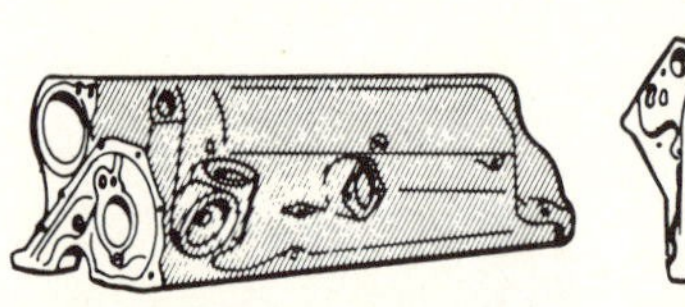

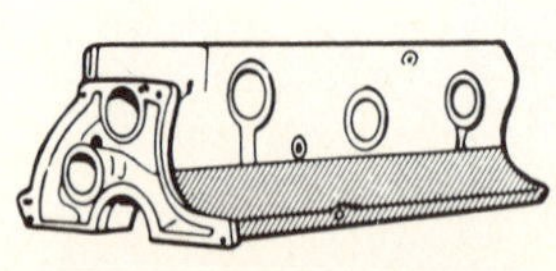

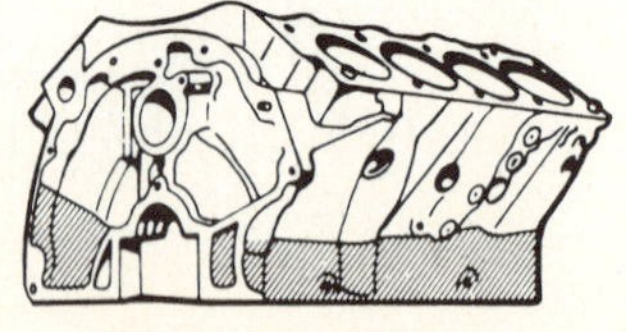

Fig. 25-41. Areas of cylinder blocks that can be repaired with epoxy. (*Ford Motor Company*)

Inspect the bore for roughness or damage that would prevent proper sealing of a new plug. If necessary, bore out the bore to take the next-size-larger plug. Before installing the new plug, coat it with the proper sealer (water resistant for cooling systems and oil resistant for oil galleries). Use the proper installation tool and proceed as follows, depending on the type of plug (Fig. 25-42).

1. *CUP-TYPE PLUG* The cup-type plug is installed with the flanged edge outward. The proper size of tool must be used. It must not contact the flange, but all driving must be against the internal cup. The flange must be brought down below the chamfered edge of the bore.

2. *EXPANSION-TYPE PLUG* The expansion-type plug is installed with the flanged edge inward, as shown. The proper tool must be used. The crowned center part must not be touched when the plug is driven in. Instead, the tool must drive against the outer part of the plug, as shown. The plug should be driven in until the top of the crown is below the chamfered edge of the bore.

⊘ 25-28 Threaded Inserts Damaged or worn threads in the block or head can often be repaired with a threaded insert. One such is the Heli-Coil (Fig. 25-43). First, drill out the worn threads. Tap the hole with the special Heli-Coil tap to make new threads. Then screw a Heli-Coil insert into the new threads, to bring the hole back to its original thread size. The original bolt can then be used in the hole.

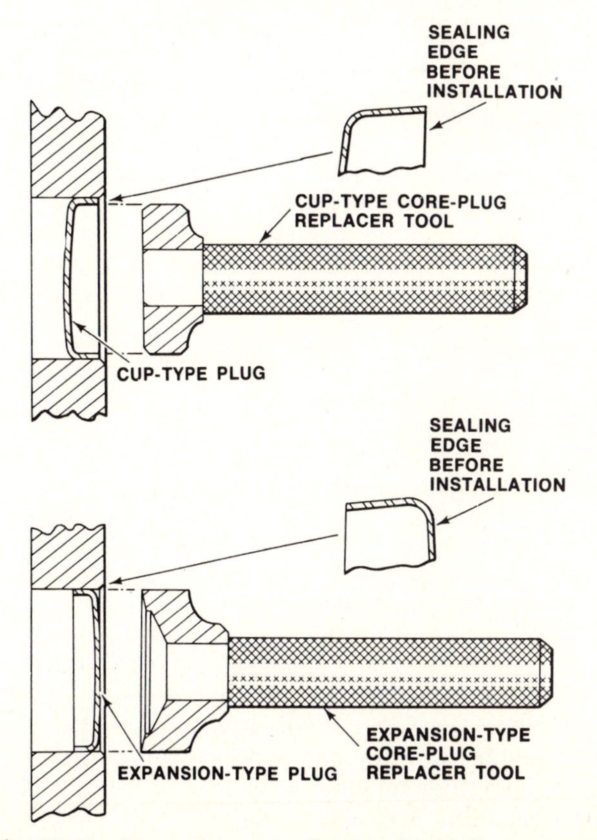

Fig. 25-42. Expansion-core plugs and installation tools. (*Ford Motor Company*)

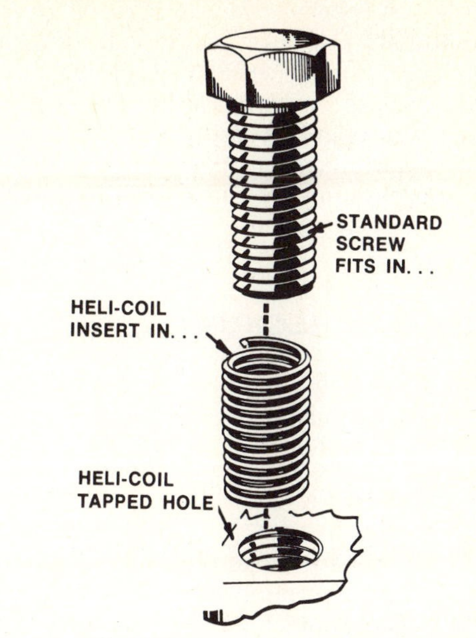

Fig. 25-43. Heli-Coil installation. (*Chrysler Corporation*)

Check Your Progress

Progress Quiz 25-2 Here is your personal checkup on cylinder service, the material we have presented in the past few pages. See how well you remember the essential points by trying the following quiz.

Completing the Sentences The sentences that follow are incomplete. After each sentence there are several words or phrases, only one of which will correctly complete the sentence. Write each sentence in your notebook, selecting the proper word or phrase to complete it correctly.

1. Taper wear in the cylinder is greatest at the: (*a*) top of the cylinder, (*b*) center of the cylinder, (*c*) bottom of the cylinder.
2. A wear spot may appear on the cylinder wall opposite the intake valve owing to: (*a*) high-speed operation, (*b*) heavy compression-ring pressure, (*c*) washing by gasoline droplets.
3. Cylinders that are oval when cold are likely to be: (*a*) more oval when hot, (*b*) less oval when hot, (*c*) about the same when hot.
4. One of the major factors causing cylinder-wall wear is the pressure of the: (*a*) piston on the walls, (*b*) compression rings on the walls, (*c*) combustion gases on the walls.
5. Two methods of finishing cylinder walls are: (*a*) grinding and honing, (*b*) reaming and boring, (*c*) roughing and finishing, (*d*) honing and boring.
6. When considerable material must be removed to clean up a cylinder bore, the cylinder should be: (*a*) bored, (*b*) honed, (*c*) ground, (*d*) reamed, (*e*) glazed.
7. When cracking the glaze, you should use a: (*a*) hammer, (*b*) wrench, (*c*) hone, (*d*) bore.
8. When boring cylinders, the main-bearing caps should be: (*a*) off, (*b*) attached lightly, (*c*) attached with normal bolt tension, (*d*) attached with heavy bolt tension.

9. After cylinders are bored or honed, they should be cleaned with either: (*a*) light oil or soapy water, (*b*) light oil or gasoline, (*c*) gasoline or kerosene.
10. Cracked blocks and scored and badly worn cylinders can sometimes be repaired by: (*a*) reboring cylinders, (*b*) rehoning cylinders, (*c*) installing cylinder sleeves.

CHAPTER 25 CHECKUP

NOTE: Since the following is a chapter review test, you should review the chapter before taking the test.

You made made wonderful progress in your study of *Automotive Engines*. If you have been able to remember most of what you have studied in the book, you are well equipped with the fundamentals of engine theory, operation, construction, and servicing. Even if you haven't been able to remember all the essential points, you have those points written in your notebook. You can skim through your notebook from time to time to refresh your memory. Of course, you can also turn back through the book and review the important points if they get hazy in your mind. See how well you remember what you have studied in this chapter by taking the following checkup test.

Completing the Sentences The sentences that follow are incomplete. After each sentence there are several words or phrases, only one of which will correctly complete the sentence. Write each sentence in your notebook, selecting the proper word or phrase to complete it correctly.

1. Crankshaft main journals should be checked for: (*a*) high spots and stretch; (*b*) taper, ridges, and out-of-roundness; (*c*) connecting-rod-bearing fit.
2. Authorities point out that any appreciable taper or out-of-roundness of the crankshaft main journals will: (*a*) cause rapid journal wear, (*b*) excessively reduce oil consumption, (*c*) shorten main-bearing life.
3. Crankshaft main-journal alignment can be checked with the crankshaft: (*a*) in the engine, (*b*) off the engine, (*c*) supported in V blocks.
4. Crankshaft main-journal taper or out-of-roundness can be checked with the crankshaft: (*a*) on or off the engine, (*b*) off the engine only, (*c*) on the engine only.
5. Precision-insert main bearings can be installed with the crankshaft on or off the engine; however, authorities agree that there is less chance for trouble if they are installed: (*a*) with the crankshaft on the engine, (*b*) with the crankshaft partly removed from the engine, (*c*) with the crankshaft off the engine.
6. Of the three methods used to check main-bearing fit, the one that is most apt to show up main-journal taper is the: (*a*) feeler-stock method, (*b*) Plastigage method, (*c*) shim-stock method.
7. When machining semifitted main bearings, not only must they be bored, but also the correct amount of crankshaft end play must be established by: (*a*) installing shims, (*b*) filing bearing caps, (*c*) facing end-thrust bearing.
8. Engine cylinder bores are likely to wear: (*a*) tapered and oval, (*b*) out of round and oval, (*c*) flat and oval, (*d*) tapered and flat.
9. When you examine a cylinder bore to decide what sort of service it requires, the more taper it has, the more likely it is that you will decide that the cylinder must be: (*a*) reamed, (*b*) bored, (*c*) honed, (*d*) glaze-cracked.
10. Cylinders may be honed: (*a*) wet or dry, (*b*) to 0.300 in [7.62 mm] oversize, (*c*) to remove 0.070 in [1.778 mm] taper.

Service Procedures In the following, you are asked to write in your notebook the servicing procedures described in this chapter. Do not copy the procedures from the book, but try to write them in your own words. Give a step-by-step account of how to do the service job asked for. This will help you remember the procedure later when you go into the engine shop.

1. Explain how to remove and replace a crankshaft.
2. List the special points to be watched when removing main-bearing caps.
3. Explain how to inspect a crankshaft for journal and crankpin wear.
4. Explain how to inspect (with shim stock and with Plastigage) and replace precision-insert main bearings.
5. Explain what causes cylinders to wear tapered.
6. Explain how to check cylinder walls with a dial indicator.
7. Explain how to hone cylinder walls.
8. Explain how to bore a cylinder.
9. Explain how to clean cylinder walls after boring or honing.
10. Explain how to remove and replace expansion plugs.

SUGGESTIONS FOR FURTHER STUDY

Keep your eyes and ears open when you are in the engine shop. You will be able to pick up much valuable information if you are alert. Notice how different jobs are done and particularly how testing and repair equipment is used. Study the operating manuals supplied by the manufacturers of the servicing equipment. For instance, the cylinder-honer and cylinder-reborer manufacturers both supply operating manuals that will prove of interest to you. Jot down in your notebook important facts that you learn from these manuals as well as information you pick up in the service shop. Jotting down these facts will help you remember them and will also preserve the information for you in permanent form.

chapter 26

WANKEL-ENGINE SERVICE

In this chapter, we cover the general servicing procedures for the Wankel engine, which was described in detail in Chap. 15. As an example, we shall use the Mazda engine. This engine, made by Toyo Kogyo of Japan, is operating in several hundred thousand RX2 and RX3 models of Mazda vehicles. You should know how this engine is serviced. For actual service work on the Mazda Wankel, you need special equipment and tools. You also need the Mazda Wankel service manual for the engine on which you are working. Here, we introduce you to the basics of Wankel-engine servicing.

⊘ **26-1 Removing the Engine** If you have had experience in removing engines from automobiles, you will have no trouble removing the Wankel engine. Follow the same general procedure used for removing the reciprocating engine. Drain the oil and the coolant; remove everything that is in the way; disconnect linkages from the carburetor, electrical wiring, and water and oil hoses; and so on. Also, on most cars, you should remove the starting motor, horn, radiator shroud, and radiator. Then attach a chain or lift cable to the two engine-lifting brackets, as shown in Fig. 26-1. Unbolt the clutch housing and put a jack or stand under the transmission. Finally, remove the nuts on the engine mounts and pull the engine forward and out.

⊘ **26-2 Disassembling the Engine** First, remove the cooling fan from the crankshaft. Then remove the alternator, intake manifold and carburetor assembly, thermal reactor, spark plugs, and oil filter. Next attach the ring-gear brake (Fig. 26-2). Remove the front-pulley mounting bolt, pulley, and its key. At the rear of the engine, straighten the tab of the lock washer securing the large flywheel nut. As shown in Fig. 26-2, use the flywheel nut wrench to remove the flywheel nut. Then use the special flywheel puller, shown in Fig. 26-3, to remove the flywheel.

Next remove the oil pan and oil strainer. Use chalk to mark and "F" and an "R" on the front and rear rotor housings. The housings must go back in their original positions.

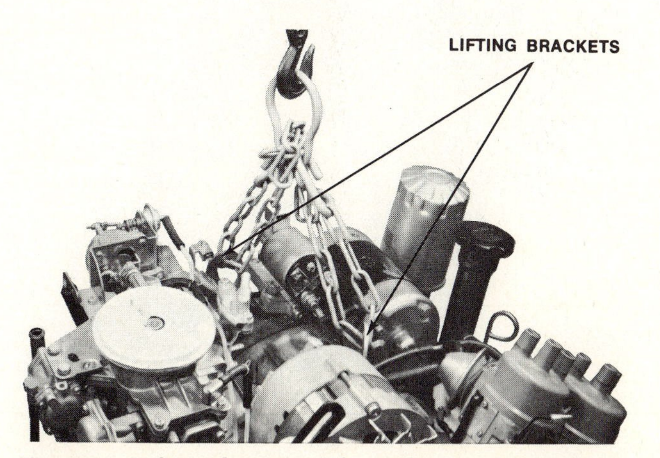

Fig. 26-1. Lifting the engine from the car. (*Toyo Kogyo Company, Limited*)

Fig. 26-2. Ring-gear brake attached and flywheel nut wrench being used to remove the flywheel nut. (*Toyo Kogyo Company, Limited*)

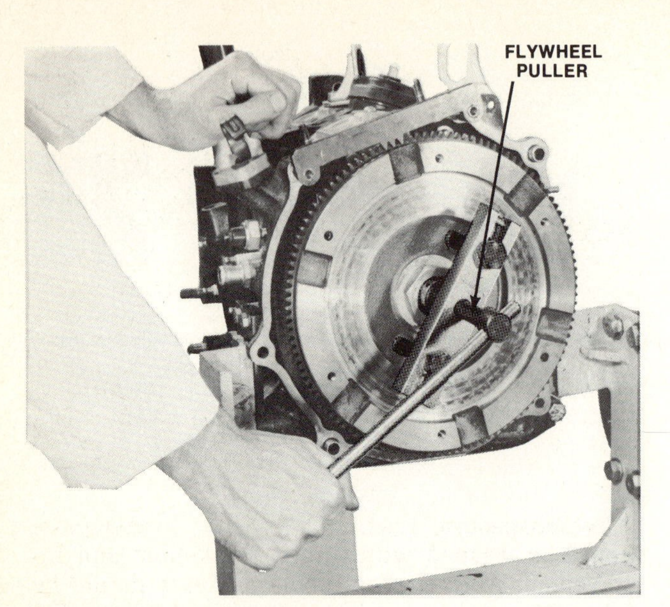

Fig. 26-3. Pulling the flywheel with a special puller. (*Toyo Kogyo Company, Limited*)

Fig. 26-4. Removing the front cover. (*Toyo Kogyo Company, Limited*)

Remove the oil-metering pump (shown in Fig. 26-24) from the right side of the front cover (Fig. 26-4). Then remove the front cover by tapping it lightly with a rubber mallet.

Remove the oil-slinger spacer and distributor drive gear from the eccentric shaft. Remove the chain adjuster and the locknut and washer for the oil-pump driven sprocket. Slide the two oil-pump sprockets with chains off their shafts at the same time. From the eccentric shaft, remove the oil-pump-sprocket key, balance weight, thrust washer, and needle bearing. Then remove the bearing housing, needle bearing, spacer, and thrust plate.

Loosen the tension bolts that hold the housings together. The tension bolts must be loosened in the order shown in Fig. 26-5. Don't take each bolt all the way out one at a time. You should go around the bolt pattern in the tension-bolt-loosening order three times before the bolts are loose enough to unscrew easily. After all the tension bolts are out, remove the front end housing by carefully lifting it off.

Now, the front rotor is exposed. Place the three corner seals and springs and the six side seals and springs from the front rotor in order in the special seal case (see Fig. 26-8).

Next, pull the tubular dowels that position the front rotor housing to the intermediate housing. Use the special dowel puller shown in Fig. 26-6. Then lift the rotor housing up just enough to get the rotor clamp into position. Now lift off the rotor housing, as shown in Fig. 26-7. The rotor clamp protects the apex seals from damage while the rotor housing is removed.

Remove the rotor clamp. Take off the three apex seals and springs from the rotor. Mark the bottom of the apex seals so that you can put them back in the same apex from which they were taken and in their original position.

CAUTION: Never mark apex seals with a punch or file! This will ruin them. Instead, use a marker, as shown in Fig. 26-8. White typewriter correction fluid can be used. The rotors are stamped "F" or "R" on the internal gear side to indicate front or rear rotor. Do not mix the rotors when putting them back into the engine.

Lift the rotor off the eccentric shaft. Be very careful not to let the seals drop from the underside of the rotor. Turn the rotor upside down and place it on a clean cloth. Remove the seals and springs from this side of the rotor.

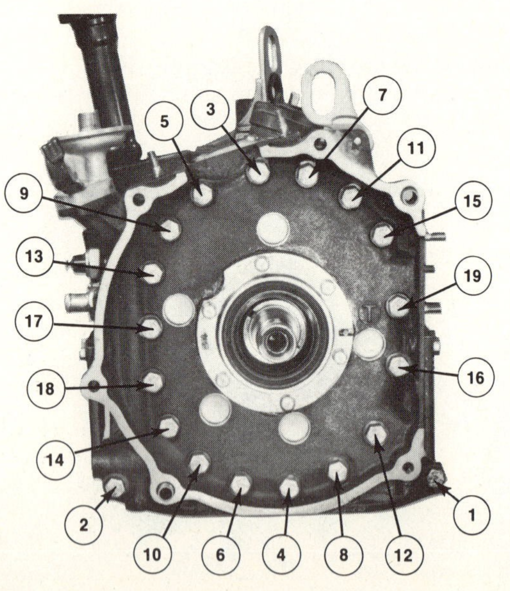

Fig. 26-5. Order in which the tension bolts must be loosened. (*Toyo Kogyo Company, Limited*)

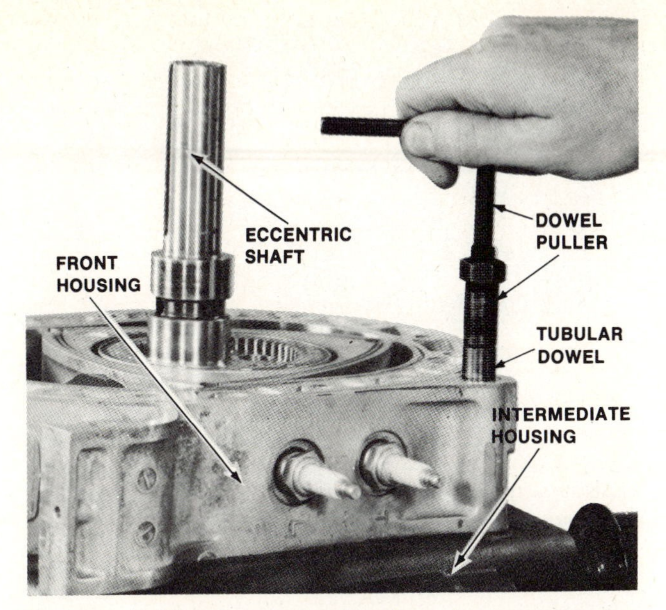

Fig. 26-6. Pulling the tubular dowels with a special puller. (*Toyo Kogyo Company, Limited*)

With the dowel puller, extract the tubular dowels from the intermediate housing. Then remove the intermediate housing. To do this, you must move the housing sideways a little as you lift it so that it clears the front rotor journal on the eccentric shaft. You can then lift off the intermediate housing. Next, remove the eccentric shaft.

Now you can remove the rear rotor housing by once again following the instructions given previously on how to remove the front rotor housing. The rear rotor assembly also is removed in the same manner as the front. Don't forget to mark the apex seals. Be sure to place each seal and spring, as it is removed, in the seal case.

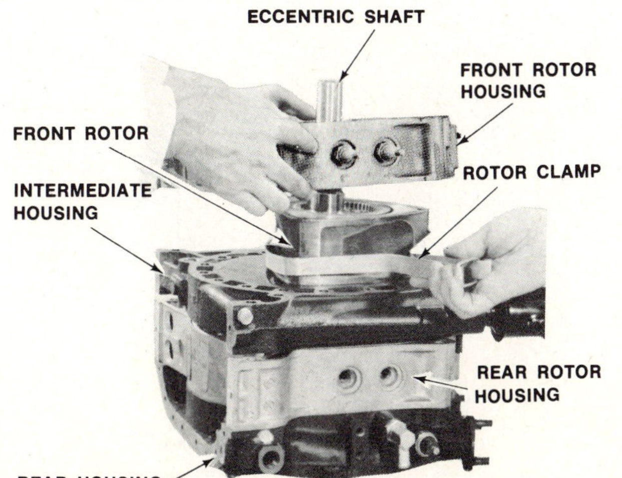

Fig. 26-7. Removing the front rotor housing. Note that the rotor clamp is in place to prevent damage to the apex seals. (*Toyo Kogyo Company, Limited*)

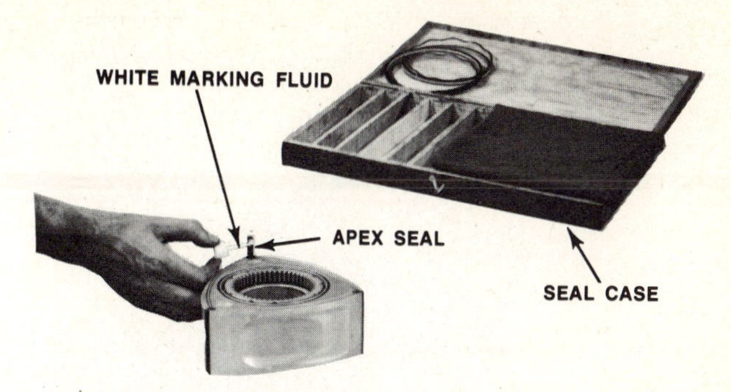

Fig. 26-8. Marking the apex seal so that it can be reinstalled in the same rotor location from which it was removed. (*Toyo Kogyo Company, Limited*)

⊘ 26-3 Inspecting Engine Parts All engine parts should be cleaned with thinner or ketone and inspected for wear or other damage. Gears should be checked for wear and chipped or broken teeth. Eccentric shaft bearings should be replaced if they show signs of wear, scratching, or flaking. Inspection of individual parts is discussed in the following sections.

⊘ 26-4 Inspecting Housing The first inspection of a cast-iron front, intermediate, or rear housing is to check for traces of gas or water leakage. With extra-fine emery paper, carefully remove all carbon deposits on the housing. (If you use a carbon scraper, be careful not to damage the mating surfaces of the housing.) To check the housing for distortion, place a precision straightedge on the housing surface. Use a feeler gauge between the housing and the straightedge, as shown in Fig. 26-9. If the distortion is greater than 0.002 inch [0.04 mm], replace the housing. Check the rotor sliding surfaces of the housing for wear using a dial indicator, as shown in Fig. 26-10. Replace the housing if the wear is more than 0.004 inch [0.10 mm].

Measure the outside diameter of the eccentric-shaft main-bearing journals. Compare the diameter of each journal with the inside diameter of each main bearing in the housings. If the clearance is

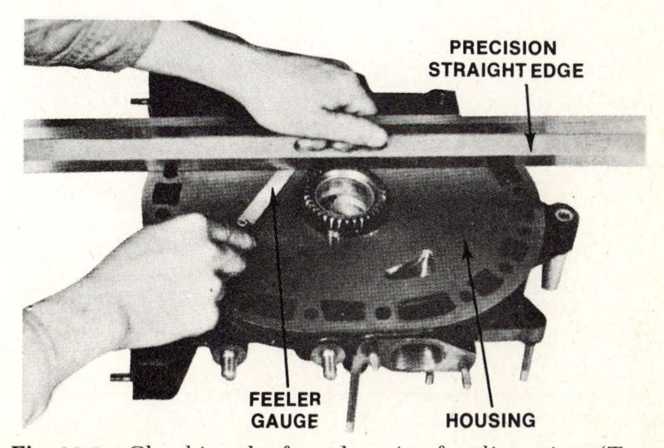

Fig. 26-9. Checking the front housing for distortion. (*Toyo Kogyo Company, Limited*)

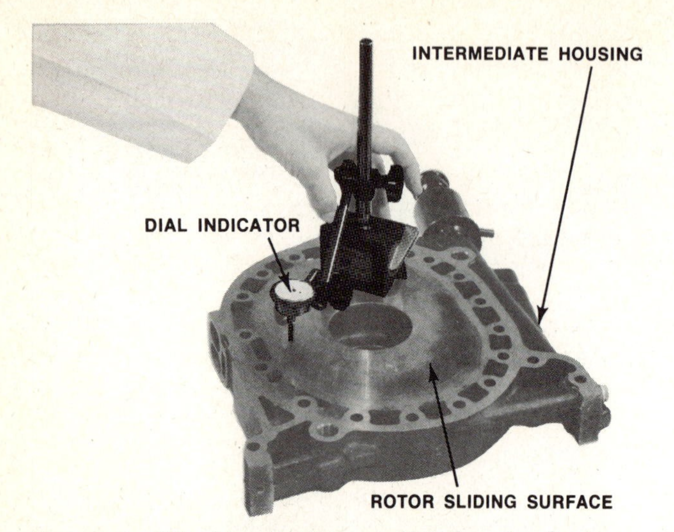

Fig. 26-10. Checking the intermediate housing for wear. (*Toyo Kogyo Company, Limited*)

more than 0.004 inch [0.10 mm], replace the bearing. To remove bearings, first remove the stationary gear and the bearing lock pin. Then press out the old bearing and press in a new one. Special tools are required for this job to prevent damage to the housing or new main bearing. Be sure the oilholes in the bearing and gear line up.

NOTE: When replacing the stationary gear in the rear housing, also check the oil seal. If the oil seal requires replacement, put a thin film of grease on the new O-ring oil seal before trying to place it in the groove in the stationary gear. This makes the O-ring seal easier to install.

Inspect the aluminum rotor housings. First check for traces of gas or water leakage along the inner margin of each side face of the rotor housing. With thinner or ketone and a cloth, clean the surface on which the apex seals ride. Never use emery cloth or a scraper on any part of the aluminum rotor housings. Check the chromium-plated interior surface of the rotor housings for flaking, cracks, or other damage. Any defect requires replacement of the rotor housing.

If the engine has overheated, measure the width of the rotor housing at various points next to the inner surface. Variations greater than 0.003 inch [0.08 mm] require replacement of the aluminum rotor housing.

⊘ 26-5 Inspecting the Rotor and Seals Note the color of the carbon on the rotor to determine the combustion conditions in the engine. In a engine operating normally, the color of the carbon on the leading side of the flank will be brown, with the color shading to black on the trailing side (Fig. 26-11). Dark streaks radiating along the rotor side surface from the side seals and corner seals indicate blow-by. In other words, gas leakage is taking place.

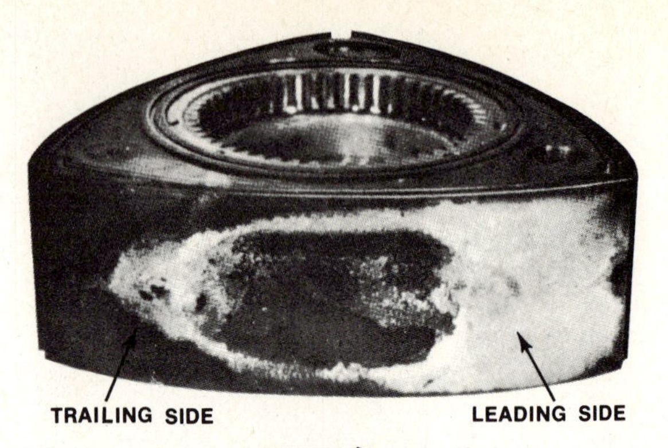

Fig. 26-11. Rotor taken from an engine operated normally. The dark areas are black; the lighter area is brown. (*Toyo Kogyo Company, Limited*)

Correction of this problem requires replacement of the seals.

Check how far the oil seals stick out above the side surfaces of the rotor. (See Fig. 26-12.) If they protrude less than 0.020 inch [0.5 mm], they must be replaced. Use a special tool, as shown in Fig. 26-13, or a small screwdriver, and carefully work the seals out. There are six parts to the oil seal on each side of the rotor: inner oil seal, outer oil seal, two O rings, and two springs.

NOTE: Every time a rotor is removed from an engine, the O rings in the oil seals must be replaced. These new O rings are required whether or not the oil seals themselves are replaced.

Clean the carbon from the rotor with a carbon scraper or emery cloth. Remove the carbon from the seal grooves, being careful not to damage the grooves. Wash the rotor in cleaner fluid, and blow the rotor dry with compressed air.

Inspect the rotor for wear or damage. If the internal gear has cracked, worn, or chipped teeth, replace the rotor. If the rotor bearing is worn, flaking, or otherwise damaged, the rotor bearing must

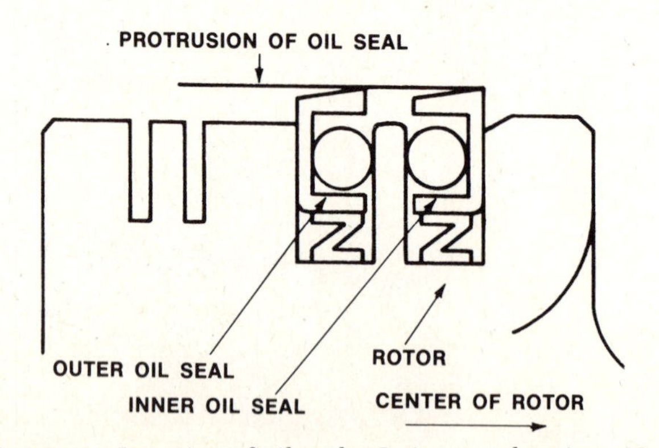

Fig. 26-12. Location of oil seals, O rings, and springs. Note that the seals stick out above the rotor surface. Also both tapers on the seals face the same way. (*Toyo Kogyo Company, Limited*)

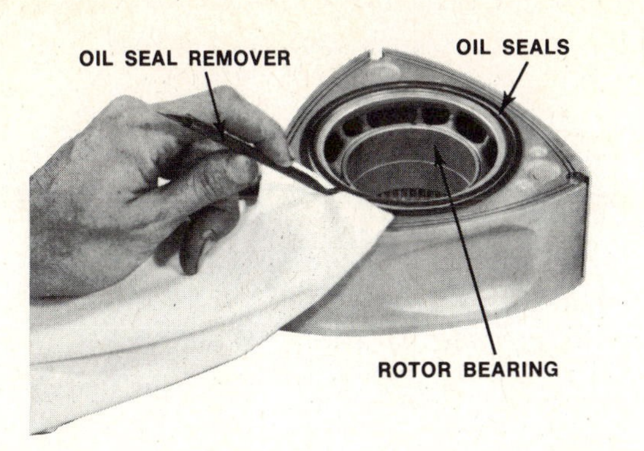

Fig. 26-13. Removing the old seal with a special tool. Note that a piece of cloth has been placed so that the rotor and other seal will not be damaged. (*Toyo Kogyo Company, Limited*)

be replaced. Check for bearing wear by comparing the inside diameter of the rotor bearing with the diameter of the rotor journal on the eccentric shaft. Replace the rotor bearing if the clearance is more than 0.004 inch [0.1 mm].

To replace the rotor bearing, install the bearing expander, as shown in Fig. 26-14. Then drill holes 0.14 inch [3.5 mm] in diameter about 0.28 inch [7 mm] deep in the three bearing locking screws.

NOTE: The rotor bearing is locked in place with three small locking screws that are staked in place with a punch to prevent loosening. These screws are in small holes that are drilled partly in the rotor and partly in the outer shell of the rotor bearing. To prevent damage to the rotor bearing when drilling, you must use the bearing expander.

Remove the bearing expander and put the rotor on a press with the internal gear facing up. Use the special bearing remover to press out the old bearing. Then press in the new bearing with the special rotor-bearing installer. Reinstall the bearing expander and drill new holes about 0.25 inch [6.35 mm] to the left or right of the old screw holes. The center of the new holes must be 0.02 inch [0.5 mm] from the rotor bore. The holes must be 0.14 inch [3.5 mm] in diameter and about 0.28 inch [7 mm] deep. Tap the holes with an M4, P-0.70-mm tap. Install new locking screws and stake them in place with a punch to prevent loosening. Then remove the bearing expander.

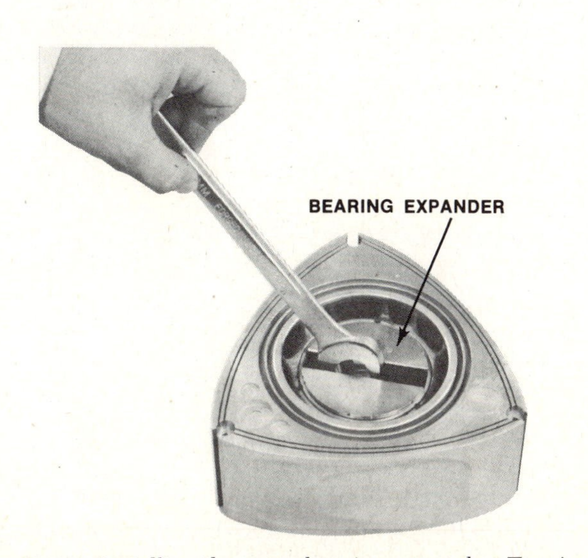

Fig. 26-14. Installing the rotor-bearing expander. Turning the bolt pushes the two halves of the expander up tight against the bearing. (*Toyo Kogyo Company, Limited*)

Handle the apex, corner, and side seals and springs with care. Apex seals are made of a special carbon material that can be damaged by careless handling. Remove all carbon deposits and wash the apex seals in cleaner. Never use emery cloth on an apex seal!

Check the apex seals for wear or other damage. Measure their height. Discard them if they measure less than 0.400 inch [10 mm]. Then insert the seal in its proper groove in the rotor. Measure the gap between the apex seal and groove with a feeler gauge (Fig. 26-15). Because the seal tends to wear unevenly, be sure to insert the feeler gauge all the way down into the groove. Next, compare the length of the apex seal with the width of the rotor housing. This determines the gap between the end of the apex seal and the side housing (Fig. 26-16). If the gap is greater than 0.006 inch [0.15 mm], replace the seal.

Measure the gap between the side seals and the rotor groove. Replace any side seal when the gap is greater than 0.004 inch [0.1 mm].

NOTE: There are four different types of side seals, as shown in Fig. 26-17: the front inner, front outer, rear inner, and rear outer seals. Be sure you install each one in its proper groove.

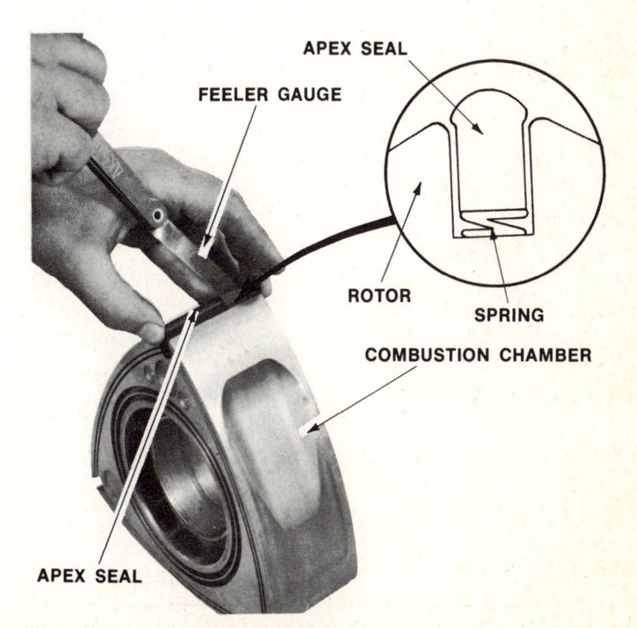

Fig. 26-15. Checking the gap between the apex seal and rotor groove. Note (*right*) how the apex seal wears on both sides. (*Toyo Kogyo Company, Limited*)

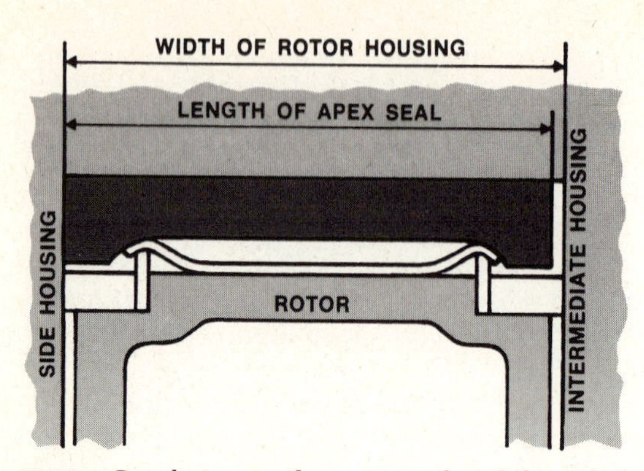

Fig. 26-16. Gap between the apex seal and the side housings. (*Toyo Kogyo Company, Limited*)

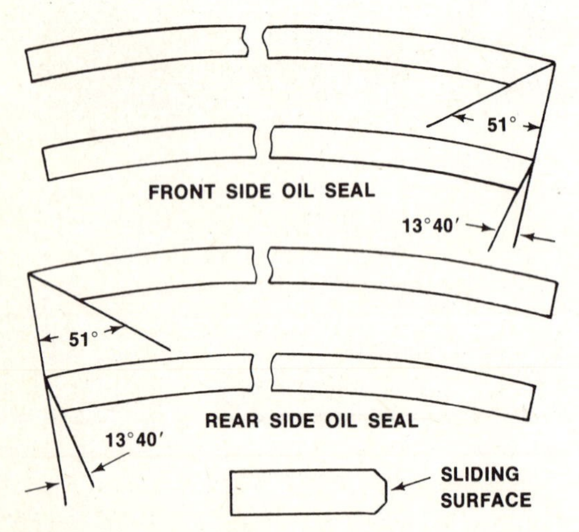

Fig. 26-17. Shapes of different side seals. Note that one end of each seal is angled, and that the sliding surface (face) of the seal is flat. (*Toyo Kogyo Company, Limited*)

Check the gap between the corner seal and the corner-seal groove (Fig. 26-18). If the gap is between 0.002 and 0.003 inch [0.048 and 0.08 mm], an oversize seal may be installed that is 0.001 inch [0.03 mm] larger than the original. If the corner seal and corner-seal groove gap exceed 0.003 inch [0.08 mm], the corner-seal groove can be rebored with a special jig and reamer. Then a new corner seal that is 0.008 inch [0.2 mm] oversize can be installed.

Check the gap between each side seal and corner seal (shown in Fig. 26-18) with both seals installed on the rotor. Insert a feeler gauge between the rear of the side seal (against the turning direction of the rotor) and the corner seal. Replace the side seal if this clearance is greater than 0.016 inch [0.4 mm].

CAUTION: Check the gap between the side seal and the corner seal when a new side seal is installed. If the gap is less than 0.002 inch [0.05 mm], with a fine file carefully file off the opposite end of the side seal along the round shape of the corner seal. The correct side-seal-to-corner-seal gap is 0.002 to 0.008 inch [0.05 to 0.15 mm].

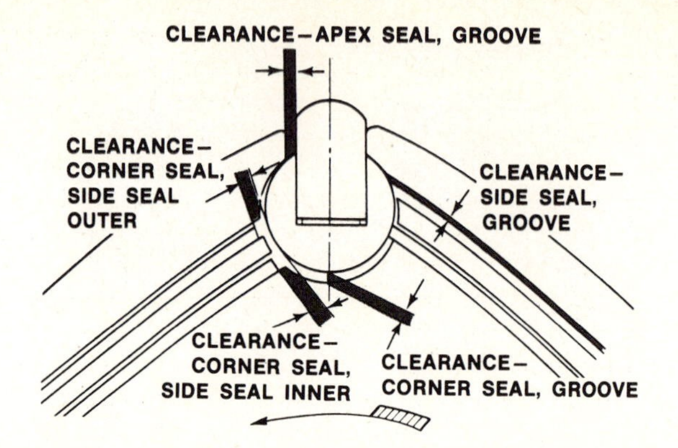

Fig. 26-18. Clearance of seals. (*Toyo Kogyo Company, Limited*)

Check each spring used with an apex seal, a corner seal, a side seal, and an oil seal for weakness, wear, or other damage. Examine carefully the sections of the springs that contact the rotor or a seal. Measure the free height of each spring to make sure that it is still strong.

⊘ 26-6 Inspecting the Eccentric Shaft Wash the shaft in cleaner and use the air hose to blow through the oil passages in the shaft. Check for cracks, scratches, wear, or other damage. Check the shaft journals for wear by measuring them with a micrometer. Mount the shaft on V blocks and check for runout with a dial indicator. A shaft with worn journals or more than 0.0008 inch [0.02 mm] run-out must be discarded.

⊘ 26-7 Assembling the Engine Put the rear rotor on a rubber pad or cloth to protect it from damage. Fit the outer and inner oil seal springs into their grooves with the gaps 180 degrees apart. Make sure the oil seals will work freely in their grooves by temporarily installing them. Then remove the seals, put new O rings in the oil seals, coat the oil seal and grooves with engine oil, and carefully work the seals with the O rings in place into the rotor grooves. Install the oil seals on both sides of the rotor.

NOTE: Be sure to install the oil seals so that the tapered side faces up, as shown in Fig. 26-12, and the flat side faces down. The bottom flat side of the oil seal has a white mark on it to help ensure proper installation.

Install the apex seals without their springs into the proper grooves in the rotor. Remember, the apex seals must go back into the same grooves from which they were removed. Also, they must be pointing in the same direction. Install the corner-seal springs and corner seals, coating these parts with engine oil. Make sure the corner seals move freely and that they stick out 0.05 to 0.06 inch [1.3 to 1.5 mm] above the rotor surface.

Fit the side-seal springs into the rotor grooves

with both ends of the springs facing up, as shown in Fig. 26-19. Install the side seals. These side seals must protrude 0.04 inch [1 mm] above the rotor and must move freely. Coat the internal gear and seals with oil. Turn the rotor over, and then place the rotor on the rear side housing, as shown in Fig. 26-20. Mesh the internal and stationary gear so that one of the rotor apexes is set to any one of the four positions shown in Fig. 26-21. Remove the apex seals, being very sure you know their proper openings and positions. Turn the rear side housing with the rotor to a horizontal position, as shown in Fig. 26-22.

To install the rear rotor housing, apply sealant to the rear side of the rear rotor housing. Cover all surfaces that will contact the rear side housing. Do not get any sealing agent into the water or oil passages! Then place new O rings and sealing rubbers on the rear rotor housing. Put a little grease on the O rings and sealing rubbers to hold them in place.

Carefully turn the rear rotor housing over. Put it into place on the rear side housing, as shown in Fig. 26-22. Coat it with oil and then install the tubular dowels through the rotor housing and into the side housing.

Install the eccentric shaft, being careful not to damage the rotor bearing or main bearing. Next, install the apex seals with springs, and the corner- and side-seal springs and seals on the upper side of the rotor. Oil the rear rotor seals and the sliding surfaces of the rear housing.

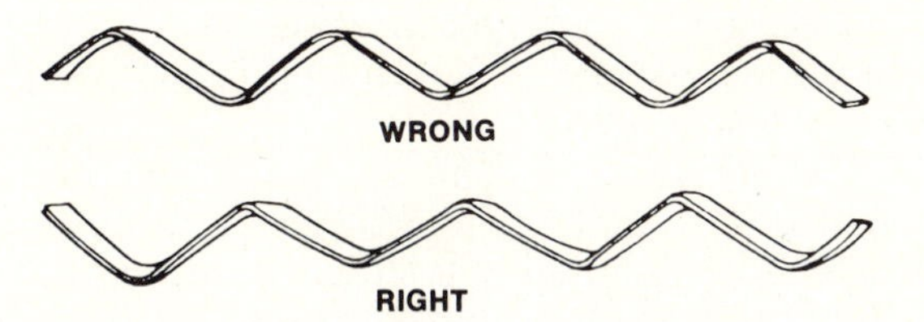

Fig. 26-19. Wrong and right ways to install the side-seal spring. (*Toyo Kogyo Company, Limited*)

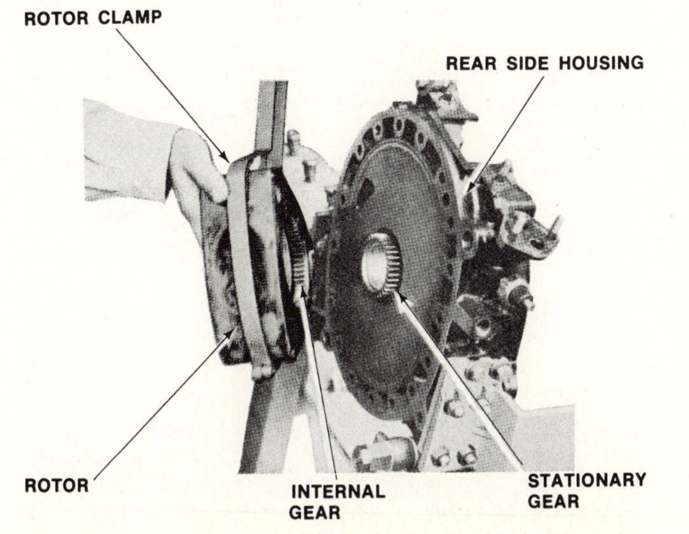

Fig. 26-20. Installing the rear rotor assembly. Note that the rear side housing is positioned upright. (*Toyo Kogyo Company, Limited*)

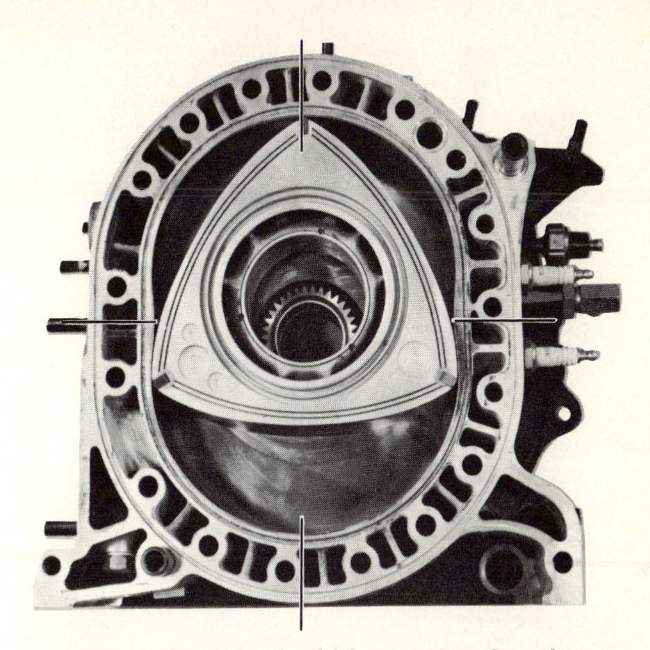

Fig. 26-21. The rotor should be positioned so that one of the rotor apexes is in one of the four places shown by the lines. (*Toyo Kogyo Company, Limited*)

Fit new O rings and sealing rubbers to the intermediate housing. Apply sealant to the mating surfaces, and install the intermediate housing. The intermediate housing must be worked sideways to get it past the upper eccentric journal on the eccentric shaft.

Install the front rotor and housing, following the procedure outlined for the rear rotor and housing. Install and tighten the tension bolts, in the order shown in Fig. 26-23. Go through the bolt-tightening order several times, tightening each bolt a little at a time until all are tightened to 22 pound-feet [3.0 kg-m].

Now check for freeness of the eccentric shaft by installing the pulley bolt temporarily and turning

Fig. 26-22. Installing the rear rotor housing. Note that the rear side housing has been turned so that it is horizontal. (*Toyo Kogyo Company, Limited*)

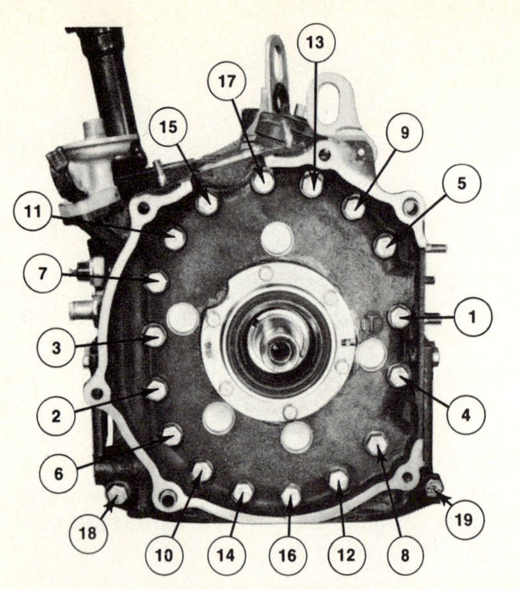

Fig. 26-23. Order in which the tension bolts should be tightened. (*Toyo Kogyo Company, Limited*)

it with a wrench. The eccentric shaft should rotate easily and smoothly.

To install the flywheel, first lubricate the oil seal in the rear housing with oil. Apply a locking agent to the threads for the flywheel lock nut on the end of the eccentric shaft. Slide the flywheel on to the eccentric shaft, making sure that the key is in the keyway. Put the lock washer in place and install the locknut. Hold the flywheel with the ring-gear brake (Fig. 26-2). Tighten the locknut to 350 pound-feet [45 kg-m]. Bend the tab of the lock washer against the locknut.

Check the eccentric-shaft end play by turning the engine so that the front side is up. Slide the thrust plate, spacer, and needle bearing onto the eccentric shaft. Oil these parts. Then install the needle-bearing housing on the front side housing, tighten the bolts, and bend the tabs of the lock washer. Now oil and install the needle bearing, thrust washer, and balance weight on the eccentric shaft. Slide the oil-pump sprockets and chain onto their shafts together. Fit the oil-pump drive-sprocket key into place. Slide the distributor drive gear and oil slinger onto the eccentric shaft.

Install the eccentric-shaft pulley with its key. Hold the flywheel with the ring-gear brake, and tighten the pulley bolt to 47 pound-feet [6.5 kg-m]. Then turn the engine to a horizontal position and measure the eccentric-shaft end play by mounting a dial indicator against the flywheel. The end play should be between 0.0016 and 0.0028 inch [0.04 inch and 0.07 mm]. If the end play is not within these limits, it can be adjusted by grinding the spacer against a flat surface covered with emery cloth. Or the spacer can be replaced. Spacers are available in different thicknesses.

Turn the engine so that the front of the engine is up. Then tighten the oil-pump drive-sprocket nut, and bend the tab up on the lock washer. A new O ring must be installed on the oil passage to the front cover. Then install the chain adjuster. Remove the pulley from the eccentric shaft. Place the front cover gasket and the front cover on the front side housing. Tighten the bolts and then oil the front cover oil seal. Reinstall the pulley on the eccentric shaft, being careful that the keyway aligns with the key. Tighten the front-pulley bolt to 47 pound-feet [6.5 kg-m]. Then turn the bottom of the engine up. Cut off any excess gasket hanging out from between the front cover and the front side housing. Install the oil strainer, and coat the mating surfaces of the oil pan to each housing with sealant. Install the oil pan along with the metal strips used for bolt-hole reinforcements. Tighten the oil-pan bolts little by little until they are torqued to 7 pound-feet [1 kg-m].

⊘ 26-8 Installing the Distributors To install the distributors, turn the engine until the white mark on the pulley aligns with the pointer on the front housing (Fig. 26-24). Alignment of the white mark and the pointer locates the front rotor at TDC on the compression stroke. Install the trailing distributor socket through the gasket. The groove on the trailing distributor drive shaft must be at an angle of about 34 degrees to the eccentric shaft. Install the leading distributor socket so that its drive shaft groove is at an inclination of about 17 degrees to the eccentric shaft. (See Fig. 26-25.)

Now install each distributor so that the end of the distributor shaft fits into the groove in the drive shaft. Make sure you install the two distributors in their correct positions. The "L" and "T" marks must match the "L" and "T" marks on the front cover. Turn the distributors in the directions shown in Fig. 26-26 until the contact points just start to open. Then tighten the distributor lock plates.

Install the other parts: alternator, intake manifold, carburetor assembly, thermal reactor, water-pump housing, spark plugs, and oil filter. You are

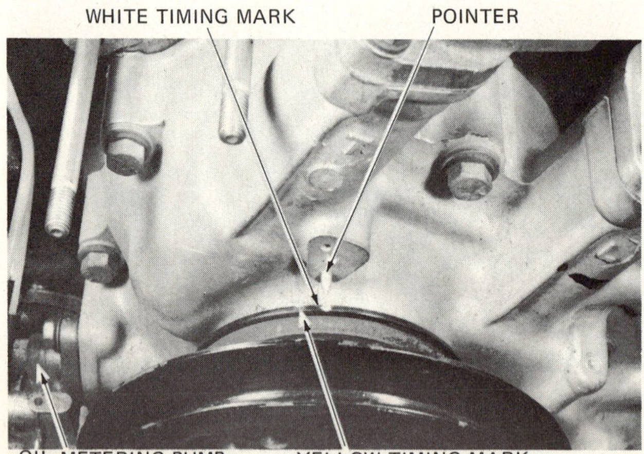

Fig. 26-24. Aligning the pulley so that the white mark is opposite the needle on the front housing. (*Toyo Kogyo Company, Limited*)

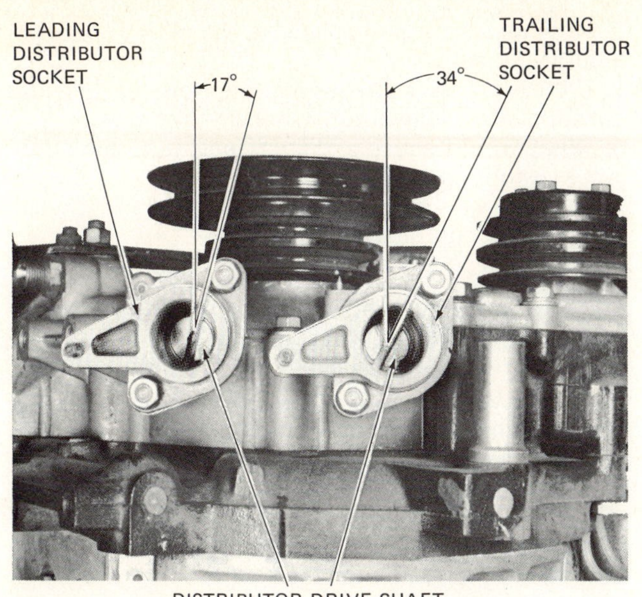

Fig. 26-25. Proper locations of the two distributor sockets. (*Toyo Kogyo Company, Limited*)

now ready to install the engine in the car by reversing the engine-removal procedure covered earlier in this chapter (⊘ 26-1).

⊘ 26-9 Engine Tune-up Certain special tools and procedures are required to tune up the Wankel engine. For example, the usual compression gauge cannot be used. As you learned in ⊘ 20-4, the compression gauge, used to check piston engines, measures the maximum pressure that will develop in a cylinder when the engine is cranked. If this gauge were used on the Wankel, it would measure the pressures of the three combustion chambers. And it would

Fig. 26-26. Direction in which to turn the two distributors to set their timing. Timing is correct when contact points just start to open. (*Toyo Kogyo Company, Limited*)

record the strongest, not the weakest, pressure. Therefore, a special strip-chart compression gauge must be used. This gauge measures the pressures as the rotor turns so that the pressures of all three combustion chambers are measured.

The failure of any one apex seal would show that the two adjoining combustion chambers are weak. Failure of a side seal would show that only one combustion chamber tests weak.

Note that you can use the cylinder leakage tester on both the Wankel and piston engines. This tester is described in Chap. 20 (⊘ 20-6). To use the tester, you apply air pressure through the spark-plug hole and then read on the tester meter how much air is leaking from the cylinder or from the combustion chamber on Wankel engines. You can pinpoint leakage points by listening at the various places where air might be leaking out.

CHAPTER 26 CHECKUP

NOTE: Since the following is a chapter review test, you should review the chapter before taking the test.

You have now learned the fundamentals of servicing Wankel engines. All Wankel engines are serviced in about the same manner. However, Wankel engines from different manufacturers will require somewhat different procedures. This means that you should always refer to the manufacturer's service manual when servicing a particular model of Wankel engine. Now see how well you remember the procedures by taking the test that follows.

Completing the Sentences The sentences that follow are incomplete. After each sentence there are several words or phrases, only one of which will correctly complete the sentence. Write each sentence in your notebook, selecting the proper word or phrase to complete it correctly.

1. To remove the flywheel nut, hold the flywheel with the: (*a*) ring-gear brake, (*b*) flywheel-nut wrench, (*c*) pulley bolts.
2. The five housings in the two-rotor Wankel are held together by: (*a*) tension bolts, (*b*) clamps, (*c*) gaskets.
3. The "F" and "R" markings on the rotors means: (*a*) forward and reverse, (*b*) front and rear, (*c*) fast and reverse.
4. Each side of each rotor has: (*a*) two side seals, (*b*) four side seals, (*c*) six side seals.
5. The bearings are locked in the rotors by: (*a*) bolts and locking nuts, (*b*) locking screws, (*c*) clamps.
6. Each rotor has: (*a*) three apex seals, (*b*) six apex seals, (*c*) one apex seal.
7. To remove the flywheel from the eccentric shaft, use: (*a*) the special flywheel puller, (*b*) the ring-gear brake, (*c*) an impact wrench.

8. In a normally operating engine, the color of the carbon on the leading sides of the rotor flanks will be: (a) black, (b) brown, (c) red.

Service Procedures In the following, you are asked to write the service procedures described in the chapter. Write the procedures in your notebook. Writing the procedures will help you remember them when you go out into the shop.

1. Describe the procedure of removing the engine from the car.
2. Explain how to remove the pulley.
3. What holds the five housings of the engine together?
4. What is the purpose of the rotor clamp?
5. Do you use a punch to mark the apex seals? Explain your answer.
6. How many apex seals are there on a rotor?
7. How many parts are there to the oil seal on each side of the rotor?
8. If there are dark streaks leading from the side and corner seals of the rotor, what do they mean?
9. What would cause a Wankel engine to have two weak adjoining combustion chambers?
10. What would cause a Wankel engine to be weak in only one combustion chamber?

SUGGESTIONS FOR FURTHER STUDY

As new Wankel engines are produced, study the shop manuals covering them. Note the similarities and differences among Wankel engines. Note especially the special service instructions. Make notes of the special procedures required and file the notes in your notebook. This will keep you up to date on new Wankel-engine developments.

GLOSSARY

This glossary of automotive terms used in the book provides a ready reference for the student. The definitions may differ somewhat from those given in a standard dictionary. They are not intended to be all-inclusive but to refresh the memory on automotive terms. More complete definitions and explanations of the terms are found in the text.

Abrasive A substance used for cutting, grinding, lapping, or polishing metal.

Accelerator The foot-operated pedal linked to the throttle valve in the carburetor.

Accelerator pump In the carburetor, a pump linked to the accelerator, which momentarily enriches the mixture when the accelerator pedal is depressed.

Additive A substance added to gasoline or oil which improves some property of the gasoline or oil.

Air cleaner A device mounted on the carburetor for filtering out dirt and dust from air being drawn into the engine.

Air-cooled engine An engine that is cooled by passage of air around the cylinders, not by passage of a liquid through water jackets.

Air-fuel mixture Name given to the air and fuel traveling to the combustion chamber after being mixed by the carburetor.

Air horn In the carburetor, the tubular passage through which the incoming air must pass.

Air-injection system A system which injects air into the exhaust manifold or thermal reactor so that the combustion of the carbon monoxide and unburned hydrocarbons in the exhaust gas can be completed.

Air pressure Atmospheric pressure of 14.7 psi (pounds per square inch) or in the metric system, 1.0355 kg/cm^2 (kilograms per square centimeter) at sea level; or the pressure of air produced by pump, by compression in engine cylinder, etc.

Alternator The device in the electric system that converts mechanical energy into electrical energy for charging the battery, etc. Also known as an ac generator, the alternator produces alternating current, which must be changed to direct current for use in the automobile.

Aluminum cylinder block An engine cylinder block cast from aluminum or aluminum alloy, and which usually has cast-iron sleeves installed for use as cylinder bores.

Antifriction bearing Name given to almost any type of ball, roller, or tapered roller bearing.

Antiknock compound An additive put into gasoline to suppress knocking, or detonation.

Arbor press Small hand-operated press used on jobs when light pressure is needed.

ATDC After top dead center.

Atmospheric pressure See "Air pressure."

Axle A cross-bar supporting a vehicle on which one or more wheels turn.

Backfiring Preexplosion of air-fuel mixture so that the explosion passes the open intake valve and flashes back through the intake manifold; also, the loud explosion of overly rich exhaust gas in the exhaust manifold.

Backlash In gearing, the clearance between meshing teeth of two gears. Generally, the amount of free motion, or lash, in a mechanical system; the amount by which the width of the tooth space exceeds the thickness of the tooth in that space.

Back pressure Pressure in the exhaust manifold; the higher the back pressure, the lower the volumetric efficiency.

Ball-peen hammer A hammer with a ball on one end of the head.

Battery An electrochemical device for storing energy in chemical form so that it can be released as electricity. A group of electric cells connected together.

BDC Bottom dead center.

Bearing The part which transmits the load to the support and, in so doing, takes the friction caused by moving parts in contact.

Bearing caps In the engine, caps held in place by bolts or nuts which, in turn, hold bearing halves in place.

Bearing crush The additional height over a full half which is purposely manufactured into each bearing half. This ensures complete contact of the bearing back with the housing bore when the engine is assembled.

Bearing oil clearance The space purposely provided between the shaft and the bearing through which lubricating oil can flow.

Bearing prelubricator A special tank attached to an air line which supplies oil at a predetermined and maintained pressure to the engine lubricating system when the engine is not operating.

Bearing roll-out tool A special tool that is a small pin with a thin head, which, when placed in the crankshaft-journal oilhole, can be used to roll-out, or roll-in, the top half of a main bearing while the crankshaft is still in place.

Bearing spin A type of bearing failure caused by lack of lubrication which overheats the bearing until it seizes on the shaft, shearing the locking lip and causing the bearing to rotate in the housing or block.

Bearing spread A purposely manufactured small extra distance across the parting faces of the bearing half in excess of the actual diameter of the housing bore.

Bell-shaped wear Deterioration of an opening (such as a brake drum) where one end is worn most so that the opening flares out like a bell.

bhp See "Brake horsepower."

Big end The crankpin end of the connecting rod.

Block See "Cylinder block."

Blow-by Leakage of unburned air-fuel mixture and some burned gases past the piston rings into the crankcase during the compression and combustion strokes.

Body The assembly of sheet-metal sections, together with windows, doors, seats, and other parts, that provides an enclosure for the passengers, engine, etc.

Bore The diameter of an engine cylinder; the diameter of any hole. Also used to describe the process of enlarging, or accurately refinishing, a hole, as to bore an engine cylinder.

Boring bar An electric motor-powered cutting tool used to machine, or bore, engine cylinders, thereby removing metal and enlarging the cylinder's bore.

Bottom dead center (BDC) The piston position at which the piston has moved to the bottom of the cylinder and the cylinder volume is at its maximum.

Brake An energy-conversion device used to retard, stop, or hold a vehicle or mechanism.

Brake drums Metal drums mounted on the car wheels which form the outer shell of the brakes; brake shoes press against the drums to slow or stop drum and wheel rotation for braking.

Brake horsepower (bhp) The power delivered by the engine which is available for driving the vehicle.

Brake shoes In drum brakes, arc-shaped metal pieces lined with a high-friction material, or brake lining, which are forced against the revolving drums to produce braking action. In disk brakes, flat metal pieces lined with brake lining which are forced against the rotor face.

Breather The opening used on engines without emission-control devices that allow air to circulate through the crankcase and thus produce crankcase ventilation.

BTDC Before top dead center.

Burr A featheredge of metal left on a part being cut with a file or other cutting tool.

Bushing A one-piece sleeve placed in a bore to serve as a bearing surface.

Butane A type of LPG liquid below 32 degrees Fahrenheit [0°C] at atmospheric pressure.

Caliper A measuring tool that can be set to measure the thickness of a block, diameter of a shaft, or bore of a hole (inside caliper). In a disk brake, a housing for pistons and brake shoes, connecting to the hydraulic system, which holds the brake shoes so that they straddle the disk.

Cam A rotating lobe or eccentric which changes rotary motion to reciprocating motion.

Cam-ground piston A piston that is ground slightly oval in shape. It becomes round as it expands with heat.

Camshaft The shaft in the engine which has a series of cams for operating the valve mechanisms. It is driven by gears or sprockets and chain from the crankshaft.

Carbon A black deposit left on engine parts by the combustion of fuel. Carbon forms on pistons, rings, valves, etc., inhibiting their action.

Carbon dioxide (CO_2) A colorless, odorless gas which results when gasoline is burned completely.

Carbon monoxide (CO) A colorless, odorless, tasteless, poisonous gas which results when gasoline is burned incompletely.

Carburetor The mixing device in the fuel system which meters gasoline into the air stream (vaporizing the gasoline as it does so) in varying proportions to suit engine operating conditions.

Catalyst A substance which can speed or slow a reaction between other substances without itself being consumed by the reaction.

Catalytic converter A mufflerlike device for use in an exhaust system that converts harmful gases in the exhaust into harmless gases by promoting a chemical reaction between a catalyst and the pollutants.

Celsius In the metric system, a temperature scale on which water boils at 100 degrees and freezes at 0 degrees; equal to a reading on a Fahrenheit thermometer of $\frac{5}{9}$(°F − 32). Also called *centigrade*.

Centigrade See "Celsius."

Centimeter (cm) A unit of linear measure in the metric system equal to approximately 0.39 inch.

Cetane number Ignition quality of diesel fuel. A high-cetane fuel ignites more easily (at lower temperature) than a low-cetane fuel.

Charcoal canister A container filled with activated charcoal used to trap gasoline vapor from the fuel tank and carburetor while the engine is off.

Chassis The assembly of mechanisms that make up the major operating part of the vehicle. It is usually assumed to include everything except the car body.

Chisel A cutting tool with a specially shaped cutting edge designed to be driven by a hammer.

Choke In the carburetor, a device used when starting a cold engine that chokes off the air flow through the air horn, producing a partial vacuum in the air horn for greater fuel delivery and a richer mixture.

Chrome-plated ring A piston compression or oil ring that has its cylinder-wall face lightly plated with hard chrome.

Clearance The space between two moving parts or between a moving and a stationary part, such as a journal and a bearing. Bearing clearance is considered to be filled with lubricating oil when the mechanism is running.

Closed-crankcase ventilating system A system in which the crankcase vapors (blow-by gases) are discharged into the engine intake system and pass through the engine cylinders rather than being discharged into the air.

Clutch In the vehicle, the mechanism in the power train that connects the engine crankshaft to, or disconnects it from, the transmission and thus the remainder of the power train.

Clutch fork In the clutch, a Y-shaped member into which is assembled the throw-out bearing.

Clutch gear See "Clutch shaft."

Clutch shaft The shaft on which the clutch is assembled, with the gear that drives the countershaft in the transmission on one end. It has external splines that can be used by a synchronizer drum to lock the clutch shaft to the main shaft for direct drive.

CO Chemical formula for carbon monoxide.

CO_2 Chemical formula for carbon dioxide.

Coated ring A piston ring having its cylinder-wall face coated with ferrous oxide, soft phosphate, or tin. This thin coating helps new rings seat by retaining oil and reducing scuffing during break-in.

Coil spring A spring made up of an elastic metal, such as steel, formed into a wire or bar and wound into a coil.

Cold welding A method of repairing cracks in metal by drilling a hole through the crack, threading the hole, and screwing in a section of threaded rod to form a seal.

Combustion Burning; in the engine, the rapid burning of the air-fuel mixture in the cylinder.

Combustion chamber The space at the top of the cylinder and in the head in which combustion of the air-fuel mixture takes place.

Compression ratio The ratio between the volume in the cylinder with the piston at BDC and the volume with the piston at TDC.

Compression rings The upper ring or rings on a piston designed to hold the compression in the cylinder and prevent blow-by.

Compression stroke The piston stroke from BDC to TDC during which both valves are closed and the air-fuel mixture is compressed.

Compression tester An instrument for testing the amount of pressure, or compression, developed in an engine cylinder during cranking.

Connecting rod In the engine, the rod that connects the crank on the crankshaft with the piston.

Connecting-rod bearing See "Rod bearing."

Connecting-rod cap The part of the connecting-rod assembly that attaches the rod to the crankpin.

Contact points In the conventional ignition system, the stationary and the movable points in the distributor which open and close the ignition primary circuit.

Coolant The liquid mixture of antifreeze and water used in the cooling system.

Cooling system In the engine, the system that removes heat by the forced circulation of coolant and thereby prevents engine overheating. It includes the water jackets, water pump, radiator, and thermostat.

Counterbored ring A piston ring, used as a compression ring, which has a counterbore on its inside diameter to promote cylinder sealing.

Countershaft The shaft in the transmission which is driven by the clutch gear; gears on the countershaft drive gears on the main shaft when the latter are shifted "into gear."

Crank A device for converting reciprocating motion into rotary motion, or vice versa.

Crankcase The lower part of the engine in which the crankshaft rotates. The upper part of the crankcase is the lower section of the cylinder block, and the lower part is made up of the oil pan.

Crankcase dilution Dilution of the lubricating oil in the oil pan by liquid gasoline seeping down the cylinder walls.

Crankcase ventilating system The system that permits air to flow through the engine crankcase when the engine is running to carry out the blow-by gases and relieve any pressure buildup.

Cranking motor See "Starting motor."

Crankpin That part of a crankshaft to which the connecting rod is attached.

Crankpin ridging A type of crankpin failure typified by deep ridges worn into the crankpin bearing surfaces.

Crankshaft The main rotating member, or shaft, of the engine with cranks to which the connecting rods are attached.

Crankshaft gauge A special type of micrometer which can measure crankshaft wear without removing the crankshaft from the block.

Crankshaft gear A gear, or sprocket, mounted on the front of the crankshaft used to drive the camshaft gear, or chain.

Cross-firing Jumping of high-voltage surge in the ignition secondary circuit to the wrong high-voltage lead so that the wrong spark plug fires. Usually caused by improper routing of the spark plug wires, by faulty insulation, or by a defective distributor cap or rotor.

Cubic centimeter (cm^3) A unit in the metric system used to measure volume; equal to approximately 0.061 cubic inch.

Cycle Any series of events which continuously repeat. In the engine, the four piston strokes (or two piston strokes) that complete the working process and produce power.

Cylinder A round hole or tubular-shaped structure in a block or casting in which a piston reciprocates. In an engine, the circular bore in the block in which the piston moves up and down.

Cylinder block The basic framework of the engine in and on which the other engine parts are attached. It includes the engine cylinders and the upper part of the crankcase.

Cylinder boring machine See "Boring bar."

Cylinder compression See "Compression tester."

Cylinder head The part that encloses the cylinder bores.

It contains the water jackets and, on I-head engines, the valves.

Cylinder hone An expandable rotating tool with abrasive fingers turned by an electric motor, used to clean and smooth the inside surface of a cylinder.

Cylinder leakage tester A type of cylinder tester that forces compressed air into the cylinder through the spark-plug hole when the valves are closed and the piston is at TDC on the compression stroke. The percentage of compressed air that leaks out is measured, and the source of the leakage accurately pinpoints the defective part.

Cylinder liner See "Cylinder sleeves."

Cylinder sleeves A replaceable sleeve, or liner, inset into the cylinder block to form the cylinder bore.

Degree $\frac{1}{360}$ of the circumference of a circle.

Detergent A chemical added to engine oil designed to help keep the internal parts of the engine clean by preventing the accumulation of deposits.

Detonation In the engine, an uncontrolled second explosion after the spark occurs, with excessively rapid burning of the compressed air-fuel mixture, resulting in a spark knock, or pinging noise.

Dial indicator A gauge that has a dial face and a needle to register movement; used to measure variations in size, movements too little to be measured conveniently by other means, etc.

Die A special tool for cutting threads on a rod.

Diesel cycle An engine cycle of events in which air alone is compressed and fuel oil is injected at the end of the compression stroke. The heat produced by compressing the air ignites the fuel oil, eliminating the need for spark plugs or a separate ignition system.

Diesel engine An engine that operates on the diesel cycle and burns oil instead of gasoline.

Dieseling A condition in which an engine continues to run after the ignition is shut off.

Differential A mechanism between axles that permits one wheel to turn at a different speed than the other while transmitting power from the drive shaft to the wheel axles.

Dipstick The oil-level indicator stick.

Direct-bonded bearing A bearing formed by pouring babbitt (bearing metal) directly into the bearing housing, and the machining of the desired size bearing diameter in that cast metal.

Disassemble To take apart.

Dispersant A chemical added to oil to prevent dirt and impurities clinging together in lumps that clog the engine lubricating system.

Displacement In an engine, the total volume of air-fuel mixture an engine is theoretically capable of drawing into all cylinders during one operating cycle. The space swept through by the piston in moving from one end of a stroke to the other.

DOHC engine Engine with double, or two, camshafts over each line of cylinders.

Drill Also called *twist drill*. A cylindrical bar with helical grooves and a point for cutting holes in material. Also refers to the device that rotates the drill.

Drive pinion A rotating shaft that transmits torque to another gear; used in the differential; also called the *clutch shaft* in the transmission.

Drive shaft An assembly of one or two universal joints connected to a hollow tube and used to transmit torque and motion. A shaft in the power train that extends from the transmission to the differential and transmits power from one to the other.

Dry friction The friction between two dry solids.

Dual carburetors An engine on which two carburetors have been mounted.

Dynamometer A device for measuring the power output, or brake horsepower, of an engine; may be an engine dynamometer, which measures power output at the flywheel, or a chassis dynamometer, which measures the power output at the drive wheels.

Eccentric A disk or offset section—of a shaft, for example—used to convert rotary to reciprocating motion.

Efficiency The ratio between the effect produced and the power expended to produce the effect; the ratio between the actual and the theoretical.

EGR system Exhaust-gas-recirculation system. It sends part of the exhaust gas back through the engine by way of the carburetor or intake manifold, which reduces the amount of NO_x that is formed.

Electric system In the automobile, the system that electrically cranks the engine for starting, furnishes high-voltage sparks to the engine cylinders to fire the compressed air-fuel charges, lights the lights, operates the heater motor, radio, and so on. It consists, in part, of the starting motor, wiring, battery, alternator, regulator, ignition distributor, and ignition coil.

Electrolyte The mixture of sulfuric acid and water used in lead-acid storage batteries. The acid enters into chemical reaction with active material in the plates to produce voltage and current.

Electronic fuel injection A fuel-injection system used for injecting gasoline into spark-ignition engines which has an electronic control system to time and meter the fuel injected.

Electronic ignition system An ignition system using transistors which does not have mechanical contact points in the distributor, but uses the distributor for distributing the secondary voltage to the spark plugs.

Emission controls A term applied to any device, or modification, added onto, or designed into, a motor vehicle for the purpose of controlling a source of air-pollution emissions.

End play As applied to the crankshaft, the amount that the crankshaft can move forward and back.

Energy The capacity or ability to do work.

Engine A machine that converts heat energy into mechanical energy. The assembly that burns fuel to produce power, sometimes referred to as the *power plant*.

Engine tune-up The procedure of checking and adjusting various engine components so that the engine is restored to top operating condition.

Ethyl See "Tetraethyl lead."

Evaporative emission-control system A system which prevents the escape of gasoline vapors from the fuel tank or carburetor float bowl to the atmosphere while the engine is off. The vapors are stored in a canister, or in the crankcase, until the engine is started.

Exhaust-gas analyzer A device for sampling the exhaust gas from an engine to determine the amounts of pollutants in the exhaust gas. Most analyzers used in the automotive shop check HC and CO, while analyzers used in testing laboratories can also check NO_x.

Exhaust-gas-recirculation system See "EGR system."

Exhaust manifold A housing with a series of connecting pipes between the exhaust ports and the exhaust pipe through which hot burned gases from the engine cylinders flow.

Exhaust stroke The piston stroke from BDC to TDC during which the exhaust valve is open so that the burned gases are forced from the cylinder.

Exhaust valve The valve which opens to allow the burned gases to exhaust from the engine cylinder during the exhaust stroke.

Expansion plug A plug that is slightly dished out and used to seal core passages in the cylinder block and cylinder head. When driven into place, it is flattened and expanded to fit tightly.

Expansion tank A tank at the top of an automobile radiator which provides room for heated coolant to expand and give off any air that may be trapped in the coolant. Also used in some fuel tanks to prevent fuel spilling from the tank because of expansion.

Fatigue failure A type of metal failure resulting from repeated stress which finally alters the character of the metal so that it cracks. In engine bearings, frequently caused by excessive idling or slow engine idle speed.

Feeler gauge Strips of metal of accurately known thicknesses used to measure clearances.

F-head engine A type of engine in which some of the valves are in the cylinder head and some in the cylinder block, giving an F-shaped appearance.

File A cutting tool with a large number of cutting edges arranged along a surface.

Filter That part in the lubricating or fuel system through which fuel, air, or oil must pass so that dust, dirt, or other contaminants are removed.

Fins Thin metal projections on an air-cooled engine cylinder and head which greatly increase the heat-radiating surfaces and help provide cooling of the engine cylinder. In a radiator, the thin metal projections, over which cooling air flows, that carry heat away from the hot coolant passages to the passing air.

Firing order The order in which the engine cylinders fire, or deliver their power strokes, beginning with No. 1 cylinder.

Flat-head engine See "L-head engine."

Float bowl In the carburetor, the reservoir from which gasoline feeds into the passing air.

Flywheel The rotating metal wheel attached to the crankshaft which helps even out the power surges from the power strokes and also serves as part of the clutch and engine-cranking system.

Flywheel ring gear The gear fitted around the flywheel that is engaged by the teeth on the starting motor drive to crank the engine.

Four cycle Short for "four-stroke cycle."

Four-stroke cycle The four piston strokes of intake, compression, power, and exhaust, which make up the complete cycle of events in the four-stroke-cycle engine.

Frame The assembly of metal structural parts and channel sections that supports the engine and body and is supported by the car wheels.

Friction The resistance to motion between two bodies in contact with each other.

Friction bearings Bearings having sliding contact between the moving surfaces. Sleeve bearings, such as those used in connecting rods, are friction bearings.

Friction horsepower (fhp) The power used up by an engine in overcoming its own internal friction; usually it increases as engine speed increases.

Fuel The substance that is burned to produce heat and create motion in an engine.

Fuel injection A system replacing the conventional carburetor which delivers fuel under pressure into the combustion chamber or into the air flow just as it enters each individual cylinder.

Fuel nozzle The tube in the carburetor through which gasoline feeds from the float bowl into the passing air. In a fuel-injection system, the tube that delivers the fuel into the air.

Fuel pump The electrical or mechanical device in the fuel system which transfers fuel from the fuel tank to the carburetor.

Fuel system In the automobile, the system that delivers to the engine cylinders the combustible mixture of vaporized fuel and air. It consists of fuel tank, lines, gauge, carburetor, fuel pump, and intake manifold.

Fuel tank The storage tank for fuel on the vehicle.

Fuel-vapor-recovery-system See "Vapor-recovery system."

Gas A state of matter, neither solid nor liquid, which has neither definite shape nor definite volume. Air is a mixture of several gases. In the automobile, the discharge from the tail pipe is called the exhaust gas. "Gas" is a slang expression used for the liquid fuel gasoline.

Gasifier section That part of a gas-turbine engine which brings in and compresses the air, mixes it with fuel, and burns the mixture in the combustor.

Gasket A flat strip, usually of cork or metal, or both, placed between two machined surfaces to provide a tight seal between them.

Gasket cement A liquid adhesive material, or sealer, used to apply gaskets; in some applications, the liquid layer of gasket cement is used as the gasket.

Gasoline A liquid blend of hydrocarbons, obtained from crude oil, used as the fuel for most automobile engines.

Gas turbine A type of internal-combustion engine with a turbine which is spun by the pressure of combustion gases flowing against the turbine blades.

Gear ratio The relative speeds at which two gears (or shafts) turn; the proportional rate of rotation.

Gears Mechanical devices to transmit power, or turning effort, from one shaft to another; gears contain teeth that interlace, or mesh, as the gears turn.

Gear-type pump A pump using a pair of matching gears that rotate; meshing of the gears forces oil (or other liquid) from between the teeth through the pump outlet.

Generator A device that converts mechanical energy into electrical energy; it can produce either ac or dc electricity. In automotive usage, the term applies to a dc generator which is now seldom used.

Glaze The mirrorlike, very smooth finish that develops on engine cylinder walls.

Glaze breaker A tool, rotated by an electric motor, used to remove the glaze from engine cylinder walls.

Goggles Special glasses worn over the eyes to protect them from flying chips, dirt, or dust.

Governor A device that governs or controls another device, usually in accordance with speed or rpm. The governor used in certain automatic transmissions is an example; it controls gear shifting in relation to car speed.

Greasy friction The friction between two solids coated with a thin film of oil.

Grinder A machine for removing metal by means of an abrasive wheel or stone.

Grinding wheel A wheel made of abrasive material used for grinding metal objects held against it.

Ground Connection of an electrical unit to the engine or frame to return the current to its source.

Guide sleeve A tubular sleeve that is put on a rod bolt when the connecting rod is removed to prevent scratching of the crankpin.

Hacksaw A special form of saw with removable blade used to saw metals.

HC Chemical formula for a hydrocarbon, such as gasoline.

Headland ring A compression ring having a cross-sectional shape of an L. Used as the top compression ring.

Heat A form of energy released by the burning of fuel.

Heat-control valve In the engine, a thermostatically operated valve in the exhaust manifold for diverting heat to the intake manifold to warm it before the engine reaches normal operating temperature.

Heat dam In a piston, a groove cut out to reduce the size of the path through which heat can travel, allowing the piston skirt to run cooler.

Heated-air system A system which uses a thermostatically controlled air cleaner to supply hot air from a stove around the exhaust manifold to the carburetor during warm-up. This improves engine operation when cold.

Heat of compression Increase of temperature brought about by compression of air or air-fuel mixture.

Heli-Coil A rethreading device to repair worn or damaged threads. It is installed in a retapped hole to bring the screw thread down to original size.

Hemispheric combustion chamber A combustion chamber having a resemblance to a hemisphere, or round ball, cut in half.

High compression A term used to refer to the increased compression ratios of modern automotive engines when compared to engines built in past years.

High-Energy Ignition (HEI) System An electronic ignition system used by General Motors that does not use contact points and has all ignition-system components contained within the distributor. It is capable of producing 35,000 volts; hence the name "High-Energy Ignition System."

Honda system A type of controlled combustion chamber for spark ignition engines which has a small chamber that surrounds the spark-plug electrodes with a rich mixture. Once the rich mixture ignites, it enters the main chamber, igniting the leaner air-fuel mixture that fills the balance of the chamber.

Hone An abrasive stone that is rotated in a bore or bushing to remove material.

Horsepower (hp) A measure of mechanical power, or the rate at which work is done. One horsepower equals 33,000 foot-pounds of work per minute.

Humidity A measure of the dampness, or moisture content, of air.

Hydraulic brakes A brake system that uses hydraulic pressure to force the brake shoes against the brake drums or rotors as the brake pedal is depressed.

Hydraulic valve lifter A valve lifter that by means of oil pressure maintains zero valve clearance so that valve noise is reduced.

Hydrocarbon (HC) A compound made of the elements hydrogen and carbon. Gasoline is a blend of hydrocarbons refined from crude oil.

Hydrometer A device used to measure specific gravity. A test instrument, consisting of a float inside a tube, which measures the specific gravity of a liquid; used to measure the specific gravity of battery electrolyte to determine the state of battery charge.

Idle speed The speed, or rpm, at which the engine runs without load when the accelerator pedal is released.

Ignition In an engine, the act of the spark in starting the combustion process in the engine cylinder.

Ignition coil That part of the ignition system which acts as a transformer to step up the battery voltage to many thousands of volts; the high-voltage surge then produces a spark at the spark-plug gap.

Ignition distributor That part of the ignition system which closes and opens the circuit to the ignition coil with correct timing and distributes to the proper spark plugs the resulting high-voltage surges from the ignition coil.

Ignition switch The switch in the ignition system which is operated with a key to open and close the ignition primary circuit.

I-head engine An overhead valve (OHV) engine with the valves in the cylinder head.

ihp See "Indicated horsepower."

Indicated horsepower (ihp) The power produced within the engine cylinders before deducting any frictional loss.

Inertia Property of objects that causes them to resist any change of speed or direction of travel.

In-line engine An engine in which all engine cylinders are in a single row, or line.

Inside micrometer A precision tool used to measure the inside diameter of a hole.

Intake manifold The part of the engine that provides a

series of passages from the carburetor to the engine cylinders through which air-fuel mixture can flow.

Intake stroke The piston stroke from TDC to BDC during which the intake valve is open and the cylinder receives a charge of air-fuel mixture.

Intake valve The valve that opens to permit air-fuel mixture to enter the cylinder on the intake stroke.

Journal The part of a rotating shaft which turns in a bearing.

Key A wedgelike metal piece, usually rectangular or semicircular, inserted in grooves to transmit torque while holding two parts in relative position; the small strip of metal with coded peaks and grooves used to operate a lock, such as in the ignition switch.

Kilogram (kg) In the metric system, a unit of weight, or mass, approximately equal to 2.2 pounds.

Kilometer (km) In the metric system, a unit of linear measure equal to 0.621 mile.

Kilowatt (kW) In the metric system, a measure of power. One horsepower equals 0.746 kilowatt.

Kinetic energy The energy of motion; the energy stored in a moving body as developed through its momentum; for example, the kinetic energy stored in a rotating flywheel.

Knock The heavy metallic sound created in an engine, which varies with engine speed, usually caused by a loose or worn bearing.

Knurl A series of ridges formed on the outer surface of a piston or on the inner surface of a valve guide which helps reduce excess clearance and hold oil for added lubrication and quieter operation.

kW See "Kilowatt."

Lapping A method of seating engine valves by which the valve is turned back and forth on the seat (no longer recommended by car manufacturers).

Leaf spring A spring made up of a single leaf, or a series of flat steel plates of graduated length assembled one on top of another, to absorb road shocks by bending, or flexing, in the middle.

L-head engine A type of engine that has the valves located in the cylinder block.

Lifter See "Valve lifter."

Line boring Setting up a special boring machine to center on the original center of the main bearing bores, and reboring the crankcase into alignment.

Liquefied petroleum gas (LPG) A hydrocarbon suitable for use as an engine fuel obtained from petroleum and natural gas; a vapor at atmospheric pressure but liquefied if put under sufficient pressure. Butane and propane are the liquefied gases most frequently used in automobile engines.

Liquid-cooled engine An engine that is cooled by the circulation of liquid coolant around the cylinders.

Liter (l) In the metric system, a measure of volume, approximately equal to 0.2642 U.S. gallon.

Lobe The projecting part, such as the rotor lobe, or the cam lobe.

Locknut A second nut turned down on a holding nut to prevent loosening.

LPG See "Liquefied petroleum gas."

Lubricating system The system in the engine that supplies moving engine parts with lubricating oil to prevent actual contact between any of the moving metal surfaces.

Lugging Low-speed, full-throttle engine operation in which the engine is heavily loaded and overworked; usually results from failure of the driver to shift to a lower gear.

Magna-Flux The process of using a special electromagnet and magnetic powder to detect cracks in iron which may be invisible to the naked eye.

Main bearings In the engine, the bearings that support the crankshaft.

Manifold vacuum The vacuum in the intake manifold that develops as a result of the vacuum in the cylinders on their intake strokes.

Master cylinder The liquid-filled cylinder in the hydraulic braking system or clutch where hydraulic pressure is developed by depression of a foot pedal.

Mechanical efficiency In an engine, the ratio between brake horsepower and indicated horsepower.

Mechanism A system of interrelated parts that make up a working assembly.

Meter (m) A unit of linear measure in the metric system equal to 39.37 inches. Also, the name given to test instruments that measure as a result of the substance being measured passing through the meter, such as an ammeter. Also, any device that measures and controls the discharge of the substance passing through it. For example, a carburetor jet is used to meter fuel flow.

Metering oil pump In a Mazda Wankel rotary engine, a plunger-type pump driven by a distributor drive gear, which meters oil to the carburetor for mixing with the gasoline. Once in the combustion chamber, this small amount of oil provides lubrication for the rotor seals.

Metering rod and jet A device, consisting of a small movable rod, which has a varied diameter, and a jet that increases or decreases fuel flow according to engine throttle opening, engine load, or a combination of both.

Micrometer A precision measuring device that measures small distances, such as crankshaft or cylinder bore diameter, or thickness of an object. Also called a *mike*.

Mike Slang term for micrometer.

Millimeter (mm) In the metric system, a unit of linear measure approximately equal to 0.039 inch.

Missing In the engine, the failure of the air-fuel mixture in a cylinder to ignite when it should.

Motor A device for converting electrical energy into mechanical energy, for example, the starting motor.

Motor vehicle Any type of self-propelled vehicle mounted on wheels or tracks.

Muffler In the exhaust, a device through which the exhaust gases must pass and which muffles the sound.

Multiple-viscosity oil An engine oil which has a low viscosity when cold (for easier cranking) and a higher viscosity when hot (to provide adequate engine lubrication.)

Mushroomed valve stem The condition that exists on a worn valve stem when the tip, or butt end, has mushroomed and metal is hanging over the valve guide. This prevents the valve from being removed from the valve guide until the mushroomed metal is removed.

Needle bearing Antifriction bearing of the roller type; the rollers are very small in diameter (needle-sized).

NO_x Chemical formula for nitrogen oxides.

NO_x control system Any type of device, or system, used to reduce the amount of NO_x produced by an engine.

Octane rating A measure of antiknock property of gasoline. The higher the octane rating, the more resistant the gasoline is to knocking, or detonation.

OHC Overhead camshaft.

Oil A liquid lubricant derived from crude oil used to provide lubrication between moving parts. In a diesel engine, oil is used for fuel.

Oil-control rings The lower ring or rings on a piston designed to prevent excessive amounts of oil from working up into the combustion chamber.

Oil cooler A small radiator through which the oil flows to lower its temperature.

Oil dilution Dilution of oil in the crankcase caused by leakage of liquid gasoline from the combustion chamber past the piston rings.

Oil filter The filter through which the crankcase oil passes to remove any impurities from the oil.

Oil-level indicator The indicator, usually called the *dipstick*, that is removed to determine the level of oil in the crankcase.

Oil pan The detachable lower part of the engine, made of sheet metal, which encloses the crankcase and acts as an oil reservoir.

Oil pump In the lubricating system, the device that delivers oil from the oil pan to the various moving engine parts.

Oil pumping Passing of oil past the piston rings into the combustion chamber because of defective rings, worn cylinder walls, etc.

Oil seal A seal placed around a rotating shaft, or other moving part, to prevent passage of oil.

Oil seal and shield Two devices used to control oil leakage past the valve stem and guide into the ports or the combustion chamber.

Oil strainer A wire mesh screen placed at the inlet end of the oil-pump pickup tube to prevent dirt and other large particles from entering the oil pump.

Orifice A small opening, or hole, into a cavity.

Oscilloscope A high-speed voltmeter which visually displays pictures of voltage variations on a televisionlike picture tube, widely used to check engine ignition systems. Also can be used to check charging systems and electronic fuel injection systems.

Otto cycle The four operations of intake, compression, power, and exhaust. Named for the inventor, Dr. Nikolaus Otto.

Overhead camshaft (OHC) engine An engine in which the camshaft is located in the cylinder head, or heads, instead of in the cylinder block.

Overhead valve (OHV) engine An engine in which the valves are mounted in the cylinder head above the combustion chamber; the camshaft is usually mounted in the cylinder block, and the valves are actuated by pushrods.

Oversquare A term applied to automotive engines which have a bore larger in diameter than the length of stroke.

Pancake engine An engine with two rows of cylinders which are opposed and on the same plane, usually set horizontally in a car, for example, Chevrolet Corvair and Volkswagen engines.

PCV Positive crankcase ventilation.

Pilot bearing A small bearing in the center of the flywheel end of the crankshaft, which carries the forward end of the clutch shaft.

Ping The sound resulting from sudden ignition of the air-fuel charge in the engine combustion chamber; characteristic sound of detonation.

Piston A movable part, fitted to a cylinder, which can receive or transmit motion as a result of pressure changes (fluid, vapor, or gas) in the cylinder.

Piston displacement The cylinder volume displaced by the piston as it moves from the bottom to the top of the cylinder during one complete stroke.

Piston pin Also called *wrist pin*. The cylindrical or tubular metal piece that attaches the piston to the connecting rod.

Piston-ring compressor A special tool used in engine overhaul work to compress the piston rings inside the piston grooves so the piston-and-rings assembly may be installed in the engine cylinder.

Piston rings Rings fitted into grooves in the piston. There are two types, compression rings for sealing the compression into the combustion chamber and oil rings to scrape excessive oil off the cylinder wall. This prevents the oil from working up into and burning in the combustion chamber.

Piston skirt The lower part of the piston below the piston-pin hole.

Piston slap Hollow, muffled, bell-like sound made by an excessively loose piston slapping the cylinder wall.

Plastic gasket compound A plastic paste in a tube which can be laid in any shape to make a gasket.

Plastigage A plastic material that comes in various sizes of wirelike lengths used to measure crankshaft main-bearing and connecting-rod bearing clearances.

Pollutant Any gas or substance in the exhaust gas from the engine or that evaporates from the fuel tank or carburetor that adds to the pollution of the atmosphere.

Poppett valve A mushroom-shaped valve, widely used in automotive engines.

Port In an engine, the valve port or opening in which the valve operates and through which air-fuel mixture or burned gases pass.

Positive crankcase ventilating (PCV) system A crankcase ventilating system in which the blow-by gas in the crankcase is returned to the intake system of the engine to be burned. This prevents the blow-by gas from escaping into the atmosphere.

Power The rate at which work is done. A common power-measuring unit is the horsepower, which is equal to 33,000 foot-pounds per minute.

Power plant The engine, or power-producing mechanism, on the vehicle.

Power section In a gas-turbine engine, the section that contains the power-turbine rotors, which, through reduction gears, turn the wheels of the vehicle.

Power steering A device that uses hydraulic pressure from a pump to multiply the driver's effort as an aid in turning the steering wheel.

Power stroke The piston stroke from TDC to BDC during which the air-fuel mixture burns and forces the piston down so that the engine produces power.

Power train The group of mechanisms that carry the rotary motion developed in the engine to the car wheels; it includes the clutch, transmission, drive shaft, differential, and axles.

Precision-insert bearings Bearings of the type that can be installed in an engine without reaming, honing, or grinding.

Precombustion chamber In some diesel engines, a separate small combustion chamber into which the fuel is injected and combustion begins.

Preignition Ignition of the air-fuel mixture in the engine cylinder (by any means) before the ignition spark occurs at the spark plug.

Preload In bearings, the amount of load originally imposed on a bearing before actual operating loads are imposed. This is done by bearing adjustments and ensures alignment and minimum looseness in the system.

Press fit A fit so tight that the pin has to be pressed into place, usually with an arbor or hydraulic press.

Pressure cap A radiator cap with valves which causes the cooling system to operate under pressure and thus at a somewhat higher and more efficient temperature.

Pressure-feed oil system A type of engine lubricating system that makes use of an oil pump to force oil through tubes and passages to the various engine parts requiring lubrication.

Pressure regulator A regulating device which operates to prevent excessive pressure from developing; in the hydraulic systems of certain automatic transmissions, a valve that opens to release oil from a line when the oil pressure attains specified maximum.

Pressure-relief valve A valve in the oil line that opens to relieve excessive pressures that the oil pump might develop.

Propane A type of LPG that is liquid below −44 degrees Fahrenheit [−42°C] at atmospheric pressure.

psi Abbreviation for pounds per square inch; often used to indicate pressure of a liquid or gas.

Puller Generally, a shop tool that permits removal of one closely fitted part from another without damage. Often contains a screw or screws which can be turned to apply gradual pressure.

Pulley A metal wheel with a V-shaped groove around the rim, which drives or is driven by a belt.

Pushrod In the I-head engine, the rod between the valve lifter and the rocker arm.

Quad carburetor A four-barrel carburetor.

Quench The space in some combustion chambers which absorbs enough heat to quench, or extinguish, the combustion flame front as it approaches a relatively cold cylinder wall. This prevents detonation of the end gas but results in hydrocarbon emissions.

Radiator In the cooling system, the device that removes heat from the coolant passing through it; it takes hot coolant from the engine and returns the coolant to the engine at a lower temperature.

Radiator-shutter system A system of engine temperature control, used mostly on trucks, that controls the amount of air flowing through the radiator by use of a shutter system.

Radius ride On a reground crankshaft, if the radius of the journal, where it comes up to the crank cheek, is not cut away enough, the journal will ride on the edge of the bearing. This contact is called *radius ride*.

RC engine A rotary combustion, or Wankel, engine.

Reamer A metal-cutting tool with a series of sharp cutting edges that remove material from a hole when the reamer is turned in it.

Rebore To bore out a cylinder larger than its original size.

Reciprocating motion Motion of an object between two limiting positions; motion back and forth, or up and down, etc.

Reed valve A type of valve used in the crankcase of some two-cycle engines. Air-fuel mixture enters the crankcase through the reed valve, which then closes as pressure builds up in the crankcase.

Refrigeration Cooling by removal of heat.

Regeneration system A system in a gas turbine that converts some of the heat that would otherwise be wasted into usuable power.

Regulator In the charging system, a device that controls alternator output to prevent excessive voltage.

Reluctor In an electronic ignition system, the metal rotor with a series of tips on it, which replaces the conventional distributor cam.

Reverse flushing A method of cleaning a radiator or engine cooling system by flushing in the direction opposite to normal coolant flow.

Ring expander A special tool used to expand piston rings for installation on the piston.

Ring gap The gap between the ends of the piston ring with the ring in place in the cylinder.

Ring grooves Grooves cut in a piston into which the piston rings are assembled.

Ring ridge Ridge left at the top of the cylinder as the cylinder wall below it is worn by piston-ring movement.

Ring-ridge remover A special tool used for removing the ring ridge from the cylinder.

Rocker arm In an I-head engine, a device that rocks on a shaft or pivots on a stud as the cam moves the pushrod, causing the valve to open.

Rod bearings In the engine, the bearings in the connecting rod in which a crankpin of the crankshaft rotates; also called *connecting-rod bearings*.

Rod big end The end of the connecting rod that attaches around the crankpin.

Rod bolts Special bolts used on the connecting rod to attach the cap.

Rod small end The end of the connecting rod through which a piston pin passes to connect the piston to the connecting rod.

Rotary Action of a part that continually rotates, or turns around.

Rotary combustion (RC) engine See "Wankel engine."

Rotor A revolving part of a machine, such as alternator rotors, disk-brake rotors, distributor rotors, and Wankel-engine rotors.

Rotor oil pump A type of oil pump using a pair of rotors, one inside the other, to produce the oil pressure required to circulate oil to engine parts.

rpm Revolutions per minute.

SA Designation of lubricating oil that is acceptable for use in engines operated under the mildest conditions.

SB Designation of lubricating oil that is acceptable for minimum-duty engines operated under mild conditions.

SC Designation of lubricating oil that meets requirements for use in gasoline engines in 1964–1967 model passenger cars and trucks.

Scored Scratched or grooved, as a cylinder wall may be scored by abrasive particles moved up and down by the piston rings.

Scraper A device in engine service to scrape carbon, etc., from engine block, pistons, etc.

Scraper ring On a piston, a type of oil-control ring designed to scrape excess oil back down the cylinder into the crankcase.

Screen A fine-mesh screen in the fuel and lubricating system that prevents large particles from entering the system.

Screwdriver A hand tool used to loosen or tighten screws.

Scuffing A type of wear of moving parts characterized by transfer of material from one to the other part and pits or grooves in the mating surfaces.

SD Designation of lubricating oil that meets requirements for use in gasoline engines in 1968–1970 model passenger cars and some trucks.

SE Designation of lubricating oil that meets requirements for use in gasoline engines in 1972 and later cars and certain 1971 model passenger cars and trucks.

Seal A material, shaped around a shaft, used to close off the operating compartment of the shaft, preventing oil leakage.

Sealer A thick, tacky compound, usually spread with a brush, which may be used as a gasket, or sealant, to seal small openings or surface irregularities.

Seat The surface upon which another part rests, such as a valve seat; also, term applied to the process of a part wearing into fit, for example, piston rings seat after a few miles of driving.

Service manual Also called *shop manual;* the annual book published by each vehicle manufacturer listing specifications and service procedures for each make and model of vehicle built.

Service ratings For lubricating oil used in engines, a designation that indicates the type of service for which the oil is best suited. See also "SA, SB, SC, SD, and SE."

Severe rings Piston rings which exert relatively high pressures against the cylinder walls, sometimes by use of an expander spring behind the ring; rings that can be used in an engine having excessive cylinder wear.

Shim A slotted strip of metal used as a spacer to adjust front-end alignment on many cars and to make small corrections in the position of body sheet metal and other parts.

Shock absorber A device placed at each vehicle wheel to regulate spring rebound and compression.

Shrink fit A tight fit of one part in another achieved by heating or cooling one part and then assembling it with the other part. If heated, the part then shrinks on cooling to provide a shrink fit. If cooled, the part expands on warming to provide the fit.

Side clearance The clearance between the sides of moving parts that do not serve as a load-carrying surface.

Single-chamber capacity In a Wankel engine, a method of comparing displacement, or size, between engines.

Slip joint In the power train, a variable-length connection that permits the drive shaft to change effective length.

Sludge Accumulation in oil pan, containing water, dirt, and oil; sludge is very viscous and tends to prevent lubrication.

Smog A term coined from *smoke* and *fog* which is applied to the foglike layer that hangs over many areas under certain atmospheric conditions. Smog is compounded from smoke, moisture, and numerous chemicals which are produced by combustion (from power plants, automotive engines, incinerators, etc.) and from numerous natural and industrial processes. The term is used generally to describe any condition of dirty air and/or fumes or smoke.

Socket wrench A wrench that fits entirely over or around the head of a bolt.

SOHC engine Engine with a single overhead camshaft.

Soldering The uniting of pieces of metal with solder, flux, and heat.

Solvent tank In the shop, a tank of cleaning fluid in which most parts are brushed and washed clean.

Spark plug The assembly, which includes a pair of electrodes and an insulator, that has the purpose of providing a spark gap in the engine cylinder.

Spark-plug heat range The distance heat must travel from the center electrode to reach the outer shell of the plug and enter the cylinder head.

Specific gravity A measure of the weight per unit volume of a liquid as compared with the weight of an equal volume of water.

Splash-feed oil system A type of engine lubricating system that depends on splashing of the oil for lubrication to moving engine parts.

Spline Slot or groove cut in a shaft or bore; a splined shaft onto which a hub, wheel, gear, etc., with matching splines in its bore is assembled so that the two must turn together.

Spring An elastic device which yields under stress or pressure but returns to its original state or position when the stress or pressure is removed; the operating component of the automotive suspension system, which absorbs the force of road shock by flexing and twisting.

Spring retainer In the valve train, the piece of metal that holds the spring in place and is itself locked in place by the valve-spring-retainer locks.

Square engine An engine having the bore and stroke of equal measurements.

Squish The action in some combustion chambers in which the last part of the compressed mixture is pushed, or squirted, out of a decreasing space between the piston and cylinder head.

Starting motor An electric motor in the electric system that cranks the engine, or turns the crankshaft, for starting.

Static friction Friction between two bodies at rest.

Steam engine An external combustion engine which works by steam generated in a boiler.

Steering gear That part of the steering system located at the lower end of the steering shaft that carries the rotary motion of the steering wheel to the car wheels for steering.

Steering system The mechanism that enables the driver to turn the wheels for changing the direction of vehicle movement.

Steering wheel The wheel at the top of the steering shaft in the driver's compartment which is used to guide, or steer, the car.

Stepped feeler gauge A feeler gauge which has a thin tip and is thicker along the rest of the gauge; a "go no-go" gauge.

Stirling engine A type of internal-combustion engine in which the piston is moved by the changing pressure of a working gas that is alternately heated and cooled.

Storage battery A lead-acid electrochemical device that changes chemical energy into electrical energy; that part of the electric system which acts as a reservoir for electric energy, storing it in chemical form.

Stratified charge In a gasoline fuel, spark ignition engine, a type of combustion chamber in which the flame starts in a small very rich pocket, or layer, of air-fuel mixture and after ignition, spreads to the leaner mixture filling the rest of the combustion chamber. The diesel engine is a stratified-charge engine.

Streamlining The shaping of an object that moves through a medium (such as air or water), or past which the medium moves, so that less energy is lost by the parting and reuniting of the medium as the object moves through it.

Stroke In an engine, the distance that the piston moves from BDC to TDC.

Stud A headless bolt that is threaded on both ends.

Stumble A term related to vehicle driveability; the tendency of an engine to falter and then catch, resulting in a noticeable stumble felt by the driver.

Sulfur oxide An acid that can be formed in small amounts as the result of a reaction between the hot exhaust gas and the catalyst in the catalytic converter.

Supercharger A device in the intake system of the engine which pressurizes the ingoing air-fuel mixture. This increases the amount of mixture delivered to the cylinders and thus increases engine output. If the supercharger is driven by the engine exhaust gas, it is called a *turbocharger*.

Surface ignition Ignition of the air-fuel mixture in the combustion chamber produced by hot metal surfaces or heated particles of carbon.

***S/V* ratio** The ratio of the surface area *S* of the combustion chamber to its volume *V*, with the piston at top dead center. Often used as a comparative indicator of hydrocarbon (HC) emission levels from an engine.

Taper A shaft or hole that gets gradually smaller toward one end. In an engine cylinder, the uneven wear which is more at the top than at the bottom.

Tappet See "Valve lifter."

TDC Top dead center.

Tel Tetraethyl lead.

Tetraethyl lead A chemical put into engine fuel which increases octane rating, or reduces knock tendency. Also called *ethyl* and *tel*.

Thermal efficiency Relationship between the power output and the energy in the fuel burned to produce the output.

Thermal reactor A chamber in the exhaust system of the engine that permits unburned hydrocarbons and carbon monoxide to react with oxygen in air pumped into the reactor so that the exhaust gases will have less of these pollutants.

Thermostat A device that operates on, or regulates, temperature changes. Several thermostats are used in engines.

Thermostatically controlled air cleaner An air cleaner which uses a thermostat to control the preheating of intake air.

Throttle valve The round disk valve in the throttle body of the carburetor that can be turned to admit more or less air, thereby controlling engine speed.

Throw-a-rod Expression used to designate an engine with a loose, knocking connecting-rod bearing, or an engine that has broken a connecting rod and shoved it through the cylinder block or pan.

Throw-out bearing In the clutch, the bearing that can be moved in to the release levers by clutch-pedal action so as to cause declutching, or a disconnection between the engine crankshaft and power train.

Thrust bearing Specifically in the engine, the main bearing that has thrust faces which prevent excessive endwise movement of the crankshaft.

Timing In the engine, refers to timing of valves, and timing of ignition, and their relation to the piston position in the cylinder.

Timing chain A chain driven by a sprocket on the crankshaft that drives the sprocket on the camshaft.

Timing gears A gear on the crankshaft that drives the camshaft by meshing with a gear on its end.

Timing light A light that is connected to the ignition system to flash each time the No. 1 spark plug fires; used for adjusting the timing of the ignition spark.

Top dead center (TDC) The piston position at which the piston has moved to the top of the cylinder and the center line of the connecting rod is parallel to the cylinder walls.

Torque Turning or twisting effort, usually measured in pound-feet [kilogram-meters].

Torque converter In an automatic transmission, a fluid coupling which incorporates a stator to permit a torque increase through the torque converter.

Torque wrench A special wrench that indicates the amount of torque being applied to a nut or bolt.

Torsional balancer See "Vibration damper."

Torsional vibration Vibration in a rotary direction that causes a twist-untwist action on a rotating shaft; a rotating shaft that repeatedly moves ahead or lags behind the remainder of the shaft; for example, the actions as a crankshaft responds to the cylinder firing impulses.

Transistor An electronic device that can be used as an electric switch; used in electronic ignition systems to replace the contact points.

Transmission The device in the power train that provides different gear ratios between the engine and rear wheels, as well as reverse.

Trouble diagnosis The detective work necessary to run down the cause of a trouble. Also implies the correction of the trouble by elimination of cause.

Tuned intake system An intake system in which the manifold has the proper length and volume to introduce a ram-jet or supercharging effect.

Tune-up The procedure of inspection, testing, and adjusting an engine and replacing any worn parts to restore the engine to its best performance.

Turbine A device that produces rotary motion as a result of gas, vapor, or hydraulic pressure. In the torque converter, the driven member.

Turbine engine An engine in which the gas pressure created by combustion is used to spin a turbine, and, through gears, move the car.

Turbocharger A supercharger driven by the engine exhaust gas.

Turbulence The state of being violently disturbed. In the engine, the rapid swirling motion imparted to the air-fuel mixture entering the cylinder.

Twist drill Drill.

Two cycle Short for "two-stroke cycle."

Two-stroke cycle The series of events taking place in a two-stroke-cycle engine, which are intake, compression, power, and exhaust all of which take place in two piston strokes. Also called two cycle.

Unit distributor An ignition distributor, used by General Motors, that uses a magnetic pickup coil and timer core instead of points and condenser. It has the ignition coil assembled in the distributor as a unit.

Vacuum An absence of air or other substance.

Vacuum advance Ignition-spark advance resulting from partial vacuum in intake manifold.

Vacuum gauge In automotive-engine service, a device that measures intake-manifold vacuum and thereby indicates actions of engine components.

Valve A device that can be opened or closed to allow or stop the flow of a liquid, gas, or vapor from one place to another.

Valve clearance The clearance between the rocker arm and the valve stem tip in an overhead valve engine; the clearance in the valve train when the valve is closed.

Valve float The condition that exists when the engine valves do not follow the cam; failure of the valves to close at the proper time.

Valve grinding Refacing a valve in a valve refacing machine.

Valve guide The cylindrical part in the cylinder block or head in which the valve is assembled and in which it moves up and down.

Valve-in-head engine An I-head engine.

Valve lash See "Valve clearance."

Valve lifter Also called *lifter, tappet, valve tappet,* and *cam follower.* A cylindrical part of the engine which rests on a cam of the camshaft and is lifted, by cam action, so that the valve is opened.

Valve-lifter foot The bottom end of the valve lifter; the part that rides on the cam lobe.

Valve overlap Number of degrees of crankshaft rotation through which both the intake and exhaust valves are open together.

Valve rack Any wood or metal container or holder which identifies and keeps the valves in order.

Valve-refacing machine A machine for removing material from the seating face of valves so that a new face appears.

Valve rotator Device used in place of the valve spring retainer; it has a built-in mechanism to rotate the valve slightly each time it opens.

Valve seat The surface in the cylinder head against which the valve face comes to rest.

Valve-seat inserts Metal rings inserted in valve seats, usually exhaust; they are of special metal more able to withstand high temperatures.

Valve-seat recession Also known as *lash loss;* the tendency for valves, in some engines run on unleaded gasoline, to contact the seat in such a way that the seat wears away, or recesses into the cylinder head.

Valve-spring retainer The device on the valve stem that holds the spring in place.

Valve-spring-retainer lock The locking device on the valve stem that locks the spring retainer in place.

Valve stem The long, thin section of the valve that fits in the valve guide.

Valve-stem seal, or shield A device placed on or surrounding the valve stem to reduce the amount of oil which can get on the stem and thereby work its way down into the combustion chamber.

Valve tappet Valve lifter.

Valve timing The timing of valve opening and closing in relation to piston position in the cylinder.

Valve train The valve-operating mechanism of an engine, from the camshaft to the valve.

Vapor lock A condition in the fuel system in which gasoline has vaporized and turned to bubbles in the fuel line or fuel pump so that fuel delivery to the carburetor is prevented or retarded.

Vapor-recovery system Recovery of the gasoline vapor escaping from the fuel tank and carburetor float bowl by an evaporative emission-control system. See "Evaporative emission-control system."

V-8 engine A type of engine with two banks of four cylinders each set at an angle to each other to form a V.

Venturi In the carburetor, the restriction in the air horn that produces the vacuum responsible for the movement of gasoline into the passing air.

VI See "Viscosity index."

Vibration A complete rapid motion back and forth; oscillation.

Vibration damper A device attached to the crankshaft of an engine which opposes crankshaft torsional vibration, that is, the twist-untwist actions of the crankshaft caused by the cylinder firing impulses. Also called *harmonic balancer.*

Viscosity The resistance to flow that a liquid has. A thick oil has greater viscosity than a thin oil.

Viscosity index A measurement used to determine how much an oil viscosity changes with heat.

Viscosity ratings Oil viscosity is rated two ways: for winter driving and for summer driving. The winter grades are SAE5W, SAE10W, and SAE20W. For summer driving, the grades are SAE20, SAE30, SAE40, and SAE50. Many oils have multiple-viscosity ratings, for example, SAE10W-30.

Viscous Thick, tending to resist flowing.

Viscous friction Friction between layers of a liquid.

Vise A gripping device for holding a piece while it is being worked on.

Volatility A measurement of the ease with which a liquid vaporizes.

Volumetric efficiency Ratio between the amount of air-fuel mixture that actually enters an engine cylinder and the theoretical amount that could enter under ideal conditions.

V-type engine Engine with two banks of cylinders set at an angle to each other to form a V.

VVR Vehicle-vapor recovery. See "Vapor-recovery system."

Wankel engine A rotary-type engine in which a three-lobe rotor turns eccentrically in an oval chamber.

Water jacket The space between inner and outer shells of the cylinder block or head through which coolant can circulate.

Water pump In the cooling system, the device that maintains circulation of the coolant between the engine water jackets and the radiator.

Wedge combustion chamber A combustion chamber resembling, in shape, a wedge.

Work The changing of the position of a body against an opposing force, measured in foot-pounds [meter-kilograms]. Product of force times the distance through which it acts.

Wrench A tool designed to tighten or loosen nuts or bolts.

Wrist pin Piston pin.

INDEX

D

E

F

G

H

I

K

L

M

O

P

Q

R

S

T

U

V

W

ANSWERS TO QUESTIONS

The answers to the questions in the progress quizzes and chapter checkups are given here. In chapter checkups, you may be asked to list parts in components, describe the purpose and operation of components, define certain terms, and so on. Obviously, no answers to these could be given here since that would mean repeating substantially the entire book. Therefore, you are asked to refer to the book to check your answer.

If you want to figure your grade on any quiz, divide the number of questions in the quiz into 100. This gives you the value of each question. For instance, suppose there are 8 questions: 8 goes into a hundred 12.5 times. Each correct answer therefore gives you 12.5 points. If you answered 6 correct out of the 8, then your grade would be 75 (6 × 12.5).

If you are not satisfied with the grade you make on a quiz or checkup, restudy the chapter or section and retake the test. This review will help you remember the important facts.

Remember, when you take a course in school, you can pass and graduate even though you make a grade of less than 100. But in the automotive shop, you must score 100 percent all the time. If you make 1 error out of 100 service jobs, for example, your average would be 99. In school that is a fine average. But in the automotive shop that one job you erred on could cause such serious trouble (a ruined engine or a wrecked car) that it would outweigh all the good jobs you performed. Therefore, always proceed carefully in performing any service job and make sure that you know exactly what you are supposed to do and how you are to do it.

CHAPTER 1

Chapter 1 Checkup
Completing the Sentences 1. (c) 2. (b) 3. (b) 4. (b) 5. (a)

CHAPTER 2

Progress Quiz 2-1
Correcting Parts Lists 1. body 2. steering wheel 3. brake shoe 4. engine support 5. transmission

Completing the Sentences 1. (c) 2. (b) 3. (b) 4. (b)

Progress Quiz 2-2
Correcting Parts Lists 1. steering wheel 2. pistons 3. tires 4. pitman arm 5. clutch

Completing the Sentences 1. (c) 2. (d) 3. (d) 4. (b) 5. (b) 6. (c) 7. (c) 8. (a) 9. (a) 10. (b)

Chapter 2 Checkup
Completing the Sentences 1. (c) 2. (a) 3. (c) 4. (c) 5. (b) 6. (c) 7. (b) 8. (b) 9. (c) 10. (a)

CHAPTER 3

Chapter 3 Checkup
Completing the Sentences 1. (b) 2. (b) 3. (b) 4. (a)

CHAPTER 4

Progress Quiz 4-1
Completing the Sentences 1. (c) 2. (b) 3. (b) 4. (a) 5. (b) 6. (a) 7. (c) 8. (a) 9. (b) 10. (a)

Progress Quiz 4-2
Completing the Sentences 1. (b) 2. (b) 3. (a) 4. (c) 5. (b) 6. (b) 7. (d) 8. (c) 9. (a) 10. (b)

Chapter 4 Checkup

Physical Properties and Their Measurements

PHYSICAL PROPERTIES	UNITS
torque	pound-feet and kilogram-meters
horsepower	foot-pounds per minute and meter-kilograms per minute
work	foot-pounds and meter-kilograms
atmospheric pressure	pounds per square inch and kilograms per square centimeter
temperature	degrees Fahrenheit and degrees Celsius

Completing the Sentences 1. (b) 2. (c) 3. (a) 4. (b) 5. (a) 6. (c) 7. (a) 8. (b) 9. (a) 10. (c)

Problems 1. 1,900 ft-lb 2. 5,500 ft-lb 3. 1 hp [0.746 kW] 4. ½ lb-ft [0.069 kg-m]

CHAPTER 5

Progress Quiz 5-1

Correcting Parts Lists 1. reverse 2. brake system 3. clutch 4. flywheel 5. differential

Completing the Sentences 1. (b) 2. (b) 3. (c) 4. (c) 5. (a) 6. (a) 7. (b) 8. (c) 9. (a) 10. (c)

Progress Quiz 5-2

Correcting Parts Lists 1. oil pump 2. carburetor motor 3. spark pump 4. float bowl

Completing the Sentences 1. (b) 2. (b) 3. (b) 4. (b) 5. (a)

Chapter 5 Checkup

Completing the Sentences 1. (c) 2. (c) 3. (d) 4. (a) 5. (c) 6. (d)

CHAPTER 6

Progress Quiz 6-1

Completing the Sentences 1. (b) 2. (d) 3. (c) 4. (b) 5. (c) 6. (c) 7. (b) 8. (b) 9. (b) 10. (a)

Progress Quiz 6-2

Completing the Sentences 1. (c) 2. (a) 3. (b) 4. (c) 5. (b) 6. (b) 7. (c) 8. (b) 9. (a)

Chapter 6 Checkup

Picking Out the Right Answer 1. (b) 2. (a) 3. (a) 4. (b) 5. (c)

CHAPTER 7

Progress Quiz 7-1

Correcting Parts Lists 1. oil pan 2. drive shaft 3. connecting rods 4. intake manifold 5. water pump

Completing the Sentences 1. (b) 2. (b) 3. (c) 4. (a) 5. (a) 6. (c) 7. (b) 8. (a) 9. (a) 10. (a)

Progress Quiz 7-2

Correcting Parts Lists 1. piston 2. piston-pin bearings 3. convertibility 4. carbon 5. depression loads

Completing the Sentences 1. (c) 2. (b) 3. (b) 4. (b) 5. (b) 6. (b) 7. (c) 8. (a)

Chapter 7 Checkup

Completing the Sentences 1. (b) 2. (c) 3. (a) 4. (b) 5. (c) 6. (b) 7. (b) 8. (b) 9. (a) 10. (c)

CHAPTER 8

Progress Quiz 8-1

Completing the Sentences 1. (a) 2. (b) 3. (a) 4. (c) 5. (a) 6. (c) 7. (c) 8. (b)

Progress Quiz 8-2

Completing the Sentences 1. (d) 2. (b) 3. (a) 4. (b) 5. (c) 6. (c) 7. (d) 8. (a) 9. (c) 10. (b)

Chapter 8 Checkup

Completing the Sentences 1. (b) 2. (c) 3. (b) 4. (a) 5. (b) 6. (b) 7. (c) 8. (a)

CHAPTER 9

Progress Quiz 9-1

Completing the Sentences 1. (a) 2. (a) 3. (c) 4. (b) 5. (c) 6. (b) 7. (b) 8. (d) 9. (c) 10. (c)

Chapter 9 Checkup

Correcting Parts Lists 1. water pump 2. skirt 3. crankshaft 4. connecting rod 5. piston ring

Completing the Sentences 1. (a) 2. (d) 3. (c) 4. (a) 5. (b) 6. (c) 7. (a) 8. (c) 9. (b) 10. (d) 11. (a) 12. (c) 13. (b) 14. (c) 15. (c)

CHAPTER 10

Completing the Sentences 1. (c) 2. (a) 3. (a) 4. (b) 5. (c) 6. (b) 7. (b) 8. (a) 9. (a) 10. (a)

CHAPTER 11

Progress Quiz 11-1
Completing the Sentences 1. (b) 2. (a) 3. (b) 4. (c) 5. (c) 6. (b) 7. (c) 8. (a) 9. (a) 10. (b)

Progress Quiz 11-2
Completing the Sentences 1. (b) 2. (a) 3. (d) 4. (b) 5. (b) 6. (a) 7. (b) 8. (a) 9. (a) 10. (a)

Chapter 11 Checkup
Completing the Sentences 1. (b) 2. (a) 3. (c) 4. (a) 5. (b) 6. (b) 7. (c) 8. (a) 9. (b) 10. (a)

Problems 1. 28.27 in^3 2. 8.5:1 3. 10:1 4. about 71% 5. 190.4 bhp 6. 203.6 ihp 7. 167 ihp 8. 81.5% 9. 800 lb-ft

CHAPTER 12

Completing the Sentences 1. (a) 2. (b) 3. (c) 4. (a) 5. (b) 6. (a) 7. (b) 8. (a) 9. (c) 10. (b)

CHAPTER 13

Progress Quiz 13-1
Completing the Sentences 1. (b) 2. (a) 3. (b) 4. (a) 5. (b) 6. (c) 7. (b) 8. (b)

Chapter 13 Checkup
Completing the Sentences 1. (b) 2. (a) 3. (b) 4. (a) 5. (b) 6. (c) 7. (b) 8. (b) 9. (a) 10. (b)

CHAPTER 14

Progress Quiz 14-1
Completing the Sentences 1. (b) 2. (a) 3. (c) 4. (b) 5. (b) 6. (b) 7. (c) 8. (b) 9. (a) 10. (c)

Progress Quiz 14-2
Completing the Sentences 1. (b) 2. (b) 3. (a) 4. (a) 5. (c) 6. (c) 7. (b) 8. (c) 9. (b) 10. (a)

CHAPTER 15

Chapter 15 Checkup
Completing the Sentences 1. (b) 2. (d) 3. (c) 4. (b) 5. (a) 6. (a) 7. (b) 8. (b) 9. (a) 10. (c)

CHAPTER 16

Chapter 16 Checkup
Completing the Sentences 1. (a) 2. (c) 3. (a) 4. (b) 5. (a) 6. (c) 7. (a) 8. (b) 9. (d) 10. (c)

CHAPTER 17

Chapter 17 Checkup
Completing the Sentences 1. (a) 2. (c) 3. (b) 4. (b) 5. (a) 6. (b) 7. (c) 8. (a) 9. (d) 10. (b)

Unscrambling the Oil Jobs
lubricates to minimize wear
lubricates to minimize power loss
cools by carrying heat from engine parts
absorbs shocks between bearings and other engine parts
seals between piston rings and cylinder walls
cleans by carrying dirt from engine parts

CHAPTER 18

Chapter 18 Checkup
Completing the Sentences 1. (b) 2. (a) 3. (b) 4. (b) 5. (c) 6. (c) 7. (b) 8. (b)

CHAPTER 19

Chapter 19 Checkup
Completing the Sentences 1. (c) 2. (b) 3. (b) 4. (b) 5. (a) 6. (a) 7. (b) 8. (a) 9. (a)

CHAPTER 20

Chapter 20 Checkup
Completing the Sentences 1. (b) 2. (a) 3. (a) 4. (c) 5. (c) 6. (a) 7. (b) 8. (b) 9. (d) 10. (b)

Unscrambling the Test Instruments

TEST INSTRUMENTS	PURPOSES
compression tester	checks cylinder compression
tachometer	checks engine speed
vacuum gauge	checks intake-manifold vacuum
combustion tester	analyzes exhaust gas
timing light	checks ignition timing

CHAPTER 21

Chapter 21 Checkup
Completing the Sentences 1. (c) 2. (c) 3. (b) 4. (a) 5. (a) 6. (d) 7. (b) 8. (c)

CHAPTER 22

Chapter 22 Checkup
Correcting Troubles Lists 1. stuck cooling-system thermostat 2. excessive carbon in engine 3. defective fuel pump 4. defective fan belt 5. excessive vacuum in intake manifold 6. vapor lock 7. run-down battery 8. incorrect idle-mixture adjustment 9. defective fuel pump 10. defective head gasket 11. overcharged battery 12. run-down battery 13. vapor lock 14. clogged air cleaner 15. run-down battery 16. spark knock 17. loose piston pin 18. broken rings 19. worn piston-pin bearings 20. misaligned rods

Completing the Sentences 1. (b) 2. (c) 3. (d) 4. (d) 5. (b) 6. (a) 7. (c) 8. (b) 9. (b) 10. (b) 11. (a) 12. (a) 13. (b) 14. (b) 15. (b) 16. (d) 17. (c) 18. (a) 19. (b) 20. (a)

CHAPTER 23

Progress Quiz 23-1
Correcting Troubles Lists 1. excessive tappet clearance 2. excessive tappet clearance 3. idle too high 4. interference angle 5. heavy valve spring

Completing the Sentences 1. (a) 2. (b) 3. (a) 4. (c) 5. (a) 6. (c) 7. (b) 8. (b) 9. (b) 10. (b)

Progress Quiz 23-2
Completing the Sentences 1. (a) 2. (b) 3. (b) 4. (a) 5. (c) 6. (c) 7. (c) 8. (a) 9. (c) 10. (c)

Chapter 23 Checkup
Completing the Sentences 1. (d) 2. (d) 3. (c) 4. (a) 5. (b) 6. (b) 7. (a) 8. (d) 9. (a) 10. (b)

CHAPTER 24

Progress Quiz 24-1
Completing the Sentences 1. (c) 2. (b) 3. (c) 4. (a) 5. (a) 6. (c) 7. (b) 8. (a) 9. (b) 10. (a)

Unscrambling Causes of Bearing Failure

BEARING FAILURES	CAUSES
scratches in bearing metal	dirt in oil
bearing metal wiped out uniformly	lack of oil
bearing metal wiped out on one side	tapered crankpin
bearing metal wiped out at center	crankpin ridged
bright spots on bearing metal	dirt under shell
craters flaked out	fatigue
bearing failure at edge	radius ride

Progress Quiz 24-2
Completing the Sentences 1. (b) 2. (d) 3. (a) 4. (c) 5. (a) 6. (a) 7. (b) 8. (c) 9. (a) 10. (b)

Chapter 24 Checkup
Completing the Sentences 1. (b) 2. (a) 3. (b) 4. (c) 5. (c) 6. (a) 7. (c) 8. (d) 9. (d) 10. (c)

CHAPTER 25

Progress Quiz 25-1
Completing the Sentences 1. (c) 2. (b) 3. (b) 4. (c) 5. (a) 6. (a) 7. (b) 8. (d) 9. (c) 10. (c)

Progress Quiz 25-2
Completing the Sentences 1. (a) 2. (c) 3. (b) 4. (b) 5. (d) 6. (a) 7. (c) 8. (c) 9. (a) 10. (c)

Chapter 25 Checkup
Completing the Sentences 1. (b) 2. (c) 3. (c) 4. (a) 5. (c) 6. (b) 7. (c) 8. (a) 9. (b) 10. (a)

CHAPTER 26

Completing the Sentences 1. (a) 2. (a) 3. (b) 4. (c) 5. (b) 6. (a) 7. (a) 8. (b)